Introductory Digital Image Processing

Prentice-Hall Series
in Geographic
Information Science

KEITH C. CLARKE,
Series Advisor

Introductory Digital Image Processing

A Remote Sensing Perspective

Third Edition

John R. Jensen

University of South Carolina

Upper Saddle River, NJ 07458

Library of Congress Cataloging-in-Publication Data

Jensen, John R.
 Introductory digital image processing: a remote sensing perspective /John R. Jensen. — 3rd ed.
    p. cm. – (Prentice Hall series in geographic information science)
Includes bibliographic references and index.
ISBN 0-13-145361-0
1. Remote sensing. 2. Image Processing — Digital techniques. I. Title. II. Series.

G70.4.J46 2005
621.36'78–dc22

2004001515

Executive Editor: Daniel E. Kaveney
Editor in Chief, Science: John Challice
Marketing Manager: Robin Farrar
Associate Editor: Amanda Griffith
Editorial Assistant: Margaret Ziegler
Vice President and Director of Production and Manufacturing, ESM: David W. Riccardi
Production Editor: Beth Lew
Media Editor: Chris Rapp
Manufacturing Manager: Trudy Pisciotti
Manufacturing Buyer: Lynda Castillo

About the Cover: Land-cover maps of North and South America produced from *Terra* MODIS 1 × 1 km data collected between November 2000 and October 2001 (courtesy of *NASA Earth Observatory*, August 13, 2002). For additional information see Friedl, M. A., McIver, D. K., Hodges, J. C. F., Zhang, X. Y., Muchoney, D., Strahler, A. H., Woodcock, C. E., Gopal, S., Schneider, A., Cooper, A., Baccini, A., Gao, F. and C. Schaaf, 2002, "Global Land Cover Mapping from MODIS: Algorithms and Early Results," *Remote Sensing of Environment*, 83(1-2):287-302.

Pearson Education Ltd., *London*
Pearson Education Australia Pty. Ltd., *Sydney*
Pearson Education Singapore, Pte. Ltd.
Pearson Education North Asia Ltd., *Hong Kong*
Pearson Education Canada, Inc., *Toronto*
Pearson Educacion de Mexico, S. A. de C. V.
Pearson Education–Japan, *Tokyo*
Pearson Education Malaysia, Pte. Ltd.

Contents

Preface

The science of remote sensing is advancing rapidly as sophisticated sensor systems obtain data with increasingly detailed spatial, spectral, temporal, and radiometric resolution. Digital image processing algorithms used to extract information from the remote sensor data continue to improve. This has resulted in an increase in our ability to extract quantitative, biophysical data (e.g., temperature, biomass, precipitation) and land-use/land-cover information from remote sensor data. Such information is becoming increasingly important as input to spatially distributed models we use to understand natural and human-modified ecosystems.

The third edition of *Introductory Digital Image Processing: A Remote Sensing Perspective* continues to focus on digital image processing of aircraft- and satellite-derived remotely sensed data for Earth resource management applications. The book was written for physical, natural, and social scientists interested in the quantitative analysis of remotely sensed data to solve real-world problems. The level of material presented assumes that the reader has completed an introductory airphoto interpretation or remote sensing course. The treatment also assumes that the reader has a reasonable background in college algebra and univariate and multivariate statistics. The text can be used in an undergraduate or graduate one-semester course on introductory digital image processing where the emphasis is on Earth resource analysis.

The following features make the third edition of *Introductory Digital Image Processing: A Remote Sensing Perspective* one of the most easily comprehended digital image processing books:

- All algorithms (except for the Fourier transform) are presented in relatively simple algebraic terms.

- Every chapter includes line drawings and digital image processing graphics designed to make complex principles easier to understand.

- The 8.5 × 10.87 in. book format renders the image processing graphics and line drawings more readable and visually informative.

- Each chapter contains a substantive reference list so the reader can obtain additional information on a topic.

- The third edition contains thirty-two color illustrations, which enhance the presentation of important image processing concepts.

- The book is organized according to the general flow or method by which digital remote sensor data are analyzed. Novices in the field can use the book as a manual as they perform the various functions associated with a remote sensing digital image processing project.

The third edition has been revised substantially. The following summary indicates significant changes in each chapter.

Chapter 1: Remote Sensing and Digital Image Processing. A revised introduction summarizes the *remote sensing process*. The various elements of the remote sensing process are reviewed, including statement of the problem, data collection (*in situ* and remote sensing), data-to-information conversion, and information presentation alternatives. A taxonomy of models used in remote sensing, geographic information systems (GIS), and environmental research is included based on the method of processing (deterministic, stochastic) or type of logic (inductive, deductive). The chapter concludes with an overview of the content of the book.

Chapter 2: Remote Sensing Data Collection. Analog (hard-copy) image digitization is presented with improved examples. Information on recent satellite and aircraft remote sensing systems is presented, including:

- Landsat Enhanced Thematic Mapper Plus (ETM$^+$)

- SPOT 4 High Resolution Visible (HRV) and SPOT 5 High Resolution Visible Infrared (HRVIR) and *Vegetation* sensors

- NASA Earth Observer (EO-1) sensors: Advanced Land Imager (ALI), Hyperion hyperspectral sensor, and LEISA atmospheric corrector

- recent NOAA Geostationary Operational Environmental Satellite (GOES) and Advanced Very High Resolution Radiometer (AVHRR) sensor systems

- ORBIMAGE, Inc., and NASA Sea-viewing Wide Field-of-View Sensor (SeaWiFS)

- Indian Remote Sensing (IRS) satellite sensors

- NASA *Terra* and *Aqua* sensors: Advanced Spaceborne Thermal Emission and Reflection Radiometer (ASTER), Multiangle Imaging Spectroradiometer (MISR), and Moderate Resolution Imaging Spectrometer (MODIS)

- high-spatial-resolution satellite remote sensing systems: IKONOS (Space Imaging), QuickBird (DigitalGlobe),

OrbView-3 (ORBIMAGE), and EROS A1 (ImageSat International)

- suborbital hyperspectral sensors such as NASA's Airborne Visible/Infrared Imaging Spectrometer (AVIRIS) and the Compact Airborne Spectrographic Imager 3 (CASI 3)

- digital frame cameras such as the EMERGE, Inc., Digital Sensor System (DSS)

- satellite photographic systems such as the Russian SPIN-2 TK-350 and KVR-1000 cameras

There is a new discussion of remote sensing data formats (band interleaved by pixel, band interleaved by line, and band sequential).

Chapter 3: Digital Image Processing Hardware and Software Considerations. The most important hardware and software considerations necessary to configure a quality remote sensing digital image processing system are updated. A historical review of the Intel, Inc., central processing unit (CPU) identifies the number of transistors and millions of instructions per second (MIPS) that could be processed through the years. New information on serial versus parallel image processing, graphical user interfaces, and the longevity of remote sensor data storage media are presented. The most important digital image processing functions found in a quality digital image processing system are updated. The functionality of numerous commercial and public domain digital image processing systems is presented along with relevant Internet addresses.

Chapter 4: Image Quality Assessment and Statistical Evaluation. This chapter provides fundamental information on univariate and multivariate statistics that are routinely extracted from remotely sensed data. It includes new information on the importance of the histogram to digital image processing, image metadata, and how to view pixel brightness values at individual locations or within geographic areas. Methods of viewing individual bands of imagery in three dimensions are examined. Two-dimensional feature space plot logic is introduced. Principles of geostatistical analysis are presented including spatial autocorrelation and the calculation of the empirical semivariogram.

Chapter 5: Initial Display Alternatives and Scientific Visualization. The concept of *scientific visualization* is introduced. Methods of visualizing data in both black-and-white and color are presented with an improved discussion of color look-up table and color space theory. New information about bitmapped images is provided. The Sheffield

Index is introduced as an alternative method for selecting the optimum bands when creating a color composite. Emphasis is placed on how to merge different types of remotely sensed data for visual display and analysis using color space transformations, including new material on the use of the chromaticity color coordinate system and the Brovey transform. The chapter concludes with a summary of the mathematics necessary to calculate distance, area, and shape measurements on rectified digital remote sensor data.

Chapter 6: Electromagnetic Radiation Principles and Radiometric Correction. This significantly revised chapter deals with radiometric correction of remote sensor data. The first half of the chapter reviews electromagnetic radiation models, atmospheric energy–matter interactions, terrain energy–matter interactions, and sensor system energy–matter interactions. Fundamental radiometric concepts are then introduced. Various methods of correcting sensor detector-induced error in remotely sensed images are presented. Remote sensing atmospheric correction is then introduced, including a discussion of when it is necessary to atmospherically correct remote sensor data. Absolute radiometric correction alternatives based on radiative transfer theory are introduced. Relative radiometric correction of atmospheric attenuation is presented. The chapter concludes with methods used to correct for the effects of terrain slope and aspect.

Chapter 7: Geometric Correction. The chapter contains new information about image offset (skew) caused by Earth rotation and how skew can be corrected. New information is introduced about the geometric effects of platform roll, pitch, and yaw during remote sensing data acquisition. The chapter then concentrates on image-to-image registration and image-to-map rectification. New graphics and discussion make clear the distinction between GIS-related input-to-output (forward) mapping logic and output-to-input (inverse) mapping logic required to resample and rectify raster remote sensor data. The chapter concludes with a new section on digital mosaicking using feathering logic.

Chapter 8: Image Enhancement. New graphics and text describe how spatial profiles (transects) and spectral profiles are extracted from multispectral and hyperspectral imagery. Piecewise linear contrast stretching is demonstrated using new examples. The use of the Fourier transform to remove striping in remote sensor data is introduced. An updated review of vegetation transformations (indices) is provided. It includes fundamental principles associated with the dominant factors controlling leaf reflectance and many newly developed indices. Texture measures based on conditional variance detection and the geostatistical semivariogram are discussed.

Chapter 9: Thematic Information Extraction: Pattern Recognition. The chapter begins with an overview of *hard* versus *fuzzy* land-use/land-cover classification logic. It then delves deeply into supervised classification. It introduces several new land-use/land-cover classification schemes. It includes a new section on nonparametric nearest-neighbor classification and a more in-depth treatment of the maximum-likelihood classification algorithm based on probability density functions. Unsupervised classification using ISODATA is made easier to understand using an additional empirical example. There is a new section on object-oriented image segmentation and how it can be used for image classification. The chapter concludes by updating methods to incorporate ancillary data into the remote sensing classification process.

Chapter 10: Information Extraction Using Artificial Intelligence. This new chapter begins by briefly reviewing the history of artificial intelligence. It then introduces the concept of an expert system, its components, the knowledge representation process, and the inference engine. Human-derived rules are then input to a rule-based expert system to extract land-cover information from remote sensor data. In a separate example, the remote sensor data are subjected to machine learning to demonstrate how the rules used in an expert system can be developed with minimal human intervention. The use of artificial neural networks in remote sensing classification is introduced. The chapter concludes with a discussion of the advantages and limitations of expert systems and artificial neural networks for information extraction.

Chapter 11: Thematic Information Extraction: Hyperspectral Image Analysis. This new chapter begins by reviewing the ways hyperspectral data are collected. It then uses an empirical case study based on AVIRIS data to introduce the general steps to extract information from hyperspectral data. Emphasis is on radiative transfer-based radiometric correction of the hyperspectral data, reducing its dimensionality, and extracting relevant endmembers. Various methods of mapping and matching are then presented including the spectral angle mapper, linear spectral unmixing, and spectroscopic library matching techniques. The chapter concludes with a summary of various narrow-band indices that can be used with hyperspectral data.

Chapter 12: Digital Change Detection. The change detection flow chart summarizes current methods. New examples of write function memory insertion, multiple-date composite image, and image algebra (image differencing) change detection are provided. New chi-square transformation and cross-correlation change detection methods are introduced.

The chapter concludes with a discussion about when it is necessary to atmospherically correct remote sensor data for change detection applications.

Chapter 13: Thematic Map Accuracy Assessment. This new chapter begins by reviewing the approaches to land-use/land-cover classification map accuracy assessment. It then summarizes the sources of error in remote sensing–derived thematic map products. Various methods of computing the sample size are introduced. Sampling designs (schemes) are discussed. The evaluation of error matrices using descriptive and discrete multivariate analytical techniques is presented. A section describes how to incorporate fuzzy information into an accuracy assessment. The chapter concludes with observations about geostatistical measures used in accuracy assessment.

Acknowledgments

The author thanks the following individuals for their support and assistance in the preparation of the third edition. Ryan Jensen contributed to the survey of digital image processing systems in Chapter 3. Kan He, Xueqiao Huang, and Brian Hadley provided insight for the radiometric correction and Fourier transform analysis sections in Chapter 6. David Vaughn contributed to the vegetation index section in Chapter 8. Jason Tullis and Xueqiao Huang contributed to the use of artificial intelligence in digital image processing in Chapter 10. Anthony M. Filippi contributed to information extraction using hyperspectral data in Chapter 11. Russ Congalton provided insight and suggestions for assessing thematic map classification accuracy in Chapter 13. Lynette Likes provided computer network support. Maria Garcia, Brian Hadley, George Raber, Jason Tullis, and David Vaughn assisted with proofreading. Finally, I would like to especially thank my wife, Marsha, for her help and unwavering encouragement.

John R. Jensen
University of South Carolina

Introductory Digital Image Processing

Remote Sensing and Digital Image Processing

1

The goal of science is to discover universal truths that are the same yesterday, today, and tomorrow. Hopefully, the knowledge obtained can be used to protect the environment and improve human quality of life. To identify these universal truths, scientists observe and make measurements about:

- the physical world (e.g., the atmosphere, water, soil, rock),

- its living inhabitants (e.g., *Homo sapiens*, flora, fauna), and

- the processes at work (e.g., mass wasting, deforestation, urban sprawl).

Scientists formulate hypotheses and then attempt to accept or reject them in a systematic, unbiased fashion. The data necessary to accept or reject a hypothesis may be collected directly in the field, often referred to as *in situ* or *in-place* data collection. This can be a time-consuming, expensive, and inaccurate process. Therefore, considerable research during the past century has gone into the development of aerial platforms (e.g., suborbital aircraft, satellites, unmanned aerial vehicles) and sensors (e.g., cameras, detectors) that can collect information some remote distance from the subject (e.g., from 10,000 m above ground level). This process is called *remote sensing* of the environment (Figure 1-1).

The remote sensor data can be stored in an analog format (e.g., a hardcopy 9 × 9 in. vertical aerial photograph) or in a digital format (e.g., Landsat Thematic Mapper imagery consists of seven registered matrices of brightness values). The analog and digital remote sensor data can be analyzed using analog (visual) and/or digital image processing techniques. The fundamental methods of analog (visual) image processing are discussed in Jensen (2000). This text introduces the fundamental concepts of digital image processing of remote sensor data to extract useful Earth resource information. First, however, it is important to review several aspects of *in situ* data collection because it is often critical to successful digital image processing (Teillet et al., 2002).

 ### *In Situ* Data Collection

One way of collecting *in situ* data about human beings and their related activities is for the scientist to go into the field or the city and directly question the persons of interest. For example, a human enumerator for the decennial census of population might go door to door, asking people questions about their age, sex, education, income, etc. These data are recorded and used to document quantitatively the demographic characteristics of the population. These *in situ* data are then used to accept or reject hypotheses associated with human activities and socioeconomic characteristics.

Conversely, a scientist might elect to place a *transducer* at the study site to make measurements. A transducer is a device that converts input energy of one form into output energy of another form. Transducers are usually placed in direct physical contact with the object of interest. Many types of transducer are available. For example, a scientist could use a thermometer to measure the temperature of the air, soil, or water; an anemometer to measure the speed of the wind; or a psychrometer to measure humidity. The data might be recorded by the transducers as an analog electrical signal with voltage variations related to the intensity of the property being measured. Often these analog signals are transformed into digital values using analog-to-digital (A-to-D) conversion procedures. *In situ* data collection using transducers relieves the scientist of monotonous data collection duties in inclement weather. Also, the scientist can distribute the transducers at important geographic locations throughout the study area, allowing the same type of measurement to be obtained at many locations at the same instant in time. Sometimes data from the transducers are telemetered electronically to a central receiving station for rapid evaluation and archiving.

Two types of *in situ* data collection often used in support of remote sensing investigations are depicted in Figure 1-2. In the first example, spectral reflectance measurements of smooth cordgrass (*Spartina alterniflora*) in Murrells Inlet, South Carolina, are being recorded in the blue, green, red and near-infrared portions of the electromagnetic spectrum (400–1100 nm) using a hand-held spectroradiometer (Figure 1-2a). Spectral reflectance measurements obtained in the field can be used to calibrate spectral reflectance measurements collected by a remote sensing system located on an aircraft or satellite. A scientist is obtaining precise *x,y,z* geographic coordinates of an *in situ* sample location using a global positioning system (GPS) in Figure 1-2b.

In Situ Data-Collection Error

Many people believe that *in situ* data are absolutely accurate because they were obtained on the ground. Unfortunately, error can also be introduced during *in situ* data collection. Sometimes the scientist is an *intrusive* agent in the field. For example, in Figure 1-2a the scientist's shadow or the ladder shadow could fall within the instantaneous-field-of-view

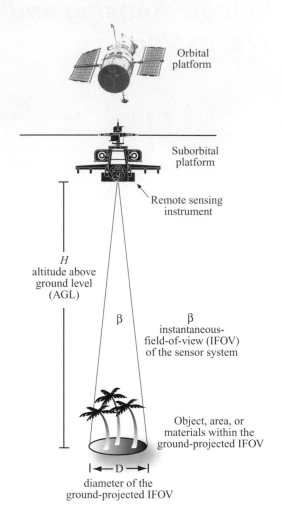

Figure 1-1   A remote sensing instrument collects information about an object or phenomenon within the instantaneous-field-of-view (IFOV) of the sensor system without being in direct physical contact with it. The sensor is usually located on a suborbital or satellite platform.

(IFOV) of the handheld spectroradiometer, causing spectral reflectance measurement error. In addition, the scientists could accidently step on the area to be measured, compacting the vegetation and soil prior to data collection. Any of these activities would result in *biased* data collection.

Scientists could also collect data in the field using biased procedures often referred to as *method-produced error*. Such error can be introduced by:

- a sampling design that does not capture all of the spatial variability of the phenomena under investigation (i.e.,

In Situ Data Collection

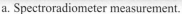

a. Spectroradiometer measurement.

b. Global positioning system (GPS) measurement.

Figure 1-2   a) A scientist is collecting *in situ* spectral reflectance measurements of smooth cordgrass (*Spartina alterniflora*) in Murrells Inlet, SC, using a handheld spectroradiometer located approximately 2 m above the mudflat surface. The *in situ* spectral reflectance measurements from 400 to 1100 nm can be used to calibrate the spectral reflectance measurements obtained from a remote sensing system onboard an aircraft or satellite. b) A scientist is obtaining *x,y,z* geographic coordinates of a site using a global positioning system (GPS) capable of accuracies within ± 50 cm.

some phenomena or geographic areas are oversampled while others are undersampled);

• operating *in situ* measurement instruments improperly; or

• using an *in situ* measurement instrument that has not been calibrated properly (or recently).

Intrusive *in situ* data collection, coupled with human method-produced error and measurement-device miscalibration, all contribute to *in situ* data-collection error. Therefore, it is a misnomer to refer to *in situ* data as *ground truth data*. Instead, we should refer to it simply as *in situ ground reference data*, and acknowledge that it also contains error.

Remote Sensing Data Collection

Fortunately, it is also possible to collect certain types of information about an object or geographic area from a distant vantage point using *remote sensing instruments* (Figure 1-1).

The American Society for Photogrammetry and Remote Sensing (ASPRS) defines *remote sensing* as (Colwell, 1983):

the measurement or acquisition of information of some property of an object or phenomenon, by a recording device that is not in physical or intimate contact with the object or phenomenon under study.

In 1988, ASPRS adopted a combined formal definition of *photogrammetry* and *remote sensing* as (Colwell, 1997):

the art, science, and technology of obtaining reliable information about physical objects and the environment, through the process of recording, measuring and interpreting imagery and digital representations of energy patterns derived from noncontact sensor systems.

Robert Green at NASA's Jet Propulsion Laboratory (JPL) suggests that the term *remote measurement* might be used instead of *remote sensing* because data obtained using hyperspectral remote sensing systems are so accurate (Robbins,

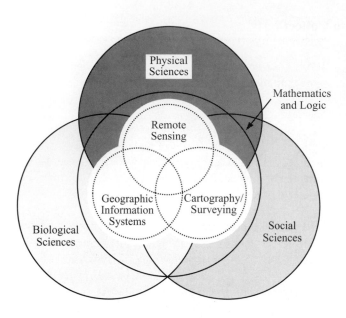

Figure 1-3 An interaction model depicting the relationships of the mapping sciences (remote sensing, geographic information systems, and cartography/surveying) as they relate to mathematics and logic and the physical, biological, and social sciences.

1999). Hyperspectral digital image processing is discussed in Chapter 11.

Observations about Remote Sensing

The following brief discussion focuses on various terms found in the formal definitions of remote sensing.

Is Remote Sensing a Science?

A *science* is defined as the broad field of human knowledge concerned with facts held together by *principles* (rules). Scientists discover and test facts and principles by the scientific method, an orderly system of solving problems. Scientists generally feel that any subject that humans can study by using the scientific method and other special rules of thinking may be called a science. The sciences include 1) *mathematics* and *logic*, 2) the *physical sciences*, such as physics and chemistry, 3) the *biological sciences*, such as botany and zoology, and 4) the *social sciences*, such as geography, sociology, and anthropology (Figure 1-3). Interestingly, some persons do not consider mathematics and logic to be sciences. But the fields of knowledge associated with mathematics and logic are such valuable *tools* for science that we

cannot ignore them. The human race's earliest questions were concerned with "how many" and "what belonged together." They struggled to count, to classify, to think systematically, and to describe exactly. In many respects, the state of development of a science is indicated by the use it makes of mathematics. A science seems to begin with simple mathematics to measure, then works toward more complex mathematics to explain.

Remote sensing is a tool or technique similar to mathematics. Using sensors to measure the amount of electromagnetic radiation (EMR) exiting an object or geographic area from a distance and then extracting valuable information from the data using mathematically and statistically based algorithms is a *scientific* activity (Fussell et al., 1986). It functions in harmony with other *spatial* data-collection techniques or tools of the *mapping sciences*, including cartography and geographic information systems (GIS) (Curran, 1987; Clarke, 2001). Dahlberg and Jensen (1986) and Fisher and Lindenberg (1989) suggest a model where there is interaction between remote sensing, cartography, and GIS; where no subdiscipline dominates; and all are recognized as having unique yet overlapping areas of knowledge and intellectual activity as they are used in physical, biological, and social science research (Figure 1-3).

Is Remote Sensing an Art?

The process of visual photo or image interpretation brings to bear not only scientific knowledge but all of the background that a person has obtained in his or her lifetime. Such learning cannot be measured, programmed, or completely understood. The synergism of combining scientific knowledge with real-world analyst experience allows the interpreter to develop heuristic rules of thumb to extract information from the imagery. Some image analysts are superior to other image analysts because they 1) understand the scientific principles better, 2) are more widely traveled and have seen many landscape objects and geographic areas, and/or 3) have the ability to synthesize scientific principles and real-world knowledge to reach logical and correct conclusions. Thus, remote sensing image interpretation is both an art and a science.

Information about an Object or Area

Sensors can be used to obtain very specific information about an object (e.g., the diameter of a cottonwood tree's crown) or the geographic extent of a phenomenon (e.g., the polygonal boundary of a cottonwood stand). The EMR reflected, emitted, or back-scattered from an object or geographic area is used as a surrogate for the actual property under investigation. The electromagnetic energy measurements must be calibrated and turned into information using visual and/or digital image processing techniques.

The Instrument (Sensor)

Remote sensing is performed using an instrument, often referred to as a *sensor*. The majority of remote sensing instruments record EMR that travels at a velocity of 3×10^8 m s^{-1} from the source, directly through the vacuum of space or indirectly by reflection or reradiation to the sensor. The EMR represents a very efficient high-speed communications link between the sensor and the remote phenomenon. In fact, we know of nothing that travels faster than the speed of light. Changes in the amount and properties of the EMR become, upon detection by the sensor, a valuable source of data for interpreting important properties of the phenomenon (e.g., temperature, color). Other types of force fields may be used in place of EMR, including sound waves (e.g., sonar). However, the majority of remotely sensed data collected for Earth resource applications are the result of sensors that record electromagnetic energy.

How Far Is Remote?

Remote sensing occurs at a distance from the object or area of interest. Interestingly, there is no clear distinction about how great this distance should be. The distance could be 1 m, 100 m, or more than 1 million meters from the object or area of interest. Much of astronomy is based on remote sensing. In fact, many of the most innovative remote sensing systems and visual and digital image processing methods were originally developed for remote sensing extraterrestrial landscapes such as the moon, Mars, Io, Saturn, Jupiter, etc. This text, however, is concerned primarily with remote sensing of the terrestrial Earth, using sensors that are placed on suborbital air-breathing aircraft or orbital satellite platforms placed in the vacuum of space.

Remote sensing and digital image processing techniques can also be used to analyze inner space. For example, an electron microscope can be used to obtain photographs of extremely small objects on the skin, in the eye, etc. An *x*-ray instrument is a remote sensing system where the skin and muscle are like the atmosphere that must be penetrated, and the interior bone or other matter is often the object of interest. Many digital image processing enhancement techniques presented in this text are well suited to the analysis of "inner space" objects.

Remote Sensing Advantages and Limitations

Remote sensing has several unique advantages as well as some limitations.

Advantages of Remote Sensing

Remote sensing is *unobtrusive* if the sensor passively records the electromagnetic energy reflected or emitted by the phenomenon of interest. Passive remote sensing does not disturb the object or area of interest.

Remote sensing devices are programmed to collect data systematically, such as within a single 9×9 in. frame of vertical aerial photography or a single line of Systeme Probatoire d'Observation de la Terre (SPOT) image data collected using a linear array. This systematic data collection can remove the sampling bias introduced in some *in situ* investigations.

Under controlled conditions, remote sensing can provide fundamental biophysical information, including *x,y* location, *z* elevation or depth, biomass, temperature, and moisture content. In this sense it is much like surveying, providing fundamental information that other sciences can use when conducting scientific investigations. However, unlike much of surveying, the remotely sensed data can be obtained systematically over very large geographic areas rather than just single-point observations. In fact, remote sensing–derived information is now critical to the successful modeling of numerous natural (e.g., water-supply estimation; eutrophication studies; nonpoint source pollution) and cultural (e.g., land-use conversion at the urban fringe; water-demand estimation; population estimation) processes (Walsh et al., 1999; Stow et al., 2003). A good example is the digital elevation model that is so important in many spatially-distributed GIS models (Clarke, 2001). Digital elevation models are now produced almost exclusively from stereoscopic imagery, light detection and ranging (LIDAR), or radio detection and ranging (RADAR) measurements.

Limitations of Remote Sensing

Remote sensing science has limitations. Perhaps the greatest limitation is that it is often oversold. *Remote sensing is not a panacea* that will provide all the information needed to conduct physical, biological, or social science research. It simply provides some spatial, spectral, and temporal information of value in a manner that we hope is efficient and economical.

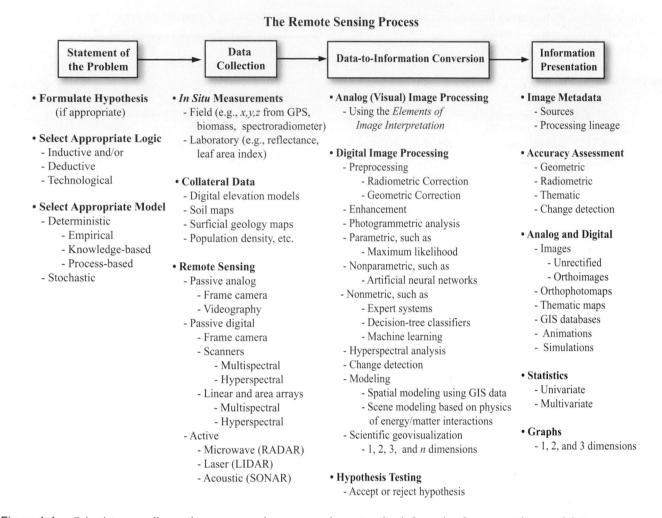

The Remote Sensing Process

Statement of the Problem	→	Data Collection	→	Data-to-Information Conversion	→	Information Presentation

• **Formulate Hypothesis**
  (if appropriate)

• **Select Appropriate Logic**
  - Inductive and/or
  - Deductive
  - Technological

• **Select Appropriate Model**
  - Deterministic
    - Empirical
    - Knowledge-based
    - Process-based
  - Stochastic

• *In Situ* **Measurements**
  - Field (e.g., *x,y,z* from GPS, biomass, spectroradiometer)
  - Laboratory (e.g., reflectance, leaf area index)

• **Collateral Data**
  - Digital elevation models
  - Soil maps
  - Surficial geology maps
  - Population density, etc.

• **Remote Sensing**
  - Passive analog
    - Frame camera
    - Videography
  - Passive digital
    - Frame camera
    - Scanners
      - Multispectral
      - Hyperspectral
    - Linear and area arrays
      - Multispectral
      - Hyperspectral
  - Active
    - Microwave (RADAR)
    - Laser (LIDAR)
    - Acoustic (SONAR)

• **Analog (Visual) Image Processing**
  - Using the *Elements of Image Interpretation*

• **Digital Image Processing**
  - Preprocessing
    - Radiometric Correction
    - Geometric Correction
  - Enhancement
  - Photogrammetric analysis
  - Parametric, such as
    - Maximum likelihood
  - Nonparametric, such as
    - Artificial neural networks
  - Nonmetric, such as
    - Expert systems
    - Decision-tree classifiers
    - Machine learning
  - Hyperspectral analysis
  - Change detection
  - Modeling
    - Spatial modeling using GIS data
    - Scene modeling based on physics of energy/matter interactions
  - Scientific geovisualization
    - 1, 2, 3, and *n* dimensions

• **Hypothesis Testing**
  - Accept or reject hypothesis

• **Image Metadata**
  - Sources
  - Processing lineage

• **Accuracy Assessment**
  - Geometric
  - Radiometric
  - Thematic
  - Change detection

• **Analog and Digital**
  - Images
    - Unrectified
    - Orthoimages
  - Orthophotomaps
  - Thematic maps
  - GIS databases
  - Animations
  - Simulations

• **Statistics**
  - Univariate
  - Multivariate

• **Graphs**
  - 1, 2, and 3 dimensions

Figure 1-4   Scientists generally use the remote sensing process when extracting information from remotely sensed data.

Human beings select the most appropriate remote sensing system to collect the data, specify the various resolutions of the remote sensor data, calibrate the sensor, select the platform that will carry the sensor, determine when the data will be collected, and specify how the data are processed. Human method-produced error may be introduced as the remote sensing instrument and mission parameters are specified.

Powerful *active* remote sensor systems that emit their own electromagnetic radiation (e.g., LIDAR, RADAR, SONAR) can be intrusive and affect the phenomenon being investigated. Additional research is required to determine how intrusive these active sensors can be.

Remote sensing instruments may become uncalibrated, resulting in uncalibrated remote sensor data. Finally, remote sensor data may be expensive to collect and analyze. Hopefully, the information extracted from the remote sensor data justifies the expense.

 The Remote Sensing Process

Urban planners (e.g., land use, transportation, utility) and natural resource managers (e.g., wetland, forest, grassland, rangeland) recognize that spatially distributed information is essential for ecological modeling and planning (Johannsen et al., 2003). Unfortunately, it is very difficult to obtain such information using *in situ* measurement for the aforementioned reasons. Therefore, public agencies and scientists have expended significant resources in developing methods to obtain the required information using remote sensing science (Goetz, 2002; Nemani et al., 2003). The remote sensing data-collection and analysis procedures used for Earth resource applications are often implemented in a systematic fashion referred to as the *remote sensing process*. The procedures in the remote sensing process are summarized here and in Figure 1-4:

- The hypothesis to be tested is defined using a specific type of logic (e.g., inductive, deductive) and an appropriate processing model (e.g., deterministic, stochastic).

- *In situ* and collateral data necessary to calibrate the remote sensor data and/or judge its geometric, radiometric, and thematic characteristics are collected.

- Remote sensor data are collected passively or actively using analog or digital remote sensing instruments, ideally at the same time as the *in situ* data.

- *In situ* and remotely sensed data are processed using a) analog image processing, b) digital image processing, c) modeling, and d) *n*-dimensional visualization.

- Metadata, processing lineage, and the accuracy of the information are provided and the results communicated using images, graphs, statistical tables, GIS databases, Spatial Decision Support Systems (SDSS), etc.

It is useful to review the characteristics of these remote sensing process procedures.

Statement of the Problem

Sometimes the general public and even children look at aerial photography or other remote sensor data and extract useful information. They typically do this without a formal hypothesis in mind. More often than not, however, they interpret the imagery incorrectly because they do not understand the nature of the remote sensing system used to collect the data or appreciate the vertical or oblique perspective of the terrain recorded in the imagery.

Scientists who use remote sensing, on the other hand, are usually trained in the *scientific method*—a way of thinking about problems and solving them. They use a formal plan that has at least five elements: 1) stating the problem, 2) forming the research hypothesis (i.e., a possible explanation), 3) observing and experimenting, 4) interpreting data, and 5) drawing conclusions. It is not necessary to follow this formal plan exactly.

The scientific method is normally used in conjunction with environmental models that are based on two primary types of logic:

- deductive logic

- inductive logic

Models based on deductive and/or inductive logic can be further subdivided according to whether they are processed *deterministically* or *stochastically*. Table 1-1 summarizes the relationship between inductive and deductive logic and deterministic and stochastic methods of processing. Sometimes information is extracted from remotely sensed images using neither deductive nor inductive logic. This is referred to as the use of *technological logic*.

Deductive Logic

When stating the remote sensing–related problem using *deductive logic,* a scientist (Skidmore, 2002):

> draws a specific conclusion from a set of general propositions (the premises). In other words, deductive reasoning proceeds from general truths or reasons (where the premises are self-evident) to a conclusion. The assumption is that the conclusion necessarily follows the premises: that is, if you accept the premises, then it would be self-contradictory to reject the conclusion (p. 10).

For example, Skidmore suggests that the simple normalized difference vegetation index (NDVI):

$$NDVI = \frac{\rho_{nir} - \rho_{red}}{\rho_{nir} + \rho_{red}} \qquad (1\text{-}1)$$

is a *deduced* relationship between land cover and the amount of near-infrared (ρ_{nir}) and red (ρ_{red}) spectral reflectance. Generally, the greater the amount of healthy green vegetation in the IFOV of the sensor, the greater the NDVI value. This relationship is *deduced* from the physiological fact that chlorophyll *a* and *b* in the palisade layer of healthy green leaves absorbs most of the incident red radiant flux while the spongy mesophyll leaf layer reflects much of the near-infrared radiant flux (refer to Chapter 8 for additional information about vegetation indices).

Inductive Logic

Problem statements based on *inductive logic* reach a conclusion from particular facts or observations that are treated as evidence. This is the type of logic used most often in physical, natural, and social science research. It usually relies heavily on statistical analysis. Building an inductive model normally involves defining the research (null) hypothesis, collecting the required data, conducting a statistical analysis, acceptance or rejection of the null hypothesis, and stating the level of confidence (probability) that should be associated with the conclusion.

Table 1-1.   A taxonomy of models used in remote sensing, GIS, and environmental science research (adapted from Skidmore, 2002).

			Models Based on the Type of Logic Used	
			Inductive	**Deductive**
			• Dominate spatial data handling (GIS, remote sensing) in environmental science • *Steps:* - define hypothesis - collect data and derive conclusion from facts - statistical analysis used to accept or reject research hypothesis - state probability of conclusion	• Develop a specific conclusion from a set of general propositions (premises) • *Steps:* - proceed from general truths (premises) to a conclusion - if the premises are accepted, then one must accept the conclusion
Models Based on the Method of Processing	**Deterministic** • Fixed output(s) for specific input(s)	**Empirical** • May be inductive or deductive • Often developed from empirically derived field measurements although rules may be encapsulated in an expert system • Can be difficult to extend through space and time	• Statistical models (e.g., bi-variate regression relationship between NDVI and biomass) • *Training* supervised classifiers • Threshold models • Rule induction • Geostatistical models	• For example, the computation of NDVI • *Classification* by supervised classification algorithms (e.g., maximum likelihood via model inversion) • Modified inductive models • Process models
		Knowledge-driven • Use rules to define relationships between dependent and independent variables	• May generate rules from training data (e.g., using C5.0) • Bayesian inference algorithm states that the conditional probability of a hypothesis occurring is a function of the evidence • Fuzzy systems	• Expert system based on - knowledge base of rules extracted from an expert - analysis by inference engine
		Process-driven • Use mathematics to describe factors controlling a process • Usually deductive • May be limited to small, simple areas • Temporally static or dynamic • May be *distributed* or *lumped*	• Modification of inductive model coefficients for local conditions using field or laboratory data	• Hydrological models • Ecological models • Atmospheric models
	Stochastic • Input is randomly varied • Output is variable		• Neural network classification: - trained based on induction - random weights may be assigned prior to first epoch (processing) • Monte Carlo simulation	• Monte Carlo simulation

Deterministic Models

Deterministic models can be based on inductive or deductive logic or both. *Deterministic models* use very specific input and result in a fixed output unlike stochastic models that may include randomly derived input and output information. There are three general types of deterministic model: empirical, knowledge-driven, and process-driven.

Empirical: An empirical model is based on *empiricism*—wherein a scientist should not accept a proposition unless he or she has witnessed the relationship or condition in the real world. Therefore, a deterministic, *empirical model* is founded on observations or measurements made in the real world or laboratory. Consequently, empirical models are usually based on data extracted from site-specific, local study areas.

For example, an analyst could harvest (clip) the amount of above-ground wetland biomass in 30 quadrats each 1×1 m. This is empirically derived wetland biomass data. He or she could then compute the remote sensing–derived NDVI value for these same 1×1 m quadrats using the deductive logic previously described. The analyst could then regress the two variables to determine their statistical correlation (e.g., Pearson product-moment correlation $r = 0.84$) and determine the level of confidence (e.g., 0.05) that we should have in the relationship. This is a classic deterministic, empirical model that makes use of both inductive and deductive logic (Table 1-1). It can be used as a *predictive* model, that is, if we know the NDVI value of a pixel then we can predict the amount of biomass present with a certain degree of confidence.

Supervised classification of remote sensor data is another example of deterministic, empirical modeling. Training data collected empirically in the field can be used to guide the selection of training site locations in the imagery. These training sites can be analyzed statistically using *inductive* logic to derive training class statistics. These training class statistics are then used in a *deductive* supervised classification algorithm (e.g., maximum likelihood) to assign each pixel to the class of which it has the highest probability of being a member (discussed in Chapter 9). Thus, each supervised classification of remotely sensed data is a combination of empirical inductive and deductive logic.

Because deterministic empirical models are usually developed based on training data from a local study area, it is often difficult to extend the deductive and/or inductive empirical model through space or time. This is a serious drawback. Hopefully, it will be minimized as more rigorous knowledge-driven or process-driven models are developed. However, it is likely that there will always be a need for empirical real-world data collection to verify the accuracy of knowledge- or process-driven models.

Knowledge-driven: A deterministic *knowledge-driven* model may be based on heuristic rules of thumb that define the relationship between dependent and independent variables. These rules may be deduced (created) by a human expert or from training data using statistical induction. In either case, the rules may be used in an expert system or decision-tree classifier (discussed in Chapter 10). The rules may be based on Boolean logic and be very precise. They may also be fuzzy or based on conditional probabilities.

Process-driven: Deterministic *process-driven* models (e.g., hydrologic, ecologic) use mathematics to describe the variables and conditions controlling a process (Skidmore, 2002). They are usually based on deductive logic. Unfortunately,

many process models output only a single value for an entire geographic area (e.g., a watershed). These are referred to as *lumped-parameter* process models. Conversely, when many elements with unique *x,y* positions within the geographic area are processed, it becomes a *spatially distributed* process model. Remote sensing and GIS are making significant contributions in the refinement and accuracy of distributed process modeling (Clarke, 2001). Unfortunately, the mathematics associated with many process-driven models were developed to run in lumped mode and do not easily incorporate the remote sensing-derived spatially distributed information (e.g., land cover) (Defries and Townshend, 1999).

Stochastic Models

Deterministic inductive and deductive models input data in a very systematic format. There is usually a fixed relationship between independent input variables and dependent output. Conversely, a stochastic model may initiate the processing by incorporating certain random values or weights. For example, neural networks (discussed in Chapter 10) routinely initialize the weights between the input and hidden layers with random values. The introduction of randomness can result in output that is useful but random. This is the nature of stochastic models.

Technological Logic

Some scientists extract new thematic information directly from remotely sensed imagery without ever explicitly using inductive or deductive logic. They are just interested in extracting information from the imagery using appropriate methods and technology. This *technological* approach is not as rigorous, but it is common in *applied remote sensing*. The approach can also generate new knowledge.

Remote sensing is used in both scientific (inductive and deductive) and technological approaches to obtain knowledge. There is debate as to how the different types of logic used in the remote sensing process and in GIScience yield new scientific knowledge (e.g., Fussell et al., 1986; Curran, 1987; Fisher and Lindenberg, 1989; Ryerson, 1989; Duggin and Robinove, 1990; Dobson, 1993; Wright et al., 1997; Skidmore, 2002).

Identification of In Situ and Remote Sensing Data Requirements

If a hypothesis is formulated using inductive and/or deductive logic, a list of variables or observations are identified that will be used during the investigation. *In situ* observation and/or remote sensing may be used to collect information on the most important variables.

In Situ Data Requirements

Scientists using remote sensing technology should be well trained in field and laboratory data-collection procedures. For example, if a scientist wants to map the surface temperature of a lake, it is usually necessary to collect some accurate empirical *in situ* lake-temperature measurements at the same time the remote sensor data are collected. The *in situ* observations may be used to 1) calibrate the remote sensor data, and/or 2) perform an unbiased accuracy assessment of the final results (Congalton and Green, 1998). Remote sensing textbooks provide some information on field and laboratory sampling techniques. The *in situ* sampling procedures, however, are learned best through formal courses in the sciences (e.g., chemistry, biology, forestry, soils, hydrology, meteorology). It is also important to know how to collect accurately socioeconomic and demographic information in urban environments (e.g., human geography, sociology).

Most *in situ* data are now collected in conjunction with global positioning system (GPS) x, y, z data (Jensen and Cowen, 1999). Scientists should know how to collect the GPS data at each *in situ* data-collection station and how to perform differential correction to obtain accurate x, y, z coordinates (Rizos, 2002).

Collateral Data Requirements

Many times *collateral* data (often called *ancillary* data), such as digital elevation models, soil maps, geology maps, political boundary files, and block population statistics, are of value in the remote sensing process. Ideally, these spatial collateral data reside in a digital GIS (Clarke, 2001).

Remote Sensing Data Requirements

Once we have a list of variables, it is useful to determine which of them can be remotely sensed. Remote sensing can provide information on two different classes of variables: *biophysical* and *hybrid*.

Biophysical Variables: Some biophysical variables can be measured directly by a remote sensing system. This means that the remotely sensed data can provide fundamental biological and/or physical (*biophysical*) information directly, generally without having to use other surrogate or ancillary data. For example, a thermal infrared remote sensing system can record the apparent temperature of a rock outcrop by measuring the radiant energy exiting its surface. Similarly, it is possible to conduct remote sensing in a very specific region of the spectrum and identify the amount of water vapor in the atmosphere. It is also possible to measure soil moisture content directly using microwave remote sensing techniques (Engman, 2000). NASA's Moderate Resolution Imaging Spectrometer (MODIS) can be used to measure absorbed photosynthetically active radiation (APAR) and leaf area index (LAI). The precise x, y location and height (z) of an object can be extracted directly from stereoscopic aerial photography, overlapping satellite imagery (e.g., SPOT), LIDAR data, or interferometric synthetic aperture radar imagery.

Table 1-2 is a list of selected biophysical variables that can be remotely sensed and useful sensors to acquire the data. Characteristics of aerial photography are discussed in Jensen (2000). Characteristics of many of these remote sensing systems are discussed in Chapter 2. Great strides have been made in remotely sensing many of these biophysical variables. They are important to the national and international effort under way to model the global environment (King, 2003).

Hybrid Variables: The second general group of variables that can be remotely sensed include *hybrid* variables, created by systematically analyzing more than one biophysical variable. For example, by remotely sensing a plant's chlorophyll absorption characteristics, temperature, and moisture content, it might be possible to model these data to detect vegetation stress, a hybrid variable. The variety of hybrid variables is large; consequently, no attempt is made to identify them. It is important to point out, however, that nominal-scale land use and land cover are hybrid variables. For example, the land cover of a particular area on an image may be derived by evaluating several of the fundamental biophysical variables at one time [e.g., object location (x, y), height (z), tone and/or color, biomass, and perhaps temperature]. So much attention has been placed on remotely sensing this hybrid *nominal*-scale variable that the *interval*- or *ratio*-scaled biophysical variables were largely neglected until the mid-1980s. Nominal-scale land-use and land-cover mapping are important capabilities of remote sensing technology and should not be minimized. Many social and physical scientists routinely use such data in their research. However, there is now a dramatic increase in the extraction of interval- and ratio-scaled biophysical data that are incorporated into quantitative models that can accept spatially distributed information (e.g., NASA, 1998; King, 2003).

Table 1-2.   Biophysical and selected hybrid variables and potential remote sensing systems used to obtain the information.

Biophysical Variable	Potential Remote Sensing System
x, y, z **Geodetic Control** *x, y* **Location from Orthocorrected Imagery**	- Global Positioning Systems (GPS) - Analog and digital stereoscopic aerial photography, Space Imaging IKONOS, DigitalGlobe QuickBird, Orbimage OrbView-3, French SPOT HRV, Landsat (Thematic Mapper, Enhanced Thematic Mapper Plus), Indian IRS-1CD, European ERS-1 and 2 microwave, Canadian RADARSAT, LIDAR
z **Elevation** - Digital Elevation Model (DEM) - Digital Bathymetric Model (DBM)	- GPS, stereoscopic aerial photography, LIDAR, SPOT, RADARSAT, IKONOS, QuickBird, OrbView-3, Shuttle Radar Topography Mission (SRTM), Interferometric Synthetic Aperture Radar (IFSAR) - SONAR, bathymetric LIDAR, stereoscopic aerial photography
Vegetation - Pigments (e.g., chlorophyll *a* and *b*) - Canopy structure and height - Biomass derived from vegetation indices - Leaf area index (LAI) - Absorbed photosynthetically active radiation (APAR) - Evapotranspiration	- Color aerial photography, Landsat ETM$^+$, IKONOS, QuickBird, OrbView-3, Orbimage SeaWiFS, Advanced Spaceborne Thermal Emission and Reflection Radiometer (ASTER), Moderate Resolution Imaging Spectrometer (MODIS), airborne hyperspectral systems (e.g., AVIRIS, HyMap, CASI) - Large-scale stereoscopic aerial photography, LIDAR, RADARSAT, IFSAR - Color-infrared (CIR) aerial photography, Landsat (TM, ETM$^+$), IKONOS, QuickBird, OrbView-3, Advanced Very High Resolution Radiometer (AVHRR), Multiangle Imaging Spectroradiometer (MISR), airborne hyperspectral systems (e.g., AVIRIS, HyMap, CASI)
Surface Temperature (land, water, atmosphere)	- ASTER, AVHRR, Geostationary Operational Environmental Satellite (GOES), Hyperion, MODIS, SeaWiFS, airborne thermal infrared sensors
Soil and Rocks - Moisture - Mineral composition - Taxonomy - Hydrothermal alteration	- ASTER, passive microwave (SSM/1), RADARSAT, MISR, ALMAZ, Landsat (TM, ETM$^+$), ERS-1 and 2, Intermap Star 3*i* - ASTER, MODIS, hyperspectral systems (e.g., AVIRIS, HyMap, CASI) - High-resolution color and CIR aerial photography, airborne hyperspectral systems (e.g., AVIRIS, HyMap, CASI) - Landsat (TM, ETM$^+$), ASTER, MODIS, airborne hyperspectral systems (e.g., AVIRIS, HyMap, CASI)
Surface Roughness	- Aerial photography, ALMAZ, ERS-1 and 2, RADARSAT, Intermap Star 3*i*, IKONOS, QuickBird, ASTER, Envisat ASAR
Atmosphere - Aerosols (e.g., optical depth) - Clouds (e.g., fraction, optical thickness) - Precipitation - Water vapor (precipitable water) - Ozone	- MISR, GOES, AVHRR, MODIS, CERES, MOPITT - GOES, AVHRR, MODIS, MISR, CERES, MOPITT, UARS - Tropical Rainfall Measurement Mission (TRMM), GOES, AVHRR, passive microwave (SSM/1) - GOES, MODIS - MODIS
Water - Color - Surface hydrology - Suspended minerals - Chlorophyll/gelbstoffe - Dissolved organic matter	- Color and CIR aerial photography, Landsat (TM, ETM$^+$), SPOT, IKONOS, QuickBird, OrbView-3, ASTER, SeaWiFS, MODIS, airborne hyperspectral systems (e.g., AVIRIS, HyMap, CASI), AVHRR, GOES, bathymetric LIDAR, MISR, CERES, Hyperion, TOPEX/POSEIDON, Envisat-1 MERIS

Table 1-2.　Biophysical and selected hybrid variables and potential remote sensing systems used to obtain the information.

Biophysical Variable	Potential Remote Sensing System
Snow and Sea Ice - Extent and characteristics	- Color and CIR aerial photography, AVHRR, GOES, Landsat (TM, ETM⁺), SPOT, RADARSAT, SeaWiFS, IKONOS, QuickBird, ASTER, MODIS
Volcanic Effects - Temperature, gases	- ASTER, MISR, Hyperion, MODIS, airborne hyperspectral systems
BRDF (bidirectional reflectance distribution function)	- MISR, MODIS, CERES

Selected Hybrid Variables	Potential Remote Sensing System
Land Use - Commercial, residential, transportation, utilities, etc. - Cadastral (property) - Tax mapping	- Very high spatial resolution panchromatic, color and /or CIR stereoscopic aerial photography, high spatial resolution satellite imagery (<1 m: IKONOS, QuickBird, OrbView-3), SPOT (2.5 m), LIDAR, high spatial resolution hyperspectral systems (e.g., AVIRIS, HyMap, CASI)
Land Cover - Agriculture, forest, urban, etc.	- Color and CIR aerial photography, Landsat (MSS, TM, ETM⁺), SPOT, ASTER, AVHRR, RADARSAT, IKONOS, QuickBird, OrbView-3, LIDAR, IFSAR, SeaWiFS, MODIS, MISR, hyperspectral systems (e.g., AVIRIS, HyMap, CASI)
Vegetation - stress	- Color and CIR aerial photography, Landsat (TM, ETM⁺), IKONOS, Quick-Bird, OrbView-3, AVHRR, SeaWiFS, MISR, MODIS, ASTER, airborne hyperspectral systems (e.g., AVIRIS, HyMap, CASI)

Remote Sensing Data Collection

Remotely sensed data are collected using passive or active remote sensing systems. *Passive* sensors record electromagnetic radiation that is reflected or emitted from the terrain (Shippert, 2004). For example, cameras and video recorders can be used to record visible and near-infrared energy reflected from the terrain. A multispectral scanner can be used to record the amount of thermal radiant flux exiting the terrain. *Active* sensors such as microwave (RADAR) or sonar bathe the terrain in machine-made electromagnetic energy and then record the amount of radiant flux scattered back toward the sensor system.

Remote sensing systems collect analog (e.g., hard-copy aerial photography or video data) and/or digital data [e.g., a matrix (raster) of brightness values obtained using a scanner, linear array, or area array]. A selected list of some of the most important remote sensing systems is presented in Table 1-3.

The amount of electromagnetic radiance, L (watts m^{-2} sr^{-1}; watts per meter squared per steradian) recorded within the IFOV of an optical remote sensing system (e.g., a picture element in a digital image) is a function of

$$L = f\,(\lambda,\, s_{x,\,y,\,z},\, t,\, \theta,\, P, \Omega) \qquad (1\text{-}2)$$

where

λ = wavelength (spectral response measured in various bands or at specific frequencies). [Wavelength (λ) and frequency (υ) may be used interchangeably based on their relationship with the speed of light (c) where $c = \lambda \times \upsilon$.];

$s_{x,y,z}$ = x, y, z location of the picture element and its size (x, y);

t = temporal information, i.e., when and how often the information was acquired,

θ = set of angles that describe the geometric relationships among the radiation source (e.g., the Sun), the terrain target of interest (e.g., a corn field), and the remote sensing system;

P = polarization of back-scattered energy recorded by the sensor; and

Table 1-3. Selected remote sensing systems and major characteristics.

Remote Sensing Systems	Blue	Green	Red	Near-infrared	Middle-infrared (SWIR)	Thermal infrared	Micro-wave	Spatial (m)	Temporal (days)
Suborbital Sensors									
Panchromatic film (black & white)		0.5 ——— 0.7 µm						Variable	Variable
Color film	0.4 ——— 0.7 µm							Variable	Variable
Color-infrared film		0.5 ——— 0.9 µm						Variable	Variable
Digital Frame Cameras (CCD)	1	1	1	1	—	—	—	0.25 – 5	Variable
ATLAS - Airborne Terrestrial Applications Sensor	0.45 ——8 bands——2.35 µm					6	—	2.5 – 25	Variable
AVIRIS - Airborne Visible Infrared Imaging Spectrometer	0.41——224 bands——2.5 µm							2.5 or 20	Variable
Intermap Star-3*i* X-band radar							1	Variable	Variable
Satellite Sensors									
NOAA-9 AVHRR LAC	—	—	1	1	—	3	—	1100	14.5/day
NOAA- K, L, M	—	—	1	1	2	2	—	1100	14.5/day
Landsat Multispectral Scanner (MSS)	—	1	1	2	—	—	—	79	16 – 18
Landsat 4 and 5 Thematic Mappers (TM)	1	1	1	1	2	1	—	30 and 120	16
Landsat 7 Enhanced TM (ETM⁺) — Multispectral	1	1	1	1	2	1	—	30 and 60	16
— Panchromatic	—	0.52 ——— 0.9 µm		—	—	—	—	15	16
SPOT 4 HRV — Multispectral	—	1	1	1	—	—	—	20	Pointable
— Panchromatic		0.51 ——— 0.73 µm		—	—	—	—	10	Pointable
GOES Series (East and West)	—	0.52 ——— 0.72 µm		—		4	—	700	0.5/hr
European Remote Sensing Satellite (ERS-1 and 2)	VV polarization C-band (5.3 GHz)						1	26 – 28	—
Canadian RADARSAT (several modes)	HH polarization C-band (5.3 GHz)						1	9 – 100	1 – 6 days
Shuttle Imaging Radar (SIR-C)	—	—	—	—	—	—	3	30	Variable
Sea-Viewing Wide Field-of-View Sensor (SeaWiFS)	3	2	1	2	—	—	—	1130	1
MODIS - Moderate Resolution Imaging Spectrometer	0.405 ——————— 36 bands ——————— 14.385 µm							250, 500, 1000	1 – 2
ASTER - Advanced Spaceborne Thermal Emission and Reflection Radiometer	0.52 — 3 bands — 0.86 µm							15	5
	1.6 – 6 bands – 2.43 µm							30	16
	8.12 – 5 bands – 11.6 µm							90	16
MISR - Multiangle Imaging SpectroRadiometer	Nine CCD cameras in four bands (440, 550, 670, 860 nm)							275 and 1100	1 – 2
NASA Topex/Poseidon — TOPEX radar altimeter	(18, 21, 37 GHz)							315,000	10
— POSEIDON single-frequency radiometer	(13.65 GHz)								
Space Imaging IKONOS — Multispectral	1	1	1	1			—	4	Pointable
— Panchromatic	0.45 ——— 0.9 µm				—	—	—	1	
Digital Globe QuickBird — Multispectral	1	1	1	1			—	2.44	Pointable
— Panchromatic	0.45 ——— 0.9 µm				—	—	—	0.61	

Ω = radiometric resolution (precision) at which the data (e.g., reflected, emitted, or back-scattered radiation) are recorded by the remote sensing system.

It is useful to briefly review characteristics of the parameters associated with Equation 1-2 and how they influence the nature of the remote sensing data collected.

Spectral Information and Resolution

Most remote sensing investigations are based on developing a deterministic relationship (i.e., a model) between the amount of electromagnetic energy reflected, emitted, or back-scattered in specific bands or frequencies and the chemical, biological, and physical characteristics of the phenomena under investigation (e.g., a corn field canopy). *Spectral resolution* is the number and dimension (size) of specific wavelength intervals (referred to as *bands* or *channels*) in the electromagnetic spectrum to which a remote sensing instrument is sensitive.

Multispectral remote sensing systems record energy in multiple bands of the electromagnetic spectrum. For example, in the 1970s and early 1980s, the Landsat Multispectral Scanners (MSS) recorded remotely sensed data of much of the Earth that is still of significant value for change detection studies. The bandwidths of the four MSS bands are displayed in Figure 1-5a (band 1 = 500 – 600 nm; band 2 = 600 – 700 nm; band 3 = 700 – 800 nm; and band 4 = 800 – 1100 nm). The nominal size of a band may be large (i.e., coarse), as with the Landsat MSS near-infrared band 4 (800 – 1100 nm) or relatively smaller (i.e., finer), as with the Landsat MSS band 3 (700 – 800 nm). Thus, Landsat MSS band 4 detectors recorded a relatively large range of reflected near-infrared radiant flux (300 nm wide) while the MSS band 3 detectors recorded a much reduced range of near-infrared radiant flux (100 nm wide).

The four multispectral bandwidths associated with the Positive Systems ADAR 5500 digital frame camera are shown for comparative purposes (Figure 1-5 a and d). The camera's bandwidths were refined to record information in more specific regions of the spectrum (band 1 = 450 – 515 nm; band 2 = 525 – 605 nm; band 3 = 640 – 690 nm; and band 4 = 750 – 900 nm). There are gaps between the spectral sensitivities of the detectors. Note that this digital camera system is also sensitive to reflected blue wavelength energy.

The aforementioned terminology is typically used to describe a sensor's *nominal spectral resolution*. Unfortu-

nately, it is difficult to create a detector that has extremely sharp bandpass boundaries such as those shown in Figure 1-5a. Rather, the more precise method of stating bandwidth is to look at the typical Gaussian shape of the detector sensitivity, such as the example shown in Figure 1-5b. The analyst then determines the Full Width at Half Maximum (FWHM). In this hypothetical example, the Landsat MSS near-infrared band 3 under investigation is sensitive to energy between 700 – 800 nm.

A *hyperspectral* remote sensing instrument acquires data in hundreds of spectral bands (Goetz, 2002). For example, the Airborne Visible and Infrared Imaging Spectrometer (AVIRIS) has 224 bands in the region from 400 to 2500 nm spaced just 10 nm apart based on the FWHM criteria (Clark, 1999). An AVIRIS hyperspectral datacube of a portion of the Savannah River Site near Aiken, SC, is shown in Figure 1-6. *Ultraspectral* remote sensing involves data collection in many hundreds of bands (Belokon, 1997).

Certain regions or spectral bands of the electromagnetic spectrum are optimum for obtaining information on biophysical parameters. The bands are normally selected to maximize the contrast between the object of interest and its background (i.e., object-to-background contrast). Careful selection of the spectral bands might improve the probability that the desired information will be extracted from the remote sensor data.

Spatial Information and Resolution

Most remote sensing studies record the spatial attributes of objects on the terrain. For example, each silver halide crystal in an aerial photograph and each picture element in a digital remote sensor image is located at a specific location in the image and associated with specific *x,y* coordinates on the ground. Once rectified to a standard map projection, the spatial information associated with each silver halide crystal or pixel is of significant value because it allows the remote sensing–derived information to be used with other spatial data in a GIS or spatial decision support system (Jensen et al., 2002).

There is a general relationship between the size of an object or area to be identified and the spatial resolution of the remote sensing system. *Spatial resolution* is a measure of the smallest angular or linear separation between two objects that can be resolved by the remote sensing system. The spatial resolution of aerial photography may be measured by 1) placing calibrated, parallel black and white lines on tarps that

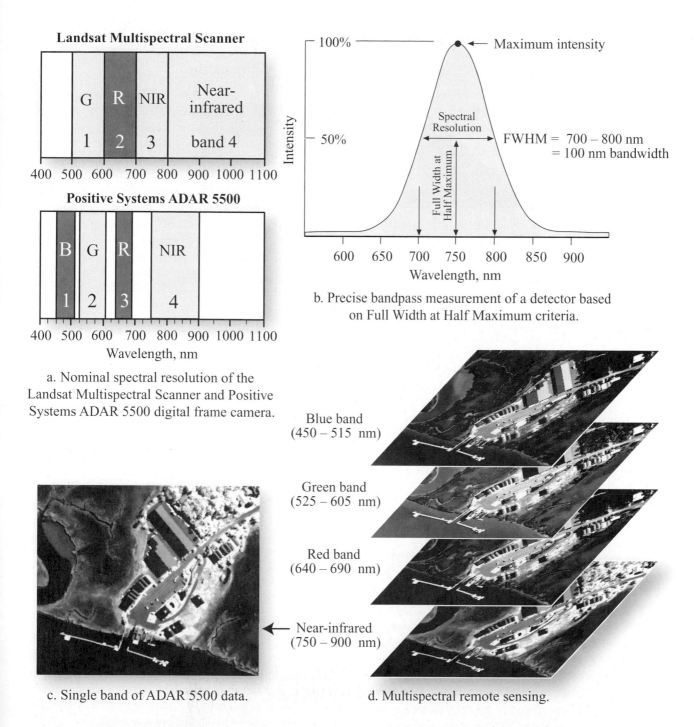

a. Nominal spectral resolution of the Landsat Multispectral Scanner and Positive Systems ADAR 5500 digital frame camera.

b. Precise bandpass measurement of a detector based on Full Width at Half Maximum criteria.

c. Single band of ADAR 5500 data.

d. Multispectral remote sensing.

Figure 1-5   a) The spectral bandwidths of the four Landsat Multispectral Scanner (MSS) bands (green, red, and two near-infrared) compared with the bandwidths of an ADAR 5500 digital frame camera. b) The true spectral bandwidth is the width of the Gaussian-shaped spectral profile at Full Width at Half Maximum (FWHM) intensity (Clark, 1999). This example has a spectral bandwidth of 100 nm between 700 and 800 nm. c) This is a 1 × 1 ft spatial resolution ADAR 5500 near-infrared image of an area in the Ace Basin in South Carolina. d) Multispectral remote sensing instruments such as the ADAR 5500 collect data in multiple bands of the electromagnetic spectrum (images acquired by Positive Systems, Inc.).

Airborne Visible/Infrared Imaging Spectrometer Datacube of the Mixed Waste Management Facility on the Savannah River Site near Aiken, SC

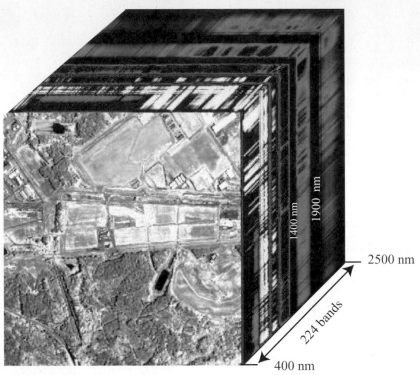

Near-infrared image on top of the datacube is just one of 224 bands at 10 nm nominal bandwidth acquired on July 26, 1999.

Figure 1-6   Hyperspectral remote sensing of an area on the Savannah River Site in South Carolina using NASA's Airborne Visible/Infrared Imaging Spectrometer (AVIRIS). The nominal spatial resolution is 3.4 × 3.4 m. The atmosphere absorbs most of the electromagnetic energy near 1400 and 1900 nm, causing the dark bands in the hyperspectral datacube (imagery collected by NASA Jet Propulsion Laboratory).

are placed in the field, 2) obtaining aerial photography of the study area, and 3) computing the number of resolvable *line pairs per millimeter* in the photography. It is also possible to determine the spatial resolution of imagery by computing its modulation transfer function, which is beyond the scope of this text (Joseph, 2000).

Many satellite remote sensing systems use optics that have a constant IFOV (Figure 1-1). Therefore, a sensor system's *nominal spatial resolution* is defined as the dimension in meters (or feet) of the ground-projected IFOV where the diameter of the circle (D) on the ground is a function of the instantaneous-field-of-view (β) times the altitude (H) of the sensor above ground level (AGL) (Figure 1-1):

$$D = \beta \times H. \tag{1-3}$$

Pixels are normally represented on computer screens and in hard-copy images as rectangles with length and width. Therefore, we typically describe a sensor system's nominal spatial resolution as being 10 × 10 m or 30 × 30 m. For example, DigitalGlobe's QuickBird has a nominal spatial resolution of 61 × 61 cm for its panchromatic band and 2.44 × 2.44 m for the four multispectral bands. The Landsat 7 Enhanced Thematic Mapper Plus (ETM^+) has a nominal spatial resolution of 15 × 15 m for its panchromatic band and 30 × 30 m for six of its multispectral bands. Generally, the smaller the nominal spatial resolution, the greater the spatial resolving power of the remote sensing system.

Imagery of Harbor Town in Hilton Head, SC, at Various Nominal Spatial Resolutions

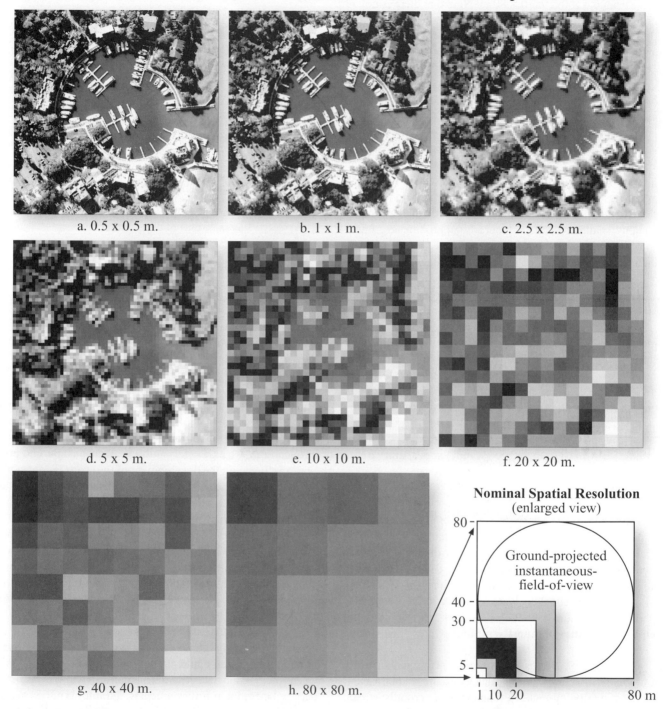

a. 0.5 x 0.5 m.   b. 1 x 1 m.   c. 2.5 x 2.5 m.

d. 5 x 5 m.   e. 10 x 10 m.   f. 20 x 20 m.

g. 40 x 40 m.   h. 80 x 80 m.

Nominal Spatial Resolution
(enlarged view)

Ground-projected instantaneous-field-of-view

Figure 1-7   Harbor Town, SC, images were obtained at a nominal spatial resolution of 0.3 × 0.3 m (approximately 1 × 1 ft) using a digital frame camera (courtesy Emerge, Inc.). The original data were resampled to derive the imagery with the simulated spatial resolutions shown.

Figure 1-7 depicts digital camera imagery of an area centered on Harbor Town in Hilton Head, SC, at spatial resolutions ranging from 0.5 × 0.5 to 80 × 80 m. Note that there is not a significant difference in the interpretability of the 0.5 × 0.5 m, 1 × 1 m, and even 2.5 x 2.5 m data. However, the urban information content decreases rapidly when using 5 × 5 m imagery and is practically useless for urban analysis at spatial resolutions >10 × 10 m. This is the reason historical Landsat MSS data (79 × 79 m) are of little value for most urban applications.

A useful heuristic rule of thumb is that in order to detect a feature, the nominal spatial resolution of the sensor system should be less than one-half the size of the feature measured in its smallest dimension. For example, if we want to identify the location of all sycamore trees in a city park, the minimum acceptable spatial resolution would be approximately one-half the diameter of the smallest sycamore tree's crown. Even this spatial resolution, however, will not guarantee success if there is no difference between the spectral response of the sycamore tree (the object) and the soil or grass surrounding it (i.e., its background).

Some sensor systems, such as LIDAR, do not completely "map" the terrain surface. Rather, the surface is "sampled" using laser pulses at some nominal time interval. The ground-projected laser pulse may be very small (e.g., 10 – 15 cm in diameter) with samples approximately every 1 to 6 m on the ground. Spatial resolution would appropriately describe the ground-projected laser pulse (e.g., 15 cm) but *sampling density* (i.e., number of points per unit area) describes the frequency of ground observations.

Because we have spatial information about the location of each pixel (*x,y*) in the matrix, it is also possible to examine the spatial relationship between a pixel and its neighbors. Therefore, the amount of spectral autocorrelation and other spatial geostatistical measurements can be determined based on the spatial information inherent in the imagery (Walsh et al., 1999). Geostatistical analysis is introduced in Chapter 4.

Temporal Information and Resolution

One of the valuable things about remote sensing science is that it obtains a record of the Earth's landscape and/or atmosphere at a unique moment in time. Multiple records of the same landscape obtained through time can be used to identify processes at work and to make predictions.

The *temporal resolution* of a remote sensing system generally refers to how often the sensor records imagery of a par-

Temporal Resolution

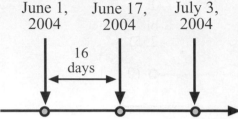

Figure 1-8   The temporal resolution of a remote sensing system refers to how often it records imagery of a particular area. This example depicts the systematic collection of data every 16 days, presumably at approximately the same time of day. Landsat Thematic Mappers 4 and 5 had 16-day revisit cycles.

ticular area. For example, the temporal resolution of the sensor system shown in Figure 1-8 is every 16 days. Ideally, the sensor obtains data repetitively to capture unique discriminating characteristics of the object under investigation (Haack et al., 1997). For example, agricultural crops have unique phenological cycles in each geographic region. To measure specific agricultural variables, it is necessary to acquire remotely sensed data at critical dates in the phenological cycle (Jensen, 2000). Analysis of multiple-date imagery provides information on how the variables are changing through time. Change information provides insight into processes influencing the development of the crop (Jensen et al., 2002). Fortunately, several satellite sensor systems such as SPOT, IKONOS, and QuickBird are pointable, meaning that they can acquire imagery off-nadir. *Nadir* is the point directly below the spacecraft. This dramatically increases the probability that imagery will be obtained during a growing season or during an emergency. However, the off-nadir oblique viewing also introduces bidirectional reflectance distribution function (BRDF) issues, discussed in the next section (Jensen, 2000).

Obtaining imagery at a high temporal resolution is very important for many applications. For example, the National Ocean and Atmospheric Administration (NOAA) Geostationary Operational Environmental Satellites (GOES) are in geostationary orbits, allowing them to obtain very high temporal resolution imagery (e.g., every half-hour). This allows meteorologists to provide hourly updates on the location of frontal systems and hurricanes and use this information along with other data to predict storm tracks.

Radiometric Resolution

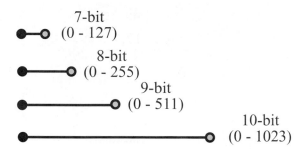

7-bit
(0 - 127)

8-bit
(0 - 255)

9-bit
(0 - 511)

10-bit
(0 - 1023)

Figure 1-9   The radiometric resolution of a remote sensing system is the sensitivity of its detectors to differences in signal strength as they record the radiant flux reflected, emitted, or back-scattered from the terrain. The energy is normally quantized during an analog-to-digital (A-to-D) conversion process to 8, 9, 10 bits or more.

Another aspect of temporal information is how many observations are recorded from a single pulse of energy that is directed at the Earth by an active sensor such as LIDAR. For example, most LIDAR sensors emit one pulse of laser energy and record multiple responses from this pulse. Measuring the time differences between multiple responses allows for the determination of object heights and terrain structure. Also, the length of time required to emit an energy signal by an active sensor is referred to as the *pulse length*. Short pulse lengths allow very precise distance (i.e., range) measurement.

Radiometric Information and Resolution

Some remote sensing systems record the reflected, emitted, or back-scattered electromagnetic radiation with more precision than other sensing systems. This is analogous to making a measurement with a ruler. If you want precisely to measure the length of an object, would you rather use a ruler with 16 or 1,024 subdivisions on it?

Radiometric resolution is defined as the sensitivity of a remote sensing detector to differences in signal strength as it records the radiant flux reflected, emitted, or back-scattered from the terrain. It defines the number of just discriminable signal levels. Therefore, radiometric resolution can have a significant impact on our ability to measure the properties of scene objects. The original Landsat 1 Multispectral Scanner launched in 1972 recorded reflected energy with a precision of 6-bits (values ranging from 0 to 63). Landsat 4 and 5 Thematic Mapper sensors launched in 1982 and 1984, respec-

tively, recorded data in 8 bits (values from 0 to 255) (Figure 1-9). Thus, the Landsat TM sensors had improved radiometric resolution (sensitivity) when compared with the original Landsat MSS. QuickBird and IKONOS sensors record information in 11 bits (values from 0 to 2,047). Several new sensor systems have 12-bit radiometric resolution (values ranging from 0 to 4,095). Radiometric resolution is sometimes referred to as the level of *quantization*. High radiometric resolution generally increases the probability that phenomena will be remotely sensed more accurately.

Polarization Information

The polarization characteristics of electromagnetic energy recorded by a remote sensing system represent an important variable that can be used in many Earth resource investigations (Curran et al., 1998). Sunlight is polarized weakly. However, when sunlight strikes a nonmetal object (e.g., grass, forest, or concrete) it becomes depolarized and the incident energy is scattered differentially. Generally, the more smooth the surface, the greater the polarization. It is possible to use polarizing filters on passive remote sensing systems (e.g., aerial cameras) to record polarized light at various angles. It is also possible to selectively send and receive polarized energy using active remote sensing systems such as RADAR (e.g., horizontal send, vertical receive - HV; vertical send, horizontal receive - VH; vertical send, vertical receive - VV; horizontal send, horizontal receive - HH). Multiple-polarized RADAR imagery is an especially useful application of polarized energy.

Angular Information

Remote sensing systems record very specific angular characteristics associated with each exposed silver halide crystal or pixel (Barnsley, 1999). The angular characteristics are a function of (See Figure 1-10a):

- the location in a three-dimensional sphere of the illumination source (e.g., the Sun for a passive system or the sensor itself in the case of RADAR, LIDAR, and SONAR) and its associated azimuth and zenith angles,

- the orientation of the terrain facet (pixel) or terrain cover (e.g., vegetation) under investigation, and

- the location of the suborbital or orbital remote sensing system and its associated azimuth and zenith angles.

There is always an angle of incidence associated with the incoming energy that illuminates the terrain and an angle of exitance from the terrain to the sensor system. This *bidirec-*

Bidirectional Reflectance Distribution Function

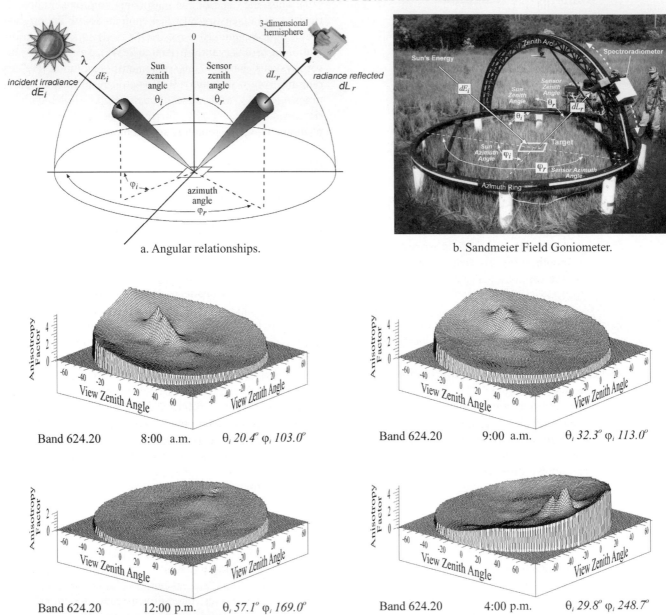

a. Angular relationships.

b. Sandmeier Field Goniometer.

Band 624.20   8:00 a.m. θ_i 20.4° φ_i 103.0°

Band 624.20   9:00 a.m. θ_i 32.3° φ_i 113.0°

Band 624.20   12:00 p.m. θ_i 57.1° φ_i 169.0°

Band 624.20   4:00 p.m. θ_i 29.8° φ_i 248.7°

c. Comparison of hourly three-dimensional plots of BRDF for smooth cordgrass (*Spartina alterniflora*) data collected at 8 a.m., 9 a.m., 12 p.m., and 4 p.m. at the boardwalk site on March 21 – 22, 2000, for band 624.20 nm.

Figure 1-10   a) The concepts and parameters of the bidirectional reflectance distribution function (BRDF). A target is bathed in irradiance (dE_i) from a specific Sun zenith and azimuth angle, and the sensor records the radiance (dL_r) exiting the target of interest at a specific azimuth and zenith angle. b) The Sandmeier Field Goniometer collecting smooth cordgrass (*Spartina alterniflora*) BRDF measurements at North Inlet, SC. Spectral measurements are made at Sun zenith angle of θ_i and Sun azimuth angle of φ_i and a sensor zenith angle of view of θ_r and sensor azimuth angle of φ_r. A GER 3700 spectroradiometer, attached to the moving sled mounted on the zenith arc, records the amount of radiance leaving the target in 704 bands at 76 angles (Sandmeier, 2000; Jensen and Schill, 2000). c) Hourly three-dimensional plots of BRDF data.

tional nature of remote sensing data collection is known to influence the spectral and polarization characteristics of the at-sensor radiance, *L,* recorded by the remote sensing system.

A *goniometer* can be used to document the changes in at-sensor radiance, *L,* caused by changing the position of the sensor and/or the source of the illumination (e.g., the Sun) (Figure 1-10b). For example, Figure 1-10c presents three-dimensional plots of smooth cordgrass (*Spartina alterniflora*) BRDF data collected at 8 am, 9 am, 12 pm, and 4 pm on March 21, 2000, for band 624.20 nm. The only thing that changed between observations was the Sun's azimuth and zenith angles. The azimuth and zenith angles of the spectroradiometer were held constant while viewing the smooth cordgrass. Ideally, the BRDF plots would be identical, suggesting that it does not matter what time of day we collect the remote sensor data because the spectral reflectance characteristics from the smooth cordgrass remain constant. It is clear that this is not the case and that the time of day influences the spectral response. The Multiangle Imaging Spectrometer (MISR) onboard the *Terra* satellite was designed to investigate the BRDF phenomena. Research continues on how to incorporate the BRDF information into the digital image processing system to improve our understanding of what is recorded in the remotely sensed imagery (Sandmeier, 2000; Jensen and Schill, 2000).

Angular information is central to the use of remote sensor data in photogrammetric applications. Stereoscopic image analysis is based on the assumption that an object on the terrain is remotely sensed from two angles. Viewing the same terrain from two vantage points introduces stereoscopic parallax, which is the foundation for all stereoscopic photogrammetric and radargrammetric analysis (Light and Jensen, 2002).

Suborbital (Airborne) Remote Sensing Systems

High-quality photogrammetric cameras mounted onboard aircraft continue to provide aerial photography for many Earth resource applications. For example, the U.S. Geological Survey's National Aerial Photography Program (NAPP) systematically collects 1:40,000-scale black-and-white or color-infrared aerial photography of much of the United States every 5 to 10 years. Some of these photogrammetric data are now being collected using digital frame cameras. In addition, sophisticated remote sensing systems are routinely mounted on aircraft to provide high spatial and spectral resolution remotely sensed data. Examples include hyperspectral sensors such as NASA's AVIRIS, the Canadian Airborne Imaging Spectrometer (CASI-3), and the Australian HyMap hyperspectral system. These sensors can collect data on

demand when disaster strikes (e.g., oil spills or floods) if cloud-cover conditions permit. There are also numerous radars, such as Intermap's Star-3*i*, that can be flown on aircraft day and night and in inclement weather. Unfortunately, suborbital remote sensor data are usually expensive to acquire per km^2. Also, atmospheric turbulence can cause the data to have severe geometric distortions that can be difficult to correct.

Current and Proposed Satellite Remote Sensing Systems

Remote sensing systems onboard satellites provide high-quality, relatively inexpensive data per km^2. For example, the European Remote Sensing satellites (ERS-1 and 2) collect 26 $\times$ 28 m spatial resolution C-band active microwave (radar) imagery of much of Earth, even through clouds. Similarly, the Canadian Space Agency RADARSAT obtains C-band active microwave imagery. The United States has progressed from multispectral scanning systems (Landsat MSS launched in 1972) to more advanced scanning systems (Landsat 7 Enhanced Thematic Mapper Plus in 1999). The Land Remote Sensing Policy Act of 1992 specified the future of satellite land remote sensing programs in the United States (Asker, 1992; Jensen, 1992). Unfortunately, Landsat 6 with its Enhanced Thematic Mapper did not achieve orbit when launched on October 5, 1993. Landsat 7 was launched on April 15, 1999, to relieve the United States' land remote sensing data gap (Henderson, 1994). Meanwhile, the French have pioneered the development of linear array remote sensing technology with the launch of SPOT satellites 1 through 5 in 1986, 1990, 1993, 1998, and 2002.

The International Geosphere–Biosphere Program (IGBP) and the United States Global Change Research Program (USGCRP) call for scientific research to describe and understand the interactive physical, chemical, and biological processes that regulate the total Earth system. Space-based remote sensing is an integral part of these research programs because it provides the only means of observing global ecosystems consistently and synoptically. The National Aeronautics and Space Administration (NASA) Earth Science Enterprise (ESE) is the name given to the coordinated plan to provide the necessary satellite platforms and instruments and an Earth Observing System Data and Information System (EOSDIS), and related scientific research for IGBP. The Earth Observing System (EOS) is a series of Earth-orbiting satellites that will provide global observations for 15 years or more. The first satellites were launched in the late 1990s. EOS is complemented by missions and instruments from international partners. For example, the Tropical Rainfall Mapping Mission (TRMM) is a joint NASA/Japanese mission.

EOS Science Plan: Asrar and Dozier (1994) conceptualized the remote sensing science conducted as part of the Earth Science Enterprise. They suggested that the Earth consists of two subsystems, 1) the physical climate, and 2) biogeochemical cycles, linked by the global hydrologic cycle, as shown in Figure 1-11.

The *physical climate* subsystem is sensitive to fluctuations in the Earth's radiation balance. Human activities have caused changes to the planet's radiative heating mechanism that rival or exceed natural change. Increases in greenhouse gases between 1765 and 1990 have caused a radiative forcing of 2.5 W m^{-2}. If this rate is sustained, it could result in global mean temperatures increasing about 0.2 to 0.5 C per decade during the next century. Volcanic eruptions and the ocean's ability to absorb heat may impact the projections. Nevertheless, the following questions are being addressed using remote sensing (Asrar and Dozier, 1994):

- How do clouds, water vapor, and aerosols in the Earth's radiation and heat budgets change with increased atmospheric greenhouse-gas concentrations?

- How do the oceans interact with the atmosphere in the transport and uptake of heat?

- How do land-surface properties such as snow and ice cover, evapotranspiration, urban/suburban land use, and vegetation influence circulation?

The Earth's *biogeochemical cycles* have also been changed by humans. Atmospheric carbon dioxide has increased by 30 percent since 1859, methane by more than 100 percent, and ozone concentrations in the stratosphere have decreased, causing increased levels of ultraviolet radiation to reach the Earth's surface. Global change research is addressing the following questions:

- What role do the oceanic and terrestrial components of the biosphere play in the changing global carbon budget?

- What are the effects on natural and managed ecosystems of increased carbon dioxide and acid deposition, shifting precipitation patterns, and changes in soil erosion, river chemistry, and atmospheric ozone concentrations?

The *hydrologic cycle* links the physical climate and biogeochemical cycles. The phase change of water between its gaseous, liquid, and solid states involves storage and release of latent heat, so it influences atmospheric circulation and globally redistributes both water and heat (Asrar and Dozier,

1994). The hydrologic cycle is the integrating process for the fluxes of water, energy, and chemical elements among components of the Earth system. Important questions to be addressed include these three:

- How will atmospheric variability, human activities, and climate change affect patterns of humidity, precipitation, evapotranspiration, and soil moisture?

- How does soil moisture vary in time and space?

- Can we predict changes in the global hydrologic cycle using present and future observation systems and models?

The EOS *Terra* satellite was launched on December 18, 1999. It contains five remote sensing instruments (MODIS, ASTER, MISR, CERES, and MOPITT) designed to address many of the research topics (King, 2003). The EOS *Aqua* satellite was launched in May, 2002. The Moderate Resolution Imaging Spectrometer (MODIS) has 36 bands from 0.405 to 14.385 μm that collect data at 250-m, 500-m, and 1-km spatial resolutions. MODIS views the entire surface of the Earth every 1 to 2 days, making observations in 36 spectral bands of land- and ocean-surface temperature, primary productivity, land-surface cover, clouds, aerosols, water vapor, temperature profiles, and fires (King, 2003).

The Advanced Spaceborne Thermal Emission and Reflection Radiometer (ASTER) has five bands in the thermal infrared region between 8 and 12 μm with 90-m pixels. It also has three broad bands between 0.5 and 0.9 μm with 15-m pixels and stereo capability, and six bands in the shortwave infrared region (1.6 – 2.5 μm) with 30-m spatial resolution. ASTER is the highest spatial resolution sensor system on the EOS *Terra* platform and provides information on surface temperature that can be used to model evapotranspiration.

The Multiangle Imaging SpectroRadiometer (MISR) has nine separate charge-coupled-device (CCD) pushbroom cameras to observe the Earth in four spectral bands and at nine view angles. It provides data on clouds, atmospheric aerosols, and multiple-angle views of the Earth's deserts, vegetation, and ice cover. The Clouds and the Earth's Radiant Energy System (CERES) consists of two scanning radiometers that measure the Earth's radiation balance and provide cloud property estimates to assess their role in radiative fluxes from the surface of the Earth to the top of the atmosphere. Finally, the Measurements of Pollution in the Troposphere (MOPITT) scanning radiometer provides information on the distribution, transport, sources, and sinks of carbon monoxide and methane in the troposphere.

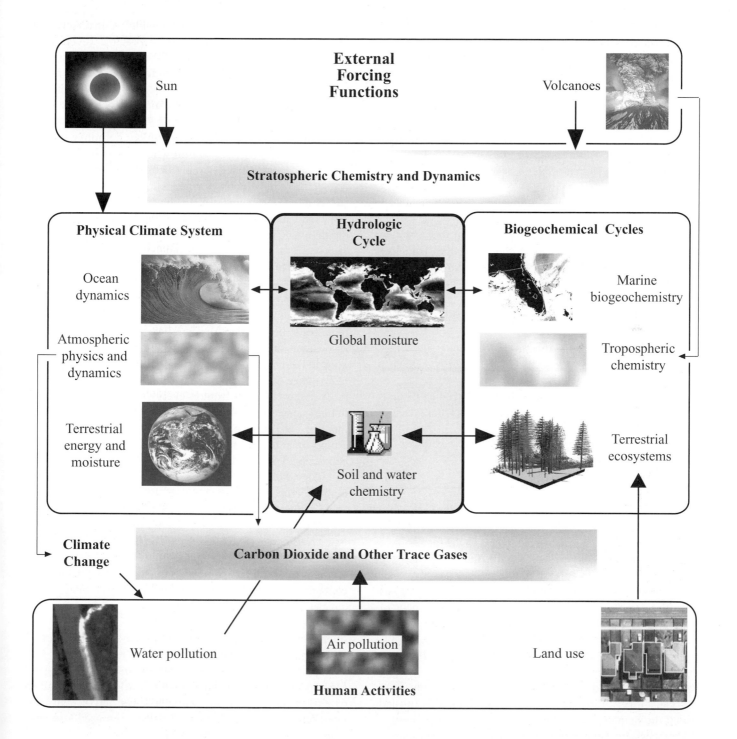

Figure 1-11   The Earth system can be subdivided into two subsystems—the physical climate system and biogeochemical cycles—that are linked by the global hydrologic cycle. Significant changes in the external forcing functions and human activities have an impact on the physical climate system, biogeochemical cycles, and the global hydrologic cycle. Examination of these subsystems and their linkages defines the critical questions that the NASA Earth Observing System (EOS) is attempting to answer (after Asrar and Dozier, 1994).

The National Polar-orbiting Operational Environmental Satellite System (NPOESS) Preparatory Project (NPP) to be launched in 2005 will extend key measurements of the EOS program in support of long-term monitoring of climate trends and global biological productivity until the NPOESS can be launched in 2008. The NPP will contain MODIS-like instruments such as the Visible Infrared Imaging Radiometer Suite (VIIRS). With a five-year design life NPP will provide data past the designed lifetimes of EOS *Terra* and *Aqua* satellites through the launch of NPOESS (King, 2003).

Commercial Vendors: Space Imaging, Inc. launched IKONOS-2 on September 24, 1999. The IKONOS-2 sensor system has a 1×1 m panchromatic band and four 4×4 m multispectral bands (Table 1-3). DigitalGlobe, Inc., launched QuickBird on October 18, 2001, with a 61×61 cm panchromatic band and four 2.44×2.44 m multispectral bands. Orbimage, Inc. launched OrbView-3 on June 26, 2003, with 1×1 m panchromatic and 4×4 m multispectral bands.

Remote Sensing Data Analysis

The analysis of remotely sensed data is performed using a variety of image processing techniques (Figure 1-12), including:

- analog (visual) image processing, and

- digital image processing.

Analog and digital analysis of remotely sensed data seek to detect and identify important phenomena in the scene. Once identified, the phenomena are usually measured, and the information is used in solving problems (Estes et al., 1983; Haack et al., 1997). Thus, both manual and digital analysis have the same general goals. However, the attainment of these goals may follow significantly different paths.

Human beings are adept at visually interpreting images produced by certain types of remote sensing devices, especially cameras. One could ask, "Why try to mimic or improve on this capability?" First, there are certain thresholds beyond which the human interpreter cannot detect "just noticeable differences" in the imagery. For example, it is commonly known that an analyst can discriminate only about nine shades of gray when interpreting continuous-tone, black-and-white aerial photography. If the data were originally recorded with 256 shades of gray, there might be more subtle information present in the image than the interpreter can extract visually. Furthermore, the interpreter brings to the task all the pressures of the day, making the interpretation

subjective and generally unrepeatable. Conversely, the results obtained by computer are repeatable (even when wrong!). Also, when it comes to keeping track of a great amount of detailed quantitative information, such as the spectral characteristics of a vegetated field throughout a growing season for crop identification purposes, the computer is very adept at storing and manipulating such tedious information and possibly making a more definitive conclusion as to what crop is being grown. This is not to say that digital image processing is superior to visual image analysis. Rather, there may be times when a digital approach is better suited to the problem at hand. Optimum results are often achieved using a synergistic combination of both visual and digital image processing (Kelly et al., 1999).

Analog (Visual) Image Processing

Human beings use the fundamental elements of image interpretation summarized in Figure 1-12, including grayscale tone, color, height (depth), size, shape, shadow, texture, site, association, and arrangement. The human mind is amazingly good at recognizing and associating these complex elements in an image or photograph because we constantly process (a) profile views of Earth features every day and (b) images seen in books, magazines, the television and the Internet. Furthermore, we are adept at bringing to bear all the knowledge in our personal background and collateral information. We then converge all this evidence to identify phenomena in images and judge their significance. Precise measurement of objects (length, area, perimeter, volume, etc.) may be performed using photogrammetric techniques applied to either monoscopic (single-photo) or stereoscopic (overlapping) images. Numerous books have been written on how to perform visual image interpretation and photogrammetric measurement. Please refer to the companion volume to review fundamental elements of image interpretation and photogrammetric principles (Jensen, 2000; Chapters 5 and 6).

Interestingly, there is a resurgence in the art and science of visual image interpretation as the digital remote sensor systems provide increasingly higher spatial resolution imagery. Many people are displaying IKONOS 1×1 m and QuickBird 61×61 cm imagery on the computer screen and then visually interpreting the data. The data are also often used as a base map in GIS projects (Clarke, 2001).

Digital Image Processing

Scientists have made significant advances in digital image processing of remotely sensed data for scientific visualization and hypothesis testing (e.g., Estes et al., 1980; Estes and Jensen, 1998; Townshend and Justice, 2002; Kraak, 2003).

Fundamental Image Analysis Tasks
• Detect, Identify, Measure
• Solve Problems

Application of the *Multi* concept
- Multispectral - Multifrequency - Multipolarization
- Multitemporal - Multiscale - Multidisciplinary

Use of *Collateral Information*
- Literature - Laboratory spectra - Dichotomous keys - Prior probabilities
- Field training sites - Field test sites - Soil maps - Surficial geology maps

Analog (Visual)
Image Processing

Digital
Image Processing

Elements of Image Interpretation	*How the Elements of Image Interpretation Are Extracted or Used in Digital Image Processing*
• Grayscale tone (black to white)	• 8- to 12-bit brightness values, or more appropriately scaled surface reflectance or emittance

• Color (red, green, blue = RGB)	• 24-bit color look-up table display
	  - Multiband RGB color composites
	  - Transforms (e.g., intensity, hue, saturation)

• Height (elevation) and depth	• Soft-copy photogrammetry, LIDAR, radargrammetry, RADAR interferometry, SONAR
• Size (length, area, perimeter, volume)	• Soft-copy photogrammetry, radargrammetry, RADAR interferometry, measurement from rectified images
• Shape	• Soft-copy photogrammetry, radargrammetry, RADAR interferometry, landscape ecology spatial statistics (metrics), object-oriented image segmentation
• Texture	• Texture transforms, geostatistical analysis (e.g., kriging), landscape ecology metrics, fractal analysis
• Pattern	• Autocorrelation, geostatistical analysis, landscape ecology metrics, fractal analysis
• Shadow	• Soft-copy photogrammetry, radargrammetry, measurement from rectified images
• Site, using convergence of evidence	• Contextual, expert system, neural network analysis
• Association, using convergence of evidence	• Contextual, expert system, neural network analysis
• Arrangement, using convergence of evidence	• Contextual, expert system, neural network analysis

Figure 1-12   Analog (visual) and digital image processing of remotely sensed data use the fundamental elements of image interpretation.

The methods are summarized in this book and others (e.g., Richards and Jia, 1999; Donnay et al., 2001; Bossler et al., 2002). Digital image processing now makes use of many elements of image interpretation using the techniques summarized in Figure 1-12. The major types of digital image processing include image preprocessing (radiometric and geometric correction), image enhancement, pattern recognition using inferential statistics, photogrammetric image processing of stereoscopic imagery, expert system (decision-tree) and neural network image analysis, hyperspectral data analysis, and change detection (Figure 1-4). Hardware and software characteristics of typical digital image processing systems are discussed in Chapter 3.

Radiometric Correction of Remote Sensor Data: Analog and digital remotely sensed imagery may contain noise or error that was introduced by the sensor system (e.g., electronic noise) or the environment (e.g., atmospheric scattering of light into the sensor's field of view). Advances have been made in our ability to remove these deleterious effects through simple image normalization techniques and more advanced absolute radiometric calibration of the data to scaled surface reflectance (for optical data). Calibrated remote sensor data allows imagery and derivative products obtained on different dates to be compared (e.g., to measure the change in leaf area index between two dates). Radiometric correction methods are discussed in Chapter 7.

Geometric Correction of Remote Sensor Data: Most analog and digital remote sensor data are now processed so that individual picture elements are in their proper planimetric positions in a standard map projection. This facilitates the use of the imagery and derivative products in GIS or spatial decision support systems. Geometric correction methods are discussed in Chapter 5.

Image Enhancement: Images can be digitally enhanced to identify subtle information in the analog or digital imagery that might otherwise be missed (Richards and Jia, 1999). Significant improvements have been made in our ability to contrast stretch and filter data to enhance low and high frequency components, edges, and texture in the imagery (e.g., Emerson et al., 1999). In addition, the remote sensor data can be linearly and nonlinearly transformed into information that is more highly correlated with real-world phenomena through principal components analysis and various vegetation indices (Townshend and Justice, 2002). Image enhancement is discussed in Chapter 6.

Photogrammetry: Significant advances have been made in the analysis of stereoscopic remote sensor data obtained from airborne or satellite platforms using computer workstations and digital image processing photogrammetric algorithms (e.g., Li, 1998; Adams and Chandler, 2002). Softcopy photogrammetric workstations can be used to extract accurate digital elevation models (DEMs) and differentially corrected orthophotography from the triangulated aerial photography or imagery (Light and Jensen, 2002). The technology is revolutionizing the way DEMs are collected and how orthophotos are produced for rural and urban–suburban applications. Fundamental photogrammetric analysis principles are discussed in Jensen (2000).

Parametric Information Extraction: Scientists attempting to extract land-cover information from remotely sensed data now routinely specify if the classification is to be:

- *hard*, with discrete mutually exclusive classes, or *fuzzy*, where the proportions of materials within pixels are extracted (Foody, 1996; Seong and Usery, 2001);

- based on individual pixels (referred to as a *per-pixel classification*) or if it will use object-oriented image segmentation algorithms that take into account not only the spectral characteristics of a pixel, but also the spectral characteristics of contextual surrounding pixels. Thus, the algorithms take into account spectral and spatial information (Herold et al., 2003; Hodgson et al., 2003; Tullis and Jensen, 2003).

Once these issues are addressed, it is a matter of determining whether to use parametric, nonparametric, and/or nonmetric classification techniques. Until recently, the maximum likelihood classification algorithm was the most widely adopted parametric classification algorithm. Unfortunately, the maximum likelihood algorithm requires normally distributed training data in *n* bands (rarely the case) for computing the class variance and covariance matrices. It is difficult to incorporate nonimage categorical data into a maximum likelihood classification. Fortunately, fuzzy maximum likelihood classification algorithms are now available (e.g., Foody, 1996). Chapter 9 introduces the fundamental principles associated with parametric information extraction.

Nonparametric Information Extraction: Nonparametric clustering algorithms, such as ISODATA, continue to be used extensively in digital image processing research. Unfortunately, such algorithms depend on how the seed training data are extracted and it is often difficult to label the clusters to turn them into information classes. For these reasons there has been a significant increase in the development and use of artificial neural networks (ANN) for remote sensing applications (e.g., Jensen et al., 1999; 2001). The ANN does not require normally distributed training data. ANN may incorporate virtually any type of spatially distributed data in the classification. The only drawback is that sometimes it is difficult to determine exactly how the ANN came up with a certain conclusion because the information is locked within the weights in the hidden layer(s). Scientists are working on ways to extract hidden information so that the rules used can be more formally stated. The ability of an ANN to *learn* should not be underestimated. Nonparametric techniques are discussed in Chapters 9 and 10.

Nonmetric Information Extraction: It is difficult to make a computer understand and use the heuristic rules of thumb and knowledge that a human expert uses when interpreting an image. Nevertheless, there has been considerable progress in the use of artificial intelligence (AI) to try to make computers do things that, at the moment, people do better. One area of AI that has great potential in remote sensing image analysis is the use of expert systems. Expert systems can be used to 1) interpret an image or 2) place all the information contained within an image in its proper context with ancillary data and extract more valuable information. Duda et al. (2001) describe various types of expert system decision-tree classifiers as *nonmetric*.

Parametric digital image classification techniques are based primarily on summary inferential statistics such as the mean, variance, and covariance matrices. Decision-tree or rule-based classifiers are not based on inferential statistics, but instead "let the data speak for itself" (Gahegan, 2003). In other words, the data retains its precision and is not dumbed down by summarizing it through means, etc. Decision-tree classifiers can process virtually any type of spatially distributed data and can incorporate prior probabilities (Defries et al., 1998; McIver and Friedl, 2002). There are three approaches to rule creation: 1) explicitly extracting knowledge and creating rules from experts, 2) implicitly extracting variables and rules using cognitive methods (Lloyd et al., 2002), and 3) empirically generating rules from observed data and automatic induction methods (Tullis and Jensen, 2003). The development of a decision tree using human-specified rules is time-consuming and difficult. However, it rewards the user with detailed information about how individual classification decisions were made (Zhang and Wang, 2003).

Ideally, computers can derive the rules from training data without human intervention. This is referred to as *machine-learning* (Huang and Jensen, 1997). The analyst identifies representative training areas. The machine learns the patterns from these training data, creates the rules, and uses them to classify the remotely sensed data. The rules are available to document how decisions were made. Chapter 10 introduces expert systems and neural network methods for extracting useful information from remote sensor data.

Hyperspectral: Special software is required to process hyperspectral data obtained by imaging spectroradiometers such as AVIRIS and MODIS. Kruse et al. (1992), Landgrebe and Biehl (2004), Digital Research Systems (2004) and others have pioneered the development of hyperspectral image analysis software. The software reduces the dimensionality of the data (number of bands) to a manageable degree, while retaining the essence of the data. Under certain conditions the software can be used to compare the remotely sensed spectral reflectance curves with a library of spectral reflectance curves. Analysts are also able to identify the type and proportion of different materials within an individual picture element (referred to as end-member spectral mixture analysis). Chapter 11 reviews hyperspectral data analysis methods.

Modeling Remote Sensing Data Using a GIS Approach: Remotely sensed data should not be analyzed in a vacuum without the benefit of collateral information such as soil maps, hydrology, and topography (Ramsey et al., 1995). For example, land-cover mapping using remotely sensed data has been significantly improved by incorporating topographic information from digital terrain models and other GIS data (e.g., Stow et al., 2003). GIS studies require timely, accurate updating of the spatially distributed variables in the database that remote sensing can provide (Clarke, 2001). Remote sensing can benefit from access to accurate ancillary information to improve classification accuracy and other types of modeling. Such synergy is critical if successful expert system and neural network analyses are to be performed (Tullis and Jensen, 2003). A framework for modeling the uncertainty between remote sensing and geographic information systems was developed by Gahegan and Ehlers (2000).

Scene Modeling: Strahler et al. (1986) describe a framework for modeling in remote sensing. Basically, a remote sensing model has three components: 1) a scene model, which specifies the form and nature of the energy and matter within the scene and their spatial and temporal order; 2) an atmospheric model, which describes the interaction between the atmosphere and the energy entering and being emitted from the scene; and 3) a sensor model, which describes the behavior of the sensor in responding to the energy fluxes incident on it and in producing the measurements that constitute the image. They suggest that the problem of scene inference, then, becomes a problem of model inversion in which the order in the scene is reconstructed from the image and the remote sensing model. For example, Woodcock et al. (1997) inverted the Li-Strahler Canopy Reflectance Model for mapping forest structure.

Basically, successful remote sensing modeling predicts how much radiant flux in certain wavelengths should exit a particular object (e.g., a conifer canopy) even without actually sensing the object. When the model's prediction is the same as the sensor's measurement, the relationship has been modeled correctly. The scientist then has a greater appreciation for energy–matter interactions in the scene and may be able to extend the logic to other regions or applications with confidence. The remote sensor data can then be used more effectively in physical deterministic models (e.g., watershed runoff, net primary productivity, and evapotranspiration models) that are so important for large ecosystem modeling. Recent work allows one to model the utility of sensors with different spatial resolutions for particular applications such as urban analysis (Collins and Woodcock, 1999).

Change Detection: Remotely sensed data obtained on multiple dates can be used to identify the type and spatial distribution of changes taking place in the landscape (Friedl et al., 2002; Zhan et al., 2002). The change information provides valuable insight into the *processes* at work. Change detection algorithms can be used on per-pixel and object-oriented (polygon) classifications. Unfortunately, there is still no universally accepted method of detecting change or of assessing the accuracy of change detection map products. Chapter 12 introduces change detection principles and methods.

Information Presentation

Information derived from remote sensor data are usually summarized as an enhanced image, image map, orthophotomap, thematic map, spatial database file, statistic, or graph (Figure 1-4). Thus, the final output products often require knowledge of remote sensing, cartography, GIS, and spatial statistics as well as the systematic science being investigated (e.g., soils, agriculture, forestry, wetland, urban studies). Scientists who understand the rules and synergistic relationships of the technologies can produce output products that communicate effectively. Conversely, those who violate fundamental rules (e.g., cartographic theory or database topology design) often produce poor output products that do not communicate effectively (Clarke, 2001).

Image maps offer scientists an alternative to line maps for many cartographic applications. Thousands of satellite image maps have been produced from Landsat MSS (1:250,000 and 1:500,000 scale), TM (1:100,000 scale) and AVHRR, and MODIS data. Image maps at scales of >1:24,000 are possible with the improved resolution of 1 x 1 m data (Light and Jensen, 2002). Because image map products can be produced for a fraction of the cost of conventional line maps, they provide the basis for a national map series oriented toward the exploration and economic development of the less-developed areas of the world, most of which have not been mapped at scales of 1:100,000 or larger.

Remote sensor data that have been geometrically rectified to a standard map projection are becoming indispensable in most sophisticated GIS databases. This is especially true of orthophotomaps, which have the metric qualities of a line map and the information content of an aerial photograph or other type of image (Jensen, 1995).

Unfortunately, *error* is introduced in the remote sensing process and must be identified and reported. Innovations in error reduction include: 1) recording the genealogy or lineage of the various operations applied to the original remote sensor data (Lanter and Veregin, 1992), 2) documenting the geometric (spatial) error and thematic (attribute) error of the individual source materials, 3) improving legend design, especially for change detection map products derived from remote sensing, and 4) precise error evaluation statistic reporting (Congalton and Green, 1998; Khorram et al., 1999). Many of these concerns have not been adequately addressed. The remote sensing and GIS community should incorporate technologies that track all error in final map and image products. This will result in more accurate information being used in the decision-making process.

 ### Earth Resource Analysis Perspective

Digital image processing is used for numerous applications, including: weapon guidance systems (e.g., the cruise missile), medical image analysis (e.g., *x*-raying a broken arm), nondestructive evaluation of machinery and products (e.g., on an assembly line), and analysis of Earth resources. *This book focuses on the art and science of applying remote sensing digital image processing for the extraction of useful Earth resource information.* Earth resource information is defined as any information concerning terrestrial vegetation, soils, minerals, rocks, water, and urban infrastructure as well as certain atmospheric characteristics. Such information may be useful for modeling the global carbon cycle, the biology and biochemistry of ecosystems, aspects of the global water and energy cycle, climate variability and prediction, atmospheric chemistry, characteristics of the solid Earth, population estimation, and monitoring land-use change and natural hazards (Paylor et al., 1999).

 ### Book Organization

This book is organized according to the *remote sensing process* (Figures 1-4 and 1-13). The analyst first defines the problem and identifies the data required to accept or reject research hypotheses (Chapter 1). If a remote sensing approach to the problem is warranted, the analyst evaluates several data acquisition alternatives, including traditional aerial photography, multispectral scanners, and linear and area array multispectral and hyperspectral remote sensing systems (Chapter 2). For example, the analyst may digitize existing aerial photography or obtain the data already in a digital format from Space Imaging, Inc., or SPOT Image Corporation. If the analysis is to be performed digitally, an appropriate digital image processing system is configured (Chapter 3). The image analysis begins by first computing

Organization of
Introductory Digital Image Processing

Chapter 1. **Remote Sensing & Digital Image Processing** • Statement of the problem • Select appropriate logic • Select appropriate model	**Chapter 8.** **Image Enhancement** • Transects, contrast, spatial filtering • Edge detection, texture • Transformations
Chapter 2. **Remote Sensing Data Collection Alternatives** • Digitize analog imagery, and/or • Collect passive and active remote sensor data	**Chapter 9.** **Information Extraction:** **Pattern Recognition** • Classification scheme • Feature selection • Supervised/unsupervised classification • Fuzzy classification • Object-oriented classification
Chapter 3. **Digital Image Processing Hardware** **and Software Considerations** • Configure software and hardware	
	Chapter 10. **Information Extraction:** **Artificial Intelligence** • Expert systems • Artificial neural networks
Chapter 4. **Image Quality Assessment and** **Statistical Evaluation** • Univariate and multivariate statistics • Geostatistics	
	Chapter 11. **Information Extraction:** **Hyperspectral Image Analysis** • Radiometric/geometric correction • Dimensionality reduction • Endmember determination • Mapping • Matched filtering
Chapter 5. **Display Alternatives and Scientific** **Visualization** • 8-bit B&W and color look-up tables • 24-bit color look-up tables	
Chapter 6. **Electromagnetic Radiation Principles and** **Radiometric Correction** • Correcting sensor-induced radiometric error • Atmospheric correction methods	**Chapter 12.** **Change Detection** • System and environmental variables • Selection of appropriate algorithm
Chapter 7. **Geometric Correction** • Image-to-image registration • Image-to-map rectification • Mosaicking	**Chapter 13.** **Accuracy Assessment** • Methods of accuracy assessment • Sampling design • Univariate/multivariate statistical analysis

Figure 1-13   This book is organized according to the remote sensing process (see Figure 1-4).

fundamental univariate and multivariate statistics of the raw digital remote sensor data (Chapter 4). The imagery is then viewed on a computer screen or output to various hard-copy devices to analyze image quality (Chapter 5). The imagery is then preprocessed to reduce environmental and/or remote sensor system distortions. This preprocessing usually includes radiometric and geometric correction (Chapters 6 and 7). Various image enhancements are then applied to the corrected data for improved visual analysis or as input to further digital image processing (Chapter 8). Thematic information may then be extracted from multispectral or hyperspectral imagery using either supervised (i.e., human-assisted) or unsupervised techniques, expert systems, or neural networks (Chapters 9 – 11). Multiple dates of imagery can be analyzed to identify change that provides insight into the processes at work (Chapter 12). Methods of assessing the accuracy of the remote sensing–derived thematic map products are summarized in Chapter 13.

 References

Adams, J. C. and J. H. Chandler, 2002, "Evaluation of Lidar and Medium Scale Photogrammetry for Detecting Soft-cliff Coastal Change," *Photogrammetric Record*, 17(99):405–418.

Asker, J. R., 1992, "Congress Considers Landsat Decommercialization Move," *Aviation Week & Space Technology,* May 11, 18–19.

Asrar, G. and J. Dozier, 1994, *EOS: Science Strategy for the Earth Observing System*, Woodbury, MA: American Institute of Physics, 342 p.

Barnsley, M., 1999, "Digital Remotely Sensed Data and Their Characteristics," in Longley, P. E., Goodchild, M. F., McGwire, D. J. and D. W. Rhind (Eds.), *Geographical Information Systems*, New York: John Wiley & Sons, 451–466.

Belokon, W. F., 1997, *Multispectral Imagery Reference Guide*, Fairfax, VA: LOGICON Geodynamics, 156 p.

Bossler, J. D., Jensen, J. R., McMaster, R. B. and C. Rizos, 2002, *Manual of Geospatial Science and Technology*, London: Taylor & Francis, 623 p.

Clark, R. N., 1999, *Spectroscopy of Rocks and Minerals, and Principles of Spectroscopy*, Denver: U.S. Geological Survey, http://speclab.cr.usgs.gov, 58 p.

Clarke, K. C., 2001, *Getting Started with Geographic Information Systems*, Upper Saddle River, NJ: Prentice-Hall, 353 p.

Collins, J. B. and C. E. Woodcock, 1999, "Geostatistical Estimation of Resolution-Dependent Variance in Remotely Sensed Images," *Photogrammetric Engineering & Remote Sensing,* 65(1):41–50.

Colwell, R. N. (Ed.), 1983, *Manual of Remote Sensing*, 2nd Ed., Bethesda: American Society for Photogrammetry & Remote Sensing.

Colwell, R. N., 1997, "History and Place of Photographic Interpretation," in *Manual of Photographic Interpretation*, 2nd Ed., W. R. Phillipson (Ed.), Bethesda: American Society for Photogrammetry & Remote Sensing, 33–48.

Congalton, R. G. and K. Green, 1998, *Assessing the Accuracy of Remotely Sensed Data*, Boca Raton, FL: Lewis Publishing, 137 p.

Curran, P. J., 1987, "Remote Sensing Methodologies and Geography," *International Journal of Remote Sensing,* 8:1255–1275.

Curran, P. J., Milton, E. J., Atkinson, P. M. and G. M. Foody, 1998, "Remote Sensing: From Data to Understanding," in P. E. Longley, S. M. Brooks, R. McDonnell, and B. Macmillan (Eds.), *Geocomputation: A Primer*, New York: John Wiley & Sons, 33–59.

Dahlberg, R. W. and J. R. Jensen, 1986, "Education for Cartography and Remote Sensing in the Service of an Information Society: The United States Case," *American Cartographer*, 13(1):51–71.

Defries, R. S. and J. R. G. Townshend, 1999, "Global Land Cover Characterization from Satellite Data: From Research to Operational Implementation?" *Global Ecology and Biogeography*, 8:367–379.

Defries, R., Hansen, M., Townshend, J. R. G. and R. Sohlberg, 1998, "Global Land Cover Classifications at 8 km Resolution: The Use of Training Data Derived from Landsat Imagery in Decision Tree Classifiers," *International Journal of Remote Sensing*, 5:3567–3586.

Digital Research, 2004, *Environment for Visualizing Images: ENVI*, www.digitalresearch.com

Dobson, J. E., 1993, "Commentary: A Conceptual Framework for Integrating Remote Sensing, Geographic Information Systems, and Geography," *Photogrammetric Engineering & Remote Sensing,* 59(10):1491–1496.

Donnay, J., Barnsley, M. J. and P. A. Longley, 2001, *Remote Sensing and Urban Analysis*, London: Taylor & Francis, 268 p.

Duda, R. O., Hart, P. E. and D. G. Stork, 2001, *Pattern Classification*, New York: John Wiley & Sons, 394–452.

Duggin, M. J. and C. J. Robinove, 1990, "Assumptions Implicit in Remote Sensing Data Acquisition and Analysis," *International Journal of Remote Sensing,* 11(10):1669–1694.

Emerson, C. W., Lam, N. and D. A. Quattrochi, 1999, "Multi-scale Fractal Analysis of Image Texture and Pattern," *Photogrammetric Engineering & Remote Sensing,* 65(1):51–61.

Engman, E. T., 2000, "Soil Moisture," in Schultz, G. A. and E. T. Engman (Eds.), *Remote Sensing in Hydrology and Water Management*, Berlin: Springer, 197–216.

Estes, J. E. and J. R. Jensen, 1998, "Development of Remote Sensing Digital Image Processing Systems and Raster GIS," *The History of Geographic Information Systems*, T. Foresman (Ed.), New York: Longman, 163–180.

Estes, J. E., Hajic, E. J. and L. R. Tinney, 1983, "Fundamentals of Image Analysis: Visible and Thermal Infrared Data," *Manual of Remote Sensing,* R. N. Colwell, (Ed.), Bethesda: American Society for Photogrammetry & Remote Sensing, 987–1125.

Estes, J. E., Jensen, J. R. and D. S. Simonett, 1980, "Impacts of Remote Sensing on U. S. Geography," *Remote Sensing of Environment*, 10(1):3–80.

Fisher, P. F. and R. E. Lindenberg, 1989, "On Distinctions among Cartography, Remote Sensing, and Geographic Information Systems," *Photogrammetric Engineering & Remote Sensing*, 55(10):1431–1434.

Foody, G. M., 1996, "Approaches for the Production and Evaluation of Fuzzy Land Cover Classifications from Remotely Sensed Data," *International Journal of Remote Sensing*, 17(7):1317–1340.

Friedl, M. A., McIver, D. K., Hodges, J. C. F., Zhang, X. Y., Muchoney, D., Strahler, A. H., Woodcock, C. E., Gopal, S., Schneider, A., Cooper, A., Baccini, A., Gao, F. and C. Schaaf, 2002, "Global Land Cover Mapping from MODIS: Algorithms and Early Results," *Remote Sensing of Environment*, 83:287–302.

Fussell, J., Rundquist, D. and J. A. Harrington, 1986, "On Defining Remote Sensing," *Photogrammetric Engineering & Remote Sensing*, 52(9):1507–1511.

Gahegan, M., 2003, "Is Inductive Machine Learning Just Another Wild Goose (or Might It Lay the Golden Egg)?" *International Journal of Geographical Information Science*, 17(1):69–92.

Gahegan, M. and M. Ehlers, 2000, "A Framework for the Modelling of Uncertainty Between Remote Sensing and Geographic Information Systems," *ISPRS Journal of Photogrammetry & Remote Sensing*, 55:176–188.

Goetz, S. J., 2002, "Recent Advances in Remote Sensing of Biophysical Variables: An Overview of the Special Issue," *Remote Sensing of Environment*, 79:145–146.

Haack, B., Guptill, S. C., Holz, R. K., Jampoler, S. M., Jensen, J. R. and R. A. Welch, 1997, "Urban Analysis and Planning," *The Manual of Photographic Interpretation*, Bethesda: American Society for Photogrammetry & Remote Sensing, 517–553.

Henderson, F., 1994, "The Landsat Program: Life After Divorce?", *Earth Observation Magazine*, April, 8.

Herold, M., Guenther, S. and K. Clarke, 2003, "Mapping Urban Areas in the Santa Barbara South Coast Using IKONOS Data and eCognition," *eCognition Application Note*, Munich: Definiens Imaging GmbH, 4(1):3.

Hodgson, M. E., Jensen, J. R., Tullis, J. A., Riordan, K. D. and C. M. Archer, 2003, "Synergistic Use of LIDAR and Color Aerial Photography for Mapping Urban Parcel Imperviousness," *Photogrammetric Engineering & Remote Sensing*, 69(9):973–980.

Huang, X. and J. R. Jensen, 1997, "A Machine Learning Approach to Automated Construction of Knowledge Bases for Image Analysis Expert Systems That Incorporate Geographic Information System Data," *Photogrammetric Engineering & Remote Sensing*, 63(10):1185–1194.

Jensen, J. R., 1992, "Testimony on S. 2297, The Land Remote Sensing Policy Act of 1992," Senate Committee on Commerce, Science, and Transportation, *Congressional Record*, (May 6):55–69.

Jensen, J. R., 1995, "Issues Involving the Creation of Digital Elevation Models and Terrain Corrected Orthoimagery Using Softcopy Photogrammetry," *Geocarto International: A Multidisciplinary Journal of Remote Sensing and GIS*, 10(1):5–21.

Jensen, J. R., 2000, *Remote Sensing of the Environment: An Earth Resource Perspective*, Upper Saddle River, NJ: Prentice-Hall, 544 p.

Jensen, J. R. and D. C. Cowen, 1999, "Remote Sensing of Urban/Suburban Infrastructure and Socioeconomic Attributes," *Photogrammetric Engineering & Remote Sensing*, 65(5):611–622.

Jensen, J. R. and S. Schill, 2000, "Bi-directional Reflectance Distribution Function (BRDF) of smooth cordgrass (*Spartina alterniflora*)," *Geocarto International: A Multidisciplinary Journal of Remote Sensing and GIS*, 15(2):21-28.

Jensen, J. R., Botchway, K., Brennan-Galvin, E., Johannsen, C., Juma, C., Mabogunje, A., Miller, R., Price, K., Reining, P., Skole, D., Stancioff, A. and D. R. F. Taylor, 2002, *Down to Earth: Geographic Information for Sustainable Development in Africa*, Washington: National Research Council, 155 p.

Jensen, J. R., Qiu, F. and K. Patterson, 2001, "A Neural Network Image Interpretation System to Extract Rural and Urban Land Use and Land Cover Information fro Remote Sensor Data," *Geocarto International: A Multidisciplinary Journal of Remote Sensing & GIS*, 16(1):19–28.

Jensen, J. R., Qiu, F. and M. Ji, 1999, "Predictive Modeling of Coniferous Forest Age Using Statistical and Artificial Neural Network Approaches Applied to Remote Sensing Data," *International Journal of Remote Sensing*, 20(14):2805–2822.

Johannsen, C. J., Petersen, G. W., Carter, P. G. and M. T. Morgan, 2003, "Remote Sensing: Changing Natural Resource Management," *Journal of Soil & Water Conservation*, 58(2):42–45.

Joseph, G., 2000, "How Well Do We Understand Earth Observation Electro-optical Sensor Parameters?" *ISPRS Journal of Photogrammetry & Remote Sensing*, 55:9–12.

Kelly, M., Estes, J. E. and K. A. Knight, 1999, "Image Interpretation Keys for Validation of Global Land-Cover Data Sets," *Photogrammetric Engineering & Remote Sensing*, 65:1041–1049.

Khorram, S., Biging, G., Chrisman, N., Colby, D., Congalton, R., Dobson, J., Ferguson, R., Goodchild, M., Jensen, J. and T. Mace, 1999, *Accuracy Assessment of Land Cover Change Detection*, Bethesda: ASPRS, 64 p.

King, M., 2003, *EOS Data Product Handbook - Volume 1*, Washington: National Aeronautics and Space Administration, 258 p.

Kraak, M., 2003, "Geovisualization Illustrated," *ISPRS Journal of Photogrammetry & Remote Sensing*, (2003):390–399.

Kruse, F. A., Lefkoff, A. B., Boardman, J. W., Heidebrecht, K. B., Shapiro, A. T., Barloon, P. J. and A. F. H. Goetz, 1992, "The Spectral Image Processing System (SIPS)—Interactive Visualization and Analysis of Imaging Spectrometer Data," *Proceedings, International Space Year Conference*, Pasadena, CA, 10 p.

Landgrebe, D. and L. Biehl, 2004, *An Introduction to MULTISPEC*, W. Lafayette, IN: Purdue University, 50 p.

Lanter, D. P. and H. Veregin, 1992, "A Research Paradigm for Propagating Error in Layer-based GIS," *Photogrammetric Engineering & Remote Sensing*, 58(6):825–833.

Li, R., 1998, "Potential of High-Resolution Satellite Imagery for National Mapping Products," *Photogrammetric Engineering & Remote Sensing*, 64(12):1165–1169.

Light, D. L. and J. R. Jensen, 2002, "Photogrammetric and Remote Sensing Considerations," in *Manual of Geospatial Science and Technology*, Bossler, J. D., Jensen, J. R., McMaster, R. B. and C. Rizos (Eds.), London: Taylor & Francis, 233–252.

Lloyd, R., Hodgson, M. E. and A. Stokes, 2002, "Visual Categorization with Aerial Photographs," *Annals of the Association of American Geographers*. 92(2):241–266.

McIver, D. K. and M. A. Friedl, 2002, "Using Prior Probabilities in Decision-tree Classification of Remotely Sensed Data," *Remote Sensing of Environment*, 81:253–261.

NASA, 1998, *Earth System Science: A Closer View*, Washington: NASA, 36 p.

Nemani, R. R., Keeling, C. D., Hashimoto, H., Jolly, W. M., Piper, S. C., Tucker, C. J., Myneni, R. B. and S. W. Running, 2003, "Climate-Driven Increases in Global Terrestrial Net Primary Production from 1982 to 1999," *Science*, 300(6):1560–1563.

Paylor, E. D., Kaye, J. A., Johnson, A. R. and N. G. Maynard, 1999, "Earth Science Enterprise: Science and Technology for Society," *Earth Observation Magazine*, 8(3):8–12.

Ramsey, R. D., Falconer, A. and J. R. Jensen, 1995, "The Relationship Between NOAA-AVHRR Normalized Difference Vegetation Index and Ecoregions in Utah," *Remote Sensing of Environment*, 53:188–198.

Richards, J. A. and X. Jia, 1999, *Remote Sensing Digital Image Analysis: An Introduction*, New York: Springer-Verlag, 363 p.

Rizos, C., 2002, "Introducing the Global Positioning System," in *Manual of Geospatial Science and Technology*, Bossler, J. D., Jensen, J. R., McMaster, R. B. and C. Rizos (Eds.), London: Taylor & Francis, 77–94.

Robbins, J., 1999, "High-Tech Camera Sees What Eyes Cannot," *New York Times*, Science Section, September 14, D5.

Ryerson, R., 1989, "Image Interpretation Concerns for the 1990s and Lessons from the Past," *Photogrammetric Engineering & Remote Sensing*, 55(10):1427–1430.

Sandmeier, S. R., 2000, "Acquisition of Bidirectional Reflectance Factor Data with Field Goniometers," *Remote Sensing of Environment*, 73:257–269.

Seong, J. C. and E. L. Usery, 2001, "Fuzzy Image Classification for Continental-Scale Multitemporal NDVI Images Using Invariant Pixels and an Image Stratification Method," *Photogrammetric Engineering & Remote Sensing*, 67(3):287–294.

Shippert, P., 2004, *Spotlight on Hyperspectral*, Boulder: Research Systems, www.geospatial_online.com/shippert, 5 p.

Skidmore, A. K., 2002, "Chapter 2: Taxonomy of Environmental Models in the Spatial Sciences," in *Environmental Modelling with GIS and Remote Sensing*, A. K. Skidmore (Ed.), London: Taylor & Francis, 8–25.

Stow, D, Coulter, L., Kaiser, J., Hope, A., Service, D., Schutte, K. and A. Walters, 2003, "Irrigated Vegetation Assessments for Urban Environments," *Photogrammetric Engineering & Remote Sensing*, 69(4):381–390.

Strahler, A. H., Woodcock, C. E. and J. A. Smith, 1986, "On the Nature of Models in Remote Sensing," *Remote Sensing of Environment*, 20:121–139.

Teillet, P. M., Gauthier, R. P., Chichagov, A. and G. Fedosejevs, 2002, "Towards Integrated Earth Sensing: Advanced Technologies for *in situ* Sensing in the Context of Earth Observation," *Canadian Journal of Remote Sensing*, 28(6):713–718.

Townshend, J. R. G. and C. O. Justice, 2002, "Towards Operational Monitoring of Terrestrial Systems by Moderate-resolution Remote Sensing," *Remote Sensing of Environment*, 83:351–359.

Tullis, J. A. and J. R. Jensen, 2003, "Expert System House Detection in High Spatial Resolution Imagery Using Size, Shape, and Context," *Geocarto International: A Multidisciplinary Journal of Remote Sensing and GIS*, 18(1):5–15.

Walsh, S. J., Evans, T. P., Welsh, W. F., Entwisle, B. and R. R. Rindfuss, 1999, "Scale-dependent Relationships Between Population and Environment in Northeastern Thailand," *Photogrammetric Engineering & Remote Sensing*, 65(1):97–105.

Woodcock, C. E., Collins, J. B., Jakabhazy, V., Li, X., Macomber, S. and Y. Wu, 1997, "Inversion of the Li-Strahler Canopy Reflectance Model for Mapping Forest Structure," *IEEE Transactions on Geoscience and Remote Sensing*, 35(2):405–414.

Wright, D. J., Goodchild, M. F. and J. D. Procter, 1997, "GIS: Tool or Science; Demystifying the Persistent Ambiguity of GIS as Tool versus Science," *The Professional Geographer*, 87(2):346–362.

Zhan, X., Sohlberg, R. A., Townshend, J. R. G., DiMiceli, C., Carrol, M. L., Eastman, J. C., Hansen, M. C. and R. S. DeFries, 2002, "Detection of Land Cover Changes Using MODIS 250 m Data," *Remote Sensing of Environment*, 83:336–350.

Zhang, Q. and J. Wang, 2003, "A Rule-based Urban Land Use Inferring Method for Fine-resolution Multispectral Imagery," *Canadian Journal of Remote Sensing*, 29(1):1–13.

Remote Sensing Data Collection 2

It is necessary to have remotely sensed imagery in a digital format in order to perform digital image processing. There are two fundamental ways to obtain digital imagery:

1. acquire remotely sensed imagery in an *analog* format (often referred to as hard-copy) and then convert it to a digital format through the process of digitization, and

2. acquire remotely sensed imagery already in a *digital* format, such as that obtained by the Landsat 7 Enhanced Thematic Mapper Plus (ETM[+]) sensor system.

This chapter reviews how different types of remotely sensed data are obtained in a digital format using these two alternatives.

 Analog (Hard-Copy) Image Digitization

Scientists or laypersons often obtain analog (hard-copy) remote sensor data that they desire to analyze using digital image processing techniques. Analog aerial photographs are, of course, ubiquitous because much of the world has been photographed many times. Occasionally scientists encounter analog thermal infrared or active microwave (RADAR) imagery. To convert analog imagery into digital imagery, the person performing the digital image processing must first understand digital image terminology.

Digital Image Terminology

Digital remote sensor data are usually stored as a matrix (array) of numbers. Each digital value is located at a specific row (i) and column (j) in the matrix (Figure 2-1). A *pixel* is defined as "a two-dimensional picture element that is the smallest nondivisible element of a digital image." Each pixel at row (i) and column (j) in the image has an original *brightness value* (BV) associated with it [some scientists use the term *digital number* (DN) value]. The dataset may consist of n individual bands (k) of multispectral or hyperspectral imagery. Thus, it is possible to identify the brightness value of a particular pixel in the dataset by specifying its row (i), column (j), and band (k) coordinate, i.e., $BV_{i,j,k}$. It is important to understand that the n bands are all geometrically registered to one another. Therefore, a road intersection in band 1 at row 4, column 4 (i.e., $BV_{4,4,1}$) should be located at the same row and column coordinate in the fourth band (i.e., $BV_{4,4,4}$).

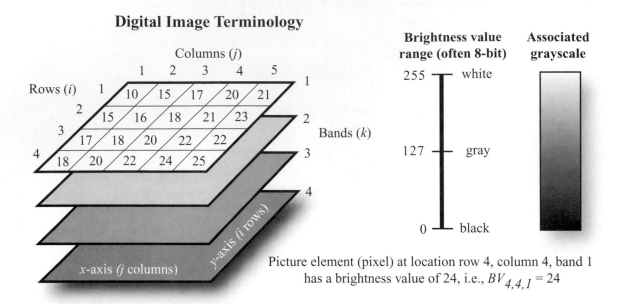

Digital Image Terminology

Picture element (pixel) at location row 4, column 4, band 1
has a brightness value of 24, i.e., $BV_{4,4,1} = 24$

Figure 2-1   Digital remote sensor data are stored in a matrix format. Picture element (pixel) brightness values *(BV)* are located at row *i*, column *j*, and band *k* in the multispectral or hyperspectral dataset. The digital remote sensor brightness values are normally stored as 8-bit bytes with values ranging from 0 to 255. However, several image digitization systems and some remote sensing systems now routinely collect 10-, 11-, or 12-bit data.

In an analog image, the brightness value *(BV)* is actually a surrogate for the *density (D)* of the light-absorbing silver or dye deposited at a specific location. The density characteristics of a negative or positive transparency film is measured using a *densitometer.* There are several types of densitometer, including flatbed and drum microdensitometers, video densitometers, and linear or area array charge-coupled-device densitometers.

Microdensitometer Digitization

The characteristics of a typical *flatbed microdensitometer* are shown in Figure 2-2. This instrument can measure the density characteristics of very small portions of a negative or positive transparency, down to just a few micrometers in size, hence the term *microdensitometer*. Basically, a known quantity of light is sent from the light source toward the receiver. If the light encounters a very dense portion of the film, very little light is transmitted to the receiver. If the light encounters a very clear portion of the film, then much of the light is transmitted to the receiver. The densitometer can output the characteristics at each *i,j* location in the photograph in terms of transmittance, opacity, or density. The ability of a portion of a developed film to pass light is called its *transmittance* ($\tau_{i,j}$). A black portion of the film may transmit no

light, while a clear portion of the film may transmit almost 100 percent of the incident light. Therefore, the transmittance at location *i,j* in the photograph is:

$$\tau_{i,j} = \frac{\text{light passing through the film}}{\text{total incident light}}. \tag{2-1}$$

There is an inverse relationship between transmittance and the opacity of an area on the film. An area in the film that is very opaque does not transmit light well. *Opacity ($O_{i,j}$)* is the reciprocal of transmittance:

$$O_{i,j} = \frac{1}{\tau_{i,j}}. \tag{2-2}$$

Transmittance and opacity are two good measures of the darkness of any portion of a developed negative. However, psychologists have found that the human visual system does not respond linearly to light stimulation, but rather we respond logarithmically. Therefore, it is common to use *density ($D_{i,j}$)*, which is the common logarithm of opacity, as our digitization measure of choice:

$$D_{i,j} = \log_{10}O_{i,j} = \log\left(\frac{1}{\tau_{i,j}}\right). \tag{2-3}$$

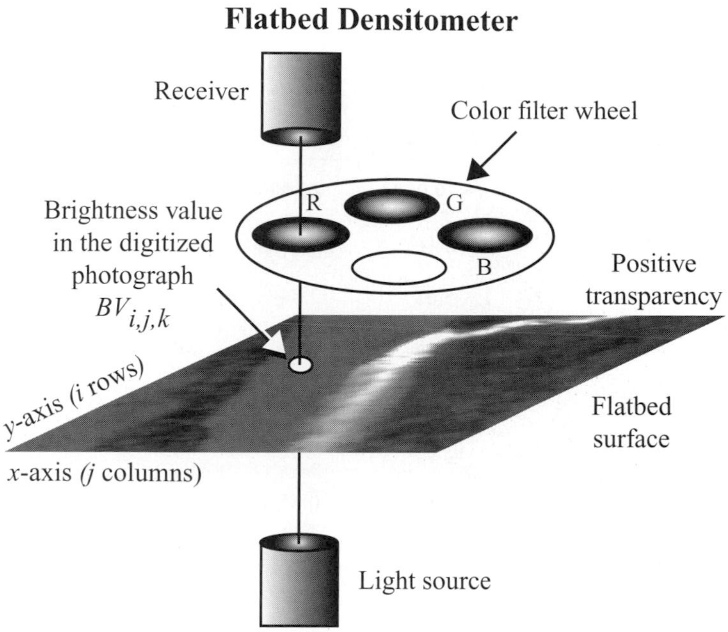

Flatbed Densitometer

Figure 2-2   Schematic of a flatbed microdensitometer. The hard-copy remotely sensed imagery (usually a positive transparency) is placed on the flatbed surface. A small light source (perhaps 10 μm in diameter) is moved mechanically across the flat imagery in the *x*-direction, emitting a constant amount of light. On the other side of the imagery, a receiver measures the amount of energy that passes through. When one line scan is complete, the light source and receiver step in the *y*-direction some Δy to scan an area contiguous and parallel to the previous scan line. The amount of energy detected by the receiver along each scan line is eventually changed from an electrical signal into a digital value through an analog-to-digital (A-to-D) conversion. After the entire image has been scanned in this fashion, a matrix of brightness values is available for digital image processing purposes. A color filter wheel may be used if the imagery has multiple dye layers that must be digitized. In this case the imagery is scanned three separate times using three different filters to separate it into its respective blue, green, and red components. The resulting three matrices should be in near-perfect registration, representing a multispectral digital dataset.

If 10 percent of the light can be transmitted through a film at a certain *i,j* location, transmittance is 1/10, opacity is 1/0.10 or 10, and density is the common logarithm of 10 or 1.0.

As previously mentioned, the amount of light recorded by the receiver in densitometric units is commonly converted into a digital brightness value ($BV_{i,j,k}$), which refers to the location in the photograph at row *i*, column *j*, and band *k*. At the end of each scan line, the light source steps in the *y*-direction some Δy to scan along a line contiguous and parallel to the previous one. As the light source is scanned across the image, the continuous output from the receiver is converted to a series of discrete numerical values on a pixel-by-pixel basis. This analog-to-digital conversion process results in a matrix of values that is usually recorded in 8-bit bytes (values ranging from 0 to 255). These data are then stored on disk or tape for future analysis.

Scanning imagery at spot sizes <12 μm may result in noisy digitized data because the spot size approaches the dimension of the film's silver halide crystals. Table 2-1 summa-

rizes the relationship between digitizer scanning spot size (IFOV) measured in dots per inch (dpi) or micrometers and the pixel ground resolution at various scales of aerial photography or imagery.

A simple black-and-white photograph has only a single band, *k = 1*. However, we may need to digitize color photography. In such circumstances, we use three specially designed filters that determine the amount of light transmitted by each of the dye layers in the film (Figure 2-2). The negative or positive transparency is scanned three times (*k = 1, 2,* and *3*), each time with a different filter. This extracts spectral information from the respective dye layers found in color and color-infrared aerial photography and results in a co-registered three-band digital dataset for subsequent image processing.

Rotating-drum optical-mechanical scanners digitize the imagery in a somewhat different fashion (Figure 2-3). The film transparency is mounted on a glass rotating drum so that it forms a portion of the drum's circumference. The light

Table 2-1.   Relationship between digitizer instantaneous-field-of-view measured in dots per inch or micrometers, and the pixel ground resolution at various scales of photography.

Digitizer Detector IFOV		Pixel Ground Resolution at Various Scales of Photography (m)					
Dots per inch	Micrometers	1:40,000	1:20,000	1:9,600	1:4,800	1:2,400	1:1,200
100	254.00	10.16	5.08	2.44	1.22	0.61	0.30
200	127.00	5.08	2.54	1.22	0.61	0.30	0.15
300	84.67	3.39	1.69	0.81	0.41	0.20	0.10
400	63.50	2.54	1.27	0.61	0.30	0.15	0.08
500	50.80	2.03	1.02	0.49	0.24	0.12	0.06
600	42.34	1.69	0.85	0.41	0.20	0.10	0.05
700	36.29	1.45	0.73	0.35	0.17	0.09	0.04
800	31.75	1.27	0.64	0.30	0.15	0.08	0.04
900	28.23	1.13	0.56	0.27	0.14	0.07	0.03
1000	25.40	1.02	0.51	0.24	0.12	0.06	0.03
1200	21.17	0.85	0.42	0.20	0.10	0.05	0.03
1500	16.94	0.67	0.34	0.16	0.08	0.04	0.02
2000	12.70	0.51	0.25	0.12	0.06	0.03	0.02
3000	8.47	0.33	0.17	0.08	0.04	0.02	0.01
4000	6.35	0.25	0.13	0.06	0.03	0.02	0.008

Useful Scanning Conversions

DPI = dots per inch;   μm = micrometers;   I = inches;   M = meters
From DPI to micrometers: μm = (2.54 / DPI)10,000
From micrometers to DPI: DPI = (2.54 / μm)10,000
From inches to meters: M = I × 0.0254
From meters to inches: I = M × 39.37

Computation of Pixel Ground Resolution

PM = pixel size in meters;   PF = pixel size in feet;   S = photo scale
Using DPI: PM = (S/DPI)/39.37   PF = (S/DPI)/12
Using micrometers: PM = (S × μm) 0.000001   PF = (S × μm) 0.00000328
For example, if a 1:6,000 scale aerial photograph is scanned at 500 dpi, the pixel size will be (6000/500)/39.37 = 0.3048 meters per pixel, or (6000/500)/12 = 1.00 foot per pixel. If a 1:9,600 scale aerial photograph is scanned at 50.8 μm, the pixel size will be (9,600 × 50.8)(0.000001) = 0.49 meters, or (9,600 × 50.8)(0.00000328) = 1.6 feet per pixel.

source is situated in the interior of the drum. The y-coordinate scanning motion is provided by the rotation of the drum. The x-coordinate is obtained by the incremental translation of the source-receiver optics after each drum revolution.

Video Digitization

It is possible to digitize hard-copy imagery by sensing it through a video camera and then performing an analog-to-

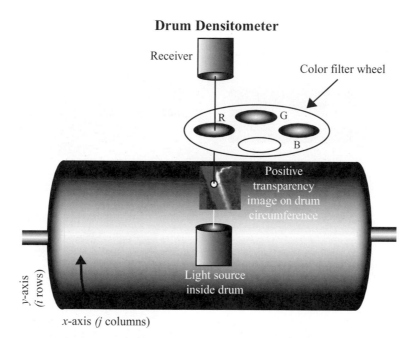

Drum Densitometer

Figure 2-3 The rotating-drum, optical-mechanical scanner works on exactly the same principle as the flatbed microdensitometer except that the remotely sensed data are mounted on a rotating drum so that they form a portion of the drum's circumference. The light source is situated in the interior of the drum, and the drum is continually rotated in the y-direction. The x-coordinate is obtained by the incremental translation of the source-receiver optics after each drum revolution. Some microdensitometers can write to film as well as digitize from film. In such cases the light source (usually a photodiode or laser) is modulated such that it exposes each picture element according to its brightness value. These are called film-writers and provide excellent hard-copy of remotely sensed data.

digital conversion on the 525 columns by 512 rows of data that are within the standard field of view (as established by the National Television System Committee). Video digitizing involves freezing and then digitizing a frame of analog video camera input. A full frame of video input can be read in approximately 1/60 sec. A high-speed analog-to-digital converter, known as a *frame grabber*, digitizes the data and stores it in frame buffer memory. The memory is then read by the host computer and the digital information is stored on disk or tape.

Video digitization of hard-copy imagery is performed very rapidly, but the results are not always useful for digital image processing purposes due to:

- differences in the radiometric sensitivity of various video cameras, and

- vignetting (light fall-off) away from the center of the image being digitized.

These characteristics can affect the spectral signatures extracted from the scene. Also, any geometric distortion in the vidicon optical system will be transferred to the digital

remote sensor data, making it difficult to edge-match between adjacent images digitized in this manner.

Linear and Area Array Charge-Coupled-Device Digitization

Advances in the personal computer industry have spurred the development of flatbed, desktop linear array digitizers based on linear array charge-coupled devices (CCDs) that can be used to digitize hard-copy negatives, paper prints, or transparencies at 300 to 6,000 pixels per inch (Figure 2-4a,b). The hard-copy photograph is placed on the glass. The digitizer optical system illuminates an entire line of the hard-copy photograph at one time with a known amount of light. A linear array of detectors records the amount of light reflected from or transmitted through the photograph along the array and performs an analog-to-digital conversion. The linear array is stepped in the y-direction, and another line of data is digitized.

It is possible to purchase desktop color digitizers for less than $300. Many digital image processing laboratories use these inexpensive desktop digitizers to convert hard-copy

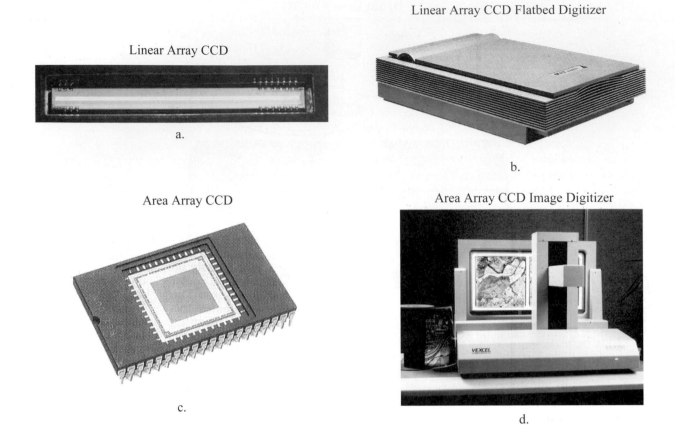

Linear Array CCD

a.

Linear Array CCD Flatbed Digitizer

b.

Area Array CCD

c.

Area Array CCD Image Digitizer

d.

Figure 2-4   a) Enlarged view of a 2,048 element charge-coupled-device (CCD) linear array. b) Inexpensive desktop linear array CCD digitizer. c) Enlarged view of a 2,048 × 2,048 area array CCD. d) An image digitizer based on area array CCD technology (courtesy Vexcel, Inc.).

remotely sensed data into a digital format. Desktop scanners provide surprising spatial precision and a reasonable characteristic curve when scanning black-and-white images. An optional device can be purchased to backlight positive-transparency film or 35-mm slides. Unfortunately, most desktop digitizers are designed for 8.5 × 14 in. originals, and most aerial photographs are 9 × 9 in. Under such conditions, the analyst must digitize the 9 × 9 in. photograph in two sections (e.g., 8.5 × 9 in. and 0.5 × 9 in.) and then digitally *mosaic* the two pieces together. The mosaicking process can introduce both geometric and radiometric error.

An area array consisting of 2,048 by 2,048 CCDs is shown in Figure 2-4c. CCD digital camera technology has been adapted specifically for remote sensing image digitization. For example, the scanner shown in Figure 2-4d digitizes from 160 dpi to 3,000 dpi (approximately 160 μm to 8.5 μm) over a 10 × 20 in. image area (254 × 508 mm). The system scans the film (ideally the original negative) as a series of

rectangular image segments, or tiles. It then illuminates and scans a *reseau grid*, which is an array of precisely located crosshatches etched into the glass of the film-carrier. The reseau grid coordinate data are used to locate the precise orientation of the CCD camera during scanning and to geometrically correct each digitized tile of the image relative to all others. Radiometric calibration algorithms are then used to compensate for uneven illumination encountered in any of the tile regions. When digitizing a color image, the scanner stops on a rectangular image section and captures that information sequentially with each of four color filters (blue, green, red, and neutral) before it moves to another section. Most other scanners digitize an entire image with one color filter and then repeat the process with the other color filters. This can result in color misregistration and loss of image radiometric quality. Area array digitizing technology has obtained geometric accuracy of ±5 μm over 23 × 23 cm images when scanned at 25 μm per pixel, and repeatability of 3 μm.

Digitized National Aerial Photography Program Data

The National Aerial Photography Program (NAPP) was initiated in 1987 as a replacement for the National High Altitude Aerial Photography (NHAP) Program. The objective of the National Aerial Photography Program is to acquire and archive photographic coverage of the coterminous United States at 1:40,000 scale using color-infrared or black-and-white film. The photography is acquired at an altitude of 20,000 ft above ground level (AGL) with a 6-in. focal-length metric camera. The aerial photography is acquired ideally on a five-year cycle, resulting in a nationwide photographic database that is readily available through the EROS Data Center in Sioux Falls, SD, or the Aerial Photography Field Office in Salt Lake City, UT.

High spatial resolution NAPP photography represents a wealth of information for on-screen photointerpretation and can become a high-resolution basemap upon which other GIS information (e.g., parcel boundaries, utility lines, tax data) may be overlaid after it is digitized and rectified to a standard map projection. Light (1993) summarized the optimum methods for converting the NAPP data into a national database of digitized photography that meets National Map Accuracy Standards. Microdensitometer scanning of the photography, using a spot size of 15 μm, preserves the 27 resolvable line-pair-per-millimeter (lp/mm) spatial resolution in the original NAPP photography. This process generally yields a digital dataset that has a ground spatial resolution of 1×1 m, depending on original scene contrast. This meets most land-cover and land-use mapping user requirements.

The digitized information can be color-separated into separate bands of information if desired. The 15-μm scanning spot size will support most digital soft-copy photogrammetry for which coordinate measurements are made using a computer and the monitor screen (Light, 1993). Because the digitized NAPP data are so useful as a high spatial resolution GIS basemap, many states have entered cost-sharing relationships with the U.S. Geological Survey and have their NAPP coverage digitized and output as digital orthophotomaps. A large amount of NAPP data has been digitized and converted into digital orthophotoquads.

Digitization Considerations

There are some basic guidelines associated with digitizing aerial photography or other types of hard-copy remote sensor data. First, the person who will be digitally processing the digitized remote sensor data should make the decision about what dpi to use (200 dpi, 1000 dpi, etc.) based on the scale of the original imagery and the desired spatial resolution (e.g., 1:40,000-scale aerial photography scanned at 1,000 dpi yields 1×1 m spatial resolution pixels; refer to Table 2-1). For example, consider the digitized NAPP photography of Three Mile Island shown in Figure 2-5. The original 1:40,000-scale panchromatic photography was scanned at 1000 to 25 dpi. The U.S. Geological Survey typically scans NAPP photography at 1000 dpi to produce digital orthophotoquads. Note that for all practical purposes it is difficult to *discern visually on the printed page* any significant difference in the quality of the digitized data until about 150 dpi. Then, the information content gradually deteriorates until at 72 to 25 dpi, individual features in the scene are very difficult to interpret. It is important to remember, however, that when the various digitized datasets are displayed on a CRT screen and magnification takes place (as simulated in the bottom nine images in Figure 2-5), it is quite evident that there is significantly more information content in the higher-resolution scans (e.g., 1000 dpi) than the lower-resolution scans (e.g., 100 dpi).

A second general principle is that when digitizing large-scale imagery, it is generally not necessary to scan at extremely high rates (e.g., 300 – 1000 dpi) to obtain visually acceptable imagery. For example, large-scale helicopter vertical photography of vehicles in a parking lot in Waikiki, Hawaii was digitized using scan resolutions from 10 to 1000 dpi (Figure 2-6). Once again, there is very little noticeable difference in the quality of the information at the high scan resolutions (e.g., 1000, 500, 300, and 200). Note however, that even the 100-dpi and possibly the 72-dpi digitized images still contain significant detailed information *on the printed page* because the original aerial photography was obtained at a very large scale.

Hopefully, there is a relationship between the brightness value ($BV_{i,j,k}$) or density ($D_{i,j}$) at any particular location in the digitized image and the energy reflected from the real-world object space ($O_{x,y}$) at the exact location. Scientists take advantage of this relationship by 1) making careful *in situ* observations in the field, such as the amount of biomass for a 1×1 m spot on the Earth located at $O_{x,y}$ and then 2) measuring the brightness value ($BV_{i,j,k}$) or density of the object at that exact location in the photograph using a *densitometer*. If enough samples are located in the field and in the photography, it may be possible to develop a correlation between the real-world object space and the image space. This is an important use of digitized aerial photography.

Vertical Aerial Photography of Three Mile Island Scanned at Various Dots-per-inch

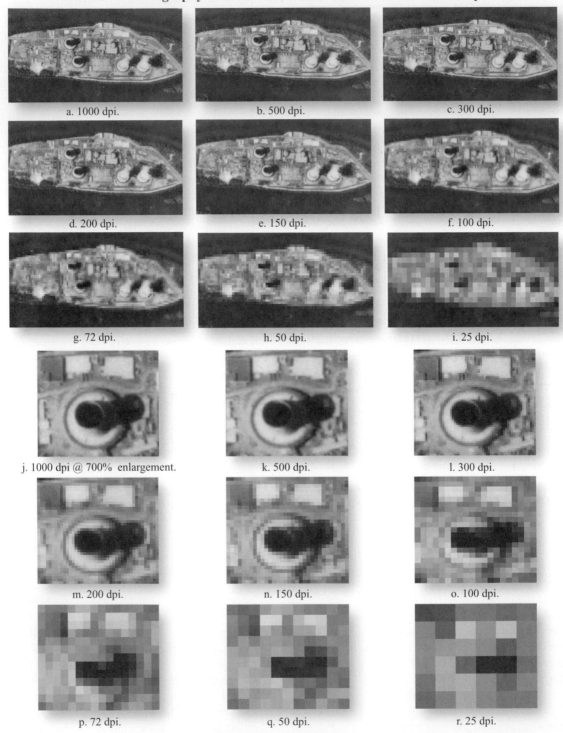

a. 1000 dpi.

b. 500 dpi.

c. 300 dpi.

d. 200 dpi.

e. 150 dpi.

f. 100 dpi.

g. 72 dpi.

h. 50 dpi.

i. 25 dpi.

j. 1000 dpi @ 700% enlargement.

k. 500 dpi.

l. 300 dpi.

m. 200 dpi.

n. 150 dpi.

o. 100 dpi.

p. 72 dpi.

q. 50 dpi.

r. 25 dpi.

Figure 2-5   National Aerial Photography Program (NAPP) photography of Three Mile Island, PA, digitized at various resolutions from 1000 to 25 dpi. The original 9 × 9 in. vertical aerial photograph was obtained on September 4, 1987 at an altitude of 20,000 ft. above ground level yielding a scale of 1:40,000. Scanning at 1000 dpi yields 1 × 1 m pixels (refer to Table 2-1). The upper nine images are printed at approximate contact scale while the lower nine are enlarged to demonstrate approximate information content (NAPP photography courtesy of the U.S. Geological Survey).

Large-scale Vertical Aerial Photography Scanned at Various Dots-per-inch

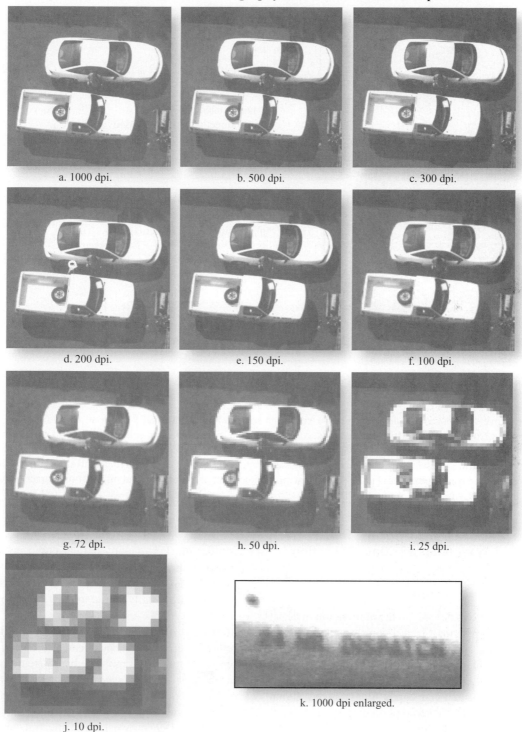

a. 1000 dpi.

b. 500 dpi.

c. 300 dpi.

d. 200 dpi.

e. 150 dpi.

f. 100 dpi.

g. 72 dpi.

h. 50 dpi.

i. 25 dpi.

j. 10 dpi.

k. 1000 dpi enlarged.

Figure 2-6   Digitized large-scale panchromatic aerial photography of vehicles in a parking lot in Waikiki, Hawaii. The original photograph was obtained from a helicopter platform and then scanned at rates from 1000 to 10 dpi. It is possible to make out the words "24 HR DISPATCH" on the vehicle when the photography is scanned at 1000 dpi and magnified.

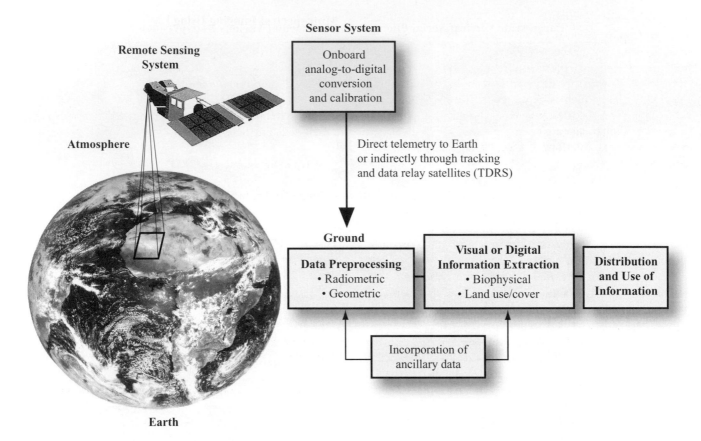

Remote Sensing System

Sensor System

Onboard analog-to-digital conversion and calibration

Atmosphere

Direct telemetry to Earth or indirectly through tracking and data relay satellites (TDRS)

Ground

Data Preprocessing
• Radiometric
• Geometric

Visual or Digital Information Extraction
• Biophysical
• Land use/cover

Distribution and Use of Information

Incorporation of ancillary data

Earth

Figure 2-7   An overview of the way digital remotely sensed data are transformed into useful information. The data recorded by the detectors are often converted from an analog electrical signal to a digital value and calibrated. Ground preprocessing removes geometric and radiometric distortions. This may involve the use of ephemeris or ancillary (collateral) data such as map *x,y* coordinates, a digital elevation model, etc. The data are then ready for visual or digital analysis to extract biophysical or land-use/land-cover information. Future sensor systems will conduct preprocessing and information extraction onboard the remote sensing system.

 Digital Remote Sensor Data Collection

The previous section was devoted to the digitization of aerial photography or other forms of hard-copy remotely sensed data. Digitized color and color-infrared aerial photography can be considered three-band multispectral datasets. Properly digitized natural color aerial photography can be converted to blue, green, and red bands of registered digital data. Digitized color-infrared aerial photography can be converted to green, red, and near-infrared bands of digital data. Although these three-band multispectral datasets are sufficient for many applications, there are times when even more spectral bands located at optimum locations throughout the electromagnetic spectrum would be useful for a specific

application. Fortunately, optical engineers have developed detectors that are sensitive to hundreds of bands in the electromagnetic spectrum. The measurements made by the detectors are usually stored in a digital format.

Multispectral remote sensing is defined as the collection of reflected, emitted, or back-scattered energy from an object or area of interest in multiple bands (regions) of the electromagnetic spectrum. *Hyperspectral remote sensing* involves data collection in hundreds of bands. *Ultraspectral remote sensing* involves data collection in many hundreds of bands (Logicon, 1997). Most multispectral and hyperspectral remote sensing systems collect data in a digital format. The remainder of this chapter introduces the characteristics of historical, current, and proposed multispectral and hyperspectral remote sensing systems.

An overview of how digital remote sensor data are turned into useful information is shown in Figure 2-7. The remote sensor system first detects electromagnetic energy that exits from the phenomena of interest and passes through the atmosphere. The energy detected is recorded as an analog electrical signal, which is usually converted into a digital value through an A-to-D conversion. If an aircraft platform is used, the digital data are simply returned to Earth. However, if a spacecraft platform is used, the digital data are telemetered to Earth receiving stations directly or indirectly via tracking and data relay satellites (TDRS). In either case, it may be necessary to perform some radiometric or geometric preprocessing of the digital remotely sensed data to improve its interpretability. The data can then be enhanced for subsequent human visual analysis or processed further using digital image processing algorithms. Biophysical, land-use, or land-cover information extracted using either a visual or digital image processing approach is distributed and used to make decisions.

There are a tremendous variety of digital multispectral and hyperspectral remote sensing systems. It is beyond the scope of this book to provide detailed information on each of them. However, it is possible to review selected remote sensing systems that are or will be of significant value for Earth resource investigations. They are organized according to the type of remote sensing technology used, as summarized in Figure 2-8, including:

Multispectral Imaging Using Discrete Detectors and Scanning Mirrors

- Landsat Multispectral Scanner (MSS)

- Landsat Thematic Mapper (TM)

- Landsat 7 Enhanced Thematic Mapper Plus (ETM$^+$)

- NOAA Geostationary Operational Environmental Satellite (GOES)

- NOAA Advanced Very High Resolution Radiometer (AVHRR)

- NASA and ORBIMAGE, Inc., Sea-viewing Wide field-of-view Sensor (SeaWiFS)

- Daedalus, Inc., Aircraft Multispectral Scanner (AMS)

- NASA Airborne Terrestrial Applications Sensor (ATLAS)

Multispectral Imaging Using Linear Arrays

- SPOT 1, 2, and 3 High Resolution Visible (HRV) sensors and SPOT 4 and 5 High Resolution Visible Infrared (HRVIR) and *Vegetation* sensor

- Indian Remote Sensing System (IRS) Linear Imaging Self-scanning Sensor (LISS)

- Space Imaging, Inc. (IKONOS)

- DigitalGlobe, Inc. (QuickBird)

- ORBIMAGE, Inc. (OrbView-3)

- ImageSat International, Inc. (EROS A1)

- NASA *Terra* Advanced Spaceborne Thermal Emission and Reflection Radiometer (ASTER)

- NASA *Terra* Multiangle Imaging Spectroradiometer (MISR)

Imaging Spectrometry Using Linear and Area Arrays

- NASA Jet Propulsion Laboratory Airborne Visible/Infrared Imaging Spectrometer (AVIRIS)

- Compact Airborne Spectrographic Imager 3(CASI 3)

- NASA *Terra* Moderate Resolution Imaging Spectrometer (MODIS)

- NASA Earth Observer (EO-1) Advanced Land Imager (ALI), Hyperion, and LEISA Atmospheric Corrector (LAC)

Digital Frame Cameras

- Emerge Spatial, Inc.

Satellite Analog and Digital Photographic Systems

- Russian SPIN-2 TK-350, and KVR-1000

- NASA Space Shuttle and International Space Station Imagery

The following discussion identifies the spatial, spectral, temporal, and radiometric characteristics of the remote sensing systems. It will be clear that certain remote sensing systems

Remote Sensing Systems Used to Collect Multispectral and Hyperspectral Imagery

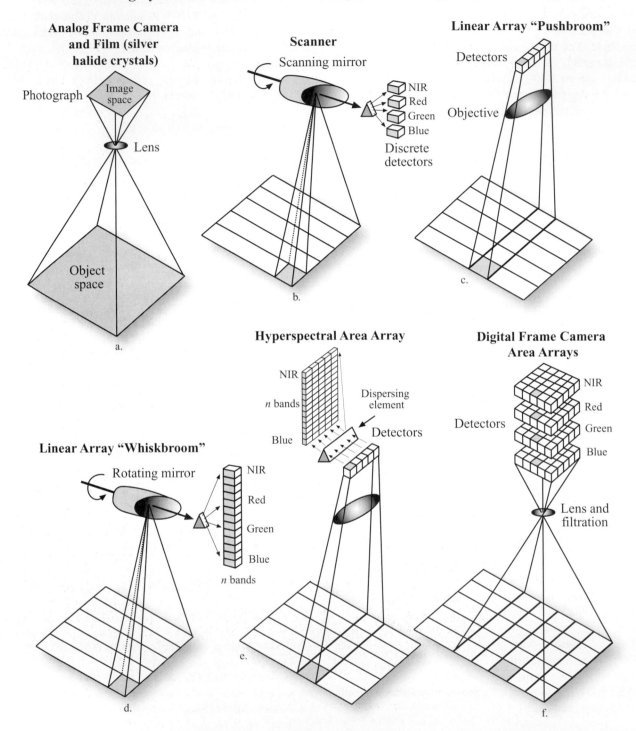

Figure 2-8   Six types of remote sensing system used for multispectral and hyperspectral data collection: a) traditional aerial photography, b) multispectral imaging using a scanning mirror and discrete detectors, c) multispectral imaging with linear arrays (often referred to as "pushbroom" technology), d) imaging with a scanning mirror and linear arrays (often referred to as "whiskbroom" technology), e) imaging spectrometry using linear and area arrays, and f) digital frame camera aerial photography based on area arrays.

Table 2-2.   Landsat Multispectral Scanner (MSS) and Landsat Thematic Mapper (TM) sensor system characteristics.

	Landsat Multispectral Scanner (MSS)			**Landsat 4 and 5 Thematic Mapper (TM)**		
Band	**Spectral Resolution (μm)**	**Radiometric Sensitivity (NEΔP)[a]**		**Band**	**Spectral Resolution (μm)**	**Radiometric Sensitivity (NEΔP)**
4[b]	0.5 – 0.6	0.57		1	0.45 – 0.52	0.8
5	0.6 – 0.7	0.57		2	0.52 – 0.60	0.5
6	0.7 – 0.8	0.65		3	0.63 – 0.69	0.5
7	0.8 – 1.1	0.70		4	0.76 – 0.90	0.5
8[c]	10.4 – 12.6	1.4K (NEΔT)		5	1.55 – 1.75	1.0
				6	10.40–12.5	0.5 (NEΔT)
				7	2.08–2.35	2.4
IFOV at nadir	79 × 79 m for bands 4 through 7 240 × 240 m for band 8			30 × 30 m for bands 1 through 5, 7 120 × 120 m for band 6		
Data rate	15 Mb/s			85 Mb/s		
Quantization levels	6 bit (values from 0 to 63)			8 bit (values from 0 to 255)		
Earth coverage	18 days Landsat 1, 2, 3 16 days Landsat 4, 5			16 days Landsat 4, 5		
Altitude	919 km			705 km		
Swath width	185 km			185 km		
Inclination	99°			98.2°		

[a] The radiometric sensitivities are the noise-equivalent reflectance differences for the reflective channels expressed as percentages (NEΔP) and temperature differences for the thermal infrared bands (NEΔT).

[b] MSS bands 4, 5, 6, and 7 were renumbered bands 1, 2, 3, and 4 on Landsats 4 and 5.

[c] MSS band 8 was present only on Landsat 3.

were designed to collect very specific types of biophysical information.

Multispectral Imaging Using Discrete Detectors and Scanning Mirrors

The collection of multispectral remote sensor data using discrete detectors and scanning mirrors has been with us since the mid-1960s. Despite the technology's age, several new remote sensing systems still use it.

Earth Resource Technology Satellites and Landsat Sensor Systems

In 1967, the National Aeronautics and Space Administration (NASA), encouraged by the U.S. Department of the Interior, initiated the Earth Resource Technology Satellite (ERTS) program. This program resulted in the deployment of five satellites carrying a variety of remote sensing systems designed primarily to acquire Earth resource information. The most noteworthy sensors were the Landsat Multispectral Scanner and the Landsat Thematic Mapper (Table 2-2). The Landsat program is the United States' oldest land-surface observation satellite system, having obtained data since

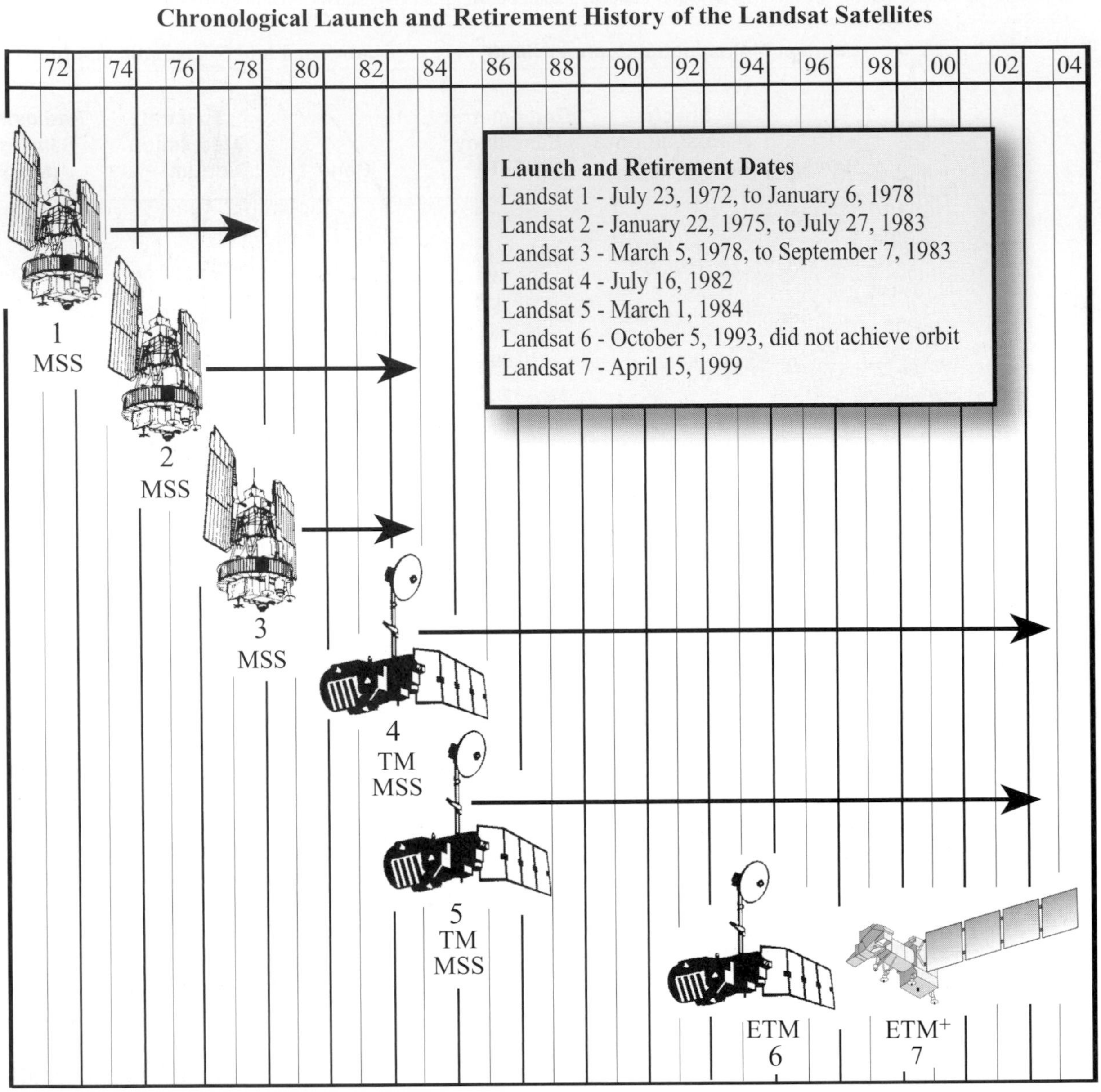

Chronological Launch and Retirement History of the Landsat Satellites

Launch and Retirement Dates
Landsat 1 - July 23, 1972, to January 6, 1978
Landsat 2 - January 22, 1975, to July 27, 1983
Landsat 3 - March 5, 1978, to September 7, 1983
Landsat 4 - July 16, 1982
Landsat 5 - March 1, 1984
Landsat 6 - October 5, 1993, did not achieve orbit
Landsat 7 - April 15, 1999

Figure 2-9   Chronological launch and retirement history of the Landsat series of satellites (1 through 7) from 1972 to 2004.

1972. It has had a tumultuous history of management and funding sources.

The chronological launch and retirement history of the satellites is shown in Figure 2-9. The ERTS-1 satellite, launched on July 23, 1972, was the first experimental system designed to test the feasibility of collecting Earth resource data by unmanned satellites. Prior to the launch of ERTS-B on January 22, 1975, NASA renamed the ERTS program *Landsat*, distinguishing it from the *Seasat* oceanographic satellite

launched on June 26, 1978. At this time, ERTS-1 was retroactively named Landsat-1 and ERTS-B became Landsat-2 at launch. Landsat-3 was launched March 5, 1978; Landsat-4 on July 16, 1982; and Landsat-5 on March 1, 1984.

The Earth Observation Satellite Company (EOSAT) obtained control of the Landsat satellites in September, 1985. Unfortunately, Landsat 6 with its Enhanced Thematic Mapper (ETM) (a 15 × 15 m panchromatic band was added) failed to achieve orbit on October 5, 1993. Landsat 7 with its

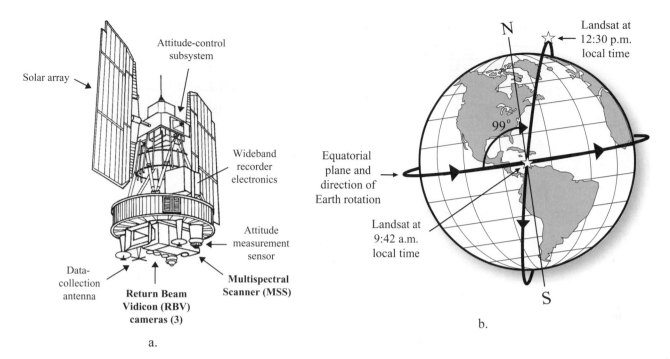

a.

b.

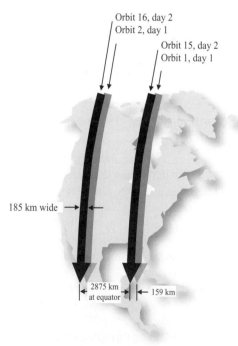

Orbit 16, day 2
Orbit 2, day 1

Orbit 15, day 2
Orbit 1, day 1

185 km wide

2875 km at equator

159 km

c.

Figure 2-10 a) Nimbus-style platform used for Landsats 1, 2, and 3 and associated sensor and telecommunication systems. b) Inclination of the Landsat orbit to maintain a Sun-synchronous orbit. c) From one orbit to the next, the position directly below the satellite moved 2875 km (1785 mi) at the equator as Earth rotated beneath it. The next day, 14 orbits later, it was approximately back to its original location, with orbit 15 displaced westward from orbit 1 by 159 km (99 mi). This is how repeat coverage of the same geographic area was obtained.

Enhanced Thematic Mapper Plus (ETM$^+$) sensor system was launched on April 15, 1999. For a detailed history of the Landsat program, refer to the *Landsat Data User Notes* published by the EROS Data Center (NOAA, 1975–1984), *Imaging Notes* published by Space Imaging, Inc., and the NASA Landsat 7 home page (NASA Landsat 7, 2004).

Landsats 1 through 3 were launched into circular orbits around Earth at a nominal altitude of 919 km (570 mi). The platform is shown in Figure 2-10a. The satellites had an orbital inclination of 99°, which made them nearly polar (Figure 2-10b) and caused them to cross the equator at an angle of approximately 9° from normal. The satellites

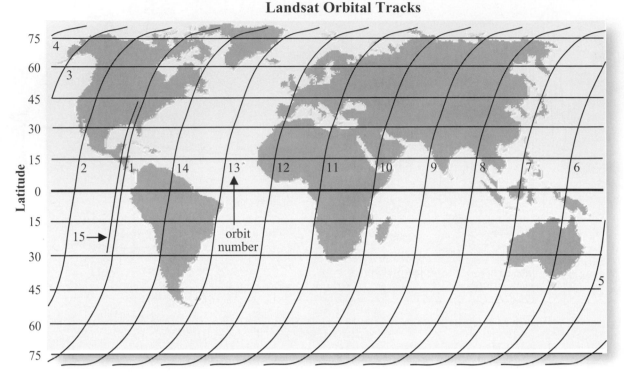

Figure 2-11   Orbital tracks of Landsat 1, 2, or 3 during a single day of coverage. The satellite crossed the equator every 103 minutes, during which time the Earth rotated a distance of 2875 km under the satellite at the equator. Every 14 orbits, 24 hours elapsed.

orbited Earth once every 103 min, resulting in 14 orbits per day (Figure 2-10c). This Sun-synchronous orbit meant that the orbital plane precessed around Earth at the same angular rate at which Earth moved around the Sun. This characteristic caused the satellites to cross the equator at approximately the same local time (9:30 to 10:00 a.m.) on the illuminated side of Earth.

Figures 2-10c and 2-11 illustrate how repeat coverage of a geographic area was acquired. From one orbit to the next, a position directly below the spacecraft moved 2875 km (1785 mi) at the equator as the Earth rotated beneath it. The next day, 14 orbits later, it was back to its original location, with orbit 15 displaced westward from orbit 1 by 159 km (99 mi) at the equator. This continued for 18 days, after which orbit 252 fell directly over orbit 1 once again. Thus, the Landsat sensor systems had the capability of observing the entire globe (except poleward of 81°) once every 18 days, or about 20 times a year. There were approximately 26 km (16 mi) of overlap between successive orbits. This overlap was a maximum at 81° north and south latitudes (about 85%) and a minimum at the equator (about 14%). This has proven useful for stereoscopic analysis applications.

The nature of the orbiting Landsat system has given rise to a Path and Row *Worldwide Reference System* (WRS) for locating and obtaining Landsat imagery for any area on Earth. The WRS has catalogued the world's landmass into 57,784 scenes. Each scene is approximately 185 km wide by 170 km long and consists of approximately 3.8 gigabits of data. Figure 2-12 depicts a small portion of the 1993 WRS map for the southeastern United States with one scene highlighted. The user locates the area of interest on the path and row map (e.g., Path 16, Row 37 is the nominal Charleston, SC, scene) and then requests information from Space Imaging, Inc., or the EROS Data Center at Sioux Falls, SD, about Landsat imagery available for this path and row. If no path and row map is available, a geographic search can be performed by specifying the longitude and latitude at the center of the area of interest or by defining an area of interest with the longitude and latitude coordinates of each corner.

In the context of this section on data acquisition, we are interested in the type of sensors carried aloft by the Landsat satellites and the nature and quality of remote sensor data provided for Earth resource investigations. The most impor-

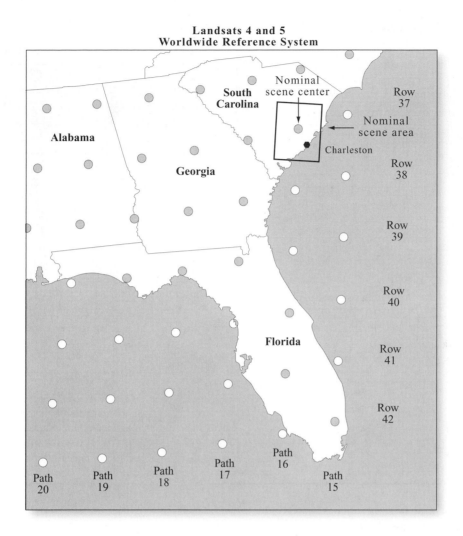

Figure 2-12   Portion of the 1993 Landsat Worldwide Reference System. Path 16, Row 37 is near Charleston, SC.

tant sensors were the Multispectral Scanner and Landsat Thematic Mapper.

Landsat Multispectral Scanner

The Landsat Multispectral Scanner was placed on Landsat satellites 1 through 5. The MSS multiple-detector array and the scanning system are shown diagrammatically in Figure 2-13a. Sensors such as the Landsat MSS (and Thematic Mapper to be discussed) are optical-mechanical systems in which a mirror scans the terrain perpendicular to the flight direction. While it scans, it focuses energy reflected or emitted from the terrain onto discrete detector elements. The detectors convert the radiant flux measured within each instantaneous-field-of-view (IFOV) in the scene into an electronic signal (Figure 2-13a). The detector elements are placed behind filters that pass broad portions of the spec-

trum. The MSS has four sets of filters and detectors, whereas the TM has seven. The primary limitation of this approach is the short residence time of the detector in each IFOV. To achieve adequate signal-to-noise ratio without sacrificing spatial resolution, such a sensor must operate in broad spectral bands of ≥ 100 nm or must use optics with unrealistically small ratios of focal-length to aperture ($f/$ stop).

The MSS scanning mirror oscillates through an angular displacement of ±5.78° off-nadir. This 11.56° field of view resulted in a swath width of approximately 185 km (115 mi) for each orbit. Six parallel detectors sensitive to four spectral bands (channels) in the electromagnetic spectrum viewed the ground simultaneously: 0.5 to 0.6 μm (green), 0.6 to 0.7 μm (red), 0.7 to 0.8 μm (reflective infrared), and 0.8 to 1.1 μm (reflective infrared). These bands were originally numbered 4, 5, 6, and 7, respectively, because a Return-Beam-

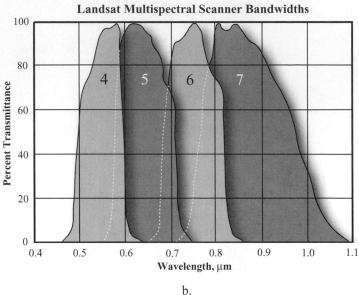

Landsat Multispectral Scanner Bandwidths

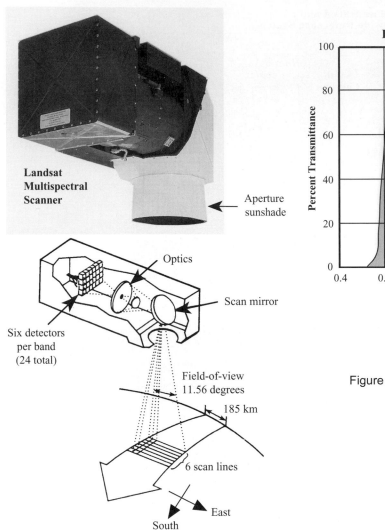

a.

Figure 2-13 a) Major components of the Landsat Multispectral Scanner system on Landsats 1 through 5 (Landsat 3 also had a thermal infrared band). A bank of 24 detectors (six for each of the four bands) measures information from Earth from an instantaneous field of view of 79 m × 79 m. b) Landsat Multispectral Scanner bandwidths. Notice that they do not end abruptly, as suggested by the usual nomenclature.

Vidicon (RBV) sensor system also onboard the satellite recorded energy in three bands labeled 1, 2, and 3.

When not viewing the Earth, the MSS detectors were exposed to internal light and Sun calibration sources. The spectral sensitivity of the bands is summarized in Table 2-2 and shown diagrammatically in Figure 2-13b. Note that there is spectral overlap between the bands.

Prior to the launch of the satellite, the engineering model of the ERTS MSS was tested by viewing the scene behind the Santa Barbara Research Center in Goleta, CA. Bands 4 and 6 (green and near-infrared, respectively) of the area are shown in Figure 2-14. Note the spatial detail present when the sensor is located only 1 to 2 km from the mountains. The

spatial resolution is much lower when the sensor is placed 919 km above Earth in orbit.

The IFOV of each detector was square and resulted in a ground resolution element of approximately 79 × 79 m (67,143 ft²). The voltage analog signal from each detector was converted to a digital value using an onboard A-to-D converter. The data were quantized to 6 bits with a range of values from 0 to 63. These data were then rescaled to 7 bits (0 to 127) for three of the four bands in subsequent ground processing (i.e., bands 4, 5, and 6 were decompressed to a range of 0 to 127). It is important to remember that the early 1970s Landsat MSS data were quantized to 6 bits when comparing MSS data collected in the late 1970s and 1980s, which were collected at 8 bits.

Terrestrial Images of Goleta, CA, Obtained Using the Landsat Multispectral Scanner

a. Landsat MSS band 4 (0.5 - 0.6 μm).

b. Landsat MSS band 6 (0.7 - 0.8 μm).

Figure 2-14   Two terrestrial images acquired by the engineering model of the Landsat MSS on March 4, 1972, at the Santa Barbara Research Center of Hughes Aircraft, Inc. The top image (a) was acquired using the MSS band 4 detectors (0.5 – 0.6 μm), and the bottom image (b) was acquired using band 6 detectors (0.7 – 0.8 μm). Note the high spatial fidelity of the images, which is possible when the terrain is close and not 919 km away.

During each scan, the voltage produced by each detector was sampled every 9.95 μs. For one detector, approximately 3300 samples were taken along a 185-km line. Thus, the IFOV of 79 m × 79 m became about 56 m on the ground between each sample (Figure 2-15). The 56 × 79 m area is called a Landsat MSS picture element. Thus, although the measurement of landscape brightness was made from a 6,241 m² area, each pixel was reformatted as if the measurement were made from a 4424 m² area (Figure 2-15). Note

the overlap of the areas from which brightness measurements were made for adjacent pixels.

The MSS scanned each line across-track from west to east as the southward orbit of the spacecraft provided the along-track progression. Each MSS scene represents a 185 × 170 km parallelogram extracted from the continuous swath of an orbit and contains approximately 10 percent overlap. A typical scene contains approximately 2340 scan lines with

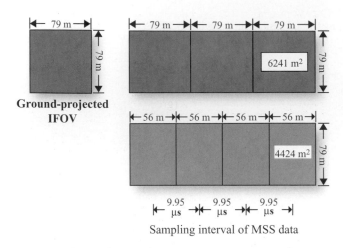

Figure 2-15 Relationship between the original 79 × 79 m ground-projected IFOV of the Landsat MSS and the rate at which it was resampled (i.e., every 9.95 μs). This resulted in picture elements (pixels) that were 56 × 79 m in dimension on tapes acquired from the EROS Data Center at Sioux Falls, SD.

about 3240 pixels per line, or about 7,581,600 pixels per channel. All four bands represent a data set of more than 30 million brightness values. Landsat MSS images provided an unprecedented ability to observe large geographic areas while viewing a single image. For example, approximately 5000 conventional vertical aerial photographs obtained at a scale of l:15,000 are required to equal the geographic coverage of a single Landsat MSS image. This allows regional terrain analysis to be performed using one data source rather than a multitude of aerial photographs.

Landsat Thematic Mapper

Landsat Thematic Mapper sensor systems were launched on July 16, 1982 (Landsat 4), and on March 1, 1984 (Landsat 5). The TM is an optical-mechanical whiskbroom sensor that records energy in the visible, reflective-infrared, middle-infrared, and thermal infrared regions of the electromagnetic spectrum. It collects multispectral imagery that has higher spatial, spectral, temporal, and radiometric resolution than the Landsat MSS. Detailed descriptions of the design and performance characteristics of the TM can be found in EOSAT (1992).

The Landsat 4 and 5 platform is shown in Figure 2-16. The Thematic Mapper sensor system configuration is shown in Figure 2-17. A telescope directs the incoming radiant flux obtained along a scan line through a scan line corrector to 1) the visible and near-infrared primary focal plane or 2) the

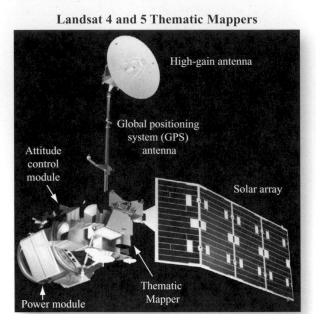

Landsat 4 and 5 Thematic Mappers

Figure 2-16 Landsat 4 and 5 platform and associated sensor and telecommunication systems.

middle-infrared and thermal infrared cooled focal plane. The detectors for the visible and near-infrared bands (1 – 4) are four staggered linear arrays, each containing l6 silicon detectors. The two middle-infrared detectors are l6 indium-antimonide cells in a staggered linear array, and the thermal infrared detector is a four-element array of mercury-cadmium-telluride cells.

Landsat TM data have a ground-projected IFOV of 30 × 30 m for bands 1 through 5 and 7. The thermal infrared band 6 has a spatial resolution of 120 × 120 m. The TM spectral bands represent important departures from the bands found on the traditional MSS, also carried onboard Landsats 4 and 5. The original MSS bandwidths were selected based on their utility for vegetation inventories and geologic studies. Conversely, the TM bands were chosen after years of analysis for their value in water penetration, discrimination of vegetation type and vigor, plant and soil moisture measurement, differentiation of clouds, snow, and ice, and identification of hydrothermal alteration in certain rock types (Table 2-3). The refined bandwidths and improved spatial resolution of the Landsat TM versus the Landsat MSS and several other sensor systems (Landsat 7 and SPOTs 1 – 4) are shown graphically in Figure 2-18. Examples of individual bands of Landsat Thematic Mapper imagery of Charleston, SC, obtained in 1994 are provided in Figure 2-19.

Table 2-3.   Characteristics of the Landsat 4 and 5 Thematic Mapper spectral bands.

Band 1: 0.45 – 0.52 μm (blue). This band provides increased penetration of waterbodies, as well as supporting analyses of land-use, soil, and vegetation characteristics. The shorter-wavelength cutoff is just below the peak transmittance of clear water, and the upper-wavelength cutoff is the limit of blue chlorophyll absorption for healthy green vegetation. Wavelengths < 0.45 μm are substantially influenced by atmospheric scattering and absorption.

Band 2: 0.52 – 0.60 μm (green). This band spans the region between the blue and red chlorophyll absorption bands and reacts to the green reflectance of healthy vegetation.

Band 3: 0.63 – 0.69 μm (red). This is the red chlorophyll absorption band of healthy green vegetation and is useful for vegetation discrimination. It is also useful for soil-boundary and geological-boundary delineations. This band may exhibit more contrast than bands 1 and 2 because of the reduced effect of atmospheric attenuation. The 0.69-μm cutoff is significant because it represents the beginning of a spectral region from 0.68 to 0.75 μm, where vegetation reflectance crossovers take place that can reduce the accuracy of vegetation investigations.

Band 4: 0.76 – 0.90 μm (near-infrared). For reasons discussed, the lower cutoff for this band was placed above 0.75 μm. This band is very responsive to the amount of vegetation biomass present. It is useful for crop identification and emphasizes soil/crop and land/water contrasts.

Band 5: 1.55 – 1.75 μm (mid-infrared). This band is sensitive to the turgidity or amount of water in plants. Such information is useful in crop drought studies and in plant vigor investigations. This is one of the few bands that can be used to discriminate among clouds, snow, and ice.

Band 6: 10.4 – 12.5 μm (thermal infrared). This band measures the amount of infrared radiant energy emitted from surfaces. The apparent temperature is a function of the emissivities and the true (kinetic) temperature of the surface. It is useful for locating geothermal activity, thermal inertia mapping for geologic investigations, vegetation classification, vegetation stress analysis, and soil moisture studies. The band often captures unique information on differences in topographic aspect in mountainous areas.

Band 7: 2.08 – 2.35 μm (mid-infrared). This is an important band for the discrimination of geologic rock formations. It has been shown to be effective for identifying zones of hydrothermal alteration in rocks.

The Landsat TM bands were selected to make maximum use of the dominant factors controlling leaf reflectance, such as leaf pigmentation, leaf and canopy structure, and moisture content, as demonstrated in Figure 2-20. Band 1 (blue) provides some water-penetration capability. Vegetation absorbs much of the incident blue, green, and red radiant flux for photosynthetic purposes; therefore, vegetated areas appear dark in TM band 1 (blue), 2 (green), and 3 (red) images, as seen in the Charleston, SC, Landsat TM data (see Figure 2-19). Vegetation reflects approximately half of the incident near-infrared radiant flux, causing it to appear bright in the band 4 (near-infrared) image. Bands 5 and 7 both provide more detail in the wetland because they are sensitive to soil and plant moisture conditions. The band 6 (thermal) image provides limited information of value.

The equatorial crossing time was 9:45 a.m. for Landsats 4 and 5 with an orbital inclination of 98.2°. The transition from an approximately 919 km orbit to a 705 km orbit for Landsats 4 and 5 disrupted the continuity of Landsats 1, 2, and 3 MSS path and row designations in the Worldwide Reference System. Consequently, a separate WRS map is required to select images obtained by Landsats 4 and 5. The lower orbit (approximately the same as the space shuttle) also increased the amount of relief displacement introduced into the imagery obtained over mountainous terrain. The new orbit also caused the period between repetitive coverage to change from 18 to 16 days for both the MSS and TM data collected by Landsats 4 and 5.

There was a substantial improvement in the level of quantization from 6 to 8 bits per pixel (Table 2-2). This, in addition to a greater number of bands and a higher spatial resolution, increased the data rate from 15 to 85 Mb/s. Ground receiving stations were modified to process the increased data flow. Based on the improvements in spectral, spatial, and radiometric resolution, Solomonson (1984) suggested that "it appears that the TM can be described as being twice as effective in providing information as the Landsat MSS. This is based on its ability to provide twice as many separable classes over a given area as the MSS, numerically provide two more independent vectors in the data or demonstrate through classical information theory that twice as much information exists in the TM data."

Efforts to move the Landsat Program into the commercial sector began under the Carter Administration in 1979 and resulted in legislation passed in 1984 that charged the National Oceanic and Atmospheric Administration to transfer the program to the private sector. The Earth Observing Satellite Company took over operation in 1985 and was given the rights to market Landsat TM data.

Landsat Thematic Mapper

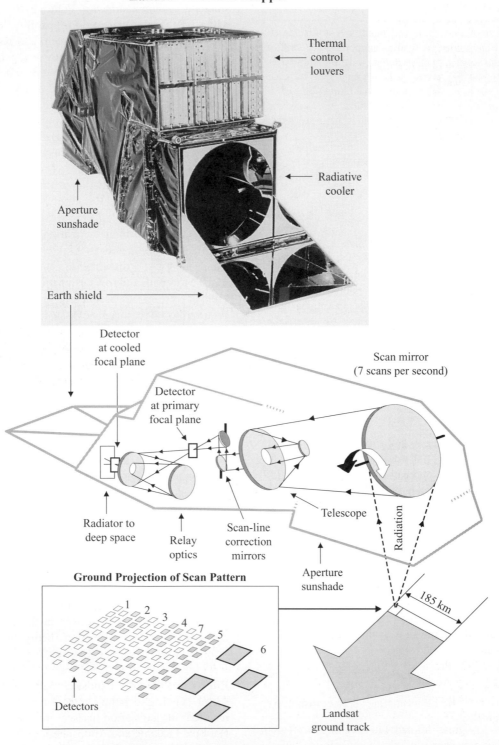

Figure 2-17   Major components of the Landsats 4 and 5 Thematic Mapper sensor system. The sensor is sensitive to the seven bands of the electromagnetic spectrum summarized in Table 2-2. Six of the seven bands have a spatial resolution of 30×30 m; the thermal infrared band has a spatial resolution of 120×120 m. The lower diagram depicts the sensor in its operational position.

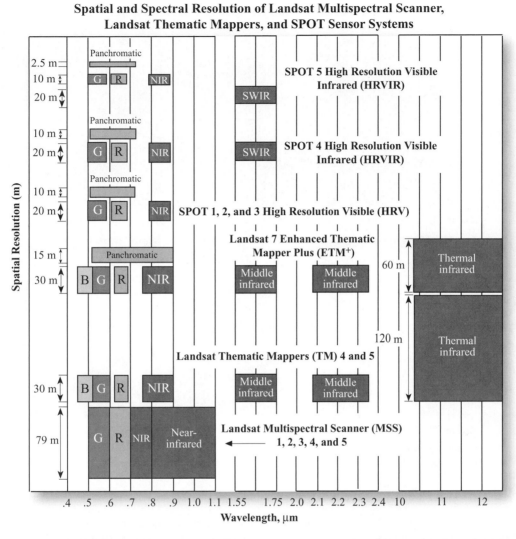

Figure 2-18 Spatial and spectral resolution of the Landsat Multispectral Scanner (MSS), Landsats 4 and 5 Thematic Mapper (TM), Landsat 7 Enhanced Thematic Mapper Plus (ETM⁺), SPOTs 1, 2, and 3 High Resolution Visible (HRV), and SPOTs 4 and 5 High Resolution Visible Infrared (HRVIR) sensor systems. The SPOTs 4 and 5 *Vegetation* sensor characteristics are not shown (it consists of four 1.15 × 1.15 km bands).

Landsat 7 Enhanced Thematic Mapper Plus

On October 28, 1992, President Clinton signed the Land Remote Sensing Policy Act of 1992 (Public Law 102-555). The law authorized the procurement of Landsat 7 and called for its launch within 5 years of the launch of Landsat 6. In parallel actions, Congress funded Landsat 7 procurement and stipulated that data from publicly funded remote sensing satellite systems like Landsat must be sold to United States government agencies and their affiliated users at the cost of fulfilling user requests. Unfortunately, Landsat 6 did not achieve orbit on October 5, 1993.

With the passage of the Land Remote Sensing Policy Act of 1992, oversight of the Landsat program began to be shifted from the commercial sector back to the federal government. NASA was responsible for the design, development, launch and on-orbit checkout of Landsat 7, and the installation and operation of the ground system. The U.S. Geological Survey (USGS) was responsible for data capture, processing, and distribution of the Landsat 7 data, mission management, and maintaining the Landsat 7 data archive.

Landsat 7 was officially integrated into NASA's Earth Observing System in 1994. It was launched on April 15,

Landsat 5 Thematic Mapper Data of Charleston, SC

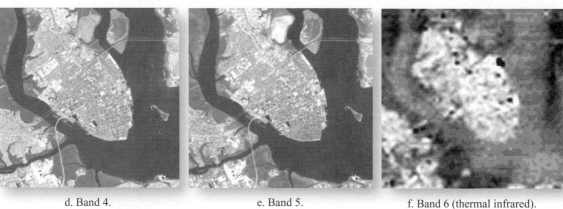

a. Band 1.   b. Band 2.   c. Band 3.

d. Band 4.   e. Band 5.   f. Band 6 (thermal infrared).

g. Band 7.

Figure 2-19   Landsat Thematic Mapper data of Charleston, SC, obtained on February 3, 1994. Bands 1 through 5 and 7 are 30 × 30 m spatial resolution. Band 6 is 120 × 120 m (courtesy Space Imaging, Inc.).

1999, from Vandenburg Air Force Base, CA, into a Sun-synchronous orbit (Figure 2-21). Landsat 7 was designed to work in harmony with NASA's EOS *Terra* satellite. It was designed to achieve three main objectives (NASA Landsat 7, 2004):

• maintain data continuity by providing data that are consistent in terms of geometry, spatial resolution, calibration, coverage characteristics, and spectral characteristics with previous Landsat data;

• generate and periodically refresh a global archive of substantially cloud-free, sunlit landmass imagery; and

• continue to make Landsat-type data available to U.S. and international users at the cost of fulfilling user requests (COFUR) and to expand the use of such data for global-change research and commercial purposes.

Landsat 7 is a three-axis stabilized platform carrying a single nadir-pointing instrument, the ETM+ (Figure 2-21). The

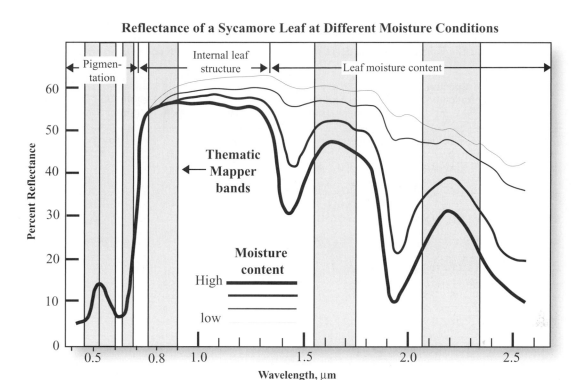

Figure 2-20   Progressive changes in percent reflectance for a sycamore leaf at varying oven-dry-weight moisture content. The dominant factors controlling leaf reflectance and the location of six of the Landsat Thematic Mapper bands are superimposed.

Landsat 7 Enhanced Thematic Mapper Plus

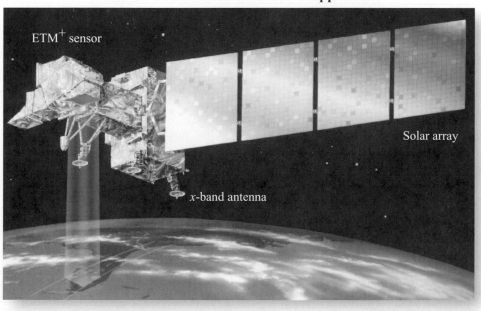

Figure 2-21   Artist's rendition of the Landsat 7 satellite with its Enhanced Thematic Mapper Plus (ETM+) sensor system (courtesy of Lockheed Martin, Inc.).

Table 2-4.   Landsat Enhanced Thematic Mapper Plus (ETM$^+$) compared with the Earth Observer (EO-1) sensors.

Landsat 7 Enhanced Thematic Mapper Plus (ETM$^+$)			EO-1 Advanced Land Imager (ALI)		
Band	Spectral Resolution (μm)	Spatial Resolution (m) at Nadir	Band	Spectral Resolution (μm)	Spatial Resolution (m) at Nadir
1	0.450 – 0.515	30 × 30	MS-1	0.433 – 0.453	30 × 30
2	0.525 – 0.605	30 × 30	MS-1	0.450 – 0.510	30 × 30
3	0.630 – 0.690	30 × 30	MS-2	0.525 – 0.605	30 × 30
4	0.750 – 0.900	30 × 30	MS-3	0.630 – 0.690	30 × 30
5	1.55 – 1.75	30 × 30	MS-4	0.775 – 0.805	30 × 30
6	10.40 – 12.50	60 × 60	MS-4'	0.845 – 0.890	30 × 30
7	2.08 – 2.35	30 × 30	MS-5'	1.20 – 1.30	30 × 30
8 (panchromatic)	0.52 – 0.90	15 × 15	MS-5	1.55 – 1.75	30 × 30
			MS-7	2.08 – 2.35	30 × 30
			Panchromatic	0.480 – 0.690	10 × 10
			EO-1 Hyperion Hyperspectral Sensor 220 bands from 0.4 to 2.4 μm at 30 × 30 m		
			EO-1 LEISA Atmospheric Corrector (LAC) 256 bands from 0.9 to 1.6 μm at 250 × 250 m		
Sensor Technology	Scanning mirror spectrometer		Advanced Land Imager is a pushbroom radiometer. Hyperion is a pushbroom spectroradiometer. LAC uses area arrays.		
Swath Width	185 km		ALI = 37 km; Hyperion = 7.5 km; LAC = 185 km		
Data Rate	250 images per day @ 31,450 km^2		—		
Revisit	16 days		16 days		
Orbit and Inclination	705 km, Sun-synchronous Inclination = 98.2° Equatorial crossing 10:00 a.m. ±15 min.		705 km, Sun-synchronous Inclination = 98.2° Equatorial crossing = Landsat 7 + 1 min		
Launch	April 15, 1999		November 21, 2000		

ETM$^+$ instrument is a derivative of the Landsat 4 and 5 Thematic Mapper sensors. Therefore, it is possible to refer to Figure 2-17 for a review of its mirror and detector design. The ETM$^+$ is based on scanning technology despite the fact that linear array "pushbroom" technology has been commercially available since the launch of the French SPOT 1 satellite in 1986. Nevertheless, the ETM$^+$ instrument is an excellent sensor with several notable improvements over its predecessors Landsat 4 and 5.

The characteristics of the Landsat 7 ETM$^+$ are summarized in Tables 2-3 and 2-4. The ETM$^+$ bands 1 through 5 and 7 are identical to those found on Landsats 4 and 5 and have the same 30 × 30 m spatial resolution. The thermal infrared band 6 has 60 × 60 m spatial resolution (instead of 120 × 120 m). Perhaps most notable is the new 15 × 15 m panchromatic band (0.52 – 0.90 μm). Landsat 7 ETM$^+$ images of San Diego, CA, are shown in Figure 2-22. An ETM$^+$ color composite of San Diego is displayed in **Color Plate 2-1.**

Landsat 7 Enhanced Thematic Mapper Plus Imagery of San Diego, CA

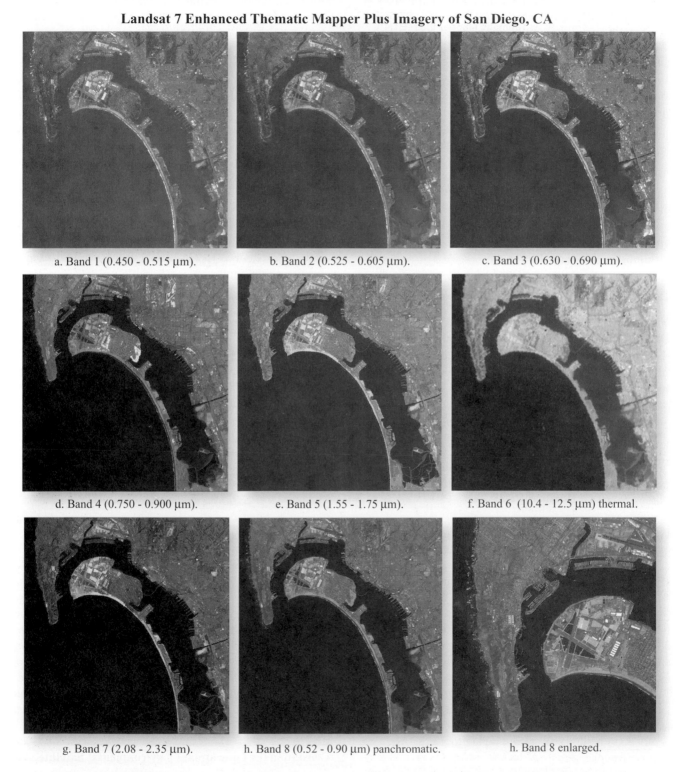

a. Band 1 (0.450 - 0.515 μm). b. Band 2 (0.525 - 0.605 μm). c. Band 3 (0.630 - 0.690 μm).

d. Band 4 (0.750 - 0.900 μm). e. Band 5 (1.55 - 1.75 μm). f. Band 6 (10.4 - 12.5 μm) thermal.

g. Band 7 (2.08 - 2.35 μm). h. Band 8 (0.52 - 0.90 μm) panchromatic. h. Band 8 enlarged.

Figure 2-22   Landsat 7 ETM$^+$ imagery of San Diego, CA, obtained on April 24, 2000. Bands 1 through 5 and 7 are 30×30 m. Thermal infrared band 6 is 60×60 m. Panchromatic band 8 is 15×15 m (courtesy of NASA).

Landsat 7 is in orbit 705 km above the Earth, collects data in a swath 185 km wide, and cannot view off-nadir. Its revisit interval is 16 days. It has a 378-gigabit solid-state recorder that can hold 42 minutes of sensor data and 29 hours of housekeeping telemetry data. The ETM$^+$ records 150 megabits of data each second. Landsat 7 can transmit data directly to ground receiving stations at the EROS Data Center in Sioux Falls, SD, or to Fairbanks, AK. Landsat 7 international data may by acquired by retransmission using TDRS satellites or by international receiving stations.

The Landsat 7 ETM$^+$ has excellent radiometric calibration, which is accomplished using partial and full aperture solar calibration. Ground look calibration is performed by acquiring images of certain Earth landmass calibration targets. Biophysical and atmospheric characteristics of these targets are well instrumented on the ground.

Data acquisition by the ETM$^+$ is directed by the mission goal of acquiring and updating periodically a global archive of daytime, substantially cloud-free images of land areas. At one time approximately 250 images were processed by the EROS Data Center each day. Unfortunately, the ETM$^+$ Scan Line Corrector (SLC) failed on May 31, 2003, resulting in imagery with significant geometric error. The SLC compensates for the forward motion of the satellite. Efforts to recover the SLC were not successful. Portions of ETM$^+$ scenes obtained after this date in SLC-off mode are usable after special processing (USGS Landsat 7, 2004).

Landsat 7 data are processed to several product levels, including (NASA Landsat 7, 2004):

• Level 0R: Reformatted, raw data including reversing the order of the reverse scan data, aligning the staggered detectors, nominal alignment of the forward and reverse scans, and metadata. The data are not geometrically corrected.

• Level 1R: Radiometrically corrected data and metadata describing calibration parameters, payload correction data, mirror scan correction data, a geolocation table, and internal calibration lamp data.

• Level 1G: Data are radiometrically and geometrically corrected. Metadata describes calibration parameters and a geolocation table. Data are resampled to a user-specified map projection.

The Level 0R data are distributed for no more than $500 per scene. The Level 1R and 1G products are more expensive. All users are charged standard prices.

Follow-on Landsat satellites may be designed based on the Earth Observer mission launched on November 21, 2000 (NASA EO-1, 2004). It was inserted into a 705-km circular, Sun-synchronous orbit at a 98.7° inclination such that it was flying in formation 1 minute behind Landsat 7 in the same ground track. This close separation enabled EO-1 to observe the same ground location (scene) through the same atmospheric region so that scene comparisons between the two satellites can be made.

EO-1 specifications are summarized in Table 2-4. It contains a linear array Advanced Land Imager (ALI) with 10 bands from 0.4 to 2.35 μm at 30×30 m spatial resolution. The Hyperion hyperspectral sensor records data in 220 bands from 0.4 to 2.4 μm at 30×30 m spatial resolution. The Linear Etalon Imaging Spectrometer Array (LEISA) Atmospheric Corrector is a 256-band hyperspectral instrument sensitive to the region from 0.9 to 1.6 μm at 250×250 m spatial resolution. It is designed to correct for water-vapor variations in the atmosphere. All three of the EO-1 land imaging instruments view all or subsegments of the Landsat 7 swath.

On September 26, 2003, NASA cancelled the implementation phase for the Landsat Data Continuity Mission. NASA is considering other options for ensuring the continuity of Landsat data consistent with the guidance contained in the Remote Sensing Policy Act of 1992 (Public Law 102-555) (USGS Landsat 7, 2004; USGS LDCM, 2004).

National Atmospheric and Oceanic Administration Multispectral Scanner Sensors

NOAA operates two series of remote sensing satellites: the Geostationary Operational Environmental Satellites (GOES) and the Polar-orbiting Operational Environmental Satellites (POES). Both are currently based on multispectral scanner technology. The U.S. National Weather Service uses data from these sensors to forecast the weather. We often see GOES images of North and South America weather patterns on the daily news. The Advanced Very High Resolution Radiometer was developed for meteorological purposes. However, global climate change research has focused attention on the use of AVHRR data to map vegetation and sea-surface characteristics..

Geostationary Operational Environmental Satellite

The GOES system is operated by the National Environmental Satellite Data and Information Service (NESDIS) of NOAA. The system was developed by NESDIS in conjunc-

Table 2-5.   NOAA Geostationary Operational Environmental Satellite (GOES) Imager sensor system characteristics.

GOES-8, 10, 12 Band	Spectral Resolution (µm)	Spatial Resolution (km)	Band Utility
1	0.52 – 0.72	1 × 1	Clouds, pollution, haze detection, and identification of severe storms
2	3.78 – 4.03	4 × 4	Fog detection, discriminates between water, clouds, snow or ice clouds during daytime, detects fires and volcanoes, nighttime sea surface temperature (SST)
3	6.47 – 7.02	8 × 8	Estimation of mid- and upper-level water vapor, detects advection, and tracks mid-level atmospheric motion
4	10.2 – 11.2	4 × 4	Cloud-drift winds, severe storms, cloud-top heights, heavy rainfall
5	11.5 – 12.5	4 × 4	Identification of low-level water vapor, SST, and dust and volcanic ash

tion with NASA. NOAA's new generation of geostationary satellites, GOES-8 East, was launched in April, 1994. GOES-9 was launched on May 23, 1995. GOES-10 West was launched April 25, 1997 (Figure 2-23a). GOES-M was launched July 23, 2001, to become GOES-12 (NOAA GOES, 2004).

The GOES system consists of several observing subsystems:

• GOES Imager (provides multispectral image data),

• GOES Sounder (provides hourly 19-channel soundings), and

• a data-collection system (DCS) that relays data from *in situ* sites at or near the Earth's surface to other locations.

The GOES spacecraft is a three-axis (x,y,z) stabilized design capable of continuously pointing the optical line of sight of the imaging and sounding radiometers toward the Earth. GOES are placed in geostationary orbits approximately 35,790 km (22,240 statute miles) above the equator. The satellites remain at a stationary point above the equator and rotate at the same speed and direction as Earth. This enables the sensors to stare at a portion of the Earth from the geosynchronous orbit and thus more frequently obtain images of clouds, monitor the Earth's surface temperature and water vapor characteristics, and sound the Earth's atmosphere for its vertical thermal and water vapor structures.

GOES-8 East is normally situated at 75° W longitude and GOES-10 West is at 135° W longitude. The geographic coverage of GOES East and GOES West is summarized in Figure 2-23a. These sensors view most of the Earth from approximately 20° W to 165° E longitude. Poleward coverage is between approximately 77° N and S latitude. GOES East and West view the contiguous 48 states, South America, and major portions of the central and eastern Pacific Ocean and the central and western Atlantic Ocean. Pacific coverage includes the Hawaiian Islands and Gulf of Alaska, the latter known to weather forecasters as "the birthplace of North American weather systems" (Loral Space Systems, 1996).

GOES Imager: The Imager is a five-channel multispectral scanner. The bandwidths and spatial resolution are summarized in Table 2-5. By means of a two-axis gimballed mirror in conjunction with a 31.1-cm (12.2-in.) diameter Cassegrain telescope, the Imager's multispectral channels can simultaneously sweep an 8-km (5-statute mile) north-to-south swath along an east-to-west/west-to-east path, at a rate of 20° (optical) per second (Loral Space Systems, 1996). The telescope concentrates both the visible and thermal radiant flux from the terrain onto a 5.3-cm secondary mirror (Figure 2-23b). Dichroic beamsplitters separate the incoming scene radiance and focus it onto 22 detectors (8 visible and 14 thermal). The visible energy passes through the initial beamsplitter and is focused onto 8 silicon visible detector elements. Each of the 8 visible detectors has an IFOV of approximately 1 × 1 km at the satellite's suborbital point on the Earth.

All thermal infrared energy is deflected to the specialized detectors in the radiative cooler. The thermal infrared energy is further separated into the 3.9-, 6.75-, 10.7-, and 12-µm channels. Each of the four infrared channels has a separate set of detectors: four-element indium-antimonide (InSb) detectors for band 2; two-element mercury-cadmium-telluride (Hg:Cd:Te) detectors for band 3; and four-element mercury-cadmium-telluride (Hg:Cd:Te) detectors for both bands 4 and 5.

GOES East and West Coverage

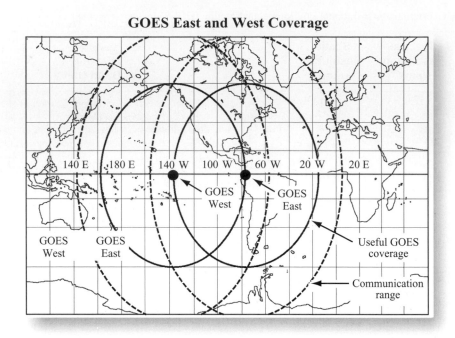

a.

GOES Imager Optical Elements

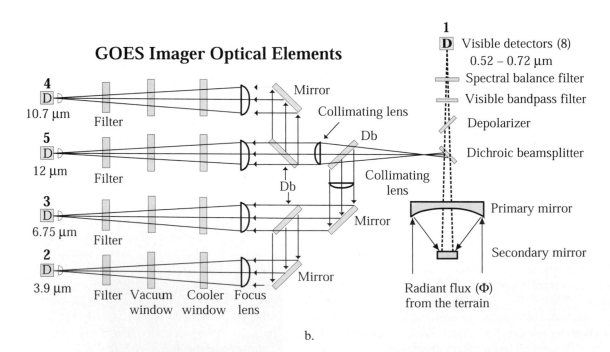

b.

Figure 2-23   a) Geographic coverage of GOES East (75° W) and GOES West (135° W). b) Radiant flux from the terrain is reflected off a scanning mirror (not shown) onto the primary and secondary mirrors. A dichroic beamsplitter separates the visible light from the thermal infrared energy. Subsequent beamsplitters separate the thermal energy into specific bands (after Loral Space Systems, 1996).

Geostationary Operational Environmental Satellite Imagery

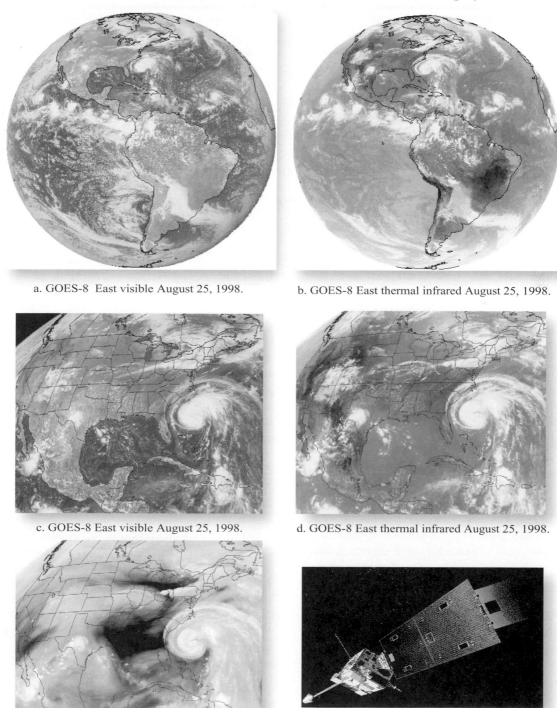

a. GOES-8 East visible August 25, 1998.

b. GOES-8 East thermal infrared August 25, 1998.

c. GOES-8 East visible August 25, 1998.

d. GOES-8 East thermal infrared August 25, 1998.

e. GOES-8 East water vapor August 25, 1998.

f. GOES-8 satellite.

Figure 2-24   a–e) Examples of GOES-8 imagery obtained on August 25, 1998. f) The GOES-8 satellite (images courtesy of NOAA).

The GOES channels have 10-bit radiometric precision. The primary utility of the visible band 1 (1 × 1 km) is in the daytime monitoring of thunderstorms, frontal systems, and tropical cyclones. Band 2 (4 × 4 km) responds to both emitted terrestrial radiation and reflected solar radiation. It is useful for identifying fog and discriminating between water and ice clouds, and between snow and clouds, and for identifying large or very intense fires. It can be used at night to track low-level clouds and monitor near-surface wind circulation. Band 3 (8 × 8 km) responds to mid- and upper-level water vapor and clouds. It is useful for identifying the jet stream, upper-level wind fields, and thunderstorms. Energy recorded by band 4 (4 × 4 km) is not absorbed to any significant degree by atmospheric gases. It is ideal for measuring cloud-top heights, identifying cloud-top features, assessing the severity of some thunderstorms, and tracking clouds and frontal systems at night. Thermal band 5 (4 × 4 km) is similar to band 4 except that this wavelength region has a unique sensitivity to low-level water vapor. GOES-8 East visible, thermal infrared, and water vapor images of Hurricane Bonnie on August 25, 1998, are shown in Figure 2-24a-e.

The Imager scans the continental United States every 15 minutes; scans most of the hemisphere from near the North Pole to approximately 20° S latitude every 30 minutes; and scans the entire hemisphere once every 3 hours in "routine" scheduling mode. Optionally, special imaging schedules are available, which allow data collection at more rapid time intervals over reduced geographic areas. During Rapid Scan Operations (RSO) and Super Rapid Scan Operations (SRSO), images are collected over increasingly reduced-area sectors at 7.5-min intervals (RSO) and at either 1-minute or 30-sec intervals (SRSO). A typical image acquired at 30-second intervals covers a rectangle of about 10° of latitude and 15° of longitude. The 1-minute SRSO collects 22 images per hour with 2 segments of 1-minute interval images, allowing for regularly scheduled 15-minute interval operational scans.

GOES Sounder: The GOES Sounder uses 1 visible and 18 infrared sounding channels to record data in a north-to-south swath across an east-to-west path. The Sounder and Imager both provide full Earth imagery, sector imagery, and area scans of local regions. The 19 bands yield the prime sounding products of vertical atmospheric temperature profiles, vertical moisture profiles, atmospheric layer mean temperature, layer mean moisture, total precipitable water, and the lifted index (a measure of stability). These products are used to augment data from the Imager to provide information on atmospheric temperature and moisture profiles, surface and cloud-top temperatures, and the distribution of atmospheric ozone (Loral Space Systems, 1996).

Advanced Very High Resolution Radiometer

The Satellite Services Branch of the National Climatic Data Center, under the auspices of NESDIS, has established a digital archive of data collected from the NOAA Polar-orbiting Operational Environmental Satellites (POES) (Kidwell, 1998). This series of satellites commenced with TIROS-N (launched in October 1978) and continued with NOAA-A (launched in March, 1983 and renamed NOAA-8) to the current NOAA-15. These Sun-synchronous polar-orbiting satellites carry the Advanced Very High Resolution Radiometer (AVHRR). Substantial progress has been made in using AVHRR data for land-cover characterization and the mapping of daytime and nighttime clouds, snow, ice, and surface temperature. Unlike the Landsat TM and Landsat 7 ETM[+] sensor systems with nadir revisit cycles of 16 days, the AVHRR sensors acquire images of the entire Earth two times each day (NOAA AVHRR, 2004). This high frequency of coverage enhances the likelihood that cloud-free observations can be obtained for specific temporal windows and makes it possible to monitor change in land-cover conditions over short periods, such as a growing season. Moreover, the moderate resolution (1.1 × 1.1 km) of the AVHRR data makes it feasible to collect, store, and process continental or global datasets. For these reasons, NASA and NOAA initiated the AVHRR Pathfinder Program to create universally available global long-term remotely sensed datasets that can be used to study global climate change.

The AVHRR satellites orbit at approximately 833 km above Earth at an inclination of 98.9° and continuously record data in a swath 2700 km wide at 1.1 × 1.1 km spatial resolution at nadir. Normally, two NOAA-series satellites are operational at one time (one odd, one even). The odd-numbered satellite typically crosses the equator at approximately 2:30 p.m. and 2:30 a.m., and the even-numbered satellite crosses the equator at 7:30 p.m. and 7:30 a.m. local time. Each satellite orbits Earth 14.1 times daily (every 102 min) and acquires complete global coverage every 24 hours. Orbital tracks of the NOAA-17 satellite on October 17, 2003, are shown in Figure 2-25. Because the number of orbits per day is not an integer, the suborbital tracks do not repeat on a daily basis, although the local solar time of the satellite's passage is essentially unchanged for any latitude. However, the satellite's orbital drift over time causes a systematic change of illumination conditions and local time of observation, which is a source of nonuniformity when analyzing multidate AVHRR data.

The AVHRR is a cross-track scanning system. The scanning rate of the AVHRR is 360 scans per minute. A total of 2,048 samples (pixels) are obtained per channel per Earth scan,

Table 2-6.   NOAA Advanced Very High Resolution Radiometer sensor system characteristics.

Band	NOAA-6, 8, 10 Spectral Resolution (μm)[a]	NOAA-7, 9, 11, 12, 13, 14 Spectral Resolution (μm)[a]	NOAA-15, 16, 17 AVHRR/3 Spectral Resolution (μm)[a]	Band Utility
1	0.580 – 0.68	0.580 – 0.68	0.580 – 0.68	Daytime cloud, snow, ice, and vegetation mapping; used to compute NDVI
2	0.725 – 1.10	0.725 – 1.10	0.725 – 1.10	Land/water interface, snow, ice, and vegetation mapping; used to compute NDVI
3	3.55 – 3.93	3.55 – 3.93	*3A*: 1.58 - 1.64 *3B*: 3.55 – 3.93	Monitor hot targets (volcanoes, forest fires), night-time cloud mapping
4	10.50 – 11.50	10.30 – 11.30	10.30 – 11.30	Day/night cloud and surface-temperature mapping
5	None	11.50 – 12.50	11.50 – 12.50	Cloud and surface temperature, day and night cloud mapping; removal of atmospheric water vapor path radiance

IFOV at nadir	1.1×1.1 km
Swath width	2700 km at nadir

[a] TIROS-N was launched on October 13, 1978; NOAA-6 on June 27, 1979; NOAA-7 on June 23, 1981; NOAA-8 on March 28, 1983; NOAA-9 on December 12, 1984; NOAA-10 on September 17, 1986; NOAA-11 on September 24, 1988; NOAA-12 on May 14, 1991; NOAA-13 on August 9, 1993; NOAA-14 on December 30, 1994; NOAA (K)-15 on May 13, 1998; NOAA (L)-16 on September 21, 2000; NOAA (M)-17 on June 24, 2002.

which spans an angle of $\pm 55.4°$ off-nadir. The IFOV of each band is approximately 1.4 milliradians leading to a resolution at the satellite subpoint of 1.1×1.1 km (Figure 2-26a). The more recent AVHRR systems have five channels (Table 2-6; Figure 2-26b).

Full resolution AVHRR data obtained at 1.1×1.1 km is called *local area coverage* (*LAC*) data. It may be resampled to 4×4 km *global area coverage* (*GAC*) data. The GAC data contains only one out of three original AVHRR lines and the data volume and resolution are further reduced by starting with the third sample along the scan line, averaging the next four samples, and skipping the next sample. The sequence of average four, skip one is continued to the end of the scan line. Some studies use GAC data while others use the full-resolution LAC data.

The AVHRR provides regional information on vegetation condition and sea-surface temperature. For example, a portion of an AVHRR image of the South Carolina coast obtained on May 13, 1993, at 3:00 p.m. is shown in Figure 2-27. Band 1 is approximately equivalent to Landsat TM band 3. Vegetated land appears in dark tones due to chlorophyll absorption of red light. Band 2 is approximately equiv-

alent to TM band 4. Vegetation reflects much of the near-infrared radiant flux, yielding bright tones, while water absorbs much of the incident energy. The land–water interface is usually quite distinct. The three thermal bands provide information about Earth's surface and water temperature. The gray scale is inverted for the thermal infrared data with cold, high clouds in black and warm land and water in lighter tones. This particular image captured a large lobe of warm Gulf Stream water.

AVHRR data are also routinely used to inventory sea surface temperature (SST). **Color Plate 2-2** is a sea surface temperature map derived from NOAA-16 AVHRR imagery obtained on October 16, 2003 (Gasparovic, 2003).

Scientists often compute a normalized difference vegetation index (NDVI) from the AVHRR data using the visible ($AVHRR_1$) and near-infrared ($AVHRR_2$) bands to map the condition of vegetation on a regional and national level. It is a simple transformation based on the following ratio:

$$NDVI = \frac{\rho_{nir} - \rho_{red}}{\rho_{nir} + \rho_{red}} = \frac{AVHRR_2 - AVHRR_1}{AVHRR_2 + AVHRR_1}. \quad (2\text{-}4)$$

NOAA 17 Overpasses on October 17, 2003

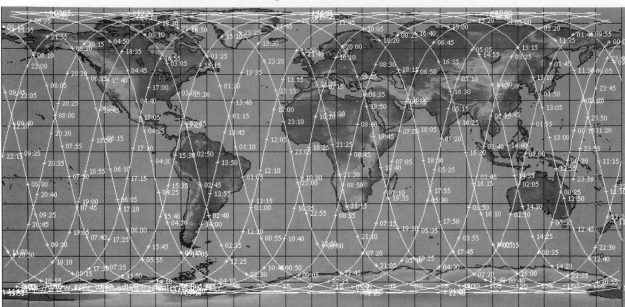

Figure 2-25   Overpasses of the NOAA-17 satellite on October 17, 2003. The AVHRR is onboard NOAA-17 (courtesy of NOAA).

The NDVI equation produces values in the range of −1.0 to 1.0, where increasing positive values indicate increasing green vegetation, and negative values indicate nonvegetated surfaces such as water, barren land, ice, and snow or clouds. To obtain the most precision, the NDVI is derived from calibrated, atmospherically corrected AVHRR channel 1 and 2 data in 16-bit precision, prior to geometric registration and sampling. The final NDVI results from −1 to 1 are normally scaled from 0 to 200. Vegetation indices are discussed in detail in Chapter 8.

NDVI data obtained from multiple dates of AVHRR data can be composited to provide summary seasonal information. The *n*-day NDVI composite is produced by examining each NDVI value pixel by pixel for each observation during the compositing period to determine the maximum value. The retention of the highest NDVI value reduces the number of cloud-contaminated pixels.

The NDVI and other vegetation indexes (refer to Chapter 8) have been used extensively with AVHRR data to monitor natural vegetation and crop condition, identify deforestation in the tropics, and monitor areas undergoing desertification and drought. For example, the U.S. Geological Survey developed the Global Land Cover Characterization dataset based primarily on the unsupervised classification of 1-km AVHRR 10-day NDVI composites. The AVHRR source

imagery dates from April 1992 through March 1993. Ancillary data sources include digital elevation data, ecoregions interpretation, and country- or regional-level vegetation and land-cover maps (USGS Global Landcover, 2004).

NOAA Global Vegetation Index products based on AVHRR data are summarized as (NOAA GVI, 2004): 1st-generation (May 1982 – April, 1985); 2nd-generation (April 1985 – present), and 3rd-generation new products (April 1985 – present).

ORBIMAGE, Inc., and NASA Sea-viewing Wide Field-of-view Sensor

The oceans cover more than two-thirds of the Earth's surface and play an important role in the global climate system. The SeaWiFS is an advanced scanning system designed specifically for ocean monitoring. The *SeaStar* satellite (OrbView-2), developed by ORBIMAGE, Inc., in conjunction with NASA, carried the SeaWiFS into orbit using a Pegasus rocket on August 1, 1997 (NASA/Orbimage SeaWiFS, 2004). The Pegasus rocket was flown aloft under the body of a modified Lockheed L-1011 and released at an altitude of about 39,000 ft, whereupon the rocket was ignited and the spacecraft was lifted into orbit. The final orbit was 705 km above the Earth. The equatorial crossing time is 12 p.m.

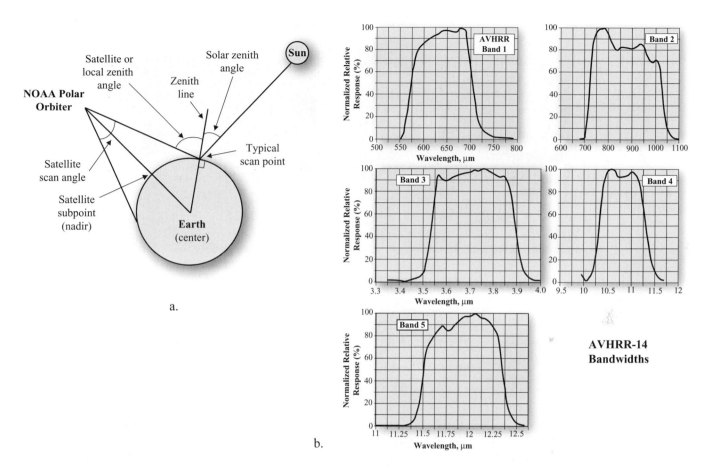

Figure 2-26   a) Relationships among the Earth, Sun, and the NOAA Polar Orbiter. The satellite subpoint lies at nadir. b) The NOAA-14 AVHRR bandwidths for bands 1 through 5.

SeaWiFS builds on all that was learned about ocean remote sensing using the Nimbus-7 satellite Coastal Zone Color Scanner (CZCS) launched in 1978. CZCS ceased operation in 1986. The SeaWiFS instrument consists of an optical scanner with a 58.3° total field of view. Incoming scene radiation is collected by a telescope and reflected onto the rotating half-angle mirror. The radiation is then relayed to dichroic beamsplitters that separate the radiation into eight wavelength intervals (Table 2-7). SeaWiFS has a spatial resolution of 1.13×1.13 km (at nadir) over a swath of 2800 km. It has a revisit time of 1 day.

SeaWiFS records energy in eight spectral bands with very narrow wavelength ranges (Table 2-7) tailored for the detection and monitoring of very specific ocean phenomena, including ocean primary production and phytoplankton processes, ocean influences on climate processes (heat storage and aerosol formation), and the cycles of carbon, sulfur, and nitrogen. In particular, SeaWiFS has specially designed

bands centered at 412 nm (to identify yellow substances through their blue wavelength absorption), at 490 nm (to increase sensitivity to chlorophyll concentration), and in the 765 and 865 nm near-infrared (to assist in the removal of atmospheric attenuation).

SeaWiFS observations help scientists understand the dynamics of ocean and coastal currents, the physics of mixing, and the relationships between ocean physics and large-scale patterns of productivity. The data fill the gaps in ocean biological observations between those of the test-bed CZCS and MODIS. SeaWiFS also provides laser water penetration depth imagery for naval operations.

Other satellites of particular value for marine remote sensing include the Japanese Marine Observation Satellites (MOS-1, 1b) launched in 1987 and 1990, respectively, and the Japanese Ocean Colour and Temperature Scanner (OCTS), launched on August 17, 1996.

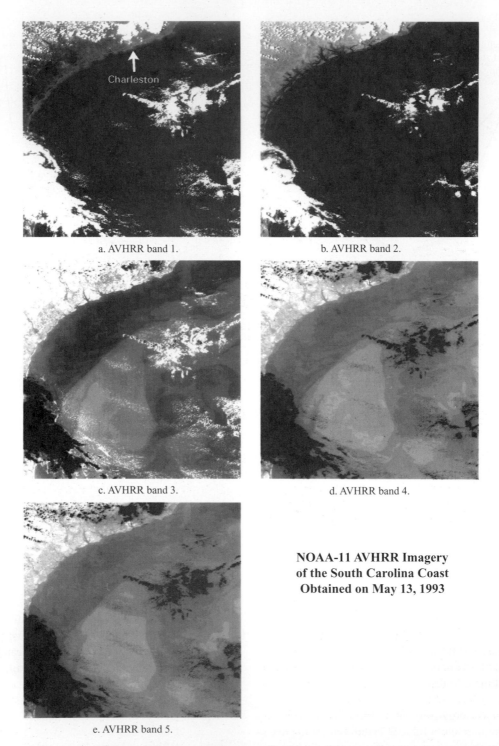

a. AVHRR band 1.

b. AVHRR band 2.

c. AVHRR band 3.

d. AVHRR band 4.

**NOAA-11 AVHRR Imagery
of the South Carolina Coast
Obtained on May 13, 1993**

e. AVHRR band 5.

Figure 2-27　Portion of a NOAA-11 AVHRR image of the South Carolina coast obtained on May 13, 1993, at 3:00 p.m. (refer to Table 2-6 for band specifications). Vegetated land appears dark in band 1 due to chlorophyll absorption of red light. Vegetation appears bright in band 2 because it reflects much of the near-infrared radiant flux. Water absorbs much of the incident energy; therefore, the land–water interface is usually distinct. The three thermal bands (3, 4, and 5) provide surface-temperature information. The grayscale is inverted with cold, high clouds in black and warm land and water in lighter tones. A large lobe of warm Gulf Stream water is easily identified (courtesy of NOAA).

Table 2-7.   Characteristics of the Sea-viewing Wide Field-of-view Sensor.

SeaWiFS Band	Bandcenter (nm)	Bandwidth (nm)	Band Utility
1	412	402 – 422	Identify yellow substances
2	443	433 – 453	Chlorophyll concentration
3	490	480 – 500	Increased sensitivity to chlorophyll concentration
4	510	500 – 520	Chlorophyll concentration
5	555	545 – 565	Gelbstoffe (yellow substance)
6	670	660 – 680	Chlorophyll concentration
7	765	745 – 785	Surface vegetation, land–water interface, atmospheric correction
8	865	845 – 885	Surface vegetation, land–water interface, atmospheric correction

Aircraft Multispectral Scanners

Orbital sensors such as the Landsat MSS, TM, and ETM$^+$ collect data on a repetitive cycle and at set spatial and spectral resolutions. Often it is necessary to acquire remotely sensed data at times that do not coincide with the scheduled satellite overpasses and at perhaps different spatial and spectral resolutions. Rapid collection and analysis of high-resolution remotely sensed data can be required for specific studies and locations. When such conditions occur or when a sensor configuration different from the Landsat or SPOT sensors is needed, agencies and companies often use a multispectral scanner placed onboard an aircraft to acquire remotely sensed data. There are several commercial and publicly available MSS that can be flown onboard aircraft, including the Daedalus and the NASA Airborne Terrestrial Applications Sensor (ATLAS).

Daedalus DS-1260, DS-1268, and Airborne Multispectral Scanner

Numerous remote sensing laboratories and government agencies in 25 countries purchased Daedalus DS-1260, DS-1268, or the Airborne Multispectral Scanner over the last 40 years (e.g., SenSyTech, 2004). These relatively expensive sensor systems have provided much of the useful high spatial and spectral resolution multispectral scanner data (including thermal infrared) for monitoring the environment. For example, the AMS records data in bands spanning the region from the ultraviolet through near-infrared (0.42 – 1.05 μm), a hot-target thermal infrared detector (3.0 – 5.5 μm), and a standard thermal infrared detector (8.5 – 12.5

μm). Table 2-8 summarizes the characteristics of the AMS sensor system.

The basic principles of operation and components of the Airborne Multispectral Scanner (AMS) are shown in Figure 2-28. Radiant flux reflected or emitted from the terrain is collected by the scanning optical system and projected onto a dichroic grating. The grating separates the reflected radiant flux from the emitted radiant flux. Energy in the reflective part of the spectrum is directed from the grating to a prism (or refraction grating) that further separates the energy into specific bands. At the same time, the emitted thermal incident energy is separated from the reflective incident energy. The electromagnetic energy is then focused onto a bank of discrete detectors situated behind the grating and the prism. The detectors that record the emitted energy are usually cooled by a dewar of liquid nitrogen or some other substance. The signals recorded by the detectors are amplified by the system electronics and recorded on a multichannel tape recorder.

The flight altitudes for aircraft MSS surveys are determined by evaluating the size of the desired ground-resolution element (or pixel) and the size of the study area. Basically, the diameter of the circular ground area viewed by the sensor, D, is a function of the instantaneous field of view, β, of the scanner and the altitude above ground level, H, where

$$D = \beta \times H. \tag{2-5}$$

For example, if the IFOV of the scanner is 2.5 mrad, the ground size of the pixel in meters is a product of the IFOV (0.0025) and the altitude AGL in meters. Table 2-9 presents

Table 2-8.   Daedalus Airborne Multispectral Scanner (AMS) and NASA Airborne Terrestrial Applications Sensor (ATLAS) characteristics.

	Airborne Multispectral Scanner (AMS)			NASA Airborne Terrestrial Applications Sensor (ATLAS)		
	Band	Spectral Resolution (μm)	Spatial Resolution (m)	Band	Spectral Resolution (μm)	Spatial Resolution (m)
	1	0.42 – 0.45	Variable,	1	0.45 – 0.52	2.5 to 25 m
	2	0.45 – 0.52	depending	2	0.52 – 0.60	depending
	3	0.52 – 0.60	upon altitude	3	0.60 – 0.63	upon altitude
	4	0.60 – 0.63	above ground	4	0.63 – 0.69	above ground
	5	0.63 – 0.69	level	5	0.69 – 0.76	level
	6	0.69 – 0.75		6	0.76 – 0.90	
	7	0.76 – 0.90		7	1.55 – 1.75	
	8	0.91 – 1.05		8	2.08 – 2.35	
	9	3.00 – 5.50		9	Removed	
	10	8.50 – 12.5		10	8.20 – 8.60	
				11	8.60 – 9.00	
				12	9.00 – 9.40	
				13	9.60 – 10.20	
				14	10.20 – 11.20	
				15	11.20 – 12.20	
IFOV	2.5 mrad			2.0 mrad		
Quantization	8 – 12 bits			8 bits		
Altitude	variable			variable		
Swath width	714 pixels			800 pixels		

Table 2-9.   Aircraft multispectral scanner flight altitude AGL and pixel size assuming an IFOV of 2.5 milliradians (mrad).

Flight Altitude AGL (m)	Pixel Size (m)
1,000	2.5
2,000	5.0
4,000	10.0
16,000	40.0
50,000	125.0

flight altitudes and corresponding pixel sizes at nadir for an IFOV of 2.5 mrad.

The following factors should be considered when collecting aircraft MSS data:

- The IFOV of the MSS optical system and the altitude AGL dictate the width of a single flight line of coverage. At lower altitudes, the high spatial resolution may be outweighed by the fact that more flight lines are required to cover the area compared to more efficient coverage at higher altitudes with larger pixels. The pixel size and the geographic extent of the area to be surveyed are considered, objectives are weighed, and a compromise is reached.

- Even single flight lines of aircraft MSS data are difficult to rectify to a standard map series because of aircraft roll, pitch, and yaw during data collection. Notches in the edge of a flight line of data indicate aircraft roll. Such data require significant human and machine resources to make the data planimetrically accurate. Several agencies place GPS units on the aircraft to obtain precise flight line coordinates, which are useful when rectifying the aircraft MSS data.

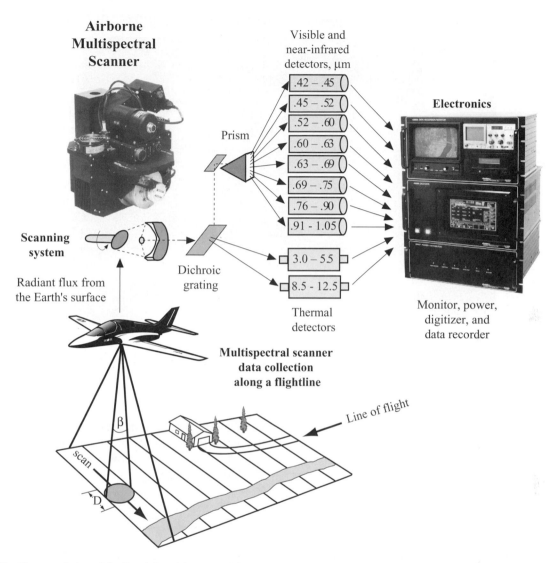

Figure 2-28 Characteristics of the Daedalus airborne multispectral scanner (AMS) and associated electronics that are carried onboard the aircraft during data collection. The diameter of the circular ground area viewed by the sensor, D, is a function of the instantaneous-field-of-view, β, of the scanner and the altitude above ground level of the aircraft, H, at the time of data collection. Radiant flux from Earth's surface is passed from the optical system onto a dichroic grate, which sends the various wavelengths of light to detectors that are continuously monitored by the sensor system electronics.

Airborne multispectral systems operating at relatively low altitudes are one of the few sensors that can acquire high spatial resolution multispectral and temperature information for a variety of environmental monitoring purposes on demand, weather permitting.

NASA Airborne Terrestrial Applications Sensor

NASA's ATLAS multispectral scanner is operated by the Stennis Space Center, MS (NASA ATLAS, 2004). ATLAS has 14 channels with a spectral range from 0.45 to 12.2 μm.

There are six visible and near-infrared bands, two short-wavelength infrared bands (identical to Thematic Mapper bands 5 and 7) and six thermal infrared bands. The band-widths are summarized in Table 2-8. The sensor has a total field of view of 72° and an IFOV of 2.0 milliradians. ATLAS is flown on a Learjet 23 aircraft from 6,000 to 41,000 ft above ground level, yielding pixels with a ground resolution of approximately 2.5 × 2.5 m to 25 × 25 m, depending upon user specifications. There are normally 800 pixels per line plus three calibration source pixels. The data are quantized to 8 bits.

Airborne Terrestrial Applications Sensor (ATLAS) Image of Sullivans Island, SC

Figure 2-29   Near-infrared band 6 (0.76 – 0.90 μm) Airborne Terrestrial Applications Sensor (ATLAS) image of a portion of the smooth cordgrass (*Spartina alterniflora*) marsh behind Sullivans Island, SC. The 2.5 × 2.5 m data were obtained on October 15, 1998.

Calibration of the thermal data is performed using two internal blackbodies. Visible and near-infrared calibration is accomplished on the ground between missions using an integrating sphere. Onboard GPS documents the location in space of each line of data collected.

The ATLAS sensor is ideal for collecting high spatial resolution data in the visible, near-infrared, middle-infrared, and thermal infrared regions so important for many commercial remote sensing applications. The Learjet is an ideal suborbital platform because of its stability in flight and its ability to travel to the study area quickly. It is particularly useful for obtaining data immediately after a disaster such as Hurricanes Hugo and Andrew. An example of 2.5 × 2.5 m ATLAS band 6 (near-infrared) data obtained in 1998 for an area adjacent to Sullivans Island, SC, is shown in Figure 2-29.

 Multispectral Imaging Using Linear Arrays

Linear array sensor systems use sensitive diodes or charge-coupled-devices to record the reflected or emitted radiance from the terrain. Linear array sensors are often referred to as *pushbroom* sensors because, like a single line of bristles in a broom, the linear array stares constantly at the ground while the spacecraft moves forward (see Figure 2-8c). The result is

usually a more accurate measurement of the reflected radiant flux because 1) there is no moving mirror and 2) the linear array detectors are able to dwell longer on a specific portion of the terrain, resulting in a more accurate representation.

SPOT Sensor Systems

The sensors onboard the TIROS and NIMBUS satellites in the 1960s provided remotely sensed imagery with ground spatial resolutions of about 1 × 1 km and were the first to reveal the potential of space as a vantage point for Earth resource observation. The multispectral Landsat MSS and TM sensor systems developed in the 1970s and 1980s provided imagery with spatial resolutions from 79 × 79 m to 30 × 30 m. The first SPOT satellite was launched on February 21, 1986 (Figure 2-30). It was developed by the French Centre National d'Etudes Spatiales (CNES) in cooperation with Belgium and Sweden. It had a spatial resolution of 10 × 10 m (panchromatic mode) and 20 × 20 m (multispectral mode), and provided several other innovations in remote sensor system design (Table 2-10). SPOT satellites 2 and 3 with identical payloads were launched on January 22, 1990, and September 25, 1993, respectively. SPOT 4 was launched on March 24, 1998, with the new *Vegetation* sensor (1 × 1 km) and a short-wavelength infrared band. SPOT 5 was launched on May 3, 2002, with visible, near-infrared, and

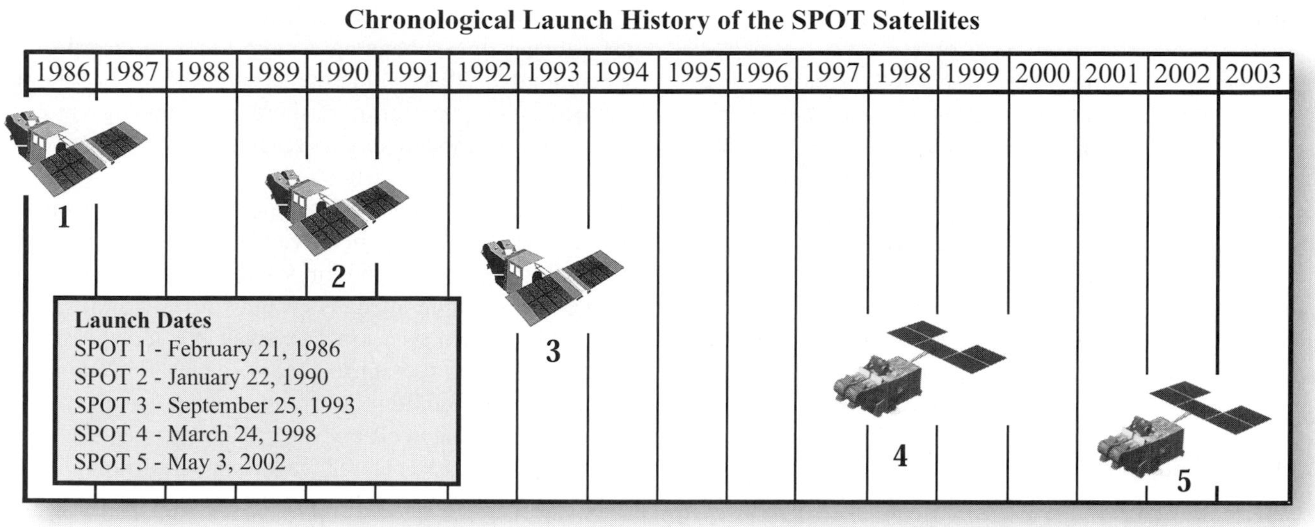

Chronological Launch History of the SPOT Satellites

| 1986 | 1987 | 1988 | 1989 | 1990 | 1991 | 1992 | 1993 | 1994 | 1995 | 1996 | 1997 | 1998 | 1999 | 2000 | 2001 | 2002 | 2003 |

Launch Dates
SPOT 1 - February 21, 1986
SPOT 2 - January 22, 1990
SPOT 3 - September 25, 1993
SPOT 4 - March 24, 1998
SPOT 5 - May 3, 2002

Figure 2-30   Chronological launch history of the SPOT satellites (SPOT, 2004).

Table 2-10.   SPOTs 1, 2, and 3 High Resolution Visible (HRV), SPOTs 4 and 5 High Resolution Visible and Infrared (HRVIR), and SPOTs 4 and 5 *Vegetation* sensor system characteristics.

SPOTs 1, 2, 3 HRV and 4 HRVIR			SPOT 5 HRVIR			SPOTs 4 and 5 *Vegetation*		
Band	Spectral Resolution (μm)	Spatial Resolution (m) at Nadir	Band	Spectral Resolution (μm)	Spatial Resolution (m) at Nadir	Band	Spectral Resolution (μm)	Spatial Resolution (km) at Nadir
1	0.50 – 0.59	20 × 20	1	0.50 – 0.59	10 × 10	0	0.43 – 0.47	1.15 × 1.15
2	0.61 – 0.68	20 × 20	2	0.61 – 0.68	10 × 10	2	0.61 – 0.68	1.15 × 1.15
3	0.79 – 0.89	20 × 20	3	0.79 – 0.89	10 × 10	3	0.78 – 0.89	1.15 × 1.15
Pan Pan (4)	0.51 – 0.73 0.61 – 0.68	10 × 10 10 × 10	Pan	0.48 – 0.71	2.5 × 2.5			
SWIR (4)	1.58 – 1.75	20 × 20	SWIR	1.58 – 1.75	20 × 20	SWIR	1.58 – 1.75	1.15 × 1.15
Sensor	Linear array pushbroom		Linear array pushbroom			Linear array pushbroom		
Swath	60 km ± 50.5°		60 km ± 27°			2250 km ± 50.5°		
Rate	25 Mb/s		50 Mb/s			50 Mb/s		
Revisit	26 days		26 days			1 day		
Orbit	822 km, Sun-synchronous Inclination = 98.7° Equatorial crossing 10:30 a.m.		822 km, Sun-synchronous Inclination = 98.7° Equatorial crossing 10:30 a.m.			822 km, Sun-synchronous Inclination = 98.7° Equatorial crossing 10:30 a.m.		

shortwave infrared (SWIR) bands at 10×10 m and a panchromatic band at 2.5×2.5 m.

Since 1986, SPOT Earth observation satellites have been a consistent, dependable source of high-resolution Earth resource information. While many countries have seen their primary Earth resource monitoring sensor systems come and go depending primarily upon politics, one could always count on SPOT Image, Inc., to provide quality imagery. Unfortunately, the imagery has always been relatively expensive, usually more than $2,000 per panchromatic *or* multispectral scene, although it has been reduced in recent years. If you desire both panchromatic and multispectral imagery of a study area, the cost often exceeds $4,000.

SPOTs 1, 2, and 3

These satellites are all identical and consist of two parts: 1) the SPOT bus, which is a standard multipurpose platform, and 2) the sensor system instruments (Figure 2-31a,b) consisting of two identical high-resolution visible (HRV) sensor systems and a package comprising two tape recorders and a telemetry transmitter. The satellites operate in a Sun-synchronous, near-polar orbit (inclination of 98.7°) at an altitude of 822 km. The satellites pass overhead at the same solar time; the local clock time varies with latitude.

The HRV sensors operate in two modes in the visible and reflective-infrared portions of the spectrum. The first is a *panchromatic* mode corresponding to observation over a broad spectral band (similar to a typical black-and-white photograph). The second is a *multispectral* (color) mode corresponding to observation in three relatively narrow spectral bands (Table 2-10). Thus, the spectral resolution of SPOTs 1 through 3 is not as good as that of the Landsat Thematic Mapper. The ground spatial resolution, however, is 10×10 m for the panchromatic band and 20×20 m for the three multispectral bands when the instruments are viewing at nadir, directly below the satellite.

Radiant flux reflected from the terrain enters the HRV via a plane mirror and is then projected onto two CCD arrays. Each CCD array consists of 6,000 detectors arranged linearly. An electron microscope view of some of the individual detectors in the linear array is shown in Figure 2-32a,b. This linear array *pushbroom* sensor images a complete line of the ground scene in the cross-track direction in one look as the sensor system progresses downtrack (refer to Figure 2-31c). This capability breaks tradition with the Landsat MSS, Landsat TM, and Landsat 7 ETM$^+$ sensors because no mechanical scanning takes place. A linear array sensor is superior because there is no mirror that must scan back and forth to collect data (mirror-scan velocity is a serious issue) and this allows the detector to literally 'stare' at the ground for a longer time, obtaining a more accurate record of the spectral radiant flux exiting the terrain. The SPOT satellites pioneered this linear array pushbroom technology in commercial Earth resource remote sensing as early as 1986.

When looking directly at the terrain beneath the sensor system, the two HRV instruments can be pointed to cover adjacent fields, each with a 60-km swath width (Figure 2-31c). In this configuration the total swath width is 117 km and the two fields overlap by 3 km. However, it is also possible to selectively point the mirror to off-nadir viewing angles through commands from the ground station. In this configuration it is possible to observe any region of interest within a 950-km-wide strip centered on the satellite ground track (i.e., the observed region may not be centered on the ground track) (Figure 2-33a). The width of the swath actually observed varies between 60 km for nadir viewing and 80 km for extreme off-nadir viewing.

If the HRV instruments were only capable of nadir viewing, the revisit frequency for any given region of the world would be 26 days. This interval is often unacceptable for the observation of phenomena evolving on time scales ranging from several days to a few weeks, especially where the cloud cover hinders the acquisition of usable data. During the 26-day period separating two successive SPOT satellite passes over a given point on Earth and taking into account the steering capability of the instruments, the point in question could be observed on seven different passes if it were on the equator and on 11 occasions if at a latitude of 45° (Figure 2-33b). A given region can be revisited on dates separated alternatively by 1 to 4 (or occasionally 5) days.

The SPOT sensors can also acquire cross-track stereoscopic pairs of images for a given geographic area (Figure 2-33c). Two observations can be made on successive days such that the two images are acquired at angles on either side of the vertical. In such cases, the ratio between the observation base (distance between the two satellite positions) and the height (satellite altitude) is approximately 0.75 at the equator and 0.50 at a latitude of 45°. SPOT data with these base-to-height ratios may be used for topographic mapping. Toutin and Beaudoin (1995) applied photogrammetric techniques to SPOT data and produced maps with a planimetric accuracy of 12 m with 90 percent confidence for well-identifiable features and an elevation accuracy for a digital elevation model of 30 m with 90 percent confidence.

SPOT 10×10 m panchromatic data are of such high geometric fidelity that they can be photointerpreted like a typical

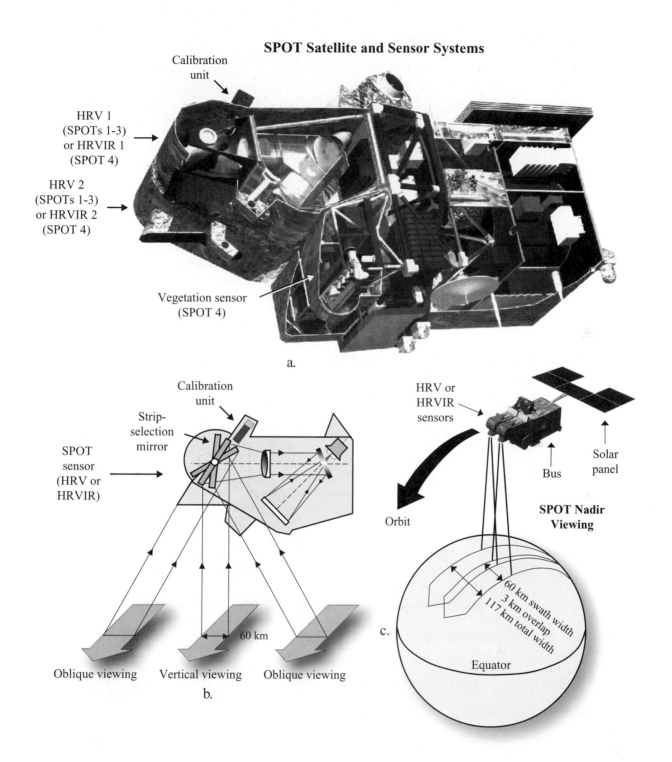

Figure 2-31   The SPOT satellites consist of the SPOT bus, which is a multipurpose platform, and the sensor system payload. Two identical high-resolution visible (HRV) sensors are found on SPOTs 1, 2, and 3 and two identical high-resolution visible infrared (HRVIR) sensors on SPOT 4. Radiant energy from the terrain enters the HRV or HRVIR via a plane mirror and is then projected onto two CCD arrays. Each CCD array consists of 6,000 detectors arranged linearly. This results in a spatial resolution of 10×10 or 20×20 m, depending on the mode in which the sensor is being used. The swath width at nadir is 60 km. The SPOT HRV and HRVIR sensors may also be pointed off-nadir to collect data. SPOT 4 carries a *Vegetation* sensor with 1.15 $\times$ 1.15 km spatial resolution and 2,250-km swath width (adapted from SPOT Image, Inc.).

SPOT Linear Array Enlargements

a.

100 μm

b.

10 μm

Figure 2-32 a) Scanning electron microscope images of the front surface of a CCD linear array like that used in the SPOT HRV sensor systems. Approximately 58 CCD detectors are visible, with rows of readout registers on both sides. b) Seven detectors of a CCD linear array are shown at higher magnification (© CNES 2004, Spot Image Corporation).

aerial photograph in many instances. For this reason, SPOT panchromatic data are often commonly registered to topographic base maps and used as orthophotomaps. Such image maps are useful in GIS databases because they contain more accurate planimetric information (e.g., new roads, subdivisions) than out-of-date 7.5-min topographic maps. The improved spatial resolution available is demonstrated in Figure 2-34, which presents a TM band 3 image and a SPOT panchromatic image of Charleston, SC.

SPOT Off-Nadir Viewing

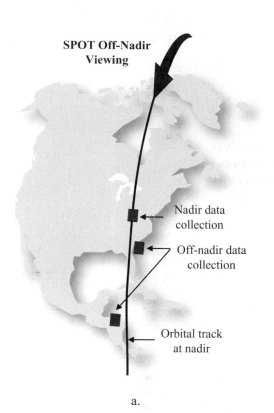

Nadir data collection

Off-nadir data collection

Orbital track at nadir

a.

SPOT Off-Nadir Revisit Capabilities

One pass on days: D + 10   D + 5   D   D - 5

60 km

Swath observed

b.

SPOT Stereoscopic Viewing Capabilities

Pass on day D   Pass on day D + 1

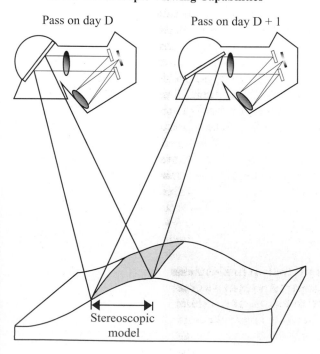

Stereoscopic model

c.

Figure 2-33   a) The SPOT HRV instruments are pointable and can be used to view areas that are not directly below the satellite (i.e., off-nadir). This is useful for collecting information in the event of a disaster, when the satellite track is not optimum for collecting stereoscopic imagery. b) During the 26-day period separating two successive SPOT satellite overpasses, a point on Earth could be observed on different passes if it is at the equator and on 11 occasions if at a latitude of 45°. A given region can be revisited on dates separated alternatively by 1, 4, and occasionally 5 days. c) Two observations can be made on successive days such that the two images are acquired at angles on either side of the vertical, resulting in stereoscopic imagery. Such imagery can be used to produce topographic and planimetric maps (adapted from SPOT Image, Inc.).

Comparison of Landsat TM (30 x 30 m) and SPOT HRV (10 x 10 m)

a. Landsat Thematic Mapper Band 3 (30 x 30 m) on February 3, 1994.

b. SPOT HRV Panchromatic Band (10 x 10 m) on January 10, 1996.

Figure 2-34   Comparison of the detail in the 30×30 m Landsat TM band 3 data and SPOT 10×10 m panchromatic data of Charleston, SC (© Spot Image Corporation, Inc.).

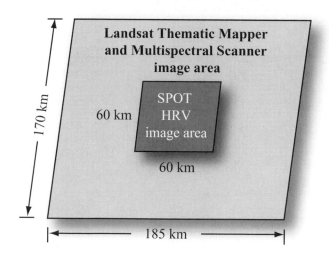

Figure 2-35 Geographic coverage of the SPOT HRV and Landsat Multispectral Scanner and Thematic Mapper remote sensing systems.

SPOT sensors collect data over a relatively small 60×60 km (3600 km^2) area compared with Landsat MSS and TM image areas of 170×185 km ($31,450$ km^2) (Figure 2-35). It takes 8.74 SPOT images to cover the same area as a single Landsat TM or MSS scene. This may be a limiting factor for extensive regional studies. However, SPOT does allow imagery to be purchased by the km^2 (e.g., for a watershed or school district) or by the linear km (e.g., along a highway).

SPOTs 4 and 5

SPOT Image, Inc., launched SPOT 4 on March 24, 1998, and SPOT 5 on May 3, 2002. Their characteristics are summarized in Table 2-10. The viewing angle can be adjusted to $\pm 27°$ off-nadir. SPOTs 4 and 5 have several notable features of significant value for Earth resource remote sensing, including:

- the addition of a short-wavelength infrared (SWIR) band ($1.58 - 1.75$ μm) for vegetation and soil moisture applications at 20×20 m;

- an independent sensor called *Vegetation* for small-scale vegetation, global change, and oceanographic studies,

- SPOT 4 has onboard registration of the spectral bands, achieved by replacing the original HRV panchromatic sensor ($0.51 - 0.73$ μm) with band 2 ($0.61 - 0.68$ μm) operating in both 10- *and* 20-m resolution mode, and

- SPOT 5 panchromatic bands ($0.48 - 0.7$ μm) can collect 2.5×2.5 m imagery. A color composite of Los Angeles, CA, incorporating 2.5×2.5 m panchromatic data is found in **Color Plate 2-3**.

Because the SPOT 4 and 5 HRV sensors are sensitive to SWIR energy, they are referred to as HRVIR 1 and HRVIR 2.

The SPOT 4 and 5 *Vegetation* sensor is independent of the HRVIR sensors. It is a multispectral electronic scanning radiometer operating at optical wavelengths with a separate objective lens and sensor for each of the four spectral bands; (blue = $0.43 - 0.47$ μm used primarily for atmospheric correction; red = $0.61 - 0.68$ μm; near-infrared = $0.78 - 0.89$ μm; and SWIR = $1.58 - 1.75$ μm). Each sensor is a 1728 CCD linear array located in the focal plane of the corresponding objective lens. The spectral resolution of the individual bands are summarized in Table 2-10. The *Vegetation* sensor has a spatial resolution of 1.15×1.15 km. The objective lenses offer a field of view of $\pm 50.5°$, which translates into a 2,250-km swath width. The *Vegetation* sensor has several important characteristics:

- multidate radiometric calibration accuracy better than 3% and absolute calibration accuracy better than 5% is superior to the AVHRR, making it more useful for repeatable global and regional vegetation surveys;

- pixel size is uniform across the entire swath width, with geometric precision better than 0.3 pixels and interband multidate registration better than 0.3 km because of the pushbroom technology;

- 10:30 a.m. equatorial crossing time versus AVHRR's 2:30 p.m. crossing time;

- a short-wavelength infrared band for improved vegetation mapping;

- it is straightforward to relate the HRVIR 10×10 m or 20×20 m data nested within the *Vegetation* 2250×2250 km swath width data; and

- individual images can be obtained or data can be summarized over a 24-hr period (called a daily synthesis), or daily synthesis data can be compiled into *n*-day syntheses.

The dependable constellation of SPOT satellites offers unrivaled ability to acquire imagery of almost any point on the globe daily.

Table 2-11.   Indian National Remote Sensing Agency (NRSA) Indian Remote Sensing (IRS) satellite characteristics.

IRS-1A and 1B			IRS-1C and 1D		
LISS-I and LISS-II Bands	Spectral Resolution (μm)	Spatial Resolution (m) at Nadir	LISS-III, Pan, and WiFS Bands	Spectral Resolution (μm)	Spatial Resolution (m) at Nadir
1	0.45 – 0.52	LISS-I @72.5 m LISS-II @36.25 m	1	-	-
2	0.52 – 0.59	LISS-I @72.5 m LISS-II @36.25 m	2	0.52 – 0.59	23.5 × 23.5
3	0.62 – 0.68	LISS-I @72.5 m LISS-II @36.25 m	3	0.62 – 0.68	23.5 × 23.5
4	0.77 – 0.86	LISS-I @72.5 m LISS-II @36.25 m	4	0.77 – 0.86	23.5 × 23.5
			5	1.55 – 1.70	70.5 × 70.5
			Pan	0.50 – 0.75	5.2 × 5.2
			WiFS 1	0.62 – 0.68	188 × 188
			WiFS 2	0.77 – 0.86	188 × 188
Sensor	Linear array pushbroom		Linear array pushbroom		
Swath width	LISS-I = 148 km; LISS-II = 146 km		LISS-III = 141 km for bands 2, 3, and 4; band 5 = 148 km Pan = 70 km; WiFS = 692 km		
Revisit	22 days at equator		LISS-III is 24 days at equator; Pan is 5 days ± 26° off-nadir viewing; WiFS is 5 days at equator		
Orbit	904 km, Sun-synchronous Inclination = 99.5° Equatorial crossing 10:26 a.m.		817 km, Sun-synchronous Inclination = 98.69° Equatorial crossing 10:30 a.m. ± 5 min		
Launch	IRS-1A on March 17, 1988 IRS-1B on August 29, 1991		IRS-1C in 1995 IRS-1D in September 1997		

Indian Remote Sensing Systems

The Indian National Remote Sensing Agency (NRSA) has launched several Indian Remote Sensing (IRS) satellites: IRS-1A on March 17, 1988, IRS-1B on August 29, 1991, IRS-1C in 1995, and IRS-1D in September 1997 (Table 2-11). IRS-P3 and IRS-P4 were launched on March 21, 1996, and May 26, 1999, respectively (NRSA, 2004). The sensors onboard the satellites use linear array sensor technology.

IRS-1A, -1B, -1C, and -1D

The IRS-1A and IRS-1B satellites acquire data with Linear Imaging Self-scanning Sensors (LISS-I and LISS-II) at spa-

tial resolutions of 72.5 × 72.5 m and 36.25 × 36.25 m, respectively (Table 2-11). The data are collected in four spectral bands that are nearly identical to the Landsat TM visible and near-infrared bands. The satellite's altitude is 904 km, the orbit is Sun-synchronous, repeat coverage is every 22 days at the equator (11-day repeat coverage with two satellites), and orbital inclination is 99.5°. The swath width is 146 to 148 km.

The IRS-1C and IRS-1D satellites carry three sensors (Table 2-11): the LISS-III multispectral sensor, a panchromatic sensor, and a Wide Field Sensor (WiFS). The LISS-III has four bands with the green, red, and near-infrared bands at 23.5 × 23.5 m spatial resolution and the short-wavelength infrared band at 70.5 × 70.5 m spatial resolution. The swath width is

Indian Remote Sensing Satellite Image of San Diego, CA

Figure 2-36   Indian Remote Sensing Satellite (IRS-1D) panchromatic image of downtown San Diego, CA. The 5.2 × 5.2 m image was re-sampled to 5 × 5 m (courtesy of Indian National Remote Sensing Agency).

141 km for bands 2, 3, and 4 and 148 km for band 5. Repeat coverage is every 24 days at the equator.

The panchromatic sensor has a spatial resolution of approximately 5.2 × 5.2 m and stereoscopic imaging capability. The panchromatic band has a 70-km swath width with repeat coverage every 24 days at the equator and a revisit time of 5 days with ±26° off-nadir viewing. An example of the 5.2 × 5.2 m panchromatic data of downtown San Diego, CA (resampled to 5 × 5 m) is shown in Figure 2-36.

The Wide Field Sensor has 188 × 188 m spatial resolution. The WiFS has two bands comparable to NOAA's AVHRR satellite (0.62 – 0.68 μm and 0.77 – 0.86 μm) with a swath width of 692 km. Repeat coverage is 5 days at the equator.

IRS-P3 and IRS-P4

The IRS-P3 WiFS is similar to the IRS-1D WiFS except for the inclusion of an additional band in the middle-infrared region (1.55 – 1.70 μm). IRS-P3 also has a Modular Opto-electronics Scanner (MOS), which collects data at three spa-

tial resolutions (MOS A, B, C = 1569 × 1395 m; 523 × 523 m; and 523 × 644 m, respectively) in three bands (MOS A, B, C = 0.755 – 0.768 μm; 0.408 – 1.01 μm; and 1.5 – 1.7 μm, respectively).

The IRS-P4 satellite is devoted to oceanographic applications based primarily on its Ocean Color Monitor (OCM) sensor, which collects data in 8 bands from 402 to 885 nm at a spatial resolution of 360 × 236 m at 12-bit radiometric resolution. The swath width is 1420 km. The IRS-P4 also carries a Multifrequency Scanning Microwave Radiometer (MSMR). IRS-P5 is scheduled for launch in the future with a 2.5 × 2.5 m panchromatic band for cartographic applications (NRSA, 2004).

Advanced Spaceborne Thermal Emission and Reflection Radiometer

ASTER is a cooperative effort between NASA and Japan's Ministry of International Trade and Industry. ASTER obtains detailed information on surface temperature, emis-

Table 2-12.   NASA Advanced Spaceborne Thermal Emission and Reflection Radiometer characteristics.

Advanced Spaceborne Thermal Emission and Reflection Radiometer (ASTER)					
Band	VNIR Spectral Resolution (μm)	Band	SWIR Spectral Resolution (μm)	Band	TIR Spectral Resolution (μm)
1 (nadir)	0.52 – 0.60	4	1.600 – 1.700	10	8.125 – 8.475
2 (nadir)	0.63 – 0.69	5	2.145 – 2.185	11	8.475 – 8.825
3 (nadir)	0.76 – 0.86	6	2.185 – 2.225	12	8.925 – 9.275
3 (backward)	0.76 – 0.86	7	2.235 – 2.285	13	10.25 – 10.95
		8	2.295 – 2.365	14	10.95 – 11.65
		9	2.360 – 2.430		
Technology (detector)	Pushbroom Si		Pushbroom PtSi:Si		Whiskbroom Hg:Cd:Te
Spatial resolution (m)	15 × 15		30 × 30		90 × 90
Swath width	60 km		60 km		60 km
Quantization	8 bits		8 bits		12 bits

sivity, reflectance, and elevation (NASA ASTER, 2004). It is the only high-spatial-resolution instrument on the *Terra* satellite. It is used in conjunction with MODIS, MISR, and CERES sensors that monitor the Earth at moderate to coarse spatial resolutions. ASTER serves as a zoom lens for the other *Terra* instruments and is particularly important for change detection and calibration/validation studies (King, 2003).

ASTER obtains data in 14 channels from the visible through the thermal infrared regions of the electromagnetic spectrum. It consists of three separate instrument subsystems. Individual bandwidths and subsystem characteristics are summarized in Table 2-12.

The VNIR detector subsystem operates in three spectral bands in the visible and near-infrared wavelength region with a spatial resolution of 15 × 15 m. It consists of two telescopes—one nadir-looking with a three-spectral-band CCD detector and another backward-looking with a single-band CCD detector. The backward-looking telescope provides a second view of the study area in band 3 for stereoscopic observations. Across-track pointing to 24° off-nadir is accomplished by rotating the entire telescope assembly. A

15 × 15 m near-infrared band 3 (0.76 – 0.86 μm) image of Pearl Harbor, HI, obtained on June 3, 2000 is shown in Figure 2-37. ASTER color composites of the island of Oahu and Pearl Harbor obtained on June 3, 2000, are found in **Color Plate 2-4**.

The SWIR subsystem operates six spectral bands in the 1.6 to 2.43 μm region through a single nadir-pointing telescope that provides 30 × 30 m spatial resolution. Cross-track pointing (± 8.55°) is accomplished by a pointing mirror.

The TIR subsystem operates in five bands in the thermal infrared region using a single, fixed-position, nadir-looking telescope with a spatial resolution of 90 × 90 m. Unlike the other subsystems, it has a whiskbroom scanning system instead of a pushbroom system. Each band uses 10 detectors in a staggered array with optical bandpass filters over each detector element. The scanning mirror functions both for scanning and cross-track pointing (±8.55°). During scanning, the mirror rotates 90° from nadir to view an internal blackbody. Multiple-date nighttime 90 × 90 m thermal infrared band 14 (10.95 – 11.65 μm) images of Pu'u O'o lava flows entering the sea at Kamokuna on the southeast side of the Island of Hawaii are shown in Figure 2-38.

ASTER Near-infrared Image of Pearl Harbor, Hawaii

Figure 2-37   *Terra* ASTER 15 × 15 m near-infrared band 3 (0.76 – 0.86 μm) image of Pearl Harbor, Hawaii, obtained on June 3, 2000 (courtesy of NASA/GSFC/MITI/ERSADC/JAROS and U.S./Japan ASTER Science Team and the California Institute of Technology).

Multiangle Imaging Spectroradiometer

The Multiangle Imaging Spectroradiometer was built by NASA's Jet Propulsion Laboratory and is one of the five *Terra* satellite instruments. The MISR instrument measures the Earth's brightness in four spectral bands, at each of nine look angles spread out in the forward and aft directions along the flight line. Spatial samples are acquired every 275 m. Over a period of 7 minutes, a 360-km wide swath of Earth comes into view at all nine angles (King, 2003; NASA MISR, 2004).

An illustration of the nine look angles is shown in Figure 2-39. The digital pushbroom sensors image the Earth at 26.1°, 45.6°, 60°, and 70.5° forward and aft of the local vertical (nadir 0°). Note that the fore and aft camera angles are the same—the cameras are arranged symmetrically about nadir. In general, large viewing angles provide enhanced sensitivity to atmospheric aerosol effects and to cloud reflectance effects, whereas more modest angles are required for land-surface viewing.

Each MISR camera sees instantaneously a single row of pixels at right angles to the ground track in a pushbroom format. It records data in four bands: blue, green, red, and near-infrared. The individual band wavelengths are identified in Figure 2-39. Each camera has four independent linear CCD arrays (one per filter), with 1,504 active pixels per linear array.

The nadir-viewing camera (labeled An in Figure 2-39) provides imagery that is less distorted by surface topographic effects than that of any other MISR camera. It also is the least affected by atmospheric scattering. It provides 1) useful reference for navigating within all the MISR imagery and 2) a base image to compare with images acquired at different angles of view. Such comparisons provide important "bidirectional reflectance distribution function, BRDF" information introduced in Chapter 1. The nadir-viewing camera also

ASTER Thermal-infrared Images of a Lava Flow on Hawaii

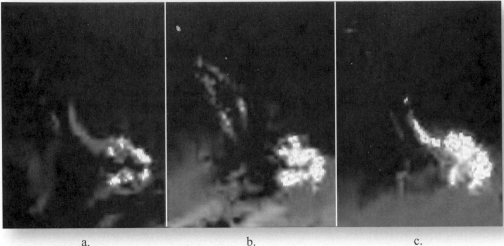

a.   b.   c.

Figure 2-38   A sequence of *Terra* ASTER nighttime 90 × 90 m thermal infrared band 14 (10.95 – 11.65 μm) images of Pu'u O'o lava flows entering the sea at Kamokuna on the southeast side of the Island of Hawaii. The images were obtained on a) May 22, 2000, b) June 30, 2000, and c) August 1, 2000. The brighter the pixel, the greater the lava temperature (courtesy of NASA/GSFC/MITI/ ERSADC/JAROS and U.S./Japan ASTER Science Team and the California Institute of Technology).

offers an opportunity to compare observations with other nadir-viewing sensors such as Landsat TM and ETM[+]. The nadir-viewing camera also facilitates calibration.

The fore and aft 26.1° view angle cameras (Af and Aa) provide useful stereoscopic information that can be of benefit for measuring topographic elevation and cloud heights. The fore and aft 45.6° view angle cameras (Bf and Ba) are positioned to be especially sensitive to atmospheric aerosol properties. The fore and aft 60° view angle cameras (Cf and Ca) provide observations looking through the atmosphere with twice the amount of air than the vertical view. This provides unique information about the hemispherical albedo of land surfaces. The fore and aft 70.5° view angle cameras (Df and Da) provide the maximum sensitivity to off-nadir effects. The scientific community is interested in obtaining quantitative information about clouds and the Earth's surface from as many angles as possible.

Very-High-Resolution Linear Array Remote Sensing Systems

In 1994, the U.S. government made a decision to allow civil commercial companies to market high spatial resolution remote sensor data (approximately 1 × 1 to 4 × 4 m). This resulted in the creation of a number of commercial consor-

tiums that have the capital necessary to create, launch, and market high spatial resolution digital remote sensor data. The most notable companies are Space Imaging, Inc., ORBIMAGE, Inc., DigitalGlobe, Inc, and ImageSat International, Inc. These companies targeted the geographic information system (GIS) and cartographic mapping markets traditionally serviced by the aerial photogrammetric industries. Some estimate the growing Earth observation industry to be 5 to 15 billion dollars a year. The new commercial remote sensing firms hope to have an impact in markets as diverse as agriculture, natural resource management, local and regional government, transportation, emergency response, mapping, and eventually an array of average consumer applications as well (Space Imaging, 2004). Warfare during the early 21st century created a significant military demand for commercial high-spatial-resolution imagery.

All commercial vendors offer an Internet on-line ordering service. All vendors offer a suite of standard and nonstandard products that can be tailored to user requirements, including the creation of digital elevation models from the remote sensor data. The commercial remote sensing companies typically price the imagery according to the type of product ordered and the amount of geographic coverage desired (km^2). The sensors used by these companies are based primarily on linear array CCD technology. The sensor system characteristics are summarized in Table 2-13.

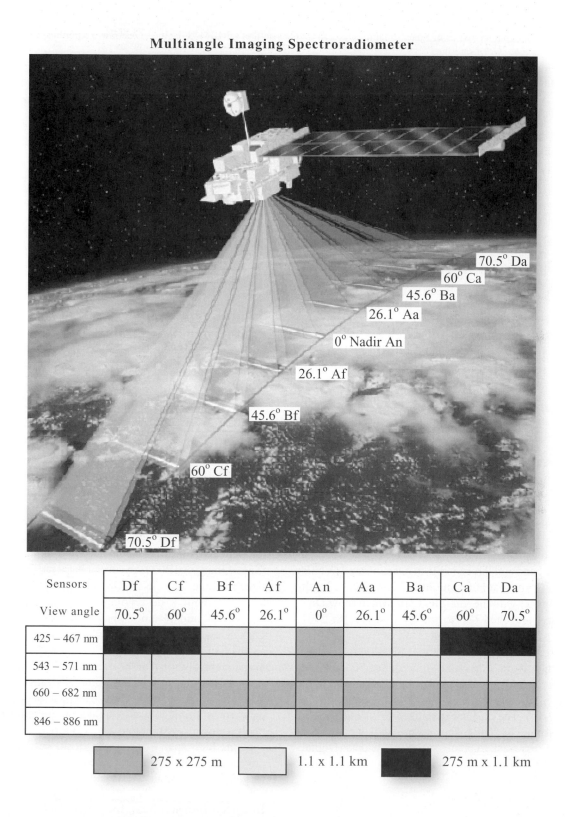

Multiangle Imaging Spectroradiometer

Sensors	Df	Cf	Bf	Af	An	Aa	Ba	Ca	Da
View angle	70.5°	60°	45.6°	26.1°	0°	26.1°	45.6°	60°	70.5°
425 – 467 nm									
543 – 571 nm									
660 – 682 nm									
846 – 886 nm									

275 x 275 m   1.1 x 1.1 km   275 m x 1.1 km

Figure 2-39   Artist's rendition of the Multiangle Imaging Spectroradiometer (MISR) on EOS *Terra*. MISR uses linear array technology to acquire imagery of the terrain in four bands at nine angles: at Nadir (0°) and at 26.1°, 45.6°, 60°, and 70.5° forward and aft of nadir (adapted from NASA Jet Propulsion Laboratory).

Table 2-13. Sensor characteristics of Space Imaging, Inc., *IKONOS* satellite; ORBIMAGE, Inc., *OrbView-3* satellite; and DigitalGlobe, Inc., *QuickBird* satellite.

	Space Imaging, Inc. *IKONOS*			ORBIMAGE, Inc. *OrbView-3*			DigitalGlobe, Inc. *QuickBird*	
Band	Spectral Resolution (μm)	Spatial Resolution (m) at Nadir	Band	Spectral Resolution (μm)	Spatial Resolution (m) at Nadir	Band	Spectral Resolution (μm)	Spatial Resolution (m) at Nadir
1	0.45 – 0.52	4 × 4	1	0.45 – 0.52	4 × 4	1	0.45 – 0.52	2.44 × 2.44
2	0.52 – 0.60	4 × 4	2	0.52 – 0.60	4 × 4	2	0.52 – 0.60	2.44 × 2.44
3	0.63 – 0.69	4 × 4	3	0.625 – 0.695	4 × 4	3	0.63 – 0.69	2.44 × 2.44
4	0.76 – 0.90	4 × 4	4	0.76 – 0.90	4 × 4	4	0.76 – 0.89	2.44 × 2.44
Pan	0.45 – 0.90	1 × 1	Pan	0.45 – 0.90	1 × 1	Pan	0.45 – 0.90	0.61 × 0.61
Sensor	Linear array pushbroom			Linear array pushbroom			Linear array pushbroom	
Swath	11 km			8 km			20 to 40 km	
Rate	25 Mb/s			50 Mb/s			50 Mb/s	
Revisit	< 3 days			< 3 days			1 to 5 days depending on latitude	
Orbit	681 km, Sun-synchronous Equatorial crossing 10–11 a.m.			470 km, Sun-synchronous Equatorial crossing 10:30 a.m.			600 km, not Sun-synchronous Equatorial crossing variable	
Launch	April 27, 1999 (failed) September 24, 1999			June 26, 2003			October 18, 2001	

DigitalGlobe, Inc., *EarlyBird*, and *QuickBird*

EarthWatch, Inc., launched *EarlyBird* in 1996 with a 3 × 3 m panchromatic band and three visible to near-infrared (VNIR) bands at 15 × 15 m spatial resolution. Unfortunately, Earth-Watch lost contact with the satellite. EarthWatch (now DigitalGlobe, Inc.) launched *QuickBird* on October 18, 2001, into a 600-km orbit. Interestingly, it is in a 66° non–Sun-synchronous orbit. Revisit times range from 1 to 5 days, depending on latitude. It has a swath width of 20 to 40 km. QuickBird has a 0.61 × 0.61 m panchromatic band and four visible/near-infrared bands at 2.44 × 2.44 m spatial resolution (Table 2-13). The data are quantized to 11 bits (brightness values from 0 to 2047). The sensor can be pointed fore and aft and across-track to obtain stereoscopic data (DigitalGlobe, 2004).

Space Imaging, Inc., *IKONOS*

Space Imaging, Inc., launched *IKONOS* on April 27, 1999. Unfortunately, the satellite never achieved orbit. Space Imaging successfully launched a second IKONOS on Sep-

tember 24, 1999. The IKONOS satellite sensor has a 1 × 1 m panchromatic band (Figure 2-40) and four multispectral visible and near-infrared bands at 4 × 4 m spatial resolution (Space Imaging, 2004). Sensor characteristics are summarized in Table 2-13. IKONOS is in a Sun-synchronous 681-km orbit, with a descending equatorial crossing time of between 10 and 11 a.m. It has both cross-track and along-track viewing instruments, which enables flexible data acquisition and frequent revisit capability: < 3 days at 1 × 1 m spatial resolution (for look angles < 26°) and 1.5 days at 4 × 4 m spatial resolution. The nominal swath width is 11 km. Data are quantized to 11 bits. The first IKONOS 1 × 1 m panchromatic image was obtained on September 30, 1999, of downtown Washington, DC (Figure 2-40). The image contains high spatial resolution detail sufficient for many city planning and earth resource investigations. IKONOS imagery of Columbia, SC, is shown in **Color Plate 2-5**.

ImageSat International, Inc., *EROS A1*

ImageSat International, Inc., successfully launched the *EROS A1* satellite using a Start-1 launcher from the Russian

a. IKONOS panchromatic 1 × 1 m image of Washington, DC.

Figure 2-40   a) IKONOS 1 × 1 m panchromatic image of Washington, DC, obtained on September 30, 1999. The Washington Monument and White House are visible. b) An enlargement of the Washington Monument (courtesy Space Imaging, Inc.).

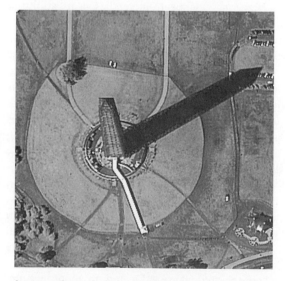

b. An enlarged view of the Washington Monument.

EROS A1 Panchromatic Image of Kamchatka Submarine Base, Russia

Figure 2-41   Panchromatic 1.8 × 1.8 m image of a portion of the Kamchatka Submarine Base on the far-eastern frontier of Russia, obtained by the EROS A1 remote sensing system on December 25, 2001 (courtesy of ImageSat, International, Inc.).

Cosmodrome in Svobodni, Siberia, on December 5, 2000 (ImageSat, 2004). The *EROS A1* was built by Israel Aircraft Industry's MBT Division and weighs only 250 kg. The satellite is in a Sun-synchronous orbit 480 km above the Earth with an equatorial crossing time of 9:45 a.m. EROS A1 is a single-band panchromatic remote sensing system. The spatial resolution is 1.8 × 1.8 m (standard) or 1 × 1 m "over-sampled" mode. An EROS A1 image of the Kamchatka Submarine Base in Russia is shown in Figure 2-41. ImageSat plans to launch EROS B and other satellites to create a constellation of commercial high-resolution satellites.

ORBIMAGE, Inc., *OrbView-3*

ORBIMAGE, Inc., launched *OrbView-3* on June 26, 2003, with 1 × 1 m panchromatic data and four visible and near-infrared multispectral bands at 4 × 4 m spatial resolution (Orbimage, 2004). OrbView-3 has a Sun-synchronous orbit at 470 km above the Earth with a 10:30 a.m. equatorial crossing time and an 8-km swath width. The sensor revisits each location on Earth in less than 3 days with an ability to turn from side to side 45°. *OrbView-3* sensor specifications are summarized in Table 2-13.

 Imaging Spectrometry Using Linear and Area Arrays

This section describes a major advance in remote sensing, *imaging spectrometry*, defined as the simultaneous acquisi-

tion of images in many relatively narrow, contiguous and/or non-contiguous spectral bands throughout the ultraviolet, visible, and infrared portions of the spectrum.

In the past, most remotely sensed data were acquired in 4 to 12 spectral bands. Imaging spectrometry makes possible the acquisition of data in hundreds of spectral bands simultaneously. Because of the very precise nature of the data acquired by imaging spectrometry, more Earth resource problems can be addressed in greater detail.

The value of an imaging spectrometer lies in its ability to provide a high-resolution reflectance spectrum for each picture element in the image. The reflectance spectrum in the region from 0.4 to 2.5 µm can be used to identify a large range of surface cover materials that cannot be identified with broadband, low-spectral-resolution imaging systems such as the Landsat MSS, TM, and SPOT. Many, although not all, surface materials have diagnostic absorption features that are only 20 to 40 nm wide. Therefore, spectral imaging systems that acquire data in contiguous 10-nm bands may produce data with sufficient resolution for the direct identification of those materials with diagnostic spectral absorption features. For example, Figure 2-42 depicts high spectral resolution crop spectra over the interval 400 to 1000 nm obtained using an Imaging Spectrometer for an agricultural area near Bakersfield, CA. The absorption spectra for the Pima and Royale cotton differ from one another from about 725 nm, where the "red edge" is located, to about 900 nm, leading to the possibility that species within the same crop type might be distinguishable (SBRC, 1994). The Landsat

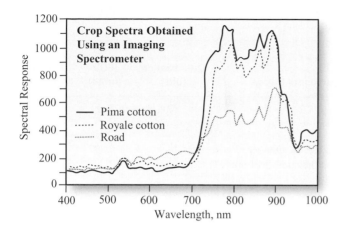

Figure 2-42 Imaging spectrometer crop spectra for Pima cotton, Royale cotton, and road surface extracted from 2 × 2 m data obtained near Bakersfield, CA.

scanners and SPOT HRV sensors, which have relatively large bandwidths, may not be able to resolve these spectral differences.

Simultaneous imaging in many contiguous spectral bands requires a new approach to remote sensor system design. One approach is to increase the residence time of a detector in each IFOV using a linear array of detector elements (Figure 2-8c). In this configuration, there is a dedicated detector element for each cross-track pixel, which increases the residence time to the interval required to move one IFOV along the flight direction. The French SPOT HRV sensor uses a linear array of detector elements. Despite the improved sensitivity of the SPOT detectors in the cross-track direction, they only record energy in three very broad green, red, and near-infrared bands. Thus, its major improvement is in the spatial domain and not in the spectral domain.

Two more practical approaches to imaging spectrometry are shown in Figures 2-8d and 8e. The whiskbroom scanner linear array approach (Figure 2-8d) is analogous to the scanner approach used for Landsat MSS and ETM$^+$, except that radiant flux from within the IFOV is passed on to a spectrometer, where it is dispersed and focused onto a linear array of detectors. Thus, each pixel is simultaneously sensed in as many spectral bands as there are detector elements in the linear array. For high spatial resolution imaging, this approach is suited only to an airborne sensor that flies slowly and when the readout time of the detector array is a small fraction of the integration time. Because of high spacecraft velocities, orbital imaging spectrometry might require the use of two-dimensional area arrays (Figure 2-8e). This elim-

inates the need for the optical scanning mechanism. In this situation, there is a dedicated column of spectral detector elements for each linear array cross-track pixel in the scene.

Thus, traditional broadband remote sensing systems such as Landsat MSS and SPOT HRV *undersample* the information available from a reflectance spectrum by making only a few measurements in spectral bands up to several hundred nanometers wide. Conversely, imaging spectrometers sample at close intervals (bands on the order of tens of nanometers wide) and have a sufficient number of spectral bands to allow construction of spectra that closely resemble those measured by laboratory instruments. Analysis of imaging spectrometer data allows extraction of a detailed spectrum for each picture element in the image (Figure 2-42). Such spectra often allow direct identification of specific materials within the IFOV of the sensor based upon their reflectance characteristics, including minerals, atmospheric gases, vegetation, snow and ice, and dissolved matter in water bodies.

Analysis of hyperspectral data often requires the use of sophisticated digital image processing software (e.g., ENVI introduced in Chapter 3). This is because it is usually necessary to calibrate (convert) the raw hyperspectral radiance data to scaled reflectance before it can be properly interpreted. This means removing the effects of atmospheric attenuation, topographic effects (slope, aspect), and any sensor anomalies. Similarly, to get the most out of the hyperspectral data it is usually necessary to use algorithms that 1) allow one to analyze a typical spectra to determine its constituent materials and 2) compare the spectra with a library of spectra obtained using handheld spectroradiometers such as that provided by the U.S. Geological Survey.

Government agencies and commercial firms have designed hundreds of imaging spectrometers capable of acquiring hyperspectral data. It is beyond the scope of this book to list them all. Only three systems are summarized: NASA JPL's Airborne Visible/Infrared Imaging Spectrometer, the commercially available Compact Airborne Spectrographic Imager 3, and NASA's Moderate Resolution Imaging Spectrometer onboard the *Terra* satellite.

Airborne Visible/Infrared Imaging Spectrometer

The first airborne imaging spectrometer (AIS) was built to test the imaging spectrometer concept with infrared area arrays (Vane and Goetz, 1993). The spectral coverage of the instrument was 1.9 to 2.1 μm in the *tree mode* and 1.2 to 2.4 μm in *rock mode* in contiguous bands that were 9.3 nm wide.

Table 2-14. Characteristics of the NASA Airborne Visible/Infrared Imaging Spectrometer (AVIRIS) and the ITRES Research, Ltd., Compact Spectrographic Imager 3 (CASI 3) hyperspectral remote sensing systems.

Sensor	Technology	Spectral Resolution (nm)	Spectral Interval (nm)	Number of Bands	Quantization (bits)	IFOV (mrad)	Total field of view (°)
AVIRIS	Whiskbroom linear array	400 – 2500	10	224	12	1.0	30°
CASI 3	Linear (1480) and area array CCD (1480 × 288)	400 – 1000	2.2	288 possible; the number of bands and the number of pixels in the across-track are programmable.	14	0.49	40.5°

To acquire data with greater spectral and spatial coverage, AVIRIS was developed at NASA's Jet Propulsion Laboratory in Pasadena, CA (Table 2-14). Using a whiskbroom scanning mirror and linear arrays of silicon (Si) and indium-antimonide (InSb) configured as in Figure 2-8d, AVIRIS acquires images in 224 bands, each 10 nm wide in the 400 to 2500 nm region (NASA AVIRIS, 2004). The sensor is typically flown onboard the NASA/ARC ER-2 aircraft at 20 km above ground level and has a 30° total field of view and an instantaneous-field-of-view of 1.0 mrad, which yields 20 × 20 m pixels. The data are recorded in 12 bits (values from 0 to 4095).

Many AVIRIS characteristics are summarized in Figure 2-43. It depicts a single band of AVIRIS imagery (band 30; 655.56 nm) obtained over the Kennedy Space Center, FL, and radiance data extracted for a single pixel of saw palmetto vegetation.

Compact Airborne Spectrographic Imager 3

ITRES Research, Ltd., of Canada markets the CASI remote sensing system introduced in 1989. CASI 3 is a pushbroom imaging spectrometer based on the use of a 1480 element across-track linear array and a 1480 × 288 area array CCD. The instrument operates over a 650-nm spectral range (400 to 1050 nm) and has a 40.5° total field of view across 1480 pixels (ITRES, 2004).

A single line of terrain 1480 pixels wide perpendicular to the flight path is sensed by the spectrometer optics (Figure 2-44). The incoming radiant flux from each pixel is then spectrally dispersed along the axis of the area array CCD so that a spectrum of energy (from blue through near-infrared) is obtained for each pixel across the swath. By repetitively reading the contents of the area array CCD as the aircraft moves along the flight path, a two-dimensional image at high spectral resolution is acquired. Since the radiant flux for all pixels in a particular swath are recorded simultaneously, spatial and spectral co-registration is assured. The across-track spatial resolution is determined by the altitude of the CASI above ground level and the IFOV, while the along-track resolution depends upon the velocity of the aircraft and the rate at which the CCD is read.

CASI 3 may be programmed to collect 14-bit data in several modes (Table 2-14) (ITRES, 2004):

- *Spatial mode* — where the full across-track resolution of 1480 pixels are obtained for up to 19 nonoverlapping spectral bands with programmable center wavelengths and bandwidths.

- *Hyperspectral mode* — up to 1480 programmable adjacent spatial pixels and up to 288 spectral bands over the full spectral range.

- *Full frame* — 1480 spatial pixels (across-track) and 288 spectral bands.

The specific bandwidths are selected according to the application (e.g., bathymetric mapping, inventorying chlorophyll *a* concentration). The result is a programmable area array remote sensing system that may be the precursor of future satellite hyperspectral sensor systems.

Moderate Resolution Imaging Spectrometer

The Moderate Resolution Imaging Spectrometer is flown on NASA's EOS *Terra* (a.m. equatorial crossing time) and *Aqua* (p.m. equatorial crossing time) satellites (Table 2-15). MODIS provides long-term observations to derive an enhanced knowledge of global dynamics and processes occurring on the surface of the Earth and in the lower atmo-

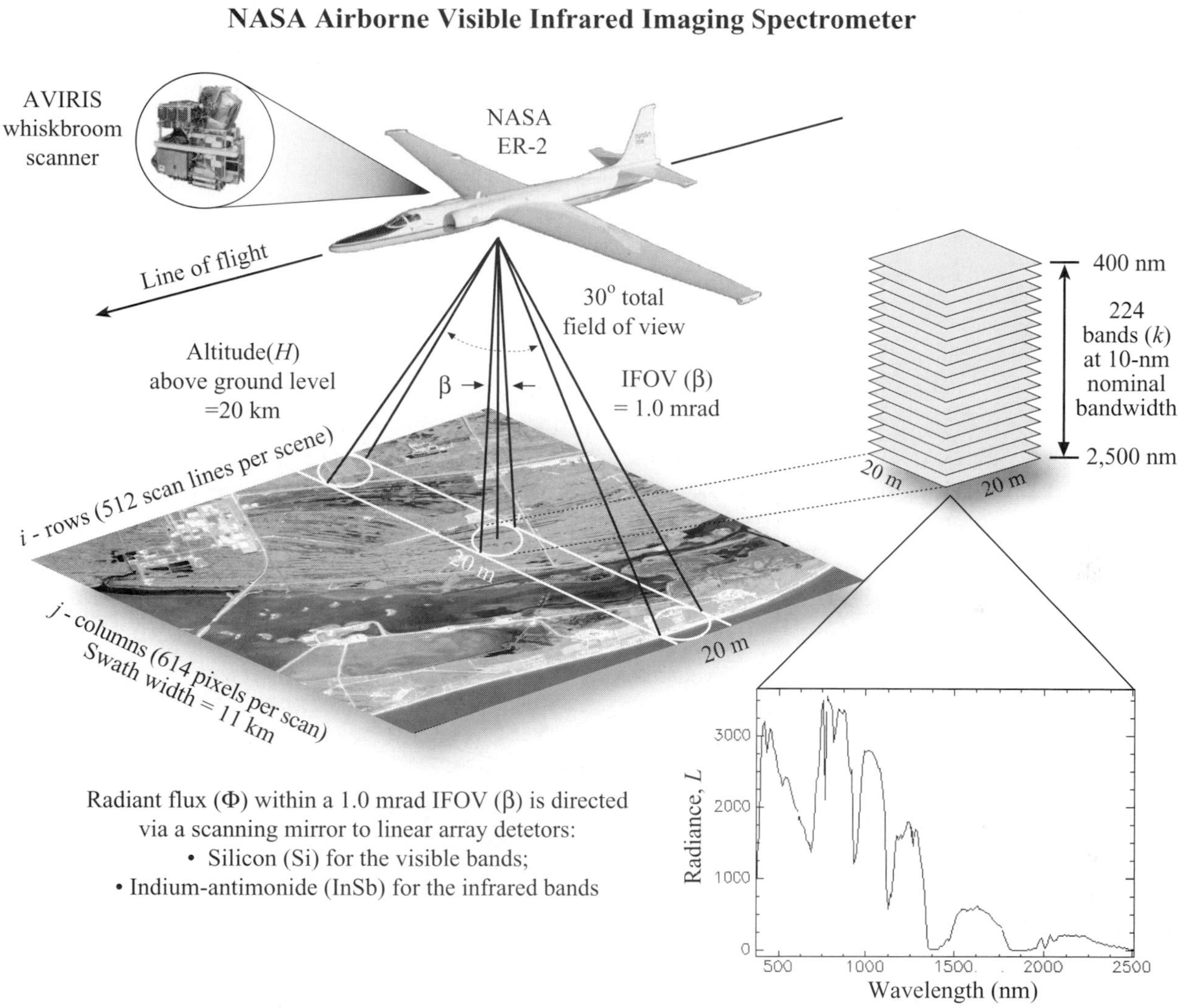

NASA Airborne Visible Infrared Imaging Spectrometer

Figure 2-43 Conceptual representation of imaging spectroscopy as implemented by the NASA Jet Propulsion Lab Airborne Visible-Infrared Imaging Spectrometer (AVIRIS). The scanner mirror focuses radiant flux onto linear arrays that contain 224 detector elements with a spectral sensitivity ranging from 400 to 2500 nm. A spectra of radiance (L) or percent reflectance can be obtained for each picture element. The AVIRIS scene was acquired over the Kennedy Space Center, FL. Only band 30 (655.56 nm) is displayed (adapted from Filippi, 1999).

sphere (King, 2003; NASA MODIS, 2004). It yields simultaneous observations of high-atmospheric (cloud cover and associated properties), oceanic (sea-surface temperature and chlorophyll), and land-surface (land-cover changes, land-surface temperature, and vegetation properties) features.

MODIS is in a 705-km Sun-synchronous orbit. It views the entire surface of the Earth every 1 to 2 days. It has a field of view of ±55° off-nadir, which yields a swath width of 2,330 km. MODIS obtains high radiometric resolution images (12-bit) of daylight-reflected solar radiation and day/night thermal emission over all regions of the globe. MODIS is a

whiskbroom scanning imaging radiometer consisting of a cross-track scan mirror, collecting optics, and a set of linear detector arrays with spectral interference filters located in four focal planes. It collects data in 36 co-registered spectral bands: 20 bands from 0.4 to 3 μm and 16 bands from 3 to 15 μm. The bandwidths and their primary uses are summarized in Table 2-15.

MODIS' coarse spatial resolution ranges from 250 × 250 m (bands 1 and 2) to 500 × 500 m (bands 3 through 7) and 1 × 1 km (bands 8 through 36). A MODIS band 4 (green: 0.545 – 0.565 μm) image of the Nile Delta, the Nile River, and

Hyperspectral Data Collection Using Linear and Area Arrays

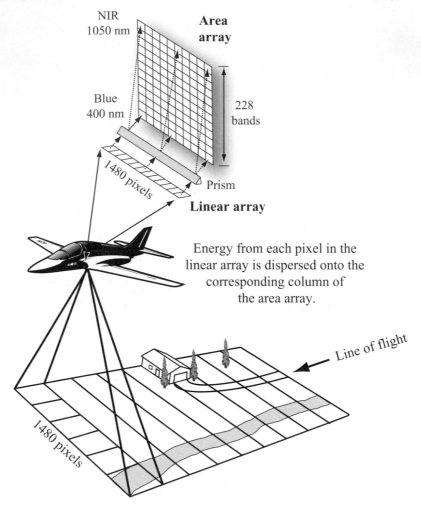

Figure 2-44   Hyperspectral data collection using a linear array pushbroom sensor that distributes the energy onto a 1480 × 228 element area array sensitive to the wavelength region from 400 to 1,050 nm.

much of the area surrounding the Red Sea is shown in Figure 2-45a. Healthy vegetation absorbs much of the incident green radiant flux, therefore, the vegetated areas are recorded in dark tones in this MODIS image. MODIS provides daylight reflection and day/night emission spectral imaging of any point on Earth at least every 2 days, with a continuous duty cycle. The swath width is 2330 km across-track. A composite image of global MODIS coverage on April 19, 2000, is shown in Figure 2-45b.

MODIS has one of the most comprehensive calibration subsystems ever flown on a remote sensing instrument. The calibration hardware includes a solar diffuser, a solar diffuser stability monitor, a spectroradiometric calibration instrument, a blackbody for thermal calibration, and a space view-

port. The calibration allows the optical data to be converted into scaled reflectance. MODIS data are being processed to create numerous global datasets, including (NASA MODIS, 2004):

• land ecosystem variables (e.g., vegetation indices, leaf area index, fraction of photosynthetically active radiation, vegetation net primary production),

• atmospheric variables (e.g., cloud fraction, cloud optical thickness, aerosol optical depth, etc.), and

• ocean variables (e.g., sea-surface temperature and chlorophyll).

Table 2-15.   Characteristics of the *Terra* satellite Moderate Resolution Imaging Spectrometer (MODIS).

Band	Spectral Resolution (μm)	Spatial Resolution	Band Utility
1	0.620 – 0.670	250 × 250 m	Land-cover classification, chlorophyll absorption, leaf-area-index mapping
2	0.841 – 0.876	250 × 250 m	
3	0.459 – 0.479	500 × 500 m	Land, cloud, and aerosol properties
4	0.545 – 0.565	500 × 500 m	
5	1.230 – 1.250	500 × 500 m	
6	1.628 – 1.652	500 × 500 m	
7	2.105 – 2.155	500 × 500 m	
8	0.405 – 0.420	1 × 1 km	Ocean color, phytoplankton, biogeochemistry
9	0.438 – 0.448	1 × 1 km	
10	0.483 – 0.493	1 × 1 km	
11	0.526 – 0.536	1 × 1 km	
12	0.546 – 0.556	1 × 1 km	
13	0.662 – 0.672	1 × 1 km	
14	0.673 – 0.683	1 × 1 km	
15	0.743 – 0.753	1 × 1 km	
16	0.862 – 0.877	1 × 1 km	
17	0.890 – 0.920	1 × 1 km	Atmospheric water vapor
18	0.931 – 0.941	1 × 1 km	
19	0.915 – 0.965	1 × 1 km	
20	3.600 – 3.840	1 × 1 km	Surface–cloud temperature
21	3.929 – 3.989	1 × 1 km	
22	3.929 – 3.989	1 × 1 km	
23	4.020 – 4.080	1 × 1 km	
24	4.433 – 4.498	1 × 1 km	Atmospheric temperature
25	4.482 – 4.549	1 × 1 km	
26	1.360 – 1.390	1 × 1 km	Cirrus clouds
27	6.535 – 6.895	1 × 1 km	Water vapor
28	7.175 – 7.475	1 × 1 km	
29	8.400 – 8.700	1 × 1 km	
30	9.580 – 9.880	1 × 1 km	Ozone
31	10.780 – 11.280	1 × 1 km	Surface–cloud temperature
32	11.770 – 12.270	1 × 1 km	
33	13.185 – 13.485	1 × 1 km	Cloud-top altitude
34	13.485 – 13.785	1 × 1 km	
35	13.785 – 14.085	1 × 1 km	
36	14.085 – 14.385	1 × 1 km	

Digital Frame Cameras

The charge-coupled-device was invented in the late 1960s by scientists at the Bell Labs. It was originally conceived as a new type of computer memory circuit, but it soon became apparent that it had many other applications, including image data collection, because of the sensitivity of silicon to

Moderate Resolution Imaging Spectrometer (MODIS) Imagery

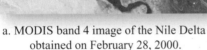

a. MODIS band 4 image of the Nile Delta
obtained on February 28, 2000.

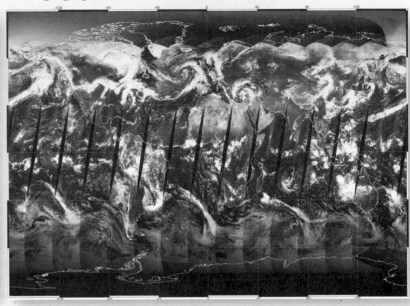

b. A composite band 1 image of global *Terra* MODIS
coverage on April 19, 2000.

Figure 2-45   *Terra* MODIS band 4 (green: 0.545 – 0.565 μm) image of the heavily vegetated Nile Delta, the Nile River, and much of the area surrounding the Red Sea. Band 4 has a spatial resolution of 500 × 500 m. b) A composite band 1 (red: 0.62 – 0.67 μm) image of the global *Terra* MODIS coverage on April 19, 2000. The MODIS swath width is 2330 km across track (courtesy NASA Goddard Space Flight Center).

light. A matrix of CCDs (referred to by Kodak, Inc., as *photosites*) represents the heart and soul of digital frame cameras that are now being used to collect remote sensor data (McGarigle, 1997).

Digital Frame Camera Data Collection

A digital frame camera draws many similarities to a regular camera. It consists of a camera body, a shutter mechanism to control the length of the exposure, and a lens to focus the incident energy onto a film or data-collection plane. However, this is where the similarity ends. Digital frame cameras use electronics rather than chemistry to capture, record, and process images. In a digital frame camera, the standard photographic silver-halide crystal emulsion at the film plane is replaced with a matrix (area array) of CCDs. There are usually thousands of light-sensitive photosites (pixels) that convert varying incident wavelengths of light into electrical signals (Kodak, 1999). An example of an area array CCD and digital frame camera is shown in Figure 2-46.

The CCD

The following is an explanation of the CCD sensor's role in the digital image capture process (Kodak, 1999):

- Mechanical shutter opens, exposing the CCD sensor to light.

- Light is converted to a charge in the CCD.

- The shutter closes, blocking the light.

- The charge is transferred to the CCD output register and converted to a signal.

- The signal is digitized and stored in computer memory.

- The stored image is processed and displayed on the camera's liquid crystal display (LCD), on a computer screen, or used to make hard-copy prints.

a. Area array.

b. Digital camera.

Figure 2-46   a) An area array CCD. b) The area array CCD is located at the film plane in a digital camera and used instead of film to record reflected light (courtesy Eastman Kodak Co.).

An image is acquired when incident light (photons) falls on the array of pixels. The energy associated with each photon is absorbed by the silicon, and a reaction takes place that creates an electron-hole charge pair (e.g., an electron). The number of electrons collected at each pixel is linearly dependent on the amount of light (photons) received per unit time and nonlinearly dependent on wavelength.

Like a traditional photographic system, the digital CCD area array captures a whole "frame" of terrain during a single exposure. The geographic area of the terrain recorded by the CCD area array is a function of 1) the dimension of the CCD array in rows and columns, 2) the focal length of the camera lens (the distance from the rear nodal point of the lens to the CCD array), and 3) the altitude of the aircraft above ground level. Moving the aircraft with the frame camera closer to the ground results in higher spatial resolution but less geographic area coverage. Increasing the actual size (dimension) of the CCD array will record more geographic area if all other variables are held constant. For example, doubling the size of the CCD array from 1000×1000 to 2000×2000 will record four times as much geographic area during the same exposure. Area arrays used for remote sensing applications typically have greater than 4000×4000 CCDs. Digital cameras based on CCD area array technology will not obtain the same resolution as traditional cameras until they contain approximately $20,000 \times 20,000$ detectors per band (Light, 1996). The general public does not have access to such technology at the present time. Fortunately, the cost of digital cameras continues to decrease.

Filtration

Because CCDs are inherently monochromatic, special filters called *color filter array* (CFA) patterns are used to capture the incident blue, green, and red photons of light. A number of CFAs have been invented, but only three are described.

The *RGB filter wheel CFA* is one of the simplest methods. A filter wheel is mounted in front of a monochromatic CCD sensor. The CCD makes three sequential exposures—one for each color. In this case, all photosites on the CCD area array capture red, blue, or green light during the appropriate exposure. This method produces true colors, but it requires three exposures. Therefore, it is suitable for still photography but not for collecting digital photography from a rapidly moving aircraft.

Three-chip cameras use three separate full-frame CCDs, each coated with a filter to make it red, green, or blue-sensitive. A beamsplitter inside the camera divides incoming energy into three distinct bands and sends the energy to the appropriate CCD. This design delivers high resolution and good color rendition of rapidly moving objects. It is the preferred method for remote sensing data collection. However, the cameras tend to be costly and bulky. Using this technology, it is also possible to send incident near-infrared light to an additional near-infrared sensitive CCD.

A *single-chip technology* filter captures all three colors with a single full-frame CCD. This is performed by placing a specially designed filter over each pixel, giving it the ability to capture red, green, and blue information. Obviously this reduces the cost and bulk of the camera since only one CCD is required. It acquires all the information instantaneously.

Timeliness

One of the most important characteristics of digital camera remote sensing is that the data are available as soon as they are collected. There is no need to send the imagery out for chemical processing and then wait for its return. If desired, the digital imagery can be downlinked electronically to the ground while the aircraft is still in the air. Furthermore, the cost of photographic processing is removed unless one wants to make hard-copies.

Anyone can point a digital camera out the window of a plane and obtain oblique aerial photography. It is quite another matter to obtain quality digital aerial photography for photogrammetric or remote digital image processing applications. This requires the use of a digital frame camera system such

Table 2-16. Emerge Digital Sensor System (DSS) characteristics.

Band	Natural Color Mode (nm)	Near-infrared Mode (nm)	Spatial Resolution (m) at Nadir
1	400 – 500	–	Variable, but usually 0.15 to 3 m
2	500 – 600	510 – 600	Variable, but usually 0.15 to 3 m
3	600 – 680	600 – 700	Variable, but usually 0.15 to 3 m
4		800 – 900	Variable, but usually 0.15 to 3 m

Sensor	Area array (4092 × 4079)
Swath	4092 pixels
Altitude	1 to 20,000 ft.
Pixel size	0.009 mm
Quantization	8- and 16-bit

as the Digital Sensor System developed by Emerge, Inc. (Light, 2001).

Emerge, Inc., Digital Sensor System

The Digital Sensor System (DSS) developed by Emerge, Inc., uses a proprietary digital camera area array configuration that acquires vertical imagery containing 4092 × 4079 pixels (Table 2-16). Each pixel in the area array is 9 × 9 microns. Users can specify color (blue, green, and red) or color-infrared (green, red, and near-infrared) multiband imagery in the spectral region from 0.4 to 0.9 μm. In near-infrared mode, the sensor has a spectral response similar to that of Kodak Aerochrome 2443 color-infrared film but with a significantly higher dynamic range. The data may be recorded at 8 or 16 bits per pixel. Emerge collects real-time differentially corrected GPS data about each digital frame of imagery. These data are used to mosaic and orthorectify the imagery using photogrammetric techniques. The pixel placement accuracy meets national map accuracy standards (Emerge, 2004). A variety of flying heights and different focal-length Nikon lenses can be used to obtain imagery

with spatial resolutions ranging from 0.15 to 3 m. Figure 2-47 depicts multispectral bands of Emerge 1 × 1 m imagery obtained over Dunkirk, NY. Such high-resolution data can be collected on demand in good weather.

 Satellite Photographic Systems

Despite the ongoing development of electro-optical remote sensing instruments, traditional optical camera systems continue to be used for space-survey purposes (Lavrov, 1997; FAS, 2004). For example, the Russian SOVINFORMSPUT-NIK SPIN-2 TK-350 and KVR-1000 cameras provide cartographic quality imagery suitable for making topographic and planimetric map products at 1:50,000 scale. Also, the U.S. space shuttle astronauts routinely obtain photography using Hasselblad and Linhof cameras.

Russian SPIN-2 TK-350 and KVR-1000 Cameras

The Russian Space Agency granted the Interbranch Association SOVINFORMSPUTNIK, Moscow, Russia, an exclusive right to use remote sensor data acquired by Russian defense satellites, to distribute these data commercially, and to produce and market value-added products. Most of the data marketed by SOVINFORMSPUTNIK are acquired by the KOMETA Space Mapping System onboard the KOSMOS series of satellites, which was designed to obtain high-spatial-resolution stereoscopic analog photography from space to produce 1:50,000-scale topographic maps. The data are also used to create digital elevation models and 2 × 2 m orthorectified images. The high-resolution image archive contains global coverage acquired since 1981. The KOMETA system includes the TK-350 Camera and the KVR-1000 Panoramic Camera (Figure 2-48).

Basically, a rocket carries the KOMETA satellite into a 220-km near-circular orbit with an orbital duration of approximately 45 days. For example, one KOMETA mission lasted from February 17, 1998, through April 3, 1998. The total film capacity of the two cameras covers approximately 10.5 million km² of land area. The entire KOMETA system is retrieved from orbit at a predetermined location. The film is then processed and digitized. SOVINFORMSPUTNIK has an agreement with Aerial Images, Inc., Raleigh, NC, and Central Trading Systems, Inc., Huntington Bay, NY, to market the data. Much of the U.S. coverage is served interactively on the Internet using Microsoft's Terra Server, a system capable of retrieving quicklooks of much of the SOVINFORMSPUTNIK KVR-1000 archive for selected

a. Green band.

b. Red band.

c. Near-infrared band.

**Digital Frame Camera
Imagery of Dunkirk, NY**

Figure 2-47   Emerge Inc., multiband digital imagery of Dunkirk, NY. The data were collected on December 12, 1998, at a spatial resolution of 1 × 1 m (courtesy Emerge, Inc.).

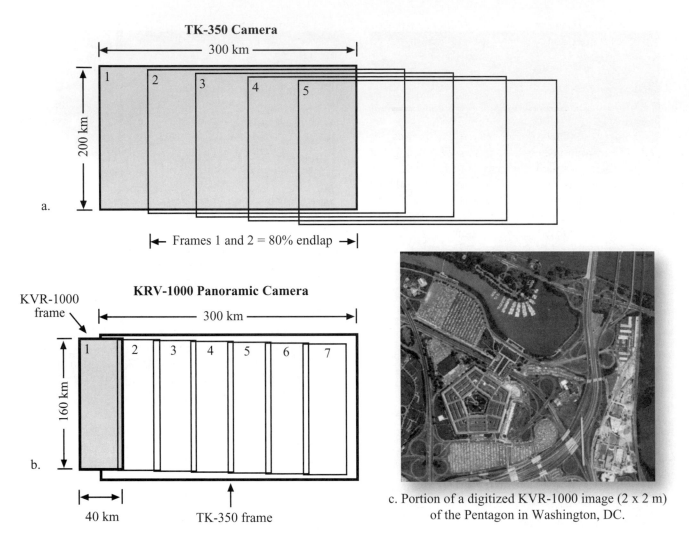

TK-350 Camera

300 km

200 km

1 2 3 4 5

a.

|← Frames 1 and 2 = 80% endlap →|

KRV-1000 Panoramic Camera

KVR-1000 frame

300 km

160 km

1 2 3 4 5 6 7

b.

|←→|
40 km   TK-350 frame

c. Portion of a digitized KVR-1000 image (2 x 2 m) of the Pentagon in Washington, DC.

Figure 2-48   a) Geographic coverage of the SPIN-2 SOVINFORMSPUTNIK TK-350 Camera system. b) Geographic coverage of the KVR-1000 Panoramic Camera. c) A small portion of a KVR-1000 frame of imagery depicting the Pentagon (courtesy of SOVINFORMSPUTNIK and Aerial Images, Inc.).

locations. The imagery can be ordered directly through the Terra Server (Microsoft, 2004).

TK-350 Camera

The TK-350 camera was developed to collect panchromatic (510 – 760 nm) stereoscopic photography with up to 80 percent endlap between frames for the extraction of topographic (elevation) data. The TK-350 camera has a 350-mm focal-length lens. Images obtained by this camera have a scale of 1:660,000. The film format is 30 × 45 cm, which encompasses a single image covering 200 × 300 km, as shown in Figure 2-48a. TK-350 images can be enlarged to 1:50,000 scale without significant loss of image quality. The area covered by a single TK-350 image is covered by seven nested

KVR-1000 images (Figure 2-48b) (Lavrov, 1997). After digitization, TK-350 imagery has a spatial resolution of approximately 10 × 10 m.

KVR-1000 Panoramic Camera

The KVR-1000 Panoramic camera has a 1000-mm focal-length lens. It records panchromatic (510 – 760 nm) photographs at approximately 1:220,000 scale. KVR-1000 images may be enlarged to 1:10,000 scale without loss of detail. The geographic coverage of individual frames is 40 × 160 km, as shown in Figure 2-48b. Note the minimal amount of endlap obtained. After digitization, the imagery is provided at 2 × 2 m spatial resolution. Ground control for rectification is achieved through an onboard system employing GPS and laser altitude control systems. This enables the imagery to be rectified even when conventional ground control is not available.

U.S. Space Shuttle Photography

NASA astronauts routinely document Earth processes during Space Transportation System (STS) missions using both analog and digital camera systems. These efforts have resulted in an impressive database of more than 400,000 Earth images. Photographic documentation of Earth processes during manned spaceflights remains the cornerstone of the Space Shuttle Observations program, as it was with the earlier *Mercury, Gemini, Apollo,* and *Skylab* Earth observations programs (Lulla and Dessinov, 2000). During the space shuttle era, more than 250 selected sites of interest to geoscientists have been identified. Data from these sites are routinely acquired during Space Shuttle missions and cataloged into a publicly accessible electronic database according to the specific mission (e.g., STS-74) or by thematic topic (NASA Shuttle Photography, 2004).

Space Shuttle Analog Cameras

The primary analog cameras used during space shuttle missions are the Hasselblad and Linhof systems. NASA-modified Hasselblad 500 EL/M 70-mm cameras are used with large film magazines, holding 100 to 130 exposures. Standard lenses include a Zeiss 50-mm CF Planar f3.5, and a Zeiss 250-mm CD Sonnar f5.6. The Aero-Technika Linhof system can be fitted with 90- and 250-mm f5.6 lenses. This system uses large-format (100×127 mm) film.

The four windows in the aft part of the Space Shuttle are typically used to obtain photography of the Earth. The windows only allow 0.4 to 0.8 µm light to pass through. This results in the use of two primary film bases in the Hasselblad and Aero-Technika Linhof camera systems, including visible color (Kodak 5017/6017 Professional Ektachrome) and color-infrared (Kodak Aerochrome 2443) films.

Space shuttle photographs are obtained at a variety of Sun angles, ranging from 1° to 80°, with the majority of pictures having Sun angles of approximately 30°. Very low Sun angle photography often provides unique topographic views of remote mountainous areas otherwise poorly mapped. Sequential photographs with different look angles can, in certain instances, provide stereoscopic coverage. Seventy-five percent of the photographs in the archive cover the regions between 28° N and 28° S latitude, providing cover-

age for many little-known tropical areas. Twenty-five percent of the images cover regions between 30° to 60° N and S latitude.

The Space Shuttle Earth Observations Project (SSEOP) photography database of the Earth Science Branch at the NASA Johnson Space Center contains the records of more than 400,000 photographs of the Earth made from space. A select set of these photographs has been digitized for downloading (http://images.jsc.nasa.gov).

Space Shuttle and Space Station Digital Photography

The International Space Station (ISS) was launched November 2, 2000. It continues the NASA tradition of Earth observation from human-tended spacecraft. The ISS U.S. Laboratory Module has a specially designed optical window with a clear aperture 50.8 cm in diameter that is perpendicular to the Earth's surface most of the time. In 2001, space shuttle astronauts began acquiring digital images using a Kodak DCS 760 camera with Nikon camera body and lenses (Kodak, 2001). Both color and monochrome digital images may be obtained. The camera uses a $3,032 \times 2,008$ astronomical grade CCD. Digital images are transmitted directly to the ground while in orbit.

An astronaut photograph of the Toquepala Copper Mine in Southern Peru taken from the International Space Station on September 22, 2003, with a Kodak DCS 760 digital camera equipped with an 400-mm lens is shown in Figure 2-49.

Digital Image Data Formats

The image analyst can order digital remote sensor data in a variety of formats. The most common formats are:

- Band Interleaved by Pixel (PIP),

- Band Interleaved by Line (BIL), and

- Band Sequential (BSQ).

To appreciate the data formats, consider a hypothetical remote sensing dataset containing just nine pixels obtained over the interface between land and water (Figure 2-50). The image consists of three bands (band 1 = green; band 2 = red; band 3 = near-infrared). The brightness value ($BV_{i,j,k}$) row, column, and band notation is provided (Figure 2-50a).

Digital Frame Camera Image of the Toquepala Copper Mine in Southern Peru Taken from the International Space Station

Figure 2-49   Astronaut photograph of the Toquepala Copper Mine in Southern Peru taken from the International Space Station on September 22, 2003, with a Kodak DCS 760 digital camera (photo #ISS007-E-15222). The open pit mine is 6.5 km across and descends more than 3,000 m into the earth. A dark line on the wall of the pit is the main access road to the bottom. Spoil dumps of material mined from the pit are arranged in tiers along the northwest lip of the pit (courtesy of NASA Earth Observatory and Dr. Kamlesh Lulla, NASA Johnson Space Center).

Band Interleaved by Pixel Format

The BIP format places the brightness values in n bands associated with each pixel in the dataset in sequential order [e.g., for a dataset containing three bands the format for the first pixel in the matrix (1,1) is 1,1,1; 1,1,2; 1,1,3]. The brightness values for pixel (1,2) are then placed in the dataset (e.g., 1,2,1; 1,2,2; and 1,2,3), and so on. An end-of-file (EOF) marker is placed at the end of the dataset (Figure 2-50b).

Band Interleaved by Line Format

The BIL format creates a file that places the brightness values in n bands associated with each line in the dataset in sequential order. For example, if there are three bands in the dataset, all of the pixels in line 1, band 1 are followed by all of the pixels in line 1, band 2, and then line 1, band 3. An EOF marker is placed at the end of the dataset (Figure 2-50c).

Remote Sensing Data Formats

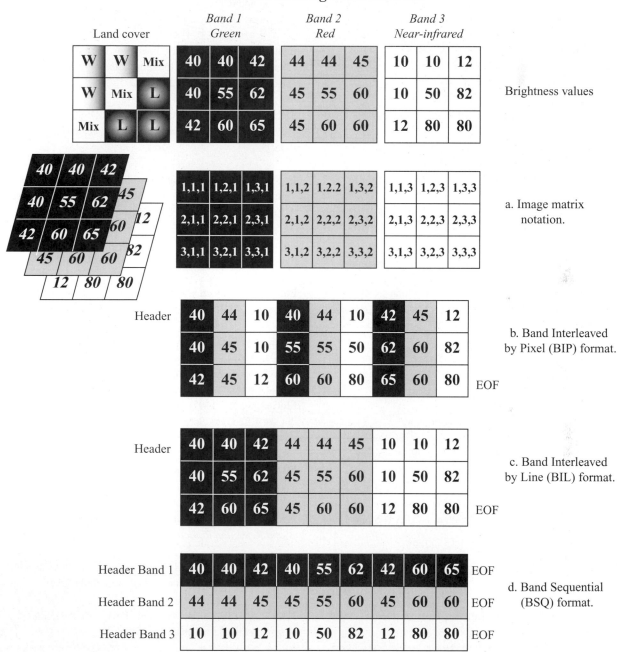

Figure 2-50   Digital image data formats.

Band Sequential Format

The BSQ format places all of the individual pixel values in each band in a separate and unique file. Each band has its own beginning header record and EOF marker (Figure 2-50d).

 Summary

The general public and scientists should perform digital image processing of remote sensor data only after they understand and appreciate exactly how the imagery was collected. This chapter provides information about the characteristics of many important digital remote sensing systems and the most often used digital data formats.

References

DigitalGlobe, 2004, *DigitalGlobe*, www.digitalglobe.com.

Emerge, 2004, *Emerge*, www.emergeweb.com.

EOSAT, 1992, *Landsat Technical Notes*, Lanham, MD: EOSAT.

FAS, 2004, *SPIN-2 Imagery*, Washington: Federation of American Scientists, www.fas.org/irp/imint/spin.htm.

Filippi, A., 1999, "Hyperspectral Image Classification Using a Batch Descending Fuzzy Learning Vector Quantization Artificial Neural Network: Vegetation Mapping at the John F. Kennedy Space Center," unpublished master's thesis, Columbia: University of South Carolina, 276 p.

Gasparovic, R. F., 2003, *Ocean Remote Sensing Program*, John Hopkins University Applied Physics Laboratory, http://fermi.jhuapl.edu/avhrr/index.html.

ImageSat, 2004, *EROS-1A*, www.imagesatintl.com/.

ITRES, 2004, *Calibrated Airborne Spectrographic Imager: CASI 3*, Canada: ITRES Research Ltd., www.itres.com.

Kidwell, K. B., 1998, *NOAA Polar Orbiter Data User's Guide: TIROS-N through NOAA-14*, Washington: NOAA U.S. Dept. of Commerce, 50 p.

King, M., 2003, *EOS Data Product Handbook - Volume 1*, Washington: NASA, 258 p.

Kodak, 1999, *CCD Technology*, Buffalo: Eastman Kodak, www.kodak.com.

Kodak, 2001, *Press Release: NASA to Launch Kodak Professional DCS 760 Digital Camera on Mission to International Space Station*, www.kodak.com/US/en/corp/pressReleases/pr200106 26-01.shtml.

Lavrov, V. N., 1997, "SPIN-2 Space Survey Photocameras for Cartographic Purposes," *Proceedings*, 4th International Conference on Remote Sensing for Marine and Coastal Environments, Orlando: ERIM, (March 17–19), 10 p.

Light, D. L., 1993, "The National Aerial Photography Program as a Geographic Information System Resource," *Photogrammetric Engineering & Remote Sensing*, 48(1):61–65.

Light, D. L., 1996, "Film Cameras or Digital Sensors? The Challenge Ahead for Aerial Imaging," *Photogrammetric Engineering & Remote Sensing*, 62(3):285–291.

Light, D. L., 2001, "An Airborne Direct Digital Imaging System," *Photogrammetric Engineering & Remote Sensing*, 67(11):1299–1305.

Liu, J. G., 2000, "Evaluation of Landsat-7 ETM+ Panchromatic Band for Image Fusion with Multispectral Bands," *Natural Resources Research*, 9(4):269–276.

Logicon Geodynamics, 1997, *Multispectral Imagery Reference Guide*, Fairfax, VA: Logicon Geodynamics, 101 p.

Loral Space Systems, 1996, *GOES I-M DataBook*, Loral Space Systems, #DRL 101-08, Revision 1; 200 p.

Lulla, K. and L. Dessinov, 2000, *Dynamic Earth Environments: Remote Sensing Observations from Shuttle-Mir Missions*, New York: John Wiley & Sons, 268 p.

McGarigle, B., 1997, "Digital Aerial Photography Becoming a Cost-Effective Solution," *Geo Info*, (March), 1–8, www.govtech.net/1997.

Microsoft, 2004, *Microsoft Terra Server*, www.terraserver.com/spin2.asp.

NASA ASTER, 2004, *Advanced Spaceborne Thermal Emission and Reflection Radiometer*, http://asterweb.jpl.nasa.gov/.

NASA ATLAS, 2004, *Airborne Terrestrial Applications Sensor*, www.earth.nasa.gov/science/suborbital/atlas2.htm.

NASA AVIRIS, 2004, *Airborne Visible Infrared Imaging Spectrometer*, http://aviris.jpl.nasa.gov/.

NASA ESE, 2004, *Earth Science Enterprise*, www.earth.nasa.gov.

NASA EO-1, 2004, *Earth Observing-1*, http://eo1.gsfc.nasa.gov/overview/eo1Overview.html.

NASA Landsat 7, 2004, *Landsat 7*, http://landsat.gsfc.nasa.gov/.

NASA MODIS, 2004, *Moderate Resolution Imaging Spectrometer*, http://modis.gsfc.nasa.gov/.

NASA MISR, 2004, *Multiangle Imaging Spectrometer*, http://www-misr.jpl.nasa.gov/.

NASA/Orbimage SeaWiFS, 2004, *Sea-viewing Wide Field-of-view Sensor*, http://daac.gsfc.nasa.gov/data/dataset/SEAWIFS/.

NASA Shuttle Photography, 2004, *NASA Space Shuttle Earth Observation Photography*, http://earth.jsc.nasa.gov/sseop/efs/.

NOAA, 1975 to 1984, *Landsat Data Users Notes, NOAA* Landsat Customer Services, Sioux Falls, SD, 57198.

NOAA AVHRR, 2004, *Advanced Very High Resolution Radiometer*, www.noaa.gov/satellites.html.

NOAA DMSP, 2004, *Defense Meteorological Satellite Program*, http://dmsp.ngdc.noaa.gov/dmsp.html.

NOAA GOES, 2004, *Geostationary Operational Environmental Satellite*, www.goes.noaa.gov/.

NOAA GVI, 2004, *Global Vegetation Index Products*, http://www2.ncdc.noaa.gov/docs/gviug/.

NOAA NPOESS, 2004, *National Polar Orbiting Operational Environmental Satellite System*, www.ipo.noaa.gov/.

NOAA NPP, 2004, *National Polar Orbiting Operational Environmental Satellite System Preparatory Project*, www.ipo.noaa.gov/Projects/npp.html.

NOAA VIIRS, 2004, *Visible/Infrared Imager/Radiometer Suite*, www.ipo.noaa.gov/Technology/viirs_summary.html.

NOAA/NASA POES, 2004, *NOAA/NASA Polar Operational Environmental Satellites (NPOES)*, http://poes.gsfc.nasa.gov/.

NRSA, 2004, *Indian Remote Sensing Program*, www.nrsa.gov.in/.

Orbimage, 2004, *OrbView 3*, www.orbimage.com.

SBRC, 1994, *Space Sensors*, Goleta, CA: Santa Barbara Research Center, 33 p.

SenSyTech, 2004, *Airborne Multispectral Scanner*, www.sensystech.com.

Solomonson, V., 1984, "Landsat 4 and 5 Status and Results from Thematic Mapper Data Analyses," *Proceedings, Machine Processing of Remotely Sensed Data*, W. Lafayette, IN: Laboratory for the Applications of Remote Sensing, 13–18.

Space Imaging, 2004, *Space Imaging*, www.spaceimaging.com.

SPOT Image, 2004, *SPOT*, www.spotimage.com/home/.

Toutin, T. and M. Beaudoin, 1995, "Real-Time Extraction of Planimetric and Altimetric Features from Digital Stereo SPOT Data Using a Digital Video Plotter," *Photogrammetric Engineering and Remote Sensing,* 61(1):63–68.

USGS Global Landcover, 2004, *Global Landcover*, http://edcwww.cr.usgs.gov/products/landcover.html.

USGS Landsat 7, 2004, *Landsat 7 Updates*, Sioux Falls, SD: USGS EROS Data Center, http://landsat7.usgs.gov/updates.php.

USGS LDCM, 2004, *Landsat Data Continuity Mission*, Sioux Falls, SD: EROS Data Center, http://ldcm.usgs.gov/.

Vane, G. and A. F. H. Goetz, 1993, "Terrestrial Imaging Spectrometry: Current Status, Future Trends," *Remote Sensing of Environment*, 44:117–126.

Digital Image Processing Hardware and Software Considerations

3

Analog (hard-copy) and digital remotely sensed data are used operationally in many Earth science, social science, and planning applications (e.g., Miller et al., 2003). Analog remotely sensed data such as positive 9 × 9 in. aerial photographs are routinely analyzed using the fundamental elements of analog (visual) image interpretation and optical instruments such as stereoscopes and zoom-transfer-scopes (ASPRS, 1997; Jensen, 2000b). Digital remote sensor data are analyzed using a *digital image processing system* that consists of computer hardware and special-purpose image processing software. This chapter describes:

- fundamental digital image processing system *hardware* characteristics,

- digital image processing *software* (computer program) requirements, and

- selected public and commercial sources of digital image processing hardware and software.

Digital image processing hardware and software are constantly improving. Therefore, an image analyst should configure a digital image processing system (hardware and software) based on the *principles* introduced in this chapter.

 Digital Image Processing System Considerations

To successfully process digital remote sensor data it is usually necessary to hire people who are trained in a systematic body of knowledge (e.g., forestry, agronomy, urban planning, geography, geology, marine science) and who have considerable knowledge in GIScience (cartography, remote sensing, and geographic information systems). They must understand the theoretical basis of remote sensing data-collection systems and the various digital image processing (and GIS) algorithms and how to properly apply the technology to their specific systematic body of knowledge. It is the quality and creativity of the people, not the hardware or the software, that dictates how useful the digital image processing will be.

Qualified image analysts select an appropriate digital image processing system that a) has a reasonable learning curve and is relatively easy to use, b) has a reputation for producing accurate results (ideally the company has ISO certification), c) will produce the desired results in an appropriate format (e.g., map products in a standard cartographic data structure compatible with most

GIS), and d) is within their department's budget. Table 3-1 summarizes some of the important factors to consider when selecting a digital image processing system. It is useful to review these factors.

Central Processing Units: Personal Computers, Workstations, and Mainframes

Many Earth resource analysis and planning projects require that large geographic areas be inventoried and monitored through time. Therefore, it is common to obtain remotely sensed data of the study area (Friedl et al., 2002). Unfortunately, the amount of digital data generated by a remote sensing system can be daunting. For example, a single SPOT 3 HRV XS multispectral 60×60 km scene consists of three bands (green, red, and near-infrared) of 8-bit remote sensor data. Each pixel is 20×20 m in dimension. Therefore, the single three-band dataset consists of 27 megabytes (Mb) of data (3000 rows $\times$ 3000 columns $\times$ 3 bands / 1,000,000 = 27 Mb). But what if a Landsat 5 Thematic Mapper 185×170 km scene was the only type of imagery available for the study area? It consists of seven bands of 30×30 m data (the thermal channel was actually 120×120 but was resampled to 30×30 m) and is approximately 244 Mb (5666 rows $\times$ 6166 columns $\times$ 7 bands / 1,000,000 = 244 Mb). More than one SPOT or Thematic Mapper scene are analyzed when change detection must take place.

Furthermore, many public agencies and Earth resource scientists are now taking advantage of even higher spatial and spectral resolution remote sensor data (Curran et al., 1998; Jensen et al., 2002). For example, a single 11×11 km Space Imaging IKONOS scene of panchromatic data consisting of 1×1 m pixels is 121 Mb (11,000 $\times$ 11,000 / 1,000,000 assuming 8-bit pixels). A single 512×512 pixel AVIRIS hyperspectral subscene contains 224 bands of 12-bit data and is 88.1 Mb ($512 \times 512 \times 1.5 \times 224$ / 1,000,000). Processing such massive remote sensor datasets requires a significant number of computations. The type of computer selected dictates how fast (efficient) the computations or operations can be performed and the precision with which they are made.

The central processing unit (CPU) is the computing part of the computer. It consists of a control unit and an arithmetic logic unit. The CPU performs:

- numerical integer and/or floating point calculations, and

- directs input and output from and to mass storage devices, color monitors, digitizers, plotters, etc.

Table 3-1.   Factors in selecting digital image processing system components.

Digital Image Processing System Considerations

- Number and speed of the computer's CPU
- Operating system (e.g., Microsoft Windows; UNIX, Linux, Macintosh)
- Amount of random access memory (RAM)
- Number of image analysts that can use the system at one time and the mode of operation (e.g., interactive or batch)
- Serial or parallel processing
- Arithmetic coprocessor or array processor
- Software compiler(s) (e.g., C^{++}, Visual Basic, Avenue, Java)
- Type of mass storage (e.g., hard disk, CD-ROM, DVD) and amount (e.g., gigabytes)
- Monitor display spatial resolution (e.g., 1024×768 pixels)
- Monitor color resolution (e.g., 24 bits of image processing video memory yields 16.7 million displayable colors)
- Input devices (e.g., optical-mechanical drum or flatbed scanners, area array digitizers)
- Output devices (e.g., CD-ROM, CD-RW, DVD-RW, film-writers, line plotters, dye sublimation printers)
- Networks (e.g., local area, wide area, Internet)
- Image processing system applications software (e.g., ERDAS Imagine, Environment for Visualizing Images, PCI Geomatica, ER MAPPER, IDRISI, ESRI Image Analyst, *e*Cognition)
- Interoperability with major GIS software

The CPU's efficiency is often measured in terms of how many millions of instructions per second (MIPS) it can process, e.g., 500 MIPS. It is also customary to describe a CPU in terms of the number of cycles it can process in 1 second measured in megahertz, e.g., 1000 Mhz (1 GHz). Manufacturers market computers with CPUs faster than 4 GHz, and this speed will continue to increase. The system bus connects the CPU with the main memory, managing data transfer and instructions between the two. Therefore, another important consideration when purchasing a computer is bus speed.

To appreciate the quality of the computers that we routinely use for digital image processing today, it is instructive to briefly review the history of the central processing unit. The ENIAC was the first computer. It was invented in 1946 and weighed approximately 30 tons. In 1968, there were only 30,000 computers in the entire world—mostly mainframes that occupied entire rooms and refrigerator-sized mini-computers. People programmed the computers using punch cards. This changed when several people who worked for Fairchild Semiconductor left to start their own business—Intel, Inc. Their first contract was to develop 12 custom inte-

Table 3-2.   Historical development of Intel central processing units for personal computers (Intel, 2004).

Central Processing Unit (Date introduced)	Clock Speed (KHz) (MHz) (GHz)	Millions of Instructions per Second (MIPS)	Bus Width (bits) or Speed (MHz)	Transistors	Address-able Memory	Significance
4004 (11/15/71)	108 KHz	0.06	4	2,300	640 bytes	First microcomputer chip; arithmetic manipulation
8008 (4/1/72)	200 KHz	0.06	8	3,500	16 Kbytes	Data/character manipulation
8080 (4/1/74)	2 MHz	0.64	8	6,000	64 Kbytes	10 × performance of the 8008
8086 (6/8/78)	10	0.75	16	29,000	1 MB	10 × performance of the 8080
8088 (6/1/79)	8	0.75	8	29,000	1 MB	Identical to 8086 except for its 8-bit external bus
80286 (2/1/82)	12	0.9 − 2.66	16	134,000	16 MB	3-6 × performance of 8086
386DX (10/17/85)	16	11.4	32	275,000	4 GB	First to process 32-bit data
486DX - 4/10/89 - 6/24/91	25 50	20	32	1.2 million 1.2 million	4 GB 4 GB	Level 1 cache on chip; first built-in math coprocessor
Pentium I - 3/22/93 - 6/2/97	60 233	100	64	3.1 million 4.5 million	4 GB	5 × performance of the 486DX; Pro had cache memory chip
Pentium II - 5/7/97 - 8/24/98	300 450	300	66 100 MHZ	7.5 million 7.5 million	64 GB	Dual independent bus, MMX; high-speed cache memory chip
Pentium III - 2/26/99 - 3/8/00	500 1 GHz	510	100 100	9.5 million 28 million	64 GB	70 new instructions; 1-GHz computer chip
Pentium 4 - 11/20/00 - 5/6/02 - 6/23/03 - 11/3/03 (Extreme)	1.4 2.8 3.2 3.2	1,700 >2,500	400 533 800 800	42 million 55 million 55 million 178 million	64 GB	0.13 μm process; Desktops and workstations for computing enthusiasts
Xeon - 5/21/01 - 6/30/03	1.7 2.8	>1,700	400 400	42 million 169 million	64 GB	Designed for dual-processor servers and workstations

grated circuits for the Busicom calculator company in Japan. Ted Hoff of Intel decided to design a single computer chip that performed all 12 functions. This was the first computer on a chip. Intel purchased the design and marketing rights to the 4004 microprocessor for $60,000 (Table 3-2). A short time later the Busicom calculator company went bankrupt.

Within a decade the microprocessor was hailed as one of the top inventions in American technological history, ranked with the light bulb, telephone, and the airplane. Table 3-2 documents the historical development of the Intel family of CPUs used in IBM-compatible personal computers (Intel, 2004).

In 1985, Gordon Moore was preparing a speech and made a memorable observation. He realized that each new chip contained roughly twice as much capacity as its predecessor and each chip was released within 18 to 24 months of the previous chip. If this trend continued, he reasoned, computing power would rise exponentially over relatively brief periods of time. *Moore's law* described a trend that has continued and is still remarkably accurate (Figure 3-1). It is the basis for many planners' performance forecasts. Since 1971, the number of transistors on a chip has increased from 2300 on the 4004 to 55 million on the Pentium 4 processor in 2003, to more than 169 million on the Xeon processor in 2003 (Table 3-2; Figure 3-1). The MIPS has also increased logarithmically.

Personal Computers

Personal computers (with 16- to 64-bit CPUs) are the workhorses of digital image processing and GIS analysis. Personal computers are based on microprocessor technology where the entire CPU is placed on a single chip. These inexpensive complex-instruction-set-computers (CISC) generally have CPUs with 32- to 64-bit registers (word size) that can compute integer arithmetic expressions at greater clock speeds and process significantly more MIPS than their 1980s – 1990s 8-bit predecessors (see Table 3-2). The 32-bit CPUs can process four 8-bit bytes at a time and 64-bit CPUs can process eight bytes at a time.

Scientists often populate digital image processing laboratories with PC-based digital image processing systems because the hardware and software are relatively inexpensive per unit and hardware maintenance is low. Educators often purchase PC-based digital image processing systems because they are able to configure numerous systems for laboratory instruction at a reasonable cost. PC maintenance agreements are also relatively inexpensive compared with those for mainframes and workstations.

The most common operating systems for personal computers are various Microsoft Windows operating systems and the Macintosh operating system. Personal computers useful for digital image processing seem to always cost approximately $2,500 with 2 GB of random access memory (RAM), a high-resolution color monitor (e.g., capable of displaying $\geq 1024 \times 768$ pixels), a reasonably sized hard disk (e.g., >300 Gb), and a rewriteable disk (e.g., CD-RW or DVD-RW). The cost of personal computers continues to decrease, and the distinction between personal computers and computer workstations is fading.

Computer Workstations

Workstations usually consist of a $\geq$64-bit reduced-instruction-set-computer (RISC) CPU that can address more random access memory than personal computers. The RISC chip is typically faster than the traditional CISC. RISC workstations application software and hardware maintenance costs are usually higher than personal computer-based image processing systems. The most common workstation operating systems are UNIX and various Microsoft Windows products.

Figure 3-2 summarizes the components of a typical digital image processing lab. These specifications will change rapidly. The computers can function independently or be networked to a file-server, as shown. Both PCs and workstations can have multiple CPUs that allow remotely sensed data to be processed in parallel and at great speed.

Mainframe Computers

Mainframe computers (with $\geq$64-bit CPU) perform calculations more rapidly than PCs or workstations and are able to support hundreds of users simultaneously, especially parallel mainframe computers such as a CRAY. This makes mainframes ideal for intensive, CPU-dependent tasks such as image registration/rectification, mosaicking multiple scenes, spatial frequency filtering, terrain rendering, classification, hyperspectral image analysis, and complex spatial GIS modeling. If desired, the output from intensive mainframe processing can be passed to a workstation or personal computer for subsequent less intensive or inexpensive processing.

Mainframe computer systems are generally expensive to purchase and maintain. Mainframe applications software is usually more expensive. In terms of computing power, mainframes are just below supercomputers, although some believe mainframes are more powerful because they can support more simultaneously running programs.

 Read-Only Memory, Random Access Memory, Serial and Parallel Processing, and Arithmetic Coprocessor

Computers have banks of memory that contain instructions that are indispensable to the successful functioning of the computer. A computer may contain a single CPU or multiple CPUs and process data serially (sequentially) or in parallel. Most CPUs now have special-purpose math coprocessors.

History of Intel Microprocessors

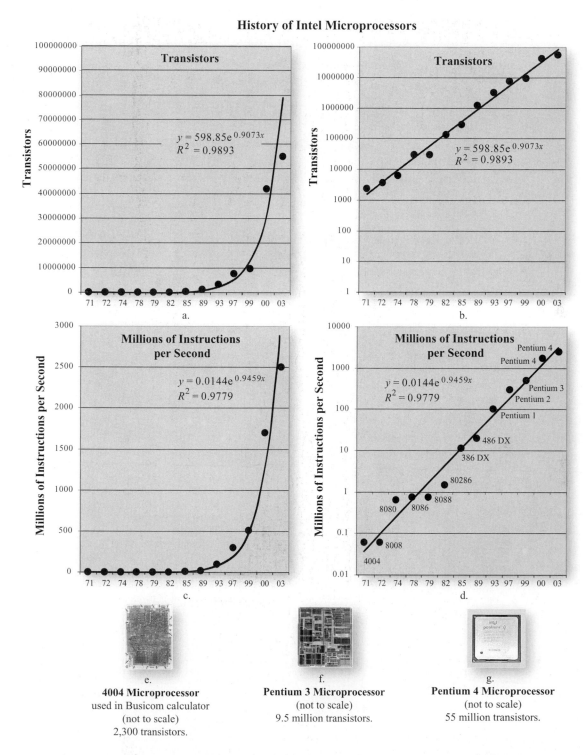

Figure 3-1   Moore's law suggests a relationship between the capacity of computer chips (e.g., number of transistors), the millions of instructions per second that can be processed (MIPS), and the 18 to 24 month release interval. a,b) Number of transistors in Intel microprocessors displayed in real numbers and exponentially. c,d) MIPS displayed in real numbers and exponentially. e) The 108-KHz 4004 Intel chip produced in 1971 had 2300 transistors and could process only 60,000 instructions per second. f) The 500-MHz Pentium 3 produced in 1999 had 9.5 million transistors and could process 510 MIPS. g) The 3.2-GHz Pentium 4 chip produced in 2003 had 55 million transistors and could process >2500 MIPS (Intel, 2004).

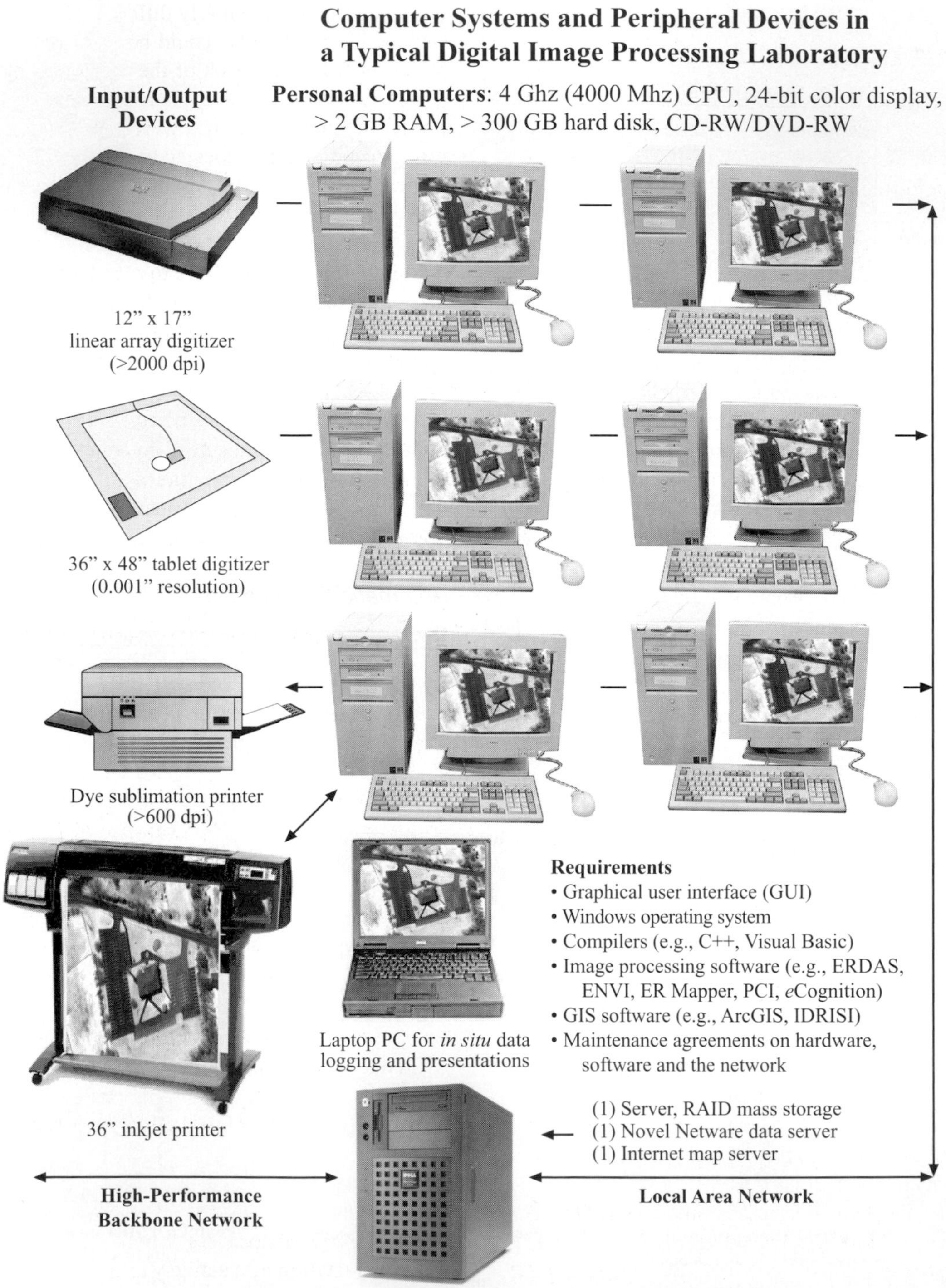

Computer Systems and Peripheral Devices in a Typical Digital Image Processing Laboratory

Input/Output Devices

Personal Computers: 4 Ghz (4000 Mhz) CPU, 24-bit color display, > 2 GB RAM, > 300 GB hard disk, CD-RW/DVD-RW

12" x 17" linear array digitizer (>2000 dpi)

36" x 48" tablet digitizer (0.001" resolution)

Dye sublimation printer (>600 dpi)

36" inkjet printer

Laptop PC for *in situ* data logging and presentations

Requirements
- Graphical user interface (GUI)
- Windows operating system
- Compilers (e.g., C++, Visual Basic)
- Image processing software (e.g., ERDAS, ENVI, ER Mapper, PCI, *e*Cognition)
- GIS software (e.g., ArcGIS, IDRISI)
- Maintenance agreements on hardware, software and the network

(1) Server, RAID mass storage
(1) Novel Netware data server
(1) Internet map server

High-Performance Backbone Network

Local Area Network

Figure 3-2　Digital image processing labs usually contain a number of complex-instruction-set-computers (CISC) and/or reduced-instruction-set-computers (RISC) and peripheral input-output devices. In this example, there are six 24-bit color desktop personal computers (PCs), one laptop, and a server. The PCs communicate locally via a local area network (LAN) and with the world via the Internet. Each 4-Ghz (4000 Mhz) PC has >2 GB random access memory, >300 GB hard disk space, and a CD RW and/or DVD-RW. Image processing software and remote sensor data can reside on each PC (increasing the speed of execution) or be served by the server, minimizing the amount of mass storage required on each PC. Scanners, digitizers, and plotters are required for input of digital remote sensor data and output of important results.

Read-Only Memory and Random Access Memory

Read-only memory (ROM) retains information even after the computer is shut down because power is supplied from a battery that must be replaced occasionally. For example, the date and time are stored in ROM after the computer is turned off. When restarted, the computer looks in the date and time ROM registers and displays the correct information. Most computers have sufficient ROM for digital image processing applications; therefore, it is not a serious consideration.

Random access memory (RAM) is the computer's primary *temporary* workspace. It requires power to maintain its content. Therefore, all of the information that is temporarily placed in RAM while the CPU is performing digital image processing must be saved to a hard disk (or other media such as a CD) before turning the computer off.

Computers should have sufficient RAM for the operating system, image processing applications software, and any remote sensor data that must be held in temporary memory while calculations are performed. Computers with 64-bit CPUs can address more RAM than 32-bit machines (see Table 3-1). Figure 3-2 depicts individual workstations with >2 GB of RAM. RAM is broken down into two types: dynamic RAM (DRAM) and static RAM (SRAM). The data stored in DRAM is updated thousands of times per second; SRAM does not need to be refreshed. SRAM is faster but is also more expensive. It seems that one can never have too much RAM for image processing applications. RAM prices continue to decline while RAM speed continues to increase.

Serial and Parallel Image Processing

It is possible to obtain PCs, workstations, and mainframe computers that have multiple CPUs that operate concurrently (Figure 3-3). Specially written parallel processing software can parse (distribute) the remote sensor data to specific CPUs to perform digital image processing. This can be much more efficient than processing the data serially. For example, consider performing a per-pixel classification on a typical 1024 row by 1024 column remote sensing dataset (Figure 3-3). In the first example, each pixel is classified by passing the spectral data to the CPU and then progressing to the next pixel. This is *serial* processing (Figure 3-3a). Conversely, suppose that instead of just one CPU we had 1024 CPUs. In this case the class of each of the 1024 pixels in the row could be determined using 1024 separate CPUs (Figure 3-3b). The *parallel* image processing would classify the line of data approximately 1024 times faster than would process-

ing it serially. In an entirely different parallel configuration, each of the 1024 CPUs could be allocated an entire row of the dataset. Finally, each of the CPUs could be allocated a separate band if desired. For example, if 224 bands of AVIRIS hyperspectral data were available, 224 of the 1024 processors could be allocated to evaluate the 224 brightness values associated with each individual pixel with 800 additional CPUs available for other tasks.

All of the CPUs do not have to reside on the same computer or even in the same city to perform parallel processing. It is possible to perform parallel processing by connecting individual computer systems via a network. This type of parallel processing requires sophisticated distributed processing software. In practice, it is difficult to parse a program so that multiple CPUs can execute different portions of the program without interfering with one another. Many vendors are developing digital image processing code that takes advantage of multiple CPUs and parallel architecture.

Arithmetic Coprocessor

An arithmetic coprocessor is a special mathematical circuit that performs high-speed floating point operations while working in harmony with the CPU. The Intel 486 processor (see Table 3-1) was the first CPU to offer a built-in math coprocessor (Intel, 2004). All of the current CPUs contain arithmetic coprocessors. If substantial resources are available, then an array processor is ideal. It consists of a bank of memory and special circuitry dedicated to performing simultaneous computations on elements of an array (matrix) of data in n dimensions. Most remotely sensed data are collected and stored as arrays of numbers so array processors are especially well suited to image enhancement and analysis operations. However, specialized software must often be written to take advantage of the array processor.

 # Mode of Operation and Interface

Image analysts can process remotely sensed data on a stand-alone workstation interactively or in batch mode. Ideally the processing takes place in an interactive environment using a well-crafted graphical user interface (GUI).

Mode of Operation

Breakthroughs in analyst *image understanding* and *scientific visualization* are generally accomplished by placing the ana-

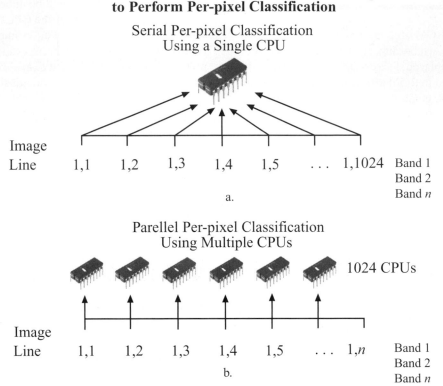

**Serial versus Parallel Digital Image Processing
to Perform Per-pixel Classification**

Figure 3-3   Digital image classification using a) *serial* and b) *parallel* image processing logic. In a serial processing environment, each pixel of each line in each band is processed sequentially using a single CPU. In a parallel processing environment, each of the CPUs could be requested to process 1) a single pixel in a line of remote sensor data (as shown), or 2) an entire line of data (not shown), or 3) an entire band of data (not shown).

lyst as intimately in the image processing loop as possible and allowing his or her intuitive capabilities to take over (Slocum, 1999). Ideally, every analyst has access to his or her own digital image processing system. Unfortunately, this is not always possible due to cost constraints. The sophisticated PC or workstation laboratory environment shown in Figure 3-2 would be ideal for six or seven people doing digital image processing. It would probably be ineffective for education or short course instruction where many analysts (e.g., ≥ 20) must be served. Fortunately, the price of personal computers continues to decline, making it practical to configure laboratory and teaching environments so that each participant has access to his or her own digital image processing system.

Graphical User Interface

One of the best scientific visualization environments for the analysis of remote sensor data takes place when the analyst communicates with the digital image processing system *interactively* using a point-and-click *graphical user interface* (Limp, 1999). Most sophisticated image processing systems are now configured with a friendly, point-and-click GUI that allows rapid display of images and the selection of important image processing functions (Chan, 2001). Several effective digital image processing graphical user interfaces include:

• ERDAS Imagine's intuitive point-and-click icons (Figure 3-4),

• Research System's Environment for Visualizing Images (ENVI) hyperspectral data analysis interface (Figure 3-5),

• ER Mapper,

• IDRISI,

• ESRI ArcGIS, and

Figure 3-4   The ERDAS Imagine graphical user interface (GUI) consists of icons that direct the user to more detailed applications (e.g., import/export functions), tool icons that perform specific functions (e.g., magnification, reduction), scroll bars on the side and bottom, and ancillary information around the image of interest such as the file name, UTM coordinates, and GRS. This is a panchromatic 1 × 1 m image of a golf course near Florence, SC, obtained by the IKONOS remote sensing system (courtesy of Space Imaging, Inc.).

• Adobe Photoshop.

Photoshop is very useful for processing photographs and images that have three or fewer bands of data.

Noninteractive, batch processing is of value for time-consuming processes such as image rectification, mosaicking, orthophoto generation, and filtering. Batch processing frees up laboratory PCs or workstations during peak demand because the jobs can be stored and executed when the computer is otherwise idle (e.g., during early morning hours). Batch processing can also be useful during peak hours because it allows the analyst to set up a series of operations

that can be executed in sequence without operator intervention. Digital image processing also can now be performed interactively over the Internet at selected sites.

Computer Operating System and Compiler(s)

The computer operating system and compiler(s) must be easy to use yet powerful enough so that analysts can program their own relatively sophisticated algorithms and experiment with them on the system. It is not wise to configure an image processing system around an unusual operating

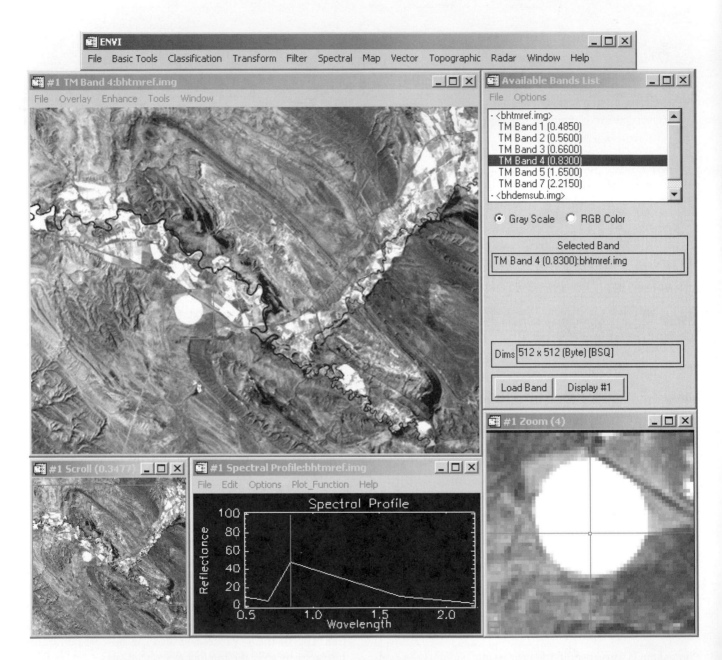

Figure 3-5　　The Environment for Visualizing Images (ENVI) graphical user interface uses a point-and-click text task bar to locate specific files and digital image processing functions (Research Systems, Inc., 2004). In this example, Landsat Thematic Mapper imagery is displayed and analyzed. The Landsat Thematic Mapper records seven bands, but only six are found in this particular dataset of the Bighorn Basin, Wyoming. Thermal infrared band 6 data are not present in the available bands list. In this example, only the band 4 (near-infrared) data are being displayed in the various views, including the scroll window in the lower left corner, which contains a much reduced version of the scene (0.3477), a portion of the entire scene at full 1× magnification, and the zoom window at 4× magnification. A cross-hair has been placed in the middle of the center pivot irrigation system in the zoom window. The spectral profile associated with this single pixel is shown over the six-band wavelength region from 0.485 to 2.215 μm. The vertical line in the spectral profile identifies where band 4 is located (0.83 μm).

system or compiler because it becomes difficult to communicate with the peripheral devices and share applications with other scientists.

Operating System

The operating system is the first program loaded into memory (RAM) when the computer is turned on. It controls all of the computer's higher-order functions. The operating system kernel resides in memory at all times. The operating system provides the user interface and controls multitasking. It handles the input and output to the hard disk and all peripheral devices such as compact disks, scanners, printers, plotters, and color displays. All digital image processing application programs must communicate with the operating system. The operating system sets the protocols for the application programs that are executed by it.

The difference between a single-user operating system and a network operating system is the latter's multi-user capability. For example, Microsoft Windows XP (home edition) and the Macintosh OS are single-user operating systems designed for one person at a desktop computer working independently. Various Microsoft Windows, UNIX, and Linux network operating systems are designed to manage multiple user requests at the same time and complex network security. The operating systems used by the most popular digital image processing systems are summarized in Table 3-5.

Compiler

A computer software compiler translates instructions programmed in a high-level language such as C^{++} or Visual Basic into machine language that the CPU can understand. A compiler usually generates assembly language first and then translates the assembly language into machine language. The compilers most often used in the development of digital image processing software are C^{++}, Assembler, and Visual Basic. Many digital image processing systems provide a toolkit that programmers can use to compile their own digital image processing algorithms (e.g., ERDAS, ER Mapper, ENVI). The toolkit consists of fundamental subroutines that perform very specific tasks such as reading a line of image data into RAM or modifying a color look-up table to change the color of a pixel (RGB) on the screen (discussed in Chapter 5).

It is often useful for remote sensing analysts to program in one of the high-level languages just listed. Very seldom will a single digital image processing software system perform all of the functions needed for a given project. Therefore, the ability to modify existing software or integrate newly developed algorithms with the existing software is important.

Storage and Archiving Considerations

Digital image processing of remote sensing and related GIS data requires substantial mass storage resources, as discussed earlier. Therefore, the mass storage media should allow relatively rapid access times, have longevity (i.e., last for a long time), and be inexpensive (Rothenberg, 1995).

Rapid Access Mass Storage

Digital remote sensor data (and other ancillary raster GIS data) are often stored in a matrix band sequential (BSQ) format in which each spectral band of imagery (or GIS data) is stored as an individual file. Each picture element of each band is typically represented in the computer by a single 8-bit byte with values from 0 to 255. The best way to make brightness values rapidly available to the computer is to place the data on a hard disk, CD-ROM, DVD, or DVD-RAM where each pixel of the data matrix may be accessed at random (*not* serially) and at great speed (e.g., within microseconds). The cost of hard disk, CD-ROM, or DVD storage per gigabyte continues to decline.

It is common for digital image processing laboratories to have gigabytes of hard-disk mass storage associated with each workstation. For example, each personal computer in the laboratory shown in Figure 3-2 has 80 GB of mass storage. Many image processing laboratories now use RAID (redundant arrays of inexpensive hard disks) technology in which two or more drives working together provide increased performance and various levels of error recovery and fault tolerance. Other storage media, such as magnetic tapes, are usually too slow for real-time image retrieval, manipulation, and storage because they do not allow random access of data. However, given their large storage capacity, they remain a cost-effective way to store data.

Companies are now developing new mass storage technologies based on atomic resolution storage (ARS), which holds the promise of storage densities of close to 1 terabit per square inch—the equivalent of nearly 50 DVDs on something the size of a credit card. The technology uses microscopic probes less than one-thousandth the width of a human hair. When the probes are brought near a conducting material, electrons write data on the surface. The same probes can detect and retrieve data and can be used to write over old data (Hewlett Packard, 2001).

Archiving Considerations: Longevity

Storing remote sensor data is no trivial matter. Significant sums of money are spent purchasing remote sensor data by commercial companies, natural resource agencies, and universities. Unfortunately, most of the time not enough attention is given to how the expensive data are stored or archived to protect the long-term investment. Figure 3-5 depicts several types of analog and digital remote sensor data mass storage devices and the average time to physical obsolescence, that is, when the media begin to deteriorate and information is lost. Interestingly, properly exposed, washed, and fixed analog black and white aerial photography negatives have considerable longevity, often more than 100 years. Color negatives with their respective dye layers have longevity, but not as much as the black-and-white negatives. Similarly, black-and-white paper prints have greater longevity than color prints (Kodak, 1995). Hard and floppy magnetic disks have relatively short longevity, often less than 20 years. Magnetic tape media (e.g., 3/4-in. tape, 8-mm tape, and 1/2-in. tape, shown in Figure 3-6) can become unreadable within 10 to 15 years if not rewound and properly stored in a cool, dry environment.

Optical disks can now be written to, read, and written over again at relatively high speeds and can store much more data than other portable media such as floppy disks. The technology used in rewriteable optical systems is magneto-optics, where data is recorded magnetically like disks and tapes, but the bits are much smaller because a laser is used to etch the bit. The laser heats the bit to 150 °C, at which temperature the bit is realigned when subjected to a magnetic field. To record new data, existing bits must first be set to zero.

Only the optical disk provides relatively long-term storage potential (>100 years). In addition, optical disks store large volumes of data on relatively small media. Advances in optical compact disc (CD) technology promise to increase the storage capacity to > 17 Gb using new rewriteable digital video disc (DVD) technology. In most remote sensing laboratories, rewritable CD-RWs or DVD-RWs have supplanted tapes as the backup system of choice. DVD drives are backwards compatible and can read data from CDs.

It is important to remember when archiving remote sensor data that sometimes it would be the loss of a) the read-write software and/or b) the read-write hardware (the drive mech-

anism and heads) that is the problem and not the digital media itself (Rothenberg, 1995; Jensen et al., 1996). Therefore, as new computers are purchased it is a good idea to set aside a single computer system that is representative of a certain computer era so that one can always read any data stored on old mass storage media.

Computer Display Spatial and Color Resolution

The display of remote sensor data on a computer screen is one of the most fundamental elements of digital image analysis (Brown and Feringa, 2003). Careful selection of the computer display characteristics will provide the optimum visual image analysis environment for the human interpreter. The two most important characteristics are computer display spatial and color resolution.

Computer Screen Display Resolution

The image processing system should be able to display at least 1024 rows by 1024 columns on the computer screen at one time. This allows larger geographic areas to be examined and places the terrain of interest in its regional context. Most Earth scientists prefer this regional perspective when performing terrain analysis using remote sensor data. Furthermore, it is disconcerting to have to analyze four 512×512 images when a single 1024×1024 display provides the information at a glance. An ideal screen display resolution is 1600×1200 pixels.

Computer Screen Color Resolution

The computer screen color resolution is the number of grayscale tones or colors (e.g., 256) that can be displayed on a CRT monitor at one time out of a palette of available colors (e.g., 16.7 million). For many applications, such as high-contrast black-and-white linework cartography, only 1 bit of color is required [i.e., either the line is black or white (0 or 1)]. For more sophisticated computer graphics for which many shades of gray or color combinations are required, up to 8 bits (or 256 colors) may be required. Most thematic mapping and GIS applications may be performed quite well by systems that display just 64 user-selectable colors out of a palette of 256 colors.

Conversely, the analysis and display of remote sensor image data may require much higher CRT screen color resolution

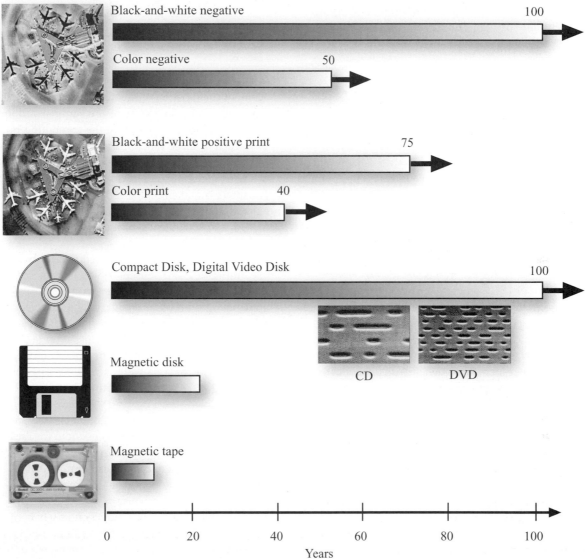

Potential Longevity of Remote Sensor Data Storage Media

Black-and-white negative — 100

Color negative — 50

Black-and-white positive print — 75

Color print — 40

Compact Disk, Digital Video Disk — 100

CD   DVD

Magnetic disk

Magnetic tape

0   20   40   60   80   100

Years

Figure 3-6   Different types of analog and digital remote sensor data mass storage devices and the average time to physical obsolescence. The lack of software or hardware to read the archived digital media is often the most serious problem.

than cartographic and GIS applications (Slocum, 1999). For example, most relatively sophisticated digital image processing systems can display a tremendous number of unique colors (e.g., 16.7 million) from a large color palette (e.g., 16.7 million). The primary reason for these color requirements is that image analysts must often display a composite of several images at one time on a CRT. This process is called *color compositing*. For example, to display a typical color-infrared image of Landsat Thematic Mapper data, it is necessary to composite three separate 8-bit images [e.g., the green band (TM 2 = 0.52 to 0.60 μm), the red band (TM 3 = 0.63 to 0.69 μm), and the reflective infrared band (TM 4 = 0.76 to 0.90 μm)]. To obtain a *true-color* composite image that provides every possible color combination for the three 8-bit images requires that 2^{24} colors (16,777,216) be available in the palette. Such true-color, direct-definition systems are relatively expensive because every pixel location must be *bitmapped*. This means that there must be a specific location in memory that keeps track of the exact blue, green, and red color value for every pixel. This requires substantial

Table 3-3.   Image processor memory required to produce various numbers of displayable colors.

Image Processor Memory (bits)	Maximum Number of Colors Displayable at One Time on the CRT Screen
1	2 (black and white)
2	4
3	8
4	16
5	32
6	64
7	128
8	256
9	512
10	1,024
11	2,048
12	4,096
13	8,192
14	16,384
15	32,768
16	65,536
17	131,072
18	262,144
24	16,777,216

amounts of computer memory which are usually collected in what is called an *image processor.* Given the availability of image processor memory, the question is: what is adequate color resolution?

Generally, 4096 carefully selected colors out of a very large palette (e.g., 16.7 million) appears to be the minimum acceptable for the creation of remote sensing color composites. This provides 12 bits of color, with 4 bits available for each of the blue, green, and red image planes (Table 3-3). For image processing applications other than compositing (e.g., black-and-white image display, color density slicing, pattern recognition classification), the 4,096 available colors and large color palette are more than adequate. However, the larger the palette and the greater the number of displayable colors at one time, the better the representation of the remote sensor data on the CRT screen for visual analysis. More information about how images are displayed using an image processor is in Chapter 5. The network configured in Figure 3-2 has six 24-bit color workstations.

Several remote sensing systems now collect data with 10-, 11-, and even 12-bit radiometric resolution with brightness values ranging from 0 to 1023, 0 to 2047, and 0 to 4095, respectively. Unfortunately, despite advances in video technology, at the present time it is necessary to generalize (i.e., dumb down) the radiometric precision of the remote sensor

data to 8 bits per pixel simply because current video display technology cannot handle the demands of the increased precision.

 Important Image Processing Functions

Many of the most important functions performed using digital image processing systems are summarized in Table 3-4. Personal computers now have the computing power to perform each of these functions. This textbook examines many of these digital image processing functions in subsequent chapters, including radiometric and geometric preprocessing (Chapters 6 and 7), enhancement (Chapter 8), information extraction (Chapters 9 through 11), change detection (Chapter 12), and accuracy assessment (Chapter 13).

It is not good for remotely sensed data to be analyzed in a vacuum (Estes and Jensen, 1998; Jensen, 2000a). Remote sensing information fulfills its promise best when used in conjunction with other ancillary data (e.g., soils, elevation, and slope) often stored in a geographic information system (GIS) (Muchoney et al., 2000; Bossler et al., 2002). Therefore, the ideal system should be able to process the digital remote sensor data as well as perform any necessary GIS processing. It is not efficient to exit the digital image processing system, log into a GIS system, perform a required GIS function, and then take the output of the procedure back into the digital image processing system for further analysis. Most integrated systems perform both digital image processing and GIS functions and consider map data as image data (or vice versa) and operate on them accordingly.

Most digital image processing systems have some limitations. For example, most systems can perform multispectral classification on a few bands of imagery, but only a few systems can perform hyperspectral analysis of hundreds of bands of imagery (e.g., Landgrebe and Biehl, 2004; Research Systems, 2004). Similarly, only a few systems can perform soft-copy analytical photogrammetric operations on overlapping stereoscopic imagery displayed on the CRT screen and generate digital orthophotographs. Also, only a few digital image processing systems incorporate expert systems or neural networks or fuzzy logic (e.g., ERDAS Imagine, ER Mapper). Finally, systems of the future should provide detailed image lineage (genealogy) information about the processing applied to each image (Lanter and Veregin, 1992). The image lineage information (metadata) is indispensable when the products derived from the analysis of remotely sensed data are subjected to intense scrutiny as in environmental litigation.

Table 3-4.   Image processing functions in quality digital image processing systems.

Preprocessing
1. *Radiometric correction* of error introduced by the sensor system electronics/optics and/or environmental effects (includes relative image-to-image normalization and absolute radiometric correction of atmospheric attenuation)
2. *Geometric correction* (image-to-image registration or image-to-map rectification)

Display and Enhancement
3. Black-and-white computer display (8-bit)
4. Color composite computer display (24-bit)
5. Magnification (zooming), reduction, roaming
6. Contrast manipulation (linear, nonlinear)
7. Color space transformations (e.g., RGB to IHS)
8. Image algebra (e.g., band ratioing, image differencing, NDVI, SAVI, Kauth-Thomas, EVI)
9. Spatial filtering (e.g., low-pass, high-pass, bandpass)
10. Edge enhancement (e.g., Kirsch, Laplacian, Sobel)
11. Principal components (standardized, nonstandardized, minimum noise fraction)
12. Texture transforms (e.g., min-max, texture spectrum, fractal dimension, geostatistical)
13. Frequency transformations (e.g., Fourier, Walsh)
14. Digital elevation models (e.g., interpolation via inverse distance weighting or kriging, analytical hill shading, calculation of slope, aspect)
15. Three-dimensional transformations (e.g., image draping over digital elevation models)
16. Image animation (e.g., movies, change detection)

Information Extraction
17. Pixel brightness value (BV_{ijk})
18. Black-and-white or color density slice
19. Transects (spatial and spectral)
20. Univariate and multivariate statistical analysis (e.g., mean, covariance, correlation, geostatistical)
21. Feature (band) selection (graphical and statistical)
22. Supervised classification (e.g., minimum distance, maximum likelihood)
23. Object-oriented image segmentation and classification
24. Incorporation of ancillary data during classification
25. Expert system image analysis including rule-based decision-tree classifiers and machine learning
26. Neural network image analysis
27. Fuzzy logic classification
28. Hyperspectral data analysis
29. Radar image processing
30. Accuracy assessment (descriptive and analytical)
31. Change detection

Photogrammetric Information Extraction
32. Soft-copy production of orthoimages
33. Soft-copy extraction of digital elevation models
34. Soft-copy extraction of planimetric detail

Table 3-4.   Image processing functions in quality digital image processing systems.

Metadata and Image/Map Lineage Documentation
35. Metadata
36. Complete image and GIS file processing history

Image and Map Cartographic Composition
37. Scaled Postscript level II output of images and maps

Geographic Information Systems
38. Raster (image)-based GIS
39. Vector (polygon)-based GIS (must allow polygon overlay)

Integrated Image Processing and GIS
40. Complete image processing systems (functions 1 through 37 plus utilities)
41. Complete image processing systems and GIS (functions 1 through 44)

Utilities
42. Network (e.g., local area network, Internet)
43. Image compression (single image, video)
44. Import and export of various file formats

Commercial and Public Digital Image Processing Systems

Commercial companies actively market digital image processing systems. Some companies provide only the software, others provide both proprietary hardware and software. Many public government agencies (e.g., NASA, NOAA, and the Bureau of Land Management) and universities (e.g., Purdue University) have developed digital image processing software (Foresman, 1998; Estes and Jensen, 1998; Miller et al., 2001; Landgrebe and Biehl, 2004). Some of these public systems are available at minimal cost. Several of the most widely used commercial and public digital image processing systems are summarized in Table 3-5.

Digital Image Processing and the National Spatial Data Infrastructure

Laypersons and scientists who use remote sensing data or share products derived from remotely sensed data should be aware of very specific spatial data standards developed by the Federal Geographic Data Committee (FGDC). The FGDC is an interagency committee of representatives from the Executive Office of the President, the Cabinet, and independent agencies. The FGDC is developing the National

Table 3-5.   Selected commercial and public digital image processing systems used for Earth resource mapping and their functions (● = significant capability; ○ = moderate capability; no symbol = little or no capability). Sources are listed in the References section. This is not an exhaustive list.

Systems	Operating System	Prepro-cessing	Display & Enhancement	Information Extraction	Soft-copy Photogram-metry	Lineage	Image/Map Cartography	GIS	IP/GIS
Commercial									
ACORN	Windows	●							
AGIS	Windows							●	○
Applied Analysis subpixel processing	Windows	●	○	●					○
ArcGIS Image and Feature Analyst	W/UNIX	●	●	●	●		●	●	●
ATCOR2	IDL	●		○					
AUTOCAD 2004	W/UNIX	●	●	○			○	●	
BAE Systems SOCET Set	W/UNIX	●	●	●	●		●		●
Blue Marble	W/UNIX	●	●				●		
EarthView	Windows	●	●	●					
*e*Cognition	Windows	●	●	●				○	○
EIDETIC Earthscope	Windows	○	○	○					
ENVI	W/U/M/IDL	●	●	●	○	○	●	○	○
DIMPLE	Mac	●	●	●			●	○	
Dragon	Windows	●	●	●					
ERDAS Imagine (Leica Geosystems)	W/UNIX	●	●	●	●	●	●	●	●
ER-Mapper	W/UNIX	●	●	●	●		●	●	●
FullPixelSearch	Mac	●	●	●					
GENASYS	W/UNIX	●	●	●			●	●	●
Global Lab Image	Windows		●	○					
GRASS	UNIX	●	●	●		●	●	●	●
IDRISI	Windows	●	●	●			●	●	●
ImagePro	Windows	●	●	●					
Intelligent Library Solution	UNIX	●	●			●	●		
Intergraph	W/UNIX	●	●	●	●	●	●	●	●

Table 3-5. Selected commercial and public digital image processing systems used for Earth resource mapping and their functions (● = significant capability; ○ = moderate capability; no symbol = little or no capability). Sources are listed in the References section. This is not an exhaustive list. (Continued)

Systems	Operating System	Prepro-cessing	Display & Enhancement	Information Extraction	Soft-copy Photogram-metry	Lineage	Image/Map Cartography	GIS	IP/GIS
MapInfo	W/UNIX		○			●	●	●	●
MrSID	W/U/M	●	●						
NOeSYS	W/Mac	○	●						
PCI Geomatica	W/UNIX/	●	●	●	●	●	●	●	●
Photoshop	W/U/M	○	●	○					
R-WEL	Windows	●	●	●	●		●		●
RemoteView	Windows	●	●						
MacSadie	Mac	●	●	●					
TNTmips	W/UNIX	●	●	●	●	●	●	●	●
OrthoView	UNIX	●	●						
VISILOG	W/UNIX	●	●	●					
Public									
C-Coast	Windows		●	●					
Cosmic VICAR-IBIS	UNIX	●	●	●			●	●	●
NOAA	UNIX	○	○						
MultiSpec	Mac/W	●	●	●					
NASA ELAS (DIPIX, Dtastar)	UNIX	●	●	●		●	●	●	●
NIH-Image	UNIX		○						

Spatial Data Infrastructure (NSDI) in cooperation with organizations from state, local and tribal governments, academics, and the private sector. The NSDI encompasses policies, standards, and procedures for organizations to cooperatively produce and share geographic data (FGDC, 2004a).

The American National Standards Institute's (ANSI) Spatial Data Transfer Standard (SDTS) is a mechanism for archiving and transferring spatial data (including metadata) between dissimilar computer systems. The SDTS specifies exchange constructs, such as format, structure, and content, for spatially referenced vector and raster (including gridded) data. Actual use of SDTS to transfer spatial data is carried out through its profiles. The FGDC *Raster Profile* standard

is of particular interest because it provides specifications for transferring spatial datasets in which features or images are represented in raster or gridded form, such as digital elevation models, digital orthophoto quarter quads (DOQQ), and digital satellite imagery (FGDC, 1999; 2004a,b).

Sources of Digital Image Processing Systems

ACORN, Atmospheric CORrection Now, www.aigllc.com/acorn/intro.asp

AGIS Software, www.agismap.com

Applied Analysis Inc., *Subpixel Processing*, www.discover-aai.com

ArcGIS Feature Analyst; www.featureanalyst.com

ATCOR2, www.geosystems.de/atcor/atcor2.html

AUTOCAD, Autodesk, Inc., usa.autodesk.com

BAE Systems SOCET Set, www.socetset.com

Blue Marble Geographics, www.bluemarblegeo.com.

C-Coast, http://coastwatch.noaa.gov/cw_ccoast.html

Cosmic, www.openchannelfoundation.org/cosmic

DIMPLE, www.process.com.au/AboutDIMPLE.shtml

Dragon, Goldin-Rudahl Systems, www.goldin-rudahl.com

EarthView, Atlantis Scientific Systems, www.pcigeomatics.com

EIDETIC Earthscope, www.eidetic.bc.ca/~eidetic/es1.htm

ENVI, Research Systems, Inc., www.rsinc.com

ELAS (DIPIX, Datastar), http://technology.ssc.nasa.gov/PDFs/SSC-00001_SS_NTTS.pdf

ERDAS Imagine, www.erdas.com

ER Mapper, www.ermapper.com

FullPixelSearch, www.themesh.com/elink13.html

Global Lab, Data Translation, 100 Locke Dr., Marlboro, MA 01752-1192

GRASS, http://grass.itc.it

IDRISI, Clarke University, www.clarklabs.org

ImagePro, www.i-cubeinc.com/software.htm

Intelligent Library System, Lockheed Martin, www.lmils.com

Intergraph, www.intergraph.com

MapInfo, www.mapinfo.com

MacSadie, www.ece.arizona.edu/~dial/base_files/software/MacSadie1.2.html

MrSID, LizardTech, www.lizardtech.com

MultiSpec, www.ece.purdue.edu/~biehl/MultiSpec/.

NIH-Image, http://rsb.info.nih.gov/nih-image

NOeSYS, www.rsinc.com/NOESYS/index.cfm

PCI, www.pcigeomatics.com

PHOTOSHOP, www.adobe.com

RemoteView, www.sensor.com/remoteview.html

R-WEL Inc., www.rwel.com

TNTmips, MicroImages, www.microimages.com

VISILOG, www.norpix.com/visilog.htm

XV image viewer, www.trilon.com/xv/

References

ASPRS, 1997, *Manual of Photographic Interpretation*, W. Philipson (Ed.), Bethesda: American Society for Photogrammetry & Remote Sensing, 689 p.

Bossler, J. D., Jensen, J. R., McMaster, R. B. and C. Rizos, 2002, *Manual of Geospatial Science and Technology*, London: Taylor & Francis, 623 p.

Brown, A. and W. Feringa, 2003, *Colour Basics for GIS Users*, London: Prentice-Hall, 171 p.

Chan, Y., 2001, "Remote Sensing and Geographic Information Systems," in *Location Theory and Decision Analysis*, Cincinnati: South-Western College Publishing, 259–338.

Curran, P. J., Milton, E. J., Atkinson, P. M. and G. M. Foody, 1998, "Remote Sensing: From Data to Understanding," Chapter 3 in *Geocomputation: A Primer*, P. A. Longley, S. M. Brooks, R. McDonnell and B. Macmillan (Eds.), New York: John Wiley & Sons, 33–59.

Estes, J. E. and J. R. Jensen, 1998, "Development of Remote Sensing Digital Image Processing and Raster GIS," in *History of Geographic Information Systems: Perspectives from the Pioneers*, T. Foresman (Ed.), Upper Saddle River, NJ: Prentice-Hall, 163–180.

FGDC, 1999, *Spatial Data Transfer Standard (SDTS): Part 5: Raster Profile and Extensions* (February), www.fgdc.gov/standards/documents/standards/sdts_pt5/srpe0299.pdf

FGDC, 2004a, *The National Spatial Data Infrastructure*, Washington: Federal Geographic Data Committee, www.fgdc.gov.

FGDC, 2004b, *Spatial Data Transfer Standard Part 5: Raster Profile with Basic Image Interchange Format (BIFF) Extension*, www.fgdc.gov/standards/status/sub4_1.html.

Foresman, T. W., 1998, *History of Geographic Information Systems: Perspectives from the Pioneers*, Upper Saddle River, NJ: Prentice-Hall, 397 p.

Friedl, M. A., McIver, D. K., Hodges, J. C. F., Zhang, X. Y., Muchoney, D., Strahler, A. H., Woodcock, C. E., Gopal, S., Schneider, A., Cooper, A., Baccini, A., Gao, F. and C. Schaaf, 2002, "Global Land Cover Mapping from MODIS: Algorithms and Early Results," *Remote Sensing of Environment*, 83:287–302.

Hewlett-Packard, 2001, *Squeezing Bits Out of Atoms: Atomic Resolution Storage*, www.hp.com/ghp/features/storage/storage1.html.

Intel Inc., 2004, *Processor Hall of Fame*, Santa Clara, CA: www.intel.com/intel/museum/25anniv/hof/hof_main.htm. Jensen, J. R., Cowen, D., Huang, X., Graves, D. and K. He, 1996, "Remote

Sensing Image Browse and Archival Systems," *Geocarto International: A Multidisciplinary Journal of Remote Sensing & GIS*, 11(2):33–42.

Jensen, J. R., 2000a, "Chapter 3: Processing Remotely Sensed Data: Hardware and Software," in *Remote Sensing in Hydrology and Water Management*, Schultz, G. A. and E. T. Engman (Eds.), Berlin: Springer, 41–63.

Jensen, J. R., 2000b, *Remote Sensing of the Environment: An Earth Resource Perspective,* Upper Saddle River, NJ: Prentice-Hall, 554 p.

Jensen, J. R., Botchway, K., Brennan-Galvin, E., Johannsen, C., Juma, C., Mabogunje, A., Miller, R., Price, K., Reining, P., Skole, D., Stancioff, A. and D. R. F. Taylor, 2002, *Down to Earth: Geographic Information for Sustainable Development in Africa*, Washington: National Research Council, 155 p.

Kodak, 1995, Correspondence with Kodak Aerial Systems Division, Rochester, NY.

Landgrebe, D. and L. Biehl, 2004, *MULTISPEC*, W. Lafayette, IN: Purdue University, www.ece.purdue.edu/~biehl/MultiSpec/.

Lanter, D. P. and H. Veregin, 1992, "A Research Paradigm for Propagating Error in Layer-based GIS," *Photogrammetric Engineering & Remote Sensing*, 58(6):825–833.

Limp, W. F., 1999, "Image Processing Software: System Selection Depends on User Needs," *GeoWorld*, May, 36–46.

Miller, R. B., Abbott, M. R., Harding, L. W., Jensen, J. R., Johannsen, C. J., Macauley, M., MacDonald, J. S. and J. S. Pearlman, 2001, *Transforming Remote Sensing Data into Information and Applications*, Washington: National Research Council, 75 p.

Miller, R. B., Abbott, M. R., Harding, L. W., Jensen, J. R., Johannsen, C. J., Macauley, M., MacDonald, J. S. and J. S. Pearlman, 2003, *Using Remote Sensing in State and Local Government: Information for Management and Decision Making*, Washington: National Research Council, 97 p.

Muchoney, D., Borak, J., Chi, H., Friedl, M., Gopal, S., Hodges, J., Morrow, N. and A. Strahler, 2000, "Application of the MODIS Global Supervised Classification Model to Vegetation and Land Cover Mapping of Central America," *International Journal of Remote Sensing*, 21(6-7):1115–1138.

Research Systems, 2004, *ENVI: Environment for Visualizing Images*, www.rsinc.com/envi/index.asp.

Rothenberg, J., 1995, "Ensuring the Longevity of Digital Documents," *Scientific American*, 272:42–47.

Slocum, T. A., 1999, *Thematic Cartography and Visualization*, Upper Saddle River, NJ: Prentice-Hall, 293 p.

Image Quality Assessment and Statistical Evaluation

4

Many remote sensing datasets contain high-quality, accurate data. Unfortunately, sometimes error (or noise) is introduced into the remote sensor data by a) the environment (e.g., atmospheric scattering), b) random or systematic malfunction of the remote sensing system (e.g., an uncalibrated detector creates striping), or c) improper airborne or ground processing of the remote sensor data prior to actual data analysis (e.g., an inaccurate analog-to-digital conversion). Therefore, the person responsible for analyzing the digital remote sensor data should first assess its quality and statistical characteristics. This is normally accomplished by:

- looking at the frequency of occurrence of individual brightness values in the image displayed in histogram format,

- viewing on a computer monitor individual pixel brightness values at specific locations or within a geographic area,

- computing fundamental univariate descriptive statistics to determine if there are unusual anomalies in the image data, and

- computing multivariate statistics to determine the amount of between-band correlation (e.g., to identify redundancy).

This chapter first reviews elements of statistical sampling theory. It then introduces the histogram and its significance to digital image processing of remote sensor data. Various methods of viewing individual pixel values and geographic areas of individual pixel values are then presented. Algorithms for the computation of univariate and multivariate statistics are introduced, including the identification of the minimum and maximum value for each band of imagery, the range, mean, standard deviation, and between-band covariance and correlation. Finally, geostatistical analysis methods are introduced that can be of value for obtaining information about spatial autocorrelation in imagery.

 Image Processing Mathematical Notation

The following notation is used to describe the mathematical operations applied to the digital remote sensor data:

i = a row (or line) in the imagery
j = a column (or sample) in the imagery
k = a band of imagery
l = another band of imagery

n = total number of picture elements (pixels) in an array

BV_{ijk} = brightness value at row i, column j, and band k

BV_{ik} = ith brightness value in band k

BV_{il} = ith brightness value in band l

min_k = minimum brightness value of band k

max_k = maximum brightness value of band k

$range_k$ = range of actual brightness values in band k

$quant_k$ = quantization level of band k (e.g., 2^8 = 0 to 255; 2^{12} = 0 to 4095)

μ_k = mean of band k

var_k = variance of band k

s_k = standard deviation of band k

$skewness_k$ = skewness of a band k distribution

$kurtosis_k$ = kurtosis of a band k distribution

cov_{kl} = covariance between pixel values in two bands, k and l

r_{kl} = correlation between pixel values in two bands, k and l

X_c = measurement vector for class c composed of brightness values (BV_{ijk}) from row i, column j, and band k

M_c = mean vector for class c

M_d = mean vector for class d

μ_{ck} = mean value of the data in class c, band k

s_{ck} = standard deviation of the data in class c, band k

v_{ckl} = covariance matrix of class c for bands k through l; shown as V_c

v_{dkl} = covariance matrix of class d for bands k through l; shown as V_d

Sampling Theory

Digital image processing is performed on only a sample of all available remote sensing information. Therefore, it is useful to review several fundamental aspects of elementary statistical sampling theory. A *population* is an infinite or finite set of elements. An infinite population would be all possible images that might be acquired of the entire Earth in 2004. All Landsat 7 Enhanced Thematic Mapper Plus (ETM$^+$) images of Charleston, SC obtained in 2004 would be a finite population.

A *sample* is a subset of the elements taken from a population used to make inferences about certain characteristics of the population. For example, we might decide to analyze a June

1, 2004 Landsat image of Charleston. If observations with certain characteristics are systematically excluded from the sample either deliberately or inadvertently (such as selecting images obtained only in the spring of the year), it is a *biased* sample. *Sampling error* is the difference between the true value of a population characteristic and the value of that characteristic inferred from a sample.

Large samples drawn randomly from natural populations usually produce a symmetrical frequency distribution, such as that shown in Figure 4-1a. Most values are clustered around a central value, and the frequency of occurrence declines away from this central point. A graph of the distribution appears bell-shaped and is called a *normal distribution*. Many statistical tests used in the analysis of remotely sensed data assume that the brightness values recorded in a scene are normally distributed. Unfortunately, remotely sensed data may *not* be normally distributed, and the analyst must be careful to identify such conditions. In such instances, nonparametric statistical theory may be preferred.

The Histogram and Its Significance to Digital Image Processing of Remote Sensor Data

The histogram is a useful graphic representation of the information content of a remotely sensed image. Histograms for each band of imagery are often displayed and analyzed in many remote sensing investigations because they provide the analyst with an appreciation of the quality of the original data (e.g., whether it is low in contrast, high in contrast, or multimodal in nature). In fact, many analysts routinely provide before (original) and after histograms of the imagery to document the effects of applying an image enhancement technique (Jahne, 1997; Gonzalez and Woods, 2002). It is instructive to review how a histogram of a single band of imagery, k, composed of i rows and j columns with a brightness value BV_{ijk} at each pixel location is constructed.

Individual bands of remote sensor data are typically quantized (digitally recorded) with brightness values ranging from 2^8 to 2^{12} (if $quant_k = 2^8$ then brightness values range from 0 to 255; 2^9 = values from 0 to 511; 2^{10} = values from 0 to 1023; 2^{11} = values from 0 to 2047; and 2^{12} = values from 0 to 4095). The majority of the remote sensor data are quantized to 8 bits, with values ranging from 0 to 255 (e.g., Landsat 5 Thematic Mapper and SPOT HRV data). Some sensor systems such as IKONOS and *Terra* MODIS obtain data with 11 bits of precision. The greater the quantization, the higher the probability that more subtle spectral reflectance (or emission) characteristics may be extracted from the imagery.

Histograms of Symmetric and Skewed Distributions

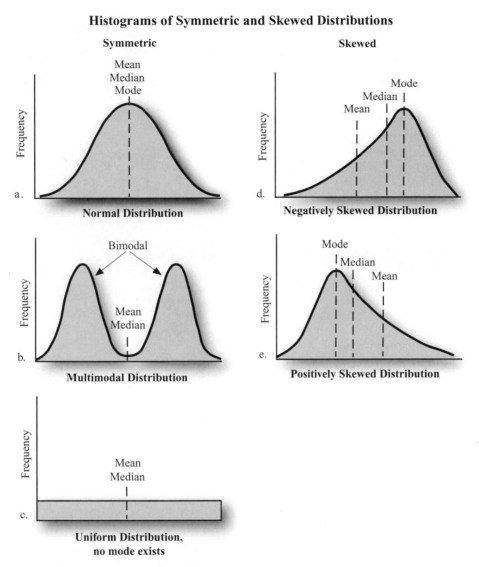

Figure 4-1   Relative position of measures of central tendency for commonly encountered frequency distributions (modified from Griffith and Amrhein, 1991).

Tabulating the frequency of occurrence of each brightness value within the image provides statistical information that can be displayed graphically in a *histogram* (Hair et al., 1998). The range of quantized values of a band of imagery, $quant_k$, is provided on the abscissa (*x* axis), while the frequency of occurrence of each of these values is displayed on the ordinate (*y* axis). For example, consider the histogram of the original brightness values for a Landsat Thematic Mapper band 4 scene of Charleston, SC (Figure 4-2). The peaks in the histogram correspond to dominant types of land cover in the image, including a) open water pixels, b) coastal wetland, and c) upland. Also, note how the Landsat Thematic Mapper band 4 data are compressed into only the lower one-third of the 0 to 255 range, suggesting that the data are rela-

tively low in contrast. If the original Landsat Thematic Mapper band 4 brightness values were displayed on a monitor screen or on the printed page they would be relatively dark and difficult to interpret. Therefore, in order to see the wealth of spectral information in the scene, the original brightness values were contrast stretched (Figure 4-2a.) Contrast stretching principles are discussed in Chapter 8.

Histograms are useful for evaluating the quality of optical daytime multispectral data and many other types of remote sensor data. For example, consider the histogram of predawn thermal infrared (8.5 to 13.5 μm) imagery of a thermal plume in the Savannah River shown in Figure 4-3. The thermal plume entered the Savannah River via Four Mile Creek,

a. Landsat Thematic Mapper Band 4 Image of Charleston, SC (contrast stretched).

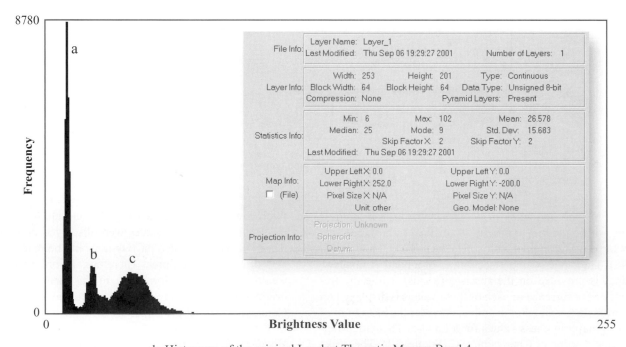

b. Histogram of the original Landsat Thematic Mapper Band 4.

Figure 4-2   a) A Landsat Thematic Mapper band 4 (near-infrared) image of Charleston, SC, obtained on November 9, 1982. b) A multi-modal histogram of the brightness values of the band 4 image. Peaks in the histogram correspond to dominant types of land cover in the image, including (a) open water pixels, (b) coastal wetland, and (c) upland. The inset provides fundamental *meta-data* about the characteristics of the Landsat Thematic Mapper subscene.

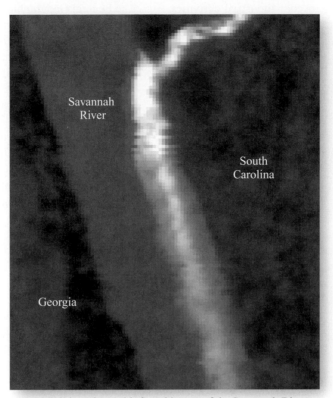

a. Nightime thermal infrared image of the Savannah River
on the South Carolina and Georgia border (contrast stretched).

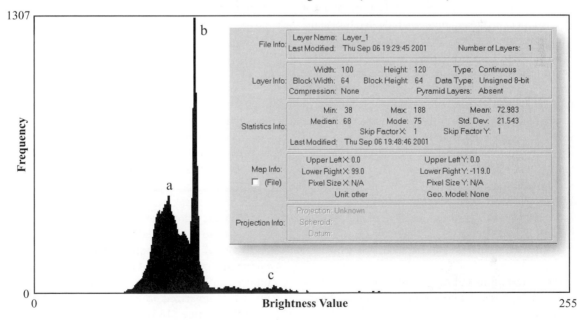

b. Histogram of the original thermal infrared data.

Figure 4-3   a) Thermal infrared image of a thermal effluent plume in the Savannah River obtained on March 28, 1981 at 4:00 a.m. b) A multimodal histogram of the brightness values in the predawn thermal-infrared image. Peaks in the histogram are associated with (a) the relatively cool temperature of the Savannah River swamp, (b) the slightly warmer temperature (12° C) of the Savannah River, and (c) the relatively hot thermal plume. The inset provides *metadata* about the characteristics of the thermal infrared image subscene.

which carried hot water used to cool industrial activities. The peaks in this histogram are associated with (a) the relatively cool temperature of the Savannah River swamp on each side of the river's natural levee, (b) the slightly warmer temperature (12° C) of the Savannah River upstream and west of the plume, and (c) the relatively hot thermal plume that progressed through the swamp and then entered the Savannah River. The north to south flow of the Savannah River caused the plume to hug the eastern bank and dissipate as it progressed away from the mouth of the creek.

When an unusually large number of pixels have the same brightness value, the traditional histogram display might not be the best way to communicate the information content of the remote sensor data. When this occurs, it might be useful to scale the frequency of occurrence (*y*-axis) according to the relative percentage of pixels within the image at each brightness level along the *x*-axis.

 ### Image Metadata

Metadata is data or information about data. Most quality digital image processing systems read, collect, and store metadata about a particular image or subimage. It is important that the image analyst have access to this metadata. In the most fundamental instance, metadata might include: the file name, date of last modification, level of quantization (e.g, 8 bits), number of rows and columns, number of bands, univariate statistics (minimum, maximum, mean, median, mode, standard deviation), georeferencing performed (if any), and pixel dimension (e.g., 5 × 5 m). Utility programs within the digital image processing system routinely provide this information. Fundamental metadata about the Charleston, SC Landsat Thematic Mapper subscene and the Savannah River thermal infrared subscene are found in Figures 4-2 and 4-3, respectively. Neither of these image has been subjected to geometric rectification as yet. Therefore, there is no metadata information concerning the projection, spheroid, or datum. This information is recorded as metadata later on in the process when geometric rectification takes place.

An ideal remote sensing metadata system would keep track of every type of processing applied to each digital image (Lanter, 1991). This 'image genealogy' or 'lineage' information can be very valuable when the remote sensor data are subjected to intense scrutiny (e.g., in a public forum) or used in litigation.

The Federal Geographic Data Committee (FGDC) has set up rigorous image metadata standards as part of the National Spatial Data Infrastructure (NSDI) (FGDC, 2004). All federal agencies that provide remote sensor data to the public are required to use the established metadata standards. For example, when anyone purchases a 1 × 1 m digital orthophotoquad from the U.S. Geological Survey produced from 1:40,000-scale National Aerial Photography Program (NAPP) data, it comes complete with detailed metadata. Image analysts should always consult the image metadata and incorporate the information in the digital image processing that takes place. Commercial remote sensing data providers are adopting FGDC metadata standards.

The histogram and metadata information help the analyst understand the content of remotely sensed data. Sometimes, however, it is very useful to look at individual brightness values at specific locations in the imagery.

 ### Viewing Individual Pixel Brightness Values at Specific Locations or within a Geographic Area

Viewing individual pixel brightness values in a remotely sensed image is one of the most useful methods for assessing the quality and information content of the data. Virtually all digital image processing systems allow the analyst to:

- use a mouse-controlled cursor (cross-hair) to identify a geographic location in the image (at a particular row and column or geographic *x,y* coordinate) and display its brightness value in *n* bands, and

- display the individual brightness values of an individual band in a matrix (raster) format.

Cursor Evaluation of Individual Pixel Brightness Values

Most people know how to control a cursor using a mouse and navigate to a desired location. For example, a cursor has been placed in the heart of the Savannah River thermal plume at row 42 (*x*) and column 20 (*y*) in Figure 4-4a. The numeric information for this location in the thermal infrared image, summarized in Figure 4-4b, includes the original brightness value (185), the color look-up table value (discussed in Chapter 5), and the number of pixels in the histogram with a value of 135 (i.e., 7). Note the directional keys that can be used to navigate to nearby pixel locations.

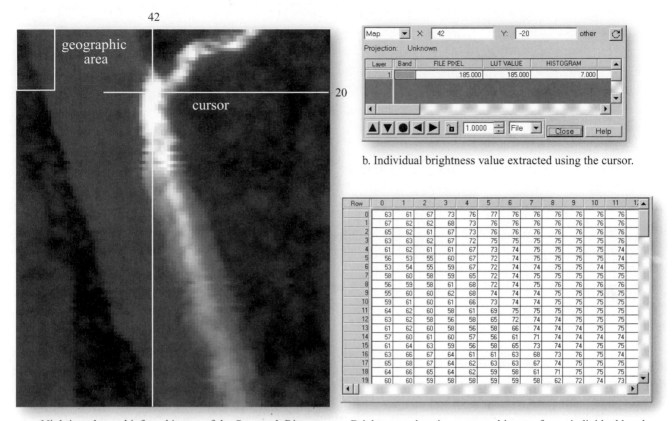

a. Nightime thermal infrared image of the Savannah River.

b. Individual brightness value extracted using the cursor.

c. Brightness values in a geographic area for an individual band.

Figure 4-4　a) Thermal infrared image of the Savannah River with a cursor located at column 42 (*x*) and row 20 (*y*). b) Spectral information located at column 42 and row 20. c) Brightness values found in the first 12 columns and first 20 rows of the thermal infrared image displayed in a matrix format (interfaces courtesy of Leica Geosystems ERDAS, Inc.).

Two- and Three-dimensional Evaluation of Pixel Brightness Values within a Geographic Area

It can become quite tedious using the cursor to evaluate the individual pixel brightness values throughout even a small geographic area. In such circumstances, it is useful to identify with the cursor a geographic area (e.g., a rectangle) and display all the pixels' values in it for a specific band or bands. For example, Figure 4-4c displays in a matrix format the brightness values found in the first 20 rows and 12 columns of the thermal infrared image. Brightness values >70 represent Savannah River pixels as well as the pixels found at the interface of the water and land. In Chapter 8 (Image Enhancement) we will perform a more rigorous examination of the data and select only those brightness values ≥ 74 as being relatively pure water pixels.

The display of individual brightness values in a matrix (raster) format is informative, but it does not convey any visual representation of the magnitude of the data within the area. Therefore, it is often useful early in the exploratory phases of a digital image processing project to extrude the value of the individual brightness values within a geographic area to create a pseudo three-dimensional display. For example, a wire-frame pseudo three-dimensional display of the individual brightness values over the entire thermal infrared image is shown in Figure 4-5b. While the wire-frame display is informative, it is usually more visually effective to drape the actual grayscale image of the individual band (or bands if a color composite is being used) over the exaggerated three-dimensional representation of the brightness values found within the specific geographic area. For example, a pseudo three-dimensional display of the thermal infrared image

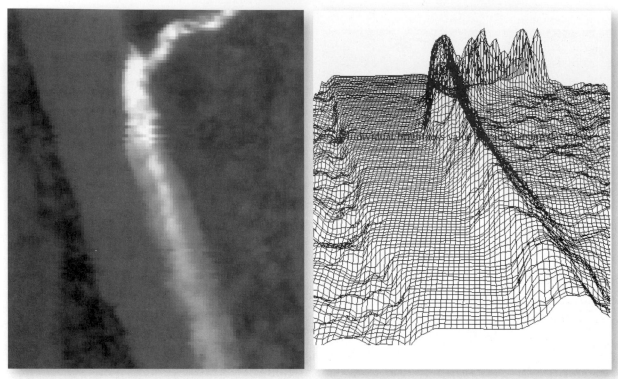

a. Thermal infrared image of the Savannah River.   b. Wire-frame pseudo three-dimensional representation of the data.

c. Thermal infrared data draped over the three-dimensional representation (5 x exaggeration).

Figure 4-5   Two- and three-dimensional views of the thermal infrared data.

viewed from two different vantage points is shown in Figure 4-5c.

The examination of histograms and the extraction of individual brightness values using a cursor or geographic area analysis methods are quite useful. However, they provide no statistical information about the remote sensor data. The following sections describe how fundamental univariate and multivariate statistics are computed for the remotely sensed imagery.

Univariate Descriptive Image Statistics

Most digital image processing systems can perform robust univariate and multivariate statistical analyses of single- and multiple-band remote sensor data. For example, image analysts have at their disposal statistical measures of central tendency and measures of dispersion that can be extracted from the imagery.

Measure of Central Tendency in Remote Sensor Data

The *mode* (e.g., see Figure 4-1a) is the value that occurs most frequently in a distribution and is usually the highest point on the curve (histogram). It is common, however, to encounter more than one mode in a remote sensing dataset, such as that shown in Figure 4-1b. The histograms of the Landsat TM image of Charleston, SC (see Figure 4-2) and the predawn thermal infrared image of the Savannah River (see Figure 4-3) have multiple modes. They are nonsymmetrical (skewed) distributions.

The *median* is the value midway in the frequency distribution (e.g., see Figure 4-1a). One-half of the area below the distribution curve is to the right of the median, and one-half is to the left. The *mean* (μ) is the arithmetic average and is defined as the sum of all brightness value observations divided by the number of observations (Freud and Wilson, 2003). It is the most commonly used measure of central tendency. The mean of a single band of imagery, μ_k, composed of n brightness values *(BV$_{ik}$)* is computed using the formula:

$$\mu_k = \frac{\sum\limits_{i=1}^{n} BV_{ik}}{n}.$$ (4-1)

The sample mean, μ_k, is an unbiased estimate of the population mean. For symmetrical distributions the sample mean tends to be closer to the population mean than any other unbiased estimate (such as the median or mode). Unfortunately, the sample mean is a poor measure of central tendency when the set of observations is skewed or contains an extreme value (outlier). As the peak (mode) becomes more extremely located to the right or left of the mean, the frequency distribution is said to be *skewed*. A frequency distribution curve (histogram) is said to be skewed in the direction of the longer tail. Therefore, if a peak (mode) falls to the right of the mean, the frequency distribution is negatively skewed. If the peak falls to the left of the mean, the frequency distribution is positively skewed (Griffith and Amrhein, 1991). Examples of positively and negatively skewed distributions are shown in Figures 4-1d and e.

Measures of Dispersion

Measures of the dispersion about the mean of a distribution also provide valuable information about the image. For example, the *range* of a band of imagery (range$_k$) is computed as the difference between the maximum (max$_k$) and minimum (min$_k$) values; that is, range$_k$ = max$_k$ - min$_k$. Unfortunately, when the minimum or maximum values are extreme or unusual observations (i.e., possibly *data blunders*), the range could be a misleading measure of dispersion. Such extreme values are not uncommon because the remote sensor data are often collected by detector systems with delicate electronics that can experience spikes in voltage and other unfortunate malfunctions. When unusual values are not encountered, the range is a very important statistic often used in image enhancement functions such as min–max contrast stretching (Chapter 8).

The *variance* of a sample is the average squared deviation of all possible observations from the sample mean. The variance of a band of imagery, var$_k$, is computed using the equation:

$$var_k = \frac{\sum\limits_{i=1}^{n} (BV_{ik} - \mu_k)^2}{n}.$$ (4-2)

The numerator of the expression, $\Sigma(BV_{ik} - \mu_k)^2$, is the corrected sum of squares (**SS**) (Davis, 2002). If the sample mean (μ_k) were actually the population mean, this would be

Areas Under the Normal Curve for Various Standard Deviations from the Mean

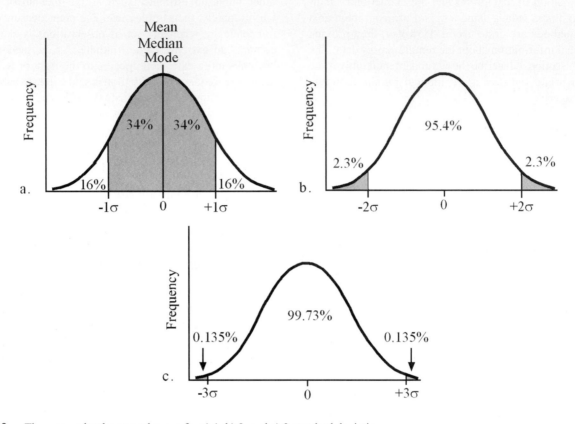

Figure 4-6   The area under the normal curve for a) 1, b) 2, and c) 3 standard deviations.

an accurate measurement of the variance. Unfortunately, there is some underestimation when variance is computed using Equation 4-2 because the sample mean (μ_k in Equation 4-1) was calculated in a manner that minimized the squared deviations about it. Therefore, the denominator of the variance equation is reduced to $n - 1$, producing a somewhat larger, unbiased estimate of the sample variance (Walpole et al., 2002):

$$\text{var}_k = \frac{\text{SS}}{n - 1}. \tag{4-3}$$

The *standard deviation* is the positive square root of the variance (Freud and Wilson, 2003). The standard deviation of the pixel brightness values in a band of imagery, s_k, is computed as:

$$s_k = \sqrt{\text{var}_k}. \tag{4-4}$$

A small standard deviation suggests that observations are clustered tightly around a central value. Conversely, a large standard deviation indicates that values are scattered widely about the mean. The total area underneath a distribution curve is equal to 1.00 (or 100%). For normal distributions, 68% of the observations lie within ±1 standard deviation of the mean, 95.4% of all observations lie within ±2 standard deviations, and 99.73% within ±3 standard deviations. The areas under the normal curve for various standard deviations are shown in Figure 4-6. The standard deviation is a statistic commonly used to perform digital image processing (e.g., linear contrast enhancement, parallelepiped classification, and error evaluation) (e.g., Schowengerdt, 1997; Richards and Jia, 1999; Russ, 2002). To interpret variance and standard deviation, analysts should not attach a significance to each numerical value but should compare one variance or standard deviation to another. The sample having the largest variance or standard deviation has the greater spread among

Table 4-1.   A sample dataset of brightness values used to demonstrate the computation of a variance–covariance matrix.

Pixel	Band 1 (green)	Band 2 (red)	Band 3 (near-infrared)	Band 4 (near-infrared)
(1,1)	130	57	180	205
(1,2)	165	35	215	255
(1,3)	100	25	135	195
(1,4)	135	50	200	220
(1,5)	145	65	205	235

the brightness values of the observations, provided all the measurements were made in the same units.

Measures of Distribution (Histogram) Asymmetry and Peak Sharpness

Sometimes it is useful to compute additional statistical measures that describe in quantitative terms various characteristics of the distribution (histogram). *Skewness* is a measure of the asymmetry of a histogram and is computed using the formula (Pratt, 1991):

$$\text{skewness}_k = \frac{\sum_{i=1}^{n} \left(\frac{BV_{ik} - \mu_k}{s_k}\right)^3}{n}. \tag{4-5}$$

A perfectly symmetric histogram has a *skewness* value of zero.

A histogram can be symmetric but have a peak that is very sharp or one that is subdued when compared with a perfectly normal distribution. A perfectly normal frequency distribution (histogram) has zero *kurtosis*. The greater the positive kurtosis value, the sharper the peak in the distribution when compared with a normal histogram. Conversely, a negative kurtosis value suggests that the peak in the histogram is less sharp than that of a normal distribution. Kurtosis is computed using the formula:

$$\text{kurtosis}_k = \left[\frac{1}{n}\sum_{i=1}^{n}\left(\frac{BV_{ik} - \mu_k}{s_k}\right)^4\right] - 3. \tag{4-6}$$

Outliers or blunders in the remotely sensed data can have a serious impact on the computation of skewness and kurtosis. Therefore, it is desirable to remove (or repair) bad data values before computing skewness or kurtosis.

 Multivariate Image Statistics

Remote sensing research is often concerned with the measurement of how much radiant flux is reflected or emitted from an object in more than one band (e.g., in red and near-infrared bands). It is useful to compute *multivariate* statistical measures such as covariance and correlation among the several bands to determine how the measurements covary. Later it will be shown that variance–covariance and correlation matrices are used in remote sensing principal components analysis (PCA), feature selection, classification and accuracy assessment (Mausel et al., 1990; Congalton and Green, 1998; Tso and Mather, 2001; Narumalani et al., 2002) (Chapters 8 through 13). For this reason, we will examine how the variance–covariance between bands is computed and then proceed to compute the correlation between bands. Although initially performed on a simple dataset consisting of just five hypothetical pixels, this example provides insight into the utility of these statistics for digital image processing purposes. Later, these statistics are computed for a seven-band Charleston, SC, Thematic Mapper scene consisting of 240×256 pixels. Note that this scene is much larger than the 120×100 pixel subscene in Figure 4-2a.

The following examples are based on an analysis of the first five pixels [(1, 1), (1, 2), (1, 3), (1, 4) and (1, 5)] in a four-band (green, red, near-infrared, near-infrared) hypothetical multispectral dataset obtained over vegetated terrain. Thus, each pixel consists of four spectral measurements (Table 4-1). Note the low brightness values in band 2 caused by plant chlorophyll absorption of red light for photosynthetic purposes. Increased reflectance of the incident near-infrared energy by the green plant results in higher brightness values in the two near-infrared bands (3 and 4). Although it is a small hypothetical sample dataset, it represents well the spectral characteristics of healthy green vegetation.

Table 4-2.   Univariate statistics for the hypothetical sample dataset.

Band	1	2	3	4
Mean (μ_k)	135.00	46.40	187.0	222.00
Standard deviation (s_k)	23.71	16.27	31.4	23.87
Variance (var_k)	562.50	264.80	1007.5	570.00
Minimum (min_k)	100.00	25.00	135.0	195.00
Maximum (max_k)	165.00	65.00	215.0	255.00
Range (BV_r)	65.00	40.00	80.0	60.00

The simple univariate statistics for such data are usually reported as shown in Table 4-2. In this example, band 2 exhibits the smallest variance (264.8) and standard deviation (16.27), the lowest brightness value (25), the smallest range of brightness values ($65 - 25 = 40$), and the lowest mean value (46.4). Conversely, band 3 has the largest variance (1007.5) and standard deviation (31.74) and the largest range of brightness values ($215 - 135 = 80$). These univariate statistics are of value but do not provide useful information concerning whether the spectral measurements in the four bands vary together or are completely independent.

Covariance in Multiple Bands of Remote Sensor Data

The different remote sensing–derived spectral measurements for each pixel often change together in a predictable fashion. If there is no relationship between the brightness value in one band and that of another for a given pixel, the values are mutually independent; that is, an increase or decrease in one band's brightness value is not accompanied by a predictable change in another band's brightness value. Because spectral measurements of individual pixels may not be independent, some measure of their mutual interaction is needed. This measure, called the *covariance*, is the joint variation of two variables about their common mean. To calculate covariance, we first compute the *corrected sum of products* (**SP**) defined by the equation (Davis, 2002):

$$\mathbf{SP}_{kl} = \sum_{i=1}^{n} (BV_{ik} - \mu_k)(BV_{il} - \mu_l). \qquad (4\text{-}7)$$

In this notation, BV_{ik} is the ith measurement of band k, and BV_{il} is the ith measurement of band l with n pixels in the

Table 4-3.   Format of a variance–covariance matrix.

	Band 1	Band 2	Band 3	Band 4
Band 1	$\mathbf{SS}_1$	$cov_{1,2}$	$cov_{1,3}$	$cov_{1,4}$
Band 2	$cov_{2,1}$	$\mathbf{SS}_2$	$cov_{2,3}$	$cov_{2,4}$
Band 3	$cov_{3,1}$	$cov_{3,2}$	$\mathbf{SS}_3$	$cov_{3,4}$
Band 4	$cov_{4,1}$	$cov_{4,2}$	$cov_{4,3}$	$\mathbf{SS}_4$

study area. The means of bands k and l are μ_k and μ_l, respectively. In our example, variable k might stand for band 1 and variable l could be band 2. It is computationally more efficient to use the following formula to arrive at the same result:

$$\mathbf{SP}_{kl} = \sum_{i=1}^{n} (BV_{ik} \times BV_{il}) - \frac{\displaystyle\sum_{i=1}^{n} BV_{ik} \sum_{i=1}^{n} BV_{il}}{n}. \qquad (4\text{-}8)$$

The quantity is called the *uncorrected sum of products*. The relationship of $\mathbf{SP}_{kl}$ to the sum of squares (**SS**) can be seen if we take k and l as being the same, that is:

$$\mathbf{SP}_{kk} = \sum_{i=1}^{n} (BV_{ik} \times BV_{ik}) - \frac{\displaystyle\sum_{i=1}^{n} BV_{ik} \sum_{i=1}^{n} BV_{ik}}{n} \qquad (4\text{-}9)$$

$$= \mathbf{SS}_k.$$

Just as simple variance was calculated by dividing the corrected sums of squares (**SS**) by $(n - 1)$, covariance is calculated by dividing **SP** by $(n - 1)$. Therefore, the covariance between brightness values in bands k and l, cov_{kl}, is equal to (Davis, 2002):

$$cov_{kl} = \frac{\mathbf{SP}_{kl}}{n - 1}. \qquad (4\text{-}10)$$

The sums of products (**SP**) and sums of squares (**SS**) can be computed for all possible combinations of the four spectral variables in Table 4-1. These data can then be arranged in a 4×4 variance–covariance matrix, as shown in Table 4-3. All elements in the matrix not on the diagonal have one duplicate (e.g., $cov_{1,2} = cov_{2,1}$ so that $cov_{kl} = cov_{lk}$).

The computation of variance for the diagonal elements of the matrix and covariance for the off-diagonal elements of the

Table 4-4. Variance–covariance matrix of the sample data.

	Band 1	Band 2	Band 3	Band 4
Band 1	562.50			
Band 2	135.00	264.80		
Band 3	718.75	275.25	1007.50	
Band 4	537.50	64.00	663.75	570.00

data is shown in Table 4-4. The manual computation of the covariance between band 1 and band 2 is shown in Table 4-5.

Correlation between Multiple Bands of Remotely Sensed Data

To estimate the degree of interrelation between variables in a manner not influenced by measurement units, *Pearson's product-moment correlation coefficient* (*r*) is commonly computed (Samuels and Witmer, 2003). For example, the correlation between two bands (*k* and *l*) of remotely sensed data, r_{kl}, is the ratio of their covariance (cov_{kl}) to the product of their standard deviations ($s_k s_l$):

$$r_{kl} = \frac{\text{cov}_{kl}}{s_k s_l}. \qquad (4\text{-}11)$$

Because the correlation coefficient is a ratio, it is a unitless number. Covariance can equal but cannot exceed the product of the standard deviation of its variables, so correlation ranges from +1 to –1. A correlation coefficient of +1 indicates a positive, perfect relationship between the brightness values in two of the bands (i.e., as one band's pixels increase in value, the other band's values also increase in a systematic fashion). Conversely, a correlation coefficient of –1 indicates that the two bands are inversely related (i.e., as brightness values in one band increase, corresponding pixels in the other band systematically decrease in value).

A continuum of less-than-perfect relationships exists between correlation coefficients of –1 and +1 (Glantz, 2001). A correlation coefficient of zero suggests that there is no linear relationship between the two bands of remote sensor data.

If we square the correlation coefficient (r_{kl}), we obtain the *sample coefficient of determination* (r^2), which expresses the proportion of the total variation in the values of "band *l*" that can be accounted for or explained by a linear relationship with the values of the random variable "band *k*." Thus, a cor-

Table 4-5. Computation of covariance between bands 1 and 2 of the sample data using Equations 4-8 and 4-10.

Band 1	(Band 1 × Band 2)	Band 2
130	7,410	57
165	5,775	35
100	2,500	25
135	6,750	50
145	9,425	65
675	31,860	232

where $\text{SP}_{1,2} = (31,860) - \dfrac{(675)(232)}{5} = 540$

$\text{cov}_{1,2} = \dfrac{540}{4} = 135$

Table 4-6. Correlation matrix of the sample data.

	Band 1	Band 2	Band 3	Band 4
Band 1	—			
Band 2	0.35	—		
Band 3	0.95	0.53	—	
Band 4	0.94	0.16	0.87	—

relation coefficient (r_{kl}) of 0.70 results in an r^2 value of 0.49, meaning that 49% of the total variation of the values of "band *l*" in the sample is accounted for by a linear relationship with values of "band *k*" (Walpole et al., 2002).

The between-band correlations are usually stored in a correlation matrix, such as the one shown in Table 4-6 that contains the between-band correlations of our sample data. Usually, only the correlation coefficients below the diagonal are displayed because the diagonal terms are 1.0 and the terms above the diagonal are duplicates.

In this hypothetical example, brightness values of band 1 are highly correlated with those of bands 3 and 4, that is, $r \geq 0.94$. A high correlation suggests substantial redundancy in the information content among these bands. Perhaps one or more of these bands could be deleted from the analysis to reduce subsequent computation. Conversely, the relatively lower correlation between band 2 and all other bands suggests that this band provides some type of unique information not found in the other bands. More sophisticated

Table 4-7.   Statistics for the Charleston, SC, Landsat Thematic Mapper scene composed of seven bands of 240 × 256 pixels each.

Band Number (μm)	1 0.45 – 0.52	2 0.52 – 0.60	3 0.63 – 0.69	4 0.76 – 0.90	5 1.55 – 1.75	7 2.08 – 2.35	6 10.4 – 12.5
Univariate Statistics							
Mean	64.80	25.60	23.70	27.30	32.40	15.00	110.60
Standard Deviation	10.05	5.84	8.30	15.76	23.85	12.45	4.21
Variance	100.93	34.14	68.83	248.40	568.84	154.92	17.78
Minimum	51.00	17.00	14.00	4.00	0.00	0.00	90.00
Maximum	242.00	115.00	131.00	105.00	193.00	128.00	130.00
Variance–Covariance Matrix							
1	100.93						
2	56.60	34.14					
3	79.43	46.71	68.83				
4	61.49	40.68	69.59	248.40			
5	134.27	85.22	141.04	330.71	568.84		
7	90.13	55.14	86.91	148.50	280.97	154.92	
6	23.72	14.33	22.92	43.62	78.91	42.65	17.78
Correlation Matrix							
1	1.00						
2	0.96	1.00					
3	0.95	0.96	1.00				
4	0.39	0.44	0.53	1.00			
5	0.56	0.61	0.71	0.88	1.00		
7	0.72	0.76	0.84	0.76	0.95	1.00	
6	0.56	0.58	0.66	0.66	0.78	0.81	1.00

methods of selecting the most useful bands for analysis are described in later sections.

The results of performing a typical statistical analysis program on the Charleston, SC, Landsat TM data are summarized in Table 4-7. Band 1 exhibits the greatest range of brightness values (from 51 to 242) due to Rayleigh and Mie atmospheric scattering of blue wavelength energy. The near- and middle-infrared bands (4, 5, and 7) all have minimums near or at zero. These values are low because much of the

Charleston scene is composed of open water, which absorbs much of the incident near- and middle-infrared radiant flux, thereby causing low reflectance in these bands. Bands 1, 2, and 3 are all highly correlated with one another ($r \geq 0.95$), indicating that there is substantial redundant spectral information in these channels. Although not to the same degree, there is also considerable redundancy among the reflective and middle-infrared bands (4, 5, and 7), as they exhibit correlations ranging from 0.66 to 0.95. Not surprisingly, the lowest correlations occur when a visible band is compared with an infrared band, especially bands 1 and 4 ($r = 0.39$). In fact, band 4 is the least redundant infrared band when compared with all three visible bands (1, 2, and 3). For this reason, TM band 4 (0.76 to 0.90 µm) is used as an example throughout much of the text. As expected, the thermal-infrared band 6 data (10.4 to 12.5 µm) are highly correlated with the middle-infrared bands (5 and 7).

Feature Space Plots

The univariate and multivariate statistics discussed provide accurate, fundamental information about the individual band statistics including how the bands covary and correlate. Sometimes, however, it is useful to examine statistical relationships graphically.

Individual bands of remotely sensed data are often referred to as *features* in the pattern recognition literature. To truly appreciate how two bands (features) in a remote sensing dataset covary and if they are correlated or not, it is often useful to produce a two-band *feature space plot*.

A two-dimensional feature space plot extracts the brightness value for every pixel in the scene in two bands and plots the frequency of occurrence in a 255 by 255 feature space (assuming 8-bit data). The greater the frequency of occurrence of unique pairs of values, the brighter the feature space pixel. For example, we know from Table 4-7 that Landsat Thematic Mapper bands 3 and 4 of the November 11, 1982, Charleston scene are not highly correlated ($r = 0.57$). A two-dimensional feature space plot of the two bands is shown in Figure 4-7. The original histograms for bands 3 and 4 are also provided. Because the two bands are not highly correlated, the cloud of points looks somewhat like a tilted cap in two-dimensional feature space. The bright areas in the plot represent pixel pairs that have a high frequency of occurrence in the images. It is usually exciting to see feature space plots such as this because it suggests that their is great information of value in these two bands. It reveals visually that the two bands do not contain much redundant information. If the two bands were highly correlated, there would be significant redundant information between the two bands and the cloud of points would appear as a relatively narrow ellipse generally trending diagonally somewhere in the feature space between 0,0 and the 255,255 coordinates (not shown).

The two-dimensional feature space plot also drives home the point that in this particular dataset, the Landsat Thematic Mapper bands 3 and 4 brightness values are seriously constrained to the lower half of the possible 256 values (i.e., band 3 has a maximum value of 131; band 4 has a maximum value of only 105). This suggests that the data in both bands will probably need to be carefully contrast stretched if humans are going to be called upon to visually extract useful information from these individual bands. More will be said about the use of feature space plots in the discussion on feature space partitioning and image classification in Chapter 9.

Geostatistical Analysis

The Earth's surface has distinct spatial properties. The brightness values in imagery constitute a record of these spatial properties. The spatial characteristics may take the form of texture or pattern, as discussed in Chapter 1. Image analysts often try to quantify the spatial texture or pattern. This requires looking at a pixel and its neighbors and trying to quantify the spatial autocorrelation relationships in the imagery. But how do we measure autocorrelation characteristics in images?

A random variable distributed in space (e.g., spectral reflectance) is said to be *regionalized* (Starck et al., 1998). We can use *geostatistical* measures to extract the spatial properties of *regionalized variables* (Curran, 1988; Woodcock et al., 1988ab). Once quantified, the regionalized variable properties can be used in many remote sensing applications such as image classification (e.g., Atkinson and Lewis, 2000; Maillard, 2003) and the allocation of spatially unbiased sampling sites during classification map accuracy assessment (e.g., Atkinson and Curran, 1995). Another application of *geostatistics* is the prediction of values at unsampled locations. For example, one could obtain many LIDAR-derived elevation values throughout a study area. Although they are very accurate, there may be small or large areas where elevation data are missing (i.e., data voids). Geostatistical interpolation techniques could be used to evaluate the spatial relationships associated with the existing data to create a new, improved systematic grid of elevation values.

Relationships among Geostatistical Analysis, Autocorrelation, and Kriging

Geostatistical techniques were pioneered by Matheron (1971) and Ghandin in the Soviet Union. Webster and his colleagues expanded the utility of regionalized variables to the study of soil science and other continuously varying data (e.g., Webster, 1985; Curran, 1988).

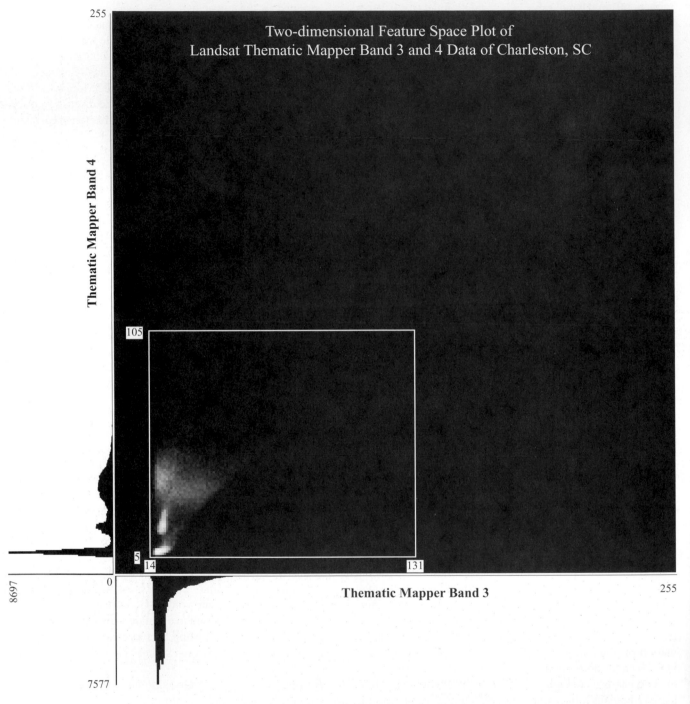

Figure 4-7   A two-dimensional feature space plot of Landsat Thematic Mapper bands 3 and 4 data of Charleston, SC, obtained on November 11, 1982. The original histograms of TM bands 3 and 4 are also provided. The minimum and maximum values within each band are bounded by a white box.The greater the frequency of occurrence of a particular pair of values in band 3 and band 4 in the original imagery, the brighter the pixel in the 255 × 255 feature space. For example, it appears that there are many pixels in the scene that have a brightness value of approximately 17 in band 3 and a brightness value of 7 in band 4. Conversely, there is at least one pixel in the scene with a brightness value of 131 in band 3 and 105 in band 4. Because there is probably only one such pair, it shows up as a dark pixel in the feature space.

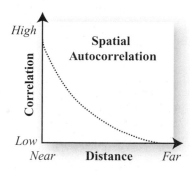

Figure 4-8   Geostatistical analysis incorporates spatial autocorrelation information in the kriging interpolation process. Phenomena that are geographically closer together are generally more highly correlated than things that are farther apart.

Geostatistics are now widely used in many fields and comprise a branch of *spatial statistics*. Originally, geostatistics was synonymous with *kriging*—a statistical version of interpolation. Kriging, named after the pioneering work of Danie Krige (1951), is a generic name for a family of least-squares linear regression algorithms that are used to estimate the value of a continuous attribute (e.g., terrain elevation or percent reflectance) at any unsampled location using only attribute data available over the study area (Lo and Yeung, 2002). However, geostatistical analysis now includes not only kriging but also the traditional deterministic spatial interpolation methods. One of the essential features of geostatistics is that the phenomenon being studied (e.g., elevation, reflectance, temperature, precipitation, a land-cover class) must be continuous across the landscape or at least capable of existing throughout the landscape.

Autocorrelation is the statistical relationship among measured points, where the correlation depends on the distance and direction that separates the locations. We know from real-world observation that spatial autocorrelation exists because we have observed generally that things that are close to one another are more alike than those farther away. As distance increases, spatial autocorrelation decreases (Figure 4-8). Kriging makes use of this spatial autocorrelation information. Kriging is similar to 'distance weighted interpolation' in that it weights the surrounding measured values to derive a prediction for each new location. However, the weights are based not only on the distance between the measured points and the point to be predicted (used in inverse distance weighting), but also on the overall spatial arrangement among the measured points (i.e., their autocorrelation). Kriging uses weights that are defined statistically from the observed data rather than *a priori*. This is the most signifi-

cant difference between deterministic (traditional) and geostatistical analysis. Traditional statistical analysis assumes the samples derived for a particular attribute are independent and not correlated in any way. Conversely, geostatistical analysis allows a scientist to compute distances between observations and to model autocorrelation as a function of distance and direction. This information is then used to refine the kriging interpolation process, hopefully, making predictions at new locations that are more accurate than those derived using traditional methods. There are numerous methods of kriging, including simple, ordinary, universal, probability, indicator, disjunctive, and multiple variable co-kriging (Johnston et al., 2001; Lo and Yeung, 2002).

The kriging process generally involves two distinct tasks:

• quantifying the spatial structure of the surrounding data points, and

• producing a prediction at a new location.

Variography is the process whereby a spatially dependent model is fit to the data and the spatial structure is quantified. To make a prediction for an unknown value at a specific location, kriging uses the fitted model from variography, the spatial data configuration, and the values of the measured sample points around the prediction location (Johnston et al., 2001).

One of the most important measurements used to understand the spatial structure of regionalized variables is the *semivariogram*, which can be used to relate the semivariance to the amount of spatial separation (and autocorrelation) between samples. The semivariance provides an unbiased description of the scale and pattern of spatial variability throughout a region. For example, if an image of a water body is examined, there may be little spatial variability (variance), which will result in a semivariogram with predictable characteristics. Conversely, a heterogeneous urban area may exhibit significant spatial variability resulting in an entirely different semivariogram.

Calculating Average Semivariance

Phenomena in the real world that are close to one another (e.g., two nearby elevation points) have a much greater likelihood of having similar values. The greater the distance between two points, the greater the likelihood that they have significantly different values. This is the underlying concept of autocorrelation. The calculation of the semivariogram makes use of this spatial separation condition, which can be

Table 4-8.   Terms and symbols in a typical empirical semivariogram (after Curran, 1988; Johnston et al., 2001; Lo and Yeung, 2002).

Term	Symbol	Definition
Lag	h	The linear (horizontal) distance that separates any two locations (i.e., a sampling pair). A lag has length (distance) and direction (orientation).
Sill	s	Maximum level of the modeled semivariogram. The value that the variogram tends to when lag distances become very large. At large lag distances, variables become uncorrelated, so the sill of the semivariogram is equal to the variance of the random variable.
Range	a	Point on the h axis where the modeled semivariogram nears a maximum. The distance beyond which there is little or no autocorrelation among variables. Places closer than the range are autocorrelated, places further apart are not.
Nugget variance	C_o	Where the modeled semivariogram intercepts the $\gamma(h)$ axis. Represents the independent error, measurement error, or microscale variation at spatial scales that are too fine to detect. The nugget effect is a discontinuity at the origin of the semivariogram model.
Partial sill	C	The sill minus the nugget variance describes the spatially dependent structural variance.

measured in the field or using remotely sensed data. This brief discussion focuses on the computation of the semivariogram using remotely sensed data although it can be computed just as easily using *in situ* field measurements.

Consider a typical remotely sensed image over a study area. Now identify the endpoints of a transect running through the scene. Twelve hypothetical individual brightness values (*BV*) found along the transect are shown in Figure 4-9a. The (*BV*) z of pixels x have been extracted at regular intervals $z(x)$, where $x = 1, 2, 3,..., n$. The relationship between a pair of pixels h intervals apart (h is referred to as the *lag distance;* Table 4-8) can be given by the average variance of the differences between all such pairs along the transect. There will be m possible pairs of observations along the transect separated by the same lag distance, h. The semivariogram $\gamma(h)$, which is a function relating one-half the squared differences between points to the directional distance between two samples, can be expressed through the relationship (Curran, 1988; Isaaks and Srivastava, 1989; Starck et al., 1998):

$$\gamma(h) = \frac{1}{2}\left[\frac{\sum_{i=1}^{m}[z(x_i) - z(x_i + h)]^2}{m}\right],$$ (4-12)

where $\gamma(h)$ is an unbiased estimate of the average semivariance of the population. The total number of possible pairs m along the transect is computed by subtracting the lag distance h from the total number of pixels present in the dataset n, that is, $m = n - h$ (Brivio and Zilioli, 2001; Johnston et al., 2001; Lo and Yeung, 2002). In practice, semivariance is

computed for pairs of observations in all directions (Maillard, 2003). Thus, directional semivariograms are derived and directional influences can be examined.

The average semivariance is a good measure of the amount of *dissimilarity* between spatially separate pixels. Generally, the larger the average semivariance $\gamma(h)$, the less similar are the pixels in an image (or the polygons if the analysis was based on ground measurement).

Empirical Semivariogram

The *semivariogram* is a plot of the average semivariance value on the y-axis (e.g., $\gamma(h)$ is expressed in brightness value units if uncalibrated remote sensor data are used) with the various lags (h) investigated on the x-axis, as shown in Figures 4-9b and 4-10. Important characteristics of the semivariogram include:

- lag distance (h) on the x-axis,

- sill (s),

- range (a),

- nugget variance (C_o), and

- spatially dependent structural variance partial sill (C).

These terms are defined in Table 4-8. A semivariogram of the autocorrelation between the twelve pixels with lags (h) ranging from 1 to 6 is shown in Figure 4-9b.

Computation of Semivariance for a Transect of Remote Sensing Brightness Values

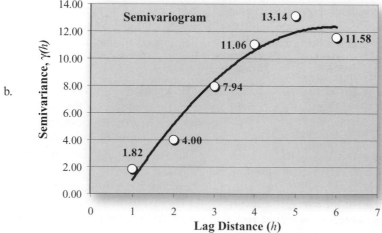

Figure 4-9   a) A hypothetical remote sensing dataset used to demonstrate the characteristics of lag distance (h) along a transect of pixels extracted from an image. b) A semivariogram of the semivariance $\gamma(h)$ characteristics found in the hypothetical dataset at various lag distances (h).

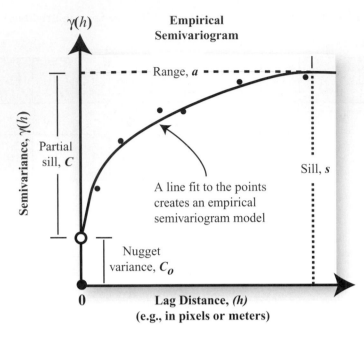

Figure 4-10   The z-values of points (e.g., pixels in an image or locations (or polygons) on the ground if collecting *in situ* data) separated by various lag distances (h) may be compared and their semivariance $\gamma(h)$ computed (adapted from Curran, 1988; Isaaks and Srivastava, 1989; Lo and Yueng, 2002). The semivariance $\gamma(h)$ at each lag distance may be displayed as a *semivariogram* with the range, sill, and nugget variance characteristics described in Table 4-8.

When spatial correlation exists, pairs of points that are close together (on the far left of the x-axis) should have less difference (be low on the y-axis). As points become farther away from each other (moving right on the x-axis), in general, the difference squared should be greater (moving up on the y-axis). The semivariogram model often flattens out at a certain lag distance from the origin. The distance where the model *first* flattens out is the *range*. The range is the distance over which the samples are spatially correlated. Some remote sensing investigations have used the range value to identify the optimum spatial resolution to discriminate the variable under investigation (e.g., Curran, 1988). The *sill* is the value on the y-axis where the range is located. It is the point of maximum variance and is the sum of the structural spatial variance and the nugget effect. The *partial sill* is the sill minus the nugget.

The semivariogram value on the y-axis should theoretically equal zero when lag (h) equals zero. However, at an infinitesimally small separation distance, the semivariogram often exhibits a *nugget* effect, which is >0. The nugget effect is attributed to measurement errors or spatial sources of variation at distances smaller than the sampling interval (or both).

A semivariogram provides information on the spatial autocorrelation of datasets. For this reason and to ensure that kriging predictions have positive kriging variances, a model is fit to the semivariogram (i.e., a continuous function or curve). This model quantifies the spatial autocorrelation in the data (Johnston et al., 2001).

The line fitted to the data points in Figures 4-9 and 4-10 is called a *semivariogram model*. The model is a line through the average semivariance versus average distance between point pairs. The user selects the functional form that best represents the distribution (e.g., spherical, circular, etc.). The coefficients of the function are empirically derived from the data.

The equations for kriging are contained in matrices and vectors that depend on the spatial autocorrelation among the measured sample locations and prediction location. The autocorrelation values come from the semivariogram model. The matrices and vectors determine the kriging weights that are assigned to each measured value in the searching neighborhood. From the kriging weights for the measured values, it is possible to calculate a prediction for the location with the unknown value (Johnston et al., 2001).

It is not necessary to work with just a) transects of image data as demonstrated in the hypothetical example, or b) with just a relatively few map observations to predict the value at

Geostatistical Analysis of Remotely Sensed Data

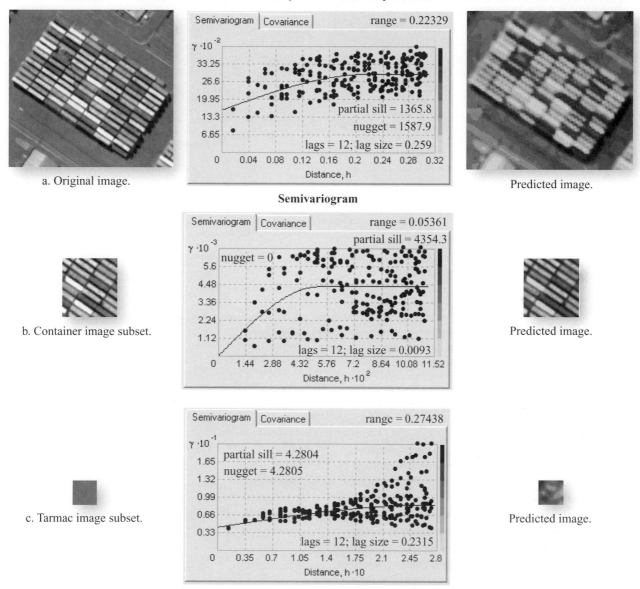

a. Original image.

Predicted image.

Semivariogram

b. Container image subset.

Predicted image.

c. Tarmac image subset.

Predicted image.

Figure 4-11   Original images, semivariograms, and predicted images. a) The green band of a 0.61 × 0.61 m image of cargo containers in the port of Hamburg, Germany, obtained on May 10, 2002 (image courtesy of DigitalGlobe, Inc.). b) A subset containing just cargo containers. c) A subset continuing just tarmac (asphalt).

unknown locations. It is also possible to compute the semivariogram of image pixels in a matrix in all directions at different lag distances. For example, consider the 0.61 × 0.61 m image of cargo containers at the port of Hamburg, Germany (Figure 4-11a). The semivariogram contains detailed information about the range, partial sill, and nugget associated with the entire raster image. A spherical geostatistical kriging algorithm was used to create the predicted image. This

was also performed for a subset containing only cargo containers (Figure 4-11b) and a subset containing only tarmac (Figure 4-11c). In each case, the geostatistical spherical kriging process did an excellent job of predicting the original surface. In addition, the semivariogram contains detailed information about the autocorrelation characteristics of the image dataset that can be used for a variety of applications including deciding the optimum resolution to identify containers or tarmac. Interestingly, Davidson and Csillag (2003) found analysis of variance techniques yielded more useful information than geostatistical analysis when measuring grassland fine-scale heterogeneity.

 ## References

Atkinson, P. M. and P. J. Curran, 1995, "Defining an Optimal Size of Support for Remote Sensing Investigations," *IEEE Transactions Geoscience Remote Sensing*, 33:768–776.

Atkinson, P. M. and P. Lewis, 2000, "Geostatistical Classification for Remote Sensing: An Introduction," *Computers & Geosciences*, 2000:361–371.

Brivio, P. A. and E. Zilioli, 2001, "Urban Pattern Characterization Through Geostatistical Analysis of Satellite Images," in J. P. Donnay, M. J. Barnsley and P. A. Longley (Eds.), *Remote Sensing and Urban Analysis*, London: Taylor & Francis, 39–53.

Congalton, R. G. and K. Green, 1998, *Assessing the Accuracy of Remotely Sensed Data: Principles and Practices*, Boca Raton, FL: Lewis, 160 p.

Curran, P. J., 1988, "The Semivariogram in Remote Sensing: An Introduction," *Remote Sensing of Environment*, 24:493–507.

Davidson, A. and F. Csillag, 2003, "A Comparison of Analysis of Variance (ANOVA) and Variograms for Characterizing Landscape Structure under a Limited Sampling Budget," *Canadian Journal of Remote Sensing*, 29(1):43–56.

Davis, J. C., 2002, *Statistics and Data Analysis in Geology*, 3rd ed., New York: John Wiley & Sons, 638 p.

Donnay, J. P., Barnsley, M. J. and P. A. Longley (Eds.), *Remote Sensing and Urban Analysis*, London: Taylor & Francis, 268 p.

FGDC, 2004, *The National Spatial Data Infrastructure*, Washington: Federal Geographic Data Committee, www.fgdc.gov.

Freud, R. J. and W. J. Wilson, 2003, *Statistical Methods*, 2nd ed., New York: Academic Press, 673 p.

Glantz, S. A., 2001, *Primer on Bio-Statistics*, 5th ed., New York: McGraw-Hill, 489 p.

Gonzalez, R. C. and R. E. Woods, 2002, *Digital Image Processing*, New York: Addison-Wesley, 793 p.

Griffith, D. A. and C. G. Armhein, 1991, *Statistical Analysis for Geographers*, Englewood Cliffs, NJ: Prentice-Hall, 75–113.

Hair, J. F., Anderson, R. E., Tatham, R. L. and W. C. Black, 1998, *Multivariate Data Analysis*, 5th Ed., Upper Saddle River, NJ: Prentice-Hall, 730 p.

Isaaks, E. H. and R. M. Srivastava, 1989, *An Introduction to Applied Geostatistics*, Oxford: Oxford University Press, 561 p.

Jahne, B., 1997, *Digital Image Processing: Concepts, Algorithms, and Scientific Applications*, New York: Springer-Verlag, 555 p.

Johnston, K., Ver Hoef, J. M., Krivoruchko, K. and N. Lucas, 2001, *Using ArcGIS Geostatistical Analyst*, Redlands, CA: Environmental Sciences Research Institute, 300 p.

Krige, D. G., 1951, *A Statistical Approach to Some Mine Valuations and Allied Problems at the Witwatersrand*, Master's Thesis, University of Witwatersrand, South Africa.

Lanter, D. P., 1991, "Design of a Lineage-based Meta-database for GIS," *Cartography and Geographic Information Systems*, 18(4):255–261.

Lo, C. P. and A. K. W. Yeung, 2002, *Concepts and Techniques of Geographic Information Systems*, Upper Saddle River, NJ: Prentice-Hall, 492 p.

Maillard, P., 2003, "Comparing Texture Analysis Methods through Classification," *Photogrammetric Engineering & Remote Sensing*, 69(4):357–367.

Matheron, G., 1971, *The Theory of Regionalized Variables and Its Applications*, Cahiers Centre de Morphologie Mathematique No. 5, Fontainebleau.

Mausel, P. W., Kamber, W. J. and J. K. Lee, 1990, "Optimum Band Selection for Supervised Classification of Multispectral Data," *Photogrammetric Engineering & Remote Sensing*, 56(1):55–60.

Narumalani, S., Hlady, J. T. and J. R. Jensen, 2002, "Information Extraction from Remotely Sensed Data," in J. D. Bossler, J. R. Jensen, R. B. McMaster and C. Rizos (Eds.), *Manual of Geospatial Science and Technology*, London: Taylor & Francis, 299–324.

Pratt, W. K., 1991, *Digital Image Processing*, 2nd ed., New York: John Wiley & Sons, 698 p.

Richards, J. A. and X. Jia, 1999, *Remote Sensing Digital Image Analysis: An Introduction*, New York: Springer-Verlag, 363 p.

Russ, J. C., 2002, *The Image Processing Handbook,* 4th ed., Boca Raton, FL: CRC Press, 744 p.

Samuels, M. L. and J. A. Witmer, 2003, *Statistics for the Life Sciences*, Upper Saddle River, NJ: Prentice-Hall, 724 p.

Schowengerdt, R. A., 1997, *Remote Sensing: Models and Methods for Image Processing*, San Diego, CA: Academic Press, 522 p.

Starck, J. L., Murtagh, F. and A. Bijaoui, 1998, *Image Processing and Data Analysis: The Multiscale Approach*, Cambridge: Cambridge University Press, 287.

Tso, B. and P. M. Mather, 2001, *Classification Methods for Remotely Sensed Data*, London: Taylor & Francis, 332 p.

Walpole, R. E., Myers, R. H., Myers, S. L. and K. Ye, 2002, *Probability and Statistics for Engineers and Scientists*, 7th ed., Upper Saddle River, NJ: Prentice-Hall, 730 p.

Webster, R., 1985, "Quantitative Spatial Analysis of Soil in the Field," *Advances in Soil Science*, 3:1–70.

Woodcock, C. E., Strahler, A. H. and D. L. B. Jupp, 1988a, "The Use of Variograms in Remote Sensing: I. Scene Models and Simulated Images," *Remote Sensing of Environment*, 25:323–348.

Woodcock, C. E., Strahler, A. H. and D. L. B. Jupp, 1988b, "The Use of Variograms in Remote Sensing: II. Real Images," *Remote Sensing of Environment*, 25:349–379.

Initial Display Alternatives and Scientific Visualization

5

Scientists interested in displaying and analyzing remotely sensed data actively participate in *scientific visualization,* defined as "visually exploring data and information in such a way as to gain understanding and insight into the data" (Earnshaw and Wiseman, 1992; Clarke et al., 2002). The difference between scientific visualization and presentation graphics is that the latter are primarily concerned with the communication of information and results that are already understood. During scientific visualization we are seeking to *understand* the data and gain insight (Harris et al., 1999; Shirley, 2002).

Scientific visualization of remotely sensed data is still in its infancy. Its origin can be traced to the simple plotting of points and lines and contour mapping (Figure 5-1). We currently have the ability to conceptualize and visualize remotely sensed images in two-dimensional space in true color as shown (two-dimensional to two-dimensional). It is also possible to drape remotely sensed data over a digital terrain model and display the synthetic three-dimensional model on a two-dimensional map or computer screen (i.e., three-dimensional to two-dimensional). If we turned this same three-dimensional model into a physical model that we could touch, it would occupy the three-dimensional to three-dimensional portion of scientific visualization mapping space. This chapter identifies the challenges and limitations associated with displaying remotely sensed data and makes suggestions about how to display and visualize the data using black-and-white and color output devices.

 Image Display Considerations

Humans are very adept at visually interpreting continuous-tone images every day as they read magazines and newspapers or watch television. Our goal is to capitalize on this talent by providing remotely sensed data in a format that can be easily visualized and interpreted to gain new insight about the Earth (Jensen and Jensen, 2002). The first problem is that the remotely sensed data collected by government agencies (e.g., NOAA AVHRR data) or private industry (e.g., Space Imaging, Inc., DigitalGlobe, Inc., and SPOT Image, Inc.) are in a digital format. How do we convert the brightness values (BVs) stored on an optical disk into an image that begins to approximate the continuous-tone photographs so familiar to humans? The answer is the creation of a brightness map, also commonly referred to as a gray-scale or color image (Bossler et al., 2002).

A *brightness map* is a computer graphic display of the brightness values, $BV_{i,j,k}$, found in digital remote sensor data (refer to Figure 2-1). Ideally, there

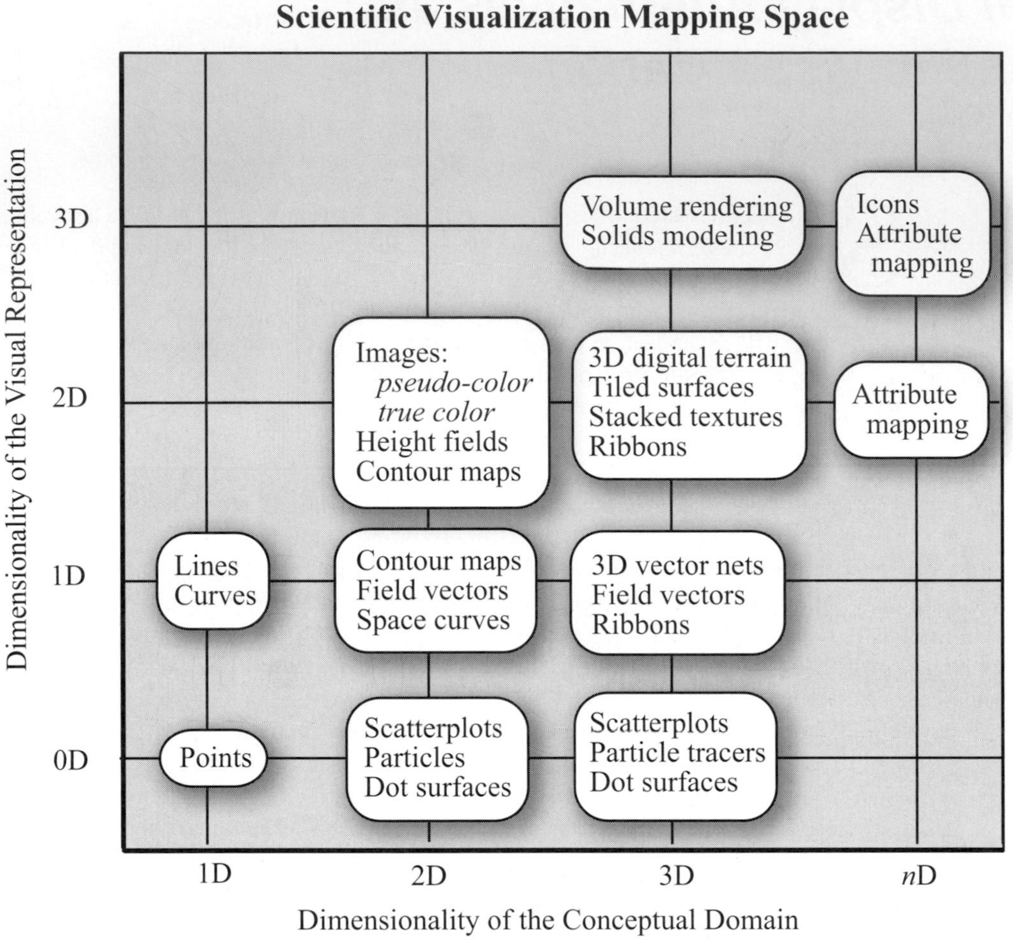

Scientific Visualization Mapping Space

Figure 5-1   Scientific visualization mapping space. The *x*-axis is the conceptual domain or how we conceive the information in our minds. The *y*-axis is the actual number of dimensions used to visually represent our conceptual ideas (after Earnshaw and Wiseman, 1992).

is a one-to-one relationship between input brightness values and the resultant intensities of the output brightness values on the display (Figure 5-2a). For example, an input *BV* of 0 would result in a very dark (black) intensity on the output brightness map, while a *BV* of 255 would produce a bright (white) intensity. All brightness values between 0 and 255 would be displayed as a continuum of grays from black to white. In such a system, an input brightness value of 127 would be displayed exactly as 127 (mid-gray) in the output image, as shown in Figure 5-2a (assuming contrast stretching does not take place). Unfortunately, it is not always easy to maintain this ideal relationship. In the past, it was common for analysts to have access to devices that displayed only a relatively small range of brightness values (e.g., <50) (Figure 5-2b). In this example, several input brightness values around *BV* 127 might be assigned the same output

brightness value of 25. When this occurs, the original remotely sensed data are generalized when displayed, and valuable information may never be seen by the image analyst. Therefore, it is important that whenever possible the one-to-one relationship between input and output brightness values be maintained.

This chapter describes the creation of remote sensing brightness maps using two fundamentally different output devices: hard-copy displays and temporary video displays. *Hard-copy displays* are based on the use of line printers, line plotters, ink-jet printers, laser printers, or film writers to produce tangible hard copies of the imagery for visual examination. *Temporary video displays* are based on the use of black-and-white or color video technology, which displays a temporary image for visual examination. The temporary image can be

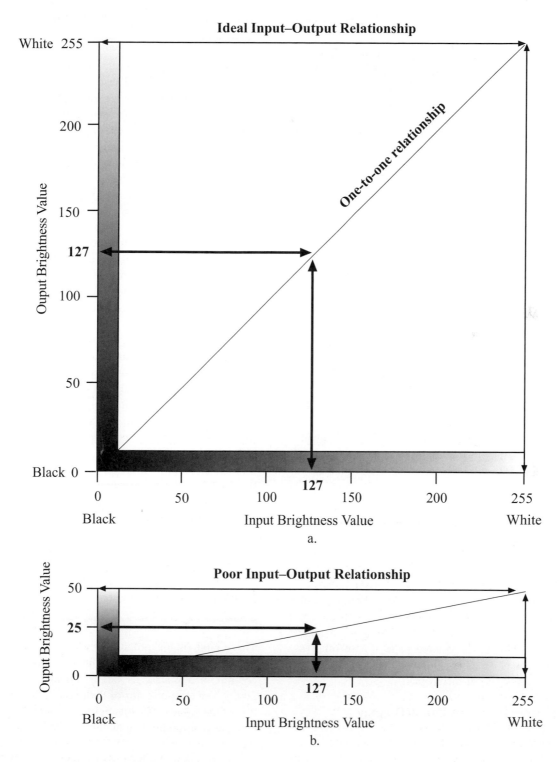

Figure 5-2 a) An ideal one-to-one relationship between 8-bit input remote sensing brightness values and the output brightness map. Most modern computer graphic and digital image processing workstation environments maintain this relationship. b) Poor situation where an analyst's output device is not capable of retaining the one-to-one relationship and must generalize the original 8-bit data down to a more manageable number of brightness map classes. In this case, the output device has only 50 classes. If the 8-bit data were uniformly distributed on the *x*-axis, a >5× reduction of information would take place using this hypothetical output device. These two examples assume that the remotely sensed data are not contrast stretched.

discarded or subsequently routed to a hard-copy device if desired (Brown and Feringa, 2003).

Black-and-White Hard-copy Image Display

Hard-copy image displays may be produced using line printer/plotters, laser printers, or ink-jet printers.

Line Printer/Plotter Brightness Maps

At one time, the 6- or 8-line per inch alphanumeric line printer was the device most often used to produce scaled hard-copy images. The input data were *density sliced* into a series of discrete class intervals in which each interval corresponded to a specific brightness value range. To produce density-sliced maps on the line printer, it was necessary to select 1) an appropriate number of class intervals, 2) the size or dimension of each class interval, and 3) the alphanumeric symbolization to be assigned to each class interval (Slocum, 1999). Sometimes, more sophisticated line plotters were available. Line plotters were programmed to give the impression of continuous tone using crossed-line shading produced by intertwining two perpendicular sets of equally spaced parallel lines (Jensen and Hodgson, 1983). Alphanumeric printers and line plotters are still used to produce hard-copy output if laser or ink-jet printers are not available.

Laser or Ink-jet Printer Brightness Maps

Most image analysts now use relatively inexpensive laser or ink-jet printers to output what appears to be continuous-tone black-and-white or color imagery. This is accomplished by the system's software, which develops a functional relationship between a pixel's input brightness value and the amount of laser toner or ink-jet ink that is applied (output) at the appropriate location on the printed page. These relatively inexpensive printers can normally apply the toner or ink-jet ink at 100 to 1200 dots per inch (dpi).

Temporary Video Image Display

The most effective display of remote sensor data is based on the use of temporary video displays that have improved brightness map display capability.

Black-and-White and Color Brightness Maps

A *video image display* of remotely sensed data can be easily modified or discarded. The central processing unit (CPU) reads the digital remote sensor data from a mass storage device (e.g., hard disk or optical disk) and transfers these data to the image processor's random access memory (RAM) frame buffer (Figure 5-3). The *image processor frame buffer* is a collection of display memory composed of *i* lines by *j* columns and *b* bits that can be accessed sequentially, line by line. Each line of digital values stored in the image processor display memory is continuously scanned by a *read mechanism*. The content of the image processor memory is read every 1/60 s, referred to as the *refresh rate* of the system. The brightness values encountered during this scanning process are passed to the color look-up table. An analyst can change the contents of the color look-up table to modify how an individual pixel eventually appears on the computer screen. The contents of the look-up table are passed to a digital-to-analog converter (DAC) that prepares an analog video signal suitable for display on the cathode-ray tube (CRT) or liquid crystal display (LCD). Thus, the analyst viewing the computer screen is actually looking at the video expression of the digital values stored in the image processor's memory. The 1/60 s refresh rate is so fast that the analyst normally does not see any significant amount of flicker on the video screen.

But how does the entire range of brightness values in a digital dataset get displayed properly? This is a function of (1) the number of bits per pixel associated with the bitmapped graphic under consideration, (2) the color coordinate system being used, and (3) the video look-up tables associated with the image processor.

Bitmapped Graphics

The digital image processing industry refers to all raster images that have a pixel brightness value at each row and column in a matrix as being *bitmapped* images (White, 2002). The tone or color of the pixel in the image is a function of the value of the bits or bytes associated with the pixel and the manipulation that takes place in a color look-up table. For example, the simplest bitmapped image is a binary image consisting of just ones (1) and zeros (0). It is useful to describe how a simple binary bitmapped image is encoded and displayed, before discussing how grayscale and color images are encoded and displayed.

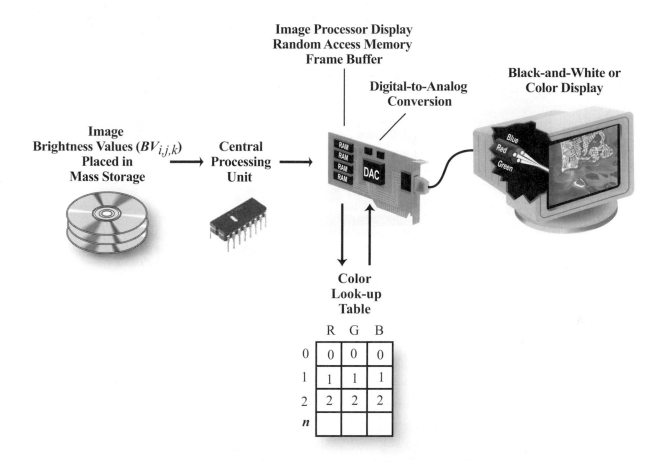

Figure 5-3   The brightness value of a picture element (pixel) is read from mass storage by the central processing unit (CPU). The digital value of the stored pixel is in its proper *i,j* location in the image processor's random access memory (RAM), often referred to as a video *frame buffer*. The brightness value is then passed through a black-and-white or color look-up table where modifications can be made. The output from the digital color look-up table is passed to a digital-to-analog converter (DAC). The output from the DAC determines the intensity of the signal for the three guns (Red, Green, and Blue) in the back of the monitor that stimulate the phosphors on a computer cathode-ray tube (CRT) at a specific *x,y* location or the transistors in a liquid crystal display (LCD).

The simple scene in Figure 5-4 consists of a house, a very large tree, and two road segments. The geographic area has been digitized into a simple raster of 1s and 0s and stored in a bitmapped format as a unique graphics file on a hard disk or other mass storage device. How do we read the contents of the bitmapped graphic file and construct a digital image from the bits and bytes stored in it? First, all raster graphic files contain a header. The first record in the header file is the *signature,* which lets the digital image processing program (application) know what type of file is present. The most common bitmapped graphics formats are BMP, TIFF, and JPEG. The next records in the header identify the number of rows (height) and columns (width) in the bitmapped dataset. The final records in the header contain a summary of the number of colors in the palette (e.g., 2) and the nature of the

colors in the original pallet (e.g., black and white). In this example, the bitmapped image contains 11 rows and 11 columns of data with a black-and-white color palette.

Based on the information in the header, the digital image processing program can extract the values of the 121 individual picture elements in the image. Note that the information content of the entire 121-element matrix can be stored in just 16 bytes (Figure 5-4). Each 8-bit byte contains 8 individual values for 8 pixels. The last 7 bits of byte 16 are not used since the header record specified that there would only be 121 useful values, not the 128 possible values (8 bits × 16 bytes = 128) in a 16-byte dataset. With all the information in hand, the program displays a binary black-and-white image of the scene with zeros in black and ones in white.

Characteristics of a Binary Bitmapped Image

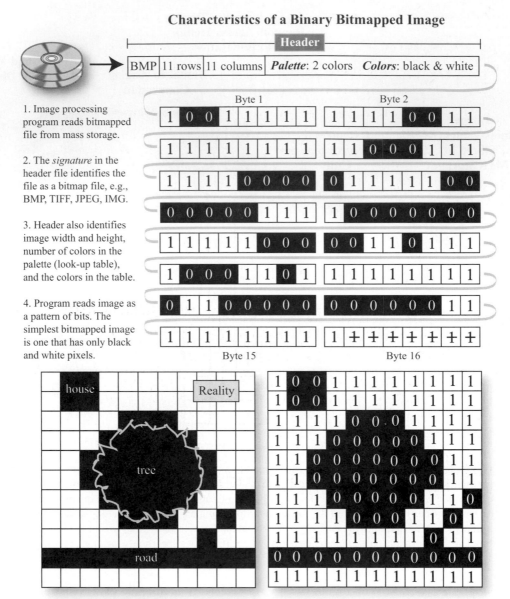

Figure 5-4   Characteristics of a binary (0 and 1), black-and-white *bitmapped* image.

Few remotely sensed images of significant value contain only binary (1-bit) information. Instead, most remote sensor data are quantized to 8 bits per pixel with the values ranging from 0 to 255 in each band. For example, consider the various bitmapped displays of the thermal plume in the Savannah River shown in Figure 5-5. As expected, the original 8-bit display contains all of the information in the original dataset with a minimum value of 38 and a maximum value of 188. The most dramatic drop-off in visual information content seems to occur when the data are displayed using 4 bits or fewer. A 4-bit display only records values that have been transformed to be within the range 0 to 15. The binary 1-bit display contains very little useful information. In effect, every pixel in the 1-bit display with a brightness value <128 has a value of 0 and every pixel ≥128 has a value of 1. It is clear from this illustration that whenever possible we

Bitmap Displays of the Savannah River Thermal Infrared Image

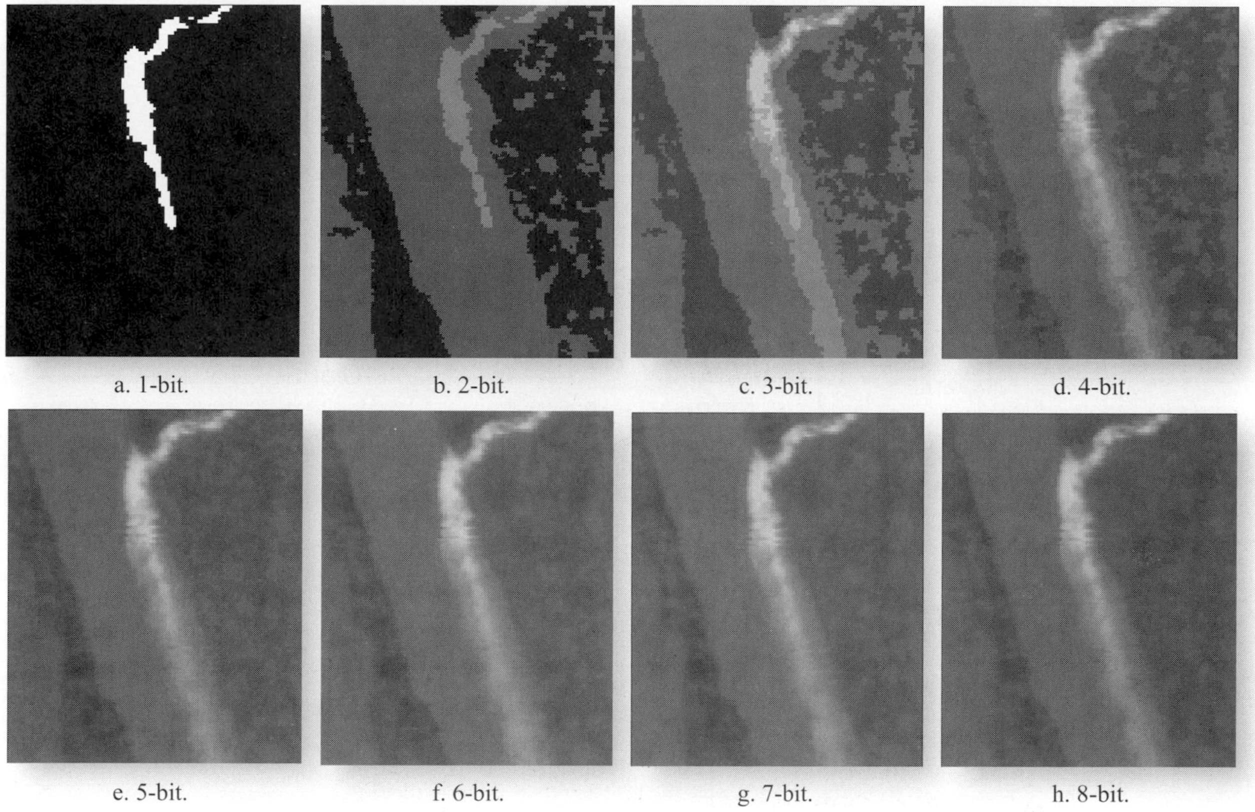

| a. 1-bit. | b. 2-bit. | c. 3-bit. | d. 4-bit. |

| e. 5-bit. | f. 6-bit. | g. 7-bit. | h. 8-bit. |

	Bits	Possible Values
original	8	0 – 255
	7	0 – 127
	6	0 – 63
	5	0 – 31
	4	0 – 15
	3	0 – 7
	2	0 – 3
	1	0 – 1

Figure 5-5   Display of the Savannah River thermal infrared data at various bitmap resolutions.

should use at least 8-bit bitmapped displays when analyzing 8-bit remotely sensed data.

The remainder of the discussion is based on the use of 8-bit imagery. However, several remote sensing systems now have 9-, 10-, 11-, and even 12-bit radiometric resolution per band. Therefore, it is important not to take this parameter for granted.

RGB Color Coordinate System

Digital remote sensor data are usually displayed using a *Red-Green-Blue (RGB) color coordinate system*, which is based on additive color theory and the three primary colors of red, green, and blue (Figure 5-6). Additive color theory is based on what happens when light is mixed, rather than

RGB Color Coordinate System

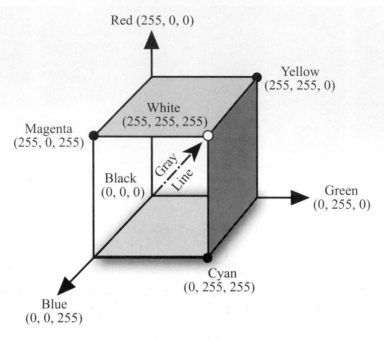

Figure 5-6   The RGB color coordinate system is based on additive color theory. If we are analyzing three 8-bit images (RGB) at one time, there is the possibility of $2^{24} = 16,777,216$ unique values. Each of these unique values lies somewhere within the three-dimensional RGB color coordinate system. Black-and-white grayscale images are located along the diagonal from 0,0,0 to 255,255,255.

when pigments are mixed using subtractive color theory. For example, in additive color theory a pixel having RGB values of 255, 255, 255 produces a bright white pixel. Conversely, we would get a dark pigment if we mixed equally high proportions of blue, green, and red paint (subtractive color theory). Using three 8-bit images and additive color theory, we can conceivably display $2^{24} = 16,777,216$ color combinations. For example, RGB brightness values of 255, 255, 0 would yield a bright yellow pixel, and RGB brightness values of 255, 0, 0 would produce a bright red pixel. RGB values of 0, 0, 0 yield a black pixel. Grays are produced along the gray line in the RGB color coordinate system (Figure 5-6) when equal proportions of blue, green, and red are encountered (e.g., an RGB of 127, 127, 127 produces a medium-gray pixel on the screen or hard-copy device).

Color Look-up Tables: 8-bit

How do we control the exact gray tone or color of the pixel on the computer screen after we have extracted a byte of remotely sensed data from the mass storage device? The gray tone or color of an individual pixel on a computer screen is controlled by the size and characteristics of a sepa-

rate block of computer memory called a *color look-up table,* which contains the exact disposition of each combination of red, green, and blue values associated with each 8-bit pixel. Evaluating the nature of an 8-bit image processor and associated color look-up table (Figure 5-7) provides insight into the way the remote sensing brightness values and color look-up table interact. Two examples of color look-up tables are provided to demonstrate how black-and-white and color density-sliced brightness maps are produced.

In Example 1 (Figure 5-7), we see that the first 256 elements of the look-up table coincide with progressively greater values of the red, green, and blue components, ranging from 0, 0, 0, which would be black on the screen, to 127, 127, 127, which would be gray, and 255, 255, 255, which is bright white. Thus, if a pixel in a single band of remotely sensed data had a brightness value of 127, the RGB value located at table entry 127 (with RGB values of 127, 127, 127) would be passed to the 8-bit DAC converter and a mid-gray pixel would be displayed on the screen. This type of logic was applied to create the black-and-white brightness map of Thematic Mapper band 4 data of Charleston, SC, shown in **Color Plate 5-1a.** This is often referred to as a true 8-bit black-and-white display because there is no

Table 5-1.   Class intervals and color look-up table values for color density slicing the Charleston, SC, Thematic Mapper band 4 scene shown in **Color Plate 5-1a** and **b**.

Color Class Interval	Visual color	Color Look-up Table Value			Brightness Value	
		Red	Green	Blue	Low	High
1	Cyan	0	255	255	0	16
	Shade of gray	17	17	17	17	17
	Shade of gray	18	18	18	18	18
	Shade of gray	19	19	19	19	19
.	.	.	.	.	.	.
.	.	.	.	.	.	.
.	.	.	.	.	.	.
	Shade of gray	58	58	58	58	58
	Shade of gray	59	59	59	59	59
2	Red	255	0	0	60	255

generalization of the original remote sensor data. This means that there is a one-to-one relationship between the 8 bits of remote sensor input data and the 8-bit color look-up table (refer to Figure 5-2a). This is the ideal mechanism for displaying remote sensor data. The same logic was applied to produce the 8-bit display of the predawn thermal infrared image of the Savannah River in **Color Plate 5-1c.**

Laypeople and scientists who analyze remotely sensed data are rarely content with the display of the fundamental gray tone (individual band) or color (using multiple bands) information contained in a dataset. They usually want to highlight in color certain brightness values in the imagery associated with important phenomena. The image analyst can accomplish this task by filling the color look-up table with very specific values (Table 5-1).

To demonstrate this point, in Example 2 (Figure 5-7), the first entry in the table, 0, is given an RGB value of 0, 255, 255. Therefore, any pixel with a value of 0 would be displayed as cyan (a bright blue-green) on the computer screen. In this way it is possible to create a special-purpose look-up table with those colors in it that are of greatest value to the analyst. This is precisely the mechanism by which a single band of remote sensor data is color *density sliced*. For example, if we wanted to highlight just the water and the most healthy vegetation found within the Landsat TM band 4 image of Charleston, we could density slice the image as shown in **Color Plate 5-1b** based on the color look-up table values summarized in Example 2 (Figure 5-7 and Table 5-1).

The color look-up table has been modified so that all pixels between 0 and 16 have RGB values of 0, 255, 255 (cyan). Color look-up table values from 60 to 255 have RGB values of 255, 0, 0 (red). All values between 17 and 59 have the normal gray-scale look-up table value; for example, a pixel with a *BV* of 17 has an RGB value 17, 17, 17, which will result in a dark gray pixel on the screen.

Entirely different brightness value class intervals were selected to density slice the predawn Savannah River thermal infrared image (**Color Plate 5-1c** and **d**). The temperature class intervals and associated color look-up table values for the Savannah River color density-sliced image are found in Table 5-2.

Color Look-up Tables: 24-bit

Much greater flexibility is provided when multiple 8-bit images can be stored and evaluated all at one time. For example, Figure 5-8 depicts the configuration of a 24-bit image processing system complete with three 8-bit banks of image processor memory and three 8-bit color look-up tables, one for each image plane. Thus, three separate 8-bit images could be stored at full resolution in the image processor memory banks (one in the red, one in the green, and one in the blue). Three separate 8-bit DACs continuously read the brightness value of a pixel in each of the red, green, and blue image planes and transform this digital value into an analog signal that can be used to modulate the intensity of

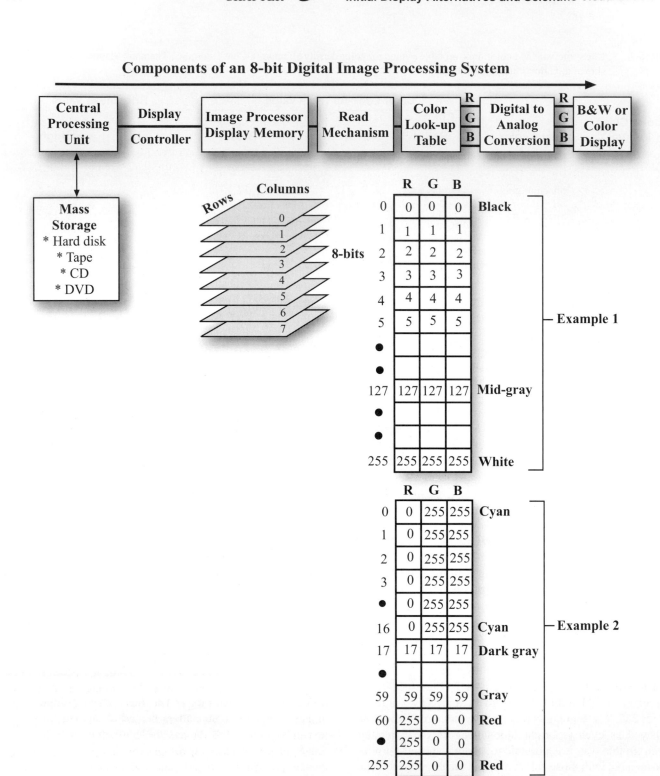

Figure 5-7   Components of an 8-bit digital image processing system. The image processor display memory is filled with the 8-bit remote sensing brightness values from a single band of imagery. These values are then manipulated for display by specifying the contents of the 256-element color look-up table. An 8-bit digital-to-analog converter (DAC) then converts the digital look-up table values into analog signals that are used to modulate the intensity of the red, green, and blue (RGB) guns that create the image on a color CRT screen.

Table 5-2. Class intervals and color look-up table values for color density slicing the predawn thermal infrared image of the thermal plume shown in **Color Plate 5-1c** and **d.**

Color Class Interval	Visual Color	Color Look-up Table Value			Apparent Temperature (°C)		Brightness Value	
		Red	Green	Blue	Low Value	Upper Value	Low	High
1. Land	Gray	127	127	127	–3.0	11.6	0	73
2. River ambient	Dark blue	0	0	120	11.8	12.2	74	76
3. +1°C	Blue	0	0	255	12.4	13.0	77	80
4. 1.2 – 2.8°C	Green	0	255	0	13.2	14.8	81	89
5. 3.0 – 5.0°C	Yellow	255	255	0	15.0	17.0	90	100
6. 5.2 – 10.0°C	Orange	255	50	0	17.2	22.0	101	125
7. 10.2 – 20.0°C	Red	255	0	0	22.2	32.0	126	176
8. >20°C	White	255	255	255	32.2	48.0	177	255

the red, green, and blue (RGB) tricolor guns on the CRT screen. For example, if pixel (1, 1) in the red image plane has a brightness value of 255, and pixel (1, 1) in both the green and blue image planes have brightness values of 0, then a bright red pixel (255, 0, 0) would be displayed at location (1, 1) on the computer screen. More than 16.7 million RGB color combinations can be produced using this configuration. Obviously, this provides a much more ideal palette of colors to choose from, all of which can be displayed on the computer screen at one time. What we have just described is the fundamental basis behind the creation of additive color composites.

Color Composites

High spectral (color) resolution is important when producing color composites. For example, if a false-color reflective infrared TM image is to be displayed accurately, each 8-bit input image (TM band 4 = near infrared; TM band 3 = red, and TM band 2 = green) must be assigned 8 bits of red, green, and blue image processor memory, respectively. In this case, the 24-bit system with three 8-bit color look-up tables would provide a true-color rendition of each pixel on the screen, as previously discussed. This is true additive color combining with no generalization taking place.

Additive color composites produced from various Landsat TM band combinations are presented in **Color Plate 5-2.**

The first is a natural color composite in which TM bands 3, 2, and 1 are placed in the red, green, and blue image processor display memory planes, respectively (**Color Plate 5-2a**). This is what the terrain would look like if the analyst were onboard the satellite platform looking down on South Carolina. A color-infrared color composite of TM bands 4, 3, and 2 (RGB) is displayed in **Color Plate 5-2b**. Healthy vegetation shows up in shades of red because photosynthesizing vegetation absorbs most of the green and red incident energy but reflects approximately half of the incident near-infrared energy (discussed in Chapter 8). Dense urban areas reflect approximately equal proportions of near-infrared, red, and green energy; therefore they appear as steel gray. Moist wetland areas appear in shades of greenish brown. The third color composite was produced using bands 4 (near-infrared), 5 (middle-infrared), and 3 (red) (**Color Plate 5-2c**). The composite provides good definition of the land–water interface. Vegetation type and condition appear in shades of brown, green, and orange. The more moist the soil, the darker it appears. **Color Plate 5-2d** was created using TM bands 7, 4, and 2 (RGB). Many analysts like this combination because vegetation is presented in familiar green tones. Also, the mid-infrared TM band 7 helps discriminate moisture content in both vegetation and soils. Urban areas appear in varying shades of magenta. Dark green areas correspond to upland forest, while greenish brown areas are wetland. But what about all the other three-band color composites that can be produced from the same Landsat TM data?

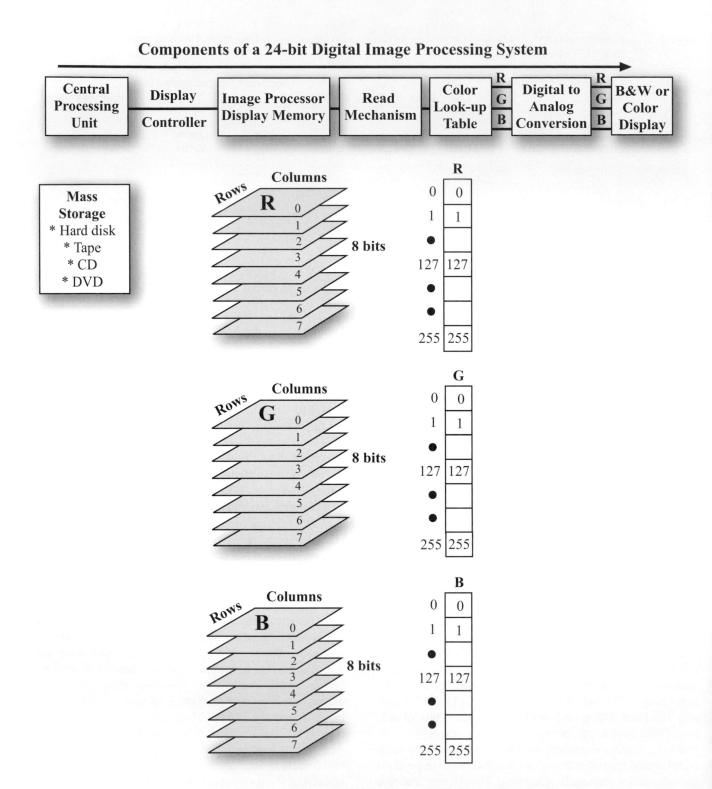

Figure 5-8   Components of a 24-bit digital image processing system. The image processor can store and continuously evaluate up to three 8-bit remotely sensed images. Three 8-bit digital-to-analog converters (DACs) scan the contents of the three 8-bit color look-up tables. The pixel color displayed on the computer screen will be just one possible combination (e.g., red = 255, 0, 0) out of a possible 16,777,216 colors that can be displayed at any time using the 24-bit image processing system.

Optimum Index Factor

Chavez et al. (1982; 1984) developed an *optimum index factor* (OIF) that ranks the 20 three-band combinations that can be made from six bands of TM data (not including the thermal-infrared band). The technique, however, is applicable to any multispectral remote sensing dataset. It is based on the amount of total variance and correlation within and between various band combinations. The algorithm used to compute the OIF for any subset of three bands is:

$$OIF = \frac{\sum_{k=1}^{3} s_k}{\sum_{j=1}^{3} Abs(r_j)}, \quad (5-1)$$

where s_k is the standard deviation for band k, and r_j is the absolute value of the correlation coefficient between any two of the three bands being evaluated. The three-band combination with the largest OIF generally has the most information (measured by variance) with the least amount of duplication (measured by correlation). Combinations within two or three rankings of each other produce similar results.

Application of the OIF criteria to the 1982 Charleston, SC, Landsat TM dataset (excluding the thermal band) resulted in 20 combinations (Table 5-3). The standard deviations and between-band correlation coefficients were obtained from Table 4-7. A three-band combination using bands 1, 4, and 5 should provide the optimum color composite, with bands 2, 4, and 5 and 3, 4, and 5 just about as good. Generally, the best three-band combinations include one of the visible bands (TM 1, 2, or 3), and one of the longer-wavelength infrared bands (TM 5 or 7), along with TM band 4. TM band 4 was present in five of the first six rankings. Such information can be used to select the most useful bands for three-band color composites. The analyst must then decide what color to assign each band (red, green, or blue) in the color composite.

Sheffield Index

Sheffield (1985) developed a statistical band selection index based on the size of the hyperspace spanned by the three bands under investigation. Sheffield suggested that the bands with the largest hypervolumes be selected. The index is based on computing the determinant of each p by p submatrix generated from the original 6×6 covariance matrix (if six bands are under investigation). The *Sheffield Index* (*SI*) is:

Table 5-3. Optimum Index Factors for six of the Charleston, SC, Landsat Thematic Mapper bands.

Rank	Combination	OIF
1	1, 4, 5	27.137
2	2, 4, 5	23.549
3	3, 4, 5	22.599
4	1, 5, 7	20.785
5	1, 4, 7	20.460
6	4, 5, 7	20.100
7	1, 3, 5	19.009
8	1, 2, 5	18.657
9	1, 3, 4	18.241
10	2, 5, 7	18.164
11	2, 3, 5	17.920
12	3, 5, 7	17.840
13	1, 2, 4	17.682
14	2, 4, 7	17.372
15	3, 4, 7	17.141
16	2, 3, 4	15.492
17	1, 3, 7	12.271
18	1, 2, 7	11.615
19	2, 3, 7	10.387
20	1, 2, 3	8.428

Six bands combined three at a time allows 20 combinations. The thermal infrared band 6 (10.4 to 12.5 μm) was not used.

For example, the OIF for Landsat TM band combination 1, 4, and 5 from Table 4-7 is:

$$\frac{10.5 + 15.76 + 23.85}{0.39 + 0.56 + 0.88} = 27.137.$$

$$SI = |Cov_{p \times p}|, \quad (5-2)$$

where $|Cov_{p \times p}|$ is the determinant of the covariance matrix of subset size p. In this case, $p = 3$ because we are trying to discover the optimum three-band combination for image display purposes. In effect, the SI is first computed from a 3×3 covariance matrix derived from just band 1, 2, and 3 data. It is then computed from a covariance matrix derived from just band 1, 2, and 4 data, etc. This process continues for all 20 possible band combinations if six bands are under investigation, as in the previous example. The band combination that results in the largest determinant is selected for image display. All of the information necessary to compute the SI is actually present in the original 6×6 covariance matrix. The Sheffield Index can be extended to datasets containing n bands. Beauchemin and Fung (2001) critiqued Chavez's Optimum Index Factor and Sheffield's Index and then suggested a normalized version of the Sheffield Index.

 Merging Remotely Sensed Data

Image analysts often merge different types of remote sensor data such as:

- SPOT 10×10 m panchromatic (PAN) data with SPOT 20×20 m multispectral (XS) data (e.g., Jensen et al., 1990; Ehlers et al., 1990)

- SPOT 10×10 m PAN data with Landsat Thematic Mapper 30×30 m data (e.g., Welch and Ehlers, 1987; Chavez and Bowell, 1988)

- Multispectral data (e.g., SPOT XS, Landsat TM, IKONOS) with active microwave (radar) and/or other data (e.g., Harris et al., 1990; Chen et al., 2003)

- Digitized aerial photography with SPOT XS or TM data (Chavez, 1986; Grasso, 1993

Merging (fusing) remotely sensed data obtained using different remote sensors must be performed carefully. All datasets to be merged must be accurately registered to one another and resampled (Chapter 6) to the same pixel size (Chavez and Bowell, 1988). Several alternatives exist for merging the data sets, including 1) simple band-substitution methods, 2) color space transformation and substitution methods using various color coordinate systems (e.g., RGB, Intensity–Hue–Saturation, Chromaticity), 3) substitution of the high spatial resolution data for principal component 1, 4) pixel-by-pixel addition of a high frequency–filtered, high spatial resolution dataset to a high spectral resolution dataset, and (5) smoothing filter-based intensity modulation.

Band Substitution

Individual SPOT bands of Marco Island, FL, are displayed in Figure 5-9. The data were geometrically rectified to a

Universal Transverse Mercator (UTM) projection and then resampled to 10×10 m pixels using bilinear interpolation (Jensen et al., 1990). Note how the 20×20 m multispectral bands (green, red, and near-infrared) appear out of focus. This is typical of 20×20 m data of urban areas. Conversely, the 10×10 m panchromatic data exhibits more detailed road network information and perimeter detail for some of the larger buildings. The merging (fusion) of the 20×20 multispectral data with the 10×10 m panchromatic data could be of value.

Color Plate 5-3a is a color-infrared color composite with the band 3 (near-infrared), band 2 (red), and band 1 (green) imagery placed in the red, green, and blue image processor display memory banks, respectively (i.e., bands 3, 2, 1 = RGB). The SPOT panchromatic data span the spectral region from 0.51 to 0.73 μm. Therefore, it is a record of both green and red energy. It can be substituted directly for either the green (SPOT 1) or red (SPOT 2) bands. **Color Plate 5-3b** is a display of the merged (fused) dataset with SPOT 3 (near-infrared), SPOT panchromatic, and SPOT 1 (green) in the RGB image processor memory planes, respectively. The result is a display that contains the spatial detail of the SPOT panchromatic data (10×10 m) and spectral detail of the 20×20 m SPOT multispectral data. This method has the advantage of not changing the radiometric qualities of any of the SPOT data.

Color Space Transformation and Substitution

All remotely sensed data presented thus far have been in the RGB color coordinate system. Other color coordinate systems may be of value when presenting remotely sensed data for visual analysis, and some of these may be used when different types of remotely sensed data are merged. Two frequently used methods are the RGB to intensity–hue–saturation (IHS) transformation (Chen et al., 2003) and the use of chromaticity coordinates (Liu, 2000a; Kulkarni, 2001).

RGB to IHS Transformation and Back Again

The *intensity–hue–saturation* color coordinate system is based on a hypothetical color sphere (Figure 5-10). The vertical axis represents *intensity* (I) which varies from black (0) to white (255) and is not associated with any color. The circumference of the sphere represents *hue* (H), which is the dominant wavelength of color. Hue values begin with 0 at the midpoint of red tones and increase counterclockwise around the circumference of the sphere to conclude with 255 adjacent to 0. *Saturation* (S) represents the purity of the color and ranges from 0 at the center of the color sphere to

SPOT Imagery of Marco Island, FL

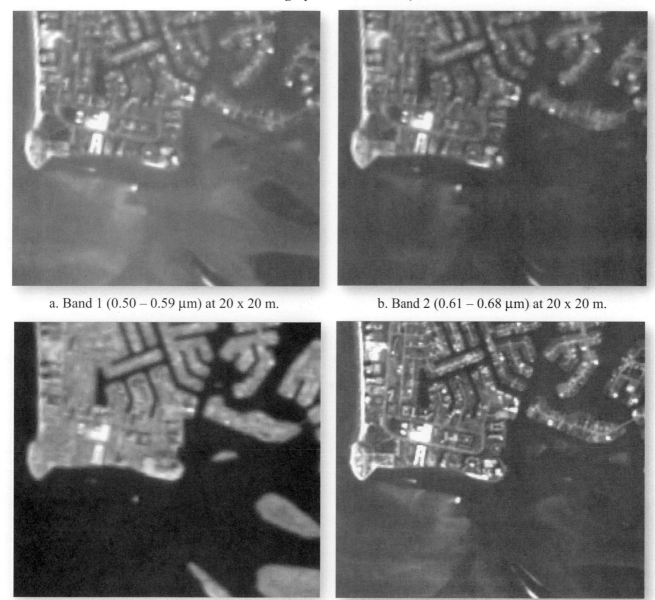

a. Band 1 (0.50 – 0.59 μm) at 20 x 20 m.

b. Band 2 (0.61 – 0.68 μm) at 20 x 20 m.

c. Band 3 (0.79 – 0.89 μm) at 20 x 20 m.

d. Panchromatic (0.51 – 0.73 μm) at 10 x 10 m.

Figure 5-9   Individual bands of SPOT 20 × 20 multispectral and 10 ×10 m panchromatic imagery used to create the normal color-infrared color composite in Color Plate 5-2a and the merged (fused) color-infrared color composite in **Color Plate 5-2b** (images © CNES 2004, Spot Image Corporation).

255 at the circumference. A saturation of 0 represents a completely impure color in which all wavelengths are equally represented and which the eye will perceive as a shade of gray that ranges from white to black depending on intensity (Sabins, 1987). Intermediate values of saturation represent pastel shades, whereas high values represent more pure, intense colors. All values used here are scaled to 8 bits, corresponding to most digital remote sensor data. Figure 5-10 highlights the location of a single pixel with IHS coordinates of 190, 0, 220 [i.e., a relatively intense (190), highly saturated (220), red (0) pixel].

Intensity, Hue, Saturation (IHS) Color Coordinate System

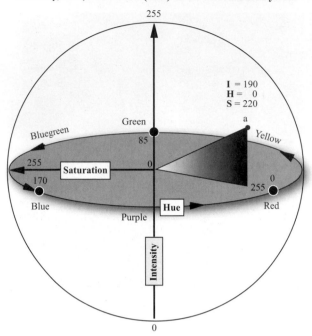

Figure 5-10   Intensity–hue–saturation (IHS) color coordinate system.

Relationship Between the RGB and IHS Color Systems

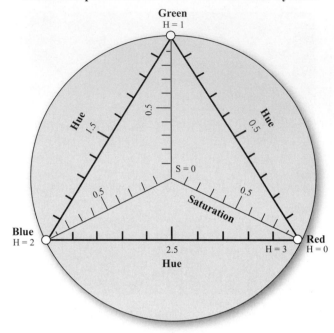

Figure 5-11   Relationship between the intensity–hue–saturation (IHS) color coordinate system and the RGB color coordinate system.

Any RGB multispectral dataset consisting of three bands may be transformed into IHS color coordinate space using an IHS transformation. This transformation is actually a limitation because many remote sensing datasets contain more than three bands (Pellemans et al., 1993). The relationship between the RGB and IHS systems is shown diagrammatically in Figure 5-11. Numerical values may be extracted from this diagram for expressing either system in terms of the other. The circle represents a horizontal section through the equatorial plane of the IHS sphere (Figure 5-10) with the intensity axis passing vertically through the plane of the diagram. The corners of the equilateral triangle are located at the position of the red, green, and blue hues. Hue changes in a counterclockwise direction around the triangle, from red (H = 0), to green (H = 1), to blue (H = 2), and again to red (H = 3). Values of saturation are 0 at the center of the triangle and increase to a maximum of 1 at the corners. Any perceived color is described by a unique set of IHS values. The IHS values can be derived from the RGB values through transformation equations (Sabins, 1987):

$$I = R + G + B, \qquad (5\text{-}3)$$

$$H = \frac{G - B}{I - 3B}, \qquad (5\text{-}4)$$

and

$$S = \frac{I - 3B}{I} \qquad (5\text{-}5)$$

for the interval $0 < H < 1$, extended to $1 < H < 3$. Pellemans et al. (1993) used different equations to compute intensity, hue, and saturation for a SPOT dataset consisting of three bands of remotely sensed data (BV_1, BV_2, and BV_3):

$$\text{Intensity} = \frac{BV_1 + BV_2 + BV_3}{3}, \qquad (5\text{-}6)$$

$$\text{Hue} = \arctan\frac{2BV_1 - BV_2 - BV_3}{\sqrt{3}(BV_2 - BV_3)} + C \qquad (5\text{-}7)$$

$$\text{where} \begin{cases} C = 0, \text{ if } BV_2 \geq BV_3 \\ C = \pi, \text{ if } BV_2 < BV_3 \end{cases},$$

and

$$\text{Saturation} =$$

$$\frac{\sqrt{6(BV_1^2 + BV_2^2 + BV_3^2 - BV_1 BV_2 - BV_1 BV_3 - BV_2 BV_3)}^{-0.5}}{3}. \qquad (5\text{-}8)$$

So what is the benefit of performing an IHS transformation? First, it may be used to improve the interpretability of multi-spectral color composites. When any three spectral bands of multispectral data are combined in the RGB system, the color composite image often lacks saturation, even when the bands have been contrast stretched. Therefore, some analysts perform an RGB-to-IHS transformation, contrast stretch the resultant saturation image, and then convert the IHS images back into RGB images using the inverse of the equations just presented. The result is usually an improved color composite.

The IHS transformation is also often used to merge multiple types of remote sensor data. The method generally involves four steps:

1. *RGB to IHS:* Three bands of lower-spatial-resolution remote sensor data in RGB color space are transformed into three bands in IHS color space.

2. *Contrast manipulation:* The high-spatial-resolution image (e.g., SPOT PAN data or digitized aerial photography) is contrast stretched so that it has approximately the same variance and mean as the intensity (I) image.

3. *Substitution:* The stretched, high-spatial-resolution image is substituted for the intensity (I) image.

4. *IHS to RGB:* The modified IHS dataset is transformed back into RGB color space using an inverse IHS transformation. The justification for replacing the intensity (I) component with the stretched higher-spatial-resolution image is that the two images have approximately the same *spectral* characteristics

Ehlers et al. (1990) used this methodology to merge SPOT 20×20 m multispectral and SPOT panchromatic 10×10 m data. The resulting multiresolution image retained the spatial resolution of the 10×10 m SPOT panchromatic data, yet provided the spectral characteristics (hue and saturation values) of the SPOT multispectral data. The enhanced detail available from merged images was found to be important for visual land-use interpretation and urban growth delineation (Ehlers et al., 1990). In a similar study, Carper et al. (1990) found that direct substitution of the panchromatic data for intensity (I) derived from the multispectral data was not ideal for visual interpretation of agricultural, forested, or heavily vegetated areas. They suggested that the original intensity value obtained in step 1 be computed using a weighted average (WA) of the SPOT panchromatic and SPOT multispectral data; that is, WA = {[(2 × SPOT Pan) +

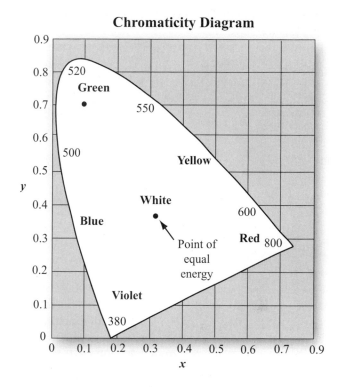

Figure 5-12   Chromaticity color coordinate system.

SPOT XS3]/3}. Chavez et al. (1991) cautioned that of all the methods used to merge multiresolution data, the IHS method distorts the spectral characteristics the most and should be used with caution if detailed radiometric analysis of the data is to be performed. Similar findings were reported by Pelle-mans et al. (1993). Chen et al. (2003) used the IHS transformation to merge radar data with hyperspectral data to enhance urban surface features. Koutsias et al. (2000) used IHS-transformed Landsat 5 Thematic Mapper data to map burned areas.

Chromaticity Color Coordinate System and the Brovey Transformation

A *chromaticity color coordinate system* can be used to specify color. A chromaticity diagram is shown in Figure 5-12. The coordinates in the chromaticity diagram represent the relative fractions of each of the primary colors (red, green, and blue) present in a given color. Since the sum of all three primaries must add to 1, we have the relationship:

$$R + G + B = 1 \tag{5-9}$$

or

$$B = 1 - (R + G). \tag{5-10}$$

Entry into the chromaticity diagram is made using the following relationships:

$$x = \frac{R}{R + G + B}, \tag{5-11}$$

$$y = \frac{G}{R + G + B}, \tag{5-12}$$

and

$$z = \frac{B}{R + G + B}, \tag{5-13}$$

where R, G, and B represent the amounts of red, green, and blue needed to form any particular color, and x, y, and z represent the corresponding *normalized* color components, also known as *trichromatic coefficients* (Kulkarni, 2001). Only x and y are required to specify the chromaticity coordinates of a color in the diagram since $x + y + z = 1$. For example, the point marked "green" in Figure 5-12 has 70% green and 10% red content. It follows from Equation 5-10 that the composition of blue is 20%.

The human eye responds to light only in the range 380 to 800 nm (Kulkarni, 2001). Therefore, the positions of the various spectrum colors from 380 to 800 nm are indicated around the boundary of the chromaticity diagram. These are the pure colors. Any point not actually on the boundary but within the diagram represents some mixture of spectrum colors. The point of equal energy shown in Figure 5-12 corresponds to equal fractions of the three primary colors and represents the CIE standard for white light. Any point located on the boundary of the chromaticity diagram is said to be completely saturated. As a point leaves the boundary and approaches the point of equal energy, more white light is added to the color and it becomes less saturated (Gonzalez and Wintz, 1977). The chromaticity diagram is useful for color mixing, because a straight-line segment joining any two points in the diagram defines all of the color variations that can be formed by combining the *two* colors additively.

The *Brovey transform* may be used to merge (fuse) images with different spatial and spectral characteristics. It is based on the chromaticity transform (Gillespie et al., 1987) and is a much simpler technique than the RGB-to-IHS transformation. The Brovey transform also can be applied to individual bands if desired. It is based on the following intensity modulation (Liu, 2000a):

$$Red_{Brovey} = \frac{R \times P}{I}, \tag{5-14}$$

$$Green_{Brovey} = \frac{G \times P}{I}, \tag{5-15}$$

$$Blue_{Brovey} = \frac{B \times P}{I}, \tag{5-16}$$

and

$$I = \frac{R + G + B}{3}, \tag{5-17}$$

where R, G, and B are the spectral band images of interest (e.g., 30×30 m Landsat ETM$^+$ bands 4, 3, and 2) to be placed in the red, green, and blue image processor memory planes, respectively, P is a coregistered band of higher spatial resolution data (e.g., 1×1 m IKONOS panchromatic data), and I = intensity. An example of the Brovey transform used to merge IKONOS 4×4 m multispectral data (bands 4, 3, and 2) with IKONOS 1×1 m panchromatic data is shown in **Color Plate 2-5f.** The merged (fused) dataset now has the spectral characteristics of the multispectral data and the spatial characteristics of the high-resolution panchromatic data.

Both the RGB-to-IHS transformation and the Brovey transform can cause color distortion if the spectral range of the intensity replacement (or modulation) image (i.e., the panchromatic band) is different from the spectral range of the three lower-resolution bands. The Brovey transform was developed to visually increase contrast in the low and high ends of an image's histogram (i.e., to provide contrast in shadows, water, and high-reflectance areas such as urban features). Consequently, the Brovey transform should not be used if preserving the original scene radiometry is important. However, it is good for producing RGB images with a higher degree of contrast in the low and high ends of the image histogram and for producing visually appealing images (ERDAS, 1999).

Principal Component Substitution

Chavez et al. (1991) used principal components analysis (discussed in Chapter 8) applied to six Landsat TM bands. The SPOT panchromatic data were contrast stretched to have approximately the same variance and average as the first principal component image. The stretched panchromatic data were substituted for the first principal component image and the data were transformed back into RGB space. The stretched panchromatic image may be substituted for the first principal component image because the first principal component image normally contains all the information that is common to all the bands input to PCA, while spectral information unique to any of the input bands is mapped to the other n principal components (Chavez and Kwarteng, 1989).

Pixel-by-Pixel Addition of High-Frequency Information

Chavez (1986), Chavez and Bowell (1988), and Chavez et al. (1991) merged both digitized National High Altitude Program photography and SPOT panchromatic data with Landsat TM data using a high-pass spatial filter applied to the high-spatial-resolution imagery. The resultant high-pass image contains high-frequency information that is related mostly to spatial characteristics of the scene. The spatial filter removes most of the spectral information. The high-pass filter results were added, pixel by pixel, to the lower-spatial-resolution TM data. This process merged the spatial information of the higher-spatial-resolution data set with the higher spectral resolution inherent in the TM dataset. Chavez et al. (1991) found that this multisensor fusion technique distorted the spectral characteristics the least.

Smoothing Filter-based Intensity Modulation Image Fusion

Liu (2000a, b) developed a *Smoothing Filter-based Intensity Modulation* (SFIM) image fusion technique based on the algorithm:

$$BV_{SFIM} = \frac{BV_{low} \times BV_{high}}{BV_{mean}} \tag{5-18}$$

where BV_{low} is a pixel from the coregistered low-resolution image, BV_{high} is a pixel from the high-resolution image, and BV_{mean} is a simulated low-resolution pixel derived from the high-resolution image using an averaging filter for a neighborhood equivalent in size to the spatial resolution of the low-resolution data. For example, suppose the high-resolution image consisted of SPOT 10×10 m panchromatic data and the low-resolution image consisted of Landsat ETM$^+$ 30×30 m data. In this case the BV_{mean} value would be the average of the nine 10×10 m pixels centered on the pixel under investigation in the high-spatial-resolution dataset. Liu (2000a) suggests that the SFIM can produce optimally fused data without altering the spectral properties of the original image if the coregistration error is minimal.

 Distance, Area, and Shape Measurement

Sometimes while viewing an image on the computer screen, an analyst desires to measure the length of an object, its area, or its shape. Most sophisticated digital image processing programs have graphical user interfaces that allow the analyst to first identify the point, line, or area feature in the imagery and then extract the type of measurement desired.

Distance Measurement

Distance is one of the most important geographic measurements extracted from remotely sensed imagery. Distance measurements in imagery are usually made using a rubber-band tool that lets the analyst identify beginning and ending vertices of a line. The x- and y-coordinates of the two vertices of the line are recorded. If the remotely sensed data has not been rectified to a standard map projection, then the x- and y-coordinates will be in simple row (i) and column (j) space. If the imagery has been geometrically rectified to a standard map projection, then the x- and y-coordinates will be in longitude and latitude or some other coordinate system. The most common map projection is the Cartesian coordinate system associated with the Universal Transverse Mercator projection with x-coordinates in meters from a standard meridian and y-coordinates in meters measured from the equator. Measurements made from rectified remote sensor data are more accurate than measurements made from unrectified imagery.

Once the coordinates of the beginning (x_1, y_1) and ending vertices (x_2, y_2) are identified, it is a relatively simple task to use the Pythagorean theorem, which states that the hypotenuse of a right triangle (c) can be computed if we know the length of the other two legs of a right triangle (a and b, respectively):

$$c^2 = a^2 + b^2 \tag{5-19}$$

$$c = \sqrt{(x_1 - x_2)^2 + (y_1 - y_2)^2}. \tag{5-20}$$

The trigonometric relationships are shown diagramatically in Figure 5-13b, where the goal is to measure the length of the longest axis of a mangrove island near Marco Island, FL, recorded on panchromatic 10×10 m SPOT imagery. In this case the SPOT panchromatic 10×10 m imagery has already been geometrically rectified to a UTM projection so the vertex coordinates are in meters. The distance between vertices 1 and 2 is computed as:

$$c = \sqrt{(431437.89 - 433346.04)^2 + (2862232.57 - 2861014.41)^2}$$

Distance and Area Measurement

a. SPOT Panchromatic 10 x 10 m image of Marco Island, FL, obtained on October 21, 1988.

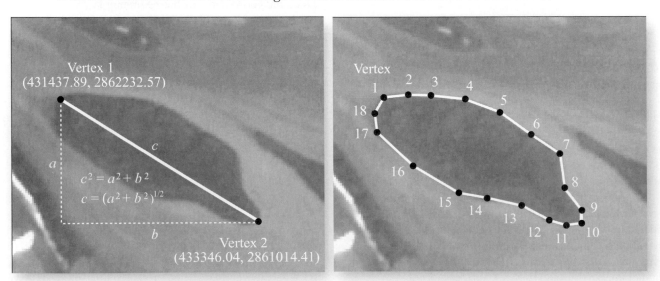

b. Computing the length of the longest dimension of a mangrove island.

c. Computing the perimeter and area of a mangrove island using 18 vertices.

Figure 5-13   a) Panchromatic 10×10 m SPOT imagery of Marco Island, FL. b) Identification of the two vertices used to compute the length of the longest axis of a mangrove island. c) Eighteen vertices used to compute the area of the mangrove island (image © CNES 2004, Spot Image Corporation).

$$c = \sqrt{3641036.422 + 1483913.786}$$

$$c = 2263.84 \text{ m}.$$

Thus, the length of the mangrove island in its longest dimension is 2263.84 m.

Area Measurement

The area of a rectangle on a remotely sensed image is computed simply by multiplying the values of its length and width, i.e., $A = l \times w$. Another elementary area computation is that of a circle, which is $A = \pi r^2$. Complications can arise, however, when the shape of the polygon varies from a rectangle or circle. In the remote sensing literature, polygons are also often referred to as *areas of interest* (AOIs).

To calculate the area of a polygon (or AOI) in remotely sensed imagery, the analyst typically uses a rubber-band tool to identify *n* vertices at unique map (x, y) or image (row and column) coordinates. The "contribution" of each point (vertex) in the polygon to the area is computed by evaluating the x-coordinate of a vertex prior to the vertex under examination (x_{i-1}) with the x-coordinate of the next vertex in the sequence (x_{i+1}) and multiplying the result by the y-coordinate (y_i) of the vertex under examination according to the following formula:

$$Area = 0.5 \sum_{i=1}^{n} y_i (x_{i+1} - x_{i-1}). \tag{5-21}$$

It is important to point out that the very first vertex in the polygon uses values from the second and the *n*th (last) vertex. The last vertex uses values from the $n-1$ vertex and the first vertex.

For example, suppose it is necessary to compute the area of a mangrove island near Marco Island, FL (Figure 5-13c). The analyst interactively identifies *n* vertices that circumscribe the relatively complex shape of the island. The more carefully the vertices are located and the greater number of vertices used during the on-screen digitizing process, the more accurate the area computation will be. In this example 18 vertices were used to outline the mangrove island. The coordinates of the 18 vertices and the contribution each makes to the computation of the area of the polygon are summarized in Table 5-4. The area of the polygon was computed using Equation 5-21. The area of the mangrove island is 1,378,510.24 m^2 or 137.85 hectares (a hectare contains 10,000 m^2).

Table 5-4. Calculation of the area of an isolated mangrove island near Marco Island, FL, from SPOT 10×10 m panchromatic imagery using 18 vertices (refer to Figure 5-13c).

Vertex	x coordinate (easting)	y coordinate (northing)	Contribution
1	431437.89	2862232.57	923041381.50
2	431670.99	2862260.02	1289705742.41
3	431888.48	2862253.68	1585974764.09
4	432225.09	2862214.60	1917483426.98
5	432558.41	2862090.40	1828875765.60
6	432864.09	2861874.74	1657025474.46
7	433137.41	2861686.32	906954245.40
8	433181.02	2861361.66	597280632.91
9	433346.15	2861140.29	472145370.66
10	433346.04	2861014.41	-433643954.12
11	433194.58	2860995.87	-899525711.49
12	433031.63	2861046.79	-1218519827.86
13	432768.68	2861185.76	-1713850270.24
14	432432.63	2861264.12	-1756730331.76
15	432154.71	2861316.30	-2027271211.71
16	431724.12	2861573.34	-2227906539.59
17	431376.15	2861899.68	-1074986757.80
18	431348.50	2862080.03	176704821.05
			2757020.48

$$Area = 0.5 \sum_{i=1}^{n} y_i (x_{i+1} - x_{i-1})$$

$$Area = 0.5 \times 2757020.48 = 1378510.24 \text{ m}^2$$

$$= \frac{1378510.24 \text{ m}^2}{10,000 \text{ m}^2} = 137.85 \text{ hectares}.$$

Polygon (AOI) area estimates extracted from rectified remote sensor data are more accurate than area estimates extracted from unrectified imagery. Unrectified imagery contains geometric error that can severely affect the accuracy of the areal estimates.

The *perimeter* of a polygon (or AOI) is simply the sum of the lengths of the individual line segments between the *n* vertices. The length of the individual line segments are computed using the Pythagorean theorem, as previously discussed. In our example, the perimeter of the mangrove island is 5168.93 m.

Shape Measurement

Polygonal areas of interest all have two-dimensional *shapes* with constant relationships of position and distance of the vertices on their perimeters. One way to measure the shape of an AOI is to compute an index that relates the real-world shape (e.g., of a mangrove island) to some regular geometric figure such as a circle or hexagon. The most commonly used shape index is *compactness,* which is based on deviations from the most compact shape, a circle. The compactness shape index assumes a value of 1.0 for a perfect circle (i.e., a circle has maximum compactness) and a value <1.0 for any less compact shape. At the opposite end of the range of the shape index is the straight line, which has no area and a shape index of zero (Earickson and Harlin, 1994).

The shape (*S*) of an area of interest may be computed using the equation

$$S = \frac{2\sqrt{(A \div \pi)}}{l}, \qquad (5\text{-}22)$$

where *l* is the length of the longest diagonal of the shape and *A* is the area. The value *l* may be computed by locating the longest diagonal in the AOI, identifying the beginning and ending vertices, and applying the Pythagorean theorem, as previously demonstrated.

The shape of the mangrove island when compared to a perfect circle is

$$S = \frac{2\sqrt{(1378510.24 \div 3.1416)}}{2263.84}$$

$$S = \frac{1324.82}{2263.84} = 0.585.$$

The shape index value of the mangrove island can be compared to the shape index of other islands to determine if there are significant differences in the shape of the islands. The

shape differences may be due to some geomorphic process at work in the intertidal zone.

 References

Beauchemin, M. and K. B. Fung, 2001, "On Statistical Band Selection for Image Visualization," *Photogrammetric Engineering & Remote Sensing*, 67(5):571–574.

Brown, A. and W. Feringa, 2003, *Colour Basics for GIS Users*, Upper Saddle River, NJ: Prentice-Hall, 171 p.

Bossler, J. D., Jensen, J. R., McMaster, R. B. and C. Rizos, 2002, *Manual of Geospatial Science and Technology*, London: Taylor & Francis, 623 p.

Carper, W. J., Kiefer, R. W. and T. M. Lillesand, 1990, "The Use of Intensity–Hue–Saturation Transformation for Merging SPOT Panchromatic and Multispectral Image Data," *Photogrammetric Engineering & Remote Sensing*, 56(4):459–467.

Chavez, P. S., 1986, "Digital Merging of Landsat TM and Digitized NHAP Data for 1:24,000 Scale Image Mapping," *Photogrammetric Engineering & Remote Sensing*, 56(2):175–180.

Chavez, P. S., and J. A. Bowell, 1988, "Comparison of the Spectral Information Content of Landsat Thematic Mapper and SPOT for Three Different Sites in the Phoenix, Arizona, Region," *Photogrammetric Engineering & Remote Sensing*, 54(12):1699–1708.

Chavez, P. L., Berlin, G. L. and L. B. Sowers, 1982, "Statistical Method for Selecting Landsat MSS Ratios," *Journal of Applied Photographic Engineering*, 8(1):23–30.

Chavez, P. S., Guptill, S. C. and J. A. Bowell, 1984, "Image Processing Techniques for Thematic Mapper Data," *Proceedings*, ASPRS Technical Papers, 2:728–742.

Chavez, P. S., and A. Y. Kwarteng, 1989, "Extracting Spectral Contrast in Landsat Thematic Mapper Image Data Using Selective Principal Component Analysis," *Photogrammetric Engineering & Remote Sensing*, 55(3):339–348.

Chavez, P. S., Sides, S. C. and J. A. Anderson, 1991, "Comparison of Three Different Methods to Merge Multiresolution and Multispectral Data: Landsat TM and SPOT Panchromatic," *Photogrammetric Engineering & Remote Sensing*, 57(3):295–303.

Chen, C. M., Hepner, G. F. and R. R. Forster, 2003, "Fusion of Hyperspectral and Radar Data Using the IHS Transformation to Enhance Urban Surface Features," *ISPRS Journal of Photogrammetry & Remote Sensing*, 58:19–30.

Clarke, K. C., Parks, B. O. and M. P. Crane, 2002, "Visualizing Environmental Data," in K. C. Clarke, B. O. Parks, and M. P. Crane, (Eds.), *Geographic Information Systems and Environmental Modeling*, Upper Saddle River: Prentice-Hall, 252–298.

Earnshaw, R. A. and N. Wiseman, 1992, *An Introduction Guide to Scientific Visualization,* New York: Springer-Verlag, 156 p.

Earickson, R. and J. Harlin, 1994, *Geographic Measurement and Quantitative Analysis*, New York: Macmillan, 350 p.

Ehlers, M., Jadkowski, M. A., Howard, R. R. and D. E. Brostuen, 1990, "Application of SPOT Data for Regional Growth Analysis and Local Planning," *Photogrammetric Engineering & Remote Sensing*, 56(2):175–180.

ERDAS, 1999, "Brovey Transform," *ERDAS Field Guide*, 5th ed., Atlanta: ERDAS, 161–162.

Gillespie, A. R., Kahle, A. B. and R. E. Walker, 1987, "Color Enhancement of Highly Correlated Images II: Channel Ratio and Chromaticity Transformation Techniques," *Remote Sensing of Environment*, 22(3):343–365.

Gonzalez, R. C. and P. Wintz, 1977, *Digital Image Processing*, Reading, MA: Addison-Wesley, 431 p.

Grasso, D. N., 1993, "Applications of the IHS Color Transformation for 1:24,000-scale Geologic Mapping: A Low Cost SPOT Alternative," *Photogrammetric Engineering & Remote Sensing*, 59(1):73–80.

Harris, J. R., Murray, R. and T. Hirose, 1990, "IHS Transform for the Integration of Radar Imagery with Other Remotely Sensed Data," *Photogrammetric Engineering & Remote Sensing*, 56(12):1631–1641.

Harris, J. R., Viljoen, D. W. and A. N. Rencz, 1999, "Integration and Visualization of Geoscience Data," Chapter 6 in Rencz, A. N. (Ed.), *Remote Sensing for the Earth Sciences*, Bethesda: American Society for Photogrammetry & Remote Sensing, 307–354.

Jensen, J. R. and M. E. Hodgson, 1983, "Remote Sensing Brightness Maps," *Photogrammetric Engineering & Remote Sensing*, 49:93–102.

Jensen, J. R. and R. R. Jensen, 2002, "Remote Sensing Digital Image Processing System Hardware and Software Considerations," in Bossler, J. D., Jensen, J. R., McMaster, R. B. and C. Rizos, (Eds.), *Manual of Geospatial Science and Technology*, London: Taylor & Francis, 623 p.

Jensen, J. R., Ramsey, E. W., Holmes, J. M., Michel, J. E., Savitsky, B. and B. A. Davis, 1990, "Environmental Sensitivity Index (ESI) Mapping for Oil Spills Using Remote Sensing and Geographic Information System Technology," *International Journal of Geographical Information Systems*, 4(2):181–201.

Koutsias, N., Karteris, M. and E. Chuvieco, 2000, "The Use of Intensity-Hue-Saturation Transformation of Landsat 5 Thematic Mapper Data for Burned Land Mapping," *Photogrammetric Engineering & Remote Sensing*, 66(7):829–839.

Kulkarni, A. D., 2001, *Computer Vision and Fuzzy-Neural Systems*, Upper Saddle River, NJ: Prentice-Hall, 504 p.

Liu, J. G., 2000a, "Evaluation of Landsat-7 ETM^+ Panchromatic Band for Image Fusion with Multispectral Bands," *Natural Resources Research*, 9(4): 269–276.

Liu, J. G., 2000b, "Smoothing Filter Based Intensity Modulation: A Spectral Preserving Image Fusion Technique for Improving Spatial Details," *International Journal of Remote Sensing*, 21(18):3461–3472.

Pellemans, A. H., Jordans, R. W. and R. Allewijn, 1993, "Merging Multispectral and Panchromatic SPOT Images with Respect to the Radiometric Properties of the Sensor," *Photogrammetric Engineering & Remote Sensing*, 59(1):81–87.

Sabins, F. F., 1987, *Remote Sensing Principles and Interpretation*, 2nd ed., San Francisco: W. H. Freeman, 251–252.

Shirley, P., 2002, *Fundamentals of Computer Graphics*, Natick, MA: A. K. Peters, 378 p.

Sheffield, C., 1985, "Selecting Band Combinations from Multispectral Data," *Photogrammetric Engineering & Remote Sensing*, 51(6):681–687.

Slocum, T. A., 1999, *Thematic Cartography and Visualization*, Upper Saddle River, NJ: Prentice-Hall, 293 p.

Welch, R. and M. Ehlers, 1987, "Merging Multiresolution SPOT HRV and Landsat TM Data," *Photogrammetric Engineering & Remote Sensing*, 53(3):301–303.

White, R., 2002, *How Computers Work,* Indianapolis: Que, 405 p.

Electromagnetic Radiation Principles and Radiometric Correction

6

The perfect remote sensing system has yet to be developed. Also, the Earth's atmosphere, land, and water are amazingly complex and do not lend themselves well to being recorded by remote sensing devices that have constraints such as spatial, spectral, temporal, and radiometric resolution. Consequently, error creeps into the data acquisition process and can degrade the quality of the remote sensor data collected (Lunetta et al., 1991; Konecny, 2003). This in turn may have an impact on the accuracy of subsequent human or machine-assisted image analysis (Lo and Yeung, 2002).

The two most common types of error encountered in remotely sensed data are radiometric and geometric. *Radiometric correction* is concerned with improving the accuracy of surface spectral reflectance, emittance, or back-scattered measurements obtained using a remote sensing system. *Geometric correction* is concerned with placing the reflected, emitted, or back-scattered measurements or derivative products in their proper planimetric (map) location so they can be associated with other spatial information in a geographic information system (GIS) or spatial decision support system (SDSS) (Jensen et al., 2002; Marakas, 2003).

Radiometric and geometric correction of remotely sensed data are normally referred to as *preprocessing* operations because they are performed prior to information extraction. Image preprocessing hopefully produces a corrected image that is as close as possible, both radiometrically and geometrically, to the true radiant energy and spatial characteristics of the study area at the time of data collection. Internal and external errors must be identified to correct the remotely sensed data:

- *Internal errors* are introduced by the remote sensing system itself. They are generally systematic (predictable) and may be identified and then corrected based on prelaunch or in-flight calibration measurements. For example, *n*-line striping in the imagery may be caused by a single detector that has become uncalibrated. In many instances, radiometric correction can adjust for detector miscalibration.

- *External errors* are usually introduced by phenomena that vary in nature through space and time. The most important external variables that can cause remote sensor data to exhibit radiometric and geometric error are the atmosphere, terrain elevation, slope, and aspect. Some external errors may be corrected by relating empirical observations made on the ground (i.e., radiometric and geometric ground control points) to sensor system measurements.

This chapter focuses on *radiometric* correction of remotely sensed data. Radiometric correction requires knowledge about electromagnetic radiation principles and what interactions take place during the remote sensing data collection process. To be exact, it also involves knowledge about the terrain slope and aspect and bidirectional reflectance characteristics of the scene. Therefore, this chapter first reviews fundamental electromagnetic radiation principles. It then discusses how these principles and relationships are used to correct for radiometric distortion in remotely sensed data caused primarily by the atmosphere and elevation. Chapter 7 discusses the geometric correction of remote sensor data.

Electromagnetic Energy Interactions

Energy recorded by remote sensing systems undergoes fundamental interactions that should be understood to properly preprocess and interpret remotely sensed data. For example, if the energy being remotely sensed comes from the Sun, the energy:

- is radiated by atomic particles at the source (the Sun),

- travels through the vacuum of space at the speed of light,

- interacts with the Earth's atmosphere,

- interacts with the Earth's surface,

- interacts with the Earth's atmosphere once again, and

- finally reaches the remote sensor, where it interacts with various optics, filters, film emulsions, or detectors.

Each of these interactions has an impact on the radiometric quality of the information stored in the remotely sensed image. It is instructive to examine each of these interactions that electromagnetic energy undergoes as it progresses from its source to the remote sensing system detector.

Conduction, Convection, and Radiation

Energy is the ability to do work. In the process of doing work, energy may be transferred from one body to another or from one place to another. The three basic ways in which energy can be transferred are conduction, convection, and radiation (Figure 6-1). Most people are familiar with *conduction,* which occurs when one body (molecule or atom) trans-

fers its kinetic energy to another by colliding with it. This is how a metal pan is heated by a hot burner on a stove. In *convection*, the kinetic energy of bodies is transferred from one place to another by physically moving the bodies. A good example is the heating of the air near the ground in the morning hours. The warmer air near the surface rises, setting up convectional currents in the atmosphere, which may produce cumulus clouds. The transfer of energy by electromagnetic *radiation* is of primary interest to remote sensing science because it is the only form of energy transfer that can take place in a vacuum such as the region between the Sun and the Earth.

Electromagnetic Radiation Models

To understand how electromagnetic radiation is created, how it propagates through space, and how it interacts with other matter, it is useful to describe the processes using two different models: the *wave* model and the *particle* model (Englert et al., 1994).

Wave Model of Electromagnetic Energy

In the 1860s, James Clerk Maxwell (1831–1879) conceptualized electromagnetic radiation (EMR) as an electromagnetic wave that travels through space at the speed of light, c, which is 3×10^8 meters per second (hereafter referred to as m s^{-1}) or 186,282.03 miles s^{-1} (Trefil and Hazen, 1995). A useful relation for quick calculations is that light travels about 1 ft per nanosecond (10^{-9} s) (Rinker, 1999). The *electromagnetic wave* consists of two fluctuating fields—one electric and the other magnetic (Figure 6-2). The two vectors are at right angles (orthogonal) to one another, and both are perpendicular to the direction of travel.

How is an electromagnetic wave created? *Electromagnetic radiation* is generated whenever an electrical charge is accelerated. The wavelength (λ) of the electromagnetic radiation depends upon the length of time that the charged particle is accelerated. Its frequency (ν) depends on the number of accelerations per second. *Wavelength* is formally defined as the mean distance between maximums (or minimums) of a roughly periodic pattern (Figures 6-2 and 6-3) and is normally measured in micrometers (μm) or nanometers (nm). *Frequency* is the number of wavelengths that pass a point per unit time. A wave that sends one crest by every second (completing one cycle) is said to have a frequency of one cycle per second, or one *hertz*, abbreviated 1 Hz. Frequently used measures of wavelength and frequency are found in Table 6-1.

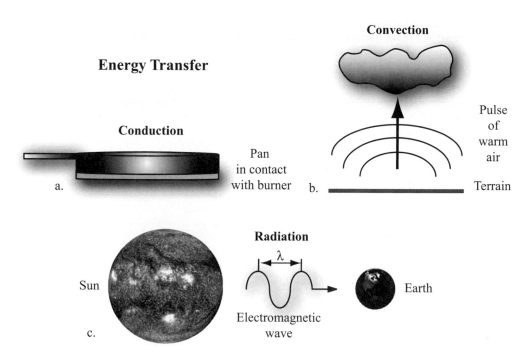

Figure 6-1 Energy may be transferred three ways: conduction, convection, and radiation. a) Energy may be conducted directly from one object to another as when a pan is in direct physical contact with a hot burner. b) The Sun bathes the Earth's surface with radiant energy causing the air near the ground to increase in temperature. The less dense air rises, creating convectional currents in the atmosphere. c) Electromagnetic energy in the form of electromagnetic waves may be transmitted through the vacuum of space from the Sun to the Earth.

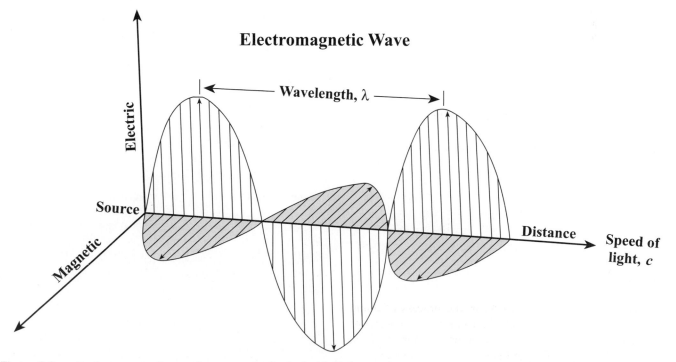

Figure 6-2 An electromagnetic wave is composed of both electric and magnetic vectors that are orthogonal (at 90° angles) to one another. The waves travel from the source at the speed of light (3×10^8 m s^{-1}).

Inverse Relationship between Wavelength and Frequency

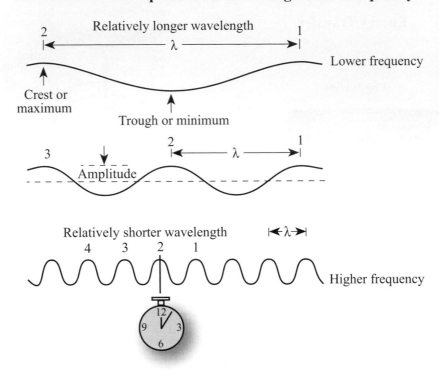

Figure 6-3   This cross-section of several electromagnetic waves illustrates the inverse relationship between wavelength (λ) and frequency (ν). The longer the wavelength, the lower the frequency; the shorter the wavelength, the higher the frequency. The amplitude of an electromagnetic wave is the height of the wave crest above the undisturbed position. Successive wave crests are numbered 1, 2, 3, and 4. An observer at the position of the clock records the number of crests that pass by in a second. This frequency is measured in cycles per second or *hertz*.

Table 6-1.   Wavelength and frequency units of measurement.

	Wavelength (λ)		Frequency (cycles per second)
kilometer (km)	1,000 m	hertz (Hz)	1
meter (m)	1.0 m	kilohertz (kHz)	$1,000 = 10^3$
centimeter (cm)	$0.01\ m = 10^{-2}\ m$	megahertz (MHz)	$1,000,000 = 10^6$
millimeter (mm)	$0.001\ m = 10^{-3}\ m$	gigahertz (GHz)	$1,000,000,000 = 10^9$
micrometer (μm)	$0.000001 = 10^{-6}\ m$		
nanometer (nm)	$0.000000001 = 10^{-9}\ m$		
angstrom (A)	$0.0000000001 = 10^{-10}\ m$		

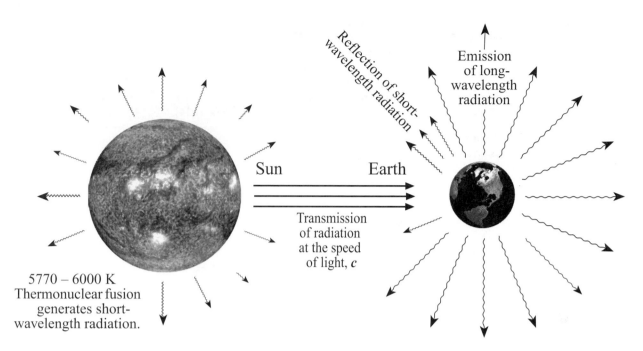

Figure 6-4   The thermonuclear fusion taking place on the surface of the Sun yields a continuous spectrum of electromagnetic energy. The 5770 – 6000 kelvin (K) temperature of this process produces a large amount of relatively short wavelength energy that travels through the vacuum of space at the speed of light. Some of this energy is intercepted by the Earth, where it interacts with the atmosphere and surface materials. The Earth reflects some of the energy directly back out to space or it may absorb the short wavelength energy and then re-emit it at a longer wavelength (after Strahler and Strahler, 1989).

The relationship between the wavelength (λ) and frequency (ν) of electromagnetic radiation is based on the following formula, where c is the speed of light (Rott, 2000):

$$c = \lambda\nu, \qquad (6-1)$$

$$\nu = \frac{c}{\lambda}, \qquad (6-2)$$

and

$$\lambda = \frac{c}{\nu}. \qquad (6-3)$$

Note that frequency is *inversely* proportional to wavelength. This relationship is shown diagrammatically in Figure 6-3, where the longer the wavelength, the lower the frequency; the shorter the wavelength, the higher the frequency. When electromagnetic radiation passes from one substance to another, the speed of light and wavelength change while the frequency remains the same.

All objects above absolute zero (–273°C or 0 K) emit electromagnetic energy, including water, soil, rock, vegetation, and the surface of the Sun. The Sun represents the initial source of most of the electromagnetic energy recorded by remote sensing systems (except RADAR, LIDAR, and SONAR) (Figure 6-4). We may think of the Sun as a 5770 – 6,000 K *blackbody* (a theoretical construct that absorbs and radiates energy at the maximum possible rate per unit area at each wavelength (λ) for a given temperature). The total emitted radiation from a blackbody (M_λ) measured in watts per m^{-2} is proportional to the fourth power of its absolute temperature (T) measured in kelvin (K). This is known as the *Stefan-Boltzmann law* and is expressed as (Rott, 2000):

$$M_\lambda = \sigma T^4 \qquad (6-4)$$

where σ is the Stefan-Boltzmann constant, 5.6697×10^{-8}W m^{-2}K^{-4}. The important thing to remember is that the amount of energy emitted by an object such as the Sun or the Earth is a function of its temperature. The greater the temperature, the greater the amount of radiant energy exiting the object. The actual amount of energy emitted by an object is computed by summing (integrating) the area under its curve (Figure 6-5). It is clear from this illustration that the total emitted radiation from the 6000 K Sun is far greater than that emitted by the 300 K Earth.

Table 6-2.   Methods of describing the color spectrum (after Nassau, 1983; 1984).

Color	Wavelength Descriptions				Energy Descriptions	
	Angstrom (A)	Nanometer (nm)	Micrometer (µm)	Frequency Hz (x 1014)	Wave Number (ψ cm⁻¹)	Electron Volt (eV)
Ultraviolet, sw	2,537	254	0.254	11.82	39,400	4.89
Ultraviolet, lw	3,660	366	0.366	8.19	27,300	3.39
Violet (limit)	4,000	400	0.40	7.50	25,000	3.10
Blue	4,500	450	0.45	6.66	22,200	2.75
Green	5,000	500	0.50	6.00	20,000	2.48
Green	5,500	550	0.55	5.45	18,200	2.25
Yellow	5,800	580	0.58	5.17	17,240	2.14
Orange	6,000	600	0.60	5.00	16,700	2.06
Red	6,500	650	0.65	4.62	15,400	1.91
Red (limit)	7,000	700	0.70	4.29	14,300	1.77
Infrared, near	10,000	1,000	1.00	3.00	10,000	1.24
Infrared, far	300,000	30,000	30.00	0.10	333	0.04

In addition to computing the total amount of energy exiting a theoretical blackbody such as the Sun, we can determine its dominant wavelength (λ_{max}) based on *Wien's displacement law*:

$$\lambda_{max} = \frac{k}{T} \qquad (6\text{-}5)$$

where k is a constant equaling 2898 µm K, and T is the absolute temperature in kelvin. Therefore, as the Sun approximates a 6000 K blackbody, its dominant wavelength (λ_{max}) is 0.48 µm:

$$0.483 \mu m = \frac{2898 \ \mu m \ K}{6000 \ K}.$$

Electromagnetic energy from the Sun travels in 8 minutes across the intervening 93 million miles (150 million km) of space to the Earth. As shown in Figure 6-5, the Earth approximates a 300 K (27 °C) blackbody and has a dominant wavelength at approximately 9.66 µm:

$$9.66 \ \mu m = \frac{2898 \ \mu m \ K}{300 \ K}.$$

Although the Sun has a dominant wavelength at 0.48 µm, it produces a continuous spectrum of electromagnetic radiation ranging from very short, extremely high-frequency gamma and cosmic waves to long, very-low-frequency radio waves (Figures 6-6 and 6-7). The Earth intercepts only a very small portion of the electromagnetic energy produced by the Sun.

As mentioned in Chapter 1, in remote sensing research we often specify a particular region of the electromagnetic spectrum (e.g., red light) by identifying a beginning and ending wavelength (or frequency) and then attaching a description. This wavelength (or frequency) interval in the electromagnetic spectrum is commonly referred to as a *band*, *channel*, or *region*. The major subdivisions of visible light are presented diagrammatically in Figure 6-7 and summarized in Table 6-2. For example, we generally think of visible light as being composed of energy in the blue (0.4 – 0.5 µm), green (0.5 – 0.6 µm), and red (0.6 – 0.7 µm) bands of the electromagnetic spectrum. Similarly, reflected near-infrared energy in the region from 0.7 to 1.3 µm is commonly used to expose black-and-white and color-infrared-sensitive film.

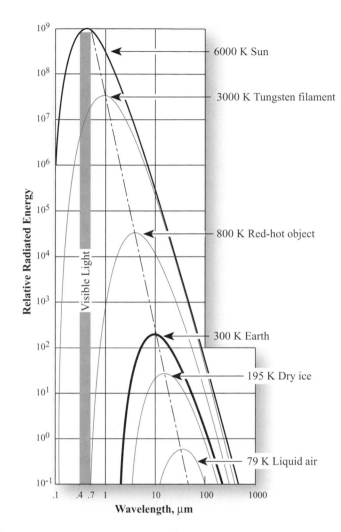

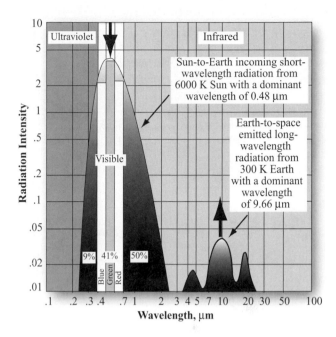

Figure 6-6   The Sun approximates a 6000 K blackbody with a dominant wavelength of about 0.48 μm. The Earth approximates a 300 K blackbody with a dominant wavelength of about 9.66 μm. The 6000 K Sun produces approximately 41% of its energy in the visible region from 0.4 to 0.7 μm (blue, green, and red light). The other 59% of the energy is in wavelengths shorter than blue light (<0.4 μm) and longer than red light (>0.7 μm). Our eyes are only sensitive to light from 0.4 to 0.7 μm (after Strahler and Strahler, 1989). Fortunately, it is possible to make remote sensor detectors sensitive to energy in these nonvisible regions of the spectrum.

Figure 6-5   Blackbody radiation curves for several objects, including the Sun and the Earth, which approximate 6000 K and 300 K blackbodies, respectively. The area under each curve may be summed to compute the total radiant energy (M_λ) exiting each object (Equation 6-4). Thus, the Sun produces more radiant exitance than the Earth because its temperature is greater. As the temperature of an object increases, its dominant wavelength (λ_{max}) shifts toward the shorter wavelengths of the spectrum.

The spectral resolution of most remote sensing systems is described in terms of bands of the electromagnetic spectrum. The band specifications for the most important remote sensing systems are summarized in Chapter 2.

Electromagnetic energy may be described not only in terms of wavelength and frequency but also in photon energy units such as joules (J) and electron volts (eV), as shown in Figure 6-7. Several of the more important mass, energy, and power conversions are summarized in Table 6-3.

The Particle Model: Radiation from Atomic Structures

The middle-infrared region (often referred to as the short wavelength infrared, SWIR) includes energy with a wavelength of 1.3 to 3 μm. The thermal infrared region has two very useful bands at 3 to 5 μm and 8 to 14 μm. The microwave portion of the spectrum consists of much longer wavelengths (1 mm – 1 m). The radio-wave portion of the spectrum may be subdivided into UHF, VHF, radio (HF), LF, and ULF frequencies.

In *Opticks* (1704), Sir Isaac Newton stated that light was a stream of particles, or corpuscles, traveling in straight lines.

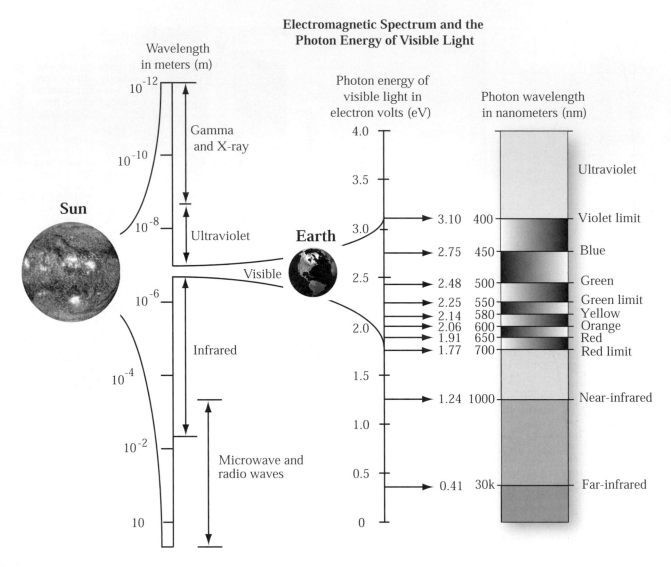

Figure 6-7 The electromagnetic spectrum and the photon energy of visible light. The Sun produces a continuous spectrum of energy from gamma rays to radio waves that continually bathe the Earth in energy. The visible portion of the spectrum may be measured using wavelength (measured in micrometers or nanometers, i.e., μm or nm) or electron volts (eV). All units are interchangeable.

He also knew that light had wavelike characteristics based on his work with glass plates. Nevertheless, during the two hundred years before 1905, light was thought of primarily as a smooth and continuous wave. Then, Albert Einstein (1879–1955) found that when light interacts with electrons, it has a different character. He concluded that when light interacts with matter, it behaves as though it is composed of many individual bodies called *photons*, which carry such particle-like properties as energy and momentum (Meadows, 1992). As a result, most physicists today would answer the question, "What is light?" by saying that light is a *particular* kind of matter (Feinberg, 1985). Thus, we sometimes describe elec-

tromagnetic energy in terms of its wavelike properties. But when the energy interacts with matter, it is useful to describe it as discrete packets of energy, or *quanta*. It is practical to review how electromagnetic energy is generated at the atomic level; this provides insight as to how light interacts with matter.

Electrons are the tiny negatively charged particles that move around the positively charged nucleus of an atom (Figure 6-8). Atoms of different substances are made up of varying numbers of electrons arranged in different ways. The interaction between the positively charged nucleus and the nega-

Table 6-3.   Mass, energy, and power conversions.

Conversion from English to SI Units

To get:	Multiply:	By:
newtons[a]	pounds	4.448
joules[b]	BTUs[c]	1055
joules	calories[d]	4.184
joules	kilowatt-hours[e]	3.6×10^6
joules	foot-pounds[f]	1.356
joules	horsepower[g]	745.7

Conversion from SI to English Units

To get:	Multiply:	By:
BTUs	joules	0.00095
calories	joules	0.2390
kilowatt-hours	joules	2.78×10^{-7}
foot-pounds	joules	0.7375
horsepower	watts	0.00134

[a]newton: force needed to accelerate a mass of 1 kg by 1 m s^{-2}
[b]joule: a force of 1 newton acting through 1 meter.
[c]British thermal unit, or BTU: energy required to raise the temperature of 1 pound of water by 1 degree Fahrenheit.
[d]calorie: energy required to raise the temperature of 1 kilogram of water by 1 degree Celsius.
[e]kilowatt-hour: 1000 joules per second for 1 hour.
[f]foot-pound: a force of 1 pound acting through 1 foot.
[g]horsepower: 550 foot-pounds per second.

tively charged electron keeps the electron in orbit. While its orbit is not explicitly fixed, each electron's motion is restricted to a definite range from the nucleus. The allowable orbital paths of electrons moving around an atom might be thought of as energy classes or levels (Figure 6-8a). In order for an electron to climb to a higher class, work must be performed. However, unless an amount of energy is available to move the electron up at least one energy level, it will accept no work. If a sufficient amount of energy is received, the electron will jump to a new level and the atom is said to be *excited* (Figure 6-8b). Once an electron is in a higher orbit, it possesses potential energy. After about 10^{-8} seconds, the

electron falls back to the atom's lowest empty energy level or orbit and gives off radiation (Figure 6-8c). The wavelength of radiation given off is a function of the amount of work done on the atom, i.e., the quantum of energy it absorbed to cause the electron to become excited and move to a higher orbit.

Electron orbits are like the rungs of a ladder. Adding energy moves the electron up the energy ladder; emitting energy moves it down. However, the energy ladder differs from an ordinary ladder in that its rungs are unevenly spaced. This means that the energy an electron needs to absorb, or to give up, in order to jump from one orbit to the next may not be the same as the energy change needed for some other step. Furthermore, an electron does not necessarily use consecutive rungs. Instead, it follows what physicists call *selection rules*. In many cases, an electron uses one sequence of rungs as it climbs the ladder and another sequence as it descends (Nassau, 1983). The energy that is left over when the electrically charged electron moves from an excited state (Figure 6-8b) to a de-excited state (Figure 6-8c) is emitted by the atom as a single packet of electromagnetic radiation, a particle-like unit of light called a *photon*. Every time an electron jumps from a higher to a lower energy level, a photon moves away at the speed of light.

Somehow an electron must disappear from its original orbit and reappear in its destination orbit without ever having to traverse any of the positions in between. This process is called a *quantum leap* or *quantum jump*. If the electron leaps from its highest excited state to the ground state in a single leap, it will emit a single photon of energy. It is also possible for the electron to leap from an excited orbit to the ground state in a series of jumps (e.g., from 4 to 2 to 1). If it takes two leaps to get to the ground state, then each of these jumps will emit photons of somewhat less energy. The energies emitted in the two jumps must sum to the total of the single large jump (Trefil and Hazen, 1995).

Niels Bohr (1885–1962) and Max Planck recognized the discrete nature of exchanges of radiant energy and proposed the *quantum theory* of electromagnetic radiation. This theory states that energy is transferred in discrete packets called quanta or photons, as discussed. The relationship between the frequency of radiation expressed by wave theory and the quantum is:

$$Q = h\nu \qquad (6\text{-}6)$$

where Q is the energy of a quantum measured in joules, h is the Planck constant (6.626×10^{-34} J s), and ν is the frequency of the radiation. Referring to Equation 6-3, we can

Creation of Light from Atomic Particles and the Photoelectric Effect

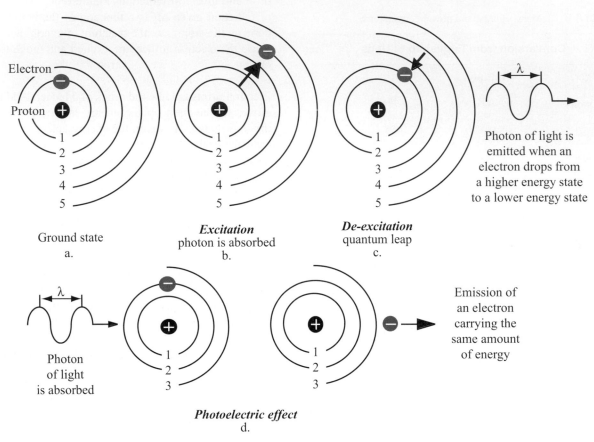

Figure 6-8   a–c) A photon of electromagnetic energy is emitted when an electron in an atom or molecule drops from a higher-energy state to a lower-energy state. The light emitted (i.e., its wavelength) is a function of the changes in the energy levels of the outer, valence electron. For example, yellow light is produced from a sodium vapor lamp in Figure 6-9. d) Matter can also be subjected to such high temperatures that electrons, which normally move in captured, nonradiating orbits, are broken free. When this happens, the atom remains with a positive charge equal to the negatively charged electron that escaped. The electron becomes a free electron, and the atom is called an ion. If another free electron fills the vacant energy level created by the free electron, then radiation from all wavelengths is produced, i.e., a continuous spectrum of energy. The intense heat at the surface of the Sun produces a continuous spectrum in this manner.

multiply the equation by h/h, or 1, without changing its value:

$$\lambda = \frac{hc}{h\nu}. \tag{6-7}$$

By substituting Q for $h\nu$ (from Equation 6-6), we can express the wavelength associated with a quantum of energy as:

$$\lambda = \frac{hc}{Q} \tag{6-8}$$

or

$$Q = \frac{hc}{\lambda}. \tag{6-9}$$

Thus, we see that the energy of a quantum is inversely proportional to its wavelength, i.e., the longer the wavelength involved, the lower its energy content. This inverse relationship is important to remote sensing because it suggests that it is more difficult to detect longer-wavelength energy being emitted at thermal infrared wavelengths than those at shorter visible wavelengths. In fact, it might be necessary to have the

**Creation of Light from Atomic Particles
in a Sodium Vapor Lamp**

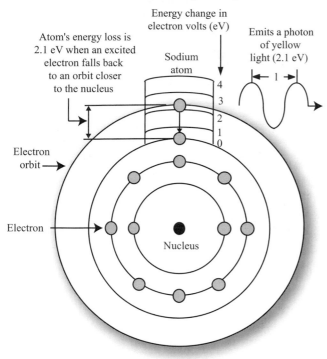

Figure 6-9   Creation of light from atomic particles in a sodium vapor lamp. After being energized by several thousand volts of electricity, the outermost electron in each energized atom of sodium vapor climbs to a high rung on the energy ladder and then returns down the ladder in a predictable fashion. The last two rungs in the descent are 2.1 eV apart. This produces a photon of yellow light, which has 2.1 eV of energy (see Table 6-2).

sensor look at or dwell longer on the parcel of ground if we are trying to measure the longer wavelength energy.

Substances have color because of differences in their energy levels and the selection rules. For example, consider energized sodium vapor that produces a bright yellow light that is used in some street lamps. When a sodium vapor lamp is turned on, several thousand volts of electricity energize the vapor. The outermost electron in each energized atom of sodium vapor climbs to a higher rung on the energy ladder and then returns down the ladder in a certain sequence of rungs, the last two of which are 2.1 eV apart (Figure 6-9). The energy released in this last leap appears as a photon of yellow light with a wavelength of 0.58 μm with 2.1 eV of energy (Nassau, 1983).

Matter can be heated to such high temperatures that electrons that normally move in captured, nonradiating orbits break free. This is called the *photoelectric effect* (Figure 6-8d). When this happens, the atom remains with a positive charge equal to the negatively charged electron that escaped. The electron becomes a free electron and the atom is called an *ion*. In the ultraviolet and visible (blue, green, and red) parts of the electromagnetic spectrum, radiation is produced by changes in the energy levels of the outer valence electrons. The wavelengths of energy produced are a function of the particular orbital levels of the electrons involved in the excitation process. If the atoms absorb enough energy to become ionized and if a free electron drops in to fill the vacant energy level, then the radiation given off is unquantized and a *continuous spectrum* is produced rather than a band or a series of bands. Every encounter of one of the free electrons with a positively charged nucleus causes rapidly changing electric and magnetic fields, so that radiation at all wavelengths is produced. The hot surface of the Sun is largely a *plasma* in which radiation of all wavelengths is produced. As previously shown in Figure 6-7, the spectrum of a plasma like the Sun is continuous.

In atoms and molecules, electron orbital changes produce the shortest-wavelength radiation, molecule vibrational motion changes produce near- and/or middle-infrared energy, and rotational motion changes produce long-wavelength infrared or microwave radiation.

 Atmospheric Energy–Matter Interactions

Radiant energy is the capacity of radiation within a spectral band to do work. Once electromagnetic radiation is generated, it is propagated through the Earth's atmosphere almost at the speed of light in a vacuum. Unlike a vacuum in which nothing happens, however, the atmosphere may affect not only the speed of radiation but also its wavelength, its intensity, and its spectral distribution. The electromagnetic radiation may also be diverted from its original direction due to refraction.

Refraction

The speed of light in a vacuum is 3×10^8 m s^{-1}. When electromagnetic radiation encounters substances of different density, like air and water, refraction may take place. *Refraction* refers to the bending of light when it passes from one medium to another of different density. Refraction occurs because the media are of differing densities and the speed of

EMR is different in each. The *index of refraction (n)* is a measure of the optical density of a substance. This index is the ratio of the speed of light in a vacuum, c, to the speed of light in a substance such as the atmosphere or water, c_n (Mulligan, 1980):

$$n = \frac{c}{c_n}. \qquad (6\text{-}10)$$

The speed of light in a substance can never reach the speed of light in a vacuum. Therefore, its index of refraction must always be greater than 1. For example, the index of refraction for the atmosphere is 1.0002926 and 1.33 for water. Light travels more slowly through water because of water's higher density.

Refraction can be described by Snell's law, which states that for a given frequency of light (we must use frequency since, unlike wavelength, it does not change when the speed of light changes), the product of the index of refraction and the sine of the angle between the ray and a line normal to the interface is constant:

$$n_1 \sin \theta_1 = n_2 \sin \theta_2. \qquad (6\text{-}11)$$

From Figure 6-10 we can see that a nonturbulent atmosphere can be thought of as a series of layers of gases, each with a slightly different density. Anytime energy is propagated through the atmosphere for any appreciable distance at any angle other than vertical, refraction occurs.

The amount of refraction is a function of the angle made with the vertical (θ), the distance involved (in the atmosphere the greater the distance, the more changes in density), and the density of the air involved (air is usually more dense near sea level). Serious errors in location due to refraction can occur in images formed from energy detected at high altitudes or at acute angles. However, these location errors are predictable by Snell's law and thus can be removed. Notice that

$$\sin \theta_2 = \frac{n_1 \sin \theta_1}{n_2}. \qquad (6\text{-}12)$$

Therefore, if one knows the index of refraction of medium n_1 and n_2 and the angle of incidence of the energy to medium n_1, it is possible to predict the amount of refraction that will take place ($\sin \theta_2$) in medium n_2 using trigonometric relationships. It is useful to note, however, that most image analysts never concern themselves with computing the index of refraction.

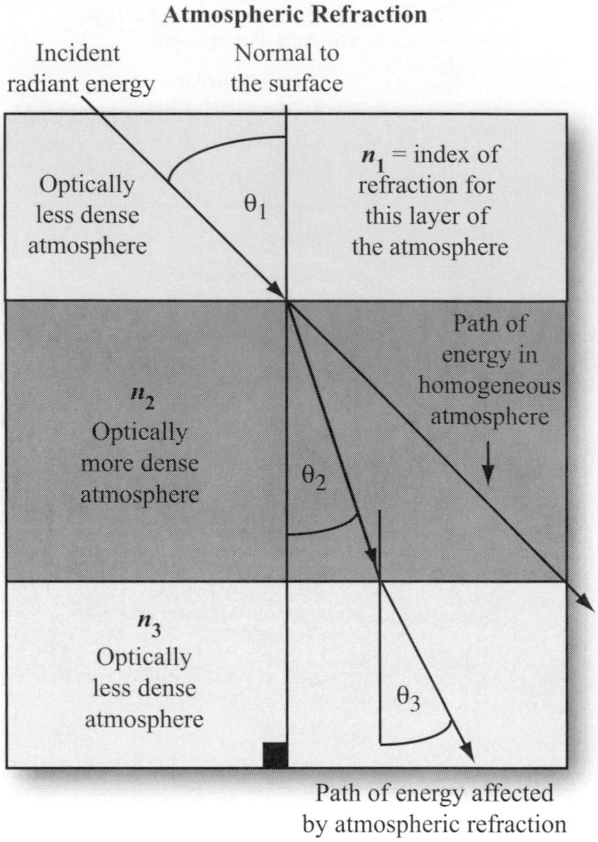

Figure 6-10   Refraction in three nonturbulent atmospheric layers. The incident energy is bent from its normal trajectory as it travels from one atmospheric layer to another. Snell's law can be used to predict how much bending will take place, based on a knowledge of the angle of incidence (θ) and the index of refraction of each atmospheric level, n_1, n_2, n_3.

Scattering

One very serious effect of the atmosphere is the scattering of radiation by atmospheric particles. *Scattering* differs from reflection in that the direction associated with scattering is unpredictable, whereas the direction of reflection (to be defined shortly) is predictable. There are essentially three types of scattering: Rayleigh, Mie, and nonselective scattering. Major subdivisions of the atmosphere and the types of molecules and aerosols found in each layer are shown in Figure 6-11 (Miller and Vermote, 2002). The relative size of the wavelength of the incident electromagnetic radiation, the diameter of the gases, water vapor, and/or dust with which the energy interacts, and the type of scattering that should occur are summarized in Figure 6-12.

Atmospheric Layers and Constituents

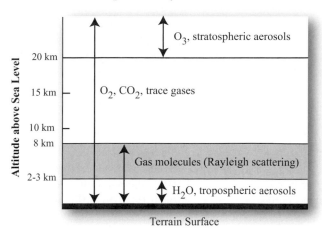

Figure 6-11   Major subdivisions of the atmosphere and the types of molecules and aerosols found in each layer (adapted from Miller and Vermote, 2002).

Atmospheric Scattering

Rayleigh Scattering

a.   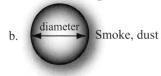   Gas molecule

Mie Scattering

b.   diameter   Smoke, dust

Nonselective Scattering

c.   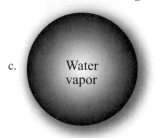   Water vapor

Photon of electromagnetic energy modeled as a wave

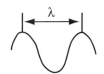

Figure 6-12   Type of scattering is a function of 1) the wavelength of the incident radiant energy and 2) the size of the gas molecule, dust particle, and/or water vapor droplet encountered.

Rayleigh scattering (often referred to as *molecular scattering*) occurs when the effective diameter of the matter (usually air molecules such as oxygen and nitrogen in the atmosphere) are many times smaller (usually < 0.1) than the wavelength of the incident electromagnetic radiation (Figure 6-12a). Rayleigh scattering is named after the English physicist Lord Rayleigh, who offered the first coherent explanation for it (Sagan, 1994). All scattering is accomplished through absorption and re-emission of radiation by atoms or molecules in the manner previously described in the section on radiation from atomic structures. It is impossible to predict the direction in which a specific atom or molecule will emit a photon, hence scattering. The energy required to excite an atom is associated with powerful short-wavelength, high-frequency radiation.

The approximate amount of Rayleigh scattering in the atmosphere in optical wavelengths (0.4 – 0.7 μm) may be computed using the Rayleigh scattering cross-section (τ_m) algorithm (Cracknell and Hayes, 1993):

$$\tau_m = \frac{8\pi^3(n^2-1)^2}{(3N^2\lambda^4)} \qquad (6\text{-}13)$$

where n = refractive index, N = number of air molecules per unit volume, and λ = wavelength. The amount of scattering is inversely related to the fourth power of the radiation's wavelength. For example, ultraviolet light at 0.3 μm is scattered approximately 16 times more than red light at 0.6 μm,

i.e., $(0.6/0.3)^4 = 16$. Blue light at 0.4 μm is scattered about 5 times more than red light at 0.6 μm, i.e., $(0.6/0.4)^4 = 5.06$. The amount of Rayleigh scattering expected throughout the visible spectrum (0.4 – 0.7 μm) is shown in Figure 6-13.

Most Rayleigh scattering by gas molecules takes place in the atmosphere 2 to 8 km above the ground (Figure 6-11). Rayleigh scattering is responsible for the blue appearance of the sky. The shorter violet and blue wavelengths are more efficiently scattered than the longer orange and red wavelengths. When we look up on a cloudless day and admire the blue sky, we are witnessing the preferential scattering of the short-wavelength sunlight. Rayleigh scattering is also responsible for red sunsets. Since the atmosphere is a thin shell of gravitationally bound gas surrounding the solid Earth, sunlight must pass through a longer slant path of air at sunset (or sunrise) than at noon. Since the violet and blue wavelengths are scattered even more during their now longer path through the air than when the Sun is overhead, what we see when we look toward the sunset is the residue—the wavelengths of sunlight that are hardly scattered away at all, especially the oranges and reds (Sagan, 1994).

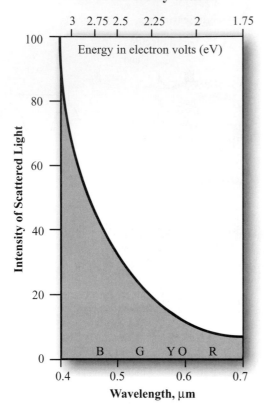

Intensitiy of Rayleigh Scattering Varies Inversely with λ^{-4}

Figure 6-13   The intensity of Rayleigh scattering varies inversely with the fourth power of the wavelength (λ^{-4}).

Mie scattering (sometimes referred to as nonmolecular or aerosol particle scattering) takes place in the lower 4.5 km of the atmosphere, where there may be many essentially spherical particles present with diameters approximately equal to the size of the wavelength of the incident energy (Figure 6-12b). The actual size of the particles may range from 0.1 to 10 times the wavelength of the incident energy. For visible light, the main scattering agents are dust and other particles ranging from a few tenths of a micrometer to several micrometers in diameter. The amount of scatter is greater than Rayleigh scatter, and the wavelengths scattered are longer. The greater the amount of smoke and dust particles in the atmospheric column, the more that violet and blue light will be scattered away and only the longer orange and red wavelength light will reach our eyes. Pollution also contributes to beautiful sunsets and sunrises.

Nonselective scattering takes place in the lowest portions of the atmosphere where there are particles >10 times the wavelength of the incident electromagnetic radiation (Figure 6-

12c). This type of scattering is nonselective, i.e., all wavelengths of light are scattered, not just blue, green, or red. Thus, the water droplets and ice crystals that make up clouds and fog banks scatter all wavelengths of visible light equally well, causing the cloud to appear white. Nonselective scattering of approximately equal proportions of blue, green, and red light always appears as white light to the casual observer. This is the reason why putting our automobile high beams on in fog only makes the problem worse as we nonselectively scatter even more light into our visual field of view.

Scattering is a very important consideration in remote sensing investigations. It can severely reduce the information content of remotely sensed data to the point that the imagery loses contrast and it becomes difficult to differentiate one object from another.

Absorption

Absorption is the process by which radiant energy is absorbed and converted into other forms of energy. The absorption of the incident radiant energy may take place in the atmosphere or on the terrain. An *absorption band* is a range of wavelengths (or frequencies) in the electromagnetic spectrum within which radiant energy is absorbed by a substance. The effects of water (H_2O), carbon dioxide (CO_2), oxygen (O_2), ozone (O_3), and nitrous oxide (N_2O) on the transmission of light through the atmosphere are summarized in Figure 6-14a. The cumulative effect of the absorption by the various constituents can cause the atmosphere to "close down" completely in certain regions of the spectrum. This is bad for remote sensing because no energy is available to be sensed. Conversely, in the visible portion of the spectrum ($0.4 - 0.7 \; \mu m$), the atmosphere does not absorb all of the incident energy but transmits it rather effectively. Portions of the spectrum that transmit radiant energy effectively are called *atmospheric windows*.

Absorption occurs when incident energy of the same frequency as the resonant frequency of an atom or molecule is absorbed, producing an excited state. If instead of reradiating a photon of the same wavelength, the energy is transformed into heat motion and is subsequently reradiated at a longer wavelength, absorption occurs. When dealing with a medium like air, absorption and scattering are frequently combined into an *extinction coefficient* (Konecny, 2003). Transmission is inversely related to the extinction coefficient times the thickness of the layer. Certain wavelengths of radiation are affected far more by absorption than by scattering. This is particularly true of infrared and wavelengths shorter than visible light. The combined effects of atmospheric absorp-

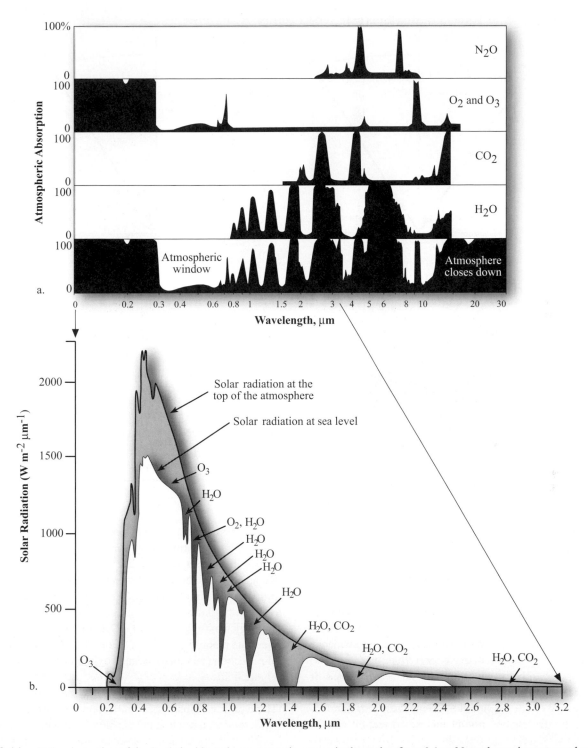

Figure 6-14 a) The absorption of the Sun's incident electromagnetic energy in the region from 0.1 to 30 μm by various atmospheric gases. The first four graphs depict the absorption characteristics of N_2O, O_2 and O_3, CO_2, and H_2O, while the final graphic depicts the cumulative result of all these constituents being in the atmosphere at one time. The atmosphere essentially "closes down" in certain portions of the spectrum while "atmospheric windows" exist in other regions that transmit incident energy effectively to the ground. It is within these windows that remote sensing systems must function. b) The combined effects of atmospheric absorption, scattering, and reflectance reduce the amount of solar irradiance reaching the Earth's surface at sea level (after Slater, 1980).

Specular versus Diffuse Reflectance

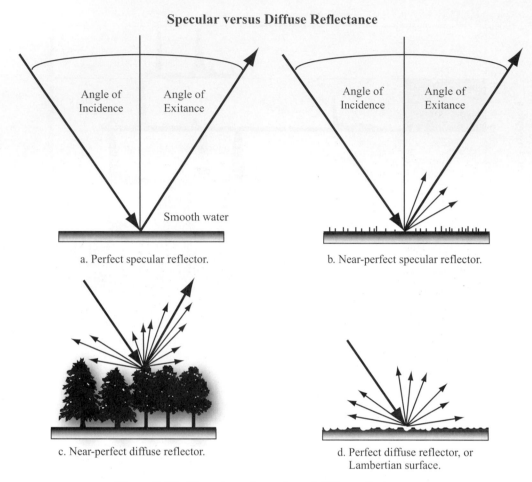

Figure 6-15   The nature of specular and diffuse reflectance.

tion, scattering, and reflectance (from cloud tops) can dramatically reduce the amount of solar radiation reaching the Earth's surface at sea level, as shown in Figure 6-14b.

Energy reaching the remote sensing system must pass through the atmosphere twice. Therefore, it is common to identify two *atmospheric transmission coefficients*: one for energy coming into the atmosphere (T_{θ_o}) at an incidence angle related to the source of the energy (e.g., the Sun) and one for the atmosphere that the Earth surface–reflected or emitted energy must pass through to reach the remote sensor system (T_{θ_v}). We will see how important these two parameters are in the discussion on atmospheric correction later in this chapter.

Reflectance

Reflectance is the process whereby radiation "bounces off" an object like the top of a cloud, a water body, or the terres-

trial Earth. Actually, the process is more complicated, involving reradiation of photons in unison by atoms or molecules in a layer approximately one-half wavelength deep. Reflection exhibits fundamental characteristics that are important in remote sensing. First, the incident radiation, the reflected radiation, and a vertical to the surface from which the angles of incidence and reflection are measured all lie in the same plane. Second, the angle of incidence and the angle of reflection (exitance) are approximately equal, as shown in Figure 6-15.

There are various types of reflecting surfaces. *Specular reflection* occurs when the surface from which the radiation is reflected is essentially smooth (i.e., the average surface-profile height is several times smaller than the wavelength of radiation striking the surface). Several features, such as calm water bodies, act like *near-perfect specular reflectors* (Figure 6-15a,b). If there are very few ripples on the surface, the incident energy will leave the water body at an angle equal and opposite to the incident energy. We know this occurs from our personal experience. If we shine a flashlight at night on a tranquil pool of water, the light will bounce off the surface and into the trees across the way at an angle equal to and opposite from the incident radiation.

If the surface has a large surface height relative to the size of the wavelength of the incident energy, the reflected rays go in many directions, depending on the orientation of the smaller reflecting surfaces. This *diffuse reflection* does not yield a mirror image, but instead produces diffused radiation (Figure 6-15c). White paper, white powders, and other materials reflect visible light in this diffuse manner. If the surface is so rough that there are no individual reflecting surfaces, then unpredictable scattering may occur. Lambert defined a perfectly diffuse surface; hence, the commonly designated *Lambertian surface* is one for which the radiant flux leaving the surface is constant for any angle of reflectance (Figure 6-15d).

A considerable amount of incident radiant flux from the Sun is reflected from the tops of clouds and other materials in the atmosphere. A substantial amount of this energy is reradiated back to space. As we shall see, the specular and diffuse reflection principles that apply to clouds also apply to the terrain.

 ## Terrain Energy–Matter Interactions

The time rate of flow of energy onto, off of, or through a surface is called *radiant flux* (Φ) and is measured in watts (W) (Table 6-4). The characteristics of the radiant flux and what happens to it as it interacts with the Earth's surface is of critical importance in remote sensing. In fact, this is the fundamental focus of much remote sensing research. By carefully monitoring the exact nature of the incoming (incident) radiant flux in selective wavelengths and how it interacts with the terrain, it is possible to learn important information about the terrain.

Radiometric quantities have been identified that allow analysts to keep a careful record of the incident and exiting radiant flux (Table 6-4). We begin with the simple *radiation budget equation,* which states that the total amount of radiant flux in specific wavelengths (λ) incident to the terrain (Φ_{i_λ}) must be accounted for by evaluating the amount of radiant flux reflected from the surface ($\Phi_{\text{reflected}_\lambda}$), the amount of radiant flux absorbed by the surface ($\Phi_{\text{absorbed}_\lambda}$), and the amount of radiant flux transmitted through the surface ($\Phi_{\text{transmitted}_\lambda}$):

$$\Phi_{i_\lambda} = \Phi_{\text{reflected}_\lambda} + \Phi_{\text{absorbed}_\lambda} + \Phi_{\text{transmitted}_\lambda}. \tag{6-14}$$

It is important to note that these radiometric quantities are based on the amount of radiant energy incident to a surface from any angle in a hemisphere (i.e., a half of a sphere).

Hemispherical Reflectance, Absorptance, and Transmittance

Hemispherical reflectance (ρ_λ) is defined as the dimensionless ratio of the radiant flux reflected from a surface to the radiant flux incident to it (Table 6-4):

$$\rho_\lambda = \frac{\Phi_{\text{reflected}_\lambda}}{\Phi_{i_\lambda}}. \tag{6-15}$$

Hemispherical transmittance (τ_λ) is defined as the dimensionless ratio of the radiant flux transmitted through a surface to the radiant flux incident to it:

$$\tau_\lambda = \frac{\Phi_{\text{transmitted}_\lambda}}{\Phi_{i_\lambda}}. \tag{6-16}$$

Hemispherical absorptance (α_λ) is defined by the dimensionless relationship:

$$\alpha_\lambda = \frac{\Phi_{\text{absorbed}_\lambda}}{\Phi_{i_\lambda}} \tag{6-17}$$

or

$$\alpha_\lambda = 1 - (r_\lambda + \tau_\lambda). \tag{6-18}$$

These definitions imply that radiant energy must be conserved, i.e., it is either returned back by reflection, transmitted through a material, or absorbed and transformed into some other form of energy inside the terrain. The net effect of absorption of radiation by most substances is that the energy is converted into heat, causing a subsequent rise in the substance's temperature.

These radiometric quantities are useful for producing general statements about the spectral reflectance, absorptance, and transmittance characteristics of terrain features. In fact, if we take the simple hemispherical reflectance equation and multiply it by 100, we obtain an expression for percent reflectance ($\rho_{\lambda_\%}$):

$$\rho_{\lambda_\%} = \frac{\Phi_{\text{reflected}_\lambda}}{\Phi_{i_\lambda}} \times 100, \tag{6-19}$$

Table 6-4.   Radiometric concepts (Colwell, 1983).

Name	Symbol	Units	Concept
Radiant energy	Q_λ	joules, J	Capacity of radiation within a specified spectral band to do work.
Radiant flux	Φ_λ	watts, W	Time rate of flow of energy onto, off of, or through a surface.
Radiant flux density at the surface			
Irradiance	E_λ	watts per square meter, W m^{-2}	Radiant flux incident upon a surface per unit area of that surface.
Radiant exitance	M_λ	watts per square meter, W m^{-2}	Radiant flux leaving a surface per unit area of that surface.
Radiance	L_λ	watts per square meter, per steradian, W m^{-2} sr^{-1}	Radiant intensity per unit of projected source area in a specified direction.
Hemispherical reflectance	ρ_λ	dimensionless	$\dfrac{\Phi_{\text{reflected}_\lambda}}{\Phi_{i_\lambda}}$
Hemispherical transmittance	τ_λ	dimensionless	$\dfrac{\Phi_{\text{transmitted}_\lambda}}{\Phi_{i_\lambda}}$
Hemispherical absorptance	α_λ	dimensionless	$\dfrac{\Phi_{\text{absorbed}_\lambda}}{\Phi_{i_\lambda}}$

which is often used in remote sensing research to describe the spectral reflectance characteristics of various phenomena. Examples of spectral percent reflectance curves for selected urban–suburban phenomena are shown in Figure 6-16. Spectral reflectance curves provide no information about the absorption and transmittance of the radiant energy. But because many of the sensor systems such as cameras and some multispectral scanners record only reflected energy, this information is still quite valuable and can form the basis for object identification and assessment. For example, it is clear from Figure 6-16 that grass reflects only approximately 15 percent of the incident red radiant energy ($0.6 - 0.7$ μm) while reflecting approximately 50 percent of the incident near-infrared radiant flux ($0.7 - 0.9$ μm). If we wanted to discriminate between grass and artificial turf, the ideal portion of the spectrum to remotely sense in would be the near-infrared region because artificial turf reflects only about 5 percent of the incident near-infrared energy. This would cause a black-and-white infrared image of the terrain to display grass in bright tones and the artificial turf in darker tones.

Hemispherical reflectance, transmittance, and absorptance radiometric quantities do not provide information about the exact amount of energy reaching a specific area on the ground from a specific direction or about the exact amount of radiant flux exiting the ground in a certain direction. Remote sensing systems can be located in space only at a single point in time, and they usually look only at a relatively small portion of the Earth at a single instant. Therefore, it is important to refine our radiometric measurement techniques so that more precise radiometric information can be extracted from the remotely sensed data. This requires the introduction of several radiometric quantities that provide progressively more precise radiometric information.

Radiant Flux Density

A flat area (e.g., 1×1 m in dimension) being bathed in radiant flux (Φ) in specific wavelengths from the Sun is shown in Figure 6-17. The amount of radiant flux intercepted divided

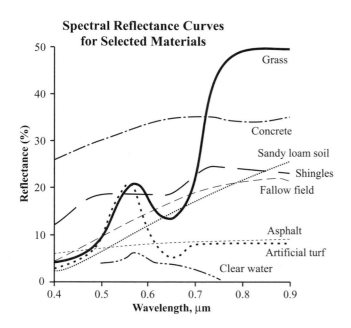

Spectral Reflectance Curves for Selected Materials

Figure 6-16   Typical spectral reflectance curves for urban–suburban phenomena in the region 0.4 – 0.9 µm.

by the area of the plane surface is the average *radiant flux density.*

Irradiance and Exitance

The amount of radiant flux incident upon a surface per unit area of that surface is called *irradiance (E_λ):*

$$E_\lambda = \frac{\Phi_\lambda}{A}. \qquad (6\text{-}20)$$

The amount of radiant flux leaving a surface per unit area of that surface is called *exitance (M_λ):*

$$M_\lambda = \frac{\Phi_\lambda}{A}. \qquad (6\text{-}21)$$

Both quantities are usually measured in watts per meter squared (W m^{-2}). Although we do not have information on the direction of either the incoming or outgoing radiant energy (i.e., the energy can come and go at any angle throughout the entire hemisphere), we have now refined the measurement to include information about the size of the study area of interest on the ground in m^2. Next we need to refine our radiometric measurement techniques to include information on what direction the radiant flux is leaving the study area.

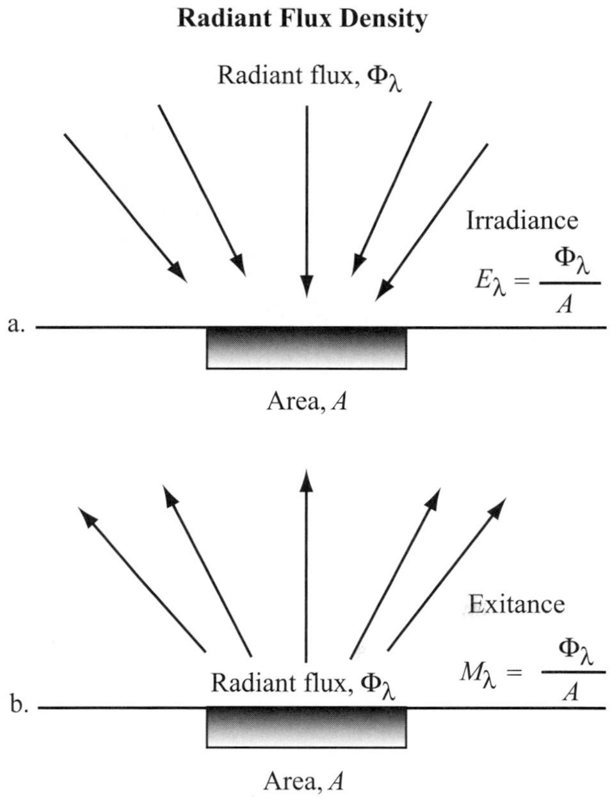

Radiant Flux Density

Radiant flux, Φ_λ

Irradiance

$$E_\lambda = \frac{\Phi_\lambda}{A}$$

a.

Area, A

Exitance

$$M_\lambda = \frac{\Phi_\lambda}{A}$$

Radiant flux, Φ_λ

b.

Area, A

Figure 6-17   The concept of radiant flux density for an area on the surface of the Earth. a) *Irradiance* is a measure of the amount of radiant flux incident upon a surface per unit area of the surface measured in watts m^{-2}. b) *Exitance* is a measure of the amount of radiant flux leaving a surface per unit area of the surface measured in watts m^{-2}.

Radiance

Radiance is the most precise remote sensing radiometric measurement. *Radiance (L_λ)* is the radiant intensity per unit of projected source area in a specified direction. It is measured in watts per meter squared per steradian (W m^{-2} sr^{-1}). The concept of radiance is best understood by evaluating Figure 6-18. First, the radiant flux leaves the projected source area in a specific direction toward the remote sensor. We are not concerned with any other radiant flux that might be leaving the source area in any other direction. We are interested only in the radiant flux in certain wavelengths (Φ_λ) leaving the projected source area (A) within a certain direction (cos θ) and solid angle (Ω) (Milman, 1999):

Radiance

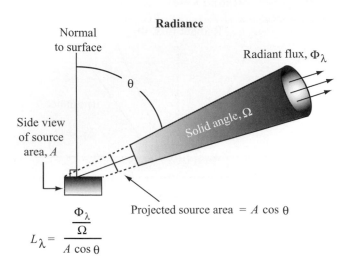

$$L_\lambda = \frac{\dfrac{\Phi_\lambda}{\Omega}}{A\cos\theta}$$

Figure 6-18 The concept of *radiance* leaving a specific projected source area on the ground, in a specific direction, and within a specific solid angle.

$$L_\lambda = \frac{\dfrac{\Phi_\lambda}{\Omega}}{A\cos\theta}. \qquad (6\text{-}22)$$

One way of visualizing the solid angle is to consider what you would see if you were in an airplane looking through a telescope at the ground. Only the energy that exited the terrain and came up to and through the telescope in a specific solid angle (measured in steradians) would be intercepted by the telescope and viewed by your eye. Therefore, the solid angle is like a three-dimensional cone (or tube) that funnels radiant flux from a specific point source on the terrain toward the sensor system. Hopefully, energy from the atmosphere or other terrain features does not become scattered into the solid angle field of view and contaminate the radiant flux from the area of interest on the ground. Unfortunately, this is not often the case because scattering in the atmosphere and from other nearby areas on the ground can contribute spurious spectral energy, which enters into the solid angle field of view.

 Energy–Matter Interactions in the Atmosphere Once Again

The radiant flux reflected or emitted from the Earth's surface once again enters the atmosphere, where it interacts with the various gases, water vapor, and particulates. Thus, atmo

spheric scattering, absorption, reflection, and refraction influence the radiant flux once again before the energy is recorded by the remote sensing system.

 Energy–Matter Interactions at the Sensor System

Finally, energy–matter interactions take place when the energy reaches the remote sensor. If an aerial camera is being used, then the radiance will interact with the camera filter, the optical glass lens, and finally the film emulsion with its light-sensitive silver halide crystals. The emulsion must then be developed and printed before an analog copy is available for analysis. Rather than storing a latent image on film, an optical–mechanical detector will digitally record the number of photons in very specific wavelength regions reaching the sensor.

 Correcting Remote Sensing System Detector Error

Ideally, the radiance recorded by a remote sensing system in various bands is an accurate representation of the radiance actually leaving the feature of interest (e.g., soil, vegetation, water, or urban land cover) on the Earth's surface. Unfortunately, noise (error) can enter the data-collection system at several points. For example, radiometric error in remotely sensed data may be introduced by the sensor system itself when the individual detectors do not function properly or are improperly calibrated (Teillet, 1986). Several of the more common remote sensing system–induced radiometric errors are:

• random bad pixels (shot noise),

• line-start/stop problems,

• line or column drop-outs,

• partial line or column drop-outs, and

• line or column striping.

Sometimes digital image processing can recover the miscalibrated spectral information and make it relatively compatible with the correctly acquired data in the scene. Unfortunately, sometimes only cosmetic adjustments can be made to compensate for the fact that no data of value were acquired.

Random Bad Pixels (Shot Noise)

Sometimes an individual detector does not record spectral data for an individual pixel. When this occurs randomly, it is called a *bad pixel*. When there are numerous random bad pixels found within the scene, it is called *shot noise* because it appears that the image was shot by a shotgun. Normally these bad pixels contain values of 0 or 255 (in 8-bit data) in one or more of the bands. Shot noise is identified and repaired using the following methodology.

It is first necessary to locate each bad pixel in the band k dataset. A simple thresholding algorithm makes a pass through the dataset and flags any pixel ($BV_{i,j,k}$) having a brightness value of zero (assuming values of 0 represent shot noise and not a real land cover such as water). Once identified, it is then possible to evaluate the eight pixels surrounding the flagged pixel, as shown below:

	col$_{j-1}$	col$_j$	col$_{j+1}$
row$_{i-1}$	BV_1	BV_2	BV_3
row$_i$	BV_8	BV_{ijk}	BV_4
row$_{i+1}$	BV_7	BV_6	BV_5

The mean of the eight surrounding pixels is computed using Equation 6-23 and the value substituted for $BV_{i,j,k}$ in the corrected image:

$$BV_{i,j,k} = \text{Int}\left(\frac{\sum_{i=1}^{8} BV_i}{8}\right). \qquad (6\text{-}23)$$

This operation is performed for every bad (shot noise) pixel in the dataset.

For example, Landsat Thematic Mapper band 7 imagery (2.08 – 2.35 μm) of the Santee Delta is shown in Figure 6-19. It contains two pixels along a bad scan line with values of zero. The eight brightness values surrounding each bad pixel are enlarged and annotated in Figure 6-19b. The Landsat TM band 7 image after shot noise removal is shown in Figure 6-19c.

Line or Column Drop-outs

An entire line containing no spectral information may be produced if an individual detector in a scanning system (e.g., Landsat MSS or Landsat 7 ETM$^+$) fails to function properly. If a detector in a linear array (e.g., SPOT XS, IRS-1C, Quick-Bird) fails to function, this can result in an entire column of data with no spectral information. The bad line or column is commonly called a *line* or *column drop-out* and contains brightness values equal to zero. For example, if one of the 16 detectors in the Landsat Thematic Mapper sensor system fails to function during scanning, this can result in a brightness value of zero for every pixel, j, in a particular line, i. This *line drop-out* would appear as a completely black line in the band, k, of imagery. This is a serious condition because there is no way to restore data that were never acquired. However, it is possible to improve the visual interpretability of the data by introducing estimated brightness values for each bad scan line.

It is first necessary to locate each bad line in the dataset. A simple thresholding algorithm makes a pass through the dataset and flags any scan line having a mean brightness value at or near zero. Once identified, it is then possible to evaluate the output for a pixel in the preceding line ($BV_{i-1,j,k}$) and succeeding line ($BV_{i+1,j,k}$) and assign the output pixel ($BV_{i,j,k}$) in the drop-out line the average of these two brightness values:

$$BV_{i,j,k} = \text{Int}\left(\frac{BV_{i-1,j,k} + BV_{i+1,j,k}}{2}\right). \qquad (6\text{-}24)$$

This is performed for every pixel in a bad scan line. The result is an image consisting of interpolated data every nth line that is more visually interpretable than one with horizontal black lines running systematically throughout the entire image. This same cosmetic digital image processing procedure can be applied to *column drop-outs* produced by a linear array remote sensing system.

Partial Line or Column Drop-outs

Occasionally an individual detector will function perfectly along a scan line and then for some unknown reason it will not function properly for n columns. Then sometimes the detector functions properly again for the remainder of the scan line. The result is a portion of a scan line with no data. This is commonly referred to as a *partial line* or *partial column drop-out problem*. This is a serious condition. It usually cannot be dealt with systematically because it occurs ran-

a. Landsat TM band 7 data of the Santee Delta with shot noise.

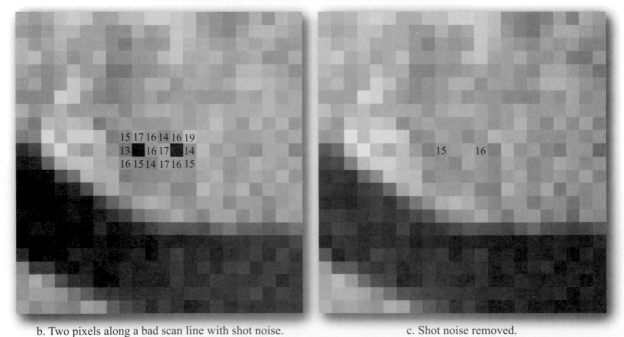

b. Two pixels along a bad scan line with shot noise.   c. Shot noise removed.

Figure 6-19   a) Landsat Thematic Mapper band 7 (2.08 – 2.35 μm) image of the Santee Delta in South Carolina. One of the 16 detectors exhibits serious striping and the absence of brightness values at pixel locations along a scan line. b) An enlarged view of the bad pixels with the brightness values of the eight surrounding pixels annotated. c) The brightness values of the bad pixels after shot noise removal. This image was not destriped.

Line-start Problems

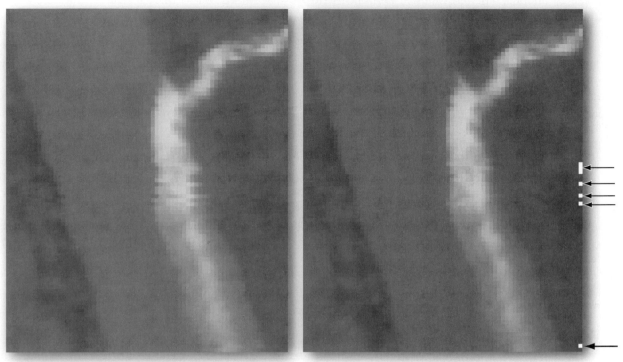

a. Predawn thermal infrared imagery of the
Savannah River with line-start problems.

b. Seven line-start problem lines were
translated one column to the left.

Figure 6-20   a) Infrared imagery of the Four Mile Creek thermal effluent plume entering the Savannah River on March 31, 1981. Seven lines with line-start problems are present. b) The result of translating each of the line-start problem lines one column to the left. The line-start problem was systematic in that offending lines only needed to be adjusted by one column. The line-start problem was unsystematic because not all of the lines in the dataset exhibited the line-start problem. Therefore, it was necessary to manually correct (using digital image processing) each of the line-start problem lines. A few additional corrections could still be made.

domly. If the portion of the image with the drop-out problem is particularly important, then the analyst must manually correct the problem pixel by pixel. The analyst must go into the dataset and compute the mean of the brightness values above and below each bad pixel and place the average in the bad pixel location. This is done for every pixel in the bad portion of the scan line.

Line-start Problems

Occasionally, scanning systems fail to collect data at the beginning of a scan line, or they place the pixel data at inappropriate locations along the scan line. For example, all of the pixels in a scan line might be systematically shifted just one pixel to the right. This is called a *line-start* problem. Also, a detector may abruptly stop collecting data somewhere along a scan and produce results similar to the line or

column drop-out previously discussed. Ideally, when data are not collected, the sensor system would be programmed to remember what was not collected and place any good data in their proper geometric locations along the scan. Unfortunately, this is not always the case. For example, the first pixel (column 1) in band k on line i (i.e., $BV_{i,1,k}$) might be improperly located at column 50 (i.e., $BV_{i,50,k}$). If the line-start problem is always associated with a horizontal bias of 50 columns, it can be corrected using a simple horizontal adjustment. However, if the amount of the line-start displacement is random, it is difficult to restore the data without extensive human interaction on a line-by-line basis. A considerable amount of MSS data collected by Landsats 2 and 3 exhibit line-start problems.

Figure 6-20a depicts predawn thermal infrared remote sensor data of the Savannah River obtained on March 31, 1981. It contains several line-start problems that are visually apparent. Seven lines with line-start problems (which contain good radiant energy information) were adjusted horizontally (to the left) just one pixel to their proper position (Figure 6-20b). Accurate linear measurements can now be made from the adjusted data, plus the image is much more visually appealing.

N-line Striping

Sometimes a detector does not fail completely, but simply goes out of radiometric adjustment. For example, a detector might record spectral measurements over a dark, deep body of water that are almost uniformly 20 brightness values greater than the other detectors for the same band. The result would be an image with systematic, noticeable lines that are brighter than adjacent lines. This is referred to as *n-line striping*. The maladjusted line contains valuable information, but should be corrected to have approximately the same radiometric scale as the data collected by the properly calibrated detectors associated with the same band.

To repair systematic *n*-line striping, it is first necessary to identify the miscalibrated scan lines in the scene. This is usually accomplished by computing a histogram of the values for each of the *n* detectors that collected data over the entire scene (ideally, this would take place over a homogeneous area, such as a body of water). If one detector's mean or median is significantly different from the others, it is probable that this detector is out of adjustment. Consequently, every line and pixel in the scene recorded by the maladjusted detector may require a *bias* (additive or subtractive) correction or a more severe *gain* (multiplicative) correction. This type of *n*-line striping correction a) adjusts all the bad scan lines so that they have approximately the same radiometric scale as the correctly collected data and b) improves the visual interpretability of the data. It looks better.

For example, Figure 6-21a presents Landsat TM band 3 (0.63 – 0.69 µm) data of the Santee Delta in South Carolina obtained in 1984. The area consists primarily of salt marsh. The image exhibits serious striping every 16 lines. A pass was made through the dataset to identify the maladjusted lines. A gain and a bias were computed and applied to each affected line. Figure 6-21b depicts the same band 3 data after destriping. The striping is noticeably subdued, although close inspection reveals some residual striping. It is usually difficult to remove all traces of striping. Interestingly, the maladjusted lines in the Santee TM band 3 image (Figure 6-21) are not the same as those encountered in the Santee TM 7 image used to demonstrate shot noise removal (Figure 6-19).

A 35-band hyperspectral 2 × 2 m spatial resolution dataset of the Mixed Waste Management Facility on the Savannah River Site in Aiken, SC, was acquired on July 31, 2002 (Figure 6-22a). The level, clay-capped hazardous waste site fields are covered with Bahia grass or Centipede grass. A horizontal spectral profile through band 5 (red; centered at 633 nm) reveals good radiometric integrity along the columns of the dataset (Figure 6-23a). Unfortunately, a vertical spectral profile through the waste site exhibits a diagnostic sawtooth pattern indicative of striping (Figure 6-23b). Using the previously described methodology, a gain and a bias were identified and applied to each offending fourth line in the dataset. The result is the radiometrically adjusted band 5 image shown in Figure 6-22c,d. Examination of a vertical spectral profile through the corrected (destriped) dataset reveals that the sawtooth striping was greatly minimized but not completely removed (Figure 6-23c). The visual appearance of the band 5 data is noticeably improved.

 Remote Sensing Atmospheric Correction

Even when the remote sensing system is functioning properly, radiometric error may be introduced into the remote sensor data. The two most important sources of environmental attenuation are 1) atmosphere attenuation caused by scattering and absorption in the atmosphere and 2) topographic attenuation. However, it is important to first consider that it may not be necessary to atmospherically correct the remote sensor data for all applications. The decision to perform an atmospheric correction is a function of the nature of the problem, the type of remote sensing data available, the amount of *in situ* historical and/or concurrent atmospheric information available, and how accurate the biophysical information to be extracted from the remote sensing data must be.

Unnecessary Atmospheric Correction

Sometimes it is possible to ignore atmospheric effects in remote sensor data completely (Cracknell and Hayes, 1993; Song et al., 2001). For example, atmospheric correction is not always necessary for certain types of classification and change detection. Theoretical analysis and empirical results indicate that only when training data from one time or place must be extended through space and/or time is atmospheric correction necessary for image classification and many types of change detection (Song et al., 2001). For example, it is not generally necessary to perform atmospheric correction on a single date of remotely sensed data that will be classified using a maximum likelihood classification algorithm (dis-

Striping

a. Landsat TM band 3 data of the Santee Delta.

b. Landsat TM band 3 data destriped.

Figure 6-21 a) Landsat Thematic Mapper band 3 (0.63 – 0.69 μm) image of the Santee Delta in South Carolina. The image exhibits serious striping. b) The TM band 3 image after destriping.

cussed in Chapter 9). As long as the training data from the image to be classified have the same relative scale (corrected or uncorrected), atmospheric correction has little effect on classification accuracy (Kawata et al., 1990; Song et al., 2001).

For example, consider land-cover classification using a single date of Landsat Thematic Mapper data. Rayleigh and other types of scattering normally add brightness to the visible bands (400 – 700 nm). Atmospheric absorption is the major factor reducing the brightness values of pixels in the near- and middle-infrared region (700 – 2400 nm). Fortunately, the Landsat TM near- and middle-infrared bandwidths were carefully chosen to minimize the effects of atmospheric absorption. Therefore, if a single date of Landsat TM data is atmospherically corrected, it is likely that the primary effect will be a simple bias adjustment applied separately to each band. This action would adjust the minimum

and maximum values of each band downward. Training class means extracted from the single-date image would change but the training class variance–covariance matrices should remain invariant. Therefore, the actual information content used in the maximum likelihood classification of the dataset would remain unchanged.

This logic also applies to certain types of change detection. For example, consider a change detection study involving two dates of imagery. If the two dates of imagery are analyzed independently using a maximum likelihood classification algorithm and the resultant classification maps are compared using post-classification change detection logic (discussed in Chapter 11), there is no need to atmospherically correct the individual dates of remote sensor data (Singh, 1989; Foody et al., 1996). Similarly, atmospheric correction is not necessary when performing multidate composite image change detection (Jensen et al., 1993) where the

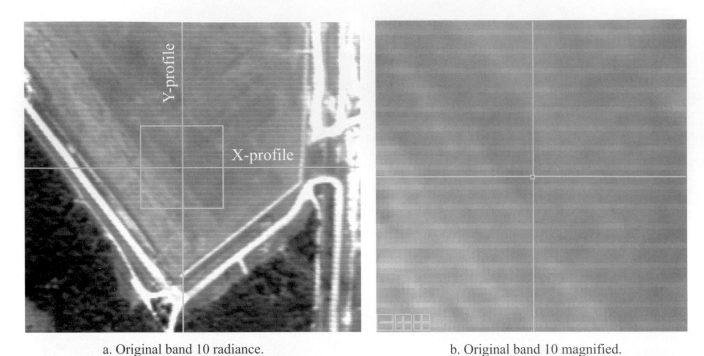

a. Original band 10 radiance.   b. Original band 10 magnified.

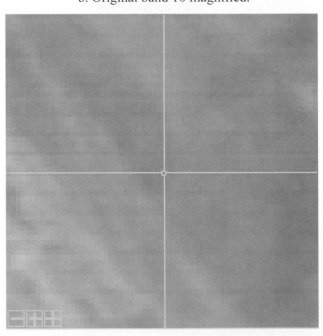

c. Destriped band 10 radiance.   d. Destriped band 10 magnified.

Figure 6-22   a) Original band 10 radiance (W m^{-2} sr^{-1}) data from a GER DAIS 3715 hyperspectral dataset of the Mixed Waste Management Facility on the Savannah River Site near Aiken, SC. The subset is focused on a clay-capped hazardous waste site covered with Bahia grass and Centipede grass. The 35-band dataset was obtained at 2×2 m spatial resolution. The radiance values along the horizontal (X) and vertical (Y) profiles are summarized in Figure 6-23. b) An enlargement of the band 10 data. c) The band 10 data after destriping. d) An enlargement of the destriped data (Jensen et al., 2003).

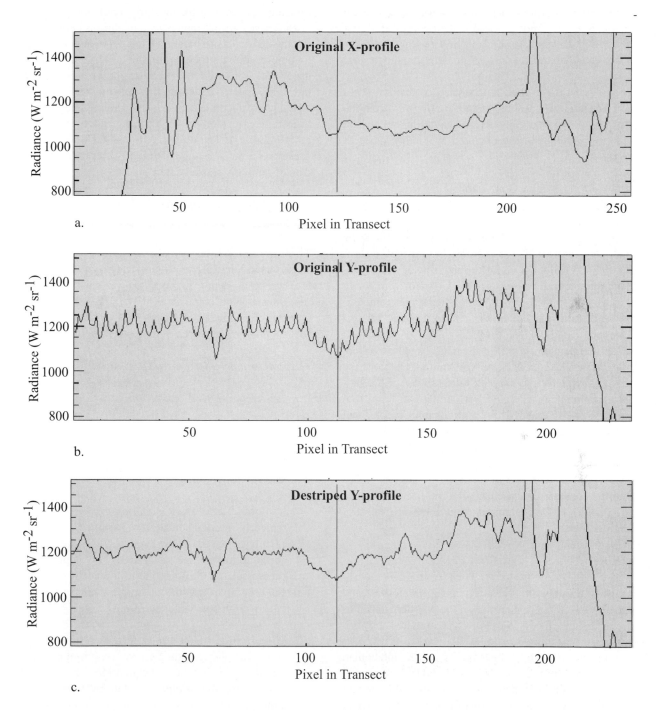

Figure 6-23   a) The radiance values along the horizontal (X) profile of the original band 10 radiance values in Figure 6-22. b) The radiance values along the vertical (Y) profile of the original band 10 radiance values in Figure 6-22. c) The radiance values along the vertical (Y) profile of the destriped band 10 radiance values. Note the reduction of the saw-toothed pattern in the destriped data (Jensen et al., 2003).

change detection algorithm identifies change classes based on an analysis of all bands on both dates being placed in a single dataset (e.g., four TM bands from date 1 and four TM bands from date 2 are placed in a single eight-band dataset).

The general principle is that atmospheric correction is not necessary as long as the training data are *extracted* from the image (or composite image) under investigation and are not imported from another image obtained at another place or time.

Necessary Atmospheric Correction

Sometimes it is essential that the remotely sensed data be atmospherically corrected. For example, it is usually necessary to atmospherically correct the remote sensor data if biophysical parameters are going to be extracted from water bodies (e.g., chlorophyll *a*, suspended sediment, temperature) or vegetation (e.g., biomass, leaf area index, chlorophyll, percent canopy closure) (Haboudane et al., 2002; Thiemann and Hermann, 2002). If the data are not corrected, the subtle differences in reflectance (or emittance) among the important constituents may be lost. Furthermore, if the biophysical measurements extracted from one image (e.g., biomass) are to be compared with the same biophysical information extracted from other images obtained on different dates, then it is usually essential that the remote sensor data be atmospherically corrected.

For example, consider the case of the normalized difference vegetation index (NDVI) derived from Landsat Thematic Mapper (TM) data:

$$NDVI = \frac{\rho_{tm4} - \rho_{tm3}}{\rho_{tm4} + \rho_{tm3}},$$

which is used routinely to measure vegetation biomass and functional health in many decision-support systems such as the Africa Famine Early Warning System and Livestock Early Warning System (Jensen et al., 2002). Erroneous NDVI estimates can result in the loss of livestock and human life. Contributions from the atmosphere to NDVI are significant and can amount to 50 percent or more over thin or broken vegetation cover. Therefore, there is significant interest in removing the deleterious effects of the atmosphere in remotely sensed data that are used to compute NDVI estimates. The simple ratio (SR) vegetation index (TM4/TM3 for Landsat TM data) is also contaminated by the atmosphere (Song et al., 2001).

Great emphasis is being placed on the development of algorithms that can extract biophysical information from remotely sensed data for local, regional, and global applications. These data are then placed in various deterministic and stochastic models (Chapter 1) and decision-support systems to monitor global processes and hopefully improve the quality of life. Consequently, great emphasis is being placed on the ability to derive measurements from the remotely sensed data that are accurate across vast distances and through time. Signature extension through space and time is becoming more important. The only way to extend signatures through space and time is to atmospherically correct each individual date of remotely sensed data.

Types of Atmospheric Correction

There are several ways to atmospherically correct remotely sensed data. Some are relatively straightforward while others are complex, being founded on physical principles and requiring a significant amount of information to function properly (Cracknell and Hayes, 1993). This discussion will focus on two major types of atmospheric correction:

- *Absolute atmospheric correction*, and

- *Relative atmospheric correction*.

There are various methods that can be used to achieve absolute or relative atmospheric correction. The following sections identify the logic, algorithms, and problems associated with each methodology.

1. It is possible to use a *model atmosphere* to correct the remotely sensed data. An assumed atmosphere is calculated based on the time of year, altitude, latitude, and longitude of the study area. This approach may be successful when atmospheric attenuation is relatively small compared with the signal from the terrain being remotely sensed (Cracknell and Hayes, 1993).

2. The use of a model atmosphere in conjunction with *in situ* atmospheric measurements acquired at the time of remote sensor data acquisition is even better. Sometimes, the *in situ* data can be provided by other atmospheric sounding instruments found onboard the sensor platform. The atmospheric model may then be fine-tuned using the local condition information. This is referred to as *absolute radiometric correction,* and an example is provided later in the chapter.

3. Minimization of atmospheric attenuation is sometimes possible using multiple looks at the same object from different vantage points (e.g., fore and aft) or by looking

at the same object using multiple bands of the spectrum. The goal is to try to have the information from the multiple looks or multiple bands cancel out the atmospheric effects. The multiple-look method suffers from the fact that the atmospheric paths for the multiple looks (e.g., fore and aft) may not be the same. Theoretically, the band-cancellation method should be capable of providing good results because it is using identical atmospheric paths for the channels that are being compared. This is called *relative radiometric correction,* and an image normalization example will be provided.

Absolute Radiometric Correction of Atmospheric Attenuation

Solar radiation is largely unaffected as it travels through the vacuum of space. When it interacts with the Earth's atmosphere, however, it is selectively scattered and absorbed. The sum of these two forms of energy loss is called *atmospheric attenuation.* Serious atmospheric attenuation may 1) make it difficult to relate hand-held *in situ* spectroradiometer measurements with remote measurements, 2) make it difficult to extend spectral signatures through space and time, and (3) have an impact on classification accuracy within a scene if atmospheric attenuation varies significantly throughout the image (Kaufman and Fraser, 1984; Cracknell and Haynes, 1993; Kaufman et al., 1997).

The general goal of *absolute radiometric correction* is to turn the digital brightness values recorded by a remote sensing system into *scaled surface reflectance* values (Du et al., 2002). These values can then be compared or used in conjunction with scaled surface reflectance values obtained anywhere else on the planet.

A considerable amount of research has been carried out to address the problem of correcting images for atmospheric effects (Du et al., 2002). These efforts have resulted in a number of atmospheric radiative transfer codes (models) that can provide realistic estimates of the effects of atmospheric scattering and absorption on satellite imagery (e.g., Alder-Golden et al., 1999; Matthew et al., 2000; Vermote et al., 1997, 2002). Once these effects have been identified for a specific date of imagery, each band and/or pixel in the scene can be adjusted to remove the effects of scattering and/or absorption. The image is then considered to be atmospherically corrected.

Unfortunately, the application of these codes to a specific scene and date also requires knowledge of both the sensor spectral profile and the atmospheric properties at the same time. Atmospheric properties are difficult to acquire even when planned. For most historic satellite data, they are not available. Even today, accurate scaled surface reflectance retrieval is not operational for the majority of satellite image sources used for land-cover change detection (Du et al., 2002). An exception is NASA's Moderate Resolution Imaging Spectroradiometer (MODIS), for which surface reflectance products are available (Justice et al., 1998). Nevertheless, we will proceed with a general discussion of the important issues associated with absolute atmospheric correction and then provide examples of how absolute radiometric correction is performed.

Target and Path Radiance

Ideally, the radiance (L) recorded by the camera or detector is a true function of the amount of radiance leaving the target terrain within the instantaneous-field-of-view (IFOV) at a specific solid angle, as previously discussed. Unfortunately, other radiant energy may enter into the field of view from various other paths and introduce confounding noise into the remote sensing process. Therefore, additional radiometric variable definitions are needed to identify the major sources and paths of this energy (Forster, 1984; Green, 2003). The variables are summarized in Table 6-5. The various paths and factors that determine the radiance reaching the remote sensor are summarized in Figure 6-24, including:

- *Path 1* contains spectral solar irradiance (E_{o_λ}) that was attenuated very little before illuminating the terrain within the IFOV. Notice in this case that we are interested in the solar irradiance from a specific solar zenith angle (θ_o) and that the amount of irradiance reaching the terrain is a function of the atmospheric transmittance at this angle (T_{θ_o}). If all of the irradiance makes it to the ground, then the atmospheric transmittance (T_{θ_o}) equals one. If none of the irradiance makes it to the ground, then the atmospheric transmittance is zero.

- *Path 2* contains spectral diffuse sky irradiance (E_{d_λ}) that never even reaches the Earth's surface (the target study area) because of scattering in the atmosphere. Unfortunately, such energy is often scattered directly into the IFOV of the sensor system. As previously discussed, Rayleigh scattering of blue light contributes much to this diffuse sky irradiance. That is why the blue band image produced by a remote sensor system is often much brighter than any of the other bands. It contains much unwanted diffuse sky irradiance that was inadvertently scattered into the IFOV of the sensor system. Therefore, if possible, we want to minimize its effects. Green (2003) refers to the

Table 6-5.   Radiometric variables used in remote sensing.

Radiometric Variables

E_o = solar irradiance at the top of the atmosphere (W m^{-2})

E_{o_λ} = spectral solar irradiance at the top of the atmosphere (W m^{-2} μm^{-1})

E_d = diffuse sky irradiance (W m^{-2})

E_{d_λ} = spectral diffuse sky irradiance (W m^{-2} μm^{-1})

E_{du_λ} = the *upward* reflectance of the atmosphere

E_{dd_λ} = the *downward* reflectance of the atmosphere

E_g = global irradiance incident on the surface (W m^{-2})

E_{g_λ} = spectral global irradiance on the surface (W m^{-2} μm^{-1})

τ = normal atmospheric optical thickness

T_θ = atmospheric transmittance at an angle θ to the zenith

θ_o = solar zenith angle

θ_v = view angle of the satellite sensor (or scan angle)

μ = cos θ

ρ_λ = surface target reflectance at a specific wavelength

ρ_{λ_n} = reflectance from a neighboring area

L_s = total radiance at the sensor (W m^{-2} sr^{-1})

L_t = total radiance from the target of interest toward the sensor (W m^{-2} sr^{-1})

L_i = intrinsic radiance of the target (W m^{-2} sr^{-1}) (i.e., what a handheld radiometer would record on the ground without intervening atmosphere)

L_p = path radiance from multiple scattering (W m^{-2} sr^{-1})

quantity as the upward reflectance of the atmosphere (E_{du_λ}).

- *Path 3* contains energy from the Sun that has undergone some Rayleigh, Mie, and/or nonselective scattering and perhaps some absorption and reemission before illuminating the study area. Thus, its spectral composition and polarization may be somewhat different from the energy that reaches the ground from path 1. Green (2003) refers to this quantity as the downward reflectance of the atmosphere (E_{dd_λ}).

- *Path 4* contains radiation that was reflected or scattered by nearby terrain (ρ_{λ_n}) covered by snow, concrete, soil,

water, and/or vegetation into the IFOV of the sensor system. The energy does not actually illuminate the study area of interest. Therefore, if possible, we would like to minimize its effects.

- *Path 5* is energy that was also reflected from nearby terrain into the atmosphere, but then scattered or reflected onto the study area.

Therefore, for a given spectral interval in the electromagnetic spectrum (e.g., λ_1 to λ_2 could be 0.6 – 0.7 μm or red light), the total solar irradiance reaching the *Earth's surface, E_{g_λ}*, is an integration of several components:

$$E_{g_\lambda} = \int_{\lambda_1}^{\lambda_2} (E_{o_\lambda} T_{\theta_o} \cos\theta_o + E_{d_\lambda}) d\lambda \quad (\text{W m}^{-2} \text{μm}^{-1}). \ (6\text{-}25)$$

It is a function of the spectral solar irradiance at the top of the atmosphere (E_{o_λ}) multiplied by the atmospheric transmittance (T_{θ_o}) at a certain solar zenith angle (θ_o) plus the contribution of spectral diffuse sky irradiance (E_{d_λ}).

Only a small amount of this irradiance is actually reflected by the terrain in the direction of the satellite sensor system. If we assume the surface of Earth is a diffuse reflector (a Lambertian surface), the total amount of radiance exiting the target study area (L_T) toward the sensor is:

$$L_T = \frac{1}{\pi} \int_{\lambda_1}^{\lambda_2} \rho_\lambda \, T_{\theta_v} (E_{o_\lambda} T_{\theta_o} \cos\theta_o + E_{d_\lambda}) d\lambda . \quad (6\text{-}26)$$

The average surface target reflectance (ρ_λ) is included because the vegetation, soil, and water within the IFOV selectively absorb some of the incident energy. Therefore, not all of the energy incident to the IFOV (E_{g_λ}) leaves the IFOV. In effect, the terrain acts like a filter, selectively absorbing certain wavelengths of light while reflecting others. Note that the energy exiting the terrain is at an angle (θ_v), requiring the use of an atmospheric transmittance factor T_{θ_v} once again.

It would be wonderful if the total radiance recorded by the sensor, L_S, equaled the radiance returned from the target study area of interest, L_T. Unfortunately, $L_S \neq L_T$ because there is some additional radiance from different *paths* that may fall within the IFOV of the sensor system detector (Figure 6-24). This is often called *path radiance*, L_P. Thus, the total radiance recorded by the sensor becomes:

$$L_S = L_T + L_P \quad (\text{W m}^{-2} \text{sr}^{-1}). \quad (6\text{-}27)$$

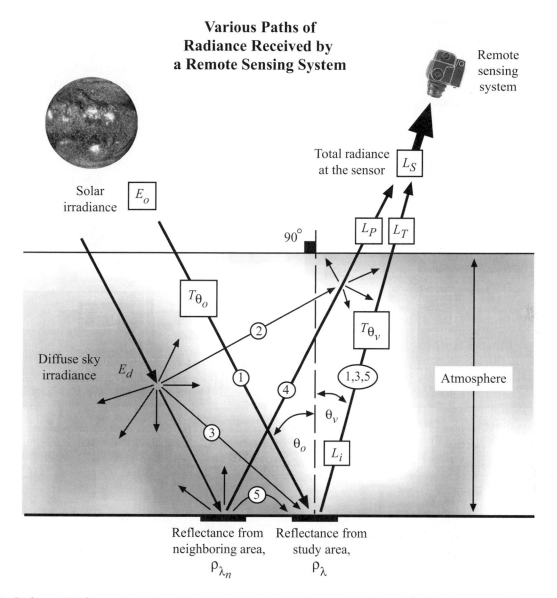

Figure 6-24 *Radiance* (L_T) from paths 1, 3, and 5 contains intrinsic valuable spectral information about the target of interest. Conversely, the *path radiance* (L_p) from paths 2 and 4 includes diffuse sky irradiance or radiance from neighboring areas on the ground. This path radiance generally introduces unwanted radiometric noise in the remotely sensed data and complicates the image interpretation process.

We see from Equation 6-27 and Figure 6-24 that the path radiance (L_P) is an intrusive (bad) component of the total amount of radiance recorded by the sensor system (L_S). It is composed of radiant energy primarily from the diffuse sky irradiance (E_d) from path 2 as well as the reflectance from nearby ground areas ρ_{λ_n} from path 4. Path radiance introduces error to the remote sensing data-collection process. It can impede our ability to obtain accurate spectral measurements.

A great deal of research has gone into developing methods to remove the contribution of path radiance (L_P). Methods for computing path radiance are summarized in Turner and Spencer (1972), Turner (1975, 1978), Forster (1984), and Richards and Jia (1999). Radiative transfer model programs such as MODTRAN, Second Simulation of the Satellite Signal in the Solar Spectrum (6S) and others may be used to predict path radiance on a particular day for a particular study area (e.g., Ontar, 1999; Alder-Golden et al., 1999; Matthew et al., 2000). Such information can then be used to remove the path radiance (L_p) contribution to the remote sensing signal (L_S).

Atmospheric Transmittance

To understand how to remove atmospheric attenuation, it is useful to review the fundamental mechanisms of atmospheric scattering and absorption related to atmospheric transmittance. In the absence of atmosphere, the transmittance of solar radiant energy to the ground would be 100%. However, because of absorption and scattering, not all of the radiant energy reaches the ground. The amount that does reach the ground, relative to that for no atmosphere, is called *transmittance.* Atmospheric transmittance (T_θ) may be computed as:

$$T_\theta = e^{-\tau/\cos\theta} \tag{6-28}$$

where τ is the normal atmospheric optical thickness and θ can represent either θ_o or θ_v in Figure 6-24 (i.e., the ability of the atmosphere to transmit radiant flux from the Sun to the target, T_{θ_o}, or the ability of the atmosphere to transmit radiant flux from the target to the sensor system, T_{θ_v}). The optical thickness of the atmosphere at certain wavelengths, $\tau(\lambda)$, equals the sum of all the attenuating coefficients, which are made up primarily of Rayleigh scattering, τ_m, Mie scattering, τ_p, and selective atmospheric absorption, τ_a:

$$\tau(\lambda) = \tau_m + \tau_p + \tau_a \tag{6-29}$$

where $\tau_a = \tau_{H_2O} + \tau_{O_2} + \tau_{O_3} + \tau_{CO_2}$.

As previously discussed, Rayleigh scattering of gas molecules occurs when the diameter of the gas molecule (d) is less than the size of the incident wavelength ($d < \lambda$) and is inversely proportional to the fourth power of the wavelength, $1/\lambda^4$. Aerosol (Mie) scattering occurs when $d = \lambda$ and is primarily a function of water vapor, dust, and other aerosol particles in the atmosphere. Selective absorption of radiant energy in the atmosphere is wavelength-dependent. Most of the optical region from 0.4 to 1.0 μm is dominated by water and ozone absorption attenuation. Atmospheric absorption by water vapor and other gases in the atmosphere mostly affects radiation of wavelengths > 0.8 μm. Thus, atmospheric scattering may add brightness to, whereas atmospheric absorption may subtract brightness from, landscape spectral measurements.

Diffuse Sky Irradiance

Path 1 in Figure 6-24 contains radiation that followed a direct path from the Sun, to the target study area within the IFOV, to the remote sensing detector. Unfortunately, some additional scene irradiance comes from scattered skylight. Path 3 in Figure 6-24 contains radiation that has undergone some scattering before illuminating the study area. Thus, its spectral composition may be somewhat different. Similarly, path 5 contains energy that was reflected from a nearby area on the ground into the atmosphere and then scattered once again onto the study area. The total diffuse sky irradiance at the pixel is E_d.

It is now possible to determine how these atmospheric effects (transmittance, diffuse sky irradiance, and path radiance) affect the radiance measured by the remote sensing system. First, however, because the bandwidths used in remote sensing are relatively narrow (e.g., 0.5 to 0.6 μm), it is possible to restate Equations 6-25 and 6-26 without the integral. For example, the total irradiance at the Earth's surface may be written:

$$E_g = E_{o\Delta\lambda} T_{\theta_o} \cos\theta_o \Delta\lambda + E_d, \tag{6-30}$$

where $E_{o\Delta\lambda}$ is the average spectral irradiance in the band interval $\Delta\lambda = \lambda_2 - \lambda_1$. The total radiance transmitted through the atmosphere toward the sensor (L_T) becomes:

$$L_T = \frac{1}{\pi}\rho T_{\theta_v}(E_{o\Delta\lambda} T_{\theta_o} \cos\theta_o \Delta\lambda + E_d). \tag{6-31}$$

The total radiance reaching the sensor then becomes:

$$L_S = \frac{1}{\pi}\rho T_{\theta_v}(E_{o\Delta\lambda} T_{\theta_o} \cos\theta_o \Delta\lambda + E_d) + L_p, \tag{6-32}$$

which may be used to relate brightness values in remotely sensed data to measured radiance, using the equation:

$$L_S = (K \times BV_{i,j,k}) + L_{min}, \tag{6-33}$$

where

K = radiance per bit of sensor count rate =
    $(L_{max} - L_{min})/C_{max}$
BV_{ijk} = brightness value of a pixel
C_{max} = maximum value in the dataset (e.g., 8-bit = 255)
L_{max} = radiance measured at detector saturation
    (W m^{-2} sr^{-1})
L_{min} = lowest radiance measured by a detector (W m^{-2} sr^{-1})

Examples of L_{min} and L_{max} for Landsat satellites 1 through 4 are found in Table 6-6.

Table 6-6.   Values of L_{min} and L_{max} for the Landsat MSS sensor system on various platforms.[a]

| | **Spectral Bands** | | | | | | | |
| | 4 | | 5 | | 6 | | 7 | |
Satellite	L_{min}	L_{max}	L_{min}	L_{max}	L_{min}	L_{max}	L_{min}	L_{max}
Landsat 1	0.0	24.8	0.0	20.0	0.0	17.6	0.0	40.0
Landsat 2 (6/22 to 7/16/75)	1.0	21.0	0.7	15.6	0.7	14.0	1.4	41.5
Landsat 2 (after 7/16/75)	0.8	26.3	0.6	17.6	0.6	15.2	1.1	39.1
Landsat 3 (3/5 to 6/1/78)	0.4	22.0	0.3	17.5	0.3	14.5	0.3	44.1
Landsat 3 (after 6/1/78)	0.4	25.9	0.3	17.9	0.3	14.9	0.3	38.3
Landsat 4	0.2	23.0	0.4	18.0	0.4	13.0	1.0	40.0

[a] Low gain mode from *Landsat Data User's Handbook* and Goddard Space Flight Center, 1982.

Example of Manual Absolute Atmospheric Correction

Radiometric correction of remotely sensed data requires the analyst to understand 1) the scattering and absorption taking place, and 2) how these affect the transmittance of radiant energy through the various paths of sky irradiance and path irradiance. Forster (1984) desired to compare two Landsat MSS scenes of Sydney, Australia, obtained on 14 December 1980 and 8 June 1980 to extract urban information. He believed the only way to extract meaningful information was to remove the atmospheric effects from each of the remote sensing datasets independently. Then spectrally stable land-cover classes from each date would be expected to cluster in approximately the same region of multispectral feature space (discussed in Chapter 9). The methods used to correct Landsat MSS band 7 data from the 14 December 1980 scene were presented based on the following pertinent information at the time of the satellite overpass:

Date	14 December 1980
Time	23.00 hours G.M.T., 9.05 hours local time
Sensor	Landsat 2 MSS
Air temperature	29°C
Relative humidity	24%
Atmospheric pressure	1004 mbar
Visibility	65 km
Altitude	30 m

Step 1. Calculate solar zenith angle (θ_o) at the time of overpass and determine $\mu_o = \cos \theta_o$:

$$\theta_o = 38°$$
$$\mu_o = 0.788.$$

Step 2. Compute the total normal optical thickness, τ. The optical thickness caused by molecular scattering, τ_m, was computed using a graph showing the atmospheric transmittance for a Rayleigh-type atmosphere as a function of wavelength. τ_{H_2O} was computed using temperature and relative humidity tables that summarize 1) the equivalent mass of liquid water (g cm^{-2}) as a function of temperature and relative humidity and 2) optical thickness of τ_{H_2O} versus equivalent mass of liquid water (g cm^{-2}) found in Forster (1984). The ozone optical thickness, τ_{O_3}, was determined to be 0.00 for band 7. The aerosol optical thickness, τ_p, for each band was computed using a graph of the aerosol optical thickness versus visual range from Turner and Spencer (1972). All these values were totaled, yielding the total normal optical thickness of the atmosphere, $\tau = 0.15$.

Step 3. The atmospheric transmittance for an angle of incidence θ can be calculated according to Equation 6-28. Landsat is for all practical purposes a nadir-viewing sensor system. Therefore, a 38° solar zenith angle yielded

$$T_{\theta_o} = e^{-0.15/\cos 38°} = 0.827$$

$$T_{\theta_v} = e^{-0.15/\cos 0°} = 0.861.$$

Step 4. Spectral solar irradiance at the top of the atmosphere varies throughout the year due to the varying distance of

Earth from the Sun. At aphelion (4 July) this distance is approximately 1.034 times that at perihelion (4 January). Using information from Slater (1980), the spectral solar irradiance at the top of the atmosphere was computed for the Landsat MSS bands, $E_o = 256$ W m^{-2}. Forster then computed total global irradiance at Earth's surface as $E_g = 186.6$ W m^{-2}.

Step 5. Forster used algorithms by Turner and Spencer (1972) that account for Rayleigh and Mie scattering and atmospheric absorption to compute path radiance:

$$L_p = 0.62 \quad \text{(W m}^{-2}\text{ sr}^{-1}\text{)}.$$

Step 6. These data were then used in Equations 6-30 and 6-32:

$$L_s = \frac{1}{\pi}\rho_{\text{band 7}}T_{\theta_v}E_g + L_p$$

$$= \frac{1}{\pi}\rho_{\text{band 7}}(0.861)(186.6) + 0.62$$

$$= 51.14\rho_{\text{band 7}} + 0.62.$$

Step 7. The Landsat 2 MSS had a band 7 brightness value range, $C_{\max}$, of 63. Therefore, from Table 6-6,

$$L_{\min} = 1.1 \quad \text{(W m}^{-2}\text{ sr}^{-1}\text{)}$$

$$L_{\max} = 39.1 \quad \text{(W m}^{-2}\text{ sr}^{-1}\text{)}$$

$$K = \frac{(L_{\max} - L_{\min})}{C_{\max}}$$

$$= \frac{(39.1 - 1.1)}{63}$$

$$= 0.603.$$

Therefore, according to Equation 6-33,
$$L_s = (K \times BV_{i,j,k}) + L_{\min}$$
$$= (0.603\, BV_{i,j,7}) + 1.1 \quad \text{(W m}^{-2}\text{ sr}^{-1}\text{)},$$

which when combined with L_s calculated in step 6 yields

$$51.14\rho_{\text{band 7}} + 0.62 = 0.603 BV_{i,j,7} + 1.1$$

$$\rho_{\text{band 7}} = 0.0118 BV_{i,j,7} + 0.0094.$$

All the brightness values in band 7 can now be transformed into percent (%) reflectance information (i.e., $\rho_\% 7 = \rho_{\text{band 7}} \times 100$). Therefore,
$$\rho_\% 7 = 1.18 BV_{i,j,7} + 0.94.$$

This can be done for each of the bands. When the same procedure is carried out on other remotely sensed data of the same geographical area, it may be possible to compare percent reflectance measurements for registered pixels acquired on different dates; that is, *absolute* scene-to-scene percent reflectance comparisons may be made. Of course, this model assumes uniform atmospheric conditions throughout each scene.

Atmospheric Correction Based on Radiative Transfer Modeling

In the previous example, manual methods were used to obtain the required input parameters necessary to characterize the scattering and absorption characteristics of the atmosphere on a specific date and time. Most current radiative transfer-based atmospheric correction algorithms can compute much of this same information if a) the user provides fundamental atmospheric characteristic information to the program or b) certain atmospheric absorption bands are present in the remote sensing dataset. For example, most radiative transfer-based atmospheric correction algorithms require that the user provide:

• latitude and longitude of the remotely sensed image scene,

• date and exact time of remote sensing data collection,

• image acquisition altitude (e.g., 20 km AGL)

• mean elevation of the scene (e.g., 200 m ASL),

• an atmospheric model (e.g., mid-latitude summer, mid-latitude winter, tropical),

• radiometrically calibrated image radiance data (i.e., the data *must* be in the form W m^2 μm^{-1} sr^{-1}),

• information about each specific band (i.e., its mean and full-width at half-maximum (FWHM)), and

• local atmospheric visibility at the time of remote sensing data collection (e.g., 10 km, obtained from a nearby airport if possible).

These parameters are then input to the atmospheric model selected (e.g., mid-latitude summer) and used to compute the absorption and scattering characteristics of the atmosphere at the instance of remote sensing data collection (InSpec, 2002). These atmospheric characteristics are then used to invert the remote sensing radiance to *scaled surface reflectance*. Many of these atmospheric correction programs derive the scattering and absorption information they require from robust atmosphere radiative transfer code such as MODTRAN 4+ (Alder-Golden et al., 1999) or Second Simulation of the Satellite Signal in the Solar Spectrum (6S) (Vermote et al., 1997; 2002).

Numerous atmospheric correction algorithms have been developed based on the use of radiative transfer principles. Several of the most important are ACORN, ATREM, FLAASH, and ATCOR. It is instructive briefly to review some of their characteristics.

- **ACORN**—**A**tmospheric **COR**rection **N**ow is based on radiative transfer equations developed by Chandrasekhar (1960). The program uses MODTRAN 4 radiative transfer code. It defines the relationship from contributions of the exo-atmospheric solar source, a homogenous plane parallel atmosphere, and the surface with respect to the radiance measured by an Earth-looking remote sensor (InSpec, 2002):

$$L_S = \frac{E_{o_\lambda}\left(E_{du_\lambda} + \dfrac{T_{\theta_o}\rho_\lambda T_{\theta_v}}{1 - E_{dd_\lambda}\rho_\lambda}\right)}{\pi}, \qquad (6\text{-}34)$$

where L_s is the total radiance measured by the sensor, E_{o_λ} is the solar spectral irradiance at the top of the atmosphere, E_{du_λ} is the *upward* reflectance of the atmosphere, T_{θ_o} is the *downward* transmittance of the atmosphere, ρ_λ is the scaled spectral reflectance of the surface, T_{θ_v} is the *upward* transmittance of the atmosphere, and E_{dd_λ} is the *downward* reflectance of the atmosphere.

When all the appropriate parameters have been derived, the scaled surface reflectance ρ_λ is computed using the equation:

$$\rho_\lambda = \frac{1}{\left[\dfrac{(E_{o_\lambda}T_{\theta_o}T_{\theta_v})/\pi}{L_s - (E_{o_\lambda}E_{du_\lambda})/\pi}\right] + E_{dd_\lambda}}. \qquad (6\text{-}35)$$

For example, consider the hyperspectral data of the Savannah River Site that has been atmospherically corrected and scaled from radiance to scaled surface reflectance. Figure 6-25a depicts the original radiance (in W m^{-2} nm^{-1} sr^{-1}) of a pixel of loblolly pine (*Pinus taeda*) in 35 bands of GER DAIS 3715 hyperspectral imagery from 509 nm to 2365 nm. Figure 6-25b is the scaled surface reflectance of the same pixel. The pine pixel now appears as it should, with a slight peak of reflectance in the green

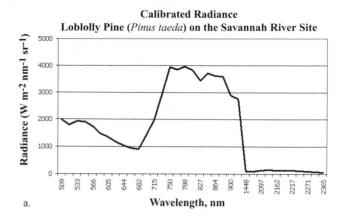

a.

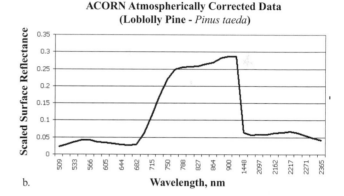

b.

Figure 6-25 a) Calibrated radiance data of a loblolly pine pixel on the Savannah River Site recorded by the GER DAIS 3715 hyperspectral remote sensing system. The 35 bands range from 509 to 2365 nm. b.) The same pixel transformed into scaled surface reflectance using the radiative transfer Atmospheric CORrection Now (ACORN) algorithm (Jensen et al., 2003).

at approximately 566 nm, chlorophyll absorption in the red centered at approximately 682 nm, and increased near-infrared reflectance (Jensen et al., 2003).

- **ATREM**—The **AT**mospheric **REM**oval program [Center for the Study of Earth from Space (CSES), University of Colorado] calculates the amount of molecular scattering (Rayleigh) present using the radiative transfer code Second Simulation of the Satellite Signal in the Solar Spectrum (6S) (Tanre et al., 1986) and a user-specified aerosol model (Research Systems, Inc., 2002). ATREM calculates the atmospheric absorption term using the Malkmus narrow-band spectral model with a user-selected standard atmospheric model (temperature, pressure, and water vapor vertical distributions) or a user-supplied atmospheric model (Gao et al., 1999). The amount of water vapor is derived on a pixel-by-pixel basis from the hyperspectral data using the 0.94-μm and the 1.14-μm water vapor bands and a three-channel ratioing technique. The derived water vapor values are then used for modeling water vapor absorption effects over the entire 400- to 2500-nm region. The result is a radiometrically corrected dataset consisting of scaled surface reflectance. ATREM is applied to hyperspectral data in Chapter 11.

- **FLAASH**—ENVI's **F**ast **L**ine-of-sight **A**tmospheric **A**nalysis of **S**pectral **H**ypercubes was developed by Spectral Sciences, Inc., in collaboration with the U.S. Air Force Research Lab (AFRL) and Spectral Information Technology Application Center personnel. FLAASH corrects images for atmospheric water vapor, oxygen, carbon dioxide, methane, ozone, and molecular and aerosol scattering using a MODTRAN 4+ radiation transfer code solution computed for each image and each pixel in the image. It also produces a column water vapor image, a cloud map, and a visibility value for the scene (Research Systems, Inc., 2003).

- **ATCOR**—The **AT**mospheric **COR**rection program was originally developed at DLR, the German Aerospace Centre. ATCOR consists of ATCOR 2 (used for flat terrain) and ATCOR 3 (used for rugged terrain, i.e., the 3 stands for 3 dimensions). ATCOR 4 is used with remotely sensed data acquired by suborbital remote sensing systems. The atmospheric correction algorithm employs the MODTRAN 4+ (Alder-Golden et al., 1999) radiative transfer code to calculate look-up tables (LUT) of the atmospheric correction functions (path radiance, atmospheric transmittance, direct and diffuse solar flux) that depend on scan angle, relative azimuth angle between scan line and solar azimuth, and terrain elevation. An example of before and after atmospheric correction is shown in Figure 6-26. Thiemann and Hermann (2002) provide a summary of how ATCOR was used to radiometrically correct hyperspectral data to extract water quality tropic parameters.

Artifact Suppression: Most robust radiative transfer-based atmospheric correction algorithms also suppress artifacts in the corrected remote sensor data (Boardman, 1998). For example, FLAASH uses an adjustable spectral "polishing" based on Adler-Golden et al. (1999). This type of spectral polishing involves applying mild adjustments to reflectance data so that the spectra appear more like spectra of real materials as recorded on the ground by a handheld spectroradiometer. ENVI's EFFORT, InSpec's ACORN, and Leica Geosystem's ATCOR all have artifact suppression capability.

Single Spectrum Enhancement: Many radiative transfer atmospheric correction programs (e.g., ATREM, ACORN, FLAASH) allow the user to incorporate *in situ* spectroradiometer measurements. This is commonly referred to as *single spectrum enhancement*. The analyst simply provides an accurate *in situ* spectral reflectance curve for a known homogeneous area (e.g., a bare soil field of kaolinite, a deep water body, or a parking lot) and also provides the geometric coordinates of this same location in the remotely sensed data. The atmospheric correction program then determines all of the traditional radiative transfer scattering and absorption adjustments, plus it attempts to make the pixel in the image have the same spectral characteristics as the *in situ* spectroradiometric data.

For example, 46 *in situ* spectroradiometric measurements were measured on the Savannah River Site Mixed Waste Management Facility (Figure 6-27a) using an Analytical Spectral Devices (ASD) spectroradiometer (Figure 6-28a). The GER DAIS 3715 hyperspectral data were radiometrically corrected using ACORN and empirical line calibration (discussed in the next section). In addition, a single spectrum enhancement was applied to both methods. A comparison between the 46 *in situ* spectroradiometer measurements for each of the 35 bands with the four atmospheric correction treatments (ELC, ACORN, ELC+SSE, ACORN+SSE) reveals that the ACORN and ELC with the single spectrum enhancement provided very accurate radiometric correction of the remote sensor data (Figure 6-27b). The DAIS 3715 sensor did not collect data in one of the optimum atmospheric absorption windows. This may be the reason why ACORN did not do as well as a stand-alone correction.

Absolute Atmospheric Correction Using Empirical Line Calibration

Absolute atmospheric correction may also be performed using *empirical line calibration (ELC)*, which forces the remote sensing image data to match *in situ* spectral reflectance measurements, hopefully obtained at approximately the same time and on the same date as the remote sensing overflight. Empirical line calibration is based on the equation (Roberts et al., 1986; Smith and Milton, 1999; InSpec, 2002):

$$BV_k = \rho_\lambda A_k + B_k, \qquad (6\text{-}36)$$

where BV_k is the digital output value for a pixel in band k, ρ_λ equals the scaled surface reflectance of the materials within the remote sensor IFOV at a specific wavelength (λ), A_k is a multiplicative term affecting the BV, and B_k is an additive term. The multiplicative term is associated primarily with atmospheric transmittance and instrumental factors, and the

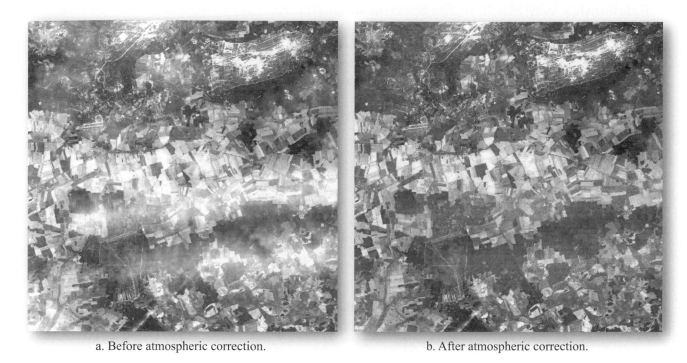

a. Before atmospheric correction.   b. After atmospheric correction.

Figure 6-26   a) Image containing substantial haze prior to atmospheric correction. b) Image after atmospheric correction using ATCOR (Courtesy Leica Geosystems and DLR, the German Aerospace Centre).

additive term deals primarily with atmospheric path radiance and instrumental offset (i.e., dark current).

To use ELC, the analyst usually selects two or more areas in the scene with different albedos (e.g., one bright target such as a sand pile and one dark target such as a deep, nonturbid water body). The areas should be as homogeneous as possible. *In situ* spectroradiometer measurements of these targets are made on the ground. The *in situ* and remote sensing–derived spectra are regressed and gain and offset values computed (Farrand et al., 1994). The gain and offset values are then applied to the remote sensor data on a band by band basis, removing atmospheric attenuation. Note that the correction is applied band by band and not pixel by pixel, as it is with ATREM, ATCOR, and ACORN.

To gain an appreciation of how empirical line calibration works, consider how GER DAIS 3715 radiance values of the Savannah River Site were converted to scaled surface reflectance values. Instead of trying to locate natural dark and light radiometric control points on the ground, dark and light 8×8 m calibration panel targets were located in the field at the time of a GER DAIS 3715 hyperspectral overflight (Figure 6-28b). Moran et al. (2001) provide guidance on how to pre-

pare and maintain reference reflectance tarps. The spatial resolution of the imagery was 2.4×2.4 m. Image pixel brightness values for the targets were extracted from flight line 05 at two locations: dark (3,682,609 N; 438,864 E) and bright (3,682,608 N; 438,855 E). *In situ* ASD spectroradiometer measurements were obtained from the calibration targets. The remote sensing and *in situ* spectral measurements for the calibration targets were then paired and input to the empirical line calibration to derive the appropriate gain and offset values to atmospherically correct the hyperspectral data. Results of the empirical line calibration are summarized in Figure 6-27b.

It is important to note that most multispectral remote sensing datasets can be calibrated using empirical line calibration. The difficulty arises when trying to locate homogeneous bright and dark targets in the study, collecting representative *in situ* spectroradiometer measurements, and extracting uncontaminated pixels of the calibration targets from the imagery. If the analyst does not have access to *in situ* spectra obtained at the time of the remote sensing overflight, it might be possible to use spectra of such fundamental materials as clear water and sand (quartz) that are stored in spectral libraries. For example, NASA's Jet Propulsion Laboratory (JPL)

Relationship between 46 *In Situ* Spectroradiometer Measurements and Pixel Spectral Reflectance after Processing Using Empirical Line Calibration and ACORN

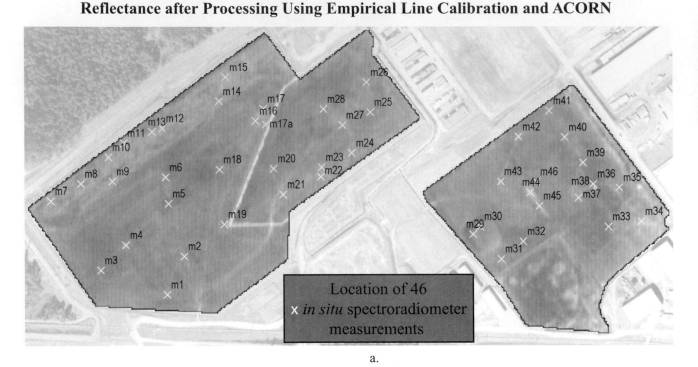

a.

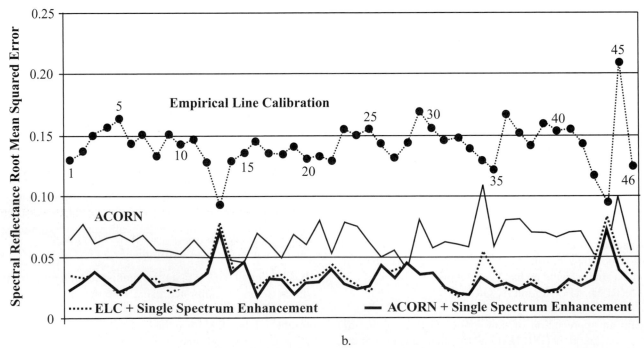

b.

Figure 6-27   a) Forty-six *in situ* ASD spectroradiometer (400 – 2500 nm) measurements were obtained on the Savannah River Site Mixed Waste Management Facility near Aiken, SC. b) The relationship between the 46 *in situ* spectroradiometer measurements and pixel spectral reflectance after processing using empirical line calibration (ELC) and ACORN. The ELC and ACORN with the single spectrum enhancements provided the most accurate atmospheric correction of the GER DAIS 3715 hyperspectral data (Jensen et al., 2003).

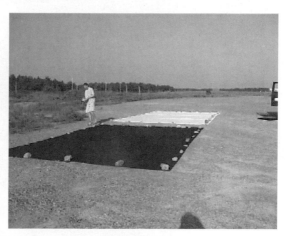

a. *In situ* spectroradiometer calibration.

b. Calibration targets.

Figure 6-28   a) Field crew taking a spectroradiometer measurement from a calibrated reflectance standard on the tripod. b) 8 × 8 m black and white calibration targets at the Savannah River Site (Jensen et al., 2003).

has a spectral library of hundreds of minerals (JPL, 2004a). The USGS has a spectral library of minerals and vegetation (Clark, 1999; USGS, 2004). The Johns Hopkins University spectral library contains spectra of a variety of materials (JPL, 2004b). Hopefully, some of these materials exist in the scene and the analyst can locate the appropriate pixel and pair the image brightness values with the library *in situ* spectroradiometer data.

To demonstrate the use of empirical line calibration applied to historical imagery, consider the Landsat Thematic Mapper dataset shown in Figure 6-29a. *In situ* spectral reflectance measurements were not obtained when the Landsat TM imagery were acquired on February 3, 1994. Therefore, how can the brightness values in this historical imagery be converted into scaled surface reflectance? The answer lies in the use of empirical line calibration, *in situ* library spectra, and the Landsat TM brightness values ($BV_{i,j,k}$). This particular scene has a substantial amount of water and beach. Therefore, *in situ* library spectroradiometer data of water (from the Johns Hopkins library) and quartz (from the NASA JPL library) were utilized. The multispectral brightness values of one pixel of water and one pixel of beach were extracted from the Landsat TM data. These were paired with the *in situ* spectral reflectance measurements in the same six wavelengths (the Landsat 6 thermal infrared channel was not used). The empirical line calibration resulted in a Landsat TM dataset scaled to scaled surface reflectance. For example, consider a pixel of healthy loblolly pine. The original brightness value response is shown in Figure 6-29b. The

scaled surface reflectance derived from the Landsat TM data is shown in Figure 6-29c. All pixels in the scene now represent scaled surface reflectance and not brightness value.

Scaled surface reflectance measurements obtained from this date of imagery can now be compared with scaled surface reflectance measurements extracted from other dates of Landsat TM imagery for purposes of monitoring biomass, etc. Of course, the use of library spectra is not as good as acquiring *in situ* spectroradiometry at selected sites in the scene at the time of the remote sensing data collection. Nevertheless, the technique is useful.

Relative Radiometric Correction of Atmospheric Attenuation

As previously discussed, absolute radiometric correction makes it possible to relate the digital counts in satellite or aircraft image data to scaled surface reflectance. Except when using empirical line calibration, this requires sensor calibration coefficients (to convert the original remote sensor data to W m^2 sr^{-1}) and an atmospheric correction algorithm usually based on radiative transfer code. Unfortunately, the application of these codes to a specific scene and date also requires knowledge of both the sensor spectral profile and the atmospheric properties at the time of remote sensor data collection (Du et al., 2002). If all this information is available, then it is possible for the atmospheric radiative transfer code to provide realistic estimates of the effects of atmo-

a. Empirical line calibrated Landsat Thematic Mapper band 4 image of Charleston, SC.

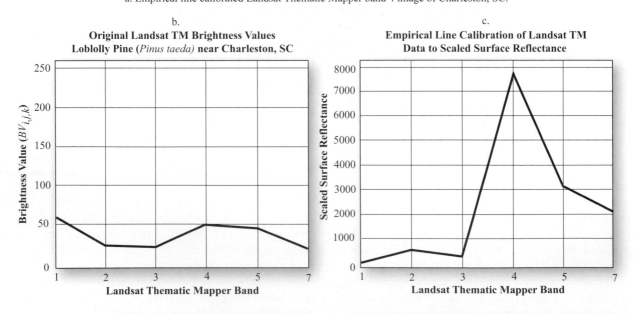

Figure 6-29   a) A Landsat Thematic Mapper image acquired on February 3, 1994, was radiometrically corrected using empirical line calibration and paired NASA JPL and Johns Hopkins University spectral library beach and water *in situ* spectroradiometer measurements and Landsat TM image brightness values ($BV_{i,j,k}$). b) A pixel of loblolly pine with its original brightness values in six bands (the TM band 6 thermal channel was not used). c) The same pixel after empirical line calibration to scaled surface reflectance. Note the correct chlorophyll absorption in the blue (band 1) and red (band 3) portions of the spectrum and the increase in near-infrared reflectance.

spheric scattering and absorption and convert the imagery to scaled surface reflectance.

Atmospheric properties are difficult to acquire even when planned for. In fact, for most historic satellite data, they are not available. Even today, accurate surface reflectance retrieval is not operational (i.e., it is not a standard archived product) for the majority of satellite image sources used for land-cover change detection (Du et al., 2002). For this reason, relative radiometric correction techniques have been developed.

Relative radiometric correction may be used 1) to *normalize* the intensities among the different bands within a single-date remotely sensed image and 2) to *normalize* the intensities of bands of remote sensor data in multiple dates of imagery to a standard scene selected by the analyst.

Single-image Normalization Using Histogram Adjustment

This simple method is based primarily on the fact that infrared data (>0.7 μm) are largely free of atmospheric scattering effects, whereas the visible region (0.4 – 0.7 μm) is strongly influenced by them. The method involves evaluating the histograms of the various bands of remotely sensed data of the desired scene. Normally, the data collected in the visible wavelengths (e.g., TM bands 1 to 3) have a higher minimum value because of the increased atmospheric scattering taking place in these wavelengths. For example, a histogram of Charleston, SC, TM data (Figure 6-30) reveals that it has a minimum of 51 and a maximum of 242. Conversely, atmospheric absorption subtracts brightness from the data recorded in the longer-wavelength intervals (e.g., TM bands 4, 5, and 7). This effect commonly causes data from the infrared bands to have minimums close to zero, even when few objects in the scene truly have a reflectance of zero (Figure 6-30).

If the histograms in Figure 6-30 are shifted to the left so that zero values appear in the data, the effects of atmospheric scattering will be somewhat minimized. This simple algorithm models the first-order effects of atmospheric scattering, or *haze*. It is based on a subtractive bias established for each spectral band. The bias may also be determined by evaluating a histogram of brightness values of a reference target such as deep water in all bands. The atmospheric effects correction algorithm is defined:

$$\text{output } BV_{i,j,k} = \text{input } BV_{i,j,k} - \text{bias},$$

where

$$\text{input } BV_{i,j,k} = \text{input pixel value at line } i \text{ and column } j \text{ of band } k$$
$$\text{output } BV_{i,j,k} = \text{the adjusted pixel value at the same location.}$$

In our example, the appropriate bias was determined for each histogram in Figure 6-30 and subtracted from the data. The histograms of the adjusted data are displayed in Figure 6-31. It was not necessary to adjust bands 5 and 7 for first-order atmospheric effects because they originally had minimums of zero.

Multiple-date Image Normalization Using Regression

Multiple-date image normalization involves selecting a base image (*b*) and then transforming the spectral characteristics of all other images obtained on different dates (e.g., those obtained on dates *b*–1, *b*–2 and/or *b*+1, *b*+2, etc.) to have approximately the same radiometric scale as the base image. It is important to remember, however, that the radiometric scale used in a relative multiple-date image normalization will most likely be simple brightness values (e.g., *BV* with a range from 0 to 255) rather than scaled surface reflectance produced when conducting an absolute radiometric correction.

Multiple-date image normalization involves the selection of *pseudo-invariant features* (PIFs), often referred to as radiometric ground control points. To be of value in the multiple-date image normalization process, the pseudo-invariant features should have the following characteristics:

- The spectral characteristics of a PIF should change very little through time, although it is acknowledged that some change is inevitable. Deep nonturbid water bodies, bare soil, large rooftops, or other homogeneous features are candidates.

- The PIF should be at approximately the same elevation as the other land in the scene. Selecting a mountaintop PIF would be of little use in estimating atmospheric conditions near sea level because most aerosols in the atmosphere occur within the lowest 1000 m.

- The PIF should normally contain only minimal amounts of vegetation. Vegetation spectral reflectance can change over time as a result of environmental stress and plant phenology. However, an extremely stable, homogeneous forest canopy imaged on near-anniversary dates might be considered.

Original Data

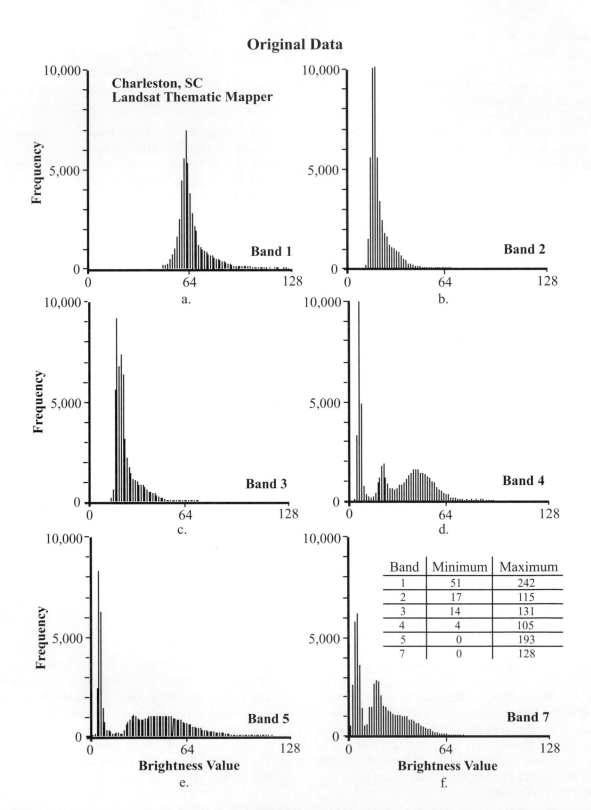

Band	Minimum	Maximum
1	51	242
2	17	115
3	14	131
4	4	105
5	0	193
7	0	128

Figure 6-30   Original histograms of six bands of the Charleston, SC, Thematic Mapper scene. Atmospheric scattering in the visible regions has increased the minimum brightness values in bands 1, 2, and 3. Generally, the shorter the wavelengths sensed by each band, the greater the offset from a brightness value of zero.

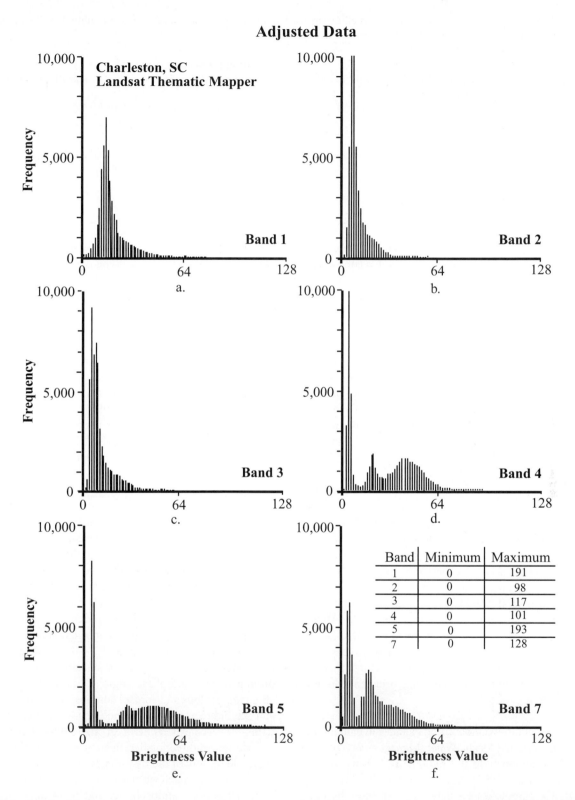

Adjusted Data

Charleston, SC
Landsat Thematic Mapper

Band 1

a.

Band 2

b.

Band 3

c.

Band 4

d.

Band	Minimum	Maximum
1	0	191
2	0	98
3	0	117
4	0	101
5	0	193
7	0	128

Band 5

e.

Band 7

f.

Frequency

Brightness Value

Figure 6-31   Result of applying a simple histogram adjustment atmospheric scattering correction to the data shown in Figure 6-30. Only the first four Thematic Mapper bands required the adjustment. This method does not correct for atmospheric absorption.

Table 6-7.   Characteristics of the remotely sensed satellite data used to inventory wetland in Water Conservation Area 2A of the South Florida Water Management District (Jensen et al., 1995).

Date	Type of Imagery	Bands Used	Nominal Instantaneous Field of View (m)	Rectification RMSE
3/22/73	Landsat MSS	1, 2, 4	79×79	±0.377
4/02/76	Landsat MSS	1, 2, 4	79×79	±0.275
10/17/82	Landsat MSS	1, 2, 4	79×79	±0.807
4/04/87	SPOT HRV	1, 2, 3	20×20	±0.675
8/10/91	SPOT HRV	1, 2, 3	20×20	±0.400

• The PIF must be in a relatively flat area so that incremental changes in Sun angle from date to date will have the same proportional increase or decrease in direct beam sunlight for all normalization targets.

Regression is used to relate the base image's PIF spectral characteristics with PIF spectral characteristics from other dates. The algorithm assumes that the pixels sampled at time $b+1$ or $b-1$ are linearly related to the pixels for the same locations on the base image (b). This implies that the spectral reflectance properties of the sampled pixels have not changed during the time interval. Thus, the key to the image regression method is the selection of quality pseudo-invariant features.

Numerous scientists have investigated the use of pseudo-invariant features to normalize multiple-date imagery. For example, Caselles and Garcia (1989), Schott et al., (1988), Conel (1990), and Hall et al. (1991) developed a *radiometric rectification technique* that corrected images of the same areas through the use of landscape elements whose reflectances were nearly constant over time. Coppin and Bauer (1994) and Jensen et al. (1995) used similar procedures to perform relative radiometric correction. Heo and FitzHugh (2000) developed different techniques to select PIFs. Du et al. (2002) developed a new procedure for the relative radiometric normalization of multitemporal satellite images where PIFs were selected objectively using principal component analysis (PCA).

A remote sensing study that identified changes in cattail distribution in the South Florida Water Management District can be used to demonstrate the radiometric normalization process (Jensen et al., 1995). Six predominantly cloud-free dates of satellite remote sensor data were collected by two sensor systems for Everglades Water Conservation Area 2A from 1973 to 1991. Landsat Multispectral Scanner (MSS) data were obtained in 1973, 1976, and 1982 and SPOT High Resolution Visible (HRV) multispectral (XS) data in 1987 and 1991. The specific date, type of imagery, bands used in the analysis, and nominal spatial resolution of the various sensor systems are summarized in Table 6-7. Twenty (20) ground control points (GCPs) were obtained and used to rectify the August 10, 1991, remote sensor data to a standard map projection. The remotely sensed data were rectified to a Universal Transverse Mercator (UTM) map projection having 20×20 m pixels using a nearest-neighbor resampling algorithm and a root-mean-square error (RMSE) of ±0.4 pixel (Rutchey and Vilchek, 1994). All other images were resampled to 20×20 m pixels using nearest-neighbor resampling and registered to the 1991 SPOT data for change detection purposes. The RMSE statistic for each image is summarized in Table 6-7.

A problem associated with using historical remotely sensed data for change detection is that the data are usually from non-anniversary dates with varying Sun angle, atmospheric, and soil moisture conditions. The multiple dates of remotely sensed data should be normalized so that these effects can be minimized or eliminated (Hall et al., 1991).

Differences in direct-beam solar radiation due to variation in Sun angle and Earth-to-Sun distance can be calculated accurately, as can variation in pixel BVs due to detector calibration differences between sensor systems. However, removal of atmospheric and phase angle effects require information about the gaseous and aerosol composition of the atmosphere and the bidirectional reflectance characteristics of elements within the scene (Eckhardt et al., 1990). Because atmospheric and bidirectional reflectance information were not available for any of the five scenes, an "empirical scene normalization" approach was used to match the detector calibration, astronomic, atmospheric, and phase angle conditions present in a reference scene. The August 10, 1991, SPOT HRV scene was selected as the base scene to which the 1973, 1976, 1982, and 1987 scenes were normalized. The 1991

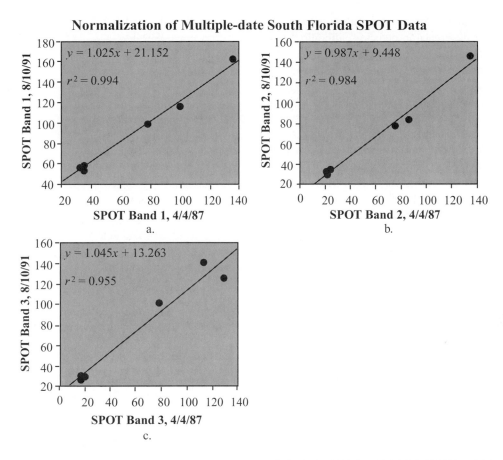

Normalization of Multiple-date South Florida SPOT Data

a.

b.

c.

Figure 6-32   a) Relationship between the same wet and dry regions found in both the 4/4/87 and 8/10/91 SPOT band 1 (green) dataset. The equation was used to normalize the 4/4/87 data to the 8/10/91 SPOT data as per methods described in Eckhardt et al. (1990). b) Relationship between wet and dry regions found in both the 4/4/87 and 8/10/91 SPOT band 2 (red) dataset. c) Relationship between wet and dry regions found in both the 4/4/87 and 8/10/91 SPOT band 3 (near-infrared) dataset (Jensen et al., 1995).

SPOT image was selected because it was the only year for which quality *in situ* ground reference data were available.

Image normalization was achieved by applying regression equations to the 1973, 1976, 1982, and 1987 imagery to predict what a given BV would be if it had been acquired under the same conditions as the 1991 reference scene. These regression equations were developed by correlating the brightness of pseudo-invariant features present in both the scene being normalized and the reference (1991) scene. PIF were assumed to be constant reflectors, so any changes in their brightness values were attributed to detector calibration, and astronomic, atmospheric, and phase angle differences. Once these variations were removed, changes in BV could be related to changes in surface conditions.

Multiple wet (water) and dry (e.g., unvegetated bare soil) PIF were found in the base year image (1991) and each of the

other dates of imagery (e.g., 1987 SPOT data). A total of 21 radiometric control points were used to normalize the 1973, 1976, 1982, and 1987 data to the 1991 SPOT data. It is useful to summarize the nature of the normalization targets used and identify adjustments that had to be made when trying to identify dry soil targets in a humid subtropical environment. Radiometric normalization targets found within the 1987 and 1991 SPOT data consisted of three wet points obtained just to the north of WCA-2A within WCA-1 and three dry points extracted from an excavated area, a dry lake area, and a limestone road area. The brightness values of the early image targets (e.g., 1987) were regressed against the brightness values of the base image targets (e.g., 1991) for each band (Figure 6-32 a, b, and c). The coefficients and intercept of the equation were used to compute a normalized 1987 SPOT dataset, which had approximately the same spectral characteristics as the 1991 SPOT data. Each regression model contained an additive component that corrected for the difference in atmospheric path radiance among dates and a multiplicative term that corrected for the difference in detector calibration, Sun angle, Earth–Sun distance, atmospheric attenuation, and phase angle between dates.

The 1982 MSS data were normalized to the 1991 data using (1) three common wet targets found within WCA-1 and (2) two dry points extracted from a bare soil excavation area in 1982, which progressed northward about 300 m (15 pixels) in the y dimension by 1991 (i.e., the x dimension was held constant). Thus, two noncommon dry radiometric control points were extracted for this date. Hall et al. (1991) suggest that the members of the radiometric control sets may not be the same pixels from image to image, in contrast to geometric control points for spatial image rectification, which are composed of identical elements in each scene. Furthermore, they suggest that "using fixed elements inevitably requires manual selection of sufficient numbers of image-to-image pairs of suitable pixels, which can be prohibitively labor intensive, particularly when several images from a number of years are being considered." Such conditions were a factor in the Everglades study.

The 1976 MSS data were normalized to the 1991 data using three wet targets located in WCA-1 and two dry points extracted along a bare soil road and a limestone bare soil area. The 1973 MSS data were normalized to the 1991 data using two wet and three dry targets. The greater the time between the base image (e.g., 1991) and the earlier year's image (e.g., 1973), the more difficult it is to locate unvegetated, dry normalization targets. For this reason, analysts sometimes use synthetic, pseudo-invariant features such as concrete, asphalt, rooftops, parking lots, and roads when normalizing historic remotely sensed data (Schott et al., 1988; Caselles and Garcia, 1989; Hall et al., 1991).

The normalization equations for each date are summarized in Table 6-8. The gain (slope) associated with the SPOT data was minimal, while the historical MSS data required significant gain and bias adjustments (because some MSS data were not originally acquired as 8-bit data). The methodology applied to all images minimized the differences in Sun angle, atmospheric effects, and soil moisture conditions between the dates. The radiometrically corrected remote sensor data were then classified and used to monitor wetland change (Jensen et al., 1995).

The ability to use remotely sensed data to classify land cover accurately is contingent on there being a robust relationship between remotely sensing brightness value (BV) and actual surface conditions. However, factors such as Sun angle, Earth–Sun distance, detector calibration differences among

Table 6-8. Equations used to normalize the radiometric characteristics of the historic remote sensor data with the August 10, 1991, SPOT XS data (Jensen et al., 1995).

Date	Band	Slope	y-intercept	r^2
3/22/73	MSS 1	1.40	31.19	0.99[a]
	2	1.01	23.49	0.98
	4	3.28	23.48	0.99
4/02/76	MSS 1	0.57	31.69	0.99
	2	0.43	21.91	0.98
	4	3.84	26.32	0.96
10/17/82	MSS 1	2.52	16.117	0.99
	2	2.142	8.488	0.99
	4	1.779	17.936	0.99
4/04/87	SPOT 1	1.025	21.152	0.99
	2	0.987	9.448	0.98
	3	1.045	13.263	0.95

[a]All regression equations were significant at the 0.001 level.

the various sensor systems, atmospheric condition, and Sun–target–sensor (phase angle) geometry will affect pixel brightness value. Image normalization reduces pixel BV variation caused by nonsurface factors, so that variations in pixel brightness value among dates may be related to actual changes in surface conditions. Normalization may allow the use of pixel classification logic developed from a base year scene to be applied to the other normalized scenes.

 Correcting for Slope and Aspect Effects

The previous sections discussed how scattering and absorption in the atmosphere can attenuate the radiant flux recorded by the sensor system. Topographic slope and aspect may also introduce radiometric distortion of the recorded signal (Gibson and Power, 2000). In some locations, the area of interest might even be in complete shadow, dramatically affecting the brightness values of the pixels involved. For these reasons, research has been directed toward the removal of topographic effects, especially in mountainous regions, on Landsat and SPOT digital multispectral data (Teillet et al., 1982; Shasby and Carneggie, 1986; Hall-Konyves, 1987;

Jones et al., 1988; Kawata et al., 1988; Leprieur et al., 1988; Civco, 1989; Meyer et al., 1993). The goal of a slope-aspect correction is to remove topographically induced illumination variation so that two objects having the same reflectance properties show the same brightness value in the image despite their different orientation to the Sun's position. If the topographic slope-aspect correction is effective, the three-dimensional impression we get when looking at a satellite image of mountainous terrain should be somewhat subdued. A good slope-aspect correction is believed to improve forest stand classification when compared to noncorrected imagery (Civco, 1989; Meyer et al., 1993).

Teillet et al. (1982) described four topographic slope-aspect correction methods: the simple cosine correction, two semi-empirical methods (the Minnaert method and the C correction), and a statistic-empirical correction. Each correction is based on *illumination,* which is defined as the cosine of the incident solar angle, thus representing the proportion of the direct solar radiation hitting a pixel. The amount of illumination is dependent on the relative orientation of the pixel toward the Sun's actual position (Figure 6-33). Each slope-aspect topographic correction method to be discussed requires a digital elevation model (DEM) of the study area. The DEM and satellite remote sensor data (e.g., Landsat TM data) must be geometrically registered and resampled to the same spatial resolution (e.g., 30×30 m pixels). The DEM is processed so that each pixel's brightness value represents the amount of illumination it should receive from the Sun. This information is then modeled using one of the four algorithms to enhance or subdue the original brightness values of the remote sensor data.

The Cosine Correction

The amount of irradiance reaching a pixel on a slope is directly proportional to the cosine of the incidence angle i, which is defined as the angle between the normal on the pixel in question and the zenith direction (Teillet et al., 1982). This assumes 1) Lambertian surfaces, 2) a constant distance between Earth and the Sun, and 3) a constant amount of solar energy illuminating Earth (somewhat unrealistic assumptions!). Only the part cos i of the total incoming irradiance, E_g, reaches the inclined pixel. It is possible to perform a simple topographic slope-aspect correction of the remote sensor data using the following cosine equation:

$$L_H = L_T \frac{\cos \theta_o}{\cos i} \qquad (6\text{-}37)$$

where

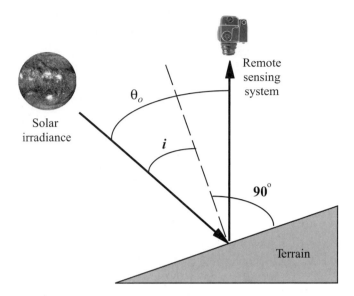

Figure 6-33   Representation of the Sun's angle of incidence, i, and the solar zenith angle, θ_o.

L_H = radiance observed for a horizontal surface (i.e., slope-aspect-corrected remote sensor data)

L_T = radiance observed over sloped terrain (i.e., the raw remote sensor data)

θ_o = Sun's zenith angle

i = Sun's incidence angle in relation to the normal on a pixel (Figure 6-33).

Unfortunately, this method models only the direct part of the irradiance that illuminates a pixel on the ground. It does not take into account diffuse skylight or light reflected from surrounding mountainsides that may illuminate the pixel in question. Consequently, weakly illuminated areas in the terrain receive a disproportionate brightening effect when the cosine correction is applied. Basically, the smaller the cos i, the greater the overcorrection is (Meyer et al., 1993). Nevertheless, several researchers have found the cosine correction of value. For example, Civco (1989) achieved good results with the technique when used in conjunction with empirically derived correction coefficients for each band of imagery.

The Minnaert Correction

Teillet et al. (1982) introduced the Minnaert correction to the basic cosine function:

$$L_H = L_T \left(\frac{\cos \theta_o}{\cos i}\right)^k \qquad (6\text{-}38)$$

where k = the Minnaert constant.

The constant varies between 0 and 1 and is a measure of the extent to which a surface is Lambertian. A perfectly Lambertian surface has $k = 1$ and represents a traditional cosine correction. Meyer et al. (1993) describe how k may be computed empirically.

A Statistical–Empirical Correction

For each pixel in the scene, it is possible to correlate (1) the predicted illumination (cos $i \times 100$) from the DEM with (2) the actual remote sensor data. For example, Meyer et al. (1993) correlated Landsat TM data of known forest stands in Switzerland with the predicted illumination from a high-resolution DEM. Any slope in the regression line suggests that a constant type of forest stand will appear differently on different terrain slopes. Conversely, by taking into account the statistical relationship in the distribution, the regression line may be rotated based on the following equation:

$$L_H = L_T - \cos(i)m - b + \overline{L_T} \qquad (6\text{-}39)$$

where

     L_H = radiance observed for a horizontal surface (i.e., slope-aspect-corrected remote sensor data)

     L_T = radiance observed over sloped terrain (i.e., the raw remote sensor data)

     $\overline{L_T}$ = average of L_T for forested pixels (according to ground reference data)

     i = Sun's incidence angle in relation to the normal on a pixel (Figure 6-33)

     m = slope of the regression line

     b = y-intercept of the regression line.

Application of this equation makes a specific object (e.g., a particular type of deciduous forest) independent of cos i and produces the same brightness values (or radiance) throughout the image for this object.

The C Correction

Teillet et al. (1982) introduced an additional adjustment to the cosine function called the c correction:

$$L_H = L_T \frac{\cos \theta_o + c}{\cos i + c} \qquad (6\text{-}40)$$

where

$c = \dfrac{b}{m}$ in the previous regression equation.

Similar to the Minnaert constant, c increases the denominator and weakens the overcorrection of faintly illuminated pixels.

The radiometric correction of topographically induced effects is improving. Civco (1989) identified some important considerations:

- The digital elevation model of the study area should have a spatial resolution comparable to the spatial resolution of the digital remote sensor data.

- When correcting for topographic effect using a Lambertian surface assumption, remotely sensed data are often overcorrected. Slopes facing away from the Sun (i.e., north-facing slopes in the northern hemisphere) appear brighter than Sun-facing slopes (south-facing slopes) of similar composition. This suggests that the ideal correction is less than the traditional cosine of the incidence angle.

- Most techniques consider only the direct solar beam contribution to irradiance and forget that the diffuse component also illuminates the local topography.

- There is often strong anisotropy of apparent reflectance, which is wavelength dependent, and solar, surface, and sensor system geometry should be considered in the modeling process (Leprieur et al., 1988).

- The amount of correction required is a function of wavelength. Particular attention must be given to middle-infrared bands, which are severely affected by the topographic effect (Kawata et al., 1988).

- It is difficult to remove the topographic effect completely from severely shadowed areas such as deep valleys (Kawata et al., 1988).

 ## References

Alder-Golden, S. M., Matthew, M. W., Bernstein, L. S., Levine, R. Y., Berk, A., Richtsmeier, S. C., Acharya, P. K., Anderson, G. P., Felde, G., Gardner, J., Hoke, M., Jeong, L. S., Pukall, B., Mello, J., Ratkowski, A. and H. H. Burke, 1999, "Atmospheric Correction for Short-wave Spectral Imagery Based on MODTRAN4," *SPIE Proceedings on Imaging Spectrometry V*, (3753), 10 p.

Boardman, J. W., 1998, "Post-ATREM Polishing of AVIRIS Apparent Reflectance Data Using EFFORT: A Lesson in Accuracy versus Precision," *Summaries of the Seventh JPL Airborne Earth Science Workshop*, Pasadena: NASA Jet Propulsion Laboratory, JPL Publication #97-21, 1:53.

Caselles, V. and M. J. L. Garcia, 1989, "An Alternative Simple Approach to Estimate Atmospheric Correction in Multitemporal Studies," *International Journal of Remote Sensing*, 10(6):1127–1134.

Civco, D. L., 1989, "Topographic Normalization of Landsat Thematic Mapper Digital Imagery," *Photogrammetric Engineering & Remote Sensing*, 55(9):1303–1309.

Clark, R. N., 1999, "Spectroscopy of Rocks and Minerals, and Principles of Spectroscopy," *Manual of Remote Sensing – Remote Sensing for the Earth Sciences*, A. N. Rencz (Ed.), New York: John Wiley & Sons, 672 p.

Colwell, R. N., (Ed.), 1983, *Manual of Remote Sensing*, 2nd ed., Falls Church, VA: American Society of Photogrammetry, 2440 p.

Conel, J. E., 1990, "Determination of Surface Reflectance and Estimates of Atmospheric Optical Depth and Single Scattering Albedo from Landsat Thematic Mapper Data," *International Journal of Remote Sensing*, 11:783–828.

Coppin, P. R. and M. E. Bauer, 1994, "Processing of Multitemporal Landsat TM Imagery to Optimize Extraction of Forest Cover Change Features," *IEEE Transactions on Geoscience and Remote Sensing*, 32:918–927.

Cracknell, A. P. and L. W. Hayes, 1993, "Atmospheric Corrections to Passive Satellite Remote Sensing Data," Chapter 8 in *Introduction to Remote Sensing*, London: Taylor & Francis, 116–158.

Crippen, R. E., 1989, "A Simple Spatial Filtering Routine for the Cosmetic Removal of Scan-line Noise from Landsat TM P-Tape Imagery," *Photogrammetric Engineering & Remote Sensing*, 55(3):327–331.

Du, Y., Teillet, P. M. and J. Cihlar, 2002, "Radiometric Normalization of Multitemporal High-resolution Satellite Images with Quality Control for Land Cover Change Detection," *Remote Sensing of Environment*, 82:123.

Eckhardt, D. W., Verdin, J. P. and G. R. Lyford, 1990, "Automated Update of an Irrigated Lands GIS Using SPOT HRV Imagery," *Photogrammetric Engineering & Remote Sensing*, 56(11):1515–1522.

Englert, B., Scully, M. O. and H. Walther, 1994, "The Duality in Matter and Light," *Scientific American*, 271(6):86–92.

Feinberg, G., 1985, "Light," in R. K. Holz (Ed.), *The Surveillant Science: Remote Sensing of the Environment*, 2nd ed., New York: John Wiley & Sons, 2–11.

Farrand, W. H., Singer, R. B. and E. Merenyi, 1994, "Retrieval of Apparent Surface Reflectance from AVIRIS Data: A Comparison of Empirical Line, Radiative Transfer, and Spectral Mixture Methods," *Remote Sensing of Environment*, 47:311–321.

Foody, G. M., 1996, "Approaches for the Production and Evaluation of Fuzzy Land Cover Classifications from Remotely Sensed Data," *International Journal of Remote Sensing*, 17(7):1317–1340.

Forster, B. C., 1984, "Derivation of Atmospheric Correction Procedures for Landsat MSS with Particular Reference to Urban Data," *International Journal of Remote Sensing*, 5:799–817.

Gao, B.C., Heidebrecht, K. B. and A. F. H. Goetz, 1999, *ATmosphere REMoval Program (ATREM) User's Guide*, Version 3.1, Boulder, CO: Center for the Study of Earth from Space, 31 p.

Gibson, P. J. and C. H. Power, 2000, *Introductory Remote Sensing: Digital Image Processing and Applications*, London: Routledge, 249 p.

Green, R. O., 2003, "Introduction to Atmospheric Correction," Chapter 2 in *ACORN Tutorial*, Boulder, CO: Analytical Imaging and Geophysics, LLC, 12–18.

Haboudane, D., Miller, J. R., Tremblaly, N., Zarco-Tajada, P. J. and L. Dextraze, 2002, "Integrated Narrow-band Vegetation Indices for Prediction of Crop Chlorophyll Content for Application to Precision Agriculture," *Remote Sensing of Environment*, 81:416–426.

Hall, F. G., Strebel, D. E., Nickeson, J. E. and S. J. Goetz, 1991, "Radiometric Rectification: Toward a Common Radiometric Response Among Multidate, Multisensor Images," *Remote Sensing of Environment*, 35:11–27.

Hall-Konyves, K., 1987, "The Topographic Effect on Landsat Data in Gentle Undulating Terrain in Southern Sweden," *International Journal of Remote Sensing*, 8(2):157–168.

Heo, J. and F. W. Fitzhugh, 2000, "A Standardized Radiometric Normalization Method for Change Detection Using Remotely Sensed Imagery," *ISPRS Journal of Photogrammetry and Remote Sensing*, 60:173–181.

InSpec, 2002, *ACORN 4.0 User's Guide*, Boulder, CO: Analytical Imaging and Geophysics, LLC, 76 p.

Jensen, J. R., Botchway, K., Brennan-Galvin, E., Johannsen, C., Juma, C., Mabogunje, A., Miller, R., Price, K., Reining, P., Skole, D., Stancioff, A. and D. R. F. Taylor, 2002, *Down to Earth: Geographic Information for Sustainable Development in Africa*, Washington: National Research Council, 155 p.

Jensen, J. R., Cowen, D., Narumalani, S., Weatherbee, O. and J. Althausen, 1993, "Evaluation of CoastWatch Change Detection Protocol in South Carolina," *Photogrammetric Engineering & Remote Sensing*, 59(6):1039–1046.

Jensen, J. R., Hadley, B. C., Tullis, J. A., Gladden, J., Nelson, S., Riley, S., Filippi, T. and M. Pendergast, 2003, *2002 Hyperspectral Analysis of Hazardous Waste Sites on the Savannah River Site*, Aiken, SC: Westinghouse Savannah River Company, WSRC-TR-2003-0025, 52 p.

Jensen, J. R., Rutchey, K., Koch, M. and S. Narumalani, 1995, "Inland Wetland Change Detection in the Everglades Water Conservation Area 2A Using a Time Series of Normalized Remotely Sensed Data," *Photogrammetric Engineering & Remote Sensing*, 61(2):199–209.

JPL, 2004a, *JPL Spectral Library*, Pasadena: Jet Propulsion Lab, http://speclib.jpl.nasa.gov.

JPL, 2004b, *Johns Hopkins University Spectral Library*, Pasadena: Jet Propulsion Lab, http://speclib.jpl.nasa.gov.

Jones, A. R., Settle, J. J. and B. K. Wyatt, 1988, "Use of Digital Terrain Data in the Interpretation of SPOT-1 HRV Multispectral Imagery," *International Journal of Remote Sensing*, 9(4):729–748.

Justice, C. O., Vermote, E., Townshend, J. R. G., Defries, R., Roy, D. P., Hall, D. K., Solomonson, V. V., Privette, J. L., Riggs, G. and A. Strahler, 1998, "The Moderate Resolution Imaging Spectroradiometer (MODIS): Land Remote Sensing for Global Change Research," *IEEE Transactions on Geoscience and Remote Sensing*, 36(4):1228–1249.

Kawata, Y., Ohtani, A., Kusaka, T. and S. Ueno, 1990, "Classification Accuracy for the MOS-1 MESSR Data Before and After the Atmospheric Correction," *IEEE Transactions on Geoscience Remote Sensing*, 28:755–760.

Kawata, Y., Ueno, S. and T. Kusaka, 1988, "Radiometric Correction for Atmospheric and Topographic Effects on Landsat MSS Images," *International Journal of Remote Sensing*, 9(4):729–748.

Kaufman, Y. J., and R. S. Fraser, 1984, "Atmospheric Effect on Classification of Finite Fields," *Remote Sensing of Environment*, 15:95–118.

Kaufman, Y. J., Wald, A. E., Remer, L. A., Gao, B. C., Li, R. R. and F. Flynn, 1997, "The MODIS 2.1-mm Channel Correlation with Visible Reflectance for Use in Remote Sensing of Aerosol," *IEEE Transactions on Geoscience and Remote Sensing*, 35:1286–1298.

Konecny, G., 2003, *Geoinformation: Remote Sensing, Photogrammetry and Geographic Information Systems*, New York: Taylor & Francis, 248 p.

Leprieur, C. E., J. M. Durand and J. L. Peyron, 1988, "Influence of Topography on Forest Reflectance Using Landsat Thematic Mapper and Digital Terrain Data," *Photogrammetric Engineering & Remote Sensing*, 54(4):491–496.

Lo, C. P. and A. K. W. Yeung, 2002, *Concepts and Techniques of Geographic Information Systems*, Upper Saddle River, NJ: Prentice-Hall, 492 p.

Lunetta, R. S., Congalton, R. G., Fenstermaker, L. K., Jensen, J. R., McGwire, K. C. and L. R. Tinney, 1991, "Remote Sensing and Geographic Information Systems Data Integration: Error Sources and Research Issues," *Photogrammetric Engineering & Remote Sensing*, 57(6):677–687.

Marakas, G. M., 2003, *Decision Support Systems in the 21st Century*, Upper Saddle River, NJ: Prentice-Hall, 611 p.

Matthew, M. W., Adler-Golden, S. M., Berk, A., Richtsmeier, S. C., Levin, R. Y., Bernstein, L. S., Acharya, P. K., Anderson, G. P., Felde, G. W., Hoke, M. P., Ratkowski, A., Burke, H. H., Kaiser, R. D. and D. P. Miller, 2000, "Status of Atmospheric Correction Using a MODTRAN4-based Algorithm," *SPIE Proc. Algorithms for Multispectral, Hyperspectral, and Ultraspectral Imagery VI*, 4049:199–207.

Meadows, J., 1992, *The Great Scientists*, New York: Oxford University Press, 248 p.

Meyer, P., Itten, K. I., Kellenberger, T., Sandmeier, S. and R. Sandmeier, 1993, "Radiometric Corrections of Topographically Induced Effects on Landsat TM Data in an Alpine Environment," *ISPRS Journal of Photogrammetry and Remote Sensing* 48(4):17–28.

Miller, S. W. and E. Vermote, 2002, *NPOESS Visible/Infrared Imager/Radiometer Suite: Algorithm Theoretical Basis Document*, Version 5, Lanham, MD: Raytheon, 83 p.

Milman, A. S., 1999, *Mathematical Principles of Remote Sensing: Making Inferences from Noisy Data*, Ann Arbor, MI: Ann Arbor Press, 37 p.

Moran, S. M., Bryant, R. B., Clarke, T. R. and J. Qi, 2001, "Deployment and Calibration of Reference Reflectance Tarps for Use with Airborne Imaging Sensors," *Photogrammetric Engineering & Remote Sensing*, 67(3):273–286.

Mulligan, J. R., 1980, *Practical Physics: The Production and Conservation of Energy*, New York: McGraw-Hill, 526 pp.

Nassau, K., 1983, *The Physics and Chemistry of Color: The Fifteen Causes of Color*, New York: John Wiley & Sons.

Nassau, K., 1984, "The Physics of Color," in *Science Year 1984*, Chicago: World Book, 126–139.

Ontar, *1999 Product Catalog: PcModWin* (commercial version of U.S. Air Force Research Lab's MODTRAN), Andover, MA: Ontar Inc., 23 p.

Research Systems, 2002, *ATREM*, Boulder, CO: Research Systems, Inc., http://www.rsinc.com.

Research Systems, 2003, *FLAASH—Fast Line-of-sight Atmospheric Analysis of Spectral Hypercubes*, Boulder, CO: Research Systems, Inc., http://www.rsinc.com/envi/flaash.asp.

Richards, J. A. and X. Jia, 1999, *Remote Sensing Digital Image Analysis*, New York: Springer-Verlag, 363 p.

Rinker, J. N., 1999, *Introduction to Spectral Remote Sensing*, Alexandria, VA: U.S. Army Topographic Engineering Center, http://www.tec.army.mil/terrain/desert/tutorial.

Roberts, D. A., Yamagushi, Y. and R. J. P. Lyon, "Comparison of Various Techniques for Calibration of AIS Data," in Vane, G. and A. F. H. Goetz (Eds.), *Proceedings, 2nd Airborne Imaging Spectrometer Data Analysis Workshop*, Pasadena: NASA Jet Propulsion Laboratory, JPL Publication #86-35, 21–30.

Rott, H., 2000, "Physical Principles and Technical Aspects of Remote Sensing," in Schultz, G. A. and E. T. Engman (Eds.), *Remote Sensing in Hydrology and Water Management*, Berlin: Springer, 16–39.

Rutchey, K. and L. Vilchek, 1994, "Development of an Everglades Vegetation Map Using a SPOT Image and the Global Positioning System," *Photogrammetric Engineering & Remote Sensing*, 60(6):767–775.

Sagan, C., 1994, *Pale Blue Dot*, New York: Random House, 429 p.

Schott, J. R., Salvaggio, C. and W. J. Wolchok, 1988, "Radiometric Scene Normalization Using Pseudoinvariant Features," *Remote Sensing of Environment*, 26:1–16.

Shasby, M., and D. Carneggie, 1986, "Vegetation and Terrain Mapping in Alaska Using Landsat MSS and Digital Terrain Data," *Photogrammetric Engineering & Remote Sensing*, 52(6):779–786.

Singh, A., 1989, "Digital Change Detection Techniques Using Remotely Sensed Data," *International Journal of Remote Sensing*, 10:989–1003.

Slater, P. N., 1980, *Remote Sensing Optics and Optical Systems*. Reading, MA: Addison-Wesley, 575 p.

Song, C., Woodcock, C. E., Soto, K. C., Lenney, M. P. and S. A. Macomber, 2001, "Classification and Change Detection Using Landsat TM Data: When and How to Correct Atmospheric Effects?" *Remote Sensing of Environment*, 75:230-244.

Smith, G. M. and E. J. Milton, 1999, "The Use of Empirical Line Method to Calibrate Remotely Sensed Data to Reflectance," *International Journal of Remote Sensing*, 20:2653-2662.

Strahler, A. N. and A. H. Strahler, 1989, *Elements of Physical Geography*, 4th ed., New York: John Wiley & Sons, 562 p.

Tanre, D., Deroo, C., Duhaut, P., Herman, M., Morcrette, J., Perbos, J. and P. Y. Deschamps, 1986, *Second Simulation of the Satellite Signal in the Solar Spectrum (6S) User's Guide,* U.S.T. de Lille, 59655 Villeneuve d'Ascq, France: Lab d'Optique Atmospherique.

Teillet, P. M., 1986, "Image Correction for Radiometric Effects in Remote Sensing," *International Journal of Remote Sensing*, 7(12):1637–1651.

Teillet, P. M., Guindon, B. and D. G. Goodenough, 1982, "On the Slope-aspect Correction of Multispectral Scanner Data," *Canadian Journal of Remote Sensing*, 8(2):84–106.

Thiemann, S. and K. Hermann, 2002, "Lake Water Quality Monitoring Using Hyperspectral Airborne Data—A Semiempirical Multisensor and Multitemporal Approach for the Mecklenburg Lake District, Germany," *Remote Sensing of Environment*, 81:228–237.

Trefil, J. and R. M. Hazen, 1995, *The Sciences: An Integrated Approach*, New York: John Wiley & Sons, 634 p.

Turner, R. E., 1975, "Signature Variations Due to Atmospheric Effects," *Proceedings*, *10th International Symposium on Remote Sensing of the Environment*, Ann Arbor, MI: Environmental Research Institute of Michigan, 671–682.

Turner, R. E., 1978. "Elimination of Atmospheric Effects from Remote Sensor Data," *Proceedings*, *12th International Symposium on Remote Sensing of Environment*, Ann Arbor, MI: Environmental Research Institute of Michigan, p. 783.

Turner, R. E. and M. M. Spencer, 1972, "Atmospheric Model for Correction of Spacecraft Data," *Proceedings*, *8th International Symposium on Remote Sensing of the Environment*, Ann Arbor, MI: Environmental Research Institute of Michigan, 895–934.

USGS, 2004, *USGS Digital Library —splib04: 9.2 to 3.0 Microns*, Washington: U.S. Geological Survey, http://speclab.cr.ugsg.gov/spectral.lib04/spectral-lib04.html.

Vermote, E., Tanre, D., Deuze, J. L., Herman, M. and J. J. Morcrette, 1997, *Second Simulation of the Satellite Signal in the Solar Spectrum (6S)*, Code 923, Washington: NASA Goddard Space Flight Center, 54 p.

Vermote, E. F., El Saleous, N. Z. and C. O. Justice, 2002, "Atmospheric Correction of MODIS Data in the Visible to Middle Infrared: First Results," *Remote Sensing of Environment*, 83:97–111.

Geometric Correction

It would be wonderful if every remotely sensed image contained data that were already in their proper geometric *x, y* locations. This would allow each image to be used as if it were a map. Unfortunately, this is not the case. Instead, it is usually necessary to *preprocess* the remotely sensed data and remove the geometric distortion so that individual picture elements (pixels) are in their proper planimetric (*x, y*) map locations. This allows remote sensing–derived information to be related to other thematic information in geographic information systems (GIS) or spatial decision support systems (SDSS) (Jensen et al., 2002; Marakas, 2003). Geometrically corrected imagery can be used to extract accurate distance, polygon area, and direction (bearing) information (Wolf, 2002; Tucker et al., 2004).

 Internal and External Geometric Error

Remotely sensed imagery typically exhibits *internal* and *external geometric error*. It is important to recognize the source of the internal and external error and whether it is *systematic* (predictable) or *nonsystematic* (random). Systematic geometric error is generally easier to identify and correct than random geometric error.

Internal Geometric Error

Internal geometric errors are generally introduced by the remote sensing system itself or in combination with Earth rotation or curvature characteristics. These distortions are often systematic (predictable) and may be identified and then corrected using prelaunch or in-flight platform ephemeris (i.e., information about the geometric characteristics of the sensor system and the Earth at the time of data acquisition). Geometric distortions in imagery that can sometimes be corrected through analysis of sensor characteristics and ephemeris data include:

- skew caused by Earth rotation effects,

- scanning system–induced variation in nominal ground resolution cell size,

- scanning system one-dimensional relief displacement, and

- scanning system tangential scale distortion.

227

Image Offset (Skew) Caused by Earth Rotation Effects

Earth-observing Sun-synchronous satellites are normally in fixed orbits that collect a path (or swath) of imagery as the satellite makes its way from the north to the south in descending mode (Figure 7-1a). Meanwhile, the Earth below rotates on its axis from west to east making one complete revolution every 24 hours. This interaction between the fixed orbital path of the remote sensing system and the Earth's rotation on its axis *skews* the geometry of the imagery collected. For example, consider just three hypothetical scans of 16 lines each obtained by the Landsat Enhanced Thematic Mapper Plus (ETM$^+$). If the data are not deskewed, they will appear in the dataset incorrectly, as shown in Figure 7-1b. While this looks correct, it is not. This matrix does not take into account Earth rotation effects. The data in this dataset are actually skewed to the *east* by a predictable amount.

Conversely, if the remotely sensed data are deskewed, then all of the pixels associated with a single scan (containing 16 lines of data) will be offset (adjusted) by the digital image processing system a certain amount to the *west* (Figure 7-1c). *Deskewing* is defined as the systematic displacement of pixels westward in a frame of imagery to correct for the interaction of the satellite sensor system's angular velocity and the Earth's surface velocity. This adjustment places all of the pixels in the scan in their proper positions relative to adjacent scans. Null values are added to the dataset to maintain the integrity of the raster (matrix) format. The amount of shift or displacement to the west is a function of the relative velocities of both the satellite and the Earth and the length of the image frame that is recorded.

Most satellite image data providers automatically deskew the data that is delivered to customers using the logic shown in Figure 7-1c. First, the surface velocity of the Earth, v_{earth}, is computed:

$$v_{earth} = \omega_{earth}\, r \cos \lambda, \qquad (7\text{-}1)$$

where r is the radius of the Earth (6.37816 Mm) and ω_{earth} is the Earth's rotational velocity (72.72 µrad s^{-1}) at a specific latitude, λ. Therefore, for Charleston, SC, located at 33°N latitude, the surface velocity of the Earth is:

$$v_{earth} = 72.72 \ \mu rad \ s^{-1} \times 6.37816 \ Mm \times 0.83867$$

$$v_{earth} = 389 \ m \ s^{-1}.$$

Next we must determine the length of time it takes for the satellite to scan a typical remote sensing frame (F) of data on the ground. In this example we will consider a 185-km frame

obtained by Landsats 1, 2, and 3 associated with the Multispectral Scanner and Landsats 4, 5, and 7 associated primarily with the Thematic Mapper and Enhanced Thematic Mapper Plus. Landsats 1, 2, and 3 have an angular velocity, $\omega_{land123}$, of 1.014 mrad s^{-1}. Landsat satellites 4, 5, and 7 have an angular velocity, $\omega_{land457}$, of 1.059 mrad s^{-1} (Williams, 2003)[1]. Therefore, a typical 185-km frame of Landsat MSS imagery (WRS1 orbit) would be scanned in:

$$s_t = \frac{L}{r \times \omega_{land123}} \qquad (7\text{-}2)$$

$$s_t = \frac{185 \ km}{(6.37816 \ Mm)(1.014 \ mrad \ s^{-1})} = 28.6 \ s \ .$$

A 185-km frame of Landsat 4, 5 and 7 imagery (WRS2 orbit) would be scanned in:

$$s_t = \frac{185 \ km}{(6.37816 \ Mm)(1.059 \ mrad \ s^{-1})} = 27.4 \ s \ .$$

Therefore, during the time that the 185-km frame of Landsat 1, 2, or MSS imagery was collected at Charleston, SC (33°N latitude), the surface of the Earth moved to the east by:

$$\Delta x_{east} = v_{earth}\, s_t \qquad (7\text{-}3)$$

$$\Delta x_{east} = 389 \ m \ s^{-1} \times 28.6 \ s = 11.12 \ km \ .$$

Similarly, during the time a 185-km frame of Landsat ETM+ imagery is collected, the surface of the Earth moves to the east by:

$$\Delta x_{east} = 389 \ m \ s^{-1} \times 27.39 \ s = 10.65 \ km \ .$$

This is approximately 6% of the 185-km frame size (e.g., 11.12/185 = 0.06; 10.65/185 = 0.057) for Landsats 1, 2, and 3 MSS and Landsat TMs 4, 5, and 7 data. Fortunately, most of the satellite imagery provided by commercial and public remote sensing data providers have already been deskewed by the appropriate amount per line scan.

The skewing described holds true for line-scan sensors such as Landsat MSS, TM, ETM$^+$ and also for pushbroom sensors

---

1. For Landsats 1, 2, and 3 (WRS1 orbit) the angular velocity is (251 paths/cycle × 2π × 1,000 mrad/path) / (18 days/cycle × 86,400 s/day) = 1.014 mrad/s. For Landsats 4, 5, and 7 (WRS2 orbit) the angular velocity is (233 paths/cycle × 2π × 1,000 mrad/path) / (16 days/cycle × 86,400 s/day) = 1.059 mrad/s.

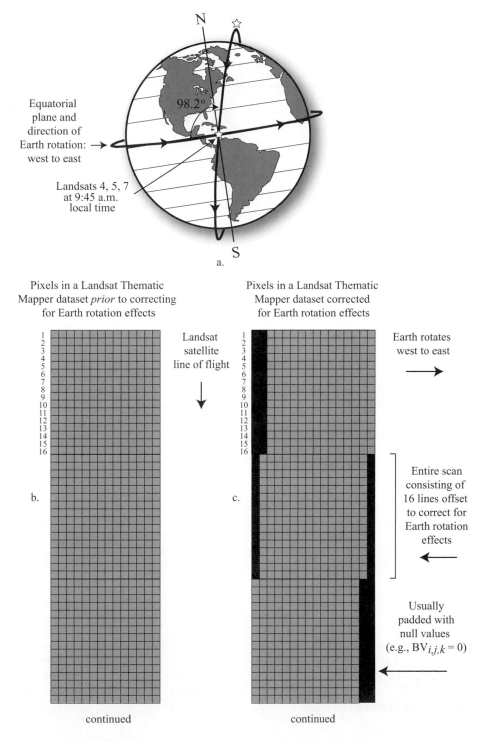

Pixels in a Landsat Thematic
Mapper dataset *prior* to correcting
for Earth rotation effects

Pixels in a Landsat Thematic
Mapper dataset corrected
for Earth rotation effects

Figure 7-1 a) Landsat satellites 4, 5, and 7 are in a Sun-synchronous orbit with an angle of inclination of 98.2°. The Earth rotates on its axis from west to east as imagery is collected. b) Pixels in three hypothetical scans (consisting of 16 lines each) of Landsat TM data. While the matrix (raster) may look correct, it actually contains systematic geometric distortion caused by the angular velocity of the satellite in its descending orbital path in conjunction with the surface velocity of the Earth as it rotates on its axis while collecting a frame of imagery. c) The result of adjusting (deskewing) the original Landsat TM data to the west to compensate for Earth rotation effects. Landsats 4, 5, and 7 use a bidirectional cross-track scanning mirror (NASA, 1998).

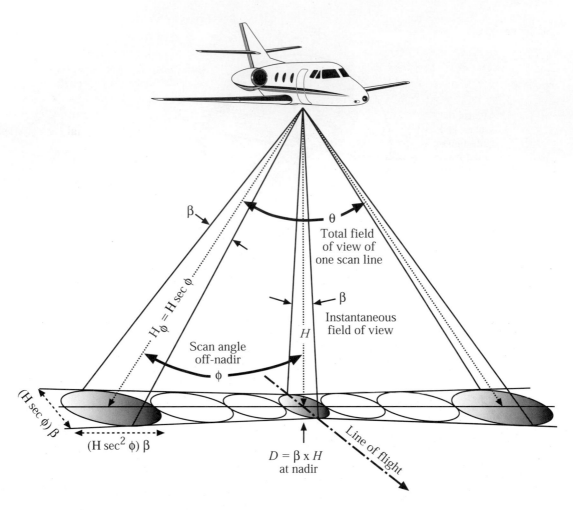

Figure 7-2   The ground resolution cell size along a single across-track scan is a function of a) the distance from the aircraft to the observation where H is the altitude of the aircraft above ground level (AGL) at nadir and $H \sec \phi$ off-nadir; b) the instantaneous-field-of-view of the sensor, β, measured in radians; and c) the scan angle off-nadir, ϕ. Pixels off-nadir have semimajor and semiminor axes (diameters) that define the resolution cell size. The total field of view of one scan line is θ. One-dimensional relief displacement and tangential scale distortion occur in the direction perpendicular to the line of flight and parallel with a line scan.

such as SPOT HRV, Digital Globe's QuickBird, and Space Imaging's IKONOS. Every fixed-orbit remote sensing system that collects data while the Earth rotates on its axis will contain frames of imagery that are skewed.

Scanning System-induced Variation in Ground Resolution Cell Size

A large amount of remote sensor data is acquired using scanning systems (e.g., Landsat 7, ASTER). Fortunately, an orbital multispectral scanning system scans through just a few degrees off-nadir as it collects data hundreds of kilometers above the Earth's surface (e.g., Landsat 7 data are col-

lected at 705 km AGL). This configuration minimizes the amount of distortion introduced by the scanning system. Conversely, a suborbital multispectral scanning system may be operating just tens of kilometers AGL with a scan field of view of perhaps 70 degrees. This introduces numerous types of geometric distortion that can be difficult to correct.

The ground swath width (*gsw*) is the length of the terrain strip remotely sensed by the system during one complete across-track sweep of the scanning mirror. It is a function of the total angular field of view of the sensor system, θ, and the altitude of the sensor system above ground level, H (Figure 7-2). It is computed:

$$gsw = \tan\left(\frac{\theta}{2}\right) \times H \times 2. \qquad (7\text{-}4)$$

For example, the ground swath width of an across-track scanning system with a 90° total field of view and an altitude above ground level of 6000 m would be 12,000 m:

$$gsw = \tan\left(\frac{90}{2}\right) \times 6000 \times 2$$

$$gsw = 1 \times 6000 \times 2$$

$$gsw = 12{,}000 \text{ m}.$$

Most scientists using across-track scanner data use only approximately the central 70 percent of the swath width (35 percent on each side of nadir) primarily because ground resolution elements have larger cell sizes the farther they are away from nadir.

The diameter of the circular ground area viewed by the sensor, D, at nadir is a function of the instantaneous-field-of-view, β, of the scanner measured in milliradians (mrad) and the altitude of the scanner above ground level, H, where $D = \beta \times H$. As the scanner's instantaneous field-of-view moves away from nadir on either side, the circle becomes an ellipsoid because the distance from the aircraft to the resolution cell is increasing, as shown in Figure 7-2. In fact, the distance from the aircraft to the resolution cell on the ground, H_ϕ, is a function of the scan angle off-nadir, ϕ, at the time of data collection and the true altitude of the aircraft, H (Lillesand et al., 2004):

$$H_\phi = H \times \sec\phi. \qquad (7\text{-}5)$$

Thus, the size of the ground-resolution cell increases as the angle increases away from nadir. The nominal (average) diameter of the elliptical resolution cell, D_ϕ, at this angular location from nadir has the dimension:

$$D_\phi = (H \times \sec\phi) \times \beta \qquad (7\text{-}6)$$

in the direction of the line of flight, and

$$D_\phi = (H \times \sec^2\phi) \times \beta \qquad (7\text{-}7)$$

in the orthogonal (perpendicular) scanning direction.

Scientists using across-track scanner data usually concern themselves only with the spatial ground resolution of the cell at nadir, D. If it is necessary to perform precise quantitative work on pixels some angle ϕ off-nadir, then it may be impor-

tant to remember that the radiant flux recorded is an integration of the radiant flux from all the surface materials in a ground-resolution cell with a constantly changing diameter. Using only the central 70 percent of the swath width reduces the impact of the larger pixels found at the extreme edges of the swath.

Scanning System One-Dimensional Relief Displacement

Truly vertical aerial photographs have a single principal point directly beneath the aircraft at nadir at the instant of exposure. This perspective geometry causes all objects that rise above the local terrain elevation to be displaced from their proper planimetric position radially outward from the principal point. For example, the four hypothetical tanks in Figure 7-3a are each 50 ft high. The greater the distance from the principal point, the greater the radial relief displacement of the top of the tank away from its base.

Images acquired using an across-track scanning system also contain relief displacement. However, instead of being radial from a single principal point, the displacement takes place in a direction that is perpendicular to the line of flight for each and every scan line, as shown in Figure 7-3b. In effect, the ground-resolution element at nadir functions like a principal point for each scan line. At nadir, the scanning system looks directly down on the tank, and it appears as a perfect circle in Figure 7-3b. The greater the height of the object above the local terrain and the greater the distance of the top of the object from nadir (i.e., the line of flight), the greater the amount of *one-dimensional relief displacement* present. One-dimensional relief displacement is introduced in both directions away from nadir for each sweep of the across-track mirror.

Although some aspects of one-dimensional relief displacement may be of utility for visual image interpretation, it seriously displaces the tops of objects projecting above the local terrain from their true planimetric position. Maps produced from such imagery contain serious planimetric errors.

Scanning System Tangential Scale Distortion

The mirror on an across-track scanning system rotates at a constant speed and typically views from 70° to 120° of terrain during a complete line scan. Of course, the amount depends on the specific sensor system. From Figure 7-2 it is clear that the terrain directly beneath the aircraft (at nadir) is closer to the aircraft than the terrain at the edges during a single sweep of the mirror. Therefore, because the mirror rotates at a constant rate, the sensor scans a shorter geographic distance at nadir than it does at the edge of the image. This rela-

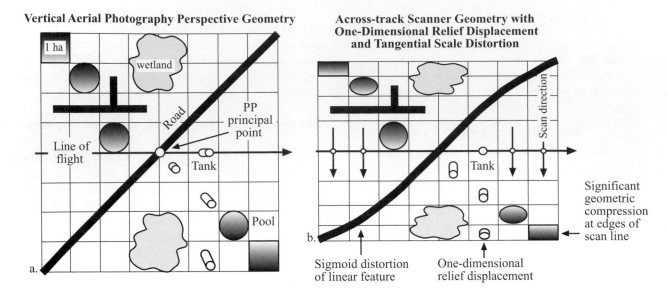

Figure 7-3   a) Hypothetical perspective geometry of a vertical aerial photograph obtained over level terrain. Four 50-ft-tall tanks are distributed throughout the landscape and experience varying degrees of radial relief displacement the farther they are from the principal point (PP). b) Across-track scanning system introduces one-dimensional relief displacement perpendicular to the line of flight and tangential scale distortion and compression the farther the object is from nadir. Linear features trending across the terrain are often recorded with *s*-shaped or sigmoid curvature characteristics due to tangential scale distortion and image compression.

tionship tends to *compress* features along an axis that is perpendicular to the line of flight. The greater the distance of the ground-resolution cell from nadir, the greater the image scale compression. This is called *tangential scale distortion*. Objects near nadir exhibit their proper shape. Objects near the edge of the flight line become compressed and their shape distorted. For example, consider the tangential geometric distortion and compression of the circular swimming pools and one hectare of land the farther they are from nadir in the hypothetical diagram (Figure 7-3b).

This tangential scale distortion and compression in the far range also causes linear features such as roads, railroads, utility right of ways, etc., to have an *s-shape* or *sigmoid distortion* when recorded on scanner imagery (Figure 7-3b). Interestingly, if the linear feature is parallel with or perpendicular to the line of flight, it does not experience sigmoid distortion.

Even single flight lines of aircraft MSS data are difficult to rectify to a standard map projection because of aircraft roll, pitch, and yaw during data collection (Jensen et al., 1988). Notches in the edge of a flight line of data are indicative of aircraft roll. Such data require significant human and machine resources to make the data planimetrically accurate. Most commercial data providers now place GPS on the air-

craft to obtain precise flight line coordinates, which are useful when rectifying the aircraft MSS data.

External Geometric Errors

External geometric errors are usually introduced by phenomena that vary in nature through space and time. The most important external variables that can cause geometric error in remote sensor data are random movements by the aircraft (or spacecraft) at the exact time of data collection, which usually involve:

- altitude changes, and/or

- attitude changes (roll, pitch, and yaw).

Altitude Changes

A remote sensing system is ideally flown at a constant altitude above ground level (AGL) resulting in imagery with a uniform scale all along the flightline. For example, a frame camera with a 12-in. focal length lens flown at 20,000 ft. AGL will yield 1:20,000-scale imagery. If the aircraft or spacecraft gradually changes its altitude along a flightline, then the scale of the imagery will change (Figure 7-4a).

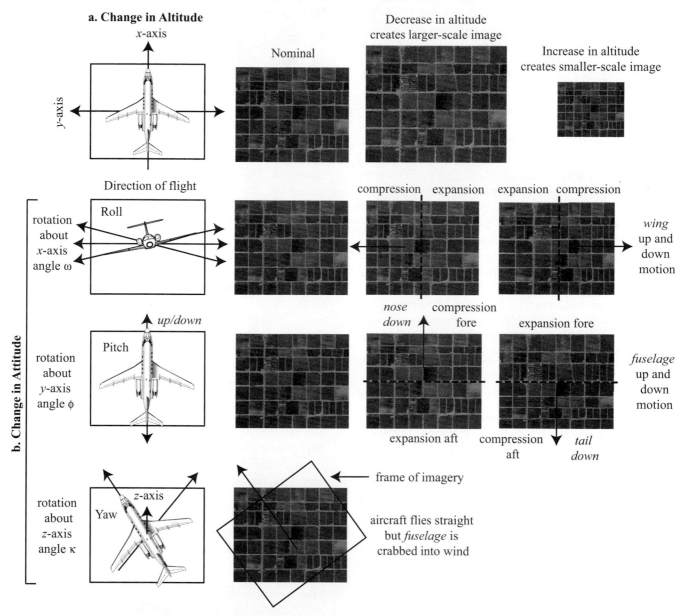

Geometric Modification of Remotely Sensed Data
Caused by Changes in Platform Altitude and Attitude

a. Change in Altitude

Figure 7-4 a) Geometric modification in imagery may be introduced by changes in the aircraft or satellite platform *altitude* above ground level (AGL) at the time of data collection. Increasing altitude results in smaller-scale imagery while decreasing altitude results in larger-scale imagery. b) Geometric modification may also be introduced by aircraft or spacecraft changes in *attitude*, including roll, pitch, and yaw. An aircraft flies in the *x*-direction. Roll occurs when the aircraft or spacecraft fuselage maintains directional stability but the wings move up or down, i.e. they rotate about the *x*-axis angle (omega: ω). Aircraft pitch occurs when the wings are stable but the fuselage nose or tail moves up or down, i.e., they rotate about the *y*-axis angle (phi: ϕ). Yaw occurs when the wings remain parallel but the fuselage is forced by wind to be oriented some angle to the left or right of the intended line of flight, i.e., it rotates about the z-axis angle (kappa: κ). Thus, the plane flies straight but all remote sensor data are displaced by κ°. Remote sensing data often are distorted due to a combination of changes in *altitude* and *attitude* (roll, pitch, and yaw).

Increasing the altitude will result in smaller-scale imagery (e.g., 1:25,000-scale). Decreasing the altitude of the sensor system will result in larger-scale imagery (e.g, 1:15,000). The same relationship holds true for digital remote sensing systems collecting imagery on a pixel by pixel basis. The diameter of the spot size on the ground (D; the nominal spatial resolution) is a function of the instantaneous-field-of-view (β) and the altitude above ground level (H) of the sensor system, i.e., $D = \beta \times H$.

It is important to remember, however, that scale changes can be introduced into the imagery even when the remote sensing system is flown at a constant elevation above ground level. This occurs when the terrain gradually increases or decreases in elevation (i.e., it moves closer to or farther away from the sensor system). For example, if the terrain surface is at 1000 ft AGL at the start of a flight line and 2000 ft AGL at the end of the flight line, then the scale of the imagery will become larger as the flight line progresses. Remote sensing platforms do not generally attempt to adjust for such gradual changes in elevation. Rather, it is acknowledged that scale changes will exist in the imagery and that the use of geometric rectification algorithms will normally be used to minimize the effects. The methods for adjusting for changes in scale using ground control points and geometric rectification coefficients will be discussed shortly.

Attitude Changes

Satellite platforms are usually stable because they are not buffeted by atmospheric turbulence or wind. Conversely, suborbital aircraft must constantly contend with atmospheric updrafts, downdrafts, head-winds, tail-winds, and cross-winds when collecting remote sensor data. Even when the remote sensing platform maintains a constant altitude AGL, it may rotate randomly about three separate axes that are commonly referred to as *roll*, *pitch*, and *yaw* (Figure 7-4b). For example, sometimes the fuselage remains horizontal, but the aircraft rolls from side to side about the x-axis (direction of flight) some $\omega°$, introducing compression and/or expansion of the imagery in the near- and far-ranges perpendicular to the line of flight (Wolf, 2002). Similarly, the aircraft may be flying in the intended direction but the nose pitches up or down a certain $\phi°$ about the y-axis. If the nose pitches down, the imagery will be compressed in the fore-direction (toward the nose of the aircraft) and expanded in the aft-direction (toward the tail). If the nose pitches up, the imagery will be compressed in the aft-direction and expanded in the fore-direction. Occasionally a remote sensing platform experiences significant headwinds (or tailwinds) that must be compensated for in order to fly in a straight direction. When this occurs, it is possible for the pilot to crab the aircraft fuselage

into the wind $\kappa°$ about the z-axis. The result is an accurate flight line but imagery that is oriented some $\kappa°$ from the intended flight line (Figure 7-4b).

High quality satellite and aircraft remote sensing systems often have gyro-stabilization equipment that, in effect, isolates the sensor system from the roll and pitch movements of the aircraft. Remote sensing systems without stabilization equipment introduce some geometric error into the remote sensing dataset through variations in roll, pitch, and yaw that can only be corrected using ground control points.

Ground Control Points

Geometric distortions introduced by sensor system attitude (roll, pitch, and yaw) and/or altitude changes can be corrected using ground control points and appropriate mathematical models (Bernstein, 1983). A *ground control point* (GCP) is a location on the surface of the Earth (e.g., a road intersection) that can be identified on the imagery and located accurately on a map. The image analyst must be able to obtain two distinct sets of coordinates associated with each GCP:

• image coordinates specified in i rows and j columns, and

• map coordinates (e.g., x, y measured in degrees of latitude and longitude, feet in a state plane coordinate system, or meters in a Universal Transverse Mercator projection).

The paired coordinates (i, j and x, y) from many GCPs (e.g., 20) can be modeled to derive geometric transformation coefficients (Buiten and Van Putten, 1997). These coefficients may then be used to geometrically rectify the remotely sensed data to a standard datum and map projection.

 Types of Geometric Correction

Most commercially available remote sensor data (e.g., from SPOT Image Inc., DigitalGlobe Inc., Space Imaging, Inc.) already have much of the systematic error removed. Unless otherwise processed, however, the unsystematic random error remains in the image, making it nonplanimetric (i.e., the pixels are not in their correct x, y planimetric map position). This section focuses on two common geometric correction procedures often used by scientists to make the digital remote sensor data of value:

• image-to-map rectification, and

Selecting Ground Control Points for Image-to-Map Rectification

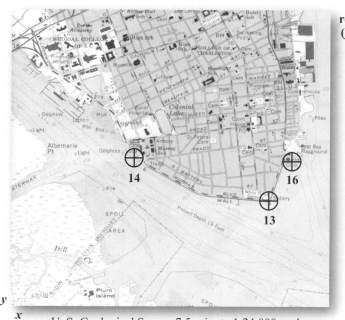

column (*x'*)

row (*y'*)

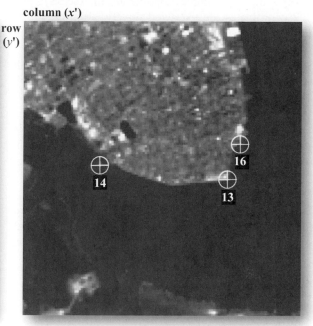

a. U. S. Geological Survey 7.5-minute 1:24,000-scale topographic map of Charleston, SC, with three ground control points identified.

b. Unrectified Landsat Thematic Mapper band 4 image obtained on November 9, 1982.

Figure 7-5   Example of *image-to-map rectification*. a) U.S. Geological Survey 7.5-minute 1:24,000-scale topographic map of Charleston, SC, with three ground control points identified (13, 14, and 16). The GCP map coordinates are measured in meters easting (*x*) and northing (*y*) in a Universal Transverse Mercator projection. b) Unrectified 11/09/82 Landsat TM band 4 image with the three ground control points identified. The image GCP coordinates are measured in rows and columns.

- image-to-image registration.

The general rule of thumb is to rectify remotely sensed data to a standard map projection whereby it may be used in conjunction with other spatial information in a GIS to solve problems. Therefore, most of the discussion will focus on image-to-map rectification.

Image-to-Map Rectification

Image-to-map rectification is the process by which the geometry of an image is made planimetric. Whenever accurate area, direction, and distance measurements are required, image-to-map geometric rectification should be performed. It may not, however, remove all the distortion caused by topographic relief displacement in images. The image-to-map rectification process normally involves selecting GCP image pixel coordinates (row and column) with their map

coordinate counterparts (e.g., meters northing and easting in a Universal Transverse Mercator map projection). For example, Figure 7-5 displays three GCPs (13, 14, and 16) easily identifiable by an image analyst in both the U.S. Geological Survey 1:24,000-scale 7.5-minute quadrangle and an unrectified Landsat TM band 4 image of Charleston, SC. It will be demonstrated how the mathematical relationship between the image coordinates and map coordinates of the selected GCPs is computed and how the image is made to fit the geometry of the map.

In the United States, there are at least four alternatives to obtaining accurate ground control point (GCP) *map* coordinate information for image-to-map rectification, including:

- hard-copy planimetric maps (e.g., U.S. Geological Survey 7.5-minute 1:24,000-scale topographic maps) where GCP coordinates are extracted using simple ruler measurements or a coordinate digitizer;

- digital planimetric maps (e.g., the U.S. Geological Survey's digital 7.5-minute topographic map series) where GCP coordinates are extracted directly from the digital map on the screen;

- digital orthophotoquads that are already geometrically rectified (e.g., U.S. Geological Survey digital orthophoto quarter quadrangles —DOQQ); and/or

- global positioning system (GPS) instruments that may be taken into the field to obtain the coordinates of objects to within ±20 cm if the GPS data are differentially corrected (Renz, 1999; Bossler et al., 2002).

GPS collection of map coordinate information to be used for image rectification is especially effective in poorly mapped regions of the world or where rapid change has made existing maps obsolete (Jensen et al., 2002).

Image-to-Image Registration

Image-to-image registration is the translation and rotation alignment process by which two images of like geometry and of the same geographic area are positioned coincident with respect to one another so that corresponding elements of the same ground area appear in the same place on the registered images (Chen and Lee, 1992). This type of geometric correction is used when it is *not* necessary to have each pixel assigned a unique *x, y* coordinate in a map projection. For example, we might want to make a cursory examination of two images obtained on different dates to see if any change has taken place. While it is possible to rectify both of the images to a standard map projection and then evaluate them (image-to-map rectification), this may not be necessary to simply identify the change that has taken place between the two images.

Hybrid Approach to Image Rectification/Registration

It is interesting that the same general image processing principles are used in both image rectification and image registration. The difference is that in image-to-map rectification the reference is a map in a standard map projection, while in image-to-image registration the reference is another image. It should be obvious that if a rectified image is used as the reference base (rather than a traditional map) any image registered to it will inherit the geometric errors existing in the reference image. Because of this characteristic, most serious Earth science remote sensing research is based on analysis of data that have been rectified to a map base. However, when conducting rigorous change detection between two or more dates of remotely sensed data, it may be useful to select a *hybrid* approach involving both image-to-map rectification and image-to-image registration (Jensen et al., 1993).

An example of the hybrid approach is demonstrated in Figure 7-6 where an October 14, 1987, Landsat TM image is being registered to a rectified November 9, 1982, Landsat TM scene. In this case, the 1982 base year image was previously rectified to a Universal Transverse Mercator map projection with 30×30 m pixels. Ground control points are being selected to register the 1987 image to the rectified 1982 base year image. It is often very difficult to locate good ground control points in remotely sensed data, especially in rural areas (e.g., forests, wetland, and water bodies). The use of the *rectified* base year image as the map allows many more common GCPs to be located in the unrectified 1987 imagery. For example, edges of water bodies and fields or the intersection of small stream segments are not usually found on a map but may be easy to identify in the rectified and unrectified imagery.

The optimum method of selecting the ground control points is to have both the rectified base year image (or reference map) and the image to be rectified on the screen at the same time (Figure 7-6). This dual display greatly simplifies GCP selection. Some image processing systems even allow the GCP selected to be reprojected onto the image to be corrected (with the appropriate transformation coefficients, to be discussed shortly) to determine the quality of the GCP point. Also, some systems allow the analyst to extract floating point row and column coordinates of GCPs (instead of just integer values) through the use of a chip extraction algorithm that zooms in and does subpixel sampling, as demonstrated in Figure 7-6. GCP subpixel row and column coordinates often improve the precision of the image-to-map rectification or image-to-image registration. Some scientists have developed methods of automatically extracting GCPs common to two images, which can be used during image-to-image registration (e.g., Chen and Lee, 1992). However, most image-to-map rectification still relies heavily on human interaction.

The following example focuses on image-to-map geometric rectification because it is the most frequently used method of removing geometric distortion from remotely sensed data.

Image-to-Map Geometric Rectification Logic

Two basic operations must be performed to geometrically rectify a remotely sensed image to a map coordinate system:

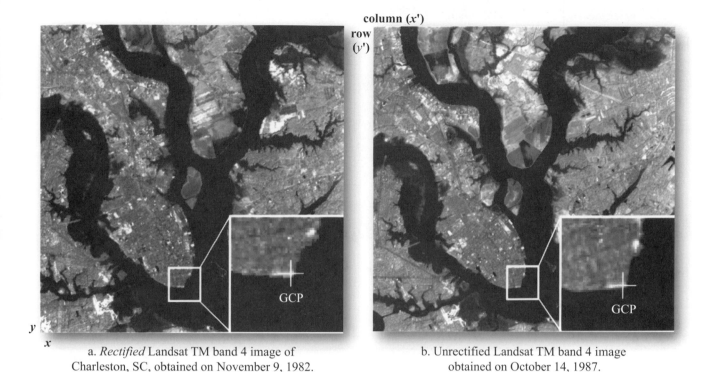

a. *Rectified* Landsat TM band 4 image of
Charleston, SC, obtained on November 9, 1982.

b. Unrectified Landsat TM band 4 image
obtained on October 14, 1987.

Figure 7-6   Example of image-to-image *hybrid* registration. a) Previously rectified Landsat TM band 4 data obtained on November 9, 1982, resampled to 30×30 m pixels using nearest-neighbor resampling logic and a UTM map projection. b) Unrectified October 14, 1987, Landsat TM band 4 data to be registered to the rectified 1982 Landsat scene.

1. The geometric relationship between the input pixel coordinates (column and row; referred to as x', y') and the associated map coordinates of this same point (x, y) must be identified (Figures 7-5 and 7-6). A number of GCP pairs are used to establish the nature of the geometric coordinate transformation that must be applied to rectify or fill every pixel in the output image (x, y) with a value from a pixel in the unrectified input image (x', y'). This process is called *spatial interpolation*.

2. Pixel brightness values must be determined. Unfortunately, there is no direct one-to-one relationship between the movement of input pixel values to output pixel locations. It will be shown that a pixel in the rectified output image often requires a value from the input pixel grid that does not fall neatly on a row-and-column coordinate. When this occurs, there must be some mechanism for determining the brightness value (BV) to be assigned to the output rectified pixel. This process is called *intensity interpolation*.

Spatial Interpolation Using Coordinate Transformations

As discussed earlier, some distortions in remotely sensed data may be removed or mitigated using techniques that model systematic orbital and sensor characteristics. Unfortunately, this does not remove error produced by changes in attitude (roll, pitch, and yaw) or altitude. Such errors are generally unsystematic and are usually removed by identifying GCPs in the original imagery and on the reference map and then mathematically modeling the geometric distortion. Image-to-map rectification requires that polynomial equations be fit to the GCP data using least-squares criteria to model the corrections directly in the image domain without explicitly identifying the source of the distortion (Novak, 1992; Bossler et al., 2002). Depending on the distortion in the imagery, the number of GCPs used, and the degree of topographic relief displacement in the area, higher-order polynomial equations may be required to geometrically correct the data. The *order* of the rectification is simply the highest exponent used in the polynomial. For example, Figure 7-7 demonstrates how different-order transformations fit a

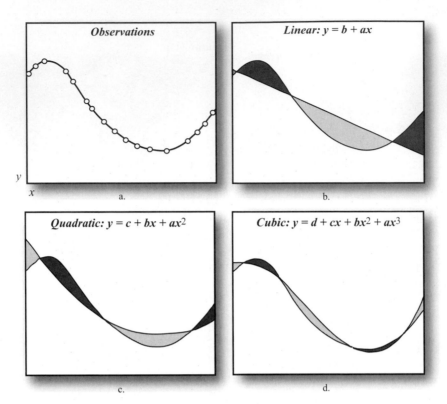

Figure 7-7   Concept of how different-order transformations fit a hypothetical surface illustrated in cross-section. a) Original observations. b) A first-order linear transformation fits a plane to the data. c) Second-order quadratic fit. d) Third-order cubic fit (Jensen et al., 1988).

hypothetical surface (Jensen et al., 1988). Generally, for moderate distortions in a relatively small area of an image (e.g., a quarter of a Landsat TM scene), a first-order, six-parameter, affine (*linear*) transformation is sufficient to rectify the imagery to a geographic frame of reference.

This type of transformation can model six kinds of distortion in the remote sensor data, including (Novak, 1992; Buiten and Van Putten, 1997):

- translation in x and y,

- scale changes in x and y,

- skew, and

- rotation.

Input-to-Output (Forward) Mapping: When all six operations are combined into a single expression it becomes:

$$x = a_0 + a_1x' + a_2y' \qquad (7-8)$$

$$y = b_0 + b_1x' + b_2y',$$

where x and y are positions in the *output*-rectified image or map, and x' and y' represent corresponding positions in the original *input* image (Figure 7-8a). These two equations can be used to perform what is commonly referred to as *input-to-output,* or *forward-mapping*. The equations function according to the logic shown in Figure 7-8a. In this example, each pixel in the *input* grid (e.g., value 15 at $x', y' = 2, 3$) is sent to an x,y location in the output image according to the six coefficients shown in Figure 7-8a.

This *forward mapping* logic works well if we are simply rectifying the location of discrete coordinates found along a linear feature such as a road in a vector map. In fact, cartographic mapping and geographic information systems typically rectify vector data using forward mapping logic. However, when we are trying to fill a rectified *output* grid (matrix) with values from an unrectified *input* image, forward mapping logic does not work well. The basic problem is that the six coefficients may require that value 15 from the x', y' location 2, 3 in the input image be located at a floating point location in the output image at $x, y = 5, 3.5$, as shown in

Landsat 7 Enhanced Thematic Mapper Plus Imagery of San Diego, CA

Path 40, Row 37

Color-infrared color composite (RGB = Landsat ETM$^+$ bands 4, 3, 2).

Color Plate 2-1 Color composite of Landsat 7 Enhanced Thematic Mapper Plus (ETM$^+$) imagery of San Diego, CA, obtained on April 24, 2000 (courtesy of NASA).

NOAA-16 Advanced Very High Resolution Radiometer Imagery

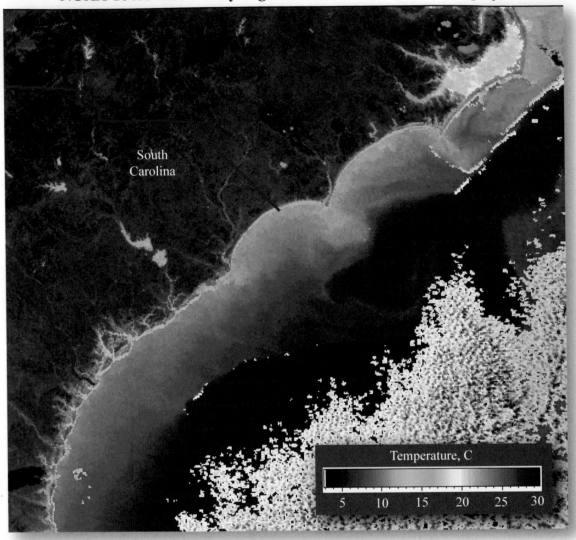

Sea-surface temperature (SST) map derived from NOAA-16 AVHRR
band 4 (10.3 - 11.3 μm) imagery obtained on October 16, 2003.

Color Plate 2-2 Sea-surface temperature (SST) map derived from NOAA-16 AVHRR thermal infrared imagery (courtesy of NOAA and the
Ocean Remote Sensing Program at Johns Hopkins University; Gasparovic, 2003).

SPOT 5 Imagery of Los Angeles, CA

SPOT 5 image of Los Angeles, CA, created by merging 2.5 x 2.5 m panchromatic data with
10 x 10 m multispectral data. The imagery is draped over a 30 x 30 m USGS digital elevation model.

Color Plate 2-3 SPOT 5 imagery of Los Angeles, CA (© CNES 2004, Spot Image Corporation).

Terra ASTER Optical Imagery of Oahu, HI

a. ASTER 15 x 15 m color composite of Oahu, HI, obtained on June 3, 2000 (RGB = bands 3, 2, 1).

b. Enlargement centered on Pearl Harbor.

Color Plate 2-4 *Terra* ASTER imagery of Oahu, HI (courtesy of NASA/GSFC/MITI/ERSADC/JAROS, U.S./Japan ASTER Science Team, and the California Institute of Technology).

IKONOS Multispectral and Panchromatic Imagery of Columbia, SC

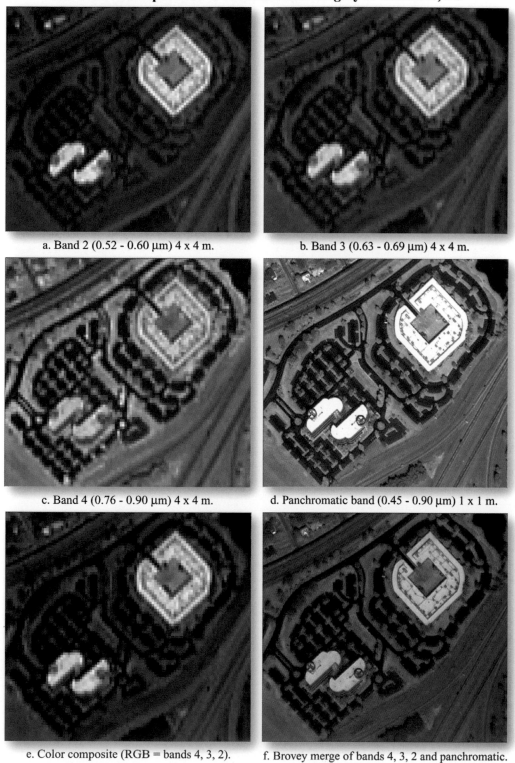

a. Band 2 (0.52 - 0.60 μm) 4 x 4 m.

b. Band 3 (0.63 - 0.69 μm) 4 x 4 m.

c. Band 4 (0.76 - 0.90 μm) 4 x 4 m.

d. Panchromatic band (0.45 - 0.90 μm) 1 x 1 m.

e. Color composite (RGB = bands 4, 3, 2).

f. Brovey merge of bands 4, 3, 2 and panchromatic.

Color Plate 2-5 IKONOS imagery of a business park in Columbia, SC. a–d) Individual 4 × 4 m multispectral bands and the 1 × 1 m panchromatic band are displayed. e) Standard color composite of IKONOS bands 4, 3, and 2. f) Color composite of a merged dataset created using a Brovey transform discussed in Chapter 5 (images courtesy of Space Imaging, Inc.).

Density Slicing Using an 8-bit Color Look-up Table

a. Landsat Thematic Mapper band 4 image of Charleston, SC, obtained on November 9, 1982.

b. Density slice based on the logic in Table 5-1.

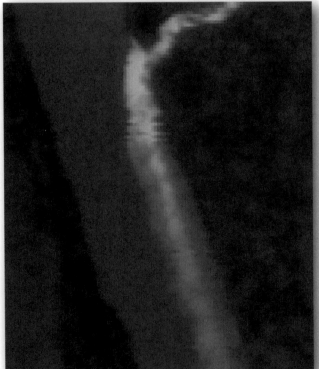

c. Predawn thermal infrared image of the Savannah River obtained on March 28, 1981.

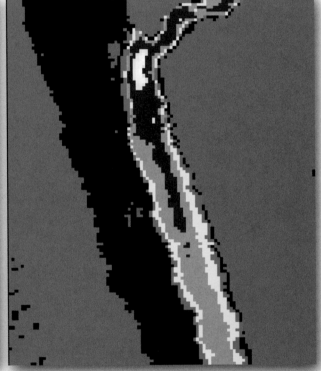

d. Density slice based on the logic in Table 5-2.

Color Plate 5-1 a) Black-and-white display of Landsat Thematic Mapper band 4 (0.76 – 0.90 μm) 30 × 30 m data of Charleston, SC. b) Color density slice using the logic summarized in Table 5-1. c) Black-and-white display of predawn thermal infrared (8.5 – 13.5 μm) imagery of the Savannah River. Each pixel is approximately 2.8 × 2.8 m on the ground. d) Color density slice using the logic summarized in Table 5-2.

Color Composites of Landsat Thematic Mapper Data of Charleston, SC, Obtained on February 3, 1994

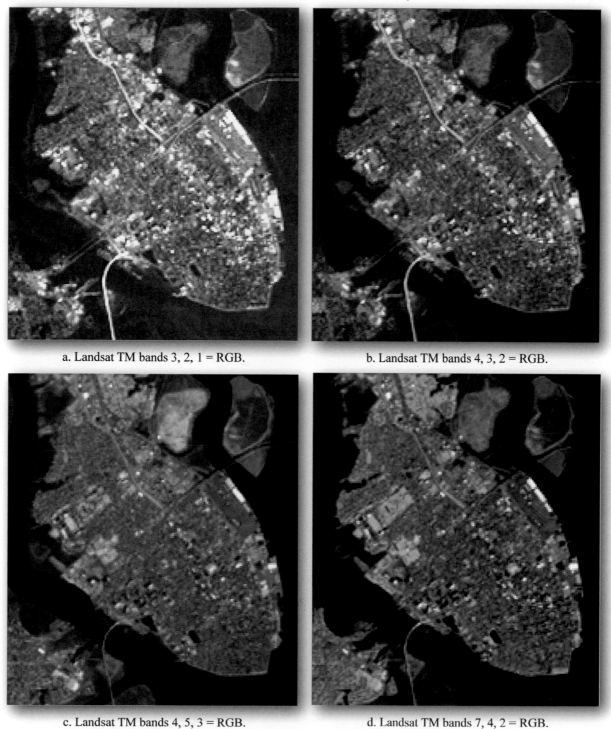

a. Landsat TM bands 3, 2, 1 = RGB.

b. Landsat TM bands 4, 3, 2 = RGB.

c. Landsat TM bands 4, 5, 3 = RGB.

d. Landsat TM bands 7, 4, 2 = RGB.

Color Plate 5-2 Color composites of Landsat Thematic Mapper data of Charleston, SC, obtained on February 3, 1994. a) Composite of Landsat TM bands 3, 2, and 1 placed in the red, green, and blue (RGB) image processor memory planes, respectively. b) TM bands 4, 3, and 2 = RGB. c) TM bands 4, 5, 3 = RGB. d) TM bands 7, 4, 2 = RGB.

**Merging (Fusion) of SPOT 20 x 20 m Multispectral
and 10 x 10 m Panchromatic Data of Marco Island, FL**

a. Color-infrared color composite of SPOT band 3 (near-infrared), 2 (red), and 1 (green) = RGB. Each band is 20 x 20 m.

b. Fused color-infrared color composite of SPOT band 3 (near-infrared), 4 (panchromatic), and 1 (green) = RGB. The panchromatic band is 10 x 10 m. The composite was created by substituting the panchromatic band for band 2 (red).

Color Plate 5-3 Merging (fusion) of SPOT multispectral data (20 × 20 m) with SPOT panchromatic data (10 × 10 m) using the band substitution method. The 20 × 20 m multispectral data were resampled to 10 × 10 m (© CNES 2004, Spot Image Corporation).

Image Mosaicking

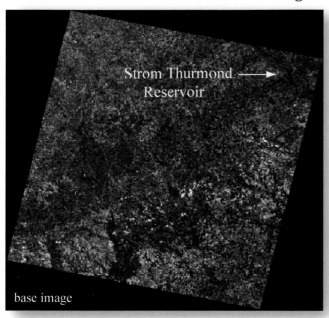

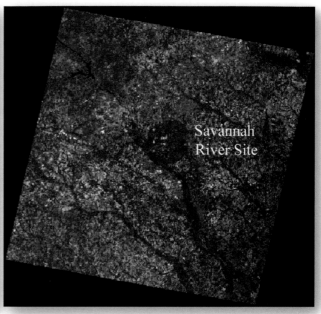

a. Rectified Landsat ETM$^+$ image of eastern Georgia obtained on October 3, 2001 (bands 4, 3, 2; Worldwide Reference System—Path 18, Row 37).

b. Rectified Landsat ETM$^+$ image of western South Carolina obtained on October 26, 2001 (bands 4, 3, 2; Worldwide Reference System—Path 17, Row 37).

c. Feathered mosaic of rectified Landsat ETM$^+$ imagery of eastern Georgia and western South Carolina.

Color Plate 7-1 Two Landsat Enhanced Thematic Mapper Plus (ETM$^+$) images are mosaicked using feathering logic.

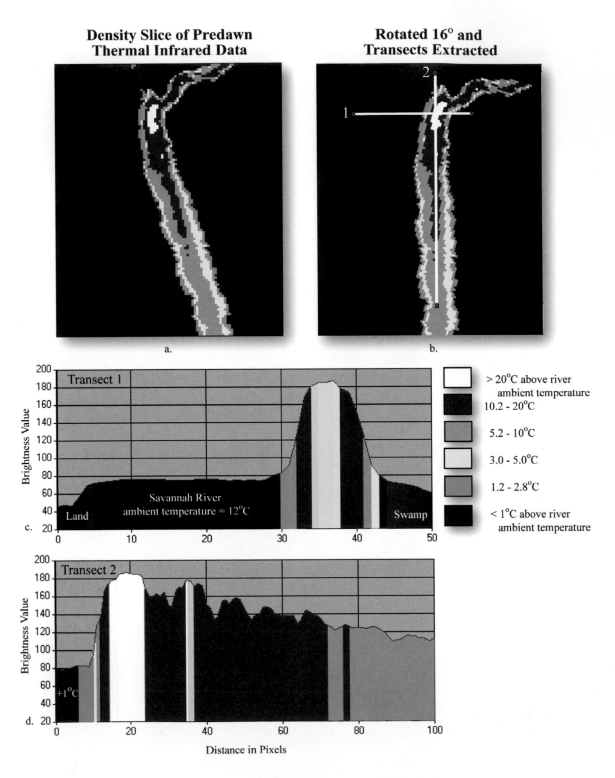

Color Plate 8-1 Transects (spatial profiles) passed through a predawn thermal infrared image of a thermal plume located in the Savannah River. a) Original image density-sliced according to the logic discussed in Table 8-1. b) Density-sliced image rotated 16° with Transects 1 and 2 displayed. c) Spatial profile of Transect 1. d) Spatial profile of Transect 2.

Spectral Profiles Extracted from SPOT 20 x 20 m Data

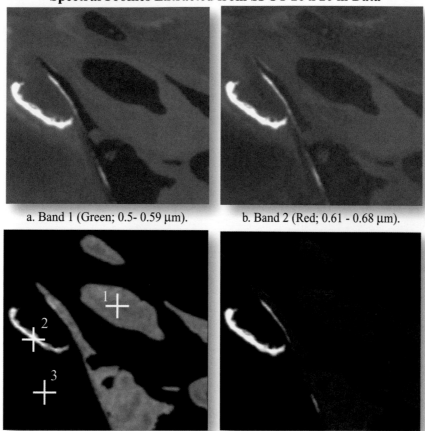

a. Band 1 (Green; 0.5- 0.59 μm).   b. Band 2 (Red; 0.61 - 0.68 μm).

c. Band 3 (Near-infrared; 0.79 - 0.89 μm).   d. Color composite (RGB = bands 3, 2, 1).

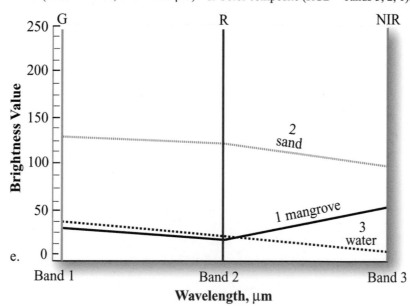

Color Plate 8-2 a–c) Three bands of SPOT 20 × 20 m multispectral data of Marco Island, FL. d) A color composite of SPOT bands 3, 2, and 1. e) Spectral profiles of mangrove, sand, and water extracted from the multispectral data (images courtesy of © SPOT Image Corporation).

Spectral Profiles Extracted from HyMap Hyperspectral Data

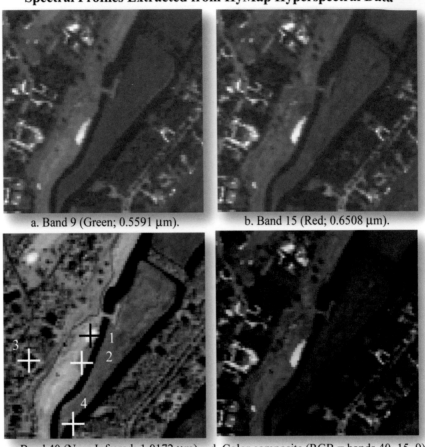

a. Band 9 (Green; 0.5591 µm).   b. Band 15 (Red; 0.6508 µm).

c. Band 40 (Near-Infrared; 1.0172 µm).   d. Color composite (RGB = bands 40, 15, 9).

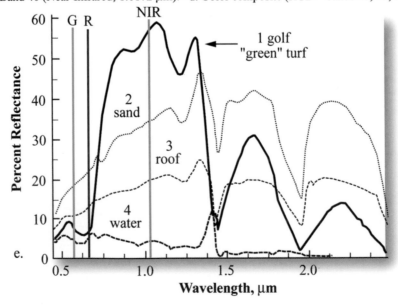

Color Plate 8-3 a–c) Three bands of HyMap hyperspectral data of the Debordiu colony near North Inlet, SC. The data were obtained at a spatial resolution of 3 × 3 m. d) Color composite of HyMap bands 40, 15, and 9. e) Spectral profiles of golf "green" turf, sand, roof, and water extracted from the 116 bands of hyperspectral data.

MODIS Enhanced Vegetation Index Map of the World

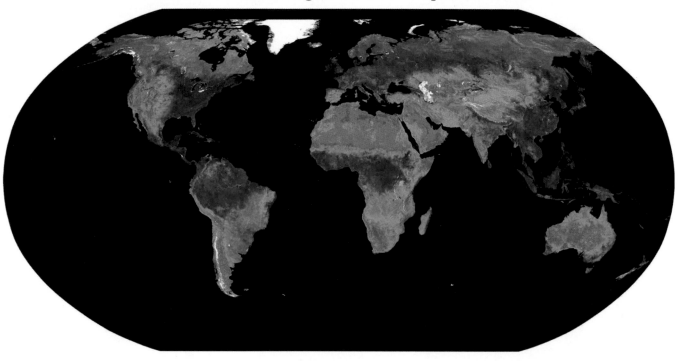

Color Plate 8-4 MODIS Enhanced Vegetation Index (EVI) map of the world obtained over a 16-day period beginning on day 193 of 2003. The greener the area, the greater the amount of biomass (courtesy Terrestrial Biophysics and Remote Sensing MODIS Team, University of Arizona and NASA).

Land Cover Map of North America Derived from *Terra* MODIS Data

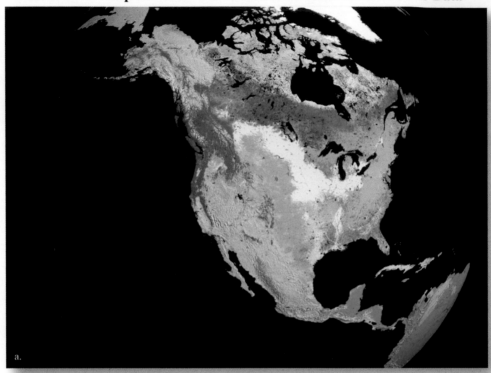

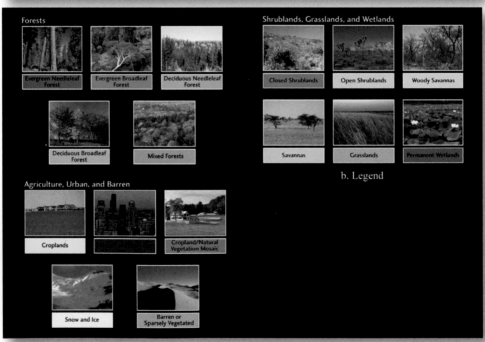

Color Plate 9-1 a) Land cover map of North America produced from *Terra* MODIS 1 × 1 km data collected between November 2000 and October 2001. b) Classification scheme legend (images courtesy of MODIS Land Cover and Land Cover Dynamics group at the Boston University Center for Remote Sensing, NASA Goddard Space Flight Center, and NASA Earth Observatory).

Minimum Distance to Means Supervised Classification of Charleston, SC, Using Landsat Thematic Mapper Data

Unsupervised Classification

a.

b.

Class	Legend
1. Residential	
2. Commercial	
3. Wetland	
4. Forest	
5. Water	

Class	Legend	Class	Legend
1. Water		11. Commercial 2	
2. Forest		12. Commercial 2	
3. Forest		13. Commercial 2	
4. Wetland		14. Residential	
5. Forest		15. Commercial 2	
6. Residential		16. Commercial 2	
7. Residential		17. Residential	
8. Park/golf		18. Residential	
9. Residential		19. Commercial 1	
10. Commercial 1		20. Commercial 2	

Color Plate 9-2 a) The results of applying a *minimum distance to means* classification algorithm to Landsat Thematic Mapper bands 4 and 5 data of Charleston, SC. Table 9-10 summarizes the number of pixels in each class. b) The results of performing an *unsupervised classification* of the Charleston, SC, Landsat Thematic Mapper imagery using bands 3, 4, and 5. Twenty spectral clusters were extracted and relabeled as information classes according to the criteria shown in Figure 9-27 and Table 9-11. Note that the unsupervised classification extracted a Park/golf class and was able to identify two classes of commercial land cover.

Information Classes Derived from an ISODATA Unsupervised Classification
Using 10 Iterations and 10 Mean Vectors of an Area Near North Inlet, SC

a. Color composite of HyMap data.

c. Classification map derived from
10 ISODATA clusters.

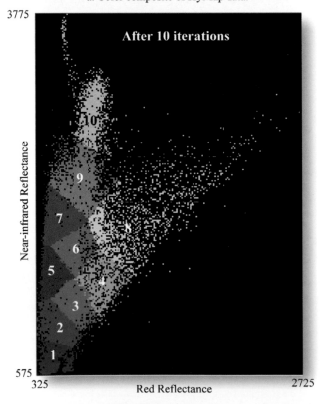

b. Final location of 10 ISODATA mean vectors after 10 iterations.

Class	Legend
1. Water	
2. Wetland	
3. Wetland	
4. Roof/asphalt	
5. Forest	
6. Wetland	
7. Forest	
8. Bare soil	
9. Forest	
10. Fairway	

Color Plate 9-3 ISODATA classification of HyMap remote sensor data of an area near North Inlet, SC. a) Color composite image. b) Location of 10 mean vectors in red and near-infrared feature space after 10 iterations. c) Classification map derived by labeling the 10 spectral classes as information classes.

Image Segmentation Based on Spectral (Green, Red, Near-infrared) and Spatial (Smoothness and Compactness) Criteria

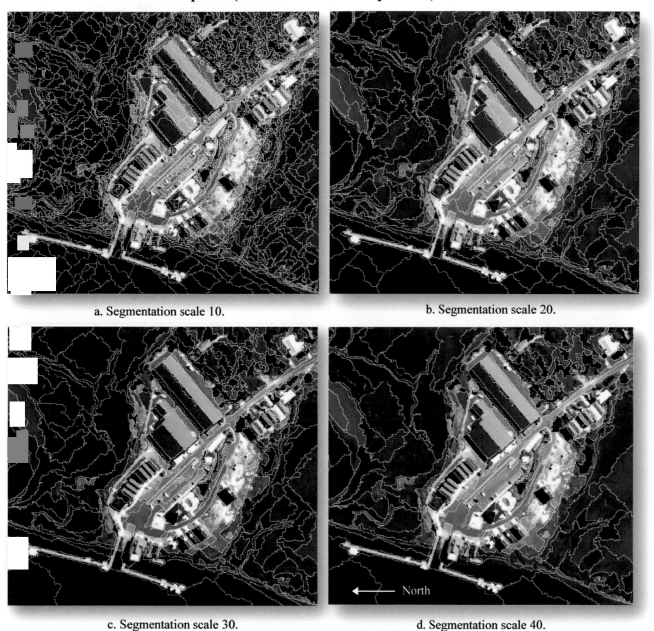

a. Segmentation scale 10.

b. Segmentation scale 20.

c. Segmentation scale 30.

d. Segmentation scale 40.

Color Plate 9-4 Color composite of ADAR 5000 imagery of a yacht harbor on Pritchard's Island, SC, in the Ace Basin obtained on September 23, 1999, at a spatial resolution of 0.7×0.7 m. Multiresolution image segmentation was performed using three bands (green, red, and near-infrared) at four segmentation scales: 10, 20, 30, and 40. The segmentation process was weighted so that spectral information (color) was more important than spatial (shape) information (weighted 0.8 to 0.2, respectively). The spatial parameter was more heavily weighted to smoothness (0.9) than compactness (0.1).

Classification Based on Image Segmentation Spectral and Spatial Criteria

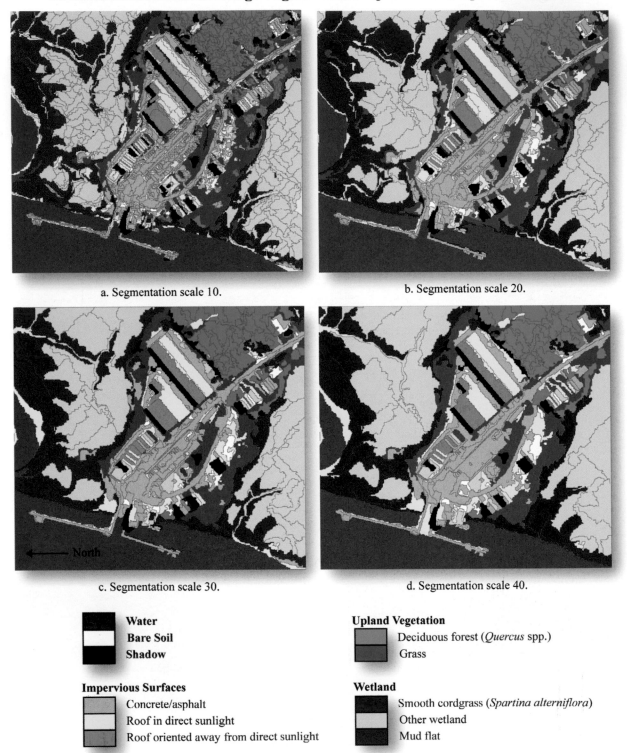

a. Segmentation scale 10.

b. Segmentation scale 20.

c. Segmentation scale 30.

d. Segmentation scale 40.

Water
Bare Soil
Shadow

Upland Vegetation
Deciduous forest (*Quercus* spp.)
Grass

Impervious Surfaces
Concrete/asphalt
Roof in direct sunlight
Roof oriented away from direct sunlight

Wetland
Smooth cordgrass (*Spartina alterniflora*)
Other wetland
Mud flat

Color Plate 9-5 Four classification maps derived from four different sets of image segmentation scale data of a yacht harbor on Pritchard's Island, SC, in the Ace Basin.

a. Landsat ETM⁺ (RGB = bands 4, 3, 2) image obtained on August 10, 1999, draped over a digital elevation model.

b. Landsat ETM⁺ (RGB = bands 5, 4, 2) image draped over a digital elevation model (azimuth = 90°).

c. Terrestrial photograph of Maple Mountain (azimuth = 113°).

d. Landsat ETM⁺ (RGB = bands 4, 3, 2) image draped over a digital elevation model viewed from the same azimuth as in part c.

e. Expert system classification of white fir.

f. Expert system classification of white fir draped over the color composite and digital elevation model.

Expert System Classification of White Fir *(Abies concolor)* on Maple Mountain in Utah County, UT

Color Plate 10-1 Expert system classification of white fir (*Abies concolor*) on Maple Mountain in Utah County, Utah. a,b) Landsat ETM⁺ color composites draped over a digital elevation model. c) Terrestrial natural color photograph. d) Maple Mountain viewed from an azimuth of 113°. e,f) Vertical and terrestrial views of the white fir classification derived using human-specified rules and an expert system.

Airborne Visible/Infrared Imaging Spectrometer Data
of the Savannah River Site Obtained July 26, 1999

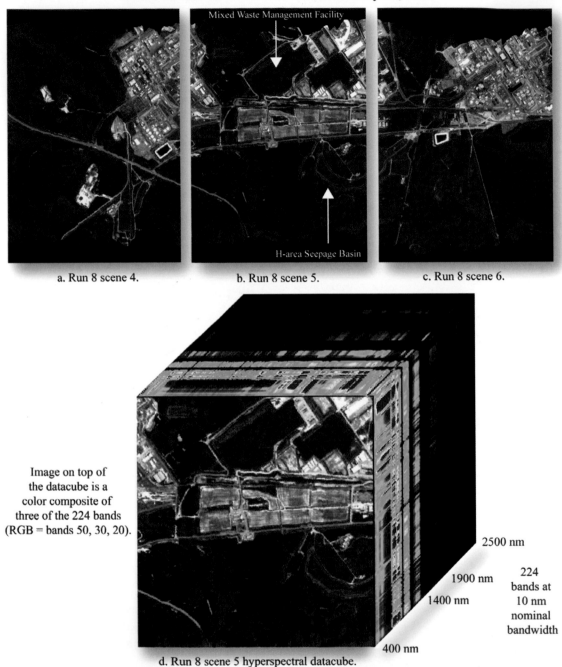

a. Run 8 scene 4.   b. Run 8 scene 5.   c. Run 8 scene 6.

Mixed Waste Management Facility

H-area Seepage Basin

Image on top of
the datacube is a
color composite of
three of the 224 bands
(RGB = bands 50, 30, 20).

2500 nm

1900 nm

1400 nm

224
bands at
10 nm
nominal
bandwidth

400 nm

d. Run 8 scene 5 hyperspectral datacube.

Color Plate 11-1 a–c) Color-infrared color composites of three AVIRIS scenes (4, 5, and 6) from run 8. The color composites represent a display of just three of the 224 AVIRIS bands (RGB = bands 50, 30, 20). The AVIRIS sensor was flown onboard a De Havilland DHC-6 Twin Otter aircraft at 11,150 ft (3.4 km) above sea level. This resulted in pixels that were 3.4 × 3.4 m. Each scene is 512 × 746 pixels in dimension. The data were preprocessed at JPL to remove fundamental geometric and radiometric errors. Pixels in the black border regions contain null values (–9999). d) Hyperspectral datacube of run 8 scene 5. All 224 bands are displayed. Dark areas in the datacube represent atmospheric absorption bands.

Masked Hyperspectral Data of the Clay-capped
Mixed Waste Management Facility and H-area Seepage Basin

a. False-color composite (RGB = bands 50, 30, 20).   b. MNF false-color composite (RGB = MNF 3, 2, 1).

Color Plate 11-2 Only the clay-capped hazardous waste sites associated with the Mixed Waste Management Facility, the H-area seepage basin, and some internal roads are included in the July 26, 1999, AVIRIS imagery. All other land cover was masked from further analysis.

**Classification Maps of Healthy and Potentially Stressed Vegetation
Created Using Image-derived Endmbers and a Spectral Angle Mapper Algorithm**

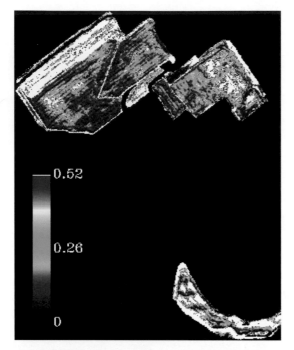

a. Healthy: derived from endmember 37.

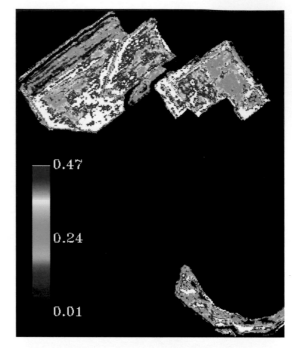

b. Potential stress: derived from endmember 15.

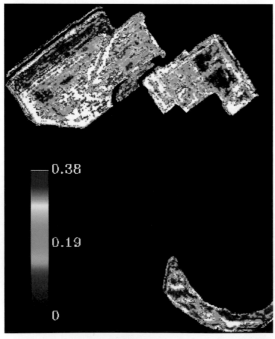

c. Potential stress: derived from endmember 25.

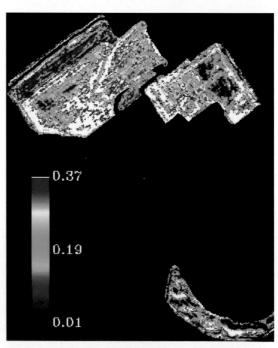

d. Potential stress: derived from endmember 36.

Color Plate 11-3 a) Healthy Bahia grass classification map derived from July 26, 1999, AVIRIS hyperspectral data and spectral angle mapper (SAM) analysis of endmember 37. b–d) Potentially stressed Bahia grass classification map derived from AVIRIS hyperspectral data and SAM analysis of endmembers 15, 25, and 36. In all cases, the darker the pixel (blue, black, etc.), the smaller the angle and the closer the match.

Hardened Classification Map of Healthy and Potentially Stressed Vegetation

Color Plate 11-4 A hardened classification map derived from July 26, 1999, AVIRIS hyperspectral data and spectral angle mapper (SAM) analysis of four endmembers. Brown areas represent healthy Bahia grass. White, red, and yellow depict areas with vegetation stress-related characteristics.

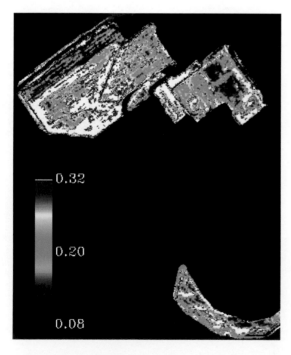

Classification Map of Potentially Stressed Vegetation Created Using Laboratory–derived Endmembers and the Spectral Angle Mapper Algorithm

Color Plate 11-5 Potentially stressed Bahia grass classification map derived from laboratory spectra with a 0.5 mg/ml copper treatment. The darker the pixel (blue, black, etc.), the smaller the angle and the closer the match.

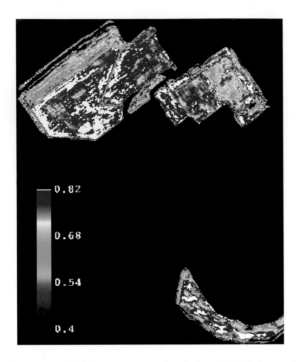

**Normalized Difference Vegetation Index
Image Derived from AVIRIS Data**

Color Plate 11-6 Normalized difference vegetation index (NDVI) image derived from AVIRIS bands 29 (0.64554 μm) and 51 (0.82593 μm), red and near-infrared, respectively. White and red pixels indicate the greatest amount of biomass. Blue, green, and orange pixels contain less biomass.

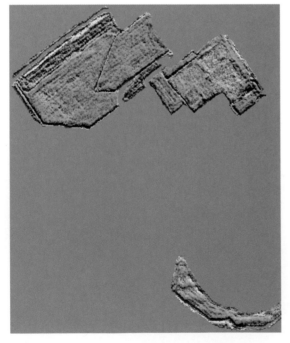

**Second-order Derivative Image
Extracted from Original AVIRIS Spectra**

Color Plate 11-7 A second-order derivative image extracted from the original AVIRIS scaled surface reflectance data (RGB = bands 42, 30, 18).

Write Function Memory Insertion Change Detection and Image Differencing Change Detection

a. Landsat ETM$^+$ data of Lake Mead, NV, obtained on May 3, 2000 (RGB = bands 4, 3, 2).

b. ASTER data obtained on April 19, 2003 (RGB = bands 3, 2, 1).

c. Write function memory insertion change detection where RGB = ASTER band 3, ETM$^+$ band 4, and ETM$^+$ band 4, respectively.

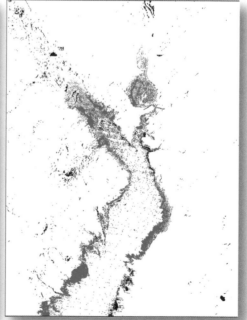

d. Image differencing change detection. Red and green change pixels are associated with thresholds located in the tails of the histogram in Figure 12-11.

Color Plate 12-1 a,b) Multiple-date images of Lake Mead, NV, collected by Landsat ETM$^+$ and ASTER. c) Write function memory insertion change detection. d) Image differencing change detection using Landsat ETM$^+$ band 4 (0.75 – 0.90 μm) May 3, 2000, data and ASTER band 3 (0.76 – 0.86 μm) April 19, 2003, data (images courtesy of NASA).

Multiple-Date Composite Image Change Detection Based on
Principal Components Viewed Using Write Function Memory Insertion

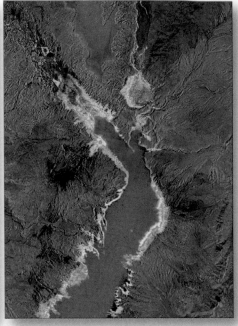

a. Write function memory insertion using
RGB = principal components 3, 2, 1.

b. Write function memory insertion using
RGB = principal components 1, 2, 3.

**Principal Component Images
Derived from a Multiple-Date
Composite Dataset Consisting of**

* Landsat ETM$^+$ data obtained on
May 3, 2000 (bands 2, 3, 4)

* ASTER data obtained on
April 19, 2003 (bands 1, 2, 3)

c. Write function memory insertion using
RGB = principal components 2, 1, 3.

Color Plate 12-2 Multiple-date composite image change detection using images that were created from a multiple-date dataset consisting of Landsat ETM$^+$ and ASTER data. Write function memory insertion is used to highlight the information content of the various principal component images. Principal component 2 contains much of the change information (original imagery courtesy of NASA).

Landsat Thematic Mapper Color-infrared Color Composites of Kittredge and Fort Moultrie, SC

a. Kittredge, SC, on November 9, 1982.

b. Kittredge, SC, on December 19, 1988.

c. Fort Moultrie, SC, on November 9, 1982.

d. Fort Moultrie, SC, on December 19, 1988.

Color Plate 12-3 a,b) Rectified Landsat Thematic Mapper data of an area near Kittredge, SC, obtained on November 9, 1982, and December 19, 1988 (RGB = bands 4, 3, 2). c,d) Rectified Landsat Thematic Mapper data of an area centered on Fort Moultrie, SC, obtained on November 9, 1982, and December 19, 1988 (RGB = bands 4, 3, 2).

Land Cover Classification of Kittredge and Fort Moultrie, SC, Using Landsat Thematic Mapper Data

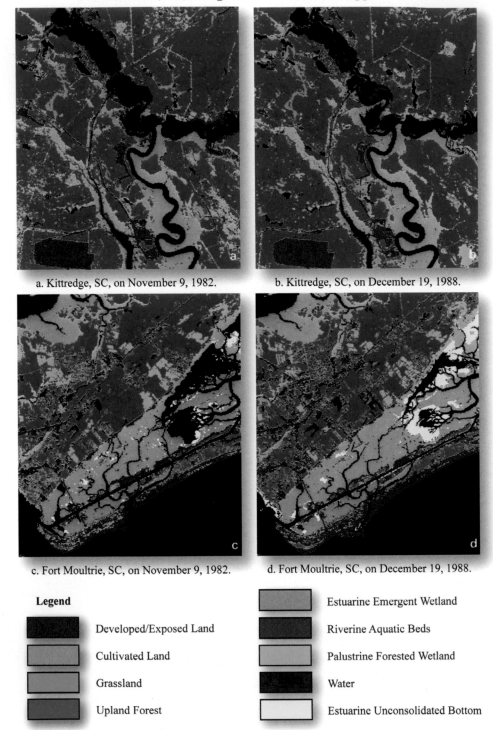

a. Kittredge, SC, on November 9, 1982.

b. Kittredge, SC, on December 19, 1988.

c. Fort Moultrie, SC, on November 9, 1982.

d. Fort Moultrie, SC, on December 19, 1988.

Legend

Developed/Exposed Land

Cultivated Land

Grassland

Upland Forest

Estuarine Emergent Wetland

Riverine Aquatic Beds

Palustrine Forested Wetland

Water

Estuarine Unconsolidated Bottom

Color Plate 12-4 a,b) Classification maps of the Kittredge, SC, study area produced from November 9, 1982, and December 19, 1988, Landsat Thematic Mapper data. c,d) Classification maps of the Fort Moultrie, SC, study area produced from the November 9, 1982, and December 19, 1988, Landsat TM data. Some barren land is included in the Developed/Exposed Land category.

From – To Change Detection Matrix

To 1988 / **From 1982**

1982 \ 1988	Developed/Exposed Land	Cultivated Land	Grassland	Upland Forest	Estuarine Emergent Wetland	Riverine Aquatic Beds	Palustrine Forested Wetland	Water	Estuarine Unconsolidated Bottom
Developed/Exposed Land	1	2	3	4	5	6	7	8	9
Cultivated Land	10	11	12	13	14	15	16	17	18
Grassland	19	20	21	22	23	24	25	26	27
Upland Forest	28	29	30	31	32	33	34	35	36
Estuarine Emergent Wetland	37	38	39	40	41	42	43	44	45
Riverine Aquatic Beds	46	47	48	49	50	51	52	53	54
Palustrine Forested Wetland	55	56	57	58	59	60	61	62	63
Water	64	65	66	67	68	69	70	71	72
Estuarine Unconsolidated Bottom	73	74	75	76	77	78	79	80	81

Color look-up table values in change detection map

	Developed/Exposed Land	Cultivated Land	Grassland	Upland Forest	Estuarine Emergent Wetland	Riverine Aquatic Beds	Palustrine Forested Wetland	Water	Estuarine Unconsolidated Bottom
Red	255	255	255	255	255	255	0	0	255
Green	0	255	255	255	163	0	255	0	255
Blue	0	255	255	255	0	255	255	255	0

- No change in land cover between dates, and not selected for display (RGB = 0, 0, 0).
- Change in land cover between dates, but not selected for display (RGB = 255, 255, 255).
- New Developed/Exposed Land (cells 10,19,28, 37,46,55,64,73) shown in red (RGB = 255, 0, 0).
- New Estuarine Unconsolidated Bottom (cells 9,18,27,36,45,54,63,72) shown in yellow (RGB = 255, 255, 0).

Color Plate 12-5 The basic elements of a change detection matrix may be used to select specific "from – to" classes for display in a *post-classification comparison* change detection map. There are ($n^2 - n$) off-diagonal possible change classes that may be displayed in the change detection map (72 in this example) although some may be highly unlikely. The colored off-diagonal cells in this particular diagram were used to produce the change maps in **Color Plate 12-6**. For example, any pixel in the 1982 map that changed to Developed/Exposed Land by 1988 is red (RGB = 255, 0, 0). Any pixel that changed into Estuarine Unconsolidated Bottom by 1988 is yellow (RGB = 255, 255, 0). Individual cells can be color-coded in the change map to identify very specific "from – to" changes (Jensen et al., 1993a).

Selected Change in Land Cover
for Kittredge and Fort Moultrie, SC

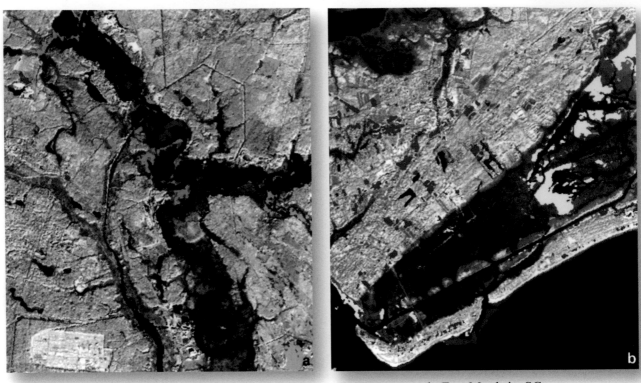

a. Kittredge, SC.

b. Fort Moultrie, SC.

Change in Land Cover from November 9, 1982, to December 19, 1988

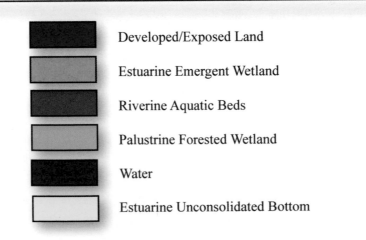

Developed/Exposed Land

Estuarine Emergent Wetland

Riverine Aquatic Beds

Palustrine Forested Wetland

Water

Estuarine Unconsolidated Bottom

Color Plate 12-6 Change detection maps of the Kittredge and Fort Moultrie, SC, study areas derived from analysis of November 11, 1982, and December 19, 1988, Landsat Thematic Mapper data. The nature of the change classes selected for display are summarized in **Color Plate 12-5**. The change information is overlaid onto the Landsat TM band 4 image of each date for orientation purposes (Jensen et al., 1993a).

Drawdown of the Aral Sea, Kazakhstan, from 1973 to 2000

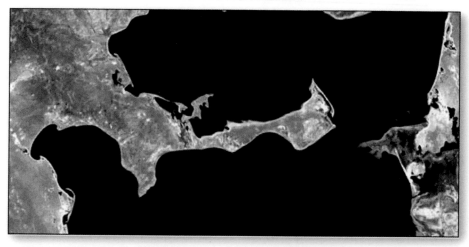

a. Landsat MSS image obtained on May 29, 1973.

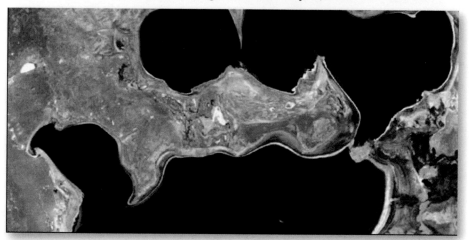

b. Landsat MSS image obtained on August 19, 1987.

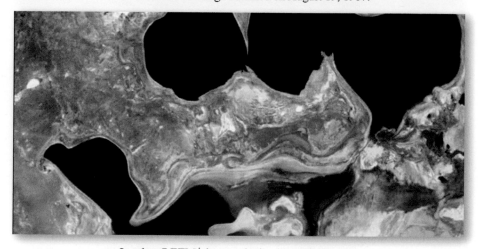

c. Landsat 7 ETM$^+$ image obtained on July 29, 2000.

Color Plate 12-7 A portion of the Aral Sea shoreline recorded by the Landsat Multispectral Scanner (MSS) and Landsat 7 Enhanced Thematic Mapper Plus (ETM$^+$) in 1973, 1987, and 2000 (courtesy of NASA Earth Observatory).

Accuracy Assessment *In Situ* Measurement Locations and Remote Sensing-derived Classification Map of the Mixed Waste Management Facility on the Savannah River Site

a. *In situ* measurement locations.

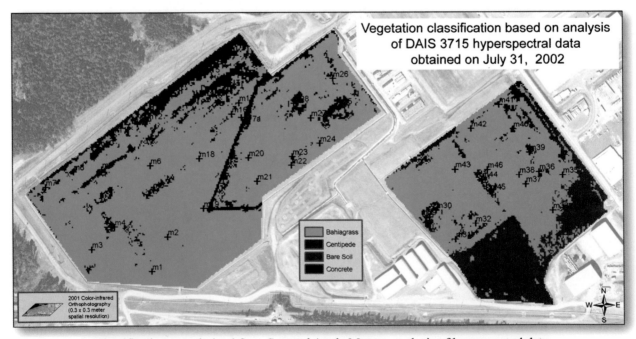

b. Classification map derived from Spectral Angle Mapper analysis of hyperspectral data.

Color Plate 13-1 Classification accuracy assessment of a land-cover map of the Mixed Waste Management Facility on the Savannah River Site in 2002 based on the analysis of DAIS 3715 hyperspectral data and *in situ* ground reference data. a) The location of Bahia grass and Centipede grass *in situ* measurement locations. b) *In situ* measurement locations within the classification map.

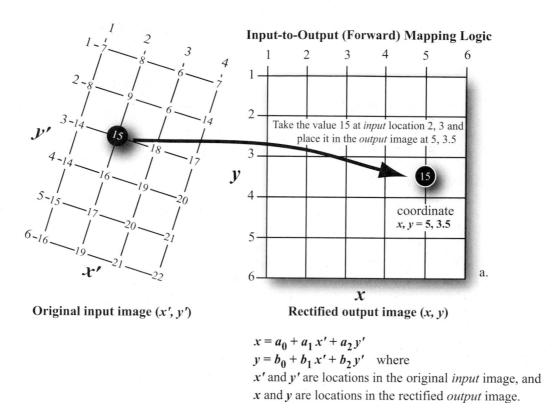

Input-to-Output (Forward) Mapping Logic

Take the value 15 at *input* location 2, 3 and place it in the *output* image at 5, 3.5

coordinate
$x, y = 5, 3.5$

a.

Original input image (x', y') Rectified output image (x, y)

$$x = a_0 + a_1 x' + a_2 y'$$
$$y = b_0 + b_1 x' + b_2 y' \quad \text{where}$$
x' and y' are locations in the original *input* image, and
x and y are locations in the rectified *output* image.

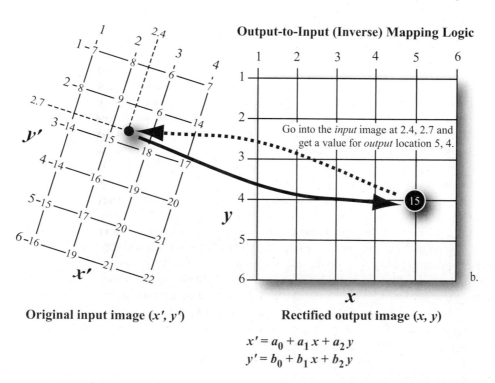

Output-to-Input (Inverse) Mapping Logic

Go into the *input* image at 2.4, 2.7 and get a value for *output* location 5, 4.

b.

Original input image (x', y') Rectified output image (x, y)

$$x' = a_0 + a_1 x + a_2 y$$
$$y' = b_0 + b_1 x + b_2 y$$

Figure 7-8   a) The logic of filling a rectified output matrix with values from an unrectified input image matrix using input-to-output (*forward*) mapping logic. b) The logic of filling a rectified output matrix with values from an unrectified input image matrix using output-to-input (*inverse*) mapping logic and nearest-neighbor resampling. Output-to-input *inverse* mapping logic is the preferred methodology because it results in a rectified output matrix with values at every pixel location.

Figure 7-8a. The output *x, y* location does not fall exactly on an integer *x* and *y* output map coordinate. *In fact, using forward mapping logic can result in output matrix pixels with no output value* (Wolberg, 1990). This is a serious condition and one that reduces the utility of the remote sensor data for useful applications. For this reason, most remotely sensed data are geometrically rectified using *output-to-input* or *inverse mapping* logic (Niblack, 1986; Richards and Jia, 1999).

Output-to-Input (Inverse) Mapping: *Output-to-input,* or *inverse mapping* logic, is based on the following two equations:

$$x' = a_0 + a_1 x + a_2 y \qquad (7\text{-}9)$$

$$y' = b_0 + b_1 x + b_2 y$$

where *x* and *y* are positions in the *output*-rectified image or map, and *x'* and *y'* represent corresponding positions in the original *input* image. If desired, this relationship can also be written in matrix notation:

$$\begin{bmatrix} x' \\ y' \end{bmatrix} = \begin{bmatrix} a_1 & a_2 \\ b_1 & b_2 \end{bmatrix} \begin{bmatrix} x \\ y \end{bmatrix} + \begin{bmatrix} a_0 \\ b_0 \end{bmatrix}. \qquad (7\text{-}10)$$

The rectified *output* matrix consisting of *x* (column) and *y* (row) coordinates is filled in the following systematic manner. Each output pixel location (e.g., *x, y* = 5, 4 in Figure 7-8b) is entered into Equation 7-9. The equation uses the six coefficients to determine where to go into the original *input* image to get a value (dashed line in Figure 7-8b). In this example, it requests a value from the floating point location *x', y'* = 2.4, 2.7. Interestingly, there is no value at this location. Nevertheless, it is possible to get a value (e.g., 15) if nearest-neighbor resampling logic is used (to be discussed shortly). The inverse mapping logic guarantees that there will be a value at every *x, y* coordinate (column and row) in the output image matrix. There will be no missing values. This procedure may seem backwards at first, but it is the only mapping function that avoids overlapping pixels and holes in the output image matrix (Schowengerdt, 1997; Konecny, 2003).

Sometimes it is difficult to gain an appreciation of exactly what impact the six coefficients in an affine (linear) transformation have on geographic coordinates. Therefore, consider Table 7-1, which demonstrates how the six coefficients (in parentheses) in a *forward* affine transformation influence the *x, y* coordinates of 16 pixels in a hypothetical 4 × 4 matrix.

Note how *translation* (shifting) in *x* and *y* is controlled by the a_0 and b_0 coefficients, respectively. *Scale changes* in *x* and *y* are controlled by the a_1 and b_2 coefficients, respectively. *Shear/rotation* in *x* and *y* are controlled by coefficients a_2 and b_1, respectively. The three examples demonstrate how modifying the coefficients causes the pixel coordinates to be translated (shifted one unit in *x* and *y*), scaled (expanded by a factor of 2 in *x* and *y*), and sheared/rotated (Brown, 1992). Please remember that this particular example is based on *forward* mapping logic.

It is possible to use higher-order polynomial transformations to rectify remotely sensed data. For example, instead of using the six-parameter affine transformation previously discussed, we could use a second-order (quadratic) polynomial:

$$x' = c_0 + c_1 x + c_2 y + c_3 xy + c_4 x^2 + c_5 y^2 \qquad (7\text{-}11)$$

$$y' = d_0 + d_1 x + d_2 y + d_3 xy + d_4 x^2 + d_5 y^2. \qquad (7\text{-}12)$$

In theory, the higher the order of the polynomial, the more closely the coefficients should model the geometric error in the original (unrectified) input image (e.g., refer to Figure 7-7 to see how the various models fit one-dimensional data) and place the pixels in their correct planimetric positions in the rectified output matrix. Higher-order polynomials often produce a more accurate fit for areas immediately surrounding ground control points. However, other geometric errors may be introduced at large distances from the GCPs (Gibson and Power, 2000). In addition, the digital image processing system time required to geometrically rectify the remote sensor data using higher-order polynomials increases because of the greater number of mathematical operations that must be performed.

A general rule of thumb is to use a first-order affine polynomial whenever possible. Select a higher-order polynomial (e.g., second or third order) only when there are serious geometric errors in the dataset. These types of error are often found in imagery obtained from suborbital aerial platforms where roll, pitch, and yaw by the aircraft introduce unsystematic, nonlinear distortions. These distortions can only be modeled using a higher-order polynomial (Buiten and Van Putten, 1997). For example, the ATLAS near-infrared image of Lake Murray shown in Figure 7-9 was geometrically rectified using a second-order polynomial. A first-order affine rectification was not sufficient to correct the significant warping in the imagery caused by the aircraft drifting off course during data collection. All pixels in this image are now in their appropriate geographic location.

Table 7-1. *Forward* mapping example of how varying the six coefficients (in parentheses) in a linear (affine) transformation can impact the coordinates of 16 pixels in a simple 4×4 matrix. The equations are of the form $x_{predict} = a_0 + a_1 x + a_2 y$ and $y_{predict} = b_0 + b_1 x + b_2 y$. Translation (shifting) in x and y is controlled by the a_0 and b_0 coefficients, respectively. Scale changes in x and y are controlled by the a_1 and b_2 coefficients, respectively. Shear/rotation in x and y are controlled by coefficients a_2 and b_1, respectively.

Original Matrix		Translation		Scale		Shear/Rotation	
$x_{predict} = (0) + (1)x + (0)y$		$x_{predict} = (1) + (1)x + (0)y$		$x_{predict} = (0) + (2)x + (0)y$		$x_{predict} = (0) + (1)x + (2)y$	
$y_{predict} = (0) + (0)x + (1)y$		$y_{predict} = (1) + (0)x + (1)y$		$y_{predict} = (0) + (0)x + (2)y$		$y_{predict} = (0) + (2)x + (1)y$	

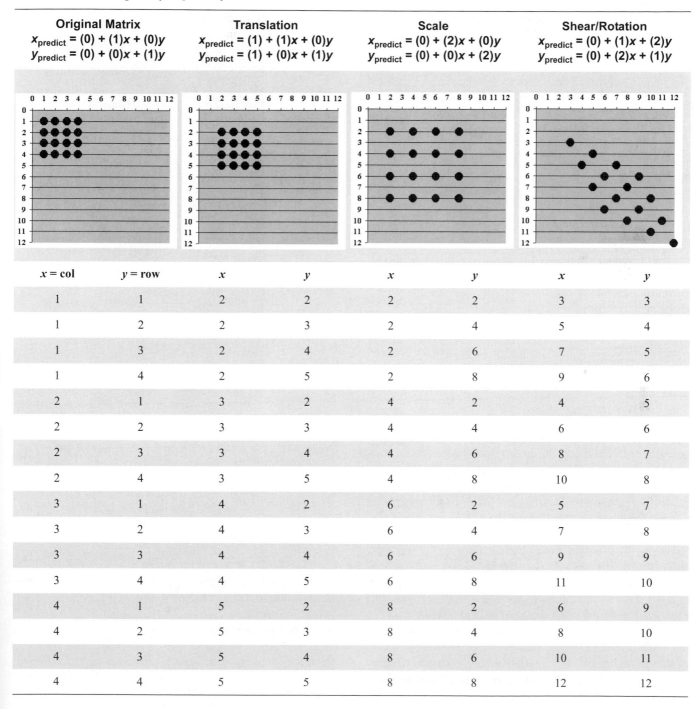

x = col	y = row	x	y	x	y	x	y
1	1	2	2	2	2	3	3
1	2	2	3	2	4	5	4
1	3	2	4	2	6	7	5
1	4	2	5	2	8	9	6
2	1	3	2	4	2	4	5
2	2	3	3	4	4	6	6
2	3	3	4	4	6	8	7
2	4	3	5	4	8	10	8
3	1	4	2	6	2	5	7
3	2	4	3	6	4	7	8
3	3	4	4	6	6	9	9
3	4	4	5	6	8	11	10
4	1	5	2	8	2	6	9
4	2	5	3	8	4	8	10
4	3	5	4	8	6	10	11
4	4	5	5	8	8	12	12

Compute the Root-Mean-Squared Error of the Inverse Mapping Function: Using the six coordinate transform coefficients that model distortions in the original scene, it is possible to use the output-to-input (inverse) mapping logic to transfer (relocate) pixel values from the original distorted image x', y' to the grid of the rectified output image, x, y.

However, before applying the coefficients to create the recti-fied output image, it is important to determine how well the six coefficients derived from the least-squares regression of the initial GCPs account for the geometric distortion in the input image. The method used most often involves the computation of the root-mean-square error (RMS$_{error}$) for each of the ground control points (Ton and Jain, 1989).

Let us consider for a moment the nature of the GCP data. We first identify a point in the image such as a road intersection. Its column and row coordinates in the original input image we will call x_{orig} and y_{orig}. The x and y position of the same road intersection is then measured from the reference map in degrees, feet, or meters. These two sets of GCP coordinates and many others selected by the analyst are used to compute the six coefficients discussed in Equation 7-9. Now, if we were to put the *map x* and *y* values for the first GCP back into Equation 7-9, with all the coefficients in place, we would get computed x' and y' values that are supposed to be the location of this point in the input image space. Ideally, x' would equal x_{orig} and y' would equal y_{orig}. Unfortunately, this is rarely the case. Any discrepancy between the values represents image geometric distortion not corrected by the six-coefficient coordinate transformation.

A simple way to measure such distortion is to compute the RMS$_{error}$ for each ground control point by using the equation:

$$RMS_{error} = \sqrt{(x' - x_{orig})^2 + (y' - y_{orig})^2}, \qquad (7\text{-}13)$$

where x_{orig} and y_{orig} are the original row and column coordi-nates of the GCP in the image and x' and y' are the computed or estimated coordinates in the original image. The square root of the squared deviations represents a measure of the accuracy of this GCP in the image. By computing RMS$_{error}$ for all GCPs, it is possible to 1) see which GCPs exhibit the greatest error, and 2) sum all the RMS$_{error}$.

Normally, the user specifies a certain amount (a threshold) of acceptable total RMS$_{error}$ (e.g., 1 pixel). If an evaluation of the total RMS$_{error}$ reveals that a given set of control points exceeds this threshold, it is common practice to 1) delete the GCP that has the greatest amount of individual error from the analysis, 2) recompute the six coefficients, and 3) recompute the RMS$_{error}$ for all points. This process continues until one of the following occurs: the total RMS$_{error}$ is less than the threshold specified or too few points remain to perform a least-squares regression to compute the coefficients. Once the acceptable RMS$_{error}$ is reached, the analyst can proceed to the intensity interpolation phase of geometric rectification,

ATLAS Imagery of Lake Murray, SC, Rectified Using a Second-order Polynomial

Figure 7-9   NASA ATLAS near-infrared image of Lake Mur-ray, SC, obtained on October 7, 1997, at a spatial resolution of 2×2 m. The image was rectified using a second-order polynomial to adjust for the signifi-cant geometric distortion in the original dataset caused by the aircraft drifting off course during data collection.

which attempts to fill an output grid (x, y) with brightness values found in the original input grid (x', y').

Intensity Interpolation

The intensity interpolation process involves the extraction of a brightness value from an x', y' location in the original (dis-torted) input image and its relocation to the appropriate x, y coordinate location in the rectified output image. This pixel-filling logic is used to produce the output image line by line, column by column. Most of the time the x' and y' coordinates to be sampled in the input image are floating point numbers

Table 7-2.   Bilinear interpolation of a weighted brightness value (BV_{wt}) at location x', y' based on the analysis of four sample points in Figure 7-8b.

Sample Point Location (column, row)	Value at Sample Point, Z	Distance from x', y' to the Sample Point, D	D_k^2	$\dfrac{Z}{D_k^2}$	$\dfrac{1}{D_k^2}$
2, 2	9	$D = \sqrt{(2.4-2)^2 + (2.7-2)^2} = 0.806$	0.65	13.85	1.539
3, 2	6	$D = \sqrt{(2.4-3)^2 + (2.7-2)^2} = 0.921$	0.85	7.06	1.176
2, 3	15	$D = \sqrt{(2.4-2)^2 + (2.7-3)^2} = 0.500$	0.25	60.00	4.000
3, 3	18	$D = \sqrt{(2.4-3)^2 + (2.7-3)^2} = 0.670$	0.45	40.00	2.222
				Σ 120.91	Σ 8.937

$$BV_{wt} = 120.91/8.937 = \mathbf{13.53}$$

(i.e., they are not integers). For example, in Figure 7-8b we see that pixel 5, 4 (x, y) in the output image is to be filled with the value from coordinates 2.4, 2.7 (x', y') in the original input image. When this occurs, there are several methods of brightness value (BV) interpolation that can be applied, including:

- nearest neighbor,

- bilinear interpolation, and

- cubic convolution.

The practice is commonly referred to as *resampling*.

Nearest-neighbor Interpolation: Using *nearest-neighbor* interpolation, the brightness value closest to the specified x', y' coordinate is assigned to the output x, y coordinate. For example, in Figure 7-8b, the output pixel 5, 4 (x, y) requests the brightness value in the original input image at location 2.4, 2.7 (x', y'). There is no value at this location. However, there are nearby values at the integer grid intersections. Distances from 2.4, 2.7 (x', y') to neighboring pixels are computed using the Pythagorean theorem. A nearest-neighbor rule would assign the output pixel (x, y) the value of 15, which is the value found at the nearest input pixel.

This is a computationally efficient procedure. It is especially liked by Earth scientists because it does not alter the pixel brightness values during resampling. It is often the very subtle changes in brightness values that make all the difference when distinguishing one type of vegetation from another, an edge associated with a geologic lineament, or different levels of turbidity, chlorophyll, or temperature in a lake. Other interpolation techniques to be discussed use averages to compute the output intensity value, often removing valuable spectral information. Therefore, nearest-neighbor resampling should be used whenever biophysical information is to be extracted from remote sensor data.

Bilinear Interpolation: First-order or *bilinear interpolation* assigns output pixel values by interpolating brightness values in two orthogonal directions in the input image. It basically fits a plane to the four pixel values nearest to the desired position (x', y') in the input image and then computes a new brightness value based on the weighted distances to these points. For example, the distances from the requested x', y' position at 2.4, 2.7 in the input image in Figure 7-8b to the closest four input pixel coordinates (2, 2; 3, 2; 2, 3; 3, 3) are computed in Table 7-2. The closer a pixel is to the desired x', y' location, the more weight it will have in the final computation of the average. The weighted average of the new brightness value (BV_{wt}) is computed using the equation:

$$\text{Bilinear}_{BV_{wt}} = \frac{\displaystyle\sum_{k=1}^{4} \frac{Z_k}{D_k^2}}{\displaystyle\sum_{k=1}^{4} \frac{1}{D_k^2}}, \tag{7-14}$$

where Z_k are the surrounding four data point values, and D_k^2 are the distances squared from the point in question (x', y') to these data points. In our example, the weighted average of

BV_{wt} is 13.53 (truncated to 13), as shown in Table 7-2. The average without weighting is 12. In many respects this method acts as a spatial moving filter that subdues extremes in brightness value throughout the output image.

Cubic convolution resampling assigns values to output pixels in much the same manner as bilinear interpolation, except that the weighted values of 16 input pixels surrounding the location of the desired x', y' pixel are used to determine the value of the output pixel. For example, the distances from the requested x', y' position at 2.4, 2.7 in the input image in Figure 7-8b to the closest 16 input pixel coordinates are computed in Table 7-3. The weighted average of the new brightness value (BV_{wt}) is computed using the equation:

$$\text{Cubic Convolution}_{BV_{wt}} = \frac{\sum_{k=1}^{16} \frac{Z_k}{D_k^2}}{\sum_{k=1}^{16} \frac{1}{D_k^2}}, \qquad (7\text{-}15)$$

where Z_k are the surrounding 16 data point values, and D_k^2 are the distances squared from the point in question (x', y') to these data points. In this example, the weighted average of BV_{wt} is 13.41 (truncated to 13), as shown in Table 7-3. The average without weighting is 12.

Example of Image-to-Map Rectification

To appreciate digital *image-to-map* rectification, it is useful to demonstrate the logic by applying it to a real dataset such as the Landsat Thematic Mapper image of Charleston, SC. The image-to-map rectification process generally involves:

- selecting an appropriate planimetric base map,

- collecting ground control points,

- determining the optimum set of geometric rectification coefficients by iteratively computing the total GCP RMS$_{error}$, and

- filling the output matrix using spatial and intensity interpolation resampling methods.

Planimetric Map Characteristics

The characteristics of the planimetric map used during the rectification process are very important and must be care-

fully selected. There are several types of map projections to be considered in this context. On an *equivalent* (equal-area) map projection, a circle of diameter n drawn at any location on the map will encompass exactly the same geographic area. This characteristic is useful if a scientist is interested in comparing land-use area, density, and so on. Unfortunately, to maintain the equal-area attribute, the shapes, angles, and scale in parts of the map may be distorted.

Conversely, a *conformal* map projection maintains correct shape and distance around a given point on the map. Because angles at each point are correct on conformal maps, the scale in every direction around any point is constant. This allows the analyst to measure distance and direction between relatively near points with good precision. For our purposes, this means that for image areas covering a few contiguous 1:24,000-scale 7.5-minute quadrangle sheets, accurate spatial measurement is possible if the data are rectified to a conformal map projection.

One of the most often used projections for rectifying remotely sensed data is the transverse Mercator projection. It is made from a normal Mercator projection by rotating the cylinder (the developable surface) so that it lies tangent along a meridian (line of longitude). The central meridians, the equator, and each line 90° from the central meridian are straight lines (Figure 7-10). The central meridian normally has a constant scale. Any lines parallel to the central meridian are lines of constant scale. The Universal Transverse Mercator projection often used by the U.S. Geological Survey in its topographic mapping program has a central scale factor of 0.9996 and is composed of 60 zones, each 6° of longitude wide, with a central meridian placed every sixth meridian beginning with 177° west. In the Charleston example, a 1979 photo-revised 7.5-minute quadrangle was selected as the base map with which to rectify the Landsat TM data (Figure 7-5a). It lies in UTM zone 17 and has a 1000-m UTM grid overlay.

Ground Control Point Collection

Twenty ground control points were located on the map and the UTM easting and northing of each point were identified (Table 7-4). The same 20 GCPs were then identified in the TM data according to their row and column coordinates (Table 7-4). The location of points 13, 14, and 16 are shown in Figure 7-5. The GCPs should be located uniformly throughout the region to be rectified and not congested into one small area simply because a) there are more easily identifiable points in that area or b) the locations are easy to get to in the field.

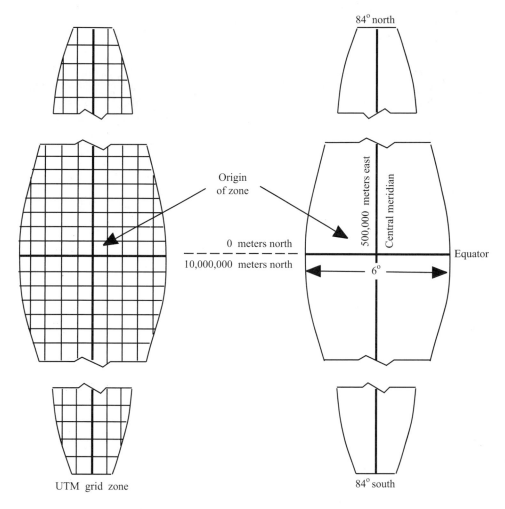

Figure 7-10   Universal Transverse Mercator (UTM) grid zone with associated parameters. This projection is often used when rectifying remote sensor data to a base map. It is found on U.S. Geological Survey 7.5- and 15-minute quadrangles.

Determine Optimum Geometric Rectification Coefficients by Evaluating GCP Total RMS$_{error}$

The 20 GCPs selected were input to the least-squares regression procedure previously discussed to identify 1) the coefficients of the coordinate transformation and 2) the individual and total RMS$_{error}$ associated with the GCPs. A threshold of 0.5 was not satisfied until 13 GCPs were deleted from the analysis. The order in which the 13 GCPs were deleted and the total RMS$_{error}$ found after each deletion are summarized in Table 7-4. The 7 GCPs finally selected that produced an acceptable RMS$_{error}$ are shown in Table 7-5. GCP 11 would have been the next point deleted if the threshold had not been satisfied. The 6 coefficients derived from the 7 suitable GCPs are found in Table 7-6. The Charleston Landsat TM scene rectified using these parameters is shown in Figure 7-6a.

It is instructive to demonstrate how the 6 coefficients were computed for this Landsat TM dataset. This was accomplished using only the final 7 GCPs, since these represent the set that produced an acceptable RMS$_{error}$ of <0.50 pixels. Remember, however, that this same operation was performed with 20 points, then 19 points, and so on, before arriving at just 7 acceptable points.

A least-squares multiple regression approach is used to compute the coefficients. Two equations are necessary. One equation computes the image y' coordinate (the dependent variable) as a function of the x and y map coordinates (representing two independent variables). The second computes the image x' coordinate as a function of the same map (x, y) coordinates. Three coefficients are determined by using each algorithm. The mathematics for computing the column coordinates (x') in the image can now be presented.

Table 7-3.   Cubic convolution interpolation of a weighted brightness value (BV_{wt}) at location x', y' based on the analysis of 16 sample points in Figure 7-8b.

Sample Point Location (column, row)	Value at Sample Point, Z	Distance from x', y' to the Sample Point, D	D_k^2	$\dfrac{Z}{D_k^2}$	$\dfrac{1}{D_k^2}$
1, 1	7	$D = \sqrt{(2.4-1)^2 + (2.7-1)^2} = 2.202$	4.85	1.443	0.206
2, 1	8	$D = \sqrt{(2.4-2)^2 + (2.7-1)^2} = 1.746$	3.05	2.623	0.328
3, 1	6	$D = \sqrt{(2.4-3)^2 + (2.7-1)^2} = 1.80$	3.24	1.852	0.309
4, 1	7	$D = \sqrt{(2.4-4)^2 + (2.7-1)^2} = 2.335$	5.45	1.284	0.183
1, 2	8	$D = \sqrt{(2.4-1)^2 + (2.7-2)^2} = 1.565$	2.45	3.265	0.408
2, 2	9	$D = \sqrt{(2.4-2)^2 + (2.7-2)^2} = 0.806$	0.65	13.85	1.539
3, 2	6	$D = \sqrt{(2.4-3)^2 + (2.7-2)^2} = 0.921$	0.85	7.06	1.176
4, 2	14	$D = \sqrt{(2.4-4)^2 + (2.7-2)^2} = 1.746$	3.05	4.59	0.328
1, 3	14	$D = \sqrt{(2.4-1)^2 + (2.7-3)^2} = 1.432$	2.05	6.829	0.488
2, 3	15	$D = \sqrt{(2.4-2)^2 + (2.7-3)^2} = 0.500$	0.25	60.00	4.000
3, 3	18	$D = \sqrt{(2.4-3)^2 + (2.7-3)^2} = 0.670$	0.45	40.00	2.222
4, 3	17	$D = \sqrt{(2.4-4)^2 + (2.7-3)^2} = 1.63$	2.65	6.415	0.377
1, 4	14	$D = \sqrt{(2.4-1)^2 + (2.7-4)^2} = 1.911$	3.65	3.836	0.274
2, 4	16	$D = \sqrt{(2.4-2)^2 + (2.7-4)^2} = 1.360$	1.85	8.649	0.541
3, 4	19	$D = \sqrt{(2.4-3)^2 + (2.7-4)^2} = 1.432$	2.05	9.268	0.488
4, 4	20	$D = \sqrt{(2.4-4)^2 + (2.7-4)^2} = 2.062$	4.25	4.706	0.235
				Σ 175.67	Σ 13.102

$$BV_{wt} = 175.67 / 13.102$$

$$BV_{wt} = \mathbf{13.41}$$

Table 7-4. Characteristics of 20 ground control points used to rectify the Charleston, SC, Landsat Thematic Mapper scene.

Point Number	Order of Points Deleted[a]	Easting on Map, X_1	Northing on Map, X_2	X' Pixel	Y' Pixel	Total RMS$_{error}$ after This Point Is Deleted
1	12	597,120	3,627,050	150	185	0.501
2	9	597,680	3,627,800	166	165	0.663
3	Kept	598,285	3,627,280	191	180	—
4	Kept	595,650	3,627,730	98	179	—
5	2	596,750	3,625,600	123	252	6.569
6	13	597,830	3,624,820	192	294	0.435
7	Kept	596,250	3,624,380	137	293	—
8	Kept	602,200	3,628,530	318	115	—
9	Kept	600,350	3,629,730	248	83	—
10	5	600,680	3,629,340	259	93	1.291
11	Kept	600,440	3,628,860	255	113	—
12	10	599,150	3,626,990	221	186	0.601
13	8	600,300	3,626,030	266	211	0.742
14	6	598,840	3,626,460	211	205	1.113
15	3	598,940	3,623,430	214	295	4.773
16	Kept	600,540	3,626,450	272	196	—
17	4	596,985	3,629,350	134	123	1.950
18	7	596,035	3,627,880	109	174	0.881
19	11	600,995	3,630,000	269	71	0.566
20	1	601,700	3,632,580	283	12	8.542
				Total RMS$_{error}$ with all 20 GCPs used:		**11.016**

[a] For example, GCP 20 was the first point deleted. After it was deleted, the total RMS$_{error}$ dropped from 11.016 to 8.542. Point 5 was the second point deleted. After it was deleted, the total RMS$_{error}$ dropped from 8.542 to 6.569.

Multiple Regression Coefficients Computation: First, we will let

Y = either the x' or y' location in the image, depending on which is being evaluated; in this example it will represent the x' values

X_1 = easting coordinate (x) of the map GCP

X_2 = northing coordinate (y) of the map GCP

It is practical here to use X_1 and X_2 instead of x and y to simplify mathematical notation.

The seven coordinates used in the computation of the coefficients are shown in Table 7-5. Notice that the independent variables (X_1 and X_2) have been adjusted (adjusted value = original value – minimum value) so that the sums of squares or sums of products do not become so large that they overwhelm the precision of the CPU being used. Note, for example, that most of the original northing UTM measurements associated with the map GCPs are already in the range of 3 million meters. The minimums subtracted in Table 7-5 are added back into the analysis at the final stage of coefficient

Table 7-5.   Information concerning the final seven ground control points used to rectify the Charleston, SC, Landsat Thematic Mapper Scene.

Point Number	Easting on Map	Adjusted Easting, X_1[a]	Northing on Map	Adjusted Northing, X_2[b]	Y[c] X′ Pixel	Y[c] Y′ Pixel
3	598,285	2,635	3,627,280	2,900	191	180
4	595,650	0	3,627,730	3,350	98	179
7	596,250	600	3,624,380	0	137	293
8	602,200	6,550	3,628,530	4,150	318	115
9	600,350	4,700	3,629,730	5,350	248	83
11	600,440	4,790	3,628,860	4,480	255	113
16	600,540	4,890	3,626,450	2,070	272	196
	Minimum = 595,650	24,165	Minimum = 3,624,380	22,300	1519	1159

[a] Adjusted easting values (X_1) used in the least-squares computation of coefficients. This is an independent variable.
[b] Adjusted northing values (X_2) used in the least-squares computation of coefficients. This is an independent variable.
[c] The dependent variable (Y) discussed in the text. In this example it was used to predict the $X′$ pixel location.

Table 7-6.   Coefficients used to rectify the Charleston, SC, Landsat Thematic Mapper scene.

$x′ = -382.2366 + 0.034187x + (-0.005481)y$

$y′ = 130,162 + (-0.005576)x + (-0.0349150)y$

where x, y are coordinates in the output image and $x′, y′$ are predicted image coordinates in the original, unrectified image.

computation. Next we will discuss the mathematics necessary to compute the $x′$ coefficients shown in Table 7-6. It is technical, but should be of value to those interested in how to compute the coefficients used in Equation 7-9.

I. Find (X^TX) and (X^TY) in deviation form:

$n = 7$, the number of control points used

A. First compute:

$$\sum_{i=1}^{n} Y_i = 1519 \qquad \sum_{i=1}^{n} X_{1i} = 24,165$$

$$\sum_{i=1}^{n} X_{2i} = 22,300 \qquad \bar{Y} = 217 \qquad \bar{X}_1 = 3452.1428$$

$$\bar{X}_2 = 3185.7142 \qquad \sum_{i=1}^{n} Y_i^2 = 366,491$$

$$\sum_{i=1}^{n} X_{1i}^2 = 119,151,925 \qquad \sum_{i=1}^{n} X_{2i}^2 = 89,832,800$$

$$\sum_{i=1}^{n} X_{1i}Y_i = 6,385,515 \qquad \sum_{i=1}^{n} X_{2i}Y_i = 5,234,140$$

$$\sum_{i=1}^{n} X_{1i}Y_{2i} = 91,550,500$$

B. Compute sums of squares:

$$1. \sum_{i=1}^{n} X_{1i}^2 - \frac{1}{n}\left(\sum_{i=1}^{n} X_{1i}\right)^2 = 119,151,925 - \frac{1}{7}(24,165)^2$$

$$= 35,730,892.8571$$

$$2. \sum_{i=1}^{n} X_{2i}^2 - \frac{1}{n}\left(\sum_{i=1}^{n} X_{2i}\right)^2 = 89,832,800 - \frac{1}{7}(22,300)^2$$

$$= 18,791,371.4286$$

3.
$$\sum_{i=1}^{n} X_{1i} X_{2i} - \frac{1}{n} \left(\sum_{i=1}^{n} X_{1i} \right) \left(\sum_{i=1}^{n} X_{2i} \right)$$

$$= 91,550,500 - \frac{1}{7}(24,165)(22,300)$$

$$= 14,567,714.2857$$

where

$$(X^T X) = \begin{Bmatrix} 35,730,892.8571 & 14,567,714.2857 \\ 14,567,714.2857 & 18,791,371.4286 \end{Bmatrix}$$

4. Covariance between Y and X_1:

$$\sum_{i=1}^{n} X_{1i} Y_i - \frac{1}{n} \left(\sum_{i=1}^{n} X_{1i} \right) \left(\sum_{i=1}^{n} Y_i \right)$$

$$= 6,385,515 - \frac{1}{7}(24,165)(1519)$$

$$= 1,141,710$$

5. Covariance between Y and X_2:

$$\sum_{i=1}^{n} X_{2i} Y_i - \frac{1}{n} \left(\sum_{i=1}^{n} X_{2i} \right) \left(\sum_{i=1}^{n} Y_i \right)$$

$$= 5,234,140 - \frac{1}{7}(22,300)(1519)$$

$$= 395,040$$

where

$$(X^T Y) = \begin{Bmatrix} 1,141,710 \\ 395,040 \end{Bmatrix}$$

II. Find the inverse of $(X^T X) = (X^T X)^{-1}$:

A. First, find the determinant of the 2×2 matrix:

$$|X^T X| = (35,730,892.8571)(18,791,371.4286)$$
$$- (14,567,714.2857)^2$$
$$= 459,214,179,643,488.9$$

B. Determine adjoint matrix of $(X^T X)$ where adjoint equals the transpose of the cofactor matrix:

$$\text{Adjoint}^* = \begin{Bmatrix} 18,791,371.4286 & -14,567,714.2857 \\ -14,567,714.2857 & 35,730,892.8571 \end{Bmatrix}$$

*Note that if

$$A = \begin{Bmatrix} a & b \\ c & d \end{Bmatrix}, \text{ then } A^{-1} = \frac{1}{\det A} \begin{Bmatrix} d & -b \\ -c & a \end{Bmatrix}.$$

C. Get $(X^T X)^{-1}$ by multiplying the adjoint of $(X^T X)$ by the det $(X^T X)$ under 1.

$$(X^T X) = \left(\frac{1}{459,214,179,643,488.9} \right)$$

$$\begin{Bmatrix} 18,791,373.7142 & -14,567,714.2857 \\ -14,567,714.2857 & 35,730,896.4285 \end{Bmatrix}$$

$$= \begin{Bmatrix} 0.41 \times 10^{-7} & -0.32 \times 10^{-7} \\ -0.32 \times 10^{-7} & 0.78 \times 10^{-7} \end{Bmatrix}$$

III. Now, compute coefficients using $a_i = (X^T X)^{-1}(X^T Y)$:

$$\begin{Bmatrix} 0.41 \times 10^{-7} & -0.32 \times 10^{-7} \\ -0.32 \times 10^{-7} & 0.78 \times 10^{-7} \end{Bmatrix} \bullet \begin{Bmatrix} 1,141,710 \\ 395,040 \end{Bmatrix} =$$

$$a_1 = (0.000000041)(1,141,710) + (-0.000000032)(395,040)$$
$$a_2 = (-0.000000032)(1,141,710) + (0.000000078)(395,040)$$

$$a_1 = 0.0341877$$
$$a_2 = -0.0054810$$

Now compute the intercept, a_0, from

$$a_0 = \bar{Y} - \sum_{i=1}^{2} a_i \bar{X}_i.$$

[The minimums of X_1 and X_2 (595,650 and 3,624,380, respectively) must be accounted for here. See Table 7-5.]

$$a_0 = 217 - [(0.0341877)(3452.1428 + 595,650) +$$

$$(-0.0054810)(3185.7142 + 3,624,380)] = -382.2366479.$$

Therefore, the equation becomes

$$Y = -382.2366479 + 0.0341877 X_1 - 0.0054810 X_2.$$

Because we actually evaluated the dependent variable x', this becomes

$$x' = -382.2366479 + 0.0341877x - 0.0054810y$$

with x and y representing the map coordinates and x' being the predicted column coordinate in the input image.

Similar procedures are required to compute the other three coefficients for the row (y') location in the input image. This would require inserting the seven y' pixel values in the equation in Table 7-5 instead of the x' pixel values just used. The X_1 and X_2 values associated with the x, y coordinates of the map remain the same.

Fill Output Matrix Using Spatial and Intensity Interpolation Resampling Methods

With the coefficients computed, it was then necessary to 1) identify the UTM coordinates of the area on the map to be rectified, 2) select the type of intensity interpolation to be performed (e.g., nearest-neighbor, bilinear, or cubic convolution), and 3) specify the desired output pixel size. A nearest-neighbor resampling algorithm was selected with a desired output pixel dimension grid of 30 × 30 m. The finer the dimension of the output grid, the greater the number of computations required to fill it. Normally, the size of the pixel is made square (e.g., 30 × 30 m) to facilitate scaling considerations when the rectified data are displayed on computer monitors and various hard-copy output devices. These procedures resulted in rectified Landsat Thematic Mapper data.

 Mosaicking

Mosaicking is the process of combining multiple images into a single seamless composite image. It is possible to mosaic unrectified individual frames or flight lines of remotely sensed data. However, it is more common to mosaic multiple images that have already been rectified to a standard map projection and datum (Figure 7-11).

Mosaicking Rectified Images

Mosaicking n rectified images requires several steps. First, the individual images should be rectified to the same map projection and datum. Ideally, rectification of the n images is performed using the same intensity interpolation resampling logic (e.g., nearest-neighbor) and pixel size (e.g., multiple

Mosaic Feathering Logic

User-specified feathering distance, e.g., 200 pixels

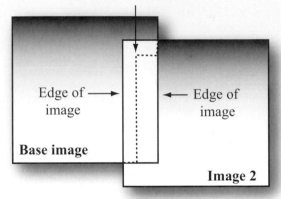

a. Cut-line feathering.

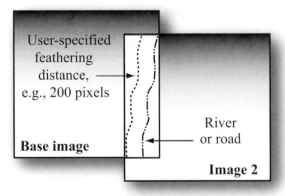

b. Edge feathering.

Figure 7-11   The visual seam between adjacent images being mosaicked may be minimized using a) cut-line feathering logic, or b) edge feathering.

Landsat TM scenes to be mosaicked are often resampled to 30 × 30 m).

Next, one of the images to be mosaicked is designated as the *base image*. The base image and image 2 will normally overlap a certain amount (e.g., 20% – 30%). A representative geographic area in the overlap region is identified. This area in the base image is contrast stretched according to user specifications. The histogram of this geographic area in the base image is extracted. The histogram from the base image is then applied to image 2 using a histogram-matching algorithm. This causes the two images to have approximately the same grayscale characteristics (Research Systems, 2000).

Image Mosaicking

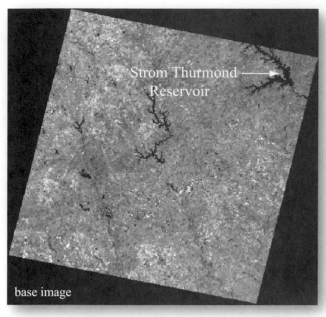

a. Rectified Landsat ETM$^+$ image of eastern Georgia obtained on October 3, 2001 (band 4; Worldwide Reference System—Path 18, Row 37).

b. Rectified Landsat ETM$^+$ image of western South Carolina obtained on October 26, 2001 (band 4; Worldwide Reference System—Path 17, Row 37).

c. Feathered mosaic of rectified Landsat ETM$^+$ imagery of eastern Georgia and western South Carolina.

Figure 7-12   Two rectified Landsat Enhanced Thematic Mapper Plus (ETM$^+$) images obtained within 23 days of one another are mosaicked using feathering logic. A color-infrared color composite is found in **Color Plate 7-1**.

It is possible to have the pixel brightness values in one scene simply dominate the pixel values in the overlapping scene. Unfortunately, this can result in noticeable seams in the final mosaic. Therefore, it is common to blend the seams between mosaicked images using *feathering* (Tucker et al., 2004). Some digital image processing systems allow the user to specific a feathering buffer distance (e.g., 200 pixels) wherein 0% of the base image is used in the blending at the edge and 100% of image 2 is used to make the output image (Figure 7-11a). At the specified distance (e.g., 200 pixels) in from the edge, 100% of the base image is used to make the output image and 0% of image 2 is used. At 100 pixels in from the edge, 50% of each image is used to make the output file.

Sometimes analysts prefer to use a linear feature such as a river or road to subdue the edge between adjacent mosaicked images. In this case, the analyst identifies a polyline in the image (using an annotation tool) and then specifies a buffer distance away from the line as before where the feathering will take place (Figure 7-11b). It is not absolutely necessary to use natural or man-made features when performing cut-line feathering. Any user-specified polyline will do.

The output file is then produced; it consists of two histogram-matched images that have been feathered at the common edge(s) of the dataset. Hopefully, the multiple image edge-match is not visible. Additional images (3, 4, etc.) to be mosaicked are histogram-matched and feathered using similar logic.

The output file of *n* mosaicked images should appear as if it were one continuous image. For example, Figure 7-12a,b depicts two rectified Landsat Enhanced Thematic Mapper ETM[+] scenes obtained over eastern Georgia and western South Carolina on October 3, 2001, and October 26, 2001, respectively. Each image was resampled using nearest-neighbor logic to 30×30 m pixels in a UTM projection. The two Landsat ETM[+] frames contained approximately 20% side-lap. Note the Strom Thurmond Reservoir in the overlap region. The two images were histogram-matched and mosaicked using edge-feathering logic (Figure 7-12c). A color-infrared color composite of the mosaicked dataset is shown in **Color Plate 7-1c**. This is a good mosaic because it is difficult to detect the seam between the two input images.

References

Bernstein, R., 1983, "Image Geometry and Rectification," Chapter 21 in R. N. Colwell, (Ed.), *Manual of Remote Sensing*, Bethesda, MD: American Society of Photogrammetry, Vol. 1, 875–881.

Bossler, J. D., Jensen, J. R., McMaster, R. B. and C. Rizos, 2002, *Manual of Geospatial Science and Technology*, London: Taylor & Francis, 623 p.

Brown, L. G., 1992, "A Survey of Image Registration Techniques," *ACM Computing Surveys*, 24(4):325–376.

Buiten, H. J. and B. Van Putten, 1997, "Quality Assessment of Remote Sensing Registration—Analysis and Testing of Control Point Residuals," *ISPRS Journal of Photogrammetry & Remote Sensing*, 52:57–73.

Chen, L., and L. Lee, 1992, "Progressive Generation of Control Frameworks for Image Registration," *Photogrammetric Engineering & Remote Sensing*, 58(9):1321–1328.

Civco, D. L., 1989, "Topographic Normalization of Landsat Thematic Mapper Digital Imagery," *Photogrammetric Engineering & Remote Sensing*, 55(9):1303–1309.

Gibson, P. J. and C. H. Power, 2000, *Introductory Remote Sensing: Digital Image Processing and Applications*, New York: Routledge, 249 p.

Jensen, J. R., Botchway, K., Brennan-Galvin, E., Johannsen, C., Juma, C., Mabogunje, A., Miller, R., Price, K., Reining, P., Skole, D., Stancioff, A. and D. R. F. Taylor, 2002, *Down to Earth: Geographic Information for Sustainable Development in Africa*, Washington: National Research Council, 155 p.

Jensen, J. R., Cowen, D., Narumalani, W., Weatherbee, O. and J. Althausen, 1993, "Evaluation of CoastWatch Change Detection Protocol in South Carolina," *Photogrammetric Engineering & Remote Sensing*, 59(6):1039–1046.

Jensen, J. R., Ramsey, E., Mackey, H. E. and M. E. Hodgson, 1988, "Thermal Modeling of Heat Dissipation in the Pen Branch Delta Using Thermal Infrared Imagery," *Geocarto International: An Interdisciplinary Journal of Remote Sensing and GIS*, 4:17–28.

Konecny, G., 2003, *Geoinformation: Remote Sensing, Photogrammetry and Geographic Information Systems*, New York: Taylor & Francis, 248 p.

Lillesand, T., Kiefer, R. W. and J. W. Chipman, 2004, *Remote Sensing and Image Interpretation*, New York: John Wiley, 368–375.

Lunetta, R. S., Congalton, R. G., Fenstermaker, L. K., Jensen, J. R., McGwire, K. C. and L. R. Tinney, 1991, "Remote Sensing and Geographic Information Systems Data Integration: Error Sources and Research Issues," *Photogrammetric Engineering & Remote Sensing*, 57(6):677–687.

Marakas, G. M., 2003, *Decision Support Systems in the 21st Century*, Upper Saddle River, NJ: Prentice-Hall, 611 p.

NASA, 1998, *Landsat 7 Initial Assessment Geometric Algorithm Theoretical Basis Document*, Washington: NASA, 177 p.

Niblack, W., 1986, *An Introduction to Digital Image Processing*, Englewood Cliffs, NJ: Prentice-Hall, 215 p.

Novak, K., 1992, "Rectification of Digital Imagery," *Photogrammetric Engineering & Remote Sensing*, 58(3):339–344.

Richards, J. R. and X. Jia, 1999, *Remote Sensing Digital Image Analysis*, New York: Springer-Verlag, 363 p.

Rinker, J. N., 1999, *Introduction to Spectral Remote Sensing*, Alexandria, VA: U.S. Army Topographic Engineering Center, http://www.tec.army.mil/terrain/desert/tutorial.

Renz, A. N. (Ed.), 1999, *Remote Sensing for the Earth Sciences*, Bethesda, MD: American Society for Photogrammetry & Remote Sensing, 707 p.

Research Systems, 2000, *The Environment for Visualizing Images Tutorials*, Boulder, CO: Research Systems, 590 p.

Schowengerdt, R. A., 1997, *Remote Sensing: Models and Methods for Image Processing*, San Diego, CA: Academic Press, 522 p.

Ton, J. and A. K. Jain, 1989, "Registering Landsat Images by Point Matching," *IEEE Transactions on Geoscience and Remote Sensing*, 27(5):642–651.

Tucker, C. J., Grant, D. M. and J. D. Dykstra, 2004, "NASA's Global Orthorectified Landsat Data Set," *Photogrammetric Engineering & Remote Sensing*, 70(3):313-322.

Welch, R. A., Remillard, M. and J. Alberts, 1992, "Integration of GPS, Remote Sensing, and GIS Techniques for Coastal Resource Management," *Photogrammetric Engineering & Remote Sensing*, 58(11):1571–1578.

Williams, D., 2003, Correspondence regarding the angular velocity of Landsat satellites 1 to 5 and 7, Greenbelt, MD: NASA Goddard Space Flight Center.

Wolberg, G., 1990, *Digital Image Warping*, New York: John Wiley–IEEE Computer Society, 340 p.

Wolf, P., 2002, *Elements of Photogrammetry*, 2nd ed., New York: McGraw-Hill, 562 p.

Image Enhancement

8

I mage enhancement algorithms are applied to remotely sensed data to improve the appearance of an image for human visual analysis or occasionally for subsequent machine analysis. There is no such thing as the ideal or best image enhancement because the results are ultimately evaluated by humans, who make subjective judgments as to whether a given image enhancement is useful. This chapter identifies a variety of image enhancement operations that have proven of value for visual analysis of remote sensor data and/or subsequent machine analysis. *Point operations* modify the brightness values of each pixel in an image dataset independent of the characteristics of neighboring pixels. *Local operations* modify the value of each pixel in the context of the brightness values of the pixels surrounding it.

Image Reduction and Magnification

Image analysts routinely view images that have been reduced in size or magnified during the image interpretation process. Image reduction techniques allow the analyst to obtain a regional perspective of the remotely sensed data. Image magnification techniques allow the analyst to zoom in and view very site-specific pixel characteristics.

Image Reduction

In the early stages of a remote sensing project it is often necessary to view the entire image in order to locate the row and column coordinates of a subimage that encompasses the study area. Most commercially available remote sensor data are composed of more than 3000 rows $\times$ 3000 columns in a number of bands. Unfortunately, most digital image processing systems only display approximately 1024 $\times$ 1024 pixels at one time. Therefore, it is useful to have a simple procedure for reducing the size of the original image dataset down to a smaller dataset that can be viewed on the screen at one time for orientation purposes. To *reduce* a digital image to just $1/m^2$ of the original data, every mth row and mth column of the imagery are systematically selected and displayed. For example, consider a rectified Landsat Enhanced Thematic Mapper image of Charleston, SC, composed of 6464 rows $\times$ 6464 columns. If every other row and every other column (i.e., $m = 2$) were selected for a single band, the entire scene could be displayed as a sampled image consisting of just 3232 rows $\times$ 3232 columns. This reduced dataset would contain only one fourth

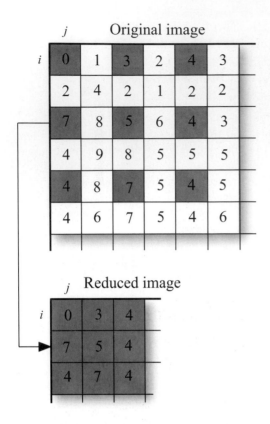

Figure 8-1　Hypothetical example of 2× image reduction achieved by sampling every other row and column of the original data. This operation results in a new image consisting of only one quarter (25%) of the original data.

(25%) of the pixels found in the original scene. The logic associated with a simple 2× integer reduction is shown in Figure 8-1.

Unfortunately, an image consisting of 3232 rows × 3232 columns is still too large to view on most screens. Therefore, it is often necessary to sample the remotely sensed data more intensively. The Charleston ETM$^+$ band 4 image displayed in Figure 8-2 was produced by sampling every eighth row and eighth column (i.e., $m = 8$) of the original imagery, often referred to as an 8× reduction. It is composed of 808 rows and 808 columns, but contains 1/64 (1.5625%) of the original data. If we compare the original data with the reduced data, there is an obvious loss of detail because so many of the pixels are not present. Therefore, we rarely interpret or digitally analyze image reductions. Instead, they are used for orienting within a scene and locating the row and column coordinates of specific *areas of interest* (AOI) that can then be extracted at full resolution for analysis.

A predawn thermal-infrared image of thermal effluent entering the Savannah River Swamp System at 4:28 a.m. on March 31, 1981, is shown in Figure 8-3. The original image consisted of 2530 rows and 1264 columns. This 2× reduction consists of 1265 rows and 632 columns and provides a regional overview of the spatial distribution of the thermal effluent.

Image Magnification

Digital image *magnification* (often referred to as *zooming*) is usually performed to improve the scale of an image for visual interpretation or, occasionally, to match the scale of another image. Just as row and column deletion is the simplest form of image reduction, row and column replication represents the simplest form of digital image magnification. To magnify a digital image by an integer factor m^2, each pixel in the original image is usually replaced by an $m \times m$ block of pixels, all with the same brightness value as the original input pixel. The logic of a 2× magnification is shown in Figure 8-4. This form of magnification is characterized by visible square tiles of pixels in the output display. Image magnifications of 1×, 2×, 3×, 4×, 6×, and 8× applied to the Charleston TM band 4 data are shown in Figure 8-5. The building shadows become more apparent as the magnification is increased. The predawn thermal-infrared data of the Savannah River are magnified 1× through 9× in Figure 8-6.

Most sophisticated digital image processing systems allow an analyst to specify floating point magnification (or reduction) factors (e.g., zoom in 2.75×). This requires that the original remote sensor data be resampled in near realtime using one of the resampling algorithms discussed in Chapter 7 (e.g., nearest-neighbor, bilinear interpolation, or cubic convolution). This is a very useful technique when the analyst is trying to obtain detailed information about the spectral reflectance or emittance characteristics of a relatively small geographic area of interest. During the training phase of a supervised classification (to be discussed in Chapter 9), it is especially useful to be able to zoom in to the raw remote sensor data at very precise floating point increments to isolate a particular field or body of water.

In addition to magnification, many digital image processing systems provide a mechanism whereby the analyst can *pan* or *roam* about a much larger geographic area (e.g., 2048 × 2048) while viewing only a portion (e.g., 512 × 512) of this area at any one time. This allows the analyst to view parts of the database much more rapidly. Panning or roaming requires additional image processor memory.

8x Reduction of Landsat Enhanced Thematic Mapper Plus (ETM+)
Imagery of Charleston, SC, Obtained on October 23, 1999 (Path 16, Row 37)

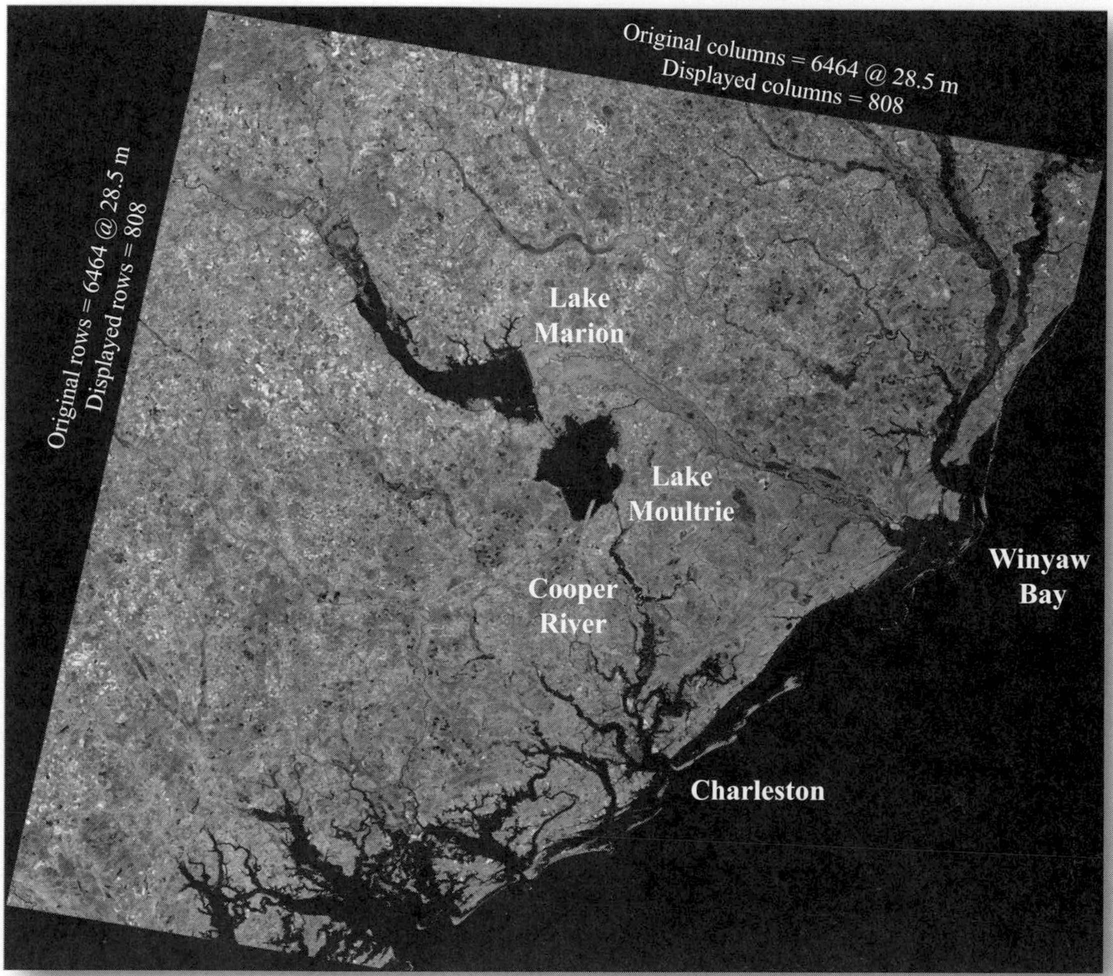

Original columns = 6464 @ 28.5 m
Displayed columns = 808

Original rows = 6464 @ 28.5 m
Displayed rows = 808

Lake Marion

Lake Moultrie

Cooper River

Winyaw Bay

Charleston

Figure 8-2   The 808 row × 808 column image represents only 1/64 of the data found in the original 6464 row × 6464 column Enhanced Thematic Mapper Plus image. It was created by sampling every eighth row and eighth column of the band 4 image.

Transects (Spatial Profiles)

The ability to extract brightness values along a user-specified *transect* (also referred to as a *spatial profile*) between two points in a single-band or multiple-band color composite image is important in many remote sensing image interpretation applications. For example, consider the three simple 50-pixel horizontal spatial profiles (A, B, and C) identified in the single-band, black-and-white thermal infrared image of the Savannah River (Figure 8-7a). In each case, the spatial profile in histogram format depicts the magnitude of the brightness value at each pixel along the 50-pixel transect. Each pixel in the dataset was 2.8 × 2.8 m.

Sometimes it is useful to view the spatial profile histogram in a single grayscale tone (or color) as shown in Figure 8-7b. Conversely, analysts often prefer to display the grayscale tone of the individual brightness values encountered along the transect as shown in Figures 8-7c, d, and e.

Each of the transects in Figure 8-7 was located along a single horizontal scan line. In this case, each of the transects was exactly 140 m long (50 pixels × 2.8 m = 140 m). But what if we wanted to extract a detailed transect across the river per-

**2x Reduction of Predawn
Thermal Infrared Scanner Data**

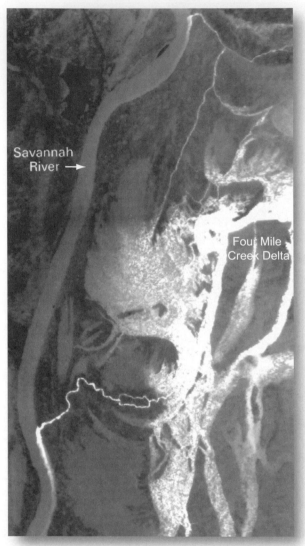

Figure 8-3   Predawn thermal infrared image of thermal effluent entering the Savannah River Swamp System
on March 31, 1981. The 2× reduction image consists of 1265 rows × 632 columns and provides a
regional overview of the spatial distribution of
the thermal effluent.

pendicular to the shoreline? Or what if we wanted to run a
transect through the heart of the plume to see how fast the
thermal plume was cooling after entering the river? In this
case it would be necessary to take into account the stair-
stepped nature of the transect and incorporate distance information derived using the Pythagorean theorem previously
discussed. Hypothetical examples of a 7-pixel diagonal

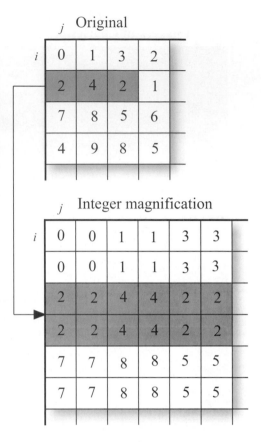

Figure 8-4   Hypothetical example of 2× image magnification
achieved by replicating every row and column in the
original image. The new image will consist of four
times as many pixels as the original scene.

transect and a 7-pixel horizontal transect are shown in Figure
8-8.

Accurate transect results can also be obtained by rotating the
image until the desired transect is aligned with either a single
line or column in the dataset. For example, consider the density-sliced image of the Savannah River thermal plume
shown in **Color Plate 8-1a** based on the color look-up table
values in Table 8-1. In this example, we are interested in
identifying a) the temperature of the thermal plume in the
river perpendicular to the shoreline using a 50-pixel transect
(1) and b) the rate of temperature decrease as the plume
progresses downstream using a 100-pixel transect (2). The
brightness values encountered along the two transects were
obtained only after the original image was geometrically
rotated 16° clockwise so that the end points of each transect
fell on the same scan line or column (**Color Plate 8-1b**). This
ensured that the number of meters in each temperature class
along each transect was accurately measured. If the analyst

Magnification of Landsat Thematic Mapper Data of Charleston, SC

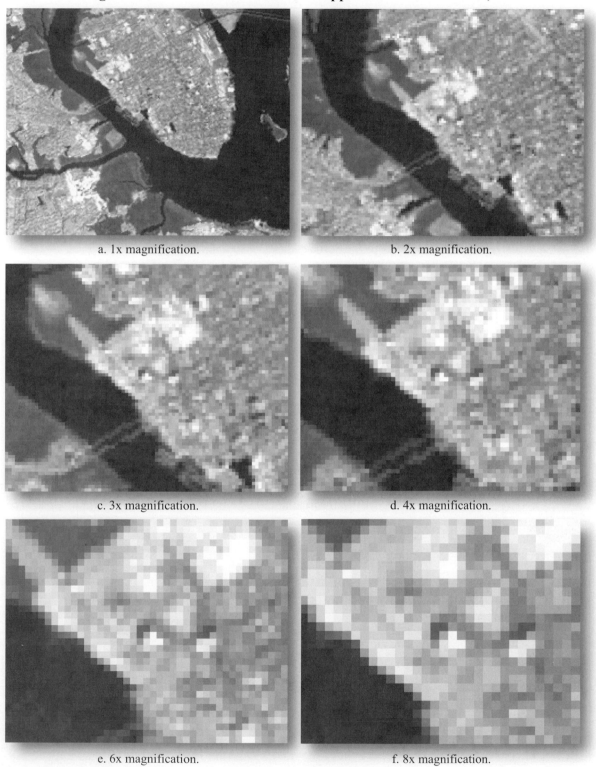

a. 1x magnification.

b. 2x magnification.

c. 3x magnification.

d. 4x magnification.

e. 6x magnification.

f. 8x magnification.

Figure 8-5   Thematic Mapper band 4 data of Charleston, SC, magnified 1×, 2×, 3×, 4×, 6×, and 8×. Note the two large buildings and their associated shadows as the magnification increases.

Magnification of Predawn Thermal Infrared Data

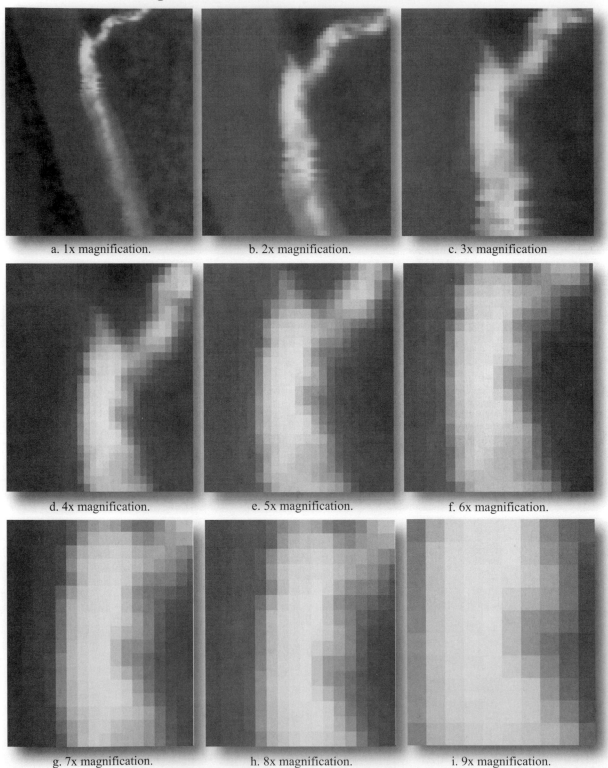

a. 1x magnification.

b. 2x magnification.

c. 3x magnification

d. 4x magnification.

e. 5x magnification.

f. 6x magnification.

g. 7x magnification.

h. 8x magnification.

i. 9x magnification.

Figure 8-6   Predawn thermal infrared data of the Savannah River magnified 1× to 9×.

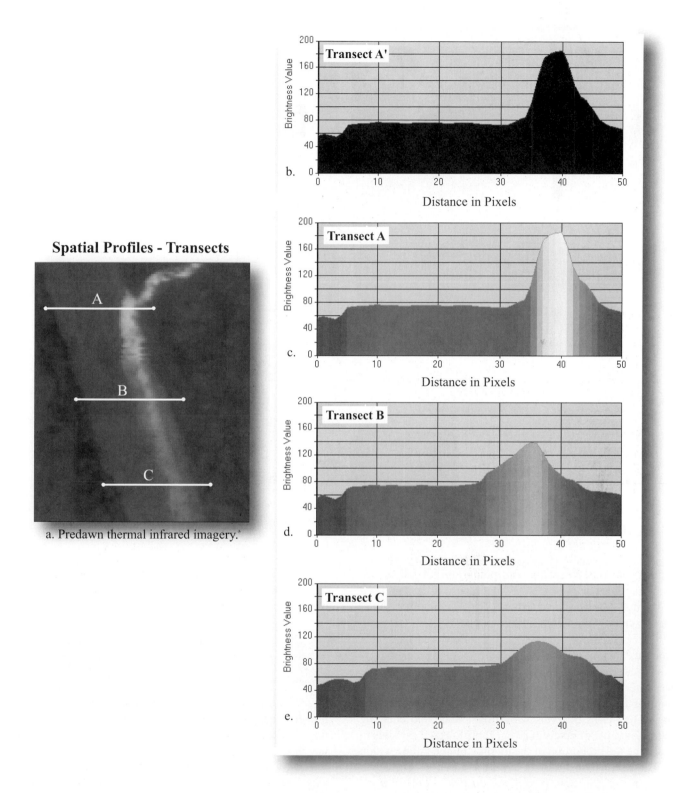

Spatial Profiles - Transects

a. Predawn thermal infrared imagery.

Figure 8-7   a) Three 50-pixel spatial profiles (transects) passed through the Savannah River predawn thermal infrared data. Each transect is 50 pixels long. Each pixel in the dataset is 2.8 × 2.8 m. b) The spatial profile data are displayed in histogram format using a single grayscale tone. c–e) The spatial profile data are displayed in histogram format according to their original brightness values.

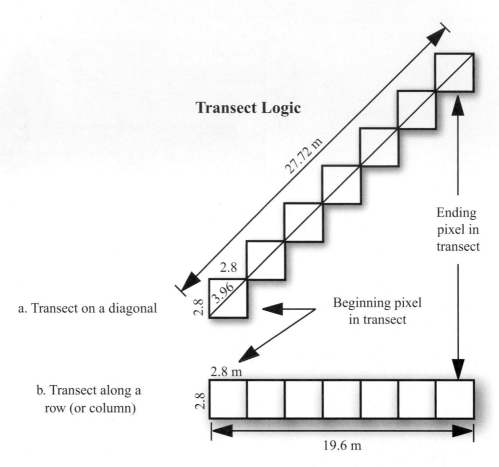

Transect Logic

a. Transect on a diagonal

27.72 m

2.8

3.96

2.8

Ending
pixel in
transect

Beginning pixel
in transect

b. Transect along a
row (or column)

2.8 m

2.8

19.6 m

Figure 8-8   a) The stair-stepped nature of a transect (spatial profile) when the beginning and ending points are not on the same line (or column). In this example, a 45° angle transect consisting of just seven 2.8 × 2.8 m pixels would be 27.72 m in length. b) A 7-pixel transect along a row (or column) of the image would only be 19.6 m long. Therefore, when diagonal transects (or spatial profiles) are extracted, it is necessary to compute their length based on the Pythagorean theorem.

extracts transects where the end points do not fall on the same scan line (or column), the hypotenuse of stair-stepped pixels must be considered instead of the simple horizontal pixel distance (Figure 8-8).

Transects 1 and 2 are shown in **Color Plate 8-1c and d**. The relationship between the original brightness values and the class intervals of the transects is provided in Table 8-1. By counting the number of pixels along a transect in specific temperature-class intervals within the plume and counting the total number of pixels of river (Table 8-1) it is possible to determine the proportion of the thermal plume falling within specific temperature-class intervals (Jensen et al., 1983; 1986). In 1981, the South Carolina Department of Health and Environmental Control (DHEC) mandated that a thermal plume could not be >2.8° above river ambient temperature for more than one third of the width of the river. Transect 1

information extracted from thermal infrared imagery (summarized in Table 8-1) can be used to determine if the plume was in compliance at that location.

Spectral Profiles

In addition to extracting pixel brightness values or percent reflectance values along selected spatial profiles (transects), it is often useful to extract the full spectrum of brightness values in *n* bands for an individual pixel. This is commonly referred to as a *spectral profile*. In a spectral profile, the *x*-axis identifies the number of the individual bands in the dataset and the *y*-axis documents the brightness value (or percent reflectance if the data have been calibrated) of the pixel under investigation for each of the bands (Figure 8-9d).

Table 8-1.   Savannah River thermal plume density slice specifications used to create **Color Plate 8-1**. The number of pixels in each class interval associated with Transect 1 are provided.

		Relationship of Class to Ambient River Temperature						
		Class 1 Dark blue (RGB = 0, 0, 120) Ambient	**Class 2** Light blue (RGB = 0, 0, 255) +1°C	**Class 3** Green (RGB = 0, 255, 0) 1.2°–2.8°C	**Class 4** Yellow (RGB = 255, 255, 0) 3.0°–5.0°C	**Class 5** Orange (RGB = 255, 50, 0) 5.2°–10°C	**Class 6** Red (RGB = 255, 0, 0) 10.2°–20°C	**Class 7** White (RGB = 255, 255, 255) >20°C
Transect[a]	**Width of River**[b]	**Brightness Value Range for Each Class Interval**						
		74 – 76	77 – 80	81 – 89	90 – 100	101 – 125	126 – 176	177 – 255
1 50 pixels @ 2.8 m	39 pixels = 109.2 m	24/67.2[c]	2/5.6	1/2.8	1/2.8	2/5.6	5/14	4/11.2

[a] The transect was 140 m long (50 pixels at 2.8 m/pixel). Transect measurements in the river were made only after the image was rotated so that the beginning and ending pixels of the transect fell on the same scan line.

[b] Includes 1 mixed pixel of land and water on each side of the river.

[c] Notation represents pixels and meters; for example, 24 pixels represent 67.2 m.

The usefulness of the spectral profile is dependent upon the quality of information in the spectral data. Analysts occasionally assume that in order to do quality remote sensing research they need a tremendous number of bands. Sometimes this is true. However, at other times just two or three optimally located bands in the electromagnetic spectrum can be sufficient to extract the desired information and solve a problem. Therefore, the goal is to have just the right number of optimally located, nonredundant spectral bands. Spectral profiles can assist the analyst by providing unique visual and quantitative information about the spectral characteristics of the phenomena under investigation and whether there are any serious problems with the spectral characteristics of the dataset.

Spectral profiles extracted from 1) mangrove, 2) sand, and 3) water locations in a 20×20 m three-band SPOT image (green, red, and near-infrared) of Marco Island, FL, are shown in Figure 8-9 and **Color Plate 8-2**. Note that these data have not been converted into percent reflectance; therefore, the y-axis is simply labeled Brightness Value. Because this dataset consists of only three bands, it is useful but not very informative. As expected, the pixel of mangrove under investigation (1) absorbs more red light than green light due to chlorophyll *a* absorption and reflects significant amounts of the incident near-infrared energy. The sandy beach (2) reflects approximately equal amounts of green, red, and near-infrared energy. As expected, the green light is reflected slightly more than the red light for the water pixel (3) while most of the incident near-infrared radiant flux is absorbed by the water, causing the brightness value to approach zero.

A second example demonstrates the information content of hyperspectral remote sensing data. The spectral profiles for several features located in HyMap hyperspectral data of the Debordiu residential colony near North Inlet, SC, are shown in Figure 8-10 and **Color Plate 8-3**. The hyperspectral dataset consists of 116 atmospherically corrected bands of data at 3×3 m spatial resolution. Note that the data have been calibrated to percent reflectance units. Only 3 of the 116 bands are displayed, including band 9 (green), band 15 (red), and band 40 (near-infrared). The spectral profile for the golf-putting green (1) exhibits all the characteristics of a well-calibrated hyperspectral vegetation pixel. There is chlorophyll absorption in the blue and red portions of the spectrum and significant reflectance throughout the near-infrared region. Reflectance in the middle-infrared bands in the region 1.55 – 1.75 µm and 2.08 – 2.35 µm is also strong. The atmospheric water absorption bands at 1.4 and 1.9 µm are evident in all the spectra. The spectral profile for sand (2) is high throughout the visible bands (blue, green, and red), causing it to appear bright white to human observers. The spectral profile for a residential rooftop (3) reflects relatively lower amounts of blue, green, and red energy throughout the visible spectrum, causing it to appear gray. Finally, water (4) absorbs more and more incident energy as we progress from the visible into the near- and middle-infrared portions of the spectrum.

Spectral Profiles Extracted from SPOT 20 x 20 m Data

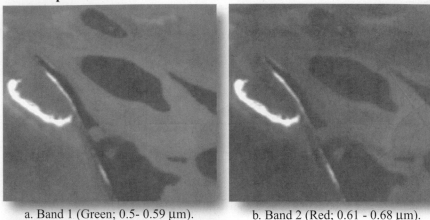

a. Band 1 (Green; 0.5- 0.59 μm).   b. Band 2 (Red; 0.61 - 0.68 μm).

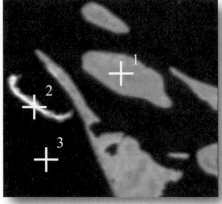

c. Band 3 (Near-infrared; 0.79 - 0.89 μm).

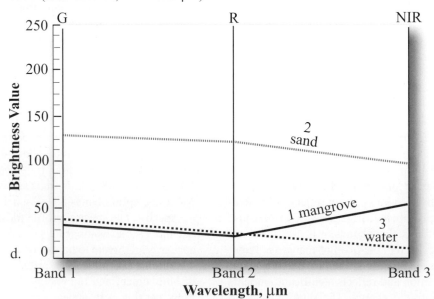

Figure 8-9   a–c) Three bands of SPOT 20 × 20 m multispectral data of Marco Island, FL. d) Spectral profiles of mangrove, sand, and water extracted from the multispectral data.

Spectral Profiles Extracted from HyMap Hyperspectral Data

a. Band 9 (Green; 0.5591 μm).

b. Band 15 (Red; 0.6508 μm).

c. Band 40 (Near-infrared; 1.0172 μm).

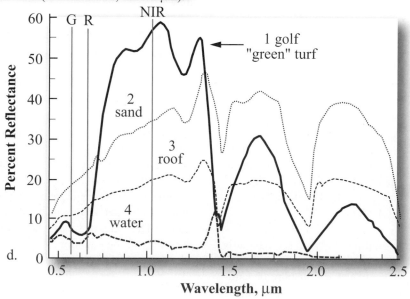

Figure 8-10   a–c) Three bands of HyMap hyperspectral data of the Debordiu colony near North Inlet, SC. The data were obtained at a spatial resolution of 3 × 3 m. d) Spectral profiles of golf "green" turf, sand, roof, and water extracted from the 116 bands of hyperspectral data.

Contrast Enhancement

Remote sensing systems record reflected and emitted radiant flux exiting from Earth's surface materials. Ideally, one material would reflect a tremendous amount of energy in a certain wavelength and another material would reflect much less energy in the same wavelength. This would result in *contrast* between the two types of material when recorded by the remote sensing system. Unfortunately, different materials often reflect similar amounts of radiant flux throughout the visible, near-infrared, and middle-infrared portions of the electromagnetic spectrum, resulting in a relatively *low-contrast* imagery. In addition, besides this obvious low-contrast characteristic of biophysical materials, there are cultural factors at work. For example, people in developing countries often construct urban areas using natural building materials (e.g., wood, sand, silt, clay) (Haack et al., 1997). This can cause urbanized areas in developing countries to have about the same reflectance characteristics as the neighboring countryside. Conversely, urban infrastructure in developed countries is usually composed of concrete, asphalt, and fertilized green vegetation. This typically causes urbanized areas in developed countries to have reflectance characteristics significantly different from the surrounding countryside.

An additional factor in the creation of low-contrast remotely sensed imagery is the sensitivity of the detectors. For example, the detectors on most remote sensing systems are designed to record a relatively wide range of scene brightness values (e.g., 0 to 255) without becoming saturated. Saturation occurs if the radiometric sensitivity of a detector is insufficient to record the full range of intensities of reflected or emitted energy emanating from the scene. The Landsat TM detectors, for example, must be sensitive to reflectance from diverse biophysical materials such as dark volcanic basalt outcrops or snow (possibly represented as BVs of 0 and 255, respectively). However, very few scenes are composed of brightness values that use the full sensitivity range of the Landsat TM detectors. Therefore, this results in relatively low-contrast imagery, with original brightness values that often range from approximately 0 to 100.

To improve the contrast of digital remotely sensed data, it is desirable to use the entire brightness range of the display medium, which is generally a video CRT display or hardcopy output device (discussed in Chapter 5). Digital methods may be more satisfactory than photographic techniques for contrast enhancement because of the precision and wide variety of processes that can be applied to the imagery. There are linear and nonlinear digital contrast-enhancement techniques.

Linear Contrast Enhancement

Contrast enhancement (also referred to as *contrast stretching*) expands the original input brightness values to make use of the total dynamic range or sensitivity of the output device. To illustrate the linear contrast-stretching process, consider the Charleston, SC, TM band 4 image produced by a sensor system whose image output levels can vary from 0 to 255. A histogram of this image is provided (Figure 8-11a). We will assume that the output device (a high-resolution black-and-white CRT) can display 256 shades of gray (i.e., $quant_k = 255$). The histogram and associated statistics of this band 4 subimage reveal that the scene is composed of brightness values ranging from a minimum of 4 (i.e., $min_4 = 4$) to a maximum value of 105 (i.e., $max_4 = 105$), with a mean of 27.3 and a standard deviation of 15.76 (refer to Table 4-7). When these data are displayed on the CRT without any contrast enhancement, we use less than one-half of the full dynamic range of brightness values that could be displayed (i.e., brightness values between 0 and 3 and between 106 and 255 are not used). The image is rather dark, low in contrast, with no distinctive bright areas (Figure 8-11a). It is difficult to visually interpret such an image. A more useful display can be produced if we expand the range of original brightness values to use the full dynamic range of the video display.

Minimum–Maximum Contrast Stretch

Linear contrast enhancement is best applied to remotely sensed images with Gaussian or near-Gaussian histograms, that is, when all the brightness values fall generally within a single, relatively narrow range of the histogram and only one mode is apparent. Unfortunately, this is rarely the case, especially for scenes that contain both land and water bodies. To perform a linear contrast enhancement, the analyst examines the image statistics and determines the minimum and maximum brightness values in band k, min_k and max_k, respectively. The output brightness value, BV_{out}, is computed according to the equation:

$$BV_{out} = \left(\frac{BV_{in} - min_k}{max_k - min_k} \right) quant_k, \qquad (8\text{-}1)$$

where BV_{in} is the original input brightness value and $quant_k$ is the maximum value of the range of brightness values that can be displayed on the CRT (e.g., 255). In the Charleston, SC,

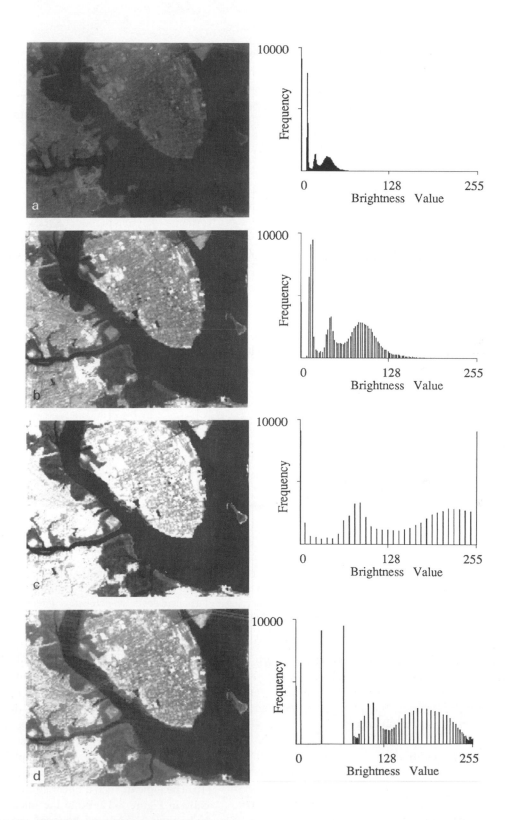

Figure 8-11   a) Original Landsat Thematic Mapper band 4 data of Charleston, SC, and its histogram. This image has *not* been contrast stretched. b) Minimum–maximum contrast stretch applied to the data and the resultant histogram. c) One standard deviation (±1σ) percentage linear contrast stretch applied to the data and the resultant histogram. d) Application of histogram equalization and the resultant histogram.

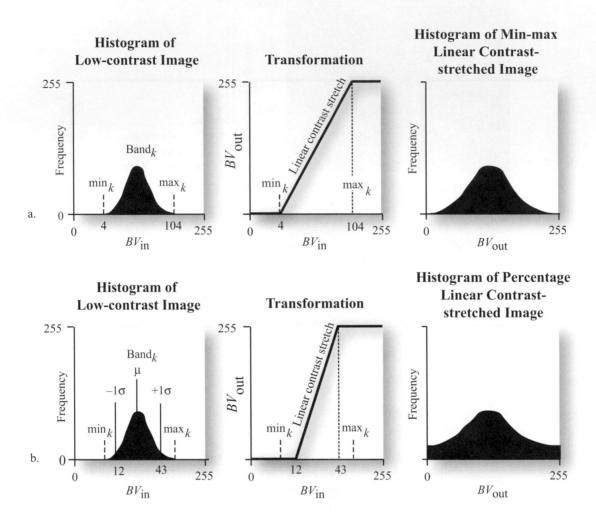

Figure 8-12   a) The result of applying a *minimum–maximum* contrast stretch to normally distributed remotely sensed data. The histograms before and after the transformation are shown. The minimum and maximum brightness values encountered in band k are $\min_k$ and $\max_k$, respectively. b) Theoretical result of applying a ±1 standard deviation *percentage linear* contrast stretch. This moves the $\min_k$ and $\max_k$ values ±34% from the mean into the tails of the distribution.

example, any pixel with a BV_{in} of 4 would now have a BV_{out} of 0, and any pixel with a BV_{in} of 105 would have a BV_{out} of 255:

$$BV_{out} = \left(\frac{4-4}{105-4}\right)255$$

$$BV_{out} = 0$$

$$BV_{out} = \left(\frac{105-4}{105-4}\right)255$$

$$BV_{out} = 255 \, .$$

The original brightness values between 5 and 104 would be linearly distributed between 0 and 255, respectively. The

application of this enhancement to the Charleston TM band 4 data is shown in Figure 8-11b. This is commonly referred to as a *minimum–maximum contrast stretch*. Most image processing systems provide for the display of a before-and-after histogram, as well as a graph of the relationship between the input brightness value (BV_{in}) and the output brightness value (BV_{out}). For example, the histogram of the min–max contrast stretch discussed is shown in Figure 8-11b. The logic of such a linear stretch is shown diagrammatically in Figure 8-12. Note the linear relationship between the brightness values of the input and output pixels and how the slope of the line becomes steeper as the minimum is increased or the maximum is decreased. The application of a minimum–maximum contrast stretch enhancement to the

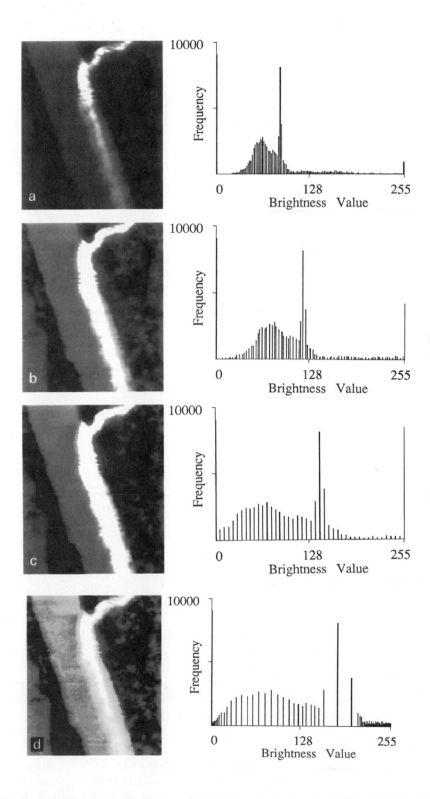

Figure 8-13 a) Original predawn thermal infrared data of the Savannah River, not contrast stretched, and its histogram. b) Minimum–maximum contrast stretch applied to the data and the resultant histogram. c) One standard deviation (±1σ) percentage linear contrast stretch applied to the data and resultant histogram. d) Application of histogram equalization and resultant histogram.

Savannah River predawn thermal infrared data is shown in Figure 8-13b.

Percentage Linear and Standard Deviation Contrast Stretching

Image analysts often specify min_k and max_k that lie a certain percentage of pixels from the mean of the histogram. This is called a *percentage linear contrast stretch*. If the percentage coincides with a standard deviation percentage, then it is called a *standard deviation contrast stretch*. For normal distributions, 68% of the observations lie within ±1 standard deviation of the mean, 95.4% of all observations lie within ±2 standard deviations, and 99.73% within ±3 standard deviations. Consider applying a ±1 standard deviation contrast stretch to the Charleston Landsat TM band 4 data. This would result in $min_k = 12$ and $max_k = 43$. All values between 12 and 43 would be linearly contrast stretched to lie within the range 0 to 255. All values between 0 and 11 are now 0, and those between 44 and 255 are set to 255. This results in more pure black-and-white pixels in the Charleston scene, dramatically increasing the contrast of the image, as shown in Figure 8-11c. The information content of the pixels that saturated at 0 and 255 is lost. The slope of a percentage linear or standard deviation contrast stretch is greater than for a simple min–max stretch (refer to Figure 8-12b).

The results of applying a ±1 standard deviation linear contrast stretch to the thermal plume data are shown in Figure 8-13c along with the histogram. The ±1 standard deviation contrast stretch effectively "burns out" the thermal plume, yet provides more detail about the temperature characteristics of vegetation on each side of the river.

Piecewise Linear Contrast Stretch

When the histogram of an image is not Gaussian (i.e., it is bimodal, trimodal, etc.), it is possible to apply a piecewise linear contrast stretch to the imagery of the type shown in Figure 8-14. Here the analyst identifies a number of linear enhancement steps that expand the brightness ranges in the modes of the histogram. In effect, this corresponds to setting up a series of min_k and max_k and using Equation 8-1 within user-selected regions of the histogram. This powerful contrast enhancement method should be used when the analyst is intimately familiar with the various modes of the histogram and what they represent in the real world. Such contrast-stretched data are rarely used in subsequent image classification.

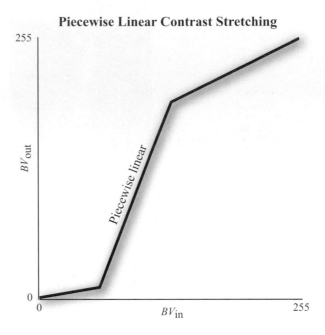

Figure 8-14 Logic of a piecewise linear contrast stretch for which selective pieces of the histogram are linearly contrast stretched. Note that the slope of the linear contrast enhancement changes.

To perform piecewise linear contrast enhancement, the analyst normally views a) the raw image and b) a display of the histogram of the raw image superimposed with an input–output line that runs diagonally from the lower left to the upper right in the display. The analyst then interactively adjusts the length and slope of n mutually exclusive contrast stretches along the input–output line.

To illustrate the process, consider the two piecewise linear contrast enhancements found in Figure 8-15. The Savannah River thermal plume is composed primarily of values from 81 to 170. A special contrast stretch to highlight just the plume is demonstrated in Figures 8-15ab (note the dotted lines). Upland and ambient Savannah River water brightness values from 0 to 80 are sent to 255 (white), values from 81 to 170 are linearly contrast stretched to have values from 0 to 255, and all values from 171 to 255 now have a value of 255 (white). Conversely, if we wanted to contrast stretch the image to highlight the spectral characteristics of the ambient Savannah River, we might use the logic shown in Figures 8-15c and d. In this example, the spike in the original histogram associated with just the ambient Savannah River water has been singled out and linearly contrast stretched to have values from 0 to 255. All the land and thermal plume pixels are set to a value of 0 and appear black (note the dotted lines).

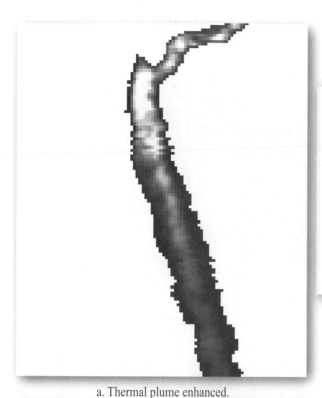

a. Thermal plume enhanced.

Piecewise Linear Contrast Stretching

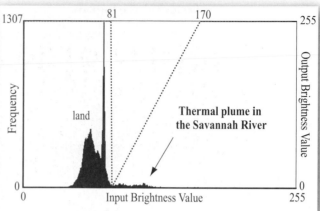

b. Enhancing the thermal plume in the Savannah River.

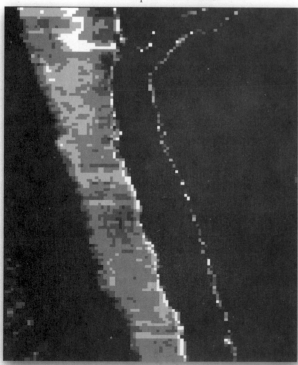

c. Savannah River enhanced.

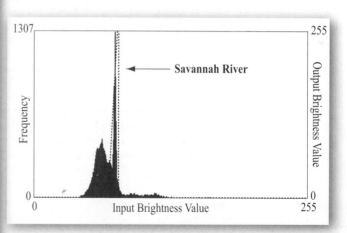

d. Enhancing the Savannah River.

Figure 8-15   a,b) Thermal infrared imagery of the Savannah River enhanced using piecewise linear contrast stretching to highlight the thermal plume at the expense of the ambient river water and the surrounding landscape. c,d) Piecewise linear contrast stretching to enhance the Savannah River at the expense of the thermal plume and the surrounding landscape.

Histogram Equalization Contrast Enhancement

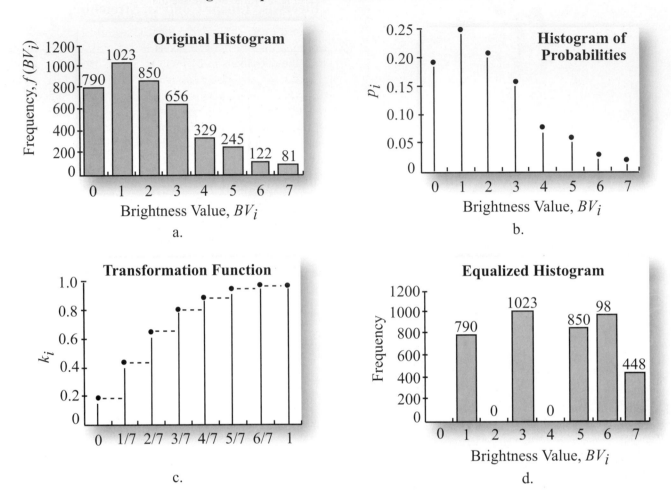

Figure 8-16 Histogram equalization process applied to hypothetical data (adapted from Gonzalez and Wintz, 1977). a) Original histogram showing the frequency of pixels in each brightness value. b) Original histogram expressed in probabilities. c) The transformation function. d) The equalized histogram showing the frequency of pixels in each brightness value.

Nonlinear Contrast Enhancement

Nonlinear contrast enhancements may also be applied. One of the most useful enhancements is *histogram equalization*. The algorithm passes through the individual bands of the dataset and assigns approximately an equal number of pixels to each of the user-specified output grayscale classes (e.g., 32, 64, 256). Histogram equalization applies the greatest contrast enhancement to the most populated range of brightness values in the image. It automatically reduces the contrast in the very light or dark parts of the image associated with the tails of a normally distributed histogram.

Histogram equalization is found in many image processing systems because it requires very little information from the analyst to implement (usually just the number of output brightness value classes desired and the bands to be equalized), yet it is often very effective. Because of its wide availability, it is instructive to review how the equalization takes place using a hypothetical dataset (Gonzalez and Wintz, 1977). For example, consider an image that is composed of 64 rows and 64 columns (4096 pixels) with the range of brightness values that each pixel can assume, $quant_k$, limited to 0 through 7 (Table 8-2). A histogram of this hypothetical image is shown in Figure 8-16a and the frequency of occurrence of the individual brightness values, $f(BV_i)$, is summa-

Table 8-2.   Statistics for a 64×64 hypothetical image with Brightness Values from 0 to 7.[a]

Brightness Value, BV_i	L_i	Frequency $f(BV_i)$	Probability[b] $p_i = f(BV_i)/n$
BV_0	0/7 = 0.00	790	0.19
BV_1	1/7 = 0.14	1023	0.25
BV_2	2/7 = 0.28	850	0.21
BV_3	3/7 = 0.42	656	0.16
BV_4	4/7 = 0.57	329	0.08
BV_5	5/7 = 0.71	245	0.06
BV_6	6/7 = 0.85	122	0.03
BV_7	7/7 = 1.00	81	0.02

[a] Source: modified from Gonzalez and Wintz, 1977.
[b] $n = 4096$ pixels.

rized in Table 8-2. For example, there are 790 pixels in the scene with a brightness value of 0 [i.e., $f(BV_0) = 790$] and 1023 pixels with a brightness value of 1 [i.e., $f(BV_1) = 1023$]. We can compute the probability of the ith brightness value, p_i, by dividing each of the frequencies, $f(BV_i)$, by the total number of pixels in the scene (i.e., $n = 4096$). Thus, the probability of encountering a pixel with a brightness value of 0 in the scene is approximately 19% [i.e., $p_0 = f(BV_i)/n = 790/4096 = 0.19$]. A plot of the probability of occurrence of each of the eight brightness values for the hypothetical scene is shown in Figure 8-16b. This particular histogram has a large number of pixels with low brightness values (0 and 1), making it a relatively low-contrast scene.

The next step is to compute a transformation function k_i for each brightness value. One way to conceptualize the histogram equalization process is to use the notation shown in Table 8-3. For each brightness value level BV_i in the $quant_k$ range of 0 to 7 of the original histogram, a new cumulative frequency value k_i is calculated:

$$k_i = \sum_{i=0}^{quant_k} \frac{f(BV_i)}{n} \qquad (8\text{-}2)$$

where the summation counts the frequency of pixels in the image with brightness values equal to or less than BV_i, and n is the total number of pixels in the entire scene (4096 in this example). The histogram equalization process iteratively compares the transformation function k_i with the original val-

ues of L_i to determine which are closest in value. The closest match is reassigned to the appropriate brightness value. For example, in Table 8-3 we see that $k_0 = 0.19$ is closest to $L_1 = 0.14$. Therefore, all pixels in BV_0 (790 of them) will be assigned to BV_1. Similarly, the 1023 pixels in BV_1 will be assigned to BV_3, the 850 pixels in BV_2 will be assigned to BV_5, the 656 pixels in BV_3 will be assigned to BV_6, the 329 pixels in BV_4 will also be assigned to BV_6, and all 448 brightness values in BV_{5-7} will be assigned to BV_7. The new image will have no pixels with brightness values of 0, 2, or 4. This is evident when evaluating the new histogram (Figure 8-16d). When analysts see such gaps in image histograms, it is usually a good indication that histogram equalization or some other operation has been applied.

Histogram-equalized versions of the Charleston TM band 4 data and the thermal plume data are found in Figure 8-11d and 8-13d, respectively. Histogram equalization is dramatically different from any other contrast enhancement because the data are redistributed according to the cumulative frequency histogram of the data, as described. Note that after histogram equalization, some pixels that originally had different values are now assigned the same value (perhaps a loss of information), while other values that were once very close together are now spread out, increasing the contrast between them. Therefore, this enhancement may improve the visibility of detail in an image, but it also alters the relationship between brightness values and image structure (Russ, 2002). For these reasons, it is not wise to extract texture or biophysical information from imagery that has been histogram equalized.

Another type of nonlinear contrast stretch involves scaling the input data *logarithmically,* as diagrammed in Figure 8-17. This enhancement has the greatest impact on the brightness values found in the darker part of the histogram. It could be reversed to enhance values in the brighter part of the histogram by scaling the input data using an inverse log function, as shown.

The selection of a contrast-enhancement algorithm depends on the nature of the original histogram and the elements of the scene that are of interest to the user. An experienced image analyst can usually identify an appropriate contrast-enhancement algorithm by examining the image histogram and then experimenting until satisfactory results are obtained. Most contrast enhancements cause some useful information to be lost. However, that which remains should be of value. Contrast enhancement is applied primarily to improve visual image analysis. It is *not* good practice to contrast stretch the original imagery and then use the enhanced imagery for computer-assisted classification, change detec-

Table 8-3.   Example of how a hypothetical 64×64 image with brightness values from 0 to 7 is histogram equalized.

Frequency, $f(BV_i)$	790	1023	850	656	329	245	122	81
Original brightness value, BV_i	0	1	2	3	4	5	6	7
$L_i = \dfrac{\text{brightness value}}{n}$	0	0.14	0.28	0.42	0.57	0.71	0.85	1.0
Cumulative frequency transformation: $k_i = \displaystyle\sum_{i=0}^{\text{quant}_k} \dfrac{f(BV_i)}{n}$	$\dfrac{790}{4096}$ $=0.19$	$\dfrac{1813}{4096}$ $=0.44$	$\dfrac{2663}{4096}$ $=0.65$	$\dfrac{3319}{4096}$ $=0.81$	$\dfrac{3648}{4096}$ $=0.89$	$\dfrac{3893}{4096}$ $=0.95$	$\dfrac{4015}{4096}$ $=0.98$	$\dfrac{4096}{4096}$ $=1.0$
Assign original BV_i class to the new class it is closest to in value.	1	3	5	6	6	7	7	7

tion, etc. Contrast stretching can distort the original pixel values, often in a nonlinear fashion.

Band Ratioing

Sometimes differences in brightness values from identical surface materials are caused by topographic slope and aspect, shadows, or seasonal changes in sunlight illumination angle and intensity. These conditions may hamper the ability of an interpreter or classification algorithm to identify correctly surface materials or land use in a remotely sensed image. Fortunately, *ratio* transformations of the remotely sensed data can, in certain instances, be applied to reduce the effects of such environmental conditions. In addition to minimizing the effects of environmental factors, ratios may also provide unique information not available in any single band that is useful for discriminating between soils and vegetation (Satterwhite, 1984).

The mathematical expression of the ratio function is:

$$BV_{i,j,r} = \frac{BV_{i,j,k}}{BV_{i,j,l}}, \qquad (8\text{-}3)$$

where $BV_{i,j,r}$ is the output ratio value for the pixel at row i, column j and $BV_{i,j,k}$ and $BV_{i,j,l}$ are the brightness values at the same location in bands k and l, respectively. Unfortunately, the computation is not always simple since $BV_{i,j} = 0$ is possible. However, there are alternatives. For example, the math-

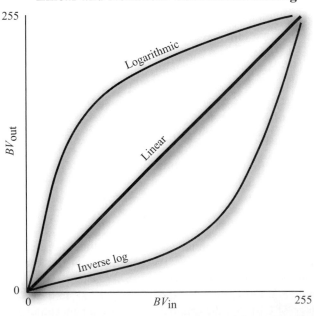

Linear and Nonlinear Contrast Stretching

Figure 8-17   Logic of a linear, nonlinear logarithmic, and inverse logarithmic contrast stretch.

ematical domain of the function is $\frac{1}{255}$ to 255 (i.e., the range of the ratio function includes all values beginning at $\frac{1}{255}$, passing through 0, and ending at 255). The way to overcome this problem is simply to give any $BV_{i,j}$ with a value of 0 the value of 1. Alternatively, some like to add a small value (e.g., 0.1) to the denominator if it equals zero.

Band Ratioing

a. Ratio of Landsat TM bands 3/4.

b. Ratio of Landsat TM bands 4/5.

c. Ratio of Landsat TM bands 4/7.

d. Ratio of Landsat TM bands 3/6.

Figure 8-18   The ratio of various Landsat TM bands of Charleston, SC.

To represent the range of the function in a linear fashion and to encode the ratio values in a standard 8-bit format (values from 0 to 255), normalizing functions are applied. Using this normalizing function, the ratio value 1 is assigned the brightness value 128. Ratio values within the range $\frac{1}{255}$ to 1 are assigned values between 1 and 128 by the function:

$$BV_{i,j,n} = \text{Int} \,[(BV_{i,j,r} \times 127) + 1]. \qquad (8\text{-}4)$$

Ratio values from 1 to 255 are assigned values within the range 128 to 255 by the function:

$$BV_{i,j,n} = \text{Int} \left(128 + \frac{BV_{i,j,r}}{2} \right). \qquad (8\text{-}5)$$

Deciding which two bands to ratio is not always a simple task. Often, the analyst simply displays various ratios and then selects the most visually appealing. The optimum index factor (OIF) and Sheffield Index (discussed in Chapter 5) can

be used to identify optimum bands for band ratioing (Chavez et al., 1984; Sheffield, 1985). Crippen (1988) recommended that all data be atmospherically corrected and free from any sensor calibration problems (e.g., a detector is out of adjustment) before it is ratioed.

The ratio of Charleston, SC, Thematic Mapper bands 3 (red) and 4 (near-infrared) is displayed in Figure 8-18a. This red/infrared ratio provides vegetation information that will be discussed in the section on vegetation indexes in this chapter. Generally, the brighter the pixel, the more vegetation present. Several other band-ratioed images are shown in Figure 8-18b–d. Generally, the lower the correlation between the bands, the greater the information content of the band-ratioed image. For example, a ratio of near-infrared (band 4) and mid-infrared (band 5) data reveals detail in the salt marsh areas. This suggests that these bands are not highly correlated and that each band provides some unique information.

Similarly, the ratio of bands 4 and 7 provides useful information. The ratio of band 3 (red) and band 6 (thermal infrared) provides detail about the water column as well as the urban structure.

Spatial Filtering

A characteristic of remotely sensed images is a parameter called *spatial frequency,* defined as the number of changes in brightness value per unit distance for any particular part of an image. If there are very few changes in brightness value over a given area in an image, this is commonly referred to as a low-frequency area. Conversely, if the brightness values change dramatically over short distances, this is an area of high-frequency detail. Because spatial frequency by its very nature describes the brightness values over a spatial *region*, it is necessary to adopt a spatial approach to extracting quantitative spatial information. This is done by looking at the local (neighboring) pixel brightness values rather than just an independent pixel value. This perspective allows the analyst to extract useful spatial frequency information from the imagery.

Spatial frequency in remotely sensed imagery may be enhanced or subdued using two different approaches. The first is *spatial convolution filtering* based primarily on the use of convolution masks. The procedure is relatively easy to understand and can be used to enhance low- and high-frequency detail, as well as edges in the imagery. Another technique is *Fourier analysis,* which mathematically separates an image into its spatial frequency components. It is then possible interactively to emphasize certain groups (or bands) of frequencies relative to others and recombine the spatial frequencies to produce an enhanced image. We first introduce the technique of spatial convolution filtering and then proceed to the more mathematically challenging Fourier analysis.

Spatial Convolution Filtering

A linear *spatial filter* is a filter for which the brightness value ($BV_{i,j}$) at location i, j in the output image is a function of some weighted average (linear combination) of brightness values located in a particular spatial pattern around the i, j location in the input image. This process of evaluating the weighted neighboring pixel values is called two-dimensional *convolution filtering* (Pratt, 2001). The procedure is often used to change the spatial frequency characteristics of an image. For example, a linear spatial filter that emphasizes high spatial frequencies may sharpen the edges within an image. A linear spatial filter that emphasizes low spatial frequencies may be used to reduce noise within an image.

Low-frequency Filtering in the Spatial Domain

Image enhancements that de-emphasize or block the high spatial frequency detail are *low-frequency* or *low-pass* filters. The simplest low-frequency filter (LFF) evaluates a particular input pixel brightness value, BV_{in}, and the pixels surrounding the input pixel, and outputs a new brightness value, BV_{out}, that is the mean of this convolution. The size of the neighborhood *convolution mask* or *kernel* (n) is usually 3×3, 5×5, 7×7, or 9×9. Examples of symmetric 3×3, 5×5, and 7×7 convolution masks are found in Figures 8-19 a, b, and d. We will constrain this discussion primarily to 3×3 convolution masks with nine coefficients, c_i, defined at the following locations:

$$\text{Mask template} = \begin{matrix} c_1 & c_2 & c_3 \\ c_4 & c_5 & c_6 \\ c_7 & c_8 & c_9 \end{matrix} \qquad (8\text{-}6)$$

For example, the coefficients in a low-frequency convolution mask might all be set equal to 1:

$$\text{Mask A} = \begin{matrix} 1 & 1 & 1 \\ 1 & 1 & 1 \\ 1 & 1 & 1 \end{matrix} \qquad (8\text{-}7)$$

The coefficients, c_i, in the mask are multiplied by the following individual brightness values (BV_i) in the input image:

$$\text{Mask template} = \begin{matrix} c_1 \times BV_1 & c_2 \times BV_2 & c_3 \times BV_3 \\ c_4 \times BV_4 & c_5 \times BV_5 & c_6 \times BV_6 \\ c_7 \times BV_7 & c_8 \times BV_8 & c_9 \times BV_9 \end{matrix} \quad (8\text{-}8)$$

where

$$\begin{aligned} BV_1 &= BV_{i-1, j-1} & BV_6 &= BV_{i, j+1} \\ BV_2 &= BV_{i-1, j} & BV_7 &= BV_{i+1, j-1} \\ BV_3 &= BV_{i-1, j+1} & BV_8 &= BV_{i+1, j} \\ BV_4 &= BV_{i, j-1} & BV_9 &= BV_{i+1, j+1} \\ BV_5 &= BV_{i, j} \end{aligned} \qquad (8\text{-}9)$$

The primary input pixel under investigation at any one time is $BV_5 = BV_{i,j}$. The convolution of Mask A (with all coefficients equal to 1) and the original data will result in a low-frequency filtered image, where

Convolution Masks of Various Sizes and Shapes

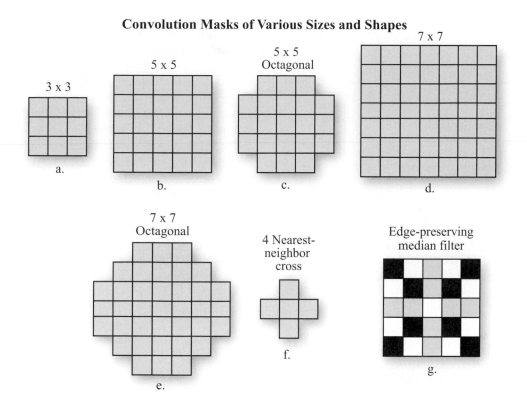

Figure 8-19   Examples of various convolution masks.

$$LFF_{5,\text{out}} = \text{Int}\frac{\displaystyle\sum_{i=1}^{n=9} c_i \times BV_i}{n} \quad (8\text{-}10)$$

$$= \text{Int}\left(\frac{BV_1 + BV_2 + BV_3 + \cdots + BV_9}{9}\right)$$

The spatial moving average then shifts to the next pixel, where the average of all nine brightness values is computed. This operation is repeated for every pixel in the input image (Figure 8-20). Such *image smoothing* is useful for removing periodic "salt and pepper" noise recorded by electronic remote sensing systems.

This simple smoothing operation will, however, blur the image, especially at the edges of objects. Blurring becomes more severe as the size of the kernel increases. To reduce blurring, unequal-weighted smoothing masks have been developed, including:

$$\text{Mask B} = \begin{matrix} 0.25 & 0.50 & 0.25 \\ 0.50 & 1.00 & 0.50 \\ 0.25 & 0.50 & 0.25 \end{matrix} \quad (8\text{-}11)$$

$$\text{Mask C} = \begin{matrix} 1.00 & 1.00 & 1.00 \\ 1.00 & 2.00 & 1.00 \\ 1.00 & 1.00 & 1.00 \end{matrix} \quad (8\text{-}12)$$

Using a 3×3 kernel can result in the low-pass image being two lines and two columns smaller than the original image. Techniques that can be applied to deal with this problem include 1) artificially extending the original image beyond its border by repeating the original border pixel brightness values, and 2) replicating the averaged brightness values near the borders, based on the image behavior within a few pixels of the border. This maintains the row and column dimension of the imagery but introduces some spurious information that should not be interpreted.

Application of a *low-pass filter* (Equation 8-7) to the Savannah River thermal infrared data and Charleston TM band 4 data are shown in Figures 8-21b and 8-24b, respectively. The smoothed thermal plume is visually appealing. Note that the lines with line-start problems discussed in Chapter 6 are subdued. The Charleston scene becomes blurred, suppressing the high-frequency detail (Figure 8-24b). Only the general trends are allowed to pass through the filter. In a heteroge-

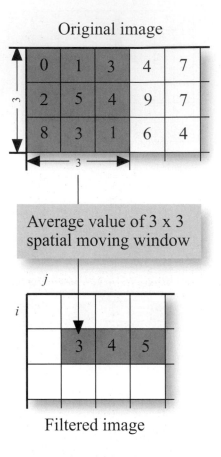

Original image

Filtered image

Figure 8-20   Result of applying low-frequency convolution mask A to hypothetical data. The nine coefficients of the 3×3 mask are all equal to 1 in this example.

neous, high-frequency urban environment, a high-frequency filter usually provides superior results.

The neighborhood ranking *median filter* is useful for removing noise in an image, especially shot noise (pixel values with no relation to the image scene). Instead of computing the average (mean) of the nine pixels in a 3×3 convolution, the median filter ranks the pixels in the neighborhood from lowest to highest and selects the median value, which is then placed in the central value of the mask (Richards and Jia, 1999). The median filter is not restricted to just a 3×3 convolution mask. Some of the more common neighborhood patterns used in median filters are shown in Figure 8-19a–f. Only the original pixel values are used in the creation of a median filter. The application of a median filter to the thermal plume and the Charleston TM band 4 data are shown in Figures 8-21c and 8-24c, respectively.

A median filter has certain advantages when compared with weighted convolution filters (Russ, 2002), including 1) it does not shift boundaries and 2) the minimal degradation to edges allows the median filter to be applied repeatedly, which allows fine detail to be erased and large regions to take on the same brightness value (often called *posterization*). The standard median filter will erase some lines in the image that are narrower than the half-width of the neighborhood and round or clip corners (Eliason and McEwen, 1990). An *edge-preserving median filter* (Nieminen et al., 1987) may be applied using the logic shown in Figure 8-19g, where 1) the median value of the black pixels is computed in a 5×5 array, 2) the median value of the gray pixels is computed, 3) these two values and the central original brightness value are ranked in ascending order, and 4) a final median value is selected to replace the central pixel. This filter preserves edges and corners.

Sometimes it is useful to apply a *minimum* or *maximum filter* to an image. Operating on one pixel at a time, these filters examine the brightness values of adjacent pixels in a user-specified region (e.g., 3×3 pixels) and replace the brightness value of the current pixel with the minimum or maximum brightness value encountered, respectively. For example, a 3×3 minimum filter was applied to the thermal data, resulting in Figure 8-21d. Note how it minimizes the width of the plume, highlighting the warm core. Conversely, a 3×3 maximum filter (Figure 8-21e) dramatically expanded the size of the plume. Such filters are for visual analysis only and should probably not be applied prior to extracting biophysical information.

Another modification to the simple averaging of neighborhood values is the *Olympic filter,* which is named after the system of scoring in Olympic events. Instead of using all nine elements in a 3×3 matrix, the highest and lowest values are dropped and the result averaged. This algorithm is useful for removing most shot noise.

Adaptive box filters are of significant value for removing noise in digital images. For example, Eliason and McEwen (1990) developed two adaptive box filters to 1) remove random bit errors (e.g., shot noise) and 2) smooth noisy data (pixels related to the image scene but with an additive or multiplicative component of noise). Both procedures rely on the computation of the standard deviation (σ) of only those pixels within a local box surrounding the central pixel (i.e., the eight values surrounding BV_5 in a 3×3 mask). The original brightness value at location BV_5 is considered a bit error if it deviates from the box mean of the eight surrounding values by more than 1.0 to 2.0σ. When this occurs, it is replaced by the box mean. It is called an adaptive filter

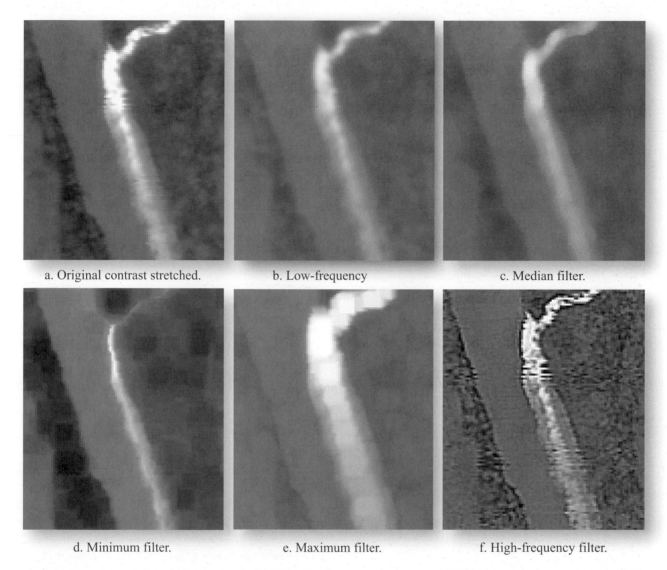

a. Original contrast stretched.   b. Low-frequency   c. Median filter.

d. Minimum filter.   e. Maximum filter.   f. High-frequency filter.

Figure 8-21   Application of various convolution masks and logic to the predawn thermal-infrared data to enhance low- and high-frequency detail: a) contrast stretched original image, b) low-frequency filter (mask A), c) median filter, d) minimum filter, e) maximum filter, and f) high-frequency filter (mask E).

because it is based on the computation of the standard deviation for each 3×3 window, rather than on the standard deviation of the entire scene. Even very minor bit errors are removed from low-variance areas, but valid data along sharp edges and corners are not replaced. Their second adaptive filter for cleaning up extremely noisy images was based on the Lee (1983) sigma filter. Lee's filter first computed the standard deviation of the entire scene. Then, each BV_5 in a 3×3 moving window was replaced by the average of only those neighboring pixels that had an intensity within a fixed σ range of the central pixel. Eliason and McEwen (1990) used the local (adaptive) σ, rather than the fixed σ

computed from the entire scene. The filter averaged only those pixels within the box that had intensities within 1.0 to 2.0σ of the central pixel. This technique effectively reduced speckle in radar images without eliminating the fine details. The two filters can be combined into a single program for processing images with both random bit errors and noisy data.

High-frequency Filtering in the Spatial Domain

High-pass filtering is applied to imagery to remove the slowly varying components and enhance the high-frequency local variations. One high-frequency filter ($HFF_{5,out}$) is computed by subtracting the output of the low-frequency filter ($LFF_{5,out}$) from twice the value of the original central pixel value, BV_5:

$$HFF_{5,out} = (2 \times BV_5) - LFF_{5,out}. \qquad (8\text{-}13)$$

Brightness values tend to be highly correlated in a nine-element window. Thus, the high-frequency filtered image will have a relatively narrow intensity histogram. This suggests that the output from most high-frequency filtered images must be contrast stretched prior to visual analysis.

High-pass filters that accentuate or sharpen edges can be produced using the following convolution masks:

$$\text{Mask D} = \begin{matrix} -1 & -1 & -1 \\ -1 & 9 & -1 \\ -1 & -1 & -1 \end{matrix} \qquad (8\text{-}14)$$

$$\text{Mask E} = \begin{matrix} 1 & -2 & 1 \\ -2 & 5 & -2 \\ 1 & -2 & 1 \end{matrix} \qquad (8\text{-}15)$$

For example, the application of high-frequency Mask E to the thermal data dramatically enhanced the line-start problem lines in the data and accentuated the spatial detail in the plume, river, and upland areas (Figure 8-21f). The application of Mask E to the Charleston scene is shown in Figure 8-24d and is much more visually interpretable than the original image. The interface between open water, wetland, and urban phenomena is easier to detect in the high-frequency image. Also, some urban structures such as individual roads and buildings are enhanced.

Edge Enhancement in the Spatial Domain

For many remote sensing Earth science applications, the most valuable information that may be derived from an image is contained in the edges surrounding various objects of interest. Edge enhancement delineates these edges and makes the shapes and details comprising the image more conspicuous and perhaps easier to analyze. Generally, what the eyes see as pictorial edges are simply sharp changes in brightness value between two adjacent pixels. The edges may be enhanced using either *linear* or *nonlinear edge enhancement* techniques.

Linear Edge Enhancement. A straightforward method of extracting edges in remotely sensed imagery is the application of a *directional first-difference* algorithm that approximates the first derivative between two adjacent pixels. The

Table 8-4.   Scale of delta versus kernel size (after Chavez and Bauer, 1982).

Delta, Δ	Smoothness/ Roughness	Kernel Size
$\leq \pm 3$	Very smooth	9×9
± 4	Smooth	
± 5	Semismooth	7×7
± 6	Smooth/rough	
± 7	Rough/smooth	5×5
± 8	Semirough	
± 9	Rough	3×3
$\geq \pm 10$	Very rough	1×1

algorithm produces the first difference of the image input in the horizontal, vertical, and diagonal directions. The algorithms for enhancing horizontal, vertical, and diagonal edges are, respectively:

Vertical: $BV_{i,j} = BV_{i,j} - BV_{i,j+1} + K. \qquad (8\text{-}16)$

Horizontal: $BV_{i,j} = BV_{i,j} - BV_{i-1,j} + K. \qquad (8\text{-}17)$

NE Diagonal: $BV_{i,j} = BV_{i,j} - BV_{i+1,j+1} + K. \qquad (8\text{-}18)$

SE Diagonal: $BV_{i,j} = BV_{i,j} - BV_{i-1,j+1} + K. \qquad (8\text{-}19)$

The result of the subtraction can be either negative or positive. Therefore, a constant K (usually 127) is added to make all values positive and centered between 0 and 255. This causes adjacent pixels with very little difference in brightness value to obtain a brightness value of around 127 and any dramatic change between adjacent pixels to migrate away from 127 in either direction. The resultant image is normally min–max contrast stretched to enhance the edges even more. It is best to make the minimum and maximum values in the contrast stretch a uniform distance from the midrange value (e.g., 127). This causes the uniform areas to appear in shades of gray, while the important edges become black or white.

It is also possible to perform edge enhancement by convolving the original data with a weighted mask or kernel, as previously discussed. Chavez and Bauer (1982) suggested that the optimum kernel size (3×3, 5×5, 7×7, etc.) typically used in edge enhancement is a function of the surface roughness and Sun-angle characteristics at the time the data were collected (Table 8-4). Taking these factors into consideration, they developed a procedure based on the "first difference in the horizontal direction" (Equation 8-17). A histogram of the differenced image reveals generally how many edges are contained in the image. The standard deviation of the first difference image is computed and multiplied by 2.3, yielding a delta value, Δ, closely associated with surface roughness (Table 8-4). The following algorithm and the information presented in Table 8-4 are then used to select the appropriate kernel size:

$$\text{kernel size} = 12 - \Delta. \tag{8-20}$$

Earth scientists may use the method to select the optimum kernel size for enhancing fine image detail without having to try several versions before selecting the appropriate kernel size for the area of interest. Once the size of the convolution mask is selected, various coefficients may be placed in the mask to enhance edges.

One of the most useful edge enhancements causes the edges to appear in a plastic shaded-relief format. This is often referred to as *embossing*. Embossed edges may be obtained by using the following convolution masks:

$$\text{Mask F} = \begin{matrix} 0 & 0 & 0 \\ 1 & 0 & -1 \\ 0 & 0 & 0 \end{matrix} \quad \text{emboss East} \tag{8-21}$$

$$\text{Mask G} = \begin{matrix} 0 & 0 & 1 \\ 0 & 0 & 0 \\ -1 & 0 & 0 \end{matrix} \quad \text{emboss NW} \tag{8-22}$$

An offset of 127 is normally added to the result and the data contrast stretched. The direction of the embossing is controlled by changing the location of the coefficients around the periphery of the mask. The plastic shaded-relief impression is pleasing to the human eye if shadows are made to fall toward the viewer. The thermal plume and Charleston TM band 4 data are embossed in Figures 8-22a and b and Figure 8-25a, respectively.

Compass gradient masks may be used to perform two-dimensional, discrete differentiation directional edge enhancement (Jain, 1989; Pratt, 2001):

$$\text{Mask H} = \begin{matrix} 1 & 1 & 1 \\ 1 & -2 & 1 \\ -1 & -1 & -1 \end{matrix} \quad \text{North} \tag{8-23}$$

$$\text{Mask I} = \begin{matrix} 1 & 1 & 1 \\ -1 & -2 & 1 \\ -1 & -1 & 1 \end{matrix} \quad \text{NE} \tag{8-24}$$

$$\text{Mask J} = \begin{matrix} -1 & 1 & 1 \\ -1 & -2 & 1 \\ -1 & 1 & 1 \end{matrix} \quad \text{East} \tag{8-25}$$

$$\text{Mask K} = \begin{matrix} -1 & -1 & 1 \\ -1 & -2 & 1 \\ 1 & 1 & 1 \end{matrix} \quad \text{SE} \tag{8-26}$$

$$\text{Mask L} = \begin{matrix} -1 & -1 & -1 \\ 1 & -2 & 1 \\ 1 & 1 & 1 \end{matrix} \quad \text{South} \tag{8-27}$$

$$\text{Mask M} = \begin{matrix} 1 & -1 & -1 \\ 1 & -2 & -1 \\ 1 & 1 & 1 \end{matrix} \quad \text{SW} \tag{8-28}$$

$$\text{Mask N} = \begin{matrix} 1 & 1 & -1 \\ 1 & -2 & -1 \\ 1 & 1 & -1 \end{matrix} \quad \text{West} \tag{8-29}$$

$$\text{Mask O} = \begin{matrix} 1 & 1 & 1 \\ 1 & -2 & -1 \\ 1 & -1 & -1 \end{matrix} \quad \text{NW} \tag{8-30}$$

The compass names suggest the slope direction of maximum response. For example, the east gradient mask produces a maximum output for horizontal brightness value changes from west to east. The gradient masks have zero weighting (i.e., the sum of the mask coefficients is zero) (Pratt, 2001). This results in no output response over regions with constant brightness values (i.e., no edges are present). Southwest and east compass gradient masks were applied to the thermal data in Figures 8-22 c and d. The southwest enhancement emphasizes the plume, while the east enhancement emphasizes the plume and the western edge of the river. A northeast compass gradient mask applied to the Charleston scene is shown in Figure 8-25b. It does a reasonable job of identifying many of the edges, although it is difficult to interpret.

Richards and Jia (1999) identified four additional 3×3 templates that may be used to detect edges in images:

$$\text{Mask P} = \begin{matrix} -1 & 0 & 1 \\ -1 & 0 & 1 \\ -1 & 0 & 1 \end{matrix} \quad \text{vertical edges} \tag{8-31}$$

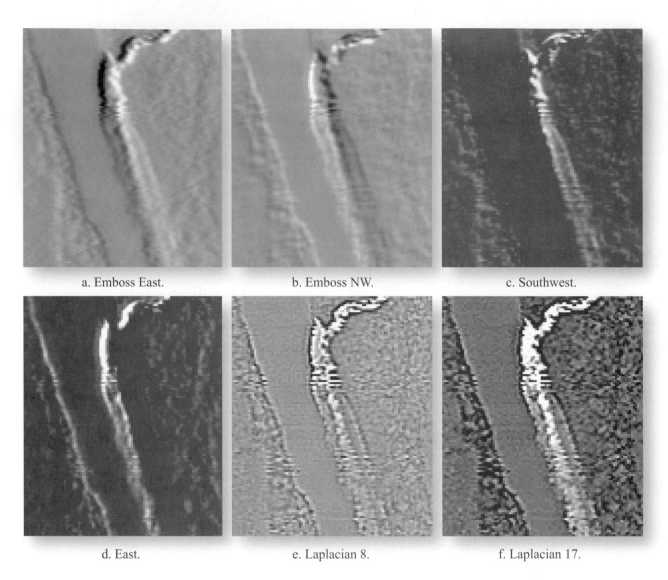

a. Emboss East.

b. Emboss NW.

c. Southwest.

d. East.

e. Laplacian 8.

f. Laplacian 17.

Figure 8-22   Application of various convolution masks and logic to the predawn thermal infrared data to enhance edges: a) embossing (Mask F), b) embossing (Mask G), c) directional filter (Mask M), d) directional filter (Mask J), e) Laplacian edge enhancement (Mask U), and f) Laplacian edge enhancement (Mask Y).

$$\text{Mask } Q = \begin{matrix} -1 & -1 & -1 \\ 0 & 0 & 0 \\ 1 & 1 & 1 \end{matrix} \qquad \text{horizontal edges} \qquad (8\text{-}32)$$

$$\text{Mask } R = \begin{matrix} 0 & 1 & 1 \\ -1 & 0 & 1 \\ -1 & -1 & 0 \end{matrix} \qquad \text{diagonal} \qquad (8\text{-}33)$$

$$\text{Mask } S = \begin{matrix} 1 & 1 & 0 \\ 1 & 0 & -1 \\ 0 & -1 & -1 \end{matrix} \qquad \text{diagonal} \qquad (8\text{-}34)$$

Laplacian convolution masks may be applied to imagery to perform edge enhancement. The Laplacian is a second derivative (as opposed to the gradient, which is a first derivative) and is invariant to rotation, meaning that it is insensitive to the direction in which the discontinuities (e.g., points, lines, edges) run. Several 3×3 Laplacian filters are shown next (Jahne, 2001; Pratt, 2001):

$$\text{Mask T} = \begin{matrix} 0 & -1 & 0 \\ -1 & 4 & -1 \\ 0 & -1 & 0 \end{matrix} \qquad (8\text{-}35)$$

$$\text{Mask U} = \begin{matrix} -1 & -1 & -1 \\ -1 & 8 & -1 \\ -1 & -1 & -1 \end{matrix} \qquad (8\text{-}36)$$

$$\text{Mask V} = \begin{matrix} 1 & -2 & 1 \\ -2 & 4 & -2 \\ 1 & -2 & 1 \end{matrix} \qquad (8\text{-}37)$$

The following operator may be used to subtract the Laplacian edges from the original image, if desired:

$$\text{Mask W} = \begin{matrix} 1 & 1 & 1 \\ 1 & -7 & 1 \\ 1 & 1 & 1 \end{matrix} \qquad (8\text{-}38)$$

Subtracting the Laplacian edge enhancement from the original image restores the overall grayscale variation, which the human viewer can comfortably interpret. It also sharpens the image by locally increasing the contrast at discontinuities (Russ, 2002).

The *Laplacian* operator generally highlights points, lines, and edges in the image and suppresses uniform and smoothly varying regions. Human vision physiological research suggests that we see objects in much the same way. Hence, the use of this operation has a more natural look than many of the other edge-enhanced images. Application of the Mask U Laplacian operator to the thermal data is shown in Figure 8-22e. Note that the Laplacian is an exceptional high-pass filter, which effectively enhances the plume and other subtle sensor noise in the image.

By itself, the Laplacian image may be difficult to interpret. Therefore, a Laplacian edge enhancement may be added back to the original image using the following mask:

$$\text{Mask X} = \begin{matrix} 0 & -1 & 0 \\ -1 & 5 & -1 \\ 0 & -1 & 0 \end{matrix} \qquad (8\text{-}39)$$

The result of applying this enhancement to the Charleston TM band 4 data is shown is Figure 8-25c. It is perhaps the best enhancement of high-frequency detail presented thus

far. Considerable detail is present in the urban structure and in the marsh. Laplacian operators do not have to be just 3×3. Below is a 5×5 Laplacian operator that adds the edge information back to the original. It is applied to the thermal plume in Figure 8-22f.

$$\text{Mask Y} = \begin{matrix} 0 & 0 & -1 & 0 & 0 \\ 0 & -1 & -2 & -1 & 0 \\ -1 & -2 & 17 & -2 & -1 \\ 0 & -1 & -2 & -1 & 0 \\ 0 & 0 & -1 & 0 & 0 \end{matrix} \qquad (8\text{-}40)$$

Numerous coefficients can be placed in the convolution masks. Usually, the analyst works interactively with the remotely sensed data, trying different coefficients and selecting those that produce the most effective results. It is also possible to combine operators for edge detection. For example, a combination of gradient and Laplacian edge operators may be superior to using either edge enhancement alone. In addition, nonlinear edge enhancements may provide superior results.

Nonlinear Edge Enhancement. Nonlinear edge enhancements are performed using nonlinear combinations of pixels. Many algorithms are applied using either 2×2 or 3×3 kernels. The *Sobel edge detector* is based on the notation of the 3×3 window previously described and is computed according to the relationship:

$$\text{Sobel}_{5,\text{out}} = \sqrt{X^2 + Y^2} \qquad (8\text{-}41)$$

where

$$X = (BV_3 + 2BV_6 + BV_9) - (BV_1 + 2BV_4 + BV_7) \quad (8\text{-}42)$$

and

$$Y = (BV_1 + 2BV_2 + BV_3) - (BV_7 + 2BV_8 + BV_9). \quad (8\text{-}43)$$

The Sobel operator may also be computed by simultaneously applying the following 3×3 templates across the image (Jain, 1989):

$$X = \begin{matrix} -1 & 0 & 1 \\ -2 & 0 & 2 \\ -1 & 0 & 1 \end{matrix} \qquad Y = \begin{matrix} 1 & 2 & 1 \\ 0 & 0 & 0 \\ -1 & -2 & -1 \end{matrix}$$

This procedure detects horizontal, vertical, and diagonal edges. A Sobel edge enhancement of the Savannah River thermal plume is shown in Figure 8-23a. It is a very effective

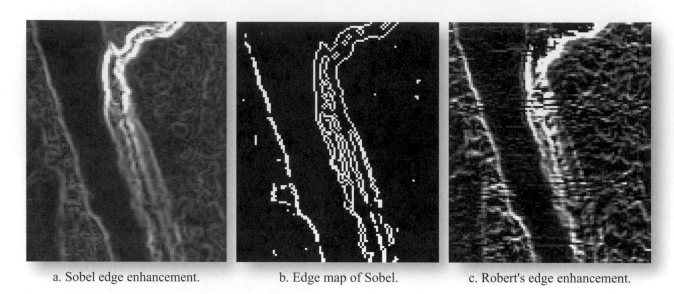

a. Sobel edge enhancement.   b. Edge map of Sobel.   c. Robert's edge enhancement.

Figure 8-23   Application of various nonlinear edge enhancements to the predawn thermal infrared data: a) Sobel, b) edge map of Sobel, and c) Robert's edge enhancement.

edge enhancement with the heart and sides of the plume surrounded by bright white lines. A Sobel edge enhancement of the Charleston scene is found in Figure 8-25d. It does an excellent job of identifying edges around rural features and large urban objects.

Each pixel in an image is declared an edge if its Sobel values exceed some user-specified threshold. Such information may be used to create *edge maps,* which often appear as white lines on a black background, or vice versa. For example, consider the edge map of the Sobel edge enhancement of the thermal data in Figure 8-23b. An edge map of the Charleston scene Sobel edge enhancement is shown in Figure 8-26a. It should be remembered that these lines are simply enhanced edges in the scene and have nothing to do with contours of equal reflectance or any other radiometric isoline.

The *Robert's edge detector* is based on the use of only four elements of a 3×3 mask. The new pixel value at pixel location $BV_{5,out}$ (refer to the 3×3 numbering scheme in Equation 8-6) is computed according to the equation (Peli and Malah, 1982):

$$\text{Roberts}_{5,\,out} = X + Y \qquad (8\text{-}44)$$

where

$$X = \left| BV_5 - BV_9 \right|$$
$$Y = \left| BV_6 - BV_8 \right|$$

The Robert's operator also may be computed by simultaneously applying the following templates across the image (Jain, 1989):

$$X = \begin{array}{ccc} 0 & 0 & 0 \\ 0 & 1 & 0 \\ 0 & 0 & -1 \end{array} \qquad Y = \begin{array}{ccc} 0 & 0 & 0 \\ 0 & 0 & 1 \\ 0 & -1 & 0 \end{array}$$

It is applied to the thermal data and the Charleston TM data in Figures 8-23c and 8-26b, respectively.

The *Kirsch nonlinear edge enhancement* calculates the gradient at pixel location $BV_{i,j}$. To apply this operator, however, it is first necessary to designate a different 3×3 window numbering scheme than used in previous discussions:

Window numbering for Kirsch =

$$\begin{array}{ccc} BV_0 & BV_1 & BV_2 \\ BV_7 & BV_{i,j} & BV_3 \\ BV_6 & BV_5 & BV_4 \end{array} \qquad (8\text{-}45)$$

The algorithm applied is (Gil et al., 1983):

$$BV_{i,j} = \max \left\{ 1, \; \max_{i=0}^{7} \left[\text{Abs}(5S_i - 3T_i) \right] \right\} \qquad (8\text{-}46)$$

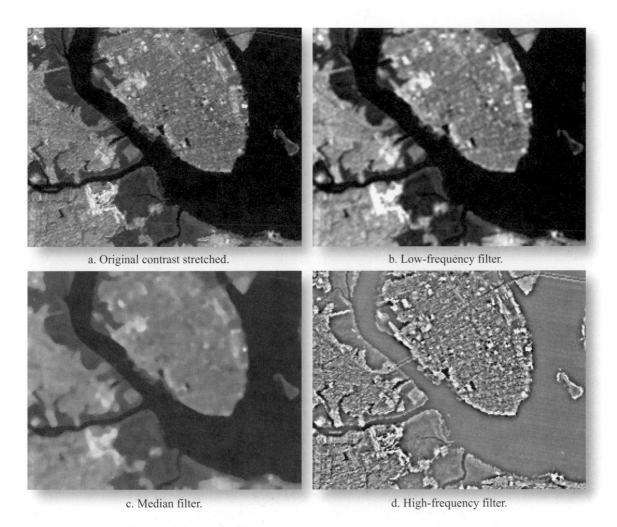

a. Original contrast stretched.

b. Low-frequency filter.

c. Median filter.

d. High-frequency filter.

Figure 8-24 Application of various convolution masks and logic to the Charleston TM band 4 data to enhance low- and high-frequency detail: a) contrast-stretched original image, b) low-frequency filter (Mask A), c) median filter, d) high-frequency filter (Mask E).

where

$$S_i = BV_i + BV_{i+1} + BV_{i+2} \qquad (8\text{-}47)$$

and

$$T_i = BV_{i+3} + BV_{i+4} + BV_{i+5} + BV_{i+6} + BV_{i+7} \ . \qquad (8\text{-}48)$$

The subscripts of BV are evaluated modulo 8, meaning that the computation moves around the perimeter of the mask in eight steps. The edge enhancement computes the maximal compass gradient magnitude about input image point $BV_{i,j}$. The value of S_i equals the sum of three adjacent pixels, while T_i equals the sum of the remaining four adjacent pixels. The input pixel value at $BV_{i,j}$ is never used in the computation.

Lines are extended edges. Analysts interested in the detection of lines as well as edges should refer to the work by Chittineni (1983). Several masks for detecting lines are:

$$\text{Mask Z} = \begin{matrix} -1 & -1 & -1 \\ 2 & 2 & 2 \\ -1 & -1 & -1 \end{matrix} \qquad \text{E–W} \qquad (8\text{-}49)$$

$$\text{Mask AA} = \begin{matrix} -1 & -1 & 2 \\ -1 & 2 & -1 \\ 2 & -1 & -1 \end{matrix} \qquad \text{NE–SW} \qquad (8\text{-}50)$$

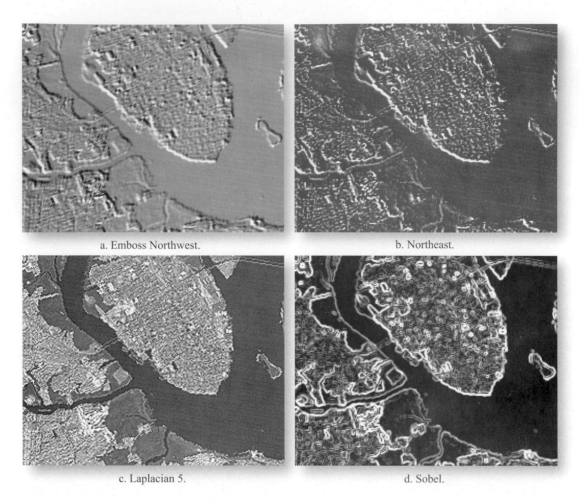

a. Emboss Northwest.   b. Northeast.

c. Laplacian 5.   d. Sobel.

Figure 8-25   Application of various convolution masks and logic to Charleston TM band 4 data to enhance edges: a) embossing (Mask F), b) northeast directional filter (Mask I), c) Laplacian edge enhancement (Mask X), and d) Sobel edge operator.

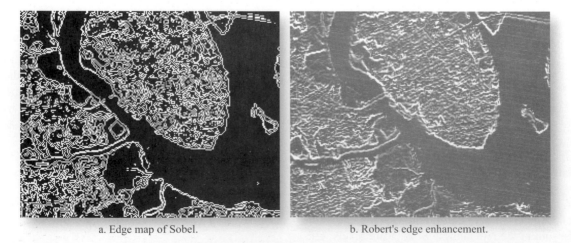

a. Edge map of Sobel.   b. Robert's edge enhancement.

Figure 8-26   Application of two nonlinear edge enhancements to Charleston TM band 4 data: a) edge map of Sobel, and b) Robert's edge enhancement.

$$\text{Mask BB} = \begin{array}{ccc} -1 & 2 & -1 \\ -1 & 2 & -1 \\ -1 & 2 & -1 \end{array} \qquad \text{N–S} \qquad (8\text{-}51)$$

$$\text{Mask CC} = \begin{array}{ccc} 2 & -1 & -1 \\ -1 & 2 & -1 \\ -1 & -1 & 2 \end{array} \qquad \text{NW–SE} \qquad (8\text{-}52)$$

Mask Z will enhance lines in the imagery that are oriented horizontally. Lines oriented at 45° will be enhanced using Mask AA and CC. Vertical lines will be enhanced using Mask BB.

The Fourier Transform

Fourier analysis is a mathematical technique for separating an image into its various spatial frequency components. First, let us consider a continuous function $f(x)$. The Fourier theorem states that any function $f(x)$ can be represented by a summation of a series of sinusoidal terms of varying spatial frequencies. These terms can be obtained by the Fourier transform of $f(x)$, which is written as:

$$F(u) = \int_{-\infty}^{\infty} f(x)e^{-2\pi iux}dx, \qquad (8\text{-}53)$$

where u is spatial frequency. This means that $F(u)$ is a frequency domain function. The spatial domain function $f(x)$ can be recovered from $F(u)$ by the inverse Fourier transform:

$$f(x) = \int_{-\infty}^{\infty} F(u)e^{2\pi iux}du. \qquad (8\text{-}54)$$

To use Fourier analysis in digital image processing, we must consider two extensions of these equations. First, both transforms can be extended from one-dimensional functions to two-dimensional functions $f(x, y)$ and $F(u, v)$. For Equation 8-53 this becomes:

$$F(u, v) = \int_{-\infty}^{\infty}\int f(x, y)e^{-2\pi i(ux + vy)}dxdy. \qquad (8\text{-}55)$$

Furthermore, we can extend both transforms to discrete functions. The two-dimensional discrete Fourier transform is written:

$$F(u, v) = \frac{1}{NM}\sum_{x=0}^{N-1}\sum_{y=0}^{M-1} f(x, y)e^{-2\pi i\left(\frac{ux}{N} + \frac{vy}{M}\right)} \qquad (8\text{-}56)$$

where N is the number of pixels in the x direction and M is the number of pixels in the y direction. Every remotely sensed image can be described as a two-dimensional discrete function. Therefore, Equation 8-56 may be used to compute the Fourier transform of an image. The image can be reconstructed using the inverse transform:

$$f(x, y) = \sum_{u=0}^{N-1}\sum_{v=0}^{M-1} F(u, v)e^{2\pi i\left(\frac{ux}{N} + \frac{vy}{M}\right)}. \qquad (8\text{-}57)$$

You are probably asking the question, "What does the $F(u, v)$ represent?" It contains the spatial frequency information of the original image $f(x, y)$ and is called the *frequency spectrum*. Note that it is a complex function because it contains i, which equals $\sqrt{-1}$. We can write any complex function as the sum of a real part and an imaginary part:

$$F(u, v) = R(u, v) + iI(u, v), \qquad (8\text{-}58)$$

which is equivalent to

$$F(u, v) = |F(u, v)|e^{i\phi(u, v)}, \qquad (8\text{-}59)$$

where $|F(u, v)|$ is a real function, and

$$|F(u, v)| = \sqrt{R(u, v)^2 + I(u, v)^2}.$$

$|F(u, v)|$ is called the magnitude of the Fourier transform and can be displayed as a two-dimensional image. It represents the magnitude and the direction of the different frequency components in the image $f(x, y)$. The variable ϕ in Equation 8-59 represents *phase* information in the image $f(x, y)$. Although we usually ignore the phase information when we display the Fourier transform, we cannot recover the original image without it.

To understand how the Fourier transform is useful in remote sensing applications, let us first consider three subimages extracted from *The Loop*, shown in Figure 8-27. The first subset includes a homogeneous, low-frequency water portion of the photograph (Figure 8-28a). Another area contains low- and medium-frequency terrain information with both horizontal and vertical linear features (Figure 8-28c). The final subset contains low- and medium-frequency terrain with some diagonal linear features (Figure 8-28e). The magnitudes of the subimages' Fourier transforms are shown in Figures 8-28b, d, and f. The Fourier magnitude images are

Loop of the Colorado River Near Moab, Utah

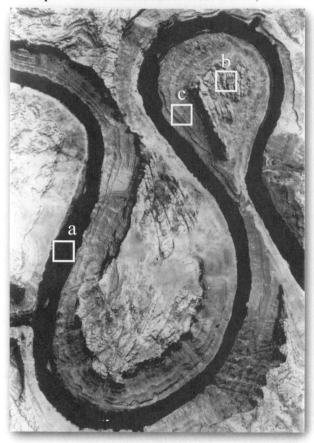

Figure 8-27   Digitized aerial photograph of The Loop on the Colorado River with three subimages identified.

symmetric about their centers, and u and v represent spatial frequency. The displayed Fourier magnitude image is usually adjusted to bring the $F(0, 0)$ to the center of the image rather than to the upper-left corner. Therefore, the intensity at the center represents the magnitude of the lowest-frequency component. The frequency increases away from the center. For example, consider the Fourier magnitude of the homogeneous water body (Figure 8-28b). The very bright values found in and around the center indicate that it is dominated by low-frequency components. In the second image, more medium-frequency components are present in addition to the background of low-frequency components. We can easily identify the high-frequency information representing the horizontal and vertical linear features in the original image (Figure 8-28d). Notice the alignment of the cloud of points in the center of the Fourier transform in Figure 8-28f. It represents the diagonal linear features trending in the NW–SE direction in the photograph.

It is important to remember that the strange-looking Fourier transformed image $F(u, v)$ contains all the information found in the original image. It provides a mechanism for analyzing and manipulating images according to their spatial frequencies. It is useful for image restoration, filtering, and radiometric correction. For example, the Fourier transform can be used to remove periodic noise in remotely sensed data. When the pattern of periodic noise is unchanged throughout the image, it is called stationary periodic noise. Striping in remotely sensed imagery is usually composed of stationary periodic noise.

When stationary periodic noise is a single-frequency sinusoidal function in the spatial domain, its Fourier transform consists of a single bright point (a peak of brightness). For example, Figure 8-29a and c displays two images of sinusoidal functions with different frequencies (which look very much like striping in remote sensor data!). Figure 8-29b and d are their Fourier transforms. The frequency and orientation of the noise can be identified by the position of the bright points. The distance from the bright points to the center of the transform (the lowest-frequency component in the image) is directly proportional to the frequency. A line connecting the bright point and the center of the transformed image is always perpendicular to the orientation of the noise lines in the original image. Striping in the remotely sensed data is usually composed of sinusoidal functions with more than one frequency in the same orientation. Therefore, the Fourier transform of such noise consists of a series of bright points lined up in the same orientation.

Because the noise information is concentrated in a point or a series of points in the frequency domain, it is relatively straightforward to identify and remove them in the frequency domain, whereas it is quite difficult to remove them in the standard spatial domain. Basically, an analyst can manually cut out these lines or points in the Fourier transform image or use a computer program to look for such noise and remove it. For example, consider the Landsat TM band 4 data of Al Jubail, Saudi Arabia, obtained on September 1, 1990 (Figure 8-30). The image contains serious stationary periodic striping, which can make the data unusable when conducting near-shore studies of suspended sediment transport. Figure 8-31 documents how a portion of the Landsat TM scene was corrected. First, a Fourier transform of the area was computed (Figure 8-31b). The analyst then modified the Fourier transform by selectively removing the points in the plot associated with the systematic striping (Figure 8-31c). This can be done manually or a special program can be written to look for and remove such systematic noise patterns in the Fourier transform image. The inverse Fourier transform was then computed, yielding a clean band 4 image, which may be

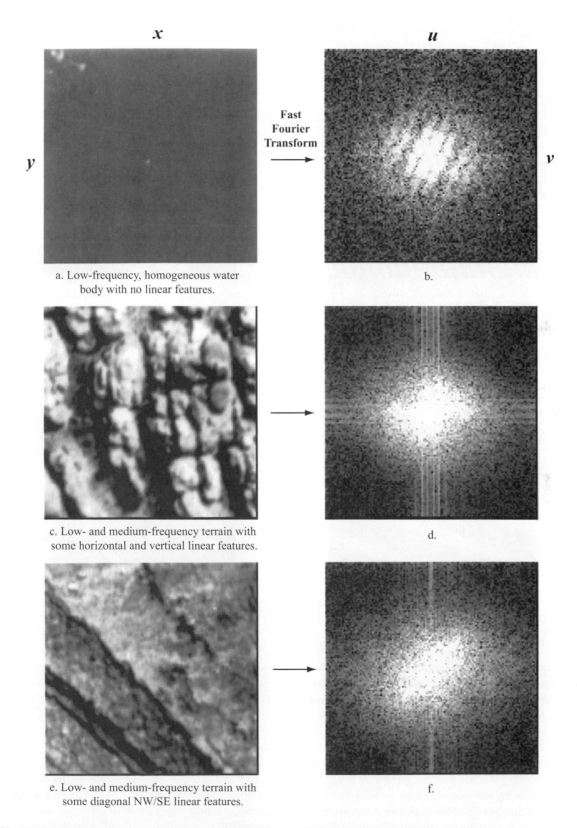

a. Low-frequency, homogeneous water body with no linear features.

b.

c. Low- and medium-frequency terrain with some horizontal and vertical linear features.

d.

e. Low- and medium-frequency terrain with some diagonal NW/SE linear features.

f.

Figure 8-28   Application of a Fourier transform to the three subimages of the Loop identified in Figure 8-27.

Stationary Periodic Noise and Its Fourier Transform

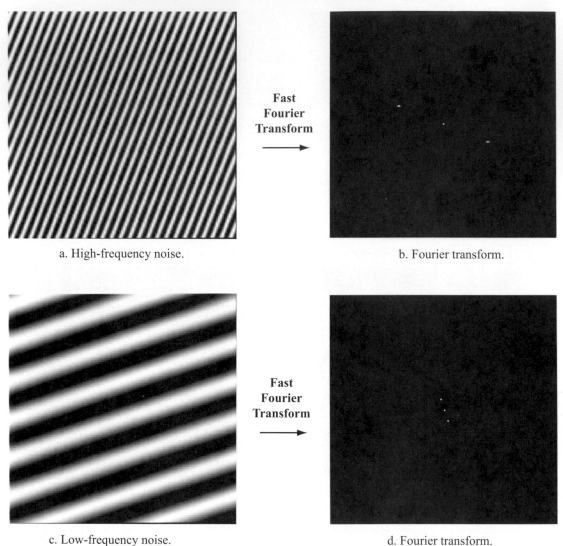

a. High-frequency noise.

Fast Fourier Transform →

b. Fourier transform.

c. Low-frequency noise.

Fast Fourier Transform →

d. Fourier transform.

Figure 8-29   Two examples of stationary periodic noise and their Fourier transforms.

more useful for biophysical analysis (Figure 8-31d). This type of noise could not be removed using a simple convolution mask. Rather, it requires access to the Fourier transform and selective editing out of the noise in the Fourier transform image.

Figure 8-32 depicts hyperspectral data of the Mixed Waste Management Facility (MWMF) on the Savannah River Site that was destriped using the same logic. A Fourier transform of a single band of data revealed the high-frequency information associated with the horizontal striping (Figure 8-32b). A cut filter was created to isolate the high-frequency informa-

tion associated with the horizontal striping (Figure 8-32c and d). The result of applying the cut filter and inverting the fast Fourier transform is shown in Figure 8-32e. The striping was reduced substantially.

Spatial Filtering in Frequency Domain

We have discussed filtering in the spatial domain using convolution filters. It can also be performed in the frequency domain. Using the Fourier transform, we can manipulate directly the frequency information of the image. The manipulation can be performed by multiplying the Fourier trans-

Landsat Thematic Mapper Data of Al Jubail, Saudi Arabia

September 1, 1990
Band 4

Figure 8-30   Landsat TM band 4 data of Al Jubail, Saudi Arabia, obtained on September 1, 1990.

form of the original image by a mask image called a frequency domain filter, which will block or weaken certain frequency components by making the values of certain parts of the frequency spectrum become smaller or even zero. Then we can compute the inverse Fourier transform of the manipulated frequency spectrum to obtain a filtered image in the spatial domain. Numerous algorithms are available for computing the Fast Fourier transform (FFT) and the inverse Fast Fourier transform (IFFT) (Russ, 2002). Spatial filtering in the frequency domain generally involves computing the FFT of the original image, multiplying the FFT of a convolution mask of the analyst's choice (e.g., a low-pass filter) with

the FFT, and inverting the resultant image with the IFFT; that is,

$$f(x, y) \text{ FFT} \rightarrow F(u, v) \rightarrow F(u, v)\ G(u, v)$$

$$\rightarrow F'(u, v) \text{ IFFT} \rightarrow f'(x, y).$$

The convolution theorem states that the convolution of two images is equivalent to the multiplication of their Fourier transformations. If

$$f'(x, y) = f(x, y)*g(x, y), \tag{8-60}$$

**Application of a Fourier Transform to Landsat Thematic Mapper Data
of an Area Near Al Jubail, Saudi Arabia to Remove Striping**

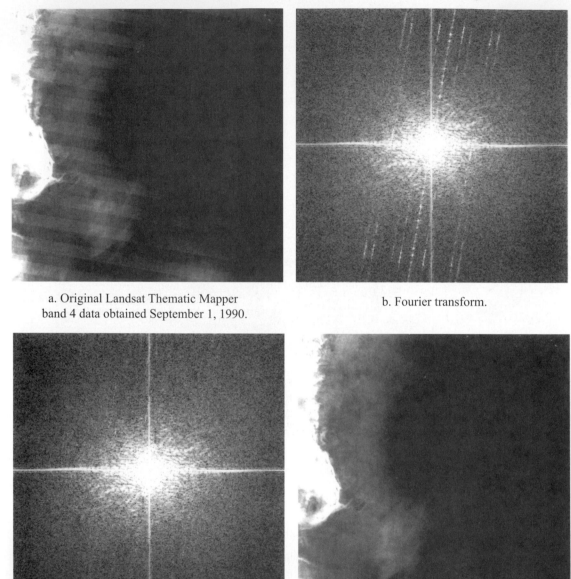

a. Original Landsat Thematic Mapper
band 4 data obtained September 1, 1990.

b. Fourier transform.

c. Interactively cleaned Fourier transform.

d. Corrected Landsat Thematic Mapper band 4 data.

Figure 8-31   Application of a Fourier transform to a portion of Landsat TM band 4 data of Al Jubail, Saudi Arabia: a) original TM band 4
data, b) Fourier transform, c) cleaned Fourier transform, and d) destriped band 4 data.

Application of a Fourier Transform to Hyperspectral Data of the Savannah River Site

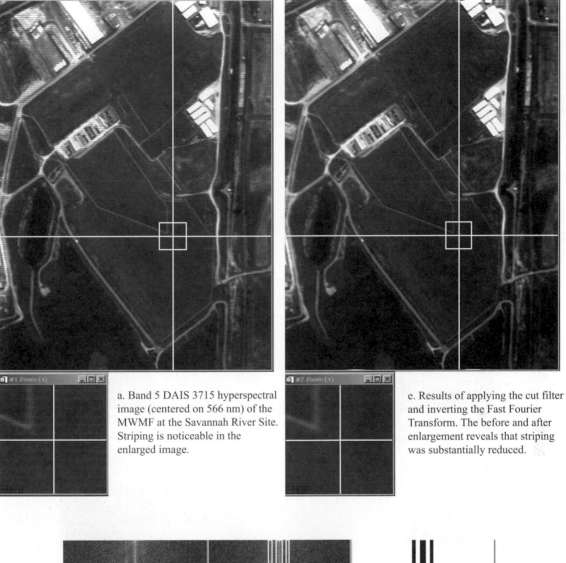

a. Band 5 DAIS 3715 hyperspectral image (centered on 566 nm) of the MWMF at the Savannah River Site. Striping is noticeable in the enlarged image.

e. Results of applying the cut filter and inverting the Fast Fourier Transform. The before and after enlargement reveals that striping was substantially reduced.

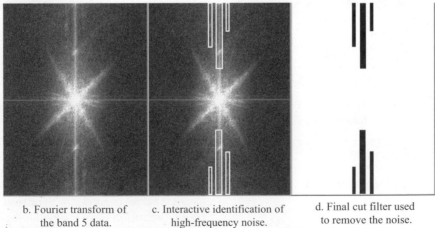

b. Fourier transform of the band 5 data.

c. Interactive identification of high-frequency noise.

d. Final cut filter used to remove the noise.

Figure 8-32   Application of a Fourier transform to a single band of hyperspectral data of the Mixed Waste Management Facility at the Savannah River Site near Aiken, SC (Jensen et al., 2003).

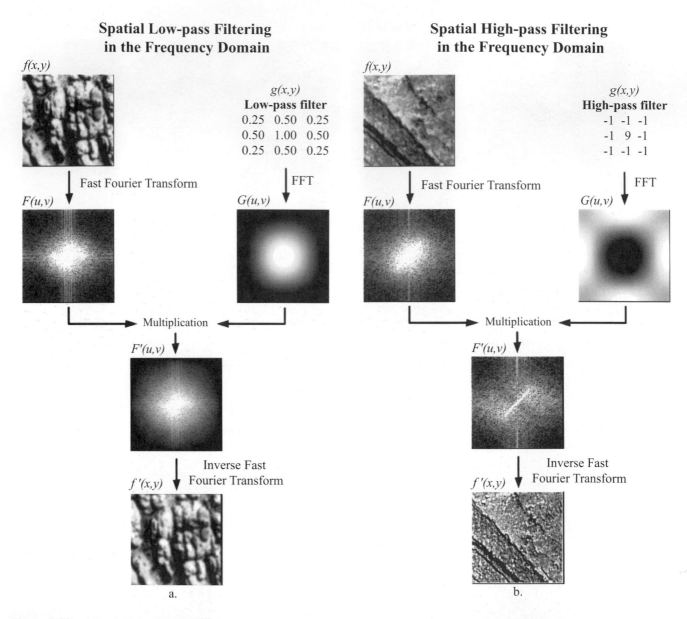

Figure 8-33   a) Spatial low-pass and b) high-pass filtering in the frequency domain using a Fourier transform.

where * represents the operation of convolution, $f(x, y)$ is the original image and $g(x,y)$ is a convolution mask filter, then

$$F'(u, v) = F(u, v)G(u, v), \qquad (8\text{-}61)$$

where F', F, and G are Fourier transforms of f', f, and g, respectively.

Two examples of such manipulation are shown in Figure 8-33a and b. A low-pass filter (Mask B) and a high-pass filter (Mask D) were used to construct the filter function $g(x, y)$ in

Figures 8-33a and b, respectively. In practice, one problem must be solved. Usually, the dimensions of $f(x, y)$ and $g(x, y)$ are different; for example, the low-pass filter in Figure 8-33a has only nine elements, while the image is composed of 128×128 pixels. Operation in the frequency domain requires that the sizes of $F(u, v)$ and $G(u, v)$ be the same. This means the sizes of f and g must be made the same because the Fourier transform of an image has the same size as the original image. The solution of this problem is to construct $g(x, y)$ by putting the convolution mask at the center of a zero-value image that has the same size as f. Note that in the Fourier

Table 8-5.   Charleston, SC Thematic Mapper scene statistics used in the principal components analysis (PCA).

Band Number:	1	2	3	4	5	7	6
μm:	0.45–0.52	0.52–0.60	0.63–0.69	0.76–0.90	1.55–1.75	2.08–2.35	10.4–12.5
Univariate Statistics							
Mean	64.80	25.60	23.70	27.30	32.40	15.00	110.60
Standard Deviation	10.05	5.84	8.30	15.76	23.85	12.45	4.21
Variance	100.93	34.14	68.83	248.40	568.84	154.92	17.78
Minimum	51	17	14	4	0	0	90
Maximum	242	115	131	105	193	128	130
Variance–Covariance Matrix							
1	100.93						
2	56.60	34.14					
3	79.43	46.71	68.83				
4	61.49	40.68	69.59	248.40			
5	134.27	85.22	141.04	330.71	568.84		
7	90.13	55.14	86.91	148.50	280.97	154.92	
6	23.72	14.33	22.92	43.62	78.91	42.65	17.78
Correlation Matrix							
1	1.00						
2	0.96	1.00					
3	0.95	0.96	1.00				
4	0.39	0.44	0.53	1.00			
5	0.56	0.61	0.71	0.88	1.00		
7	0.72	0.76	0.84	0.76	0.95	1.00	
6	0.56	0.58	0.66	0.66	0.78	0.81	1.00

transforms of the two convolution masks the low-pass convolution mask has a bright center (Figure 8-33a), while the high-pass filter has a dark center (Figure 8-33b). The multiplication of Fourier transforms $F(u, v)$ and $G(u, v)$ results in a new Fourier transform, $F'(u, v)$. Computing the Inverse Fast Fourier transformation yields $f'(x, y)$, a filtered version of the original image. Thus, spatial filtering can be performed in both the spatial and frequency domain.

As demonstrated, filtering in the frequency domain involves one multiplication and two transformations. For general applications, convolution in the spatial domain may be more cost-effective. Only when the size of $g(x, y)$ is very large does the Fourier method become cost-effective. However, with the frequency domain method we can also do some filtering that is not easy to do in the spatial domain. We may construct a frequency domain filter $G(u,v)$ specifically designed to remove certain frequency components in the image. Numerous articles describe how to construct frequency filters (Pan and Chang, 1992; Khan, 1992). Watson (1993) describes how the two-dimensional FFT may be applied to image mosaicking, enlargement, and registration.

Principal Components Analysis

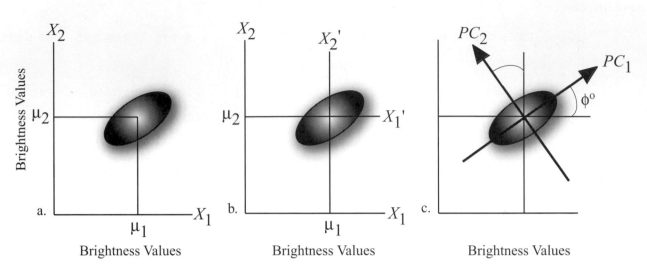

Figure 8-34 Diagrammatic representation of the spatial relationship between the first two principal components. a) Scatterplot of data points collected from two remotely sensed bands labeled X_1 and X_2 with the means of the distribution labeled μ_1 and μ_2. b) A new coordinate system is created by shifting the axes to an X' system. The values for the new data points are found by the relationship $X_1' = X_1 - \mu_1$ and $X_2' = X_2 - \mu_2$. c) The X' axis system is then rotated about its origin (μ_1, μ_2) so that PC_1 is projected through the semi-major axis of the distribution of points and the variance of PC_1 is a maximum. PC_2 must be perpendicular to PC_1. The PC axes are the principal components of this two-dimensional data space. Principal component 1 often accounts for > 90% of the variance, and principal component 2 accounts for 2% to 10%, etc.

Principal Components Analysis

Principal components analysis (often called PCA, or Karhunen–Loeve analysis) has proven to be of value in the analysis of multispectral and hyperspectral remotely sensed data (Zhao and Maclean, 2000; Mitternicht and Zinck, 2003). Principal components analysis is a technique that transforms the original remotely sensed dataset into a substantially smaller and easier to interpret set of uncorrelated variables that represents most of the information present in the original dataset. Principal components are derived from the original data such that the first principal component accounts for the maximum proportion of the variance of the original dataset, and subsequent orthogonal components account for the maximum proportion of the remaining variance (Holden and LeDrew, 1998, Zhao and Maclean, 2000). The ability to reduce the *dimensionality* (i.e., the number of bands in the dataset that must be analyzed to produce usable results) from *n* to just a few bands is an important economic consideration, especially if the potential information that can be recovered from the transformed data is just as good as the original remote sensor data.

A form of PCA may also be useful for reducing the dimensionality of hyperspectral datasets. For example, many scientists use a *minimum noise fraction* (MNF) procedure, which is a cascaded principal components analysis for data compression and noise reduction of hyperspectral data (Jensen et al., 2003). Chapter 11 provides an example of the use of the MNF procedure.

To perform principal component analysis we apply a transformation to a *correlated* set of multispectral data. For example, the Charleston TM scene is a likely candidate since bands 1, 2, and 3 are highly correlated, as are bands 5 and 7 (Table 8-5). The application of the transformation to the correlated remote sensor data will result in another *uncorrelated* multispectral dataset that has certain ordered variance properties (Singh and Harrison, 1985; Carr, 1998). This transformation is conceptualized by considering the two-dimensional distribution of pixel values obtained in two TM bands, which we will label X_1 and X_2. A scatterplot of all the brightness values associated with each pixel in each band is shown in Figure 8-34a, along with the location of the respective means, μ_1 and μ_2. The spread or variance of the distribution of points is an indication of the correlation and quality of information associated with both bands. If all the data points were clustered in an extremely tight zone in the two-dimensional space, these data would probably provide very little information.

The initial measurement coordinate axes (X_1 and X_2) may not be the best arrangement in multispectral feature space to analyze the remote sensor data associated with these two bands. The goal is to use principal components analysis to *translate* and/or *rotate* the original axes so that the original brightness values on axes X_1 and X_2 are redistributed (reprojected) onto a new set of axes or dimensions, X'_1 and X'_2 (Wang, 1993). For example, the best *translation* for the original data points from X_1 to X'_1 and from X_2 to X'_2 coordinate systems might be the simple relationship $X'_1 = X_1 - \mu_1$ and $X'_2 = X_2 - \mu_2$. Thus, the origin of the new coordinate system (X'_1 and X'_2) now lies at the location of both means in the original scatter of points (Figure 8-34b).

The X' coordinate system might then be *rotated* about its new origin (μ_1, μ_2) in the new coordinate system some ϕ degrees so that the first axis X'_1 is associated with the maximum amount of variance in the scatter of points (Figure 8-34c). This new axis is called the first *principal component* ($PC_1 = \lambda_1$). The second principal component ($PC_2 = \lambda_2$) is perpendicular (orthogonal) to PC_1. Thus, the major and minor axes of the ellipsoid of points in bands X_1 and X_2 are called the principal components. The third, fourth, fifth, and so on, components contain decreasing amounts of the variance found in the dataset.

To transform (reproject) the original data on the X_1 and X_2 axes onto the PC_1 and PC_2 axes, we must obtain certain transformation coefficients that we can apply in a linear fashion to the original pixel values. The linear transformation required is derived from the covariance matrix of the original dataset. Thus, this is a data-dependent process with each new remote sensing dataset yielding different transformation coefficients.

The transformation is computed from the original spectral statistics, as follows (Short, 1982):

1. The $n \times n$ *covariance matrix, Cov*, of the *n*-dimensional remote sensing dataset to be transformed is computed (Table 8-5). Use of the covariance matrix results in a nonstandardized PCA, whereas use of the correlation matrix results in a standardized PCA (Eastman and Fulk, 1993; Carr, 1998).

The eigenvalues, $E = [\lambda_{1,1}, \lambda_{2,2}, \lambda_{3,3}, ..., \lambda_{n,n}]$, and eigenvectors, $EV = [a_{kp} ...$ for $k = 1$ to n bands, and $p = 1$ to n components], of the covariance matrix are computed such that:

$$E$$

$$EV \cdot Cov \cdot EV^T = \begin{bmatrix} \lambda_{1,1} & 0 & 0 & 0 & 0 & 0 & 0 \\ 0 & \lambda_{2,2} & 0 & 0 & 0 & 0 & 0 \\ 0 & 0 & \lambda_{3,3} & 0 & 0 & 0 & 0 \\ 0 & 0 & 0 & \lambda_{4,4} & 0 & 0 & 0 \\ 0 & 0 & 0 & 0 & \lambda_{5,5} & 0 & 0 \\ 0 & 0 & 0 & 0 & 0 & \lambda_{6,6} & 0 \\ 0 & 0 & 0 & 0 & 0 & 0 & \lambda_{n,n} \end{bmatrix} \quad (8\text{-}62)$$

where EV^T is the transpose of the eigenvector matrix, EV, and E is a diagonal covariance matrix whose elements λ_{ii}, called *eigenvalues*, are the variances of the *p*th *principal components*, where $p = 1$ to n components. The nondiagonal eigenvalues, λ_{ij}, are equal to zero and therefore can be ignored. The number of nonzero eigenvalues in an $n \times n$ covariance matrix always equals *n*, the number of bands examined. The eigenvalues are often called *components* (i.e., eigenvalue 1 may be referred to as principal component 1). Eigenvalues and eigenvectors were computed for the Charleston, SC, Thematic Mapper scene (Tables 8-6 and 8-7).

The eigenvalues contain important information. For example, it is possible to determine the percent of total variance explained by each of the principal components, $\%_p$, using the equation:

$$\%_p = \frac{\text{eigenvalue } \lambda_p \times 100}{\displaystyle\sum_{p=1}^{n} \text{eigenvalue } \lambda_p}, \quad (8\text{-}63)$$

where λ_p is the *p*th eigenvalue out of the possible *n* eigenvalues. For example, the first principal component (eigenvalue λ_1) of the Charleston TM scene accounts for 84.68% of the variance in the entire multispectral dataset (Table 8-6). Component 2 accounts for 10.99% of the remaining variance. Cumulatively, these first two principal components account for 95.67% of the variance. The third component accounts for another 3.15%, bringing the total to 98.82% of the variance explained by the first three components (Table 8-6). Thus, the seven-band TM dataset of Charleston might be compressed into just three new principal component images (or bands) that explain 98.82% of the variance.

But what do these new components represent? For example, what does component 1 stand for? By computing the correlation of each band *k* with each component *p*, it is possible to determine how each band "loads" or is associated with each principal component. The equation is:

Table 8-6.   Eigenvalues computed for the covariance matrix.

	Component p						
	1	**2**	**3**	**4**	**5**	**6**	**7**
Eigenvalues, λ_p	1010.92	131.20	37.60	6.73	3.95	2.17	1.24
Difference	879.72	93.59	30.88	2.77	1.77	.93	--
Total Variance = 1193.81							

Percent of total variance in the data explained by each component:

Computed as $\%_p = \dfrac{\text{eigenvalue } \lambda_p \times 100}{\displaystyle\sum_{p=1}^{7} \text{eigenvalue } \lambda_p}$.

For example,

$$\sum_{p=1}^{7} \lambda_p = 1010.92 + 131.20 + 37.60 + 6.73 + 3.95 + 2.17 + 1.24 = 1193.81 .$$

Percentage of variance explained by first component $= \dfrac{1010.92 \times 100}{1193.81} = 84.68$.

	1	**2**	**3**	**4**	**5**	**6**	**7**
Percentage	84.68	10.99	3.15	0.56	0.33	0.18	0.10
Cumulative	84.68	95.67	98.82	99.38	99.71	99.89	100.00

Table 8-7.   Eigenvectors (a_{kp}) (factor scores) computed for the covariance matrix found in Table 8-5.

	Component p						
	1	**2**	**3**	**4**	**5**	**6**	**7**
band$_k$ 1	0.205	0.637	0.327	−0.054	0.249	−0.611	−0.079
2	0.127	0.342	0.169	−0.077	0.012	0.396	0.821
3	0.204	0.428	0.159	−0.076	−0.075	0.649	−0.562
4	0.443	−0.471	0.739	0.107	−0.153	−0.019	−0.004
5	0.742	−0.177	−0.437	−0.300	0.370	0.007	0.011
7	0.376	0.197	−0.309	−0.312	−0.769	−0.181	0.051
6	0.106	0.033	−0.080	0.887	0.424	0.122	0.005

$$R_{kp} = \frac{a_{kp} \times \sqrt{\lambda_p}}{\sqrt{\text{Var}_k}} \qquad (8\text{-}64)$$

where

a_{kp} = eigenvector for band k and component p
λ_p = pth eigenvalue

Var_k = variance of band k in the covariance matrix.

This computation results in a new $n \times n$ matrix (Table 8-8) filled with *factor loadings*. For example, the highest correlations (i.e., factor loadings) for principal component 1 were for bands 4, 5, and 7 (0.894, 0.989, and 0.961, respectively; Table 8-8). This suggests that this component is a near- and

Table 8-8.   Degree of correlation, R_{kp}, between each band k and each principal component p.

Computed: $R_{kp} = \dfrac{a_{kp} \times \sqrt{\lambda_p}}{\sqrt{\mathrm{Var}_k}}$.

For example:

$R_{1,1} = \dfrac{0.205 \times \sqrt{1010.92}}{\sqrt{100.93}} = \dfrac{0.205 \times 31.795}{10.046} = 0.649$

$R_{5,1} = \dfrac{0.742 \times \sqrt{1010.92}}{\sqrt{568.84}} = \dfrac{0.742 \times 31.795}{23.85} = 0.989$

$R_{2,2} = \dfrac{0.342 \times \sqrt{131.20}}{\sqrt{34.14}} = \dfrac{0.342 \times 11.45}{5.842} = 0.670.$

| | Component p | | | | | | |
	1	2	3	4	5	6	7
Band$_k$ 1	0.649	0.726	0.199	−0.014	0.049	−0.089	−0.008
2	0.694	0.670	0.178	−0.034	0.004	0.099	0.157
3	0.785	0.592	0.118	−0.023	−0.018	0.115	−0.075
4	0.894	−0.342	0.287	0.017	−0.019	−0.002	−0.000
5	0.989	−0.084	−0.112	−0.032	0.030	0.000	0.000
7	0.961	0.181	−0.152	0.065	−0.122	−0.021	0.004
6	0.799	0.089	−0.116	0.545	0.200	0.042	0.001

middle-infrared reflectance band. This makes sense because the golf courses and other vegetation are particularly bright in this image (Figure 8-35). Conversely, principal component 2 has high loadings only in the visible bands 1, 2, and 3 (0.726, 0.670, and 0.592), and vegetation is noticeably darker in this image. This is a visible spectrum component. Component 3 loads heavily on the near-infrared (0.287) and appears to provide some unique vegetation information. Component 4 accounts for little of the variance but is easy to label since it loads heavily (0.545) on the thermal infrared band 6. Components 5, 6, and 7 provide no useful information and contain most of the systematic noise. They account for very little of the variance and should probably not be used further.

Now that we understand what information each component contributes, it is useful to see how the principal component images are created. To do this, it is necessary to first identify the original brightness values ($BV_{i,j,k}$) associated with a given pixel. In this case we will evaluate the first pixel in a hypothetical image at row 1, column 1 for each of seven bands. We will represent this as the vector X, such that

$$X = \begin{bmatrix} BV_{1,1,1} = 20 \\ BV_{1,1,2} = 30 \\ BV_{1,1,3} = 22 \\ BV_{1,1,4} = 60 \\ BV_{1,1,5} = 70 \\ BV_{1,1,7} = 62 \\ BV_{1,1,6} = 50 \end{bmatrix} \tag{8-65}$$

We will now apply the appropriate transformation to this data such that it is projected onto the first principal component's axes. In this way we will find out what the new brightness value ($newBV_{i,j,p}$) will be for this component, p. It is computed according to the formula:

$$newBV_{i,j,p} = \sum_{k=1}^{n} a_{kp} BV_{i,j,k}, \tag{8-66}$$

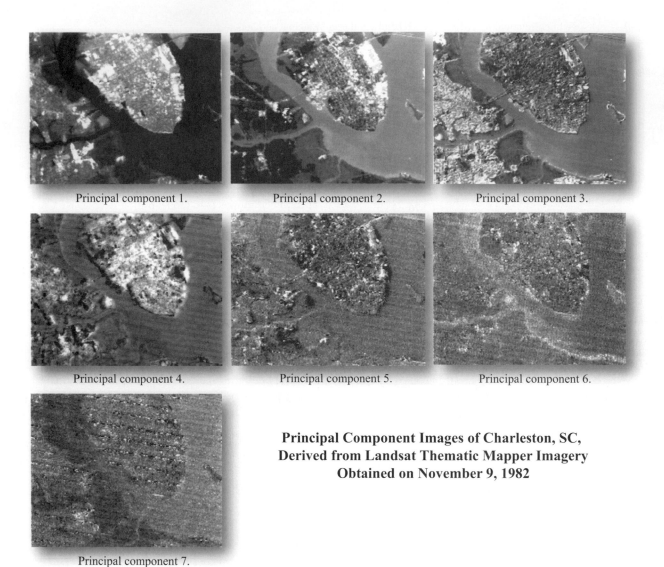

Principal component 1.   Principal component 2.   Principal component 3.

Principal component 4.   Principal component 5.   Principal component 6.

Principal component 7.

**Principal Component Images of Charleston, SC,
Derived from Landsat Thematic Mapper Imagery
Obtained on November 9, 1982**

Figure 8-35   Seven principal component images of the Charleston Thematic Mapper data computed using all seven bands. Component 1 consists of both near- and middle-infrared information (bands 4, 5, and 7). Component 2 contains primarily visible light information (bands 1, 2, and 3). Component 3 contains primarily near-infrared information. Component 4 consists of the thermal infrared information contributed by band 6. Thus, the seven-band TM data can be reduced in dimension to just four principal components (1, 2, 3, and 4), which account for 99.38% of the variance.

where a_{kp} = eigenvectors, $BV_{i,j,k}$ = brightness value in band k for the pixel at row i, column j, and n = number of bands. In our hypothetical example, this yields:

$$newBV_{1,1,1} = a_{1,1}(BV_{1,1,1}) + a_{2,1}(BV_{1,1,2}) +$$

$$a_{3,1}(BV_{1,1,3}) + a_{4,1}(BV_{1,1,4}) + a_{5,1}(BV_{1,1,5}) +$$

$$a_{6,1}(BV_{1,1,7}) + a_{7,1}(BV_{1,1,6})$$

$$= 0.205(20) + 0.127(30) + 0.204(22) + 0.443(60) +$$

$$0.742(70) + 0.376(62) + 0.106(50)$$

$$= 119.53$$

This pseudo-measurement is a linear combination of original brightness values and factor scores (eigenvectors). The new brightness value for row 1, column 1 in principal component 1 after truncation to an integer is new $BV_{1,1,1} = 119$.

This procedure takes place for every pixel in the original remote sensor image data to produce the principal component 1 image dataset. Then p is incremented by 1 and principal component 2 is created pixel by pixel. This is the method used to produce the principal component images shown in Figure 8-35. If desired, any two or three of the principal components can be placed in the blue, green, and/or red image planes to create a principal component color composite. These displays often depict more subtle differences in color shading and distribution than traditional color-infrared color composite images.

If components 1, 2, and 3 account for most of the variance in the dataset, perhaps the original seven bands of TM data can be set aside, and the remainder of the image enhancement or classification can be performed using just these three principal component images. This greatly reduces the amount of data to be analyzed and completely bypasses the expensive and time-consuming process of feature selection so often necessary when classifying remotely sensed data (discussed in Chapter 9).

Eastman and Fulk (1993) suggest that *standardized PCA* (based on the computation of eigenvalues from correlation matrices) is superior to nonstandardized PCA (computed from covariance matrices) when analyzing change in multitemporal image datasets. Standardized PCA forces each band to have equal weight in the derivation of the new component images and is identical to converting all image values to standard scores (by subtracting the mean and dividing by the standard deviation) and computing nonstandardized PCA of the results. Eastman and Fulk (1993) processed 36 monthly AVHRR-derived normalized difference vegetation index (NDVI) images of Africa for the years 1986 to 1988. They found the first component was always highly correlated with NDVI regardless of season, while the second, third, and fourth components related to seasonal changes in NDVI. Almeida-Filho and Shimabukuro (2002) used PCA to identify degraded areas in the Amazon. PCA was used to create Wisconsin's statewide land-cover map as part of the U.S. Geological Survey's Gap Analysis Program (Reese et al., 2002). Mitternicht and Zinck (2003) found PCA valuable for delineating saline from nonsaline soils in remote sensor data.

There are other uses for principal components analysis. For example, Gillespie (1992) used PCA to perform decorrela-

tion contrast stretching of multispectral thermal infrared data. Du et al. (2002) used PCA to find linear relationships among multitemporal images of the same area. The information was then used to radiometrically normalize the data for change detection purposes.

 Vegetation Transformations (Indices)

The goal of global agriculture production and the grain sector of most economies is to feed 6 billion people (Kogan, 2001). In addition, one of the primary interests of Earth-observing systems is to study the role of terrestrial vegetation in large-scale global processes (e.g., the carbon cycle) with the goal of understanding how the Earth functions as a system. Both these endeavors require an understanding of the global distribution of vegetation types as well as their biophysical and structural properties and spatial/temporal variations (TBRS, 2003).

Biophysical measurements that document vegetation type, productivity, and functional health are needed for land resource management, combating deforestation and desertification, and promoting sustainable agriculture and rural development (Jensen et al., 2002). Table 1-1 identifies numerous variables that may be sensed if we want to monitor vegetation type, condition, and change through time. Fortunately, we now have more sensitive remote sensing systems (e.g., MODIS, MISR, ASTER, IKONOS, QuickBird, airborne hyperspectral sensors) and improved vegetation indices to monitor the health and productivity of vegetated ecosystems. Scores of vegetation indices have been developed that function as surrogates for important biophysical vegetation parameters. They may be applied to local problems or to global land cover assessments.

Before presenting the various vegetation indices, however, it is important to understand how the indices are related to various leaf physiological properties. Therefore, we will first discuss the dominant factors controlling leaf reflectance. This will help us appreciate why the linear combination of certain bands can be used in vegetation indices as a surrogate for leaf and/or canopy biophysical properties.

Dominant Factors Controlling Leaf Reflectance

The dominant factors controlling leaf reflectance in the region from 0.35 to 2.6 μm are summarized in Figure 8-36 (Gausmann et al., 1969; Peterson and Running, 1989).

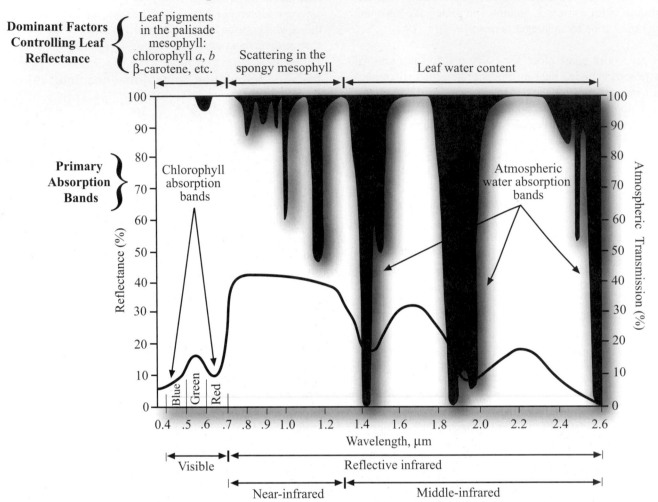

Figure 8-36   Spectral reflectance characteristics of healthy, green vegetation for the wavelength interval 0.4 – 2.6 μm. The dominant factors controlling leaf reflectance are the various leaf pigments in the palisade mesophyll (e.g., chlorophyll *a* and *b,* and β-carotene), the scattering of near-infrared energy in the spongy mesophyll, and the amount of water in the plant. The primary chlorophyll absorption bands occur at 0.43 – 0.45 μm and 0.65 – 0.66 μm in the visible region. The primary water atmospheric absorption bands occur at 0.97, 1.19, 1.45, 1.94, and 2.7 μm.

Visible Light Interaction with Pigments in the Palisade Mesophyll Cells

Photosynthesis is an energy-storing process that takes place in leaves and other green parts of plants in the presence of light. The photosynthetic process is:

$$6CO_2 + 6H_2O + \text{light energy} \rightarrow C_6H_{12}O_6 + 6O_2 . \quad (8\text{-}67)$$

Sunlight provides the energy that powers photosynthesis. The light energy is stored in a simple sugar molecule (glu-

cose) that is produced from carbon dioxide (CO_2) present in the air and water (H_2O) absorbed by the plant primarily through the root system. When the carbon dioxide and the water are combined and form a sugar molecule ($C_6H_{12}O_6$) in a chloroplast, oxygen gas (O_2) is released as a by-product. The oxygen diffuses out into the atmosphere.

The photosynthetic process begins when sunlight strikes *chloroplasts*, small bodies in the leaf that contain a green substance called chlorophyll. It is the process of food-making via photosynthesis that determines how a leaf and the

Hypothetical Leaf Cross-section

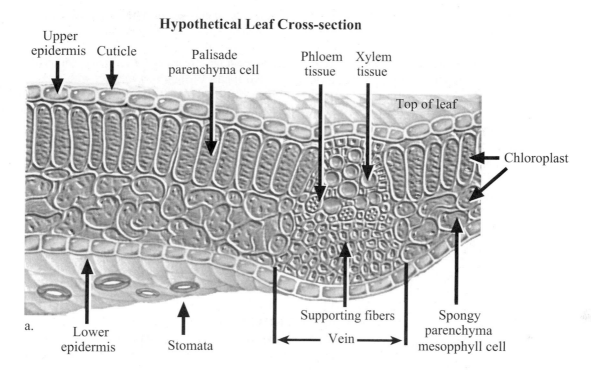

a.

Actual Leaf Cross-section

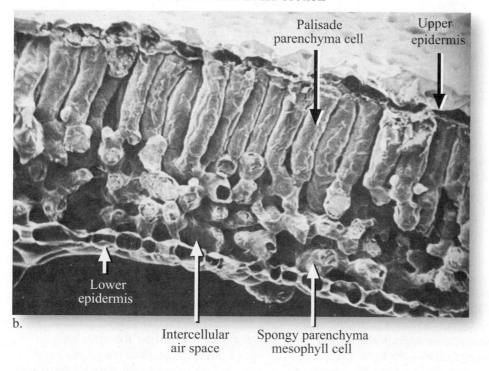

b.

Figure 8-37   a) Hypothetical cross-section of a healthy green leaf showing both the top and underside of the leaf. The chlorophyll pigments in the palisade parenchyma cells have a significant impact on the absorption and reflectance of visible light (blue, green, and red), while the spongy parenchyma mesophyll cells have a significant impact on the absorption and reflectance of near-infrared incident energy. b) Electron microscope image of a green leaf.

associated plant canopy components appear on remotely sensed images.

The leaf is the primary photosynthesizing organ. A cross-section of a typical green leaf is shown in Figure 8-37. The cell structure of leaves is highly variable depending upon species and environmental conditions during growth. Carbon dioxide enters the leaf from the atmosphere through tiny pores called *stomata* or *stoma,* located primarily on the underside of the leaf on the *lower epidermis.* Each stomata is surrounded by *guard cells* that swell or contract. When they swell, the stomata pore opens and allows carbon dioxide to enter the leaf. A typical sunflower leaf might have 2 million stomata, but they make up only about 1% of the leaf's surface area. Usually, there are more stomata on the bottom of a leaf; however, on some leaves the stomata are evenly distributed on both the upper and lower epidermis.

The top layer of leaf *upper epidermis* cells has a *cuticular* surface that diffuses, but reflects very little, light. It is usually only 3 to 5 μm thick with cell dimensions of approximately $18 \times 15 \times 20$ μm. It is useful to think of it as a waxy, translucent material similar to the cuticle at the top of your fingernail. Leaves of many plants that grow in bright sunlight have a thick cuticle that can filter out some light and guard against excessive plant water loss. Conversely, some plants such as ferns and some shrubs on the forest floor must survive in shaded conditions. The leaves of many of these plants have a thin cuticle so that the plant can collect as much of the dim sunlight as possible for photosynthesis.

Many leaves in direct sunlight have hairs growing out of the upper (and lower) epidermis, causing them to feel fuzzy. These hairs can be beneficial, as they reduce the intensity of the incident sunlight to the plant. Nevertheless, much of the visible and near-infrared wavelength energy is transmitted through the cuticle and upper epidermis to the palisade parenchyma mesophyll cells and spongy parenchyma mesophyll cells below.

Photosynthesis occurs inside the typical green leaf in two kinds of food-making cells—*palisade parenchyma* and *spongy parenchyma* mesophyll cells. Most leaves have a distinct layer of long palisade parenchyma cells in the upper part of the mesophyll and more irregularly shaped, loosely arranged spongy parenchyma cells in the lower part of the mesophyll. The palisade cells tend to form in the portion of the mesophyll toward the side from which the light enters the leaf. In most horizontal (planophile) leaves the palisade cells are toward the upper surface, but in leaves that grow nearly vertical (erectophile), the palisade cells may form from both sides. In some leaves the elongated palisade cells are absent

and only spongy parenchyma cells will exist within the mesophyll.

The cellular structure of the leaf is large compared to the wavelengths of light that interact with it. Palisade cells are typically $15 \times 15 \times 60$ μm, while spongy mesophyll parenchyma cells are smaller. The palisade parenchyma mesophyll plant cells contain chloroplasts with chlorophyll pigments.

The chloroplasts are generally 5 to 8 μm in diameter and about 1 μm wide. As many as 50 chloroplasts may be present in each parenchyma cell. Within the chloroplasts are long slender *grana* strands (not shown) within which the chlorophyll is actually located (approximately 0.5 μm long and 0.05 μm in diameter). The chloroplasts are generally more abundant toward the upper side of the leaf in the palisade cells and hence account for the darker green appearance of the upper leaf surface compared with the bottom lighter surface.

A molecule, when struck by a photon of light, reflects some of the energy or it can absorb the energy and thus enter into a higher energy or excited state (refer to Chapter 6). Each molecule absorbs or reflects its own characteristic wavelengths of light. Molecules in a typical green plant have evolved to absorb wavelengths of light in the visible region of the spectrum (0.35 – 0.70 μm) very well and are called *pigments.* An *absorption spectrum* for a particular pigment describes the wavelengths at which it can absorb light and enter into an excited state. Figure 8-38a presents the absorption spectrum of pure chlorophyll pigments in solution. Chlorophyll *a* and *b* are the most important plant pigments absorbing blue and red light: chlorophyll *a* at wavelengths of 0.43 and 0.66 μm and chlorophyll *b* at wavelengths of 0.45 and 0.65 μm (Farabee, 1997). A relative lack of absorption in the wavelengths between the two chlorophyll absorption bands produces a trough in the absorption efficiency at approximately 0.54 μm in the green portion of the electromagnetic spectrum (Figure 8-38a). Thus, it is the relatively lower absorption of green wavelength light (compared to blue and red light) by the leaf that causes healthy green foliage to appear green to our human eyes.

There are other pigments in the palisade mesophyll cells that are usually masked by the abundance of chlorophyll pigments. For example, there are yellow *carotenes* and pale yellow *xanthophyll* pigments, with strong absorption primarily in the blue wavelength region. The β-carotene absorption spectra is shown in Figure 8-38b with its strong absorption band centered at about 0.45 μm. *Phycoerythrin* pigments may also be present in the leaf and absorb predominantly in the green region centered at about 0.55 μm, allowing blue

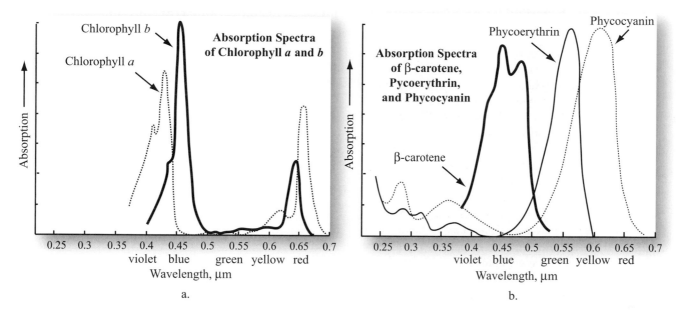

Figure 8-38   Spectral reflectance characteristics of healthy, green vegetation for the wavelength interval 0.4 – 2.6 μm. The dominant factors controlling leaf reflectance are the various leaf pigments in the palisade mesophyll (e.g., chlorophyll *a* and *b,* and β-carotene), the scattering of near-infrared energy in the spongy mesophyll, and the amount of water in the plant. The primary chlorophyll absorption bands occur at 0.43 – 0.45 μm and 0.65 – 0.66 μm in the visible region. The primary water atmospheric absorption bands occur at 0.97, 1.19, 1.45, 1.94, and 2.7 μm.

and red light to be reflected. *Phycocyanin* pigments absorb primarily in the green and red regions centered at about 0.62 μm, allowing much of the blue and some of the green light (i.e., the combination produces cyan) to be reflected (Figure 8-38b). Because chlorophyll *a* and *b* chloroplasts are also present and have a similar absorption band in this blue region, they tend to dominate and mask the effect of the other pigments. When a plant undergoes senescence in the fall or encounters stress, the chlorophyll pigment may disappear, allowing the carotenes and other pigments to become dominant. For example, in the fall, chlorophyll production ceases, causing the yellow coloration of the carotenes and other specific pigments in the tree foliage to become more visible to our eyes. In addition, some trees produce great quantities of *anthocyanin* in the fall, causing the leaves to appear bright red.

The two optimum spectral regions for sensing the chlorophyll absorption characteristics of a leaf are believed to be 0.45 to 0.52 μm and 0.63 to 0.69 μm (Figure 8-38a). The former region is characterized by strong absorption by carotenoids and chlorophylls, whereas the latter is characterized by strong chlorophyll absorption. Remote sensing of chlorophyll absorption within a canopy represents a fundamental biophysical variable useful for many biogeographical inves-

tigations. The absorption characteristics of plant canopies may be coupled with other remotely sensed data to identify vegetation stress, yield, and other hybrid variables. Thus, many remote sensing studies are concerned with monitoring what happens to the *photosynthetically active radiation* (PAR) as it interacts with individual leaves and/or the plant canopy. The use of high-spectral-resolution imaging spectrometers are particularly useful for measuring the absorption and reflectance characteristics of the photosynthetically active radiation.

It is important to understand the physiology of the plants under investigation and especially their pigmentation characteristics so that we can appreciate how a typical plant will appear when chlorophyll absorption starts to decrease, either due to seasonal senescence or environmental stress. As demonstrated, when a plant is under stress and/or chlorophyll production decreases, the lack of chlorophyll pigmentation typically causes the plant to absorb less in the chlorophyll absorption bands. Such plants have a much higher reflectance, particularly in the green and red portion of the spectrum, and therefore may appear yellowish or *chlorotic*. In fact, Carter (1993) suggests that increased reflectance in the visible spectrum is the most consistent leaf reflectance response to plant stress. Infrared reflectance responds consis-

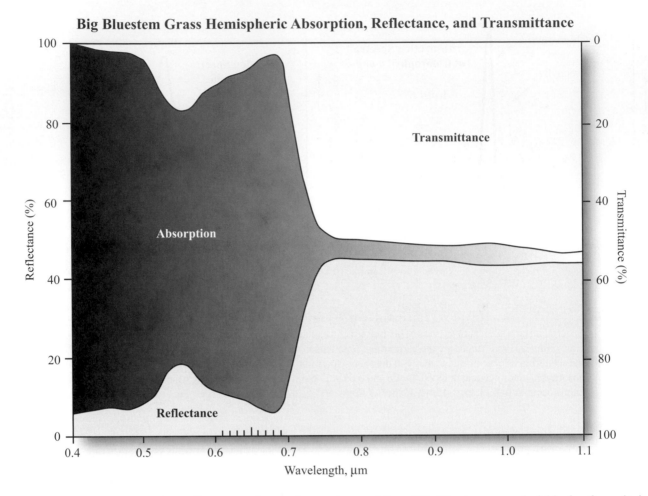

Figure 8-39 Hemispheric absorption, reflectance, and transmittance characteristics of Big Bluestem grass adaxial leaf surfaces obtained in the laboratory using a spectroradiometer. The reflectance and transmittance curves are almost mirror images of one another throughout the visible and near-infrared portions of the electromagnetic spectrum. The blue and red chlorophyll in plants absorbs much of the incident energy in the visible portion of the spectrum (0.4 – 0.7 µm) (after Walter-Shea and Biehl, 1990). Imaging spectrometers such as AVIRIS are capable of identifying small changes in the absorption and reflection characteristics of plants because the sensors often have channels that are only 10 nm apart, i.e., there could be 10 channels in the region from 0.6 to 0.7 µm (600 – 700 nm).

tently only when stress has developed sufficiently to cause severe leaf dehydration (to be demonstrated shortly).

Leaf spectral reflectance is most likely to indicate plant stress first in the sensitive 535 to 640 and 685 to 700 nm visible light wavelength ranges. Increased reflectance near 700 nm represents the often reported "blue shift of the red edge," i.e., the shift toward shorter wavelengths of the red-infrared transition curve that occurs in stressed plants when reflectance is plotted versus wavelength (Cibula and Carter, 1992). The shift toward shorter wavelengths in the region from 650 – 700 nm is particularly evident. Remote sensing within these spectrally narrow ranges may provide improved capability to

detect plant stress not only in individual leaves but for whole plants and perhaps for densely vegetated canopies (Carter, 1993; Carter et al., 1996).

Normal color film is sensitive to blue, green, and red wavelength energy. Color-infrared film is sensitive to green, red, and near-infrared energy after minus-blue (yellow) filtration. Therefore, even the most simple camera with color or color-infrared film and appropriate bandpass filtration (i.e., only certain wavelengths of light are allowed to pass) can be used to remotely sense differences in spectral reflectance caused by the pigments in the palisade parenchyma layer of cells in a typical leaf. However, to detect very subtle spectral reflectance differences in the relatively narrow bands suggested by Cibula and Carter (1992) and Carter et al. (1996), it may be necessary to use a high-spectral-resolution imaging spectro-radiometer that has very narrow bandwidths.

Near-Infrared Energy Interaction Within the Spongy Mesophyll Cells

In a typical healthy green leaf, the near-infrared reflectance increases dramatically in the region from 0.7 to 1.2 µm (Figure 8-39). Healthy green leaves absorb radiant energy very efficiently in the blue and red portions of the spectrum where incident light is required for photosynthesis. But immediately to the long-wavelength side of the red chlorophyll absorption band, why does the reflectance and transmittance of plant leaves increase so dramatically, causing the absorptance to fall to low values? This condition occurs throughout the near-infrared wavelength range where the direct sunlight incident on plants has the bulk of its energy. If plants absorbed this energy with the same efficiency as they do in the visible region, they could become much too warm and the proteins would be irreversibly denatured. As a result, plants have adapted so they do not use this massive amount of near-infrared energy and simply reflect it or transmit it through to underlying leaves or the ground.

The spongy mesophyll layer in a typical green leaf controls the amount of near-infrared energy that is reflected. The spongy mesophyll layer typically lies below the palisade mesophyll layer and is composed of many cells and intercellular air spaces, as shown in Figure 8-37. It is here that the oxygen and carbon dioxide exchange takes place for photosynthesis and respiration. In the near-infrared region, healthy green vegetation is generally characterized by high reflectance (40% – 60%), high transmittance (40% – 60%) through the leaf onto underlying leaves, and relatively low absorptance (5% – 10%). A healthy green leaf's reflectance and transmittance spectra throughout the visible and near-infrared portion of the spectrum are almost mirror images of one another.

The high diffuse reflectance of the near-infrared (0.7 – 1.2 µm) energy from plant leaves is due to the internal scattering at the cell wall–air interfaces within the leaf (Gausmann et al., 1969; Peterson and Running, 1989). A water vapor absorption band exists at 0.92 to 0.98 µm; consequently, the optimum spectral region for remote sensing in the near-infrared region is believed to be from 0.74 to 0.90 µm (Tucker, 1979).

The main reasons that healthy plant canopies reflect so much near-infrared energy are:

- the leaf already reflects 40% to 60% of the incident near-infrared energy from the spongy mesophyll (Figure 8-40), and

- the remaining 45% to 50% of the energy penetrates (i.e., is transmitted) through the leaf and can be reflected once again by leaves below it.

This phenomenon is called *leaf additive reflectance.* For example, consider the reflectance and transmission characteristics of the hypothetical two-layer plant canopy shown in Figure 8-40. Assume that leaf 1 reflects 50 percent of the incident near-infrared energy back into the atmosphere and that the remaining 50 percent of the near-infrared energy is transmitted through leaf 1 onto leaf 2. The transmitted energy then falls on leaf 2 where 50 percent again is transmitted (25 percent of the original) and 50 percent is reflected. The reflected energy then passes back through leaf 1 which allows half of that energy (or 12.5 percent of the original) to be transmitted and half reflected. The resulting total energy exiting leaf 1 in this two-layer example is 62.5 percent of the incident energy. Therefore, the greater the number of leaf layers in a healthy, mature canopy, theoretically the greater the infrared reflectance. Conversely, if the canopy is composed of only a single, sparse leaf layer, then the near-infrared reflectance will not be as great because the energy that is transmitted through the leaf layer may be absorbed by the ground cover beneath.

It is important to point out that changes in the near-infrared spectral properties of healthy green vegetation may provide information about plant senescence and stress. Photosynthesizing green vegetation typically exhibits strong chlorophyll absorption in the blue and red wavelength regions, an understandable increase in green reflectance, and approximately 76 percent reflectance in the near-infrared region. After a certain point, near-infrared reflectance may decrease as the leaves senesce. However, if the leaves were to dry out significantly during senescence, we would expect to see much higher reflectance in the near-infrared region.

Scientists have known since the 1960s that a *direct* relationship exists between response in the near-infrared region and various biomass measurements. Conversely, it has been shown that an *inverse* relationship exists between the response in the visible region, particularly the red, and plant biomass. The best way to appreciate this is to plot all of the pixels in a typical remote sensing scene in red and near-infrared reflectance space. For example, Figure 8-41a depicts where pixels in a typical agricultural scene are located in red and near-infrared multispectral feature space (i.e., within the gray area). Dry bare-soil fields and wet bare-soil fields in the

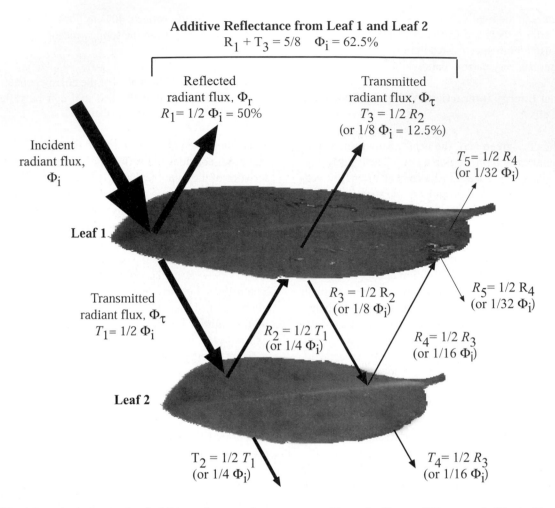

Additive Reflectance from Leaf 1 and Leaf 2

$R_1 + T_3 = 5/8$ $\Phi_i = 62.5\%$

Reflected radiant flux, Φ_r
$R_1 = 1/2\ \Phi_i = 50\%$

Transmitted radiant flux, Φ_τ
$T_3 = 1/2\ R_2$
(or $1/8\ \Phi_i = 12.5\%$)

Incident radiant flux, Φ_i

$T_5 = 1/2\ R_4$
(or $1/32\ \Phi_i$)

Leaf 1

Transmitted radiant flux, Φ_τ
$T_1 = 1/2\ \Phi_i$

$R_3 = 1/2\ R_2$
(or $1/8\ \Phi_i$)

$R_5 = 1/2\ R_4$
(or $1/32\ \Phi_i$)

$R_2 = 1/2\ T_1$
(or $1/4\ \Phi_i$)

$R_4 = 1/2\ R_3$
(or $1/16\ \Phi_i$)

Leaf 2

$T_2 = 1/2\ T_1$
(or $1/4\ \Phi_i$)

$T_4 = 1/2\ R_3$
(or $1/16\ \Phi_i$)

Figure 8-40   A hypothetical example of additive reflectance from a canopy with two leaf layers. Fifty percent of the incident radiant flux, Φ_i, to leaf 1 is reflected (R_1) and the other 50 percent is transmitted onto leaf 2 (T_1). Fifty percent of the radiant flux incident to leaf 2 is transmitted through leaf 2 (T_2); the other 50 percent is reflected toward the base of leaf 1 (R_2). Fifty percent of the energy incident at the base of leaf 1 is transmitted through it (T_3) while the remaining 50 percent (R_3) is reflected toward leaf 2 once again. At this point, an additional 12.5 percent (1/8) reflectance has been contributed by leaf 2, bringing the total reflected radiant flux to 62.5 percent. However, to be even more accurate, one would have to also take into account the amount of energy reflected from the base of leaf 1 (R_3) onto leaf 2, and the amount reflected from leaf 2 (R_4) and eventually transmitted through leaf 1 once again (T_5). This process would continue.

scene are located at opposite ends of the soil line. This means that wet bare soil would have very low red and near-infrared reflectance. Conversely, dry bare soil would probably have high red and near-infrared reflectance. As a vegetation canopy matures, it reflects more near-infrared energy while absorbing more red radiant flux for photosynthetic purposes. This causes the spectral reflectance of the pixel to move in a perpendicular direction away from the soil line. As biomass increases and as the plant canopy cover increases, the field's location in the red and near-infrared spectral space moves farther away from the soil line.

Figure 8-41b demonstrates how one agricultural pixel might move about in the red and near-infrared spectral space during a typical growing season. If the field was prepared properly, it would probably be located in the moist bare-soil region of the soil line with low red and near-infrared reflectance at the beginning of the growing season. After the crop emerges, it would depart from the soil line, eventually reaching complete canopy closure. At this point the reflected near-infrared radiant flux would be high and the red reflectance would be low. After harvesting, the pixel would probably be found once again on the soil line but perhaps in a drier condition.

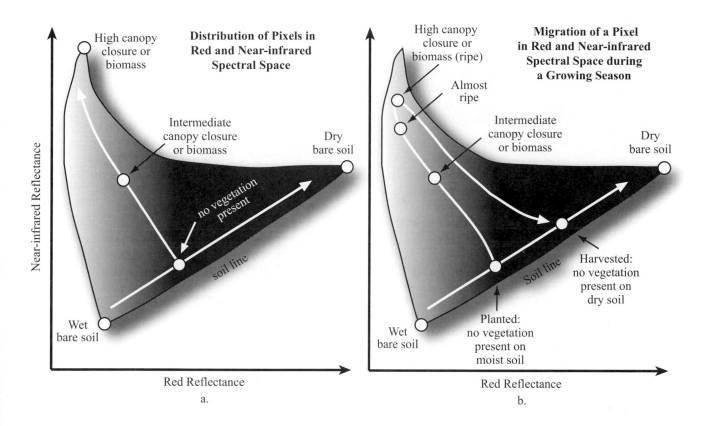

Figure 8-41   a) The distribution of all the pixels in a scene in red and near-infrared multispectral space is found in the gray shaded area. Wet and moist bare-soil fields are located along the soil line. The greater the biomass and/or crop canopy closure, the greater the near-infrared reflectance and the lower the red reflectance. This condition moves the pixel's spectral location a perpendicular direction away from the soil line. b) The migration of a single vegetated agricultural pixel in red and near-infrared multispectral space during a growing season is shown. After the crop emerges, it departs from the soil line, eventually reaching complete canopy closure. After harvesting, the pixel will be found on the soil line, but perhaps in a drier soil condition.

The relationship between red and near-infrared canopy reflectance has resulted in the development of numerous remote sensing vegetation indices and biomass-estimating techniques that utilize multiple measurements in the visible and near-infrared region (e.g., Lyon et al., 1998). The result is a linear combination that may be more highly correlated with biomass than either red or near-infrared measurement alone. Several of these algorithms are summarized in subsequent sections.

Middle-infrared Energy Interaction with Water in the Spongy Mesophyll

Plants require water to grow. A leaf obtains water that was absorbed by the plant's roots. The water travels from the roots, up the stem, and enters the leaf through the *petiole*. Veins carry water to the cells in the leaf. If a plant is watered so that it contains as much water as it can possibly hold at a given time, it is said to be fully *turgid*. Much of the water is found in the spongy mesophyll portion of the plant. If we forget to water the plant or rainfall decreases, the plant will contain an amount of water that is less than it can hold. This is called its *relative turgidity*. It would be useful to have a remote sensing instrument that was sensitive to how much water was in a plant leaf. Remote sensing in the middle-infrared, thermal infrared, and passive microwave portion of the electromagnetic spectrum can provide such information to a limited extent.

Liquid water in the atmosphere creates five major absorption bands in the near-infrared through middle-infrared portions of the electromagnetic spectrum at 0.97, 1.19, 1.45, 1.94, and 2.7 μm (Figure 8-36). The fundamental vibrational water-absorption band at 2.7 μm is the strongest in this part of the spectrum (there is also one in the thermal infrared region at 6.27 μm). However, there is also a strong relationship between the reflectance in the middle-infrared region from 1.3 to 2.5 μm and the amount of water present in the leaves of a plant canopy. Water in plants absorbs incident energy between the absorption bands with increasing strength at longer wavelengths. In these middle-infrared wavelengths, vegetation reflectance peaks at about 1.6 and 2.2 μm, between the major atmospheric water absorption bands (Figure 8-36).

Water is a good absorber of middle-infrared energy, so the greater the turgidity of the leaves, the lower the middle-infrared reflectance. Conversely, as the moisture content of leaves decreases, reflectance in the middle-infrared region increases substantially. As the amount of plant water in the intercellular air spaces decreases, the incident middle-infrared energy becomes more intensely scattered at the interface of the intercellular walls, resulting in greater middle-infrared reflectance from the leaf. For example, consider the spectral reflectance of magnolia leaf samples at five different moisture conditions displayed over the wavelength interval from 0.4 to 2.5 μm (Figure 8-42). The middle-infrared wavelength intervals from about 1.5 to 1.8 μm and from 2.1 to 2.3 μm appear to be more sensitive to changes in the moisture content of the plants than the visible or near-infrared portions of the spectrum (i.e., the y-axis distance between the spectral reflectance curves is greater as the moisture content decreases). Also note that substantive changes in the visible reflectance curves (0.4 – 0.7 μm) did not begin to appear until the plant moisture in the leaves decreased to about 50 percent. When the relative water content of the plant decreases to 50 percent, almost any portion of the visible, near- and middle-infrared regions might provide some valuable spectral reflectance information.

Leaf reflectance in the middle-infrared region is inversely related to the absorptance of a layer of water approximately 1 mm in depth (Carter, 1991). The degree to which incident solar energy in the middle-infrared region is absorbed by vegetation is a function of the total amount of water present in the leaf and the leaf thickness. If proper choices of sensors and spectral bands are made, it is possible to monitor the relative turgidity in plants.

Most optical remote sensing systems (except radar) are generally constrained to function in the wavelength intervals from 0.3 to 1.3, 1.5 to 1.8, and 2.0 to 2.6 μm due to the strong atmospheric water absorption bands at 1.45, 1.94, and 2.7 μm. Fortunately, as demonstrated in Figure 8-36, there is a strong "carry-over" sensitivity to water content in the 1.5 to 1.8 and 2.0 to 2.6 μm regions adjacent to the major water absorption bands. This is the reason that the Landsat Thematic Mapper (4 and 5) and Landsat 7 Enhanced Thematic

Mapper Plus (ETM⁺) were made sensitive to two bands in this region: band 5 (1.55 – 1.75 μm) and band 7 (2.08 – 2.35 μm). The 1.55 to 1.75 μm middle-infrared band has consistently demonstrated a sensitivity to canopy moisture content. For example, Pierce et al. (1990) found that this band and vegetation indices produced using it were correlated with canopy water stress in coniferous forests.

Much of the water in a plant is lost via transpiration. *Transpiration* occurs as the Sun warms the water inside the leaf, causing some of the water to change its state to water vapor, which escapes through the stomata. The following are several important functions that transpiration performs:

- It cools the inside of the leaf because the escaping water vapor contains heat.

- It keeps water flowing up from the roots, through the stem, to the leaves.

- It ensures a steady supply of dissolved minerals from the soil.

As molecules of water vapor at the top of the leaf in the tree are lost to transpiration, the entire column of water is pulled upward. Plants lose a considerable amount of water through transpiration each day. For example, a single corn plant can lose about 4 quarts (3.8 liters) of water on a very hot day. If the roots of the plant cannot replace this water, the leaves wilt, photosynthesis stops, and the plant dies. Thus, monitoring the moisture content of plant canopies, which is correlated with rates of transpiration, can provide valuable information about the health of a crop or stand of vegetation. Thermal infrared and passive microwave remote sensing have provided valuable plant canopy evapotranspiration information.

The most practical application of plant moisture information is the regional assessment of crop water conditions for irrigation scheduling, stress assessment, and yield modeling for agriculture, rangeland, and forestry management.

Vegetation Indices

Since the 1960s, scientists have extracted and modeled various vegetation biophysical variables using remotely sensed data. Much of this effort has involved the use of *vegetation indices*—dimensionless, radiometric measures that indicate relative abundance and activity of green vegetation, including leaf area index (LAI), percentage green cover, chlorophyll content, green biomass, and absorbed photosynthetic-

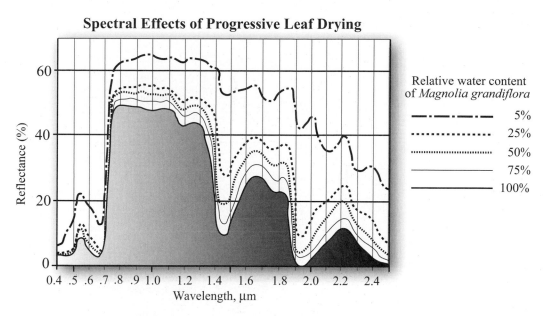

Figure 8-42 Reflectance response of a single magnolia leaf (*Magnolia grandiflora*) to decreased relative water content. As moisture content decreased, reflectance increased throughout the 0.4 to 2.5 μm region. However, the greatest increase occurred in the middle-infrared region from 1.3 to 2.5 μm (after Carter, 1991).

ally active radiation (APAR). A vegetation index should (Running et al., 1994; Huete and Justice, 1999):

- maximize sensitivity to plant biophysical parameters, preferably with a linear response in order that sensitivity be available for a wide range of vegetation conditions, and to facilitate validation and calibration of the index;

- normalize or model external effects such as Sun angle, viewing angle, and the atmosphere for consistent spatial and temporal comparisons;

- normalize internal effects such as canopy background variations, including topography (slope and aspect), soil variations, and differences in senesced or woody vegetation (nonphotosynthetic canopy components); and

- be coupled to some specific measurable biophysical parameter such as biomass, LAI, or APAR as part of the validation effort and quality control.

There are many vegetation indices. A few of the most widely adopted are summarized in Table 8-9. Many are functionally equivalent (redundant) in information content (Perry and Lautenschlager, 1984) and some provide unique biophysical information (Qi et al., 1995). It is useful to review the historical development of the main indices and provide information about recent advances in index development. Detailed

summaries are found in Running et al. (1994) and Lyon et al. (1998). Many of the indices make use of the inverse relationship between red and near-infrared reflectance associated with healthy green vegetation (Figure 8-43a).

Simple Ratio — SR

Cohen (1991) suggests that the first true vegetation index was the *Simple Ratio* (SR), which is the ratio of red reflected radiant flux (ρ_{red}) to near-infrared radiant flux (ρ_{nir}) as described in Birth and McVey (1968):

$$SR = \frac{\rho_{red}}{\rho_{nir}}. \tag{8-68}$$

Normalized Difference Vegetation Index — NDVI

Rouse et al. (1974) developed the generic *Normalized Difference Vegetation Index* (NDVI):

$$NDVI = \frac{\rho_{nir} - \rho_{red}}{\rho_{nir} + \rho_{red}}. \tag{8-69}$$

The NDVI is functionally equivalent to the simple ratio; that is, there is no scatter in an SR vs. NDVI plot and each SR value has a fixed NDVI value. When we plot the mean NDVI and SR values for various biomes we find that the NDVI

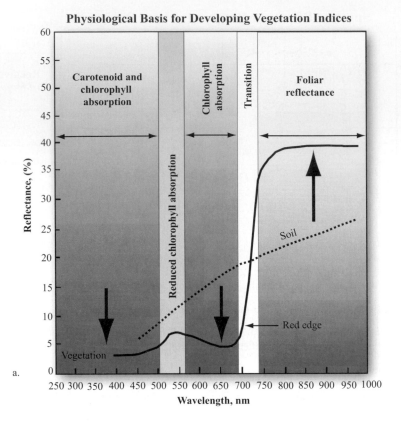

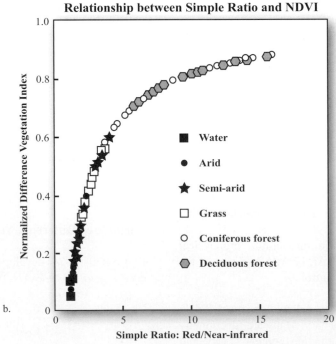

Figure 8-43   a) The physiological basis for developing vegetation indices. Typical spectral reflectance characteristics for healthy green grass and bare dry soil for the wavelength interval from 250 to 1,000 nm. b) The NDVI is a normalized ratio of the near-infrared and red bands. The NDVI is functionally equivalent to and is a nonlinear transform of the simple ratio (adapted from Huete et al., 2002b).

Table 8-9.   Selected remote sensing vegetation indices.

Vegetation Index	Equation	References
Simple Ratio (SR)	$SR = \dfrac{\rho_{red}}{\rho_{nir}}$	Birth and McVey, 1968 Colombo et al., 2003
Normalized Difference Vegetation Index (NDVI)	$NDVI = \dfrac{\rho_{nir} - \rho_{red}}{\rho_{nir} + \rho_{red}}$	Rouse et al., 1974 Deering et al., 1975 Huete et al., 2002a
Kauth-Thomas Transformation Brightness Greenness Yellow stuff Non-such Brightness Greenness Wetness	*Landsat Multispectral Scanner (MSS)* $B = 0.332\,MSS1 + 0.603\,MSS2 + 0.675\,MSS3 + 0.262\,MSS4$ $G = -0.283\,MSS1 - 0.660\,MSS2 + 0.577\,MSS3 + 0.388\,MSS4$ $Y = -0.899\,MSS1 + 0.428\,MSS2 + 0.076\,MSS3 - 0.041\,MSS4$ $N = -0.016\,MSS1 + 0.131\,MSS2 - 0.452\,MSS3 + 0.882\,MSS4$ *Landsat Thematic Mapper (TM)* $B = 0.2909\,TM1 + 0.2493\,TM2 + 0.4806\,TM3 +$ $\quad 0.5568\,TM4 + 0.4438\,TM5 + 0.1706\,TM7$ $G = -0.2728\,TM1 - 0.2174\,TM2 - 0.5508\,TM3 +$ $\quad 0.7221\,TM4 + 0.0733\,TM5 - 0.1648\,TM7$ $W = 0.1446\,TM1 + 0.1761\,TM2 + 0.3322\,TM3 +$ $\quad 0.3396\,TM4 - 0.6210\,TM5 - 0.4186\,TM7$	Kauth and Thomas, 1976 Kauth et al., 1979 Crist and Kauth, 1986 Lunetta et al., 2002 Price et al., 2002 Rogan et al., 2002
Infrared Index (II)	$II = \dfrac{NIR_{TM4} - MidIR_{TM5}}{NIR_{TM4} + MidIR_{TM5}}$	Hardisky et al., 1983
Perpendicular Vegetation Index (PVI)	$PVI = \sqrt{\left(0.355\,MSS4 - 0.149\,MSS2\right)^2 + \left(0.355\,MSS2 - 0.852\,MSS4\right)^2}$	Richardson and Wiegand, 1977
Greenness Above Bare Soil (GRABS)	$GRABS = G - 0.09178\,B + 5.58959$	Hay et al., 1979
Moisture Stress Index (MSI)	$MSI = \dfrac{MidIR_{TM5}}{NIR_{TM4}}$	Rock et al., 1986
Leaf Relative Water Content Index (LWCI)	$LWCI = \dfrac{-\log[1 - (NIR_{TM4} - MidIR_{TM5_{ft}})]}{-\log[1 - NIR_{TM4} - MidIR_{TM5_{ft}}]}$	Hunt et al., 1987
MidIR Index	$MidIR = \dfrac{MidIR_{TM5}}{NIR_{TM7}}$	Musick and Pelletier, 1988
Soil Adjusted Vegetation Index (SAVI) and Modified SAVI (MSAVI)	$SAVI = \dfrac{(1 + L)(\rho_{nir} - \rho_{red})}{\rho_{nir} + \rho_{red} + L}$	Huete, 1988; Huete and Liu, 1994; Running et al., 1994; Qi et al., 1995
Atmospherically Resistant Vegetation Index (ARVI)	$ARVI = \left(\dfrac{\rho^{*}_{nir} - \rho^{*}_{rb}}{\rho^{*}_{nir} + \rho^{*}_{rb}}\right)$	Kaufman and Tanre, 1992; Huete and Liu, 1994
Soil and Atmospherically Resistant Vegetation Index (SARVI)	$SARVI = \dfrac{\rho^{*}_{nir} - \rho^{*}_{rb}}{\rho^{*}_{nir} + \rho^{*}_{rb} + L}$	Huete and Liu, 1994; Running et al., 1994

Table 8-9.   Selected remote sensing vegetation indices.

Vegetation Index	Equation	References
Enhanced Vegetation Index (EVI)	$EVI = G\dfrac{\rho^*_{nir} - \rho^*_{red}}{\rho^*_{nir} + C_1\rho^*_{red} - C_2\rho^*_{blue} + L}(1+L)$	Huete et al., 1997 Huete and Justice, 1999 Huete et al., 2002a
New Vegetation Index (NVI)	$NVI = \dfrac{\rho_{777} - \rho_{747}}{\rho_{673}}$	Gupta et al., 2001
Aerosol Free Vegetation Index (AFRI)	$AFRI_{1.6\mu m} = \dfrac{(\rho_{nir} - 0.66\rho_{1.6\mu m})}{(\rho_{nir} - 0.66\rho_{1.6\mu m})}$ $AFRI_{2.1\mu m} = \dfrac{(\rho_{nir} - 0.5\rho_{2.1\mu m})}{(\rho_{nir} + 0.5\rho_{2.1\mu m})}$	Karnieli et al., 2001
Triangular Vegetation Index (TVI)	$TVI = 0.5(120(\rho_{nir} - \rho_{green})) - 200(\rho_{red} - \rho_{green})$	Broge and Leblanc, 2000
Reduced Simple Ratio (RSR)	$RSR = \dfrac{\rho_{nir}}{\rho_{red}}\left(1 - \dfrac{\rho_{swir} - \rho_{swirmin}}{\rho_{swirmax} + \rho_{swirmin}}\right)$	Chen et al., 2002
Ratio TCARI /OSAVI	$TCARI = 3\left[(\rho_{700} - \rho_{670}) - 0.2(\rho_{700} - \rho_{550})\left(\dfrac{\rho_{700}}{\rho_{670}}\right)\right]$ $OSAVI = \dfrac{(1+0.16)(\rho_{800} - \rho_{670})}{(\rho_{800} + \rho_{670} + 0.16)}$ $\dfrac{TCARI}{OSAVI}$	Kim et al., 1994 Rondeaux et al., 1996 Daughtry et al., 2000 Haboudane et al., 2002
Visible Atmospherically Resistant Index (VARI)	$VARI_{green} = \dfrac{\rho_{green} - \rho_{red}}{\rho_{green} + \rho_{red} - \rho_{blue}}$	Gitelson et al., 2002
Normalized Difference Built-up Index (NDBI)	$NDBI = \dfrac{MidIR_{TM5} - NIR_{TM4}}{MidIR_{TM5} + NIR_{TM4}}$ $built\text{-}up_{area} = NDBI - NDVI$	Zha et al., 2003

approximates a nonlinear transform of the simple ratio (Figure 8-43b) (Huete et al., 2002b). The NDVI is an important vegetation index because:

- Seasonal and inter-annual changes in vegetation growth and activity can be monitored.

- The ratioing reduces many forms of *multiplicative* noise (Sun illumination differences, cloud shadows, some atmospheric attenuation, some topographic variations) present in multiple bands of multiple-date imagery.

Disadvantages of the NDVI include (Huete et al., 2002a):

- The ratio-based index is nonlinear and can be influenced by *additive* noise effects such as atmospheric path radiance (Chapter 6).

- It exhibits scaling problems with saturated signals often encountered in high-biomass conditions. For example, Figure 8-43b reveals that the NDVI dynamic range is stretched in favor of low-biomass conditions and compressed in high-biomass, forested regions. The opposite is true for the Simple Ratio, in which most of the dynamic range encompasses the high-biomass forests with little variation reserved for the lower-biomass regions (grassland, semi-arid, and arid biomes).

**NDVI Image of Charleston, SC,
Derived from Landsat Thematic Mapper Data**

Figure 8-44 Normalized difference vegetation index (NDVI) image of Charleston, SC derived using Landsat Thematic Mapper bands 3 and 4.

• It is very sensitive to canopy background variations (e.g., soil visible through the canopy). NDVI values are particularly high with darker-canopy backgrounds.

The NDVI index was widely adopted and applied to the original Landsat MSS digital remote sensor data. Deering et al. (1975) added 0.5 to the NDVI to avoid negative values and took the square root of the result to stabilize the variance. This index is referred to as the *Transformed Vegetation Index* (TVI). These three indices (SR, NDVI, and TVI) respond to changes in the amount of green biomass. Many scientists continue to use the NDVI (e.g., Miura et al., 2001; Trishcenko et al., 2002). Two of the standard MODIS land products are 16-day composite NDVI datasets of the world at a spatial resolution of 500 m and 1 km (Huete et al., 2002a). An NDVI image derived from Charleston Landsat Thematic Mapper dataset is shown in Figure 8-44.

Kauth-Thomas Tasseled Cap Transformation

Kauth and Thomas (1976) produced an orthogonal transformation of the original Landsat MSS data space to a new four-dimensional feature space. It was called the *tasseled cap* or *Kauth-Thomas transformation*. It created four new axes: the soil brightness index (B), greenness vegetation index (G), yellow stuff index (Y), and non-such (N). The names attached

to the new axes indicate the characteristics the indices were intended to measure. The coefficients are (Kauth et al., 1979):

$$B = 0.332\,MSS1 + 0.603\,MSS2 + 0.675\,MSS3 + 0.262\,MSS4 \qquad (8\text{-}70)$$

$$G = -0.283\,MSS1 - 0.660\,MSS2 + 0.577\,MSS3 + 0.388\,MSS4 \qquad (8\text{-}71)$$

$$Y = -0.899\,MSS1 + 0.428\,MSS2 + 0.076\,MSS3 - 0.041\,MSS4 \qquad (8\text{-}72)$$

$$N = -0.016\,MSS1 + 0.131\,MSS2 - 0.452\,MSS3 + 0.882\,MSS4 . \qquad (8\text{-}73)$$

Crist et al. (1986) derived the visible, near-infrared, and middle-infrared coefficients for transforming Landsat Thematic Mapper imagery into brightness, greenness, and wetness variables:

$$B = 0.2909\,TM1 + 0.2493\,TM2 + 0.4806\,TM3 + \qquad (8\text{-}74)$$
$$0.5568\,TM4 + 0.4438\,TM5 + 0.1706\,TM7$$

$$G = -0.2728\,TM1 - 0.2174\,TM2 - 0.5508\,TM3 + \qquad (8\text{-}75)$$
$$0.7221\,TM4 + 0.0733\,TM5 - 0.1648\,TM7$$

$$W = 0.1446\,TM1 + 0.1761\,TM2 + 0.3322\,TM3 + \qquad (8\text{-}76)$$
$$0.3396\,TM4 - 0.6210\,TM5 - 0.4186\,TM7 .$$

It is called the tasseled cap transformation because of its cap shape (Figures 8-45a–c). Crist and Cicone (1984) identified a third component that is related to soil features, including moisture status (Figure 8-45d). Thus, an important source of soil information is available through the inclusion of the middle-infrared bands of the Thematic Mapper (Price et al., 2002).

The Charleston Landsat TM scene is transformed into brightness, greenness, and wetness (moisture) content images in Figure 8-46 based on the use of the TM tasseled cap coefficients (Equations 8-74 through 8-76). Urbanized areas are particularly evident in the brightness image. The greater the biomass, the brighter the pixel value in the greenness image. The wetness image provides subtle information concerning the moisture status of the wetland environment. As expected, the greater the moisture content, the brighter the response. As expected, water bodies are bright. Crist (1984) and Crist and Kauth (1986) identified the fourth tasseled cap parameter as being haze.

The tasseled cap transformation is a global vegetation index. Theoretically, it can be used anywhere in the world to disaggregate the amount of soil brightness, vegetation, and moisture content in individual pixels in a Landsat MSS or Thematic Mapper image. Practically, however, it is better to compute the coefficients based on local conditions. Jackson

Characteristics of the Kauth-Thomas Tasseled Cap Transformation

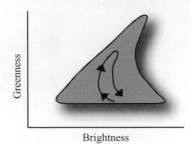

a. Crop development in the tasseled cap brightness-greenness transformation.

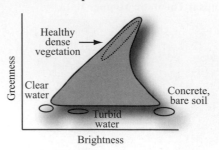

b. Location of land cover when plotted in the brightness-greenness spectral space.

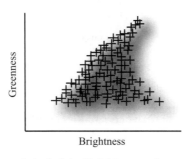

c. Actual plot of brightness and greenness values for an agricultural area. The shape of the distribution looks like a cap.

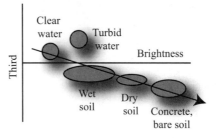

d. Approximate direction of moisture variation in the plane of soils. The arrow points in the direction of less moisture.

Figure 8-45   Characteristics of the Kauth-Thomas tasseled cap transformation (adapted from Crist, 1983; Crist and Cicone, 1984; Crist and Kauth, 1986). The tasseled cap transformation is applied correctly if the tasseled cap's base is parallel with the brightness axis.

(1983) provides a computer program for this purpose. The Kauth-Thomas tasseled cap transformation continues to be widely used (e.g., Lunetta et al., 2002; Price et al., 2002; Rogan et al., 2002).

Infrared Index — II

Hardisky et al. (1983) found that an *Infrared Index* (II), based on Landsat TM near- and middle-infrared bands,

$$ II = \frac{NIR_{TM4} - MidIR_{TM5}}{NIR_{TM4} + MidIR_{TM5}}, \tag{8-77} $$

was more sensitive to changes in plant biomass and water stress than NDVI for wetland studies.

Perpendicular Vegetation Index — PVI

Richardson and Wiegand (1977) used the perpendicular distance to the "soil line" as an indicator of plant development. The "soil line," a two-dimensional analog of the Kauth-Tho-

mas soil brightness index, was estimated by linear regression. The *Perpendicular Vegetation Index* (PVI) based on MSS band 4 data was:

$$ PVI = \sqrt{\left(0.355\,MSS4 - 0.149\,MSS2\right)^2 + \left(0.355\,MSS2 - 0.852\,MSS4\right)^2}\,. \tag{8-78} $$

Greenness Above Bare Soil — GRABS

Hay et al. (1979) proposed a vegetation index called *Greenness Above Bare Soil* (GRABS):

$$ GRABS = G - 0.09178B + 5.58959\,. \tag{8-79} $$

The calculations were made using the Kauth-Thomas transformation greenness (*G*) and soil brightness index (*B*) applied to Sun-angle and haze-corrected Landsat MSS data. The resulting index is similar to the Kauth-Thomas greenness vegetation index since the contribution of soil brightness index (*B*) is less than 10 percent of GVI.

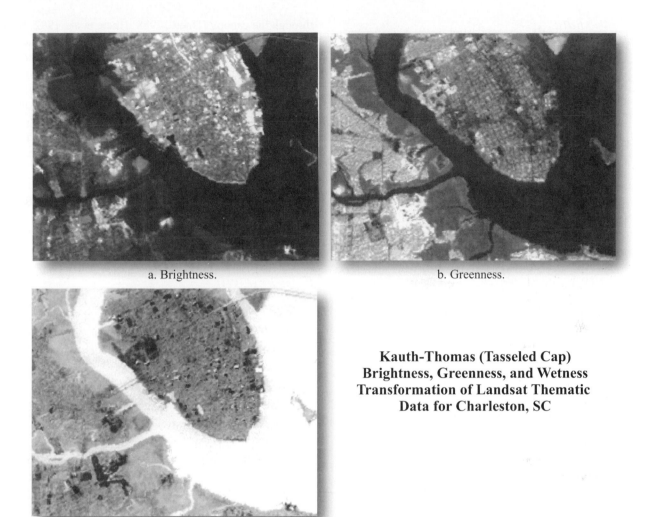

a. Brightness.

b. Greenness.

Kauth-Thomas (Tasseled Cap)
Brightness, Greenness, and Wetness
Transformation of Landsat Thematic
Data for Charleston, SC

c. Wetness.

Figure 8-46   Brightness, greenness, and wetness images derived by applying Kauth-Thomas tasseled cap transformation coefficients to the Charleston, SC Thematic Mapper data (6 bands).

Moisture Stress Index — MSI

Rock et al. (1986) utilized a *Moisture Stress Index* (MSI):

$$\text{MSI} = \frac{MidIR_{TM5}}{NIR_{TM4}} \qquad (8\text{-}80)$$

based on the Landsat Thematic Mapper near-infrared and middle-infrared bands.

Leaf Water Content Index — LWCI

Hunt et al. (1987) developed the *Leaf Water Content Index* (LWCI) to assess water stress in leaves:

$$\text{LWCI} = \frac{-\log[1 - (NIR_{TM4} - MidIR_{TM5})]}{-\log[1 - (NIR_{TM4_{ft}} - MidIR_{TM5_{ft}})]}, \qquad (8\text{-}81)$$

where *ft* represents reflectance in the specified bands when leaves are at their maximum relative water content (fully turgid; RWC) defined as:

$$\text{RWC} = \frac{\text{field weight} - \text{oven dry weight}}{\text{turgid weight} - \text{oven dry weight}} \times 100 . \quad (8\text{-}82)$$

MidIR Index

Musick and Pelletier (1988) demonstrated a strong correlation between the *MidIR Index* and soil moisture:

$$\text{MidIR} = \frac{MidIR_{TM5}}{NIR_{TM7}} . \qquad (8\text{-}83)$$

Soil Adjusted Vegetation Index — SAVI

The utility of the normalized difference vegetation index and related indices for satellite and airborne assessment of the Earth's vegetation cover has been demonstrated for almost three decades. The time series analysis of seasonal NDVI data have provided a method of estimating net primary production over varying biome types, of monitoring phenological patterns of the Earth's vegetated surface, and of assessing the length of the growing season and dry-down periods (Huete and Liu, 1994; Ramsey et al., 1995). For example, global vegetation analysis was initially based on linearly regressing NDVI values (derived from AVHRR, Landsat MSS, Landsat TM, and SPOT HRV data) with *in situ* measurements of LAI, APAR, percent cover, and/or biomass. This empirical approach revolutionized global science land-cover biophysical analysis in just one decade (Running et al., 1994). Unfortunately, studies have found that the empirically derived NDVI products can be unstable, varying with soil color and moisture conditions, bidirectional reflectance distribution function (BRDF) effects (refer to Chapter 1), atmospheric conditions, and the presence of dead material in the canopy itself (Qi et al., 1995). For example, Goward et al. (1991) found errors of $\pm 50\%$ in NDVI images used for global vegetation studies derived from the NOAA Global Vegetation Index product. What is needed are globally accurate NDVI-related products that do not need to be calibrated by *in situ* measurement within each geographic area yet will remain constant under changing atmospheric and soil background conditions (Huete and Justice, 1999).

Therefore, emphasis has been given to the development of improved vegetation indices that take advantage of calibrated sensor systems such as the Moderate Resolution Imaging Spectrometer (MODIS) (Running et al., 1994). Although the NDVI has been shown to be useful in estimating vegetation properties, many important external and internal influences restrict its global utility. The improved indices typically incorporate a soil background and/or atmospheric adjustment factor.

The *Soil Adjusted Vegetation Index* (SAVI) is:

$$\text{SAVI} = \frac{(1+L)(\rho_{nir} - \rho_{red})}{\rho_{nir} + \rho_{red} + L} \qquad (8\text{-}84)$$

where L is a canopy background adjustment factor that accounts for differential red and near-infrared extinction

through the canopy (Huete, 1988; Huete et al., 1992; Karnieli, et al., 2001). An L value of 0.5 in reflectance space was found to minimize soil brightness variations and eliminate the need for additional calibration for different soils (Huete and Liu, 1994). The utility of SAVI for minimizing the soil "noise" inherent in the NDVI has been corroborated in several studies (Bausch, 1993). Qi et al. (1995) developed a modified SAVI, called MSAVI, that uses an iterative, continuous L function to optimize soil adjustment and increase the dynamic range of SAVI.

Atmospherically Resistant Vegetation Index — ARVI

SAVI was made less sensitive to atmospheric effects by normalizing the radiance in the blue, red, and near-infrared bands. This became the *Atmospherically Resistant Vegetation Index* (ARVI):

$$\text{ARVI} = \left(\frac{\rho^{*}_{nir} - \rho^{*}_{rb}}{\rho^{*}_{nir} + \rho^{*}_{rb}} \right) \qquad (8\text{-}85)$$

where

$$\rho^{*}_{rb} = \rho^{*}_{red} - \gamma(\rho^{*}_{blue} - \rho^{*}_{red}) . \qquad (8\text{-}86)$$

The technique requires prior correction for molecular scattering and ozone absorption of the blue, red, and near-infrared remote sensor data, hence the term p^{*}. ARVI uses the difference in the radiance between the blue channel and the red channel to correct the radiance in the red channel and thus reduce atmospheric effects. Unless the aerosol model is known a priori, gamma (γ) is normally equal to 1.0 to minimize atmospheric effects. Kaufman and Tanre (1992) provide guidelines where different gammas might be used over continental, maritime, desert (e.g., the Sahel), or heavily vegetated areas.

Soil and Atmospherically Resistant Vegetation Index — SARVI

Huete and Liu (1994) integrated the L function from SAVI and the blue-band normalization in ARVI to derive a *Soil and Atmospherically Resistant Vegetation Index* (SARVI) that corrects for both soil and atmospheric noise, as would a modified SARVI (MSARVI):

$$\text{SARVI} = \frac{\rho^{*}_{nir} - \rho^{*}_{rb}}{\rho^{*}_{nir} + \rho^{*}_{rb} + L} \qquad (8\text{-}87)$$

and

$$\text{MSARVI} = \frac{2\rho^*_{nir} + 1 - \sqrt{\left(2\rho^*_{nir} + 1\right)^2 - \gamma(\rho^*_{nir} - \rho^*_{rb})}}{2}. \qquad (8\text{-}88)$$

Huete and Liu (1994) performed a sensitivity analysis on the original NDVI and improved vegetation indices (SAVI, ARVI, SARVI, MSARVI) and drew these conclusions:

- If there were a total atmospheric correction then there would mainly be "soil noise," and the SAVI and MSARVI would be the best equations to use and the NDVI and ARVI would be the worst.

- If there were a partial atmospheric correction to remove the Rayleigh and ozone components, then the best vegetation indices would be the SARVI and MSARVI, with the NDVI and ARVI being the worst.

- If there were no atmospheric correction at all (i.e., no Rayleigh, ozone, or aerosol correction), the SARVI would become slightly worse but still would have the least overall noise. The NDVI and ARVI would have the most noise and error.

Aerosol Free Vegetation Index — AFRI

Karnieli et al. (2001) found that under clear sky conditions the spectral bands centered on 1.6 and 2.1 μm are highly correlated with visible spectral bands centered on blue (0.469 μm), green (0.555 μm), and red (0.645 μm). Empirical linear relationships such as $\rho_{0.469\mu m} = 0.25\rho_{2.1\mu m}$, $\rho_{0.555\mu m} = 0.33\rho_{2.1\mu m}$ and $\rho_{0.645\mu m} = 0.66\rho_{1.6\mu m}$ were found to be statistically significant. Therefore, based on these and other relationships, two *Aerosol Free Vegetation Indices* (AFRI) were developed:

$$\text{AFRI}_{1.6\mu m} = \frac{(\rho_{nir} - 0.66\rho_{1.6\mu m})}{(\rho_{nir} + 0.66\rho_{1.6\mu m})} \text{ and} \qquad (8\text{-}89)$$

$$\text{AFRI}_{2.1\mu m} = \frac{(\rho_{nir} - 0.5\rho_{2.1\mu m})}{(\rho_{nir} + 0.5\rho_{2.1\mu m})}. \qquad (8\text{-}90)$$

Under clear sky conditions, the AFRIs (especially $\text{AFRI}_{2.1\mu m}$) result in values very similar to NDVI. However, if the atmospheric column contains smoke or sulfates then the AFRIs are superior to NDVIs. This is because the electromagnetic energy centered at 1.6 and 2.1 μm is able to penetrate the atmospheric column better than red wavelength energy used in the NDVI. Therefore, the AFRIs have a major application in assessing vegetation in the presence of smoke, anthropogenic pollution, and volcanic plumes. Limited success is expected in the case of dust due to presence of larger

particles that are similar in size to the wavelength, and therefore not transparent at 2.1 μm (Kaufman et al., 2000). The AFRIs can be implemented using any sensor that incorporates bands centered on 1.6 and/or 2.1 μm such as the Landsat TM, and ETM⁺, MODIS, ASTER, and the Japanese Earth Resource Satellite-Optical System (JERS-OPS), SPOT 4-Vegetation, and IRS-1C/D.

Enhanced Vegetation Index — EVI

The MODIS Land Discipline Group developed the *Enhanced Vegetation Index* (EVI) for use with MODIS data:

$$\text{EVI} = G\frac{\rho^*_{nir} - \rho^*_{red}}{\rho^*_{nir} + C_1\rho^*_{red} - C_2\rho^*_{blue} + L}(1 + L). \qquad (8\text{-}91)$$

The EVI is a modified NDVI with a soil adjustment factor, L, and two coefficients, C_1 and C_2, which describe the use of the blue band in correction of the red band for atmospheric aerosol scattering. The coefficients, C_1, C_2, and L, are empirically determined as 6.0, 7.5, and 1.0, respectively. G is a gain factor set to 2.5. This algorithm has improved sensitivity to high-biomass regions and has improved vegetation monitoring through a decoupling of the canopy background signal and a reduction in atmospheric influences (Huete et al., 1997; Huete and Justice, 1999; Huete et al., 2002). A comparison of MODIS-derived NDVI values and MODIS-derived EVI values for dense vegetation is shown in Figure 8-47 (Didan, 2002). The EVI has improved sensitivity to high biomass. Two MODIS vegetation indices, the MODIS NDVI and the EVI, are produced at 1×1 km and 500×500 m resolutions (and occasionally at 250×250 m) using 16-day compositing periods. A black-and-white MODIS EVI map of the Earth obtained starting on day 193 of 2003 is shown in Figure 8-48. A color version is found in **Color Plate 8-4**.

Triangular Vegetation Index — TVI

Broge and Leblanc (2000) developed a *Triangular Vegetation Index* (TVI), which describes the radiative energy absorbed by pigments as a function of the relative difference between red and near-infrared reflectance in conjunction with the magnitude of reflectance in the green region, where the light absorption by chlorophyll *a* and *b* is relatively insignificant. The index is calculated as the area of the triangle defined by the green peak, the chlorophyll absorption minimum, and the near-infrared shoulder in spectral space. It is based on the fact that both chlorophyll absorption causing a decrease of red reflectance and leaf tissue abundance causing increased near-infrared reflectance will increase the total area of the triangle. The TVI index

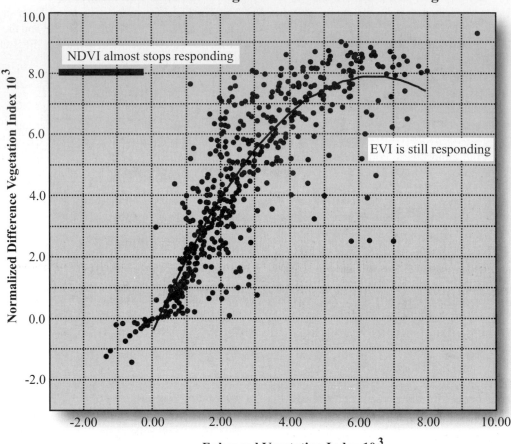

Figure 8-47 A comparison of the sensitivity of the MODIS-derived Normalized Difference Vegetation Index (NDVI) values and MODIS Enhanced Vegetation Index (EVI) values for dense vegetation (Didan, 2002).

encompasses the area spanned by the triangle ABC with the coordinates given in spectral space:

$$TVI = 0.5(120(\rho_{nir} - \rho_{green})) - 200(\rho_{red} - \rho_{green}) \quad (8\text{-}92)$$

where ρ_{green}, ρ_{red}, and ρ_{nir} are the reflectances centered at 0.55 μm, 0.67 μm, and 0.75 μm, respectively.

Reduced Simple Ratio — RSR

Chen et al. (2002) modified the simple ratio algorithm to include information from the short-wavelength infrared (SWIR) band found in the SPOT *Vegetation* sensor. They used the sensor to map the leaf area index of Canada. The *Reduced Simple Ratio* (RSR) is:

$$RSR = \frac{\rho_{nir}}{\rho_{red}} \left(1 - \frac{\rho_{swir} - \rho_{swirmin}}{\rho_{swirmax} + \rho_{swirmin}} \right) \quad (8\text{-}93)$$

where $\rho_{swirmin}$ and $\rho_{swirmax}$ are the minimum and maximum SWIR reflectance found in each image and defined as the 1% minimum and maximum cut-off points in the histograms of SWIR reflectance in a scene. The major advantages of RSR over SR are a) the difference between land-cover types is much reduced so that the accuracy for leaf area index retrieval for mixed cover types can be improved or a single LAI algorithm can be developed without resorting to coregistered land-cover maps as the first approximation, and b) the background (understory, litter, and soil) influence is suppressed using RSR because the SWIR band is most sensitive to the amount of vegetation containing liquid water in the background (Chen et al., 2002).

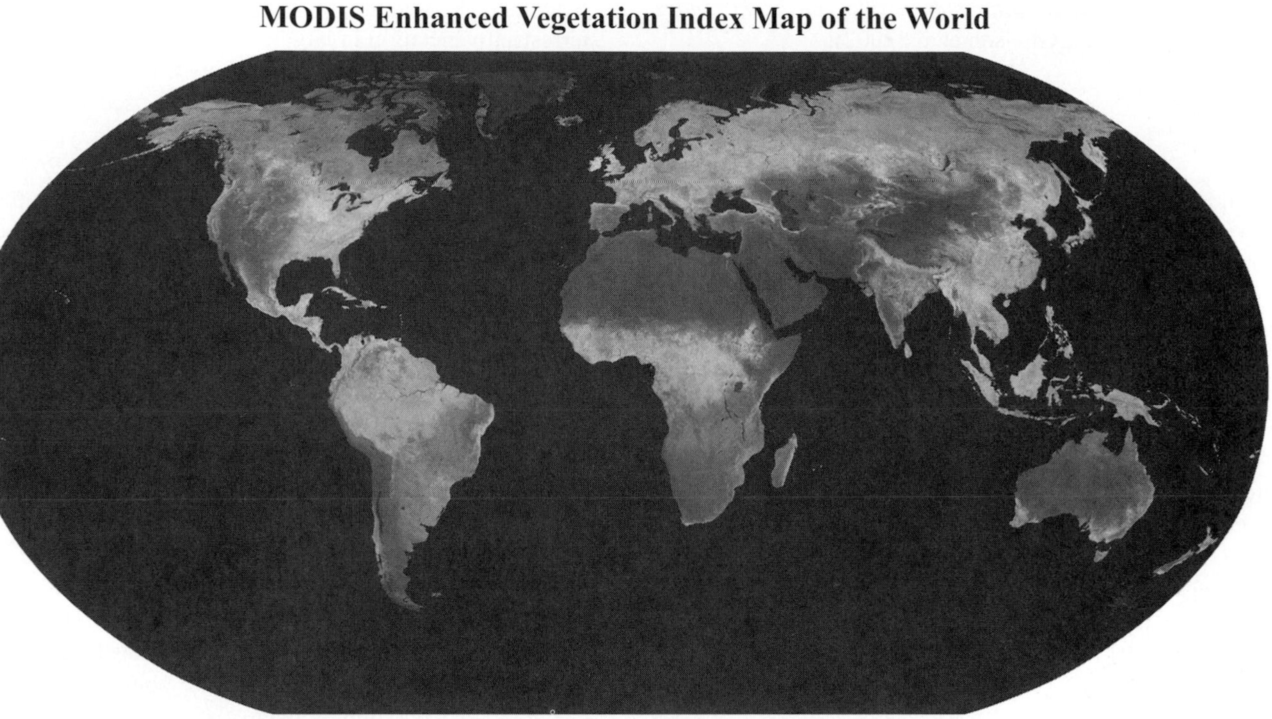

MODIS Enhanced Vegetation Index Map of the World

Figure 8-48   MODIS Enhanced Vegetation Index (EVI) map of the world obtained over a 16-day period beginning on day 193 of 2003. The brighter the area, the greater the amount of biomass (courtesy Terrestrial Biophysics and Remote Sensing MODIS Team, University of Arizona and NASA).

Chlorophyll Absorption in Reflectance Index — CARI; Modified Chlorophyll Absorption in Reflectance Index — MTCARI; Optimized Soil-Adjusted Vegetation Index — OSAVI; Ratio TCARI/OSAVI

Many scientists are interested in the amount of chlorophyll in vegetation (e.g., Daughtry et al., 2000). Kim et al. (1994) developed the *Chlorophyll Absorption in Reflectance Index* (CARI). It was modified to become the *Transformed Absorption in Reflectance Index* (TCARI):

$$TCARI = 3\left[(\rho_{700} - \rho_{670}) - 0.2(\rho_{700} - \rho_{550})\left(\frac{\rho_{700}}{\rho_{670}}\right)\right]. \quad (8\text{-}94)$$

It uses bands corresponding to the minimum absorption of the photosynthetic pigments, centered at 550 and 700 nm, in conjunction with the chlorophyll *a* maximum absorption band, around 670 nm. The choice of 700 nm is due to its location at the boundary between the region where vegetation reflectance is dominated by pigment absorption and the beginning of the red edge portion where vegetation structural characteristics (i.e., the spongy mesophyll) have more influence on reflectance (Kim et al., 1994).

Unfortunately, TCARI is still sensitive to the underlying soil reflectance properties, particularly for vegetation with a low leaf area index. Therefore, Daughtry et al. (2000) proposed that TCARI be combined with a soil line vegetation index like the *Optimized Soil-Adjusted Vegetation Index* (OSAVI) (Rondeaux et al., 1996):

$$OSAVI = \frac{(1 + 0.16)(\rho_{800} - \rho_{670})}{(\rho_{800} + \rho_{670} + 0.16)}. \quad (8\text{-}95)$$

The ratio became:

$$\frac{TCARI}{OSAVI} \quad (8\text{-}96)$$

and is highly correlated with vegetation chlorophyll content (Haboudane et al., 2002).

Visible Atmospherically Resistant Index — VARI

Many resource managers would like vegetation fraction information (e.g., 60%) (Rundquist, 2002). Building upon the Atmospherically Resistant Vegetation Index, scientists developed the *Visible Atmospherically Resistant Index* (VARI) computed as (Gitelson et al., 2002):

$$VARI_{green} = \frac{\rho_{green} - \rho_{red}}{\rho_{green} + \rho_{red} - \rho_{blue}}.$$ (8-97)

The index was minimally sensitive to atmospheric effects, allowing estimation of vegetation fraction with an error of <10% in a wide range of atmospheric optical thickness.

Normalized Difference Built-up Index — NDBI

Many professionals working on urban/suburban problems are interested in monitoring the spatial distribution and growth of urban built-up areas. These data can be used for watershed runoff prediction and other planning applications. Zha et al. (2003) calculated a *Normalized Difference Built-up Index* (NDBI):

$$NDBI = \frac{MidIR_{TM5} - NIR_{TM4}}{MidIR_{TM5} + NIR_{TM4}}$$ (8-98)

$$Built\text{-}up_{area} = NDBI - NDVI$$ (8-99)

Equation 8-98 is identical to the Hardisky et al. (1983) Infrared Index (II) (Equation 8-77). Zha et al. (2003), however, used the NDBI in conjunction with a traditional NDVI. This resulted in an output image that contained only built-up and barren pixels having positive values while all other covers had a value of 0 or –254. The technique was reported to be 92% accurate.

New Vegetation Index — NVI

The near-infrared bands found on Landsat TM, NOAA AVHRR, and the Indian Remote Sensing Linear Imaging Self Scanning (LISS) sensor are in the region 770 to 860, 760 to 900, and 725 to 1100 nm, respectively. Unfortunately, these near-infrared regions include water vapor absorption bands. Therefore, to improve the biomass monitoring capability of the NDVI, Gupta et al. (2001) chose to exclude water vapor absorption bands in the creation of a *New Vegetation Index* (NVI):

$$NVI = \frac{\rho_{777} - \rho_{747}}{\rho_{673}}$$ (8-100)

where ρ_{777}, ρ_{747}, and ρ_{673} are the reflectances centered at 777, 747, and 673 nm, respectively, in 3- to 10-nm bandwidth hyperspectral data. Other narrow-band vegetation indices are found in Chapter 11.

Scientists throughout the world are studying the role of terrestrial vegetation in large-scale global processes. This is necessary in order to understand how the Earth functions as a system. Vegetation indices are being used to accurately inventory the global distribution of vegetation types as well as their biophysical (e.g., LAI, biomass, APAR) and structural (e.g., percent canopy closure) properties. Monitoring these characteristics through space and time will provide valuable information for understanding the Earth as a system.

 Texture Transformations

When humans visually interpret remotely sensed imagery, they synergistically take into account context, edges, texture, and tonal variation or color. Conversely, most digital image processing classification algorithms are based only on the use of the spectral (tonal) information (i.e., brightness values). Thus, it is not surprising that there has been considerable activity in trying to incorporate some of these other characteristics into digital image classification procedures.

A *discrete tonal feature* is a connected set of pixels that all have the same or almost the same gray shade (brightness value). When a small area of the image (e.g., a 3 × 3 area) has little variation of discrete tonal features, the dominant property of that area is a gray shade. Conversely, when a small area has a wide variation of discrete tonal features, the dominant property of that area is *texture*. Most researchers trying to incorporate texture into the classification process have attempted to create a new texture image that can then be used as another feature or band in the classification process. Thus, each new pixel of the texture image has a brightness value that represents the texture at that location (i.e., $BV_{i,j,\text{texture}}$).

There are several standard approaches to automatic texture classification, including texture features based on first- and second-order gray-level statistics and on the Fourier power spectrum and measures based on fractals. Several studies have concluded that the use of the Fourier transform for texture analysis generally yields poor results (Weszka et al., 1976; Gong et al., 1992; Maillard, 2003).

First-order Statistics in the Spatial Domain

One class of picture properties that can be used for texture synthesis is first-order statistics of local areas such as means (averages), variance, standard deviation, and entropy (Hsu,

1978; Gong et al., 1992, Ferro and Warner, 2002; Mumby and Edwards, 2002). Typical algorithms include:

$$\text{AVE} = \frac{1}{w} \sum_{i=0}^{quant_k} i \times f_i \qquad (8\text{-}101)$$

$$\text{STD} = \sqrt{\frac{1}{w} \sum_{i=0}^{quant_k} (i - \text{AVE})^2 \times f_i} \qquad (8\text{-}102)$$

$$\text{ENT}_1 = \sum_{i=0}^{quant_k} \frac{f_i}{w} \ln \frac{f_i}{w} \qquad (8\text{-}103)$$

where

f_i = frequency of gray level i occurring in a pixel window

$quant_k$ = quantization level of band k (e.g., 2^8 = 0 to 255)

w = total number of pixels in a window.

The pixel windows typically range from 3×3 to 5×5 to 7×7. Application of a standard deviation texture transformation to the Charleston, S.C., data is shown in Figure 8-49. The brighter the pixel is, the greater the heterogeneity (more coarse the texture) within the window of interest.

Numerous scientists have evaluated these and other texture transformations. Irons and Petersen (1981) applied 11 of Hsu's (1978) local texture transforms to Landsat MSS data using a 3×3 moving window (mean, variance, skewness, kurtosis, range, Pearson's second coefficient of skewness, absolute value of mean norm length differences, mean of squared norm length differences, maximum of squared norm length differences, mean Euclidean distance, and maximum Euclidean distance). Gong et al. (1992) used two of Hsu's measures [gray-level average (AVE) and standard deviation (STD)] and developed a third, entropy (ENT$_1$), as shown in Equations 8-101 through 8-103. They found that the standard deviation measure was the best of the statistical texture features, but it was not as effective as the brightness value spatial-dependency co-occurrence matrix measures, to be discussed shortly. Mumby and Edwards (2002) extracted the variance from IKONOS imagery using a 5×5 pixel window. They found that when texture features were used in conjunction with spectral data the thematic map accuracy increased for medium and fine levels of coral reef habitat. Ferro and Warner (2002) used a variance texture measure and found that land-cover class separability increased when texture was used in addition to spectral information, and that texture separability increased with larger windows.

Standard Deviation Texture Image of Charleston, SC, Derived from Landsat Thematic Mapper Data

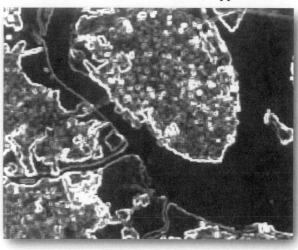

Figure 8-49   Standard deviation texture transformation applied to the Charleston, SC, Landsat TM band 4 data.

Conditional Variance Detection

Zhang (2001) developed a texture measure to identify tree features in high-spatial-resolution imagery. It consists of two components: 1) directional variance detection and 2) standard local variance measurement, as previously discussed. The spatial logic of the directional variance detection is shown in Figure 8-50. Directional variance detection is used to determine whether the central pixel within the moving spatial window (e.g., 7×7) is located in a 'treed' area. If the central pixel is in a treed area, the local variance calculation is then carried out to highlight the pixel in an output file. Otherwise, the local variance calculation is avoided to suppress the pixel. To effectively detect edges of other objects and separate them from treed areas, the size of the window for directional variance detection should be larger (e.g., 7×7) than the window for local variance calculation (e.g., 3×3).

The directional variance detector measures the pixel variances along the shaded pixels in Figure 8-50 on each side of the central pixel in four directions using the equation:

$$\text{DVar} = \frac{1}{n} \sum_{i=-n}^{n-1} [f(i,j) - \overline{f(i,j)}]^2 \qquad (8\text{-}104)$$

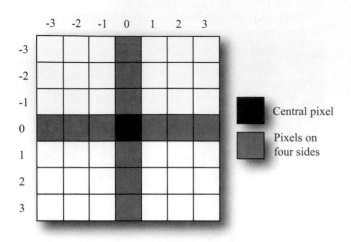

Figure 8-50   Window used to compute directional variance (adapted from Zhang, 2001). It does not have to be 7×7.

with $\overline{f(i,j)} = \dfrac{1}{n} \displaystyle\sum_{i=-n}^{n-1} f(i,j)$

and

$i < 0,\; j = 0$ for the upper side

$i \geq 0,\; j = 0$ for the lower side

$i = 0,\; j < 0$ for the left side

$i \geq 0,\; j \geq 0$ for the right side

where DVar is directional variance, $f(i,j)$ is the value of the pixel located at the ith row and jth column in the spatial moving window (Figure 8-50), and n is the pixel count on each side of the central pixel.

If the variance on one of the four sides is less than a user-specified threshold, it is concluded that there is a homogenous area on this side or the central line is along a straight edge. In this case, the central pixel is regarded as a non-tree pixel. It is assigned a lower value and the local variance calculation is not carried out. If, on the other hand, the variance in one of the directions exceeds the threshold, then the calculation of the local variance (e.g., within a 3×3 window) is carried out and the central pixel receives a higher value. Zhang (2001) found that the algorithm detected tree features more accurately than common gray-

level co-occurrence matrix-derived texture algorithms, to be discussed below.

Min–Max Texture Operator

Briggs and Nellis (1991) developed a *min–max texture operator* based on the analysis of the brightness values found within the following five-element spatial moving window:

$$
\begin{array}{ccc}
 & A & \\
B & C & D \\
 & E &
\end{array}
$$

where

$$texture_C = brightest_{A,\,B,\,C,\,D,\,E} - darkest_{A,\,B,\,C,\,D,\,E}. \quad (8\text{-}105)$$

They found that the min–max texture features and NDVI transformations of seven SPOT HRV scenes provided accurate information on the seasonal variation and heterogeneity of a portion of the tallgrass Konza Prairie Research Natural Area in Kansas.

Second-order Statistics in the Spatial Domain

A suite of very useful texture measures was originally developed by Haralick and associates (Haralick et al., 1973; Haralick and Shanmugan, 1974; Haralick, 1979; Haralick and Fu, 1983; Haralick, 1986). The higher-order set of texture measures is based on brightness value spatial-dependency gray-level co-occurrence matrices (GLCM). The GLCM-derived texture transformations have been widely adopted by the remote sensing community and are often used as an additional feature in multispectral classification (e.g., Schowengerdt, 1997; Clausi and Jernigan, 1998; Franklin et al., 2001; Maillard, 2003). Clausi (2002) summarizes their application as features in the interpretation of synthetic aperture radar (SAR) sea-ice imagery.

But how are these higher-order texture measures computed? If $c = (\Delta x, \Delta y)$ is considered a vector in the (x, y) image plane, for any such vector and for any image $f(x, y)$ it is possible to compute the joint probability density of the pairs of brightness values that occur at pairs of points separated by c. If the brightness values in the image can take upon themselves any value from 0 to the highest quantization level in the image (e.g., $quant_k = 255$), this joint density takes the form of an array h_c, where $h_c(i, j)$ is the probability that the pairs of brightness values (i, j) occur at separation c. This array h_c is $quant_k$ by $quant_k$ in size. It is easy to compute the h_c array for $f(x, y)$, where Δx and Δy are integers by simply counting the number of times each pair of brightness values occurs at separation c (Δx and Δy) in the image. For example, consider the following image that has just five lines and five columns and contains brightness values ranging from only 0 to 3:

$$\text{Original image} = \begin{array}{ccccc} 0 & 1 & 1 & 2 & 3 \\ 0 & 0 & 2 & 3 & 3 \\ 0 & 1 & 2 & 2 & 3 \\ 1 & 2 & 3 & 2 & 2 \\ 2 & 2 & 3 & 3 & 2 \end{array}$$

If (Δx and Δy) = (1, 0), then these numbers are represented by the brightness value spatial-dependency matrix h_c:

$$h_c = \begin{array}{c|cccc} & 0 & 1 & 2 & 3 \\ \hline 0 & 1 & 2 & 1 & 0 \\ 1 & 0 & 1 & 3 & 0 \\ 2 & 0 & 0 & 3 & 5 \\ 3 & 0 & 0 & 2 & 2 \end{array}$$

where the entry in row i and column j of this matrix is the number of times brightness value i occurs to the left of brightness value j. For example, brightness value 1 is to the left of brightness value 2 a total of three times in this simplified image [i.e., $h_c(1, 2) = 3$]. It is assumed that all textural information is contained in the brightness value spatial-dependency matrices that are developed for angles of 0°, 45°, 90°, and 135° (Figure 8-51). Generally, the greater the number found in the diagonal of the gray-level co-occurrence matrices, the more homogeneous the texture is for that part of the image being analyzed.

There are a variety of measures that can be used to extract useful textural information from the h_c matrices. Four of the more widely used include the angular second moment (ASM), contrast (CON), correlation (COR), entropy (ENT$_2$) and homogeneity (HOM) (Haralick, 1986; Gong et al., 1992; Zhang, 2001; Herold et al., 2003; Maillard, 2003):

$$\text{ASM} = \sum_{i=0}^{quant_k} \sum_{j=0}^{quant_k} h_c(i,j)^2 \tag{8-106}$$

$$\text{CON} = \sum_{i=0}^{quant_k} \sum_{j=0}^{quant_k} (i-j)^2 \times h_C(i,j)^2 \tag{8-107}$$

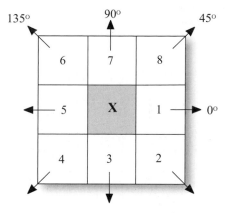

Figure 8-51 The eight nearest neighbors of pixel X according to angle ϕ used in the creation of spatial-dependency matrices for the measurement of image texture.

$$\text{COR} = \sum_{i=0}^{quant_k} \sum_{j=0}^{quant_k} \frac{(i-\mu)(j-\mu)h_c(i,j)^2}{\sigma^2} \tag{8-108}$$

$$\text{ENT}_2 = \sum_{i=0}^{quant_k} \sum_{j=0}^{quant_k} h_c(i,j) \times \log[h_c(i,j)] \tag{8-109}$$

$$\text{HOM} = \sum_{i=0}^{quant_k} \sum_{j=0}^{quant_k} \frac{1}{1+(i-j)^2} \cdot h_c(i,j) \tag{8-110}$$

where

$quant_k$ = quantization level of band k (e.g., $2^8 = 0$ to 255)
$h_c(i, j)$ = the (i, j)th entry in one of the angular brightness value spatial-dependency matrices,

and

$$\mu = \sum_{i=0}^{quant_k} \sum_{j=0}^{quant_k} i \times h_c(i,j) \tag{8-111}$$

$$\sigma^2 = \sum_{i=0}^{quant_k} \sum_{j=0}^{quant_k} (i-\mu)^2 \times h_c(i,j). \tag{8-112}$$

During computation, four brightness value spatial-dependency matrices (0°, 45°, 90°, and 135°) are derived for each pixel based on neighboring pixel values. The average of these four measures is normally output as the texture value for the pixel under consideration.

To use a GLCM-derived texture measure, the analyst usually has to make several important decisions, including (Franklin et al., 2001):

- the texture measure(s),

- window size (e.g., $3 \times 3, 5 \times 5, 32 \times 32$)

- input channel (i.e., the spectral channel used to extract texture information)

- quantization level of the input data (e.g., 8-bit, 6-bit, 4-bit) used to produce the output texture image, and

- the spatial component (i.e., the interpixel distance and angle used during co-occurrence computation).

It is possible to use any GLCM-derived texture measure alone. However, Clausi (2002) found that there was a preferred subset of statistics (contrast, correlation, and entropy) that was superior to the use of any single statistic or using the entire sent of statistics. Gong et al. (1992) found that 3×3 and 5×5 windows were generally superior to larger windows. Conversely, Ferro and Warner (2002) found that land-cover texture separability increased with larger windows. Several scientists suggest that it is a good idea to reduce the quantization level of the input data (e.g., from 8-bit data with values from 0 to 255 to 5-bit data with values from 0 to 31) when creating the texture images so that the spatial-dependency matrices to be computed for each pixel do not become too large (Clausi, 2002).

Jensen and Toll (1982) reported on the use of Haralick's angular second moment (ASM) for use as an additional feature in the supervised classification of remotely sensed data obtained at the urban fringe and in urban change detection mapping. They found it improved the classification when used as an additional feature in the multispectral classification. Gong et al. (1992) found that ASM, CON, and COR provided more valuable texture information than the first-order statistical texture measures. Peddle and Franklin (1991) used GLCM texture measures and found that the spatial co-occurrence matrices contain important textural information that improved the discrimination of classes with internal heterogeneity and structural/geomorphometric patterns. Franklin et al. (2001) found that GLCM texture measures increased the classification accuracy of forest species. Herold et al. (2003) investigated a variety of landscape ecology spatial metrics and GLCM texture measures to classify urban land use in IKONOS high spatial resolution imagery.

Texture Units as Elements of a Texture Spectrum

Wang and He (1990) computed texture based on an analysis of the eight possible clockwise ways of ordering the 3×3 matrix of pixel values shown in Figure 8-52a. This represents a set containing nine elements $V = \{V_0, V_1, ..., V_8\}$, with V_0 representing the brightness value of the central pixel and V_i the intensity of the neighboring pixel i. The corresponding *texture unit* is a set containing eight elements, $\text{TU} = \{E_1, E_2, ..., E_8\}$, where E_i is computed:

for $i = 1, 2, ..., 8$

$$E_i = 0 \text{ if } V_i < V_0$$

$$E_i = 1 \text{ if } V_i = V_0, \qquad (8\text{-}113)$$

$$E_i = 2 \text{ if } V_i > V_0$$

and the element E_i occupies the same position as pixel i. Because each element of TU has one of three possible values, the combination of all eight elements results in $3^8 = 6561$ possible texture units. There is no unique way to label and order the 6561 texture units. Therefore, the texture unit of a 3×3 neighborhood of pixels (Figure 8-52b–d) is computed:

$$N_{\text{TU}} = \sum_{i=1}^{8} 3^{i-1} E_i \qquad (8\text{-}114)$$

where E_i is the ith element of the texture unit set $\text{TU} = \{E_1, E_2, ..., E_8\}$. The first element, E_i, may take any one of the eight possible positions from a through h in Figure 8-52a. An example of transforming a 3×3 neighborhood of image brightness values into a texture unit (TU) and a texture unit number (N_{TU}) using the ordering method starting at a is shown in Figure 8-52. In this example, the texture unit number, N_{TU}, for the central pixel has a value of 6095. The eight brightness values in the hypothetical neighborhood are very diverse (that is, there is a lot of heterogeneity in this small region of the image); therefore, it is not surprising that the central pixel has such a high texture unit number. Eight separate texture unit numbers could be calculated for this central pixel based on the eight ways of ordering shown in Figure 8-52a. The eight N_{TU} values could then be averaged to obtain a mean N_{TU} value for the central pixel.

The possible texture unit values, which range from 0 to 6560, describe the local texture of a pixel in relationship to its eight neighbors. The frequency of occurrence of all the pixel tex-

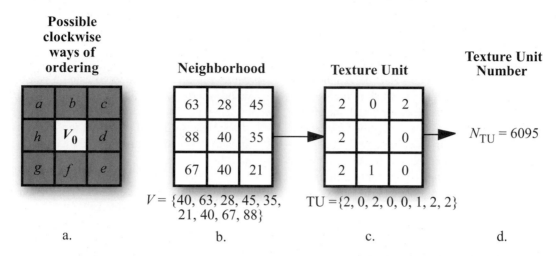

Possible clockwise ways of ordering — **Neighborhood** — **Texture Unit** — **Texture Unit Number**

$V = \{40, 63, 28, 45, 35, 21, 40, 67, 88\}$

$TU = \{2, 0, 2, 0, 0, 1, 2, 2\}$

$N_{TU} = 6095$

a. b. c. d.

Figure 8-52   How a 3×3 neighborhood of brightness values is transformed into a texture unit number (N_{TU}), which has values ranging from 0 to 6560. a) Possible clockwise ways of ordering the eight elements of the texture unit. The first element E_i in Equation 8-114 may take any of the eight positions from a through h. In this example, the ordering position begins at a. b) Brightness values found in the 3×3 neighborhood. This is a very heterogeneous group of pixels and should result in a high texture unit number. c) Transformation of the neighborhood brightness values into a texture unit. d) Computation of the texture unit number based on Equation 8-114 (values range from 0 to 6560). It is possible to compute eight separate texture unit numbers from this neighborhood and then take the mean (adapted from Wang and He, 1990; He and Wang, 1990).

ture unit numbers over a whole image is called the *texture spectrum*. It may be viewed in a graph with the range of possible texture unit numbers (N_{TU}) on the x-axis (values from 0 to 6560) and the frequency of occurrence on the y-axis (Figure 8-53). Each image (or subimage) should have a unique texture spectrum if its texture is truly different from other images (or subimages).

Wang and He (1990) developed algorithms for extracting textural features from the texture spectrum of an image, including black–white symmetry, geometric symmetry, and degree of direction. Only the geometric symmetry measure is presented here. For a given texture spectrum, let $S_j(i)$ be the occurrence frequency of the texture unit numbered i in the texture spectrum under the ordering way j, where $i = 0, 1, 2, ..., 6560$ and $j = 1, 2, 3, ..., 8$ (the ordering ways $a, b, c, ..., h$ are, respectively, represented by $j = 1, 2, 3, ..., 8$). Geometric symmetry (GS) for a given image (or subimage) is:

$$GS = \left[1 - \frac{1}{4} \sum_{j=1}^{4} \frac{\sum_{i=0}^{6560} |S_j(i) - S_{j+4}(i)|}{2 \times \sum_{i=0}^{6560} S_j(i)} \right] \times 100. \qquad (8\text{-}115)$$

GS values are normalized from 0 to 100 and measure the symmetry between the spectra under the ordering ways a and

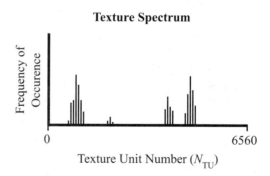

Texture Spectrum

Frequency of Occurrence (y-axis)

Texture Unit Number (N_{TU}) (x-axis, 0 to 6560)

Figure 8-53   Hypothetical texture spectrum derived from an image or subimage (adapted from Wang and He, 1990).

e, b and f, c and g, and d and h for a given image. This measure provides information on the shape regularity of images (Wang and He, 1990). A high value means the texture spectrum will remain approximately the same even if the image is rotated 180°. The degree of direction measure provides information about the orientation characteristics of images.

Fractal Dimension as a Measure of Spatial Complexity or Texture

From classical Euclidean geometry we know that the dimension of a curve is 1, a plane is 2, and a cube is 3. This is called topological dimension, D, and is characterized by integer values. A difficulty arises when analyzing complex spatial phenomena that are not simple lines, planes, or cubes. To overcome this limitation, the concept of fractional dimension was first formulated by mathematicians Hausdorff and Beiscovitch. Mandelbrot (1977, 1982) renamed it the *fractal* dimension and defined fractals as "a set for which the Hausdorff–Beiscovitch dimension strictly exceeds the topological dimension." This allows the complexity of natural real-world forms and phenomena to be measured. In fractal geometry, the dimension D of a complex *line* (e.g., a contour line) will have a fractal dimension value anywhere between the topological dimension of 1 and the Euclidean dimension of 2, while a complex *surface* (e.g., mountainous terrain or the brightness values associated with a remotely sensed image) will have a fractal dimension value of between 2 and 3 (Lam, 1990; Lam and DeCola, 1993).

Mandelbrot's fractal dimension is based on the concept of *self-similarity*, where a line or surface is composed of copies of itself that display increasing levels of detail at an enlarged scale. Fractals use self-similarity to define D (Peitgen et al., 1992). Many curves and surfaces are statistically self-similar, meaning that each portion can be considered as a reduced-scale image of the whole. Fractals have been used by a diverse set of scientists to measure the complexity of terrain lines and surfaces (e.g., Goodchild, 1980; Shelberg et al., 1983; Mark and Aronson, 1984; Roy et al., 1987). For example, Mark and Aronson (1984) used the dividers method to measure the lengths of contour lines on topographic maps using calipers with a known separation. By measuring the total number of caliper increments required to measure each line and the distance of the caliper separation, they were able to determine the length of the line. Measuring the same contour line again with a smaller caliper separation resulted in a new estimate of the length of the line. When the log of the caliper separation was plotted on the x-axis against the log of the total length of the contour line on the y-axis, an increase in the total length occurred with every decrease in caliper separation. This increase in length relates to the self-similarity of Mandelbrot's fractal dimension.

The D value can be determined by a linear regression of this plot:

$$\log L = C + B \log G, \tag{8-116}$$

where L is the length of the curve, G is the step size (caliper separation), B is the slope of the regression, and C is a constant. The actual fractal dimension D of a line is

$$D = 1 - B. \tag{8-117}$$

The fractal dimension of a surface (e.g. digital terrain model or remotely sensed image) is (Lam, 1990)

$$D = 2 - B. \tag{8-118}$$

Fractals can also be computed using a cell or box counting method where (Klinkenberg and Goodchild, 1992):

- the surface (a digital elevation model or remotely sensed image) is sliced by a horizontal plane at a given elevation or brightness value;

- cells with values above or on the plane are coded black, while those below are coded white;

- a count is made of the number of adjacencies (i.e., where a black cell occurs next to a white cell); and

- the cell size is increased (by replacing four cells with one cell that is assigned the average value of the four original cells) and the process repeated until the cell size is equal to the size of the array.

The surface fractal dimension is derived from the slope of the best-fitting line from the graphs showing the log of the average number of boundary cells plotted against the log of the cell size. The fractal dimensions of surfaces have also been computed using variograms (Chapter 4), of either the whole surface or certain profiles extracted from the surface (e.g. Roy et al., 1987). Klinkenberg and Goodchild (1992) review the dividers method, the cell or box counting method, and the variogram methods to estimate D.

Lines with a D value between 1.1 and 1.3 often look very much like real curves (e.g., coastlines) while surfaces with D values of from 2.1 to 2.3 look like real topographic surfaces. This is part of the reason they have been used so extensively in the generation of computer graphic images of terrain for flight simulators.

Lam (1990) computed the fractal dimension for three different landscapes in Louisiana using seven bands of Landsat TM data using the box counting method. An average fractal dimension for each study area was computed by taking the mean of all the individual band fractal dimensions. The urban landscape had the highest average fractal dimension ($D = 2.609$), the coastal region was second ($D = 2.597$), and the rural area was third ($D = 2.539$). DeCola (1989) calculated fractal dimensions for classification maps derived from Landsat TM data of Vermont using a box method.

These results suggest that the fractal dimension obtained over relatively small neighborhoods may eventually be useful as a texture or complexity measure, which may be of value in hard or fuzzy supervised or unsupervised classifications. Lam (1990) points out that factors such as striping noise, sun elevation angle, and atmospheric effect may affect the brightness values and therefore the fractal dimension statistic.

Although texture features have been increasingly incorporated into multispectral classifications, no single algorithm that combines efficiency and effectiveness has yet to be widely adopted. Also, the texture features derived for one type of application (e.g., land-use classification at the urban fringe) are not necessarily useful when applied to another geographic problem (e.g., identification of selected geomorphic classes). Finally, some parameters central to the computation of the texture features are still derived empirically (e.g., the size of the window or the location of certain thresholds). This makes it difficult to compare and contrast studies when so many variables in the creation of the texture features are not held constant.

Texture Statistics Based on the Semi-variogram

Several authors have investigated the use of the semi-variogram discussed in Chapter 4 to derive texture information (e.g., Woodcock et al., 1988: Lark, 1996). Maillard (2003) identified the problems associated with using different variogram models (e.g., spherical, exponential, and sinusoidal) and other criteria. He developed a variogram texture operator that:

- uses a rather large window to cover larger distance lags (up to 32 pixels),

- is rotation-invariant and preserves anisotrophy, and

- incorporates the mean square-root pair difference function (SRPD) as a semi-variance estimator.

Maillard compared the variogram, GLCM, and Fourier-based texture measures and concluded that the variogram and GLCM texture measures were generally superior.

 ## References

Almeida-Filho, R. and Y. E. Shimabukuro, 2002, "Digital Processing of a Landsat TM Time Series for Mapping and Monitoring Degraded Areas Caused by Independent Gold Miners, Roraima State, Brazilian Amazon," *Remote Sensing of Environment*, 79:42–50.

Bausch, W. C., 1993, "Soil Background Effects on Reflectance-based Crop Coefficients for Corn," *Remote Sensing of Environment*, 46:1–10.

Birth, G. S. and G. McVey, 1968, "Measuring the Color of Growing Turf with a Reflectance Spectroradiometer," *Agronomy Journal*, 60:640–643.

Briggs, J. M. and M. D. Nellis, 1991, "Seasonal Variation of Heterogeneity in the Tallgrass Prairie: A Quantitative Measure Using Remote Sensing," *Photogrammetric Engineering & Remote Sensing*, 57(4):407–411.

Broge, N. H. and E. Leblanc, 2000, "Comparing Prediction Power and Stability of Broadband and Hyperspectral Vegetation Indices for Estimation of Green Leaf Area Index and Canopy Chlorophyll Density," *Remote Sensing of Environment*, 76:156–172.

Carr, J. R., 1998, "A Visual Basic Program for Principal Components Transformation of Digital Images," *Computers & Geosciences*, 24(3):209–281.

Carter, G. A., 1991, "Primary and Secondary Effects of the Water Content on the Spectral Reflectance of Leaves," *American Journal of Botany*, 78(7):916–924.

Carter, G. A., 1993, "Responses of Leaf Spectral Reflectance to Plant Stress," *American Journal of Botany*, 80(3):231–243.

Carter, G. A., W. G. Cibula and R. L. Miller, 1996, "Narrow-band Reflectance Imagery Compared with Thermal Imagery for Early Detection of Plant Stress," *Journal of Plant Physiology*, 148:515–522.

Chavez, P. C. and B. Bauer, 1982, "An Automatic Optimum Kernel-size Selection Technique for Edge Enhancement," *Remote Sensing of Environment*, 12:23–38.

Chavez, P. C., Guptill, S. C., and J. A. Bowell, 1984, "Image Processing Techniques for Thematic Mapper Data," *Proceedings*, American Society of Photogrammetry, 2:728–743.

Chen, J. M., Pavlic, G., Brown, L., Cihlar, J., Leblanc, S. G., White, H. P., Hall, R. J., Peddle, D. R., King, D. J., Trofymow, J. A., Swift, E., Van der Sanden, J., and P. K. Pellikka, 2002, "Derivation and Validation of Canada-wide Coarse-resolution Leaf Area Index Maps Using High-resolution Satellite Imagery and Ground Measurements," *Remote Sensing of Environment*, 80:165–184.

Chittineni, C. B., 1983, "Edge and Line Detection in Multidimensional Noisy Imagery Data," *IEEE Transactions on Geoscience and Remote Sensing*, GE-21:163–174.

Cibula, W. G. and G. A. Carter, 1992, "Identification of a Far-Red Reflectance Response to *Ectomycorrhizae* in Slash Pine," *International Journal of Remote Sensing*, 13(5):925–932.

Clausi, D. A., 2002, "An Analysis of Co-occurrence Texture Statistics as a Function of Grey Level Quantization," *Canadian Journal of Remote Sensing*, 28(1):45–62.

Clausi, D. A. and M. E. Jernigan, 1998, "A Fast Method to Determine Co-occurrence Texture Features," *IEEE Transactions on Geoscience and Remote Sensing*, 36(1):298–300.

Cohen, W. B., 1991, "Response of Vegetation Indices to Changes in Three Measures of Leaf Water Stress," *Photogrammetric Engineering & Remote Sensing*, 57(2):195–202.

Colombo, R., Bellingeri, D., Fasolini, D. and C. M. Marino, 2003, "Retrieval of Leaf Area Index in Different Vegetation Types Using High Resolution Satellite Data," *Remote Sensing of Environment*, 86:120–131.

Crippen, R. E., 1988, "The Dangers of Underestimating the Importance of Data Adjustments in Band Ratioing," *International Journal of Remote Sensing*, 9(4):767–776.

Crist, E. P., 1984, "Comparison of Coincident Landsat-4 MSS and TM Data over an Agricultural Region," *Proceedings, American Society for Photogrammetry & Remote Sensing*, 2:508–517.

Crist, E. P. and R. C. Cicone, 1984, "Application of the Tasseled Cap Concept to Simulated Thematic Mapper Data," *Photogrammetric Engineering & Remote Sensing*, 50:343–352.

Crist, E. P. and R. J. Kauth, 1986, "The Tasseled Cap De-mystified," *Photogrammetric Engineering & Remote Sensing*, 52(1):81–86.

Daughtry, C. S. T., Walthall, C. L., Kim, M. S., Brown de Colstoun, E. and J. E. McMurtrey III, 2000, "Estimating Corn Leaf Chlorophyll Concentration from Leaf and Canopy Reflectance," *Remote Sensing of Environment*, 74:229–239.

DeCola, L., 1989, "Fractal Analysis of a Classified Landsat Scene," *Photogrammetric Engineering & Remote Sensing*, 55(5):601–610.

Deering, D. W., Rouse, J. W., Haas, R. H. and J. A. Schell, 1975, "Measuring Forage Production of Grazing Units from Landsat MSS Data," *Proceedings, 10th International Symposium on Remote Sensing of Environment*, 2:1169–1178.

Didan, K., 2002, *MODIS Vegetation Index Production Algorithms*, MODIS Vegetation Workshop, Missoula, Montana, July 15–18; Terrestrial Biophysics and Remote Sensing (TBRS) MODIS Team, Tucson: University of Arizona, www.ntsg.umt.edu/MODISCon/index.html.

Du, Y., Teillet, P. M. and J. Cihlar, 2002, "Radiometric Normalization of Multitemporal High-resolution Satellite Images with Quality Control for Land Cover Change Detection," *Remote Sensing of Environment*, 82:123–134.

Eastman, J. R. and M. Fulk, 1993, "Long Sequence Time Series Evaluation Using Standardized Principal Components," *Photogrammetric Engineering & Remote Sensing*, 59(6):991–996.

Eliason, E. M. and A. S. McEwen, 1990, "Adaptive Box Filters for Removal of Random Noise from Digital Images," *Photogrammetric Engineering & Remote Sensing*, 56(4):453–458.

Farabee, M. J., 1997, *Photosynthesis*, http://gened.emc.maricopa.edu/bio/bio181/BIOBK/BioBookPS.html.

Ferro, C. J. S. and T. A. Warner, 2002, "Scale and Texture Digital Image Classification," *Photogrammetric Engineering & Remote Sensing*, 68(1):51–63.

Franklin, S. E., Maudie, A. J. and M. B. Lavigne, 2001, "Using Spatial Co-occurrence Texture to Increase Forest Structure and Species Composition Classification Accuracy," *Photogrammetric Engineering & Remote Sensing*, 67(7):849–855.

Gausmann, H. W., Allen, W. A. and R. Cardenas, 1969, "Reflectance of Cotton Leaves and their Structure," *Remote Sensing of Environment*, 1:110–122.

Gil, B., Mitiche, A. and J. K. Aggarwal, 1983, "Experiments in Combining Intensity and Range Edge Maps," *Computer Vision, Graphics, and Image Processing*, 21:395–411.

Gillespie, A. R., 1992, "Enhancement of Multispectral Thermal Infrared Images: Decorrelation Contrast Stretching," *Remote Sensing of Environment*, 42:147–155.

Gitelson, A. A., Kaufman, Y. J., Stark, R. and D. Rundquist, 2002, "Novel Algorithms for Remote Estimation of Vegetation Fraction," *Remote Sensing of Environment*, 80:76–87.

Gong, P., Marceau, D. J. and P. J. Howarth, 1992, "A Comparison of Spatial Feature Extraction Algorithms for Land-Use Classification with SPOT HRV Data," *Remote Sensing of Environment,* 40:137–151.

Gonzalez, R. C. and P. Wintz, 1977, *Digital Image Processing.* Reading, MA: Addison-Wesley, 431 p.

Goodchild, M. F., 1980, "Fractals and the Accuracy of Geographical Measures," *Mathematical Geology,* 12:85–98.

Goward, S. N., Markham, B., Dye, D. G., Dulaney, W. and J. Yang, 1991, "Normalized Difference Vegetation Index Measurements from the AVHRR," *Remote Sensing of Environment,* 35:257–277.

Gupta, R. K., Vijayan, D. and T. S. Prasad, 2001, "New Hyperspectral Vegetation Characterization Parameters," *Advances in Space Research,* 28(1):201–206.

Haack, B., Guptill, S. C., Holz, R. K., Jampoler, S. M., Jensen, J. R. and R. A. Welch, 1997, "Urban Analysis and Planning," Chapter 15 in W. R. Phillipson (Ed.), *Manual of Photographic Interpretation,* 2nd ed., Bethesda, MD: American Society for Photogrammetry & Remote Sensing, 517–553.

Haboudane, D., Miller, J. R., Tremblay, N., Zarco-Tejada, P. J. and L. Dextraze, 2002, "Integrated Narrow-band Vegetation Indices for Prediction of Crop Chlorophyll Content for Application to Precision Agriculture," *Remote Sensing of Environment,* 81:416–426.

Haralick, R. M., 1979, "Statistical and Structural Approaches to Texture," *Proceedings of the IEEE,* 67:786–804.

Haralick, R. M., 1986, "Statistical Image Texture Analysis," T. Y. Young and K. S. Fu (Eds.), *Handbook of Pattern Recognition and Image Processing,* New York: Academic Press, 247–280.

Haralick, R. M. and K. Fu, 1983, "Pattern Recognition and Classification," Chapter 18 in R. N. Colwell (Ed.), *Manual of Remote Sensing,* Falls Church, VA: American Society of Photogrammetry, 793–805.

Haralick, R. M. and K. S. Shanmugam, 1974, "Combined Spectral and Spatial Processing of ERTS Imagery Data," *Remote Sensing of Environment,* 3:3–13.

Haralick, R. M., Shanmugan, K. and I. Dinstein, 1973, "Texture Feature for Image Classification," *IEEE Transactions Systems, Man and Cybernetics,* SMC-3:610–621.

Hardisky, M. A., Klemas, V. and R. M. Smart, 1983, "The Influence of Soil Salinity, Growth Form, and Leaf Moisture on the Spectral Radiance of *Spartina alterniflora* Canopies," *Photogrammetric Engineering & Remote Sensing,* 49(1):77–83.

Hay, C. M., Kuretz, C. A., Odenweller, J. B., Scheffner, E. J. and B. Wood, 1979, *Development of AI Procedures for Dealing with the Effects of Episodal Events on Crop Temporal Spectral Response,* AGRISTARS Report SR-B9-00434.

He, D. C. and L. Wang, 1990, "Texture Unit, Texture Spectrum, and Texture Analysis," *IEEE Transactions on Geoscience and Remote Sensing,* 28(4):509–512.

Herold, M., Liu, XiaoHang, L. and K. C. Clarke, 2003, "Spatial Metrics and Image Texture for Mapping Urban Land Use," *Photogrammetric Engineering & Remote Sensing,* 69(9):991–1001.

Holden, H. and E. LeDrew, 1998, "Spectral Discrimination of Healthy and Non-healthy Corals Based on Cluster Analysis, Principal Components Analysis, and Derivative Spectroscopy," *Remote Sensing of Environment,* 65:217–224.

Hsu, S., 1978, "Texture-tone Analysis for Automated Land Use Mapping," *Photogrammetric Engineering & Remote Sensing,* 44:1393–1404.

Huete, A. R., 1988, "A Soil-adjusted Vegetation Index (SAVI)," *Remote Sensing of Environment,* 25:295–309.

Huete, A. R. and C. Justice, 1999, *MODIS Vegetation Index (MOD 13) Algorithm Theoretical Basis Document,* Greenbelt: NASA Goddard Space Flight Center, http://modarch.gsfc.nasa.gov/MODIS/LAND/#vegetation-indices, 129 p.

Huete, A. R., Didan, K., Miura, T., Rodriguez, E. P., Gao, X. and G. Ferreira, 2002a, "Overview of the Radiometric and Biophysical Performance of the MODIS Vegetation Indices," *Remote Sensing of Environment,* 83:195–213.

Huete, A. R., Didan, K, and Y. Yin, 2002b, *MODIS Vegetation Workshop,* Missoula, Montana, July 15–18; Terrestrial Biophysics and Remote Sensing (TBRS) MODIS Team, University of Arizona, http://utam.geophys.utah.edu/ebooks/gg527/modis/ndvi.html.

Huete, A. R., Hua, G., Qi, J., Chehbouni A. and W. J. Van Leeuwem, 1992, "Normalization of Multidirectional Red and Near-infrared Reflectances with the SAVI," *Remote Sensing of Environment,* 40:1–20.

Huete, A. R. and H. Q. Liu, 1994, "An Error and Sensitivity Analysis of the Atmospheric and Soil-Correcting Variants of the

NDVI for the MODIS-EOS," *IEEE Transactions on Geoscience and Remote Sensing*, 32(4):897–905.

Huete, A. R., Liu, H. Q., Batchily, K. and W. J. van Leeuwen, 1997, "A Comparison of Vegetation Indices Over a Global Set of TM Images for EOS-MODIS," *Remote Sensing of Environment*, 59:440–451.

Hunt, E. R., Rock, B. N. and P. S. Nobel, 1987, "Measurement of Leaf Relative Water Content by Infrared Reflectance," *Remote Sensing of Environment*, 22:429–435.

Irons, J. R. and G. W. Petersen, 1981, "Texture Transforms of Remote Sensing Data," *Remote Sensing of Environment*, 11:359–370.

Jackson, R. D., 1983, "Spectral Indices in *n*-Space," *Remote Sensing of Environment*, 13:409–421.

Jahne, B., 2001, *Digital Image Processing*, New York: Springer-Verlag, 572 p.

Jain, A. K., 1989, *Fundamentals of Digital Image Processing*, Englewood Cliffs, NJ: Prentice-Hall, 342–357.

Jensen, J. R., Botchway, K., Brennan-Galvin, E., Johannsen, C., Juma, C., Mabogunje, A., Miller, R., Price, K., Reining, P., Skole, D., Stancioff, A. and D. R. F. Taylor, 2002, *Down to Earth: Geographic Information for Sustainable Development in Africa*, Washington: National Research Council, 155 p.

Jensen, J. R., Hadley, B. C., Tullis, J. A., Gladden, J., Nelson, E., Riley, S., Filippi, T. and M. Pendergast, 2003, *Hyperspectral Analysis of Hazardous Waste Sites on the Savannah River Site in 2002*, Westinghouse Savannah River Company: Aiken, WSRC-TR-2003-00275, 52 p.

Jensen, J. R., Hodgson, M. E., Christensen, E., Mackey, H. E., Tinney, L. R. and R. Sharitz, 1986, "Remote Sensing Inland Wetlands: A Multispectral Approach," *Photogrammetric Engineering & Remote Sensing*, 52(2):87–100.

Jensen, J. R., Pace, P. J. and E. J. Christensen, 1983, "Remote Sensing Temperature Mapping: The Thermal Plume Example," *American Cartographer*, 10:111–127.

Jensen, J. R., Lin, H., Yang, X., Ramsey, E., Davis, B. and C. Thoemke, 1991, "Measurement of Mangrove Characteristics in Southwest Florida Using SPOT Multispectral Data," *Geocarto International*, 2:13–21.

Jensen, J. R. and D. L. Toll, 1982, "Detecting Residential Land-Use Developments at the Urban Fringe," *Photogrammetric Engineering & Remote Sensing*, 48(4):629–643.

Karnieli, A., Kaufman, Y. J., Remer, L. and A. Wald, 2001, "AFRI: Aerosol Free Vegetation Index," *Remote Sensing of Environment*, 77:10–21.

Kaufman, Y. J., Karnieli, A. and D. Tanre, 2000, "Detection of Dust Over Deserts Using Satellite Data in the Solar Wavelengths," *IEEE Transactions on Geoscience and Remote Sensing*, 38:525–531.

Kaufman, Y. J. and D. Tanre, 1992, "Atmospherically Resistant Vegetation Index (ARVI) for EOS-MODIS," *IEEE Transactions on Geoscience and Remote Sensing*, 30(2):261–270.

Kauth, R. J. and G. S. Thomas, 1976, "The Tasseled Cap—A Graphic Description of the Spectral-Temporal Development of Agricultural Crops as Seen by Landsat," *Proceedings, Symposium on Machine Processing of Remotely Sensed Data*, West Lafayette, IN: LARS, 41–51.

Kauth, R. J., P. F. Lambeck, W. Richardson, G. S. Thomas and A. P. Pentland, 1979, "Feature Extraction Applied to Agricultural Crops as Seen by Landsat," *Proceedings, LACIE Symposium*, Houston: NASA, 705–721.

Khan, M. A., 1992, "Analysis of Edge Enhancement Operators and their Application to SPOT Data," *International Journal of Remote Sensing*, 13:3189–3203.

Kim, M., Daughtry, C. S., Chappelle, E. W., McMurtrey III, J. E. and C. L. Walthall, 1994, "The Use of High Spectral Resolution Bands for Estimating Absorbed Photosynthetically Active Radiation (APAR)," *Proceedings, 6th Symposium on Physical Measurements and Signatures in Remote Sensing*, January 17–21, Val D'Isere, France, 299–306.

Klinkenberg, B. and M. Goodchild, 1992, "The Fractal Properties of Topography: A Comparison of Methods," *Earth Surface Processes and Landforms*, 17:217–234.

Kogan, F. N., 2001, "Operational Space Technology for Global Vegetation Assessment," *Bulletin of the American Meteorological Society*, 82(9):1949–1964.

Lam, N. S., 1990, "Description and Measurement of Landsat TM Images Using Fractals," *Photogrammetric Engineering and Remote Sensing*, 56(2):187–195.

Lam, N. S. and L. DeCola, 1993, *Fractals in Geography*, Englewood Cliffs, NJ: Prentice-Hall, 308 p.

Lark, R. J., 1996, "Geostatistical Description of Texture on an Aerial Photograph for Discriminating Classes of Land Cover," *International Journal of Remote Sensing*, 17(11):2115–2133.

Lee, J. S., 1983, "Digital Image Smoothing and the Sigma Filter," *Computer Vision, Graphics, and Image Processing*, 24:255–269.

Li, Y. and J. Peng, 2003, "Remote Sensing Texture Analysis Using Multi-parameter and Multi-scale Features," *Photogrammetric Engineering & Remote Sensing*, 69(4):351–355.

Lunetta, R. S., Ediriwickrema, J., Johnson, D. M., Lyon, J. G. and A. McKerrow, 2002, "Impacts of Vegetation Dynamics on the Identification of Land-cover Change in a Biologically Complex Community in North Carolina, USA," *Remote Sensing of Environment*, 82:258–270.

Lyon, J. G., Yuan, D., Lunetta, R. and C. Elvidge, 1998, "A Change Detection Experiment Using Vegetation Indices," *Photogrammetric Engineering & Remote Sensing*, 64(2):143–150.

Maillard, P., 2003, "Comparing Texture Analysis Methods through Classification," *Photogrammetric Engineering & Remote Sensing*, 69(4):357–367.

Mandelbrot, B. B., 1977, *Fractals: Form, Chance and Dimension*, San Francisco: W. H. Freeman.

Mandelbrot, B. B., 1982, *The Fractal Geometry of Nature*, San Francisco: W. H. Freeman.

Mark, D. M. and P. B. Aronson, 1984, "Scale-dependent Fractal Dimensions of Topographic Surfaces: An Empirical Investigation, with Applications in Geomorphology and Computer Mapping," *Mathematical Geology*, 11:671–684.

Mitternicht, G. I. and J. A. Zinck, 2003, "Remote Sensing of Soil Salinity: Potentials and Constraints," *Remote Sensing of Environment*, 85:1–20.

Miura, T., Huete, A. F., Yoshioka, H. and B. N. Holben, 2001, "An Error and Sensitivity Analysis of Atmospheric Resistant Vegetation Indices Derived from Dark Target-based Atmospheric Correction," *Remote Sensing of Environment*, 78:284–298.

Mumby, P. J. and A. J. Edwards, 2002, "Mapping Marine Environments with IKONOS Imagery: Enhanced Spatial Resolution Can Deliver Greater Thematic Accuracy," *Remote Sensing of Environment*, 82:248–257.

Nieminen, A., Heinonen, P. and Y. Nuevo, 1987, "A New Class of Detail Preserving Filters for Image Processing," *IEEE Transactions in Pattern Analysis & Machine Intelligence*, 9:74–90.

Pan, J. and C. Chang, 1992, "Destriping of Landsat MSS Images by Filtering Techniques," *Photogrammetric Engineering & Remote Sensing*, 58(10):1417–1423.

Peddle, D. R. and S. E. Franklin, 1991, "Image Texture Processing and Data Integration for Surface Pattern Discrimination," *Photogrammetric Engineering & Remote Sensing*, 57(4):413–420.

Peterson, D. L. and S. W. Running, 1989, "Applications in Forest Science and Management," in *Theory and Applications of Optical Remote Sensing*, New York: John Wiley & Sons, 4210–4273.

Peitgen, H., Jurgens, H. and D. Saupe, 1992, *Fractals for the Classroom*, New York: Springer-Verlag, 450 p.

Peli, T. and D. Malah, 1982, "A Study of Edge Detection Algorithms," *Computer Graphics and Image Processing*, 20:1–21.

Perry, C. R. and L. F. Lautenschlager, 1984, "Functional Equivalence of Spectral Vegetation Indices," *Remote Sensing of Environment*, 14:169–182.

Pierce, L. L., Running, S. W. and G. A. Riggs, 1990, "Remote Detection of Canopy Water Stress in Coniferous Forests Using NS001 Thematic Mapper Simulator and the Thermal Infrared Multispectral Scanner," *Photogrammetric Engineering & Remote Sensing*, 56(5):5710–586.

Pratt, W. K., 2001, *Digital Image Processing*, New York: John Wiley, 656 p.

Price, K. P., Guo, X. and J. M. Stiles, 2002, "Optimal Landsat TM Band Combinations and Vegetation Indices for Discrimination of Six Grassland Types in Eastern Kansas," *International Journal of Remote Sensing*, 23:5031–5042.

Qi, J., Cabot, F., Moran, M. S. and G. Dedieu, 1995, "Biophysical Parameter Estimations Using Multidirectional Spectral Measurements," *Remote Sensing of Environment*, 54:71–83.

Ramsey, R. D., Falconer, A. and J. R. Jensen, 1995, "The Relationship between NOAA-AVHRR NDVI and Ecoregions in Utah," *Remote Sensing of Environment*, 3:188–198.

Reese, H. M., Lillesand, T. M., Nagel, D. E., Stewart, J. S., Goldmann, R. A., Simmons, T. E., Chipman, J. W. and P. A. Tessar, 2002, "Statewide Land Cover Derived from Multiseasonal Landsat TM Data: A Retrospective of the WISCLAND Project," *Remote Sensing of Environment*, 82:224–237.

Richards, J. A. and X. Jia, 1999, *Remote Sensing Digital Image Analysis*, New York: Springer-Verlag, 363 p.

Richardson, A. J. and C. L. Wiegand, 1977, "Distinguishing Vegetation from Soil Background Information," *Remote Sensing of Environment*, 8:307–312.

Rock, B. N., Vogelmann, J. E., Williams, D. L., Vogelmann A. F. and T. Hoshizaki, 1986, "Remote Detection of Forest Damage," *BioScience*, 36:439.

Rogan, J., Franklin, J. and D. A. Roberts, 2002, "A Comparison of Methods for Monitoring Multitemporal Vegetation Change Using Thematic Mapper Imagery," *Remote Sensing of Environment*, 80:143–156.

Rondeaux, G., Steven, M. and F. Baret, 1996, "Optimization of Soil-adjusted Vegetation Indices," *Remote Sensing of Environment*, 55:95–107.

Rouse, J. W., Haas, R. H., Schell, J. A. and D. W. Deering, 1974, "Monitoring Vegetation Systems in the Great Plains with ERTS, *Proceedings*, *3rd Earth Resource Technology Satellite (ERTS) Symposium*, Vol. 1, 48–62.

Roy, A. G., Gravel, G. and C. Gauthier, 1987, "Measuring the Dimension of Surfaces: A Review and Appraisal of Different Methods," *Proceedings*, *8th International Symposium on Computer-Assisted Cartography*, Baltimore, MD, 68–77.

Rundquist, B. C, 2002, "The Influence of Canopy Green Vegetation Fraction on Spectral Measurements over Native Tallgrass Prairie," *Remote Sensing of Environment*, 81:129–135.

Running, S. W., Justice, C. O., Solomonson, V., Hall, D., Barker, J., Kaufmann, Y. J., Strahler, A. H., Huete, A. R., Muller, J. P., Vanderbilt, V., Wan, Z. M., Teillet, P. and D. Carneggie, 1994, "Terrestrial Remote Sensing Science and Algorithms Planned for EOS/MODIS," *International Journal of Remote Sensing*, 15(17):3587–3620.

Russ, J. C., 2002, *The Image Processing Handbook*, Boca Raton, FL: CRC Press, 744 p.

Satterwhite, M. B., 1984, "Discriminating Vegetation and Soils Using Landsat MSS and Thematic Mapper Bands and Band Ratios," *Proceedings, American Society for Photogrammetry & Remote Sensing*, 2:479–485.

Schowengerdt, R. A., 1997, *Remote Sensing: Models and Methods for Image Processing*, San Diego: Academic Press, 522 p.

Sheffield, C., 1985, "Selecting Band Combinations from Multispectral Data," *Photogrammetric Engineering & Remote Sensing*, 51(6):681–687.

Shelberg, M. C., Lam, S. N. and H. Moellering, 1983, "Measuring the Fractal Dimensions of Surfaces," *Proceedings*, *6th International Symposium on Automated Cartography*, Ottawa, 2:319–328.

Short, N., 1982, "Principles of Computer Processing of Landsat Data," Appendix A in *Landsat Tutorial Workbook*, Publication #1078, Washington: NASA, 421–453.

Singh, A. and A. Harrison, 1985, "Standardized Principal Components," *International Journal of Remote Sensing*, 6:883–896.

TBRS, 2003, *Enhanced Vegetation Index*, Terrestrial Biophysics and Remote Sensing Lab, University of Arizona, http://tbrs.arizona.edu/project/MODIS/evi.php.

Trishchenko, A. P., Cihlar, J. and Z. Li, 2002, "Effects of Spectral Response Function on Surface Reflectance and NDVI Measured with Moderate Resolution Satellite Sensors," *Remote Sensing of Environment*, 81:1–18.

Tucker, C. J., 1979, "Red and Photographic Infrared Linear Combinations for Monitoring Vegetation," *Remote Sensing of Environment*, 8:127–150.

Walter-Shea, E. A. and L. L. Biehl, 1990, "Measuring Vegetation Spectral Properties," *Remote Sensing Reviews*, 5(1):179–205.

Wang, F., 1993, "A Knowledge-based Vision System for Detecting Land Changes at Urban Fringes," *IEEE Transactions on Geoscience and Remote Sensing*, 31(1):136–145.

Wang, L. and D. C. He, 1990, "A New Statistical Approach for Texture Analysis," *Photogrammetric Engineering & Remote Sensing*, 56(1):61–66.

Watson, K., 1993, "Processing Remote Sensing Images Using the 2-D FFT—Noise Reduction and Other Applications," *Geophysics*, 58(6):835–852.

Weszka, J., Dyer, C. and A. Rosenfeld, 1976, "A Comparative Study of Texture Measures for Terrain Classification," *IEEE Transactions on Systems, Man and Cybernetics*, SMC-6:269–285.

Woodcock, C. E., Strahler, A. H. and D. L. B. Jupp, 1988, "The Use of Variogram in Remote Sensing and Simulated Image, II: Real Digital Images," *Remote Sensing of Environment*, 25:349–379.

Zha, Y., Gao, J. and S. Ni, 2003, "Use of Normalized Difference Built-up Index in Automatically Mapping Urban Areas from TM Imagery," *International Journal of Remote Sensing*, 24(3):583–594.

Zhang, Y., 2001, "Texture-Integrated Classification of Urban Treed Areas in High-resolution Color-infrared Imagery," *Photogrammetric Engineering & Remote Sensing*, 67(12):1359–1365.

Zhao, G. and A. L Maclean, 2000, "A Comparison of Canonical Discriminant Analysis and Principal Component Analysis for Spectral Transformation," *Photogrammetric Engineering & Remote Sensing*, 66(7):841–847.

Thematic Information Extraction: Pattern Recognition

9

Remotely sensed data of the Earth (or other extraterrestrial bodies) may be analyzed to extract thematic information. Note that *data* are transformed into *information*. Remote sensing has increasingly been used as a source of information for characterizing land-use and land-cover change at local, regional, and global scales (Townshend and Justice, 2002; Lunetta and Lyons, 2003). Land-use/land-cover classification based on statistical pattern recognition techniques applied to multispectral remote sensor data is one of the most often used methods of information extraction (Narumalani et al., 2002). This procedure assumes that imagery of a specific geographic area is collected in multiple bands of the electromagnetic spectrum and that the images are in good geometric registration. General steps required to extract land-cover information from digital multispectral remote sensor data are summarized in Figure 9-1.

The actual multispectral classification may be performed using a variety of methods, including (Figures 9-1 and 9-2):

- algorithms based on *parametric* and *nonparametric* statistics that use ratio- and interval-scaled data and *nonmetric* methods that can also incorporate nominal scale data (Duda et al., 2001),

- the use of *supervised* or *unsupervised* classification logic,

- the use of *hard* or *soft (fuzzy) set classification* logic to create hard or fuzzy thematic output products,

- the use of *per-pixel* or *object-oriented classification* logic, and

- *hybrid* approaches.

Parametric methods such as maximum likelihood classification and unsupervised clustering assume normally distributed remote sensor data and knowledge about the forms of the underlying class density functions (Duda et al., 2001). *Nonparametric* methods such as nearest-neighbor classifiers, fuzzy classifiers, and neural networks may be applied to remote sensor data that are not normally distributed and without the assumption that the forms of the underlying densities are known (e.g., Friedl et al., 2002; Liu et al., 2002; Qui and Jensen, 2004). *Nonmetric* methods such as rule-based decision tree classifiers can operate on both real-valued data (e.g., reflectance values from 0 to

100%) and nominal scaled data (e.g., class 1 = forest; class 2 = agriculture) (see Tullis and Jensen, 2003; Stow et al., 2003).

In a *supervised classification,* the identity and location of some of the land-cover types (e.g., urban, agriculture, or wetland) are known a priori through a combination of field-work, interpretation of aerial photography, map analysis, and personal experience (Hodgson et al., 2003). The analyst attempts to locate specific sites in the remotely sensed data that represent homogeneous examples of these known land-cover types. These areas are commonly referred to as *training sites* because the spectral characteristics of these known areas are used to train the classification algorithm for eventual land-cover mapping of the remainder of the image. Multivariate statistical parameters (means, standard deviations, covariance matrices, correlation matrices, etc.) are calculated for each training site. Every pixel both within and outside the training sites is then evaluated and assigned to the class of which it has the highest likelihood of being a member.

In an *unsupervised classification*, the identities of land-cover types to be specified as classes within a scene are not generally known a priori because ground reference information is lacking or surface features within the scene are not well defined. The computer is required to group pixels with similar spectral characteristics into unique clusters according to some statistically determined criteria (Duda et al., 2001). The analyst then relabels and combines the spectral clusters into information classes.

Supervised and unsupervised classification algorithms typically use *hard classification* logic to produce a classification map that consists of hard, discrete categories (e.g., forest, agriculture) (Figure 9-2a). Conversely, it is also possible to use *fuzzy set classification* logic, which takes into account the heterogeneous and imprecise nature of the real world (Figure 9-2b). Fuzzy classification produces thematic output products that contain fuzzy information. Fuzzy classification is based on the fact that remote sensing detectors record the reflected or emitted radiant flux from heterogeneous mixtures of biophysical materials such as soil, water, and vegetation found within the IFOV (Foody, 1996; 2000; Karaska et al., 1997). The land-cover classes found within the IFOV (pixel) often grade into one another without sharp, hard boundaries. Thus, reality is actually very imprecise and heterogeneous; that is, it is fuzzy (Ji and Jensen, 1999). Instead of being assigned to just a single class out of *m* possible classes, each pixel in a fuzzy classification has *m* membership grade values that describe the proportion of the *m* land cover types found within the pixel (e.g., 10% bare soil, 10%

scrub-shrub, 80% forest) (Figure 9-2b). This information may be used to extract more precise land-cover information, especially concerning the makeup of mixed pixels (Ji and Jensen, 1996; Foody, 2002).

In the past, most digital image classification was based on processing the entire scene pixel by pixel. This is commonly referred to as *per-pixel classification* (Blaschke and Strobl, 2001). *Object-oriented classification* techniques allow the analyst to decompose the scene into many relatively homogenous image objects (referred to as patches or segments) using a multiresolution image segmentation process (Baatz et. al., 2001). The various statistical characteristics of these homogeneous image objects in the scene are then subjected to traditional statistical or fuzzy logic classification. Object-oriented classification based on image segmentation is often used for the analysis of high-spatial-resolution imagery (e.g., 1×1 m Space Imaging IKONOS and 0.61×0.61 m Digital Globe QuickBird).

No pattern classification method is inherently superior to any other. The nature of the classification problem, the biophysical characteristics of the study area, the distribution of the remotely sensed data (e.g., normally distributed), and a priori knowledge determine which classification algorithm will yield useful results. Duda et al. (2001) provide sound advice: "We should have a healthy skepticism regarding studies that purport to demonstrate the overall superiority of a particular learning or recognition algorithm."

This chapter introduces some of the common fundamental statistical pattern recognition techniques used to extract land-cover information from multispectral remote sensor data including supervised and unsupervised classification using parametric and nonparametric techniques. Fundamental characteristics of fuzzy classification are introduced. Finally, object-oriented image segmentation and classification principles are presented. Nonmetric methods including neural networks and expert system decision-tree classifiers are introduced in Chapter 10 (Jensen et al., 2000, 2001). The algorithms required to extract useful information from hyperspectral imagery are discussed in Chapter 11.

Supervised Classification

Useful thematic information may be obtained using supervised classification algorithms if the general steps summarized in Figure 9-1 are understood and applied. The analyst first specifies the geographic region of interest (ROI) on which to test hypotheses. The classes of interest to be exam-

General Steps Used to Extract Thematic Land-Cover Information from Digital Remote Sensor Data

State the nature of the land-cover classification problem.
* Specify the geographic region of interest.
* Define the classes of interest.
* Determine if it is to be a hard or fuzzy classification.
* Determine if it is to be a per-pixel or object-oriented classification.

Acquire appropriate remote sensing and initial ground reference data.
* Select remotely sensed data based on the following criteria:
  - Remote sensing system considerations
    - Spatial, spectral, temporal, and radiometric resolution
  - Environmental considerations
    - Atmospheric, soil moisture, phenological cycle, etc.
* Obtain initial ground reference data based on:
  - A priori knowledge of the study area

Process remote sensor data to extract thematic information.
* Radiometric correction (or normalization) (Chapter 6).
* Geometric correction (Chapter 7).
* Select appropriate image classification logic:
  - Parametric (e.g., maximum likelihood, clustering)
  - Nonparametric (e.g., nearest-neighbor, neural network)
  - Nonmetric (e.g., rule-based decision-tree classifier)
* Select appropriate image classification algorithm:
  - Supervised, e.g.,
    - Parallelepiped, minimum distance, maximum likelihood
    - Others (hyperspectral matched filtering, spectral angle mapper – Chapter 11)
  - Unsupervised, e.g.,
    - Chain method, multiple-pass ISODATA
    - Others (fuzzy c-means)
  - Hybrid involving artificial intelligence (Chapter 10)
    - Expert system decision-tree, neural network
* Extract data from initial training sites (if required).
* Select the most appropriate bands using feature selection criteria:
  - Graphical (e.g., cospectral plots)
  - Statistical (e.g., transformed divergence, TM-distance)
* Extract training statistics and rules based on:
  - Final band selection (if required), and/or
  - Machine-learning (Chapter 10)
* Extract thematic information:
  - For each pixel or for each segmented image object (supervised)
  - Label pixels or image objects (unsupervised)

Perform accuracy assessment (Chapter 13).
* Select method:
  - Qualitative confidence-building
  - Statistical measurement
* Determine number of samples required by class.
* Select sampling scheme.
* Obtain ground reference test information.
* Create and analyze error matrix:
  - Univariate and multivariate statistical analysis.

Accept or reject previously stated hypothesis.
Distribute results if accuracy is acceptable.

Figure 9-1   The general steps used to extract thematic land-cover information from digital remote sensor data.

Classification of Remotely Sensed Data Based on Hard versus Fuzzy Logic

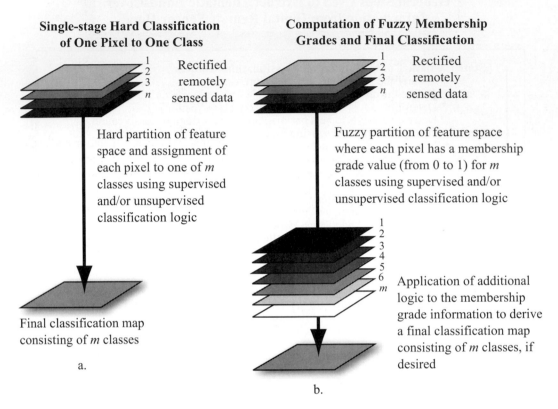

Single-stage Hard Classification of One Pixel to One Class

1
2
3
n
Rectified remotely sensed data

Hard partition of feature space and assignment of each pixel to one of m classes using supervised and/or unsupervised classification logic

Final classification map consisting of m classes

a.

Computation of Fuzzy Membership Grades and Final Classification

1
2
3
n
Rectified remotely sensed data

Fuzzy partition of feature space where each pixel has a membership grade value (from 0 to 1) for m classes using supervised and/or unsupervised classification logic

1
2
3
4
5
6
m

Application of additional logic to the membership grade information to derive a final classification map consisting of m classes, if desired

b.

Figure 9-2   Relationship between a traditional single-stage hard classification using supervised or unsupervised classification logic and a classification based on the use of fuzzy logic.

ined are then carefully defined in a classification scheme. The classes of interest normally dictate whether the analyst should produce hard or fuzzy output products and whether per-pixel or object-oriented classification logic should be used (Green and Congalton, 2003). Next, the analyst obtains the appropriate digital remote sensor data, keeping in mind both sensor system and environmental constraints. Ideally ground reference information is obtained at the same time as the remote sensing data acquisition. The remote sensor data are radiometrically and geometrically corrected as discussed in previous chapters. An appropriate classification algorithm is then selected and initial training data are collected (if necessary). Feature (band) selection is performed to determine the multispectral bands optimal for discriminating one training class from another. Additional training data are collected (if necessary) and the classification algorithm is applied, yielding a classification map. A rigorous accuracy assessment (error evaluation) is then performed (see Chapter 13). The classification maps and associated statistics are then distributed to colleagues and agencies when the results are acceptable.

Land-use and Land-cover Classification Schemes

Land cover refers to the type of material present on the landscape (e.g., water, sand, crops, forest, wetland, human-made materials such as asphalt). *Land use* refers to what people do on the land surface (e.g., agriculture, commerce, settlement). The pace, magnitude, and scale of human alterations of the Earth's land surface are unprecedented in human history. Consequently, land-cover and land-use data are central to such United Nations' *Agenda 21* issues as combating deforestation, managing sustainable settlement growth, and protecting the quality and supply of water resources (DeFries and Townshend, 1999; Jensen et al., 2002). In light of the human impacts on the landscape, there is a need to establish baseline datasets against which changes in land cover and land use can be assessed (Lunetta and Elvidge, 1998).

The International Geosphere-Biosphere Programme (IGBP) and the International Human Dimensions of Global Environmental Change Programme (IHDP) suggest that:

over the coming decades, the global effects of land use and cover change may be as significant, or more so, than those associated with potential climate change. Unlike climate change per se, land use and cover change are known and undisputed aspects of global environmental change. These changes and their impacts are with us now, ranging from potential climate warming to land degradation and biodiversity loss and from food production to spread of infectious diseases.

Land-cover data have proved especially valuable for predicting the distribution of both individual species and species assemblages across broad areas that could not otherwise be surveyed. Various predictive models have gained currency as the availability and accuracy of land-cover datasets have improved. For example, remote sensing–derived land-cover information is used extensively in the Gap Analysis Program (GAP), which is the largest species distribution modeling effort. The goal is to develop detailed maps of habitat preferences for target species and monitor plant phenology (Kerr and Ostrovsky, 2003).

All classes of interest must be selected and defined carefully to classify remotely sensed data successfully into land-use and/or land-cover information (Lunetta et al., 1991; Congalton and Green, 1999). This requires the use of a *classification scheme* containing taxonomically correct definitions of classes of information that are organized according to logical criteria. If a hard classification is to be performed, then the classes in the classification system should normally be:

- mutually exclusive,

- exhaustive, and

- hierarchical.

Mutually exclusive means that there is no taxonomic overlap (or fuzziness) of any classes (i.e., deciduous forest and evergreen forest are distinct classes). *Exhaustive* means that all land-cover classes present in the landscape are accounted for and none have been omitted. *Hierarchical* means that sublevel classes (e.g., single-family residential, multiple-family residential) may be hierarchically combined into a higher-level category (e.g., residential) that makes sense. This allows simplified thematic maps to be produced when required.

It is also important for the analyst to realize that there is a fundamental difference between information classes and spectral classes. *Information classes* are those that human

beings define. Conversely, *spectral classes* are those that are inherent in the remote sensor data and must be identified and then labeled by the analyst. For example, in a remotely sensed image of an urban area there is likely to be single-family residential housing. A relatively coarse spatial resolution remote sensor such as SPOT (20×20 m) might be able to record a few relatively pure pixels of vegetation and a few pure pixels of concrete/asphalt road or asphalt shingles. However, it is more likely that in this residential area the pixel brightness values will be a function of the reflectance from mixtures of vegetation and asphalt/concrete. Few planners or administrators want to see a map labeled with classes like (1) concrete, (2) asphalt, (3) vegetation, and (4) mixture of vegetation and concrete/asphalt (unless they are interested in a map of impervious surfaces). Rather, they typically prefer the analyst to relabel the mixture class as single-family residential. The analyst should do this only if in fact there is a good association between the mixture class and single-family residential housing. Thus, we see that an analyst must often translate *spectral classes* into *information classes* to satisfy bureaucratic requirements. An analyst should understand well the spatial and spectral characteristics of the sensor system and be able to relate these system parameters to the types and proportions of materials found within the scene and within pixel IFOVs. If these parameters are understood, spectral classes often can be thoughtfully relabeled as information classes.

Certain hard classification schemes have been developed that can readily incorporate land-use and/or land-cover data obtained by interpreting remotely sensed data. Only a few will be discussed here, including:

- American Planning Association *Land-Based Classification System* which is oriented toward detailed land-use classification (Figure 9-3);

- United States Geological Survey *Land-Use/Land-Cover Classification System for Use with Remote Sensor Data* and its adaptation for the U.S. National Land Cover Dataset and the NOAA Coastal Change Analysis Program (C-CAP);

- U.S. Department of the Interior Fish & Wildlife Service, *Classification of Wetlands and Deepwater Habitats of the United States*;

- U.S. *National Vegetation and Classification System*;

- International Geosphere-Biosphere Program *IGBP Land Cover Classification System* modified for the creation of MODIS land-cover products.

American Planning Association Land-Based Classification System

Table 1: Activity

Parcel ID	Activity	Description
10-5-100	2100	shopping
10-5-100	2200	restaurant
10-5-100	6600	social, religious assembly
10-5-100	2100	furniture
10-5-100	5210	vehicular parking
etc.	etc.	etc.

Table 2: Function

Function	Description
2110	retail sales and services
2510	full-service restaurant
6620	religious institutions
2121	furniture
5200	parking facilities
etc.	etc.

Table 3: Structure

Parcel ID	Structure	Description
10-5-100	2500	mall, shopping center

Figure 9-3   The American Planning Association developed the *Land-Based Classification System* (LBCS) that contains detailed definitions of urban/suburban land use. The system incorporates information derived *in situ* and using remote sensing techniques. This is an oblique aerial photograph of a mall in Ontario, CA. Hypothetical activity and structure codes associated with this large parcel are identified. Site development and ownership information attribute tables are not shown (concept courtesy American Planning Association).

American Planning Association *Land-Based Classification Standard*

Few classification schemes attempt to classify land use. In fact, most explicitly state that they are concerned only with land-cover information. Therefore, if the user is most interested in extracting urban/suburban land-use information from relatively high-spatial-resolution remote sensor data, then the most practical and comprehensive hierarchical classification system is the *Land-Based Classification Standard* (LBCS) developed by the American Planning Association (2004a). This standard updates the *Standard Land Use Coding Manual* (SLUCM) (Urban Renewal Administration, 1965), which is cross-referenced with the *1987 Standard*

Industrial Classification (SIC) *Manual* (Bureau of the Budget, 1987) and the updated *North American Industrial Classification Standard* (NAICS).

The LBCS provides the following conversions of different land-use classification systems:

- SLUCM to LBCS,

- SIC to LBCS,

- NAICS to LBCS, and

- Department of Defense Air Force Land-use Compatibility Zones Coding to LBCS.

The LBCS requires input from *in situ* surveys, aerial photography, and satellite remote sensor data to obtain information at the parcel level on the following five characteristics: *activity, function, site development, structure,* and *ownership* (American Planning Association, 2004b). The system provides a unique code and description for almost every commercial and industrial land-use activity. Examples of aerial photography annotated with LBCS coding are found at American Planning Association (2004c). An example of a mall in Ontario, CA, is shown in Figure 9-3 along with the associated LBCS activity, function, and structure codes.

The LBCS is always under development. Users are encouraged to keep abreast of the LBCS and to use it for very intensive urban/suburban studies that require detailed commercial and industrial land-use classification codes. The LBCS does not provide information on land-cover or vegetation characteristics in the urban environment; it relies on the Federal Geographic Data Committee standards on this topic.

U.S. Geological Survey *Land-Use/Land-Cover Classification System for Use with Remote Sensor Data*

The U.S. Geological Survey's *Land-Use/Land-Cover Classification System for Use with Remote Sensor Data* (Anderson et al., 1976) is primarily a resource-oriented land-cover classification system in contrast with people or activity land-use classification systems such as the APA's *Land-Based Classification System*. The USGS rationale is that "although there is an obvious need for an urban-oriented land-use classification system, there is also a need for a resource-oriented classification system whose primary emphasis would be the remaining 95 percent of the United States land area." The USGS system addresses this need with eight of the nine original Level I categories that treat land area that is not in urban

Table 9-1. U.S. Geological Survey *Land-Use/Land-Cover Classification System for Use with Remote Sensor Data* (Anderson et al., 1976).

Classification Level

1 Urban or Built-up Land
11 Residential
12 Commercial and Services
13 Industrial
14 Transportation, Communications, and Utilities
15 Industrial and Commercial Complexes
16 Mixed Urban or Built-up
17 Other Urban or Built-up Land

2 Agricultural Land
21 Cropland and Pasture
22 Orchards, Groves, Vineyards, Nurseries, and Ornamental Horticultural Areas
23 Confined Feeding Operations
24 Other Agricultural Land

3 Rangeland
31 Herbaceous Rangeland
32 Shrub–Brushland Rangeland
33 Mixed Rangeland

4 Forest Land
41 Deciduous Forest Land
42 Evergreen Forest Land
43 Mixed Forest Land

5 Water
51 Streams and Canals
52 Lakes
53 Reservoirs
54 Bays and Estuaries

6 Wetland
61 Forested Wetland
62 Nonforested Wetland

7 Barren Land
71 Dry Salt Flats
72 Beaches
73 Sandy Areas Other Than Beaches
74 Bare Exposed Rock
75 Strip Mines, Quarries, and Gravel Pits
76 Transitional Areas
77 Mixed Barren Land

8 Tundra
81 Shrub and Brush Tundra
82 Herbaceous Tundra
83 Bare Ground Tundra
84 Wet Tundra
85 Mixed Tundra

9 Perennial Snow or Ice
91 Perennial Snowfields
92 Glaciers

Table 9-2.   Four Levels of the U.S. Geological Survey *Land-Use/Land-Cover Classification System for Use with Remote Sensor Data* and the type of remotely sensed data typically used to provide the information (Anderson et al., 1976; Jensen and Cowen, 1999).

Classification Level	Typical Data Characteristics
I	Satellite imagery such as NOAA AVHRR (1.1×1.1 km), MODIS (250×250 m; 500×500 m), Landsat MSS (79×79 m), Landsat Thematic Mapper (30×30 m), and SPOT XS (20×20 m).
II	Satellite imagery such as SPOT HRV multispectral (10×10 m) and Indian IRS 1-C panchromatic (5×5 m). High-altitude aerial photography acquired at scales smaller than 1:80,000.
III	Satellite imagery with 1×1 m to 2.5×2.5 m nominal spatial resolution. Medium-altitude aerial photography at scales from 1:20,000 to 1:80,000.
IV	Satellite imagery with $\leq 1 \times 1$ m nominal spatial resolution (e.g., QuickBird, IKONOS). Low-altitude aerial photography at scales from 1:4,000 to 1:20,000 scale.

Table 9-3.   U.S. Geological Survey *Land-Use/Land-Cover Classification System for Use with Remote Sensor Data* modified for the National Land Cover Dataset and NOAA Coastal Change Analysis Program (NOAA, 2004).

Classification Level
1 **Water**
11 Open Water
12 Perennial Ice/Snow
2 **Developed**
21 Low-Intensity Residential
22 High-Intensity Residential
23 Commercial/Industrial/Transportation
3 **Barren**
31 Bare Rock/Sand/Clay
32 Quarries/Strip Mines/Gravel Pits
33 Transitional
4 **Forested Upland**
41 Deciduous Forest
42 Evergreen Forest
43 Mixed Forest
5 **Shrubland**
51 Shrubland
6 **Non-Natural Woody**
61 Orchards/Vineyards, Other
7 **Herbaceous Upland Natural/Seminatural Vegetation**
71 Grasslands/Herbaceous
8 **Herbaceous Planted/Cultivated**
81 Pasture/Hay
82 Row Crops
83 Small Grains
84 Fallow
85 Urban/Recreation
86 Grasses
9 **Wetland**
91 Woody Wetlands
92 Emergent Herbaceous Wetlands

or built-up categories (Table 9-1). The system is designed to be driven primarily by the interpretation of remote sensor data obtained at various scales and resolutions (Table 9-2) and not data collected *in situ*. The classification system continues to be modified to support a variety of land-cover mapping activities by the U.S. Geological Survey, Environmental Protection Agency, NOAA Coastal Services Center, and others.

For example, the U.S. Geological Survey's Land-Cover Characterization Program (LCCP) was initiated in 1995 to address national and international requirements for land-cover data. The program supports:

• *Global Land-Cover Characterization*: The goal, to develop a global 1-km land-cover characteristics database, was met in 1997. Current efforts focus on revising the database using input from users around the world.

• *National Land-Cover Characterization*: Continuous production of a U.S. National Land-Cover Dataset

(NLCD) based upon 30-m Landsat Thematic Mapper or other data.

Table 9-3 documents how the *Land-Use/Land-Cover Classification System for Use with Remote Sensor Data* has been modified to include 22 classes. Many of the Level II classes are best derived using aerial photography. It is not appropriate to attempt to derive some of these classes using Landsat TM data due to issues of spatial resolution and interpretability. Thus, no attempt was made to derive classes that were extremely difficult or impractical to obtain using Landsat

TM data, such as the Level III urban classes. In addition, some Anderson Level II classes were consolidated into a single NLCD class (USGS, 2004).

The same classes are also used by the NOAA Coastal Services Center in the creation of the Coastal Change Analysis Program (C-CAP) products. Baseline C-CAP products include land cover for the most current date available, a 5-year retrospective land-cover product, and a product that illustrates the changes between the two dates (Jensen et al., 1993; Dobson et al., 1995; NOAA, 2004).

The Multi-Resolution Land Characteristics (MRLC) consortium group of federal agencies originally joined forces in 1992 to purchase Landsat imagery for the United States and to develop a land-cover dataset. One of the results of this consortium was the September, 2000 completion of the conterminous National Land-Cover Dataset from circa 1992 Landsat TM data at Anderson et al. (1976) Level II thematic detail. Beginning in 1999, a second-generation consortium was formed to create a new Landsat image and land-cover database for the nation called MRLC 2001, designed to meet the need of federal agencies for nationally consistent satellite remote sensing and land-cover data. The MRLC 2001 image database covers 1780 Landsat Thematic Mapper paths and rows. The second goal of the MRLC consortium is the creation of an updated National Land-Cover Database covering all 50 states and Puerto Rico. To be compatible with the 1992 NLCD, MRLC 2001 will build on and modify the land-cover classification system in Table 9-3.

U.S. Department of the Interior Fish & Wildlife Service
Classification of Wetlands and Deepwater Habitats of the United States

The conterminous United States continues to lose inland and coastal wetland to agricultural, residential, and commercial land-use development. The U.S. Department of the Interior Fish & Wildlife Service is responsible for mapping and inventorying wetland in the United States. Therefore, they developed a wetland classification system that incorporates information extracted from remote sensor data and *in situ* measurement (Cowardin et al., 1979).

The Cowardin system describes ecological taxa, arranges them in a system useful to resource managers, and provides uniformity of concepts and terms. Wetlands are classified based on plant characteristics, soils, and frequency of flooding. Ecologically related areas of deep water, traditionally not considered wetlands, are included in the classification as

deep-water habitats. Five systems form the highest level of the classification hierarchy: marine, estuarine, riverine, lacustrine, and palustrine (Figure 9-4). Marine and estuarine systems each have two subsystems: subtidal and intertidal. The riverine system has four subsystems: tidal, lower perennial, upper perennial, and intermittent. The lacustrine has two, littoral and limnetic, and the palustrine has no subsystem. Within the subsystems, classes are based on substrate material and flooding regime or on vegetative life form. The same classes may appear under one or more of the systems or subsystems. The distinguishing features of the riverine system are shown in Figure 9-5. This was the first nationally recognized wetland classification scheme.

The Cowardin system was adopted as the National Vegetation Classification Standard for wetlands mapping and inventory by the Wetlands Subcommittee of the Federal Geographic Data Committee (FGDC, 1996). The Cowardin wetland classification system is the most practical scheme to use if you are going to extract wetland information from remotely sensed data and share the information with others interested in wetland-related problems.

U.S. *National Vegetation Classification System*

The Vegetation Subcommittee of the Federal Geographic Data Committee has endorsed the *National Vegetation Classification System* (NVCS) which produces uniform vegetation resource data at the national level (FGDC Vegetation Subcommittee, 1997; FGDC, 2004). The NVCS uses a systematic approach to classifying a continuum of natural, existing vegetation. The combined physiognomic-floristic hierarchy uses both qualitative and quantitative data appropriate for conservation and mapping at various scales. As can be seen in Table 9-4, physiognomic characteristics include the more general and less precise levels of taxonomy, whereas the floristic characteristics are found in the more specific levels of taxonomy.

Currently, standards for physiognomic classification are well developed and included in the NVCS (ESA, 2004a). Floristic standards have been only partially developed. The Ecological Society of America (ESA) Panel on Vegetation Classification is developing its recommendations for the floristic standards titled "An Initiative for a Standardized Classification of Vegetation in the United States" (ESA, 2004b). The *National Vegetation Classification System* is a good place to begin if a scientist is interested in extracting detailed vegetation information from remote sensor data and placing it in an ecologically sound vegetation classification system.

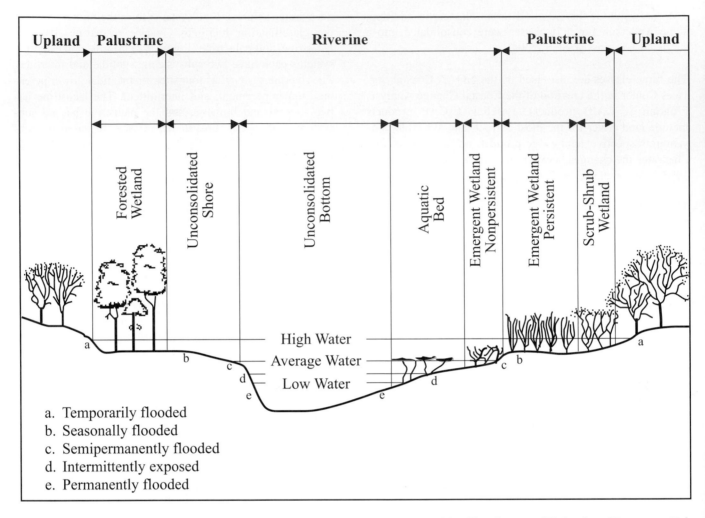

Figure 9-4   Distinguishing features and examples of habitats in the riverine system of the *Classification of Wetlands and Deepwater Habitats of the United States* (Cowardin et al., 1979).

International Geosphere-Biosphere Program *IGBP Land-Cover Classification System* Modified for the Creation of MODIS Land-Cover Products

If a scientist is interested in inventorying land cover at the regional, national, and global scale, then the modified International Geosphere-Biosphere Program *Land-Cover Classification System* may be appropriate (Loveland et al., 1999). For example, the Moderate Resolution Imaging Spectroradiometer (MODIS) of NASA's Earth Observing System (EOS) is providing global land-surface information at spatial resolutions of 250 to 1,000 m. There are approximately 44 standard MODIS-derived data products that scientists are using to study global change. For example, the MODIS Land Science Team is producing a global land-cover change product at 1-km (0.6 mile) resolution to depict broad-scale land-cover changes.

The land-cover type and land-cover change parameters are produced at 1-km resolution on a quarterly basis. The land-cover parameter identifies 17 categories of land-cover following the IGBP global vegetation database (Table 9-5), which defines nine classes of natural vegetation, three classes of developed lands, two classes of mosaic lands, and three classes of nonvegetated lands (snow/ice, bare soil/rocks, water) (Strahler, 1999; NASA GSFC, 2004). The first global land-cover map based on MODIS data was distributed in August, 2002. The 1-km resolution maps are based on MODIS images collected between November, 2000, and October, 2001 (NASA Earth Observatory, 2002).

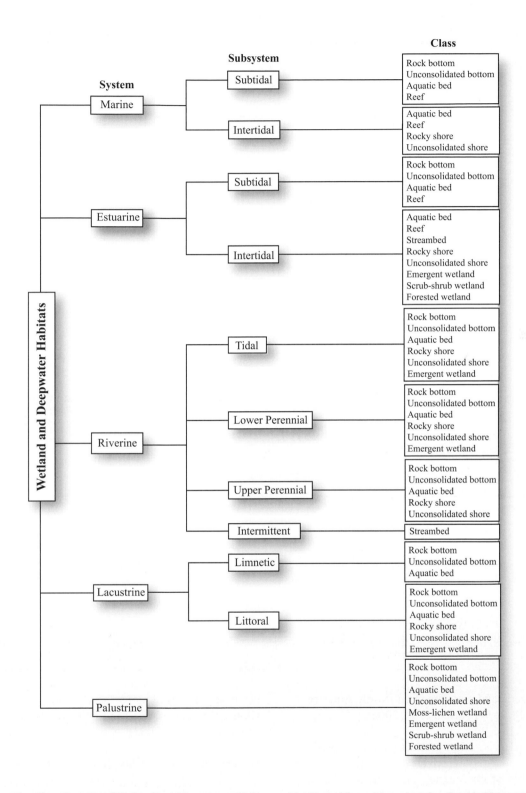

Figure 9-5   The *Classification of Wetlands and Deepwater Habitats of the United States* hierarchy of wetland habitat systems, subsystems, and classes (Cowardin et al., 1979). The palustrine system does not include deepwater habitats. The Cowardin system is the National Vegetation Classification Standard for wetlands mapping and inventory (Wetlands Subcommittee of the Federal Geographic Data Committee; FGDC, 2004).

Table 9-4.   *National Vegetation Classification System* logic, levels, and primary basis for classification (FGDC Vegetation Committee, 1997; FGDC, 2004).

Logic

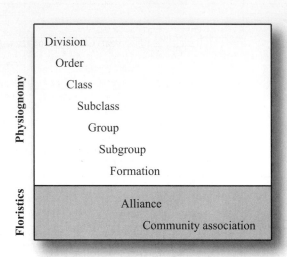

Level	Primary Basis for Classification	Example
Order	Dominant vegetation	Tree dominated
Class	Growth form and structure of vegetation	Woodland
Subclass	Growth form, e.g., leaf phenology	Deciduous woodland
Group	Leaf types, corresponding to climate	Cold-deciduous woodland
Subgroup	Relative human impact (natural/seminatural, or cultural)	Natural/seminatural
Formation	Additional physiognomic and environmental factors, including hydrology	Temporarily flooded cold-deciduous woodland
Alliance	Dominant/diagnostic species of uppermost or dominant stratum	*Populus deltoides* temporarily flooded woodland alliance
Community association	Additional dominant/diagnostic species from any stratum	*Populus deltoides* (*Salix Amygdaloides/ Salix exigua*) woodland

A MODIS-derived land-cover map of North America is shown in **Color Plate 9-1** (Friedl et al., 2002).

In addition to the basic 1-km product, summary products containing proportions of land covers and change characteristics are available at one-quarter, one-half, and 1-degree resolutions. These products are prepared independently from the 250-m Land-Cover Change Product and use a different algorithm (EROS, 2003). The University of Maryland provides a product at 250 m resolution to depict land-cover changes caused by anthropogenic activities, which generally occur at finer resolutions than 1 km (Townshend, 1999). Global plant phenology information may also become available (Zhang et al., 2003).

Observations about Classification Schemes

Geographical information (including remote sensor data) is often imprecise. For example, there is usually a gradual transition at the interface of forests and rangeland, yet many of the aforementioned classification schemes insist on a hard boundary between the classes at this transition zone. The

Table 9-5. *IGBP Land-Cover Classification System* modified for the creation of MODIS Land-Cover Products (Strahler, 1999; Loveland et al., 1999; Friedl et al., 2002; NASA Earth Observatory, 2002).

Natural Vegetation	
Forest	
Evergreen Needleleaf	Dominated by woody vegetation with a cover >60% and height > 2 m. Almost all trees remain green all year. Canopy is never without green foliage.
Evergreen Broadleaf	Dominated by woody vegetation with a cover >60% and height > 2 m. Most trees and shrubs remain green year round. Canopy is never without green foliage.
Deciduous Needleleaf	Dominated by woody vegetation with a cover >60% and height > 2 m. Seasonal needleleaf communities with an annual cycle of leaf-on and leaf-off periods.
Deciduous Broadleaf	Dominated by woody vegetation with a cover >60% and height > 2 m. Broadleaf tree communities with an annual cycle of leaf-on and leaf-off periods.
Mixed Forest	Dominated by trees with a cover >60% and height > 2 m. Communities with interspersed mixtures or mosaics of the other four forest types. None of the forest types exceeds 60% of landscape.
Shrubland, Grasslands, and Wetland	
Closed Shrublands	Woody vegetation <2 m tall and with shrub canopy cover >60%. The shrub foliage can be either evergreen or deciduous.
Open Shrublands	Woody vegetation <2 m tall and with shrub canopy cover between 10% and 60%. The shrub foliage can be either evergreen or deciduous.
Woody Savannas	Herbaceous and other understory systems, and with forest canopy cover between 30% and 60%. The forest cover height >2 m.
Savannas	Herbaceous and other understory systems, and with forest canopy cover between 10% and 30%. The forest cover height >2 m.
Grasslands	Lands with herbaceous types of cover. Tree and shrub cover is <10%.
Permanent Wetlands	Lands with a permanent mixture of water and herbaceous or woody vegetation. Vegetation can be present in salt, brackish, or fresh water.
Developed and Mosaic Lands	
Agriculture	
Croplands	Lands covered with temporary crops followed by harvest and a bare-soil period (e.g., single and multiple cropping systems). Perennial woody crops are classified as the appropriate forest or shrub land-cover type.
Cropland/Natural Vegetation Mosaic	
Cropland/Natural Vegetation Mosaic	Mosaic of croplands, forests, shrubland, and grasslands in which no one component comprises > 60% of the landscape.
Urban	
Built-up	Land covered by buildings and other human-made structures.
Nonvegetated Lands	
Barren	
Barren or Sparsely Vegetated	Exposed soil, sand, rocks, or snow and ≤10% vegetated cover during the year.
Snow and Ice	
Snow and Ice	Lands under snow/ice cover throughout the year.
Water	
Water	Oceans, seas, lakes, reservoirs, and rivers. Can be fresh or salt water.

schemes should contain fuzzy definitions because the thematic information they contain is fuzzy (Wang, 1990a). Fuzzy classification schemes are not currently standardized. They are typically developed by individual researchers for site-specific projects. The fuzzy classification systems developed may not be transferable to other environments. Therefore, we tend to see the use of existing hard classification schemes, which are rigid, based on a priori knowledge, and generally difficult to use. They continue to be widely employed because they are scientifically based and different

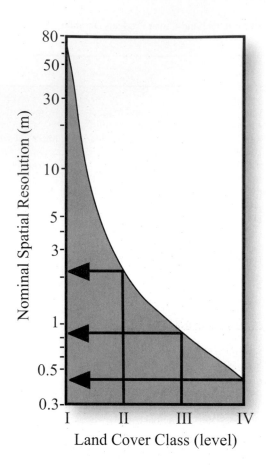

Figure 9-6   Nominal spatial resolution requirements as a function of the mapping requirements for Levels I to IV land-cover classes in the United States (based on Anderson et al., 1976). Note the dramatic increase in spatial resolution required to map Level II classes (adapted from Welch, 1982; Jensen and Cowen, 1999).

individuals using the same classification system can compare results.

This brings us to another important consideration. If a reputable classification system already exists, it is foolish to develop an entirely new system that will probably be used only by ourselves. It is better to adopt or modify existing nationally or internationally recognized classification systems. This allows us to interpret the significance of our classification results in light of other studies and makes it easier to share data.

Finally, it should be reiterated that there is a general relationship between the level of detail in a classification scheme and the spatial resolution of remote sensor systems used to

provide information. Welch (1982) summarizes this relationship for mapping urban/suburban land use and land cover in the United States (Figure 9-6). A similar relationship exists when mapping vegetation (Botkin et al., 1984). For example, the sensor systems and spatial resolutions useful for discriminating vegetation from a global to an *in situ* perspective are summarized in Figure 9-7. This suggests that the level of detail in the desired classification system dictates the spatial resolution of the remote sensor data that should be used. Of course, the spectral resolution of the remote sensing system is also an important consideration, especially when inventorying vegetation, water, ice, snow, soil, and rock.

Training Site Selection and Statistics Extraction

An image analyst may select *training sites* within the image that are representative of the land-cover or land-use classes of interest after the classification scheme is adopted. The training data should be of value if the environment from which they were obtained is relatively homogeneous. For example, if all the soils in a grassland region are composed of well-drained sandy-loam soil, then it is likely that grassland training data collected throughout the region would be representative. However, if the soil conditions change dramatically across the study area (e.g., one-half of the region has a perched water table with moist near-surface soil), it is likely that grassland training data acquired in the dry-soil part of the study area will *not* be representative of the spectral conditions for grassland found in the moist-soil portion of the study area. Thus, we have a *geographic signature extension* problem, meaning that it may not be possible to extend our grassland remote sensing training data through *x, y* space.

The easiest way to remedy this situation is to apply *geographical stratification* during the preliminary stages of a project. At this time all significant environmental factors that contribute to geographic signature extension problems should be identified, such as differences in soil type, water turbidity, crop species (e.g., two strains of wheat), unusual soil moisture conditions possibly caused by a thunderstorm that did not uniformly deposit its precipitation, scattered patches of atmospheric haze, and so on. Such environmental conditions should be carefully annotated on the imagery and the selection of training sites made based on the geographic stratification of these data. In such cases, it may be necessary to train the classifier over relatively short geographic distances. Each individual stratum may have to be classified separately. The final classification map of the entire region will then be a composite of the individual stratum classifica-

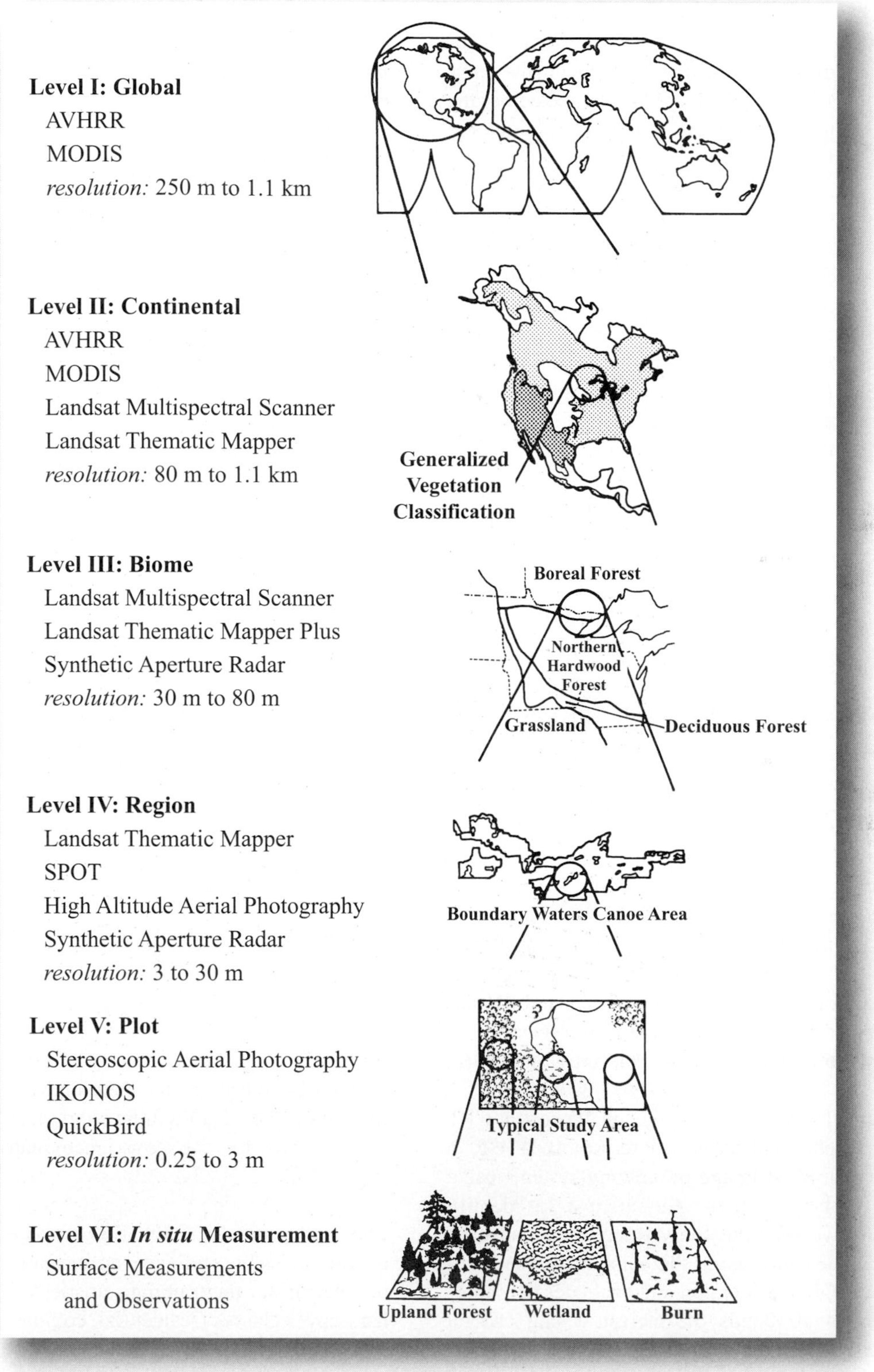

Level I: Global
 AVHRR
 MODIS
 resolution: 250 m to 1.1 km

Level II: Continental
 AVHRR
 MODIS
 Landsat Multispectral Scanner
 Landsat Thematic Mapper
 resolution: 80 m to 1.1 km

**Generalized
Vegetation
Classification**

Level III: Biome
 Landsat Multispectral Scanner
 Landsat Thematic Mapper Plus
 Synthetic Aperture Radar
 resolution: 30 m to 80 m

Boreal Forest

**Northern
Hardwood
Forest**

Grassland **Deciduous Forest**

Level IV: Region
 Landsat Thematic Mapper
 SPOT
 High Altitude Aerial Photography
 Synthetic Aperture Radar
 resolution: 3 to 30 m

Boundary Waters Canoe Area

Level V: Plot
 Stereoscopic Aerial Photography
 IKONOS
 QuickBird
 resolution: 0.25 to 3 m

Typical Study Area

Level VI: *In situ* Measurement
 Surface Measurements
  and Observations

Upland Forest **Wetland** **Burn**

Figure 9-7   Relationship between the level of detail required and the spatial resolution of representative remote sensing systems for vegetation inventories.

tions. However, if environmental conditions are homogeneous or can be held constant (e.g., through band ratioing or atmospheric correction), it may be possible to extend signatures vast distances in space, significantly reducing the training cost and effort. Additional research is required before the concept of geographic and temporal (through time) signature extension is fully understood.

Once spatial and temporal signature extension factors have been considered, the analyst selects representative training sites for each class and collects the spectral statistics for each pixel found within each training site. Each site is usually composed of many pixels. The general rule is that if training data are being extracted from n bands then >$10n$ pixels of training data are collected for each class. This is sufficient to compute the variance–covariance matrices required by some classification algorithms.

There are a number of ways to collect the *training site* data, including:

- collection of *in situ* information such as tree type, height, percent canopy closure, and diameter-at-breast-height (dbh) measurements,

- on-screen selection of polygonal training data, and/or

- on-screen seeding of training data.

Ideally, each training site is visited in the field and its perimeter and/or centroid coordinates are measured directly using a global positioning system (GPS). When the U.S. government has *selective availability* "off," the differentially corrected GPS x,y coordinates should be within ±1 m of their planimetric position. The GPS-derived x,y coordinates of a training site (e.g., a polygon encompassing a stand of oak) may then be input directly to the image processing system and used to extract training class (e.g., oak forest) statistics.

The analyst may also view the image on the color CRT screen and select polygonal areas of interest (AOI) (e.g., a stand of oak forest). Most image processing systems use a "rubber band" tool that allows the analyst to identify detailed AOIs. Conversely, the analyst may seed a specific location in the image space using the cursor. The seed program begins at a single x, y location and evaluates neighboring pixel values in all bands of interest. Using criteria specified by the analyst, the seed algorithm expands outward like an amoeba as long as it finds pixels with spectral characteristics similar to the original seed pixel. This is a

very effective way of collecting homogeneous training information.

Each pixel in each training site associated with a particular class (c) is represented by a *measurement vector, X_c:*

$$X_c = \begin{bmatrix} BV_{ij1} \\ BV_{ij2} \\ BV_{ij3} \\ \vdots \\ BV_{ijk} \end{bmatrix} \tag{9-1}$$

where $BV_{i,j,k}$ is the brightness value for the i,jth pixel in band k. The brightness values for each pixel in each band in each training class can then be analyzed statistically to yield a mean measurement vector, M_c, for each class:

$$M_c = \begin{bmatrix} \mu_{c1} \\ \mu_{c2} \\ \mu_{c3} \\ \vdots \\ \mu_{ck} \end{bmatrix} \tag{9-2}$$

where μ_{ck} represents the mean value of the data obtained for class c in band k. The raw measurement vector can also be analyzed to yield the covariance matrix for each class c:

$$V_c = V_{ckl} = \begin{bmatrix} Cov_{c11} & Cov_{c12} & \dots & Cov_{c1n} \\ Cov_{c21} & Cov_{c22} & \dots & Cov_{c2n} \\ \vdots & \vdots & \vdots & \vdots \\ Cov_{cn1} & Cov_{cn2} & \dots & Cov_{cnn} \end{bmatrix} \tag{9-3}$$

where Cov_{ckl} is the covariance of class c between bands k through l. For brevity, the notation for the covariance matrix for class c (i.e., V_{ckl}) will be shortened to just V_c. The same will be true for the covariance matrix of class d (i.e., $V_{dkl} = V_d$).

The mean, standard deviation, variance, minimum value, maximum value, variance–covariance matrix, and correlation matrix for the training statistics of five Charleston, SC, land-cover classes (residential, commercial, wetland, forest, and water) are listed in Table 9-6. These represent fundamental information on the spectral characteristics of the five classes.

Table 9-6. Univariate and multivariate training statistics for the five land-cover classes using six bands of Landsat Thematic Mapper data of Charleston, SC.

a. Statistics for Residential

	Band 1	Band 2	Band 3	Band 4	Band 5	Band 7
Univariate Statistics						
Mean	70.6	28.8	29.8	36.7	55.7	28.2
Std. dev.	6.90	3.96	5.65	4.53	10.72	6.70
Variance	47.6	15.7	31.9	20.6	114.9	44.9
Minimum	59	22	19	26	32	16
Maximum	91	41	45	52	84	48
Variance–Covariance Matrix						
1	47.65					
2	24.76	15.70				
3	35.71	20.34	31.91			
4	12.45	8.27	12.01	20.56		
5	34.71	23.79	38.81	22.30	114.89	
7	30.46	18.70	30.86	12.99	60.63	44.92
Correlation Matrix						
1	1.00					
2	0.91	1.00				
3	0.92	0.91	1.00			
4	0.40	0.46	0.47	1.00		
5	0.47	0.56	0.64	0.46	1.00	
7	0.66	0.70	0.82	0.43	0.84	1.00

b. Statistics for Commercial

	Band 1	Band 2	Band 3	Band 4	Band 5	Band 7
Univariate Statistics						
Mean	112.4	53.3	63.5	54.8	77.4	45.6
Std. dev.	5.77	4.55	3.95	3.88	11.16	7.56
Variance	33.3	20.7	15.6	15.0	124.6	57.2
Minimum	103	43	56	47	57	32
Maximum	124	59	72	62	98	57

b. Statistics for Commercial (Continued)

	Band 1	Band 2	Band 3	Band 4	Band 5	Band 7
Variance–Covariance Matrix						
1	33.29					
2	11.76	20.71				
3	19.13	11.42	15.61			
4	19.60	12.77	14.26	15.03		
5	−16.62	15.84	2.39	0.94	124.63	
7	−4.58	17.15	6.94	5.76	68.81	57.16
Correlation Matrix						
1	1.00					
2	0.45	1.00				
3	0.84	0.64	1.00			
4	0.88	0.72	0.93	1.00		
5	−0.26	0.31	0.05	0.02	1.00	
7	−0.10	0.50	0.23	0.20	0.82	1.00

c. Statistics for Wetland

	Band 1	Band 2	Band 3	Band 4	Band 5	Band 7
Univariate Statistics						
Mean	59.0	21.6	19.7	20.2	28.2	12.2
Std. dev.	1.61	0.71	0.80	1.88	4.31	1.60
Variance	2.6	0.5	0.6	3.5	18.6	2.6
Minimum	54	20	18	17	20	9
Maximum	63	25	21	25	35	16
Variance–Covariance Matrix						
1	2.59					
2	0.14	0.50				
3	0.22	0.15	0.63			
4	−0.64	0.17	0.60	3.54		
5	−1.20	0.28	0.93	5.93	18.61	
7	−0.32	0.17	0.40	1.72	4.53	2.55

c. Statistics for Wetland (Continued)

	Band 1	Band 2	Band 3	Band 4	Band 5	Band 7
Correlation Matrix						
1	1.00					
2	0.12	1.00				
3	0.17	0.26	1.00			
4	−0.21	0.12	0.40	1.00		
5	−0.17	0.09	0.27	0.73	1.00	
7	−0.13	0.15	0.32	0.57	0.66	1.00

d. Statistics for Forest

	Band 1	Band 2	Band 3	Band 4	Band 5	Band 7
Univariate Statistics						
Mean	57.5	21.7	19.0	39.1	35.5	12.5
Std. dev.	2.21	1.39	1.40	5.11	6.41	2.97
Variance	4.9	1.9	1.9	26.1	41.1	8.8
Minimum	53	20	17	25	22	8
Maximum	63	28	24	48	54	22
Variance–Covariance Matrix						
1	4.89					
2	1.91	1.93				
3	2.05	1.54	1.95			
4	5.29	3.95	4.06	26.08		
5	9.89	5.30	5.66	13.80	41.13	
7	4.63	2.34	2.22	3.22	16.59	8.84
Correlation Matrix						
1	1.00					
2	0.62	1.00				
3	0.66	0.80	1.00			
4	0.47	0.56	0.57	1.00		
5	0.70	0.59	0.63	0.42	1.00	
7	0.70	0.57	0.53	0.21	0.87	1.00

e. Statistics for Water

	Band 1	Band 2	Band 3	Band 4	Band 5	Band 7
Univariate Statistics						
Mean	61.5	23.2	18.3	9.3	5.2	2.7
Std. dev.	1.31	0.66	0.72	0.56	0.71	1.01
Variance	1.7	0.4	0.5	0.3	0.5	1.0
Minimum	58	22	17	8	4	0
Maximum	65	25	20	10	7	5
Variance–Covariance Matrix						
1	1.72					
2	0.06	0.43				
3	0.12	0.19	0.51			
4	0.09	0.05	0.05	0.32		
5	−0.26	−0.05	−0.11	−0.07	0.51	
7	−0.21	−0.05	−0.03	−0.07	0.05	1.03
Correlation Matrix						
1	1.00					
2	0.07	1.00				
3	0.13	0.40	1.00			
4	0.12	0.14	0.11	1.00		
5	−0.28	−0.10	−0.21	−0.17	1.00	
7	−0.16	−0.08	−0.04	−0.11	0.07	1.00

Sometimes the manual selection of polygons results in the collection of training data with multiple modes in a training class histogram. This suggests that there are at least two types of land cover within the training area. This condition is not good when we are attempting to discriminate among individual classes. Therefore, it is a good practice to discard multimodal training data and retrain on specific parts of the polygon of interest until unimodal histograms are derived per class.

Positive spatial *autocorrelation* exists among pixels that are contiguous or close together (Gong and Howarth, 1992). This means that adjacent pixels have a high probability of having similar brightness values. Training data collected from autocorrelated data tend to have reduced variance, which may be caused more by the way the sensor is collect-

ing the data than by actual field conditions (e.g., most detectors dwell on an individual pixel for a very short time and may smear spectral information from one pixel to an adjacent pixel). The ideal situation is to collect training data within a region using every *n*th pixel or some other sampling criterion. The goal is to get non-autocorrelated training data. Unfortunately, most digital image processing systems do not provide this option in training data–collection modules.

Selecting the Optimum Bands for Image Classification: Feature Selection

Once the training statistics have been systematically collected from each band for each class of interest, a judgment must be made to determine the bands that are most effective in discriminating each class from all others. This process is commonly called *feature selection* (Duda et al., 2001). The goal is to delete from the analysis the bands that provide redundant spectral information. In this way the *dimensionality* (i.e., the number of bands to be processed) in the dataset may be reduced. This process minimizes the cost of the digital image classification process (but should not affect the accuracy). Feature selection may involve both statistical and graphical analysis to determine the degree of between-class separability in the remote sensor training data. Using statistical methods, combinations of bands are normally ranked according to their potential ability to discriminate each class from all others using *n* bands at a time (Beauchemin and Fung, 2001). Statistical measures such as divergence will be discussed shortly.

Why use graphical methods of feature selection if statistical techniques provide the information necessary to select the most appropriate bands for classification? The reason is simple. An analyst may base a decision solely on the statistic, yet never obtain a fundamental understanding of the spectral nature of the data being analyzed. In effect, without ever visualizing where the spectral measurements cluster in *n*-dimensional feature space, each new supervised classification finds the analyst beginning anew, relying totally on the abstract statistical analysis. Many practitioners of remote sensing are by necessity very graphically literate; that is, they can readily interpret maps and graphs. Therefore, a graphic display of the statistical data is useful and often necessary for a thorough analysis of multispectral training data and feature selection. Several graphic feature selection methods have been developed for this purpose.

Graphic Methods of Feature Selection

Bar graph spectral plots were one of the first simple feature selection aids where the mean ±1 standard deviation (σ) is displayed in a bar graph format for each band (Figure 9-8). This provides an effective visual presentation of the degree of between-class separability for one band at a time. In the example, band 3 is useful only for discriminating between water (class 1) and all other classes. Bands 1 and 2 appear to provide good separability between most of the classes (with the possible exception of classes 5 and 6). The display provides no information on how well any two bands would perform.

Cospectral mean vector plots may be used to present statistical information about at least two bands at one time. Hodgson and Plews (1989) provided several methods for displaying the mean vectors for each class in two- and three-dimensional feature space. For example, in Figure 9-9a we

see 49 mean vectors derived from Charleston, SC, TM data arrayed in two-dimensional feature space (bands 3 and 4). Theoretically, the greater the distance is between numbers in the feature space distribution, the greater the potential for accurate between-class discrimination. Using this method, only two bands of data can be analyzed at one time. Therefore, they devised an alternative method whereby the size of the numeral depicts the location of information in a third dimension of feature space (Figure 9-9b). For example, Figure 9-9c depicts the same 49 mean vectors in simulated three-dimensional feature space (bands 2, 3, and 4). Normal viewing of the *trispectral mean vector plot* looks down the *z* axis; thus the *z* axis is not seen. Scaling of the numeral size is performed by linear scaling:

$$\text{Size} = \frac{BV_{ck}}{\text{quant}_k} * \text{MaxSize} \qquad (9\text{-}4)$$

where

Size = the numeral size in feature space,
BV_{ck} = brightness value in class *c* for band *k* depicted by the *z* axis,
quant_k = quantization level of band *k* (e.g., 0 to 255), and
MaxSize = maximum numeral size.

Size and *MaxSize* are in the units of the output device (e.g., inches for a pen plotter or pixels for a raster display). By depicting cluster labels farther from the viewer with smaller numeric labels, the relative proximity of the means in the third band can be visually interpreted in the trispectral plot. It is also possible to make the thickness of the lines used to construct the numeric labels proportional to the distance from the viewer, adding a second depth perception visual cue (Figure 9-9b).

Feature space plots in two dimensions depict the distribution of all the pixels in the scene using two bands at a time (Figure 9-9). Such plots are often used as a backdrop for the display of various graphic feature selection methods. A typical plot usually consists of a 256 × 256 matrix (0 to 255 in the *x*-axis and 0 to 255 in the *y*-axis), which is filled with values in the following manner. Let us suppose that the first pixel in the entire dataset has a brightness value of 50 in band 1 and a value of 30 in band 3. A value of 1 is placed at location 50, 30 in the feature space plot matrix. If the next pixel in the dataset also has brightness values of 50 and 30 in bands 1 and 3, the value of this cell in the feature space matrix is incremented by 1, becoming 2. This logic is applied to each pixel in the scene. The brighter the pixel is in the feature space plot display, the greater the number of pixels having the same values in the two bands of interest.

Bar Graph Spectral Plots

```
          Brightness Value                              Brightness Value
        1    2    3    4    5    6                    1    2    3    4    5    6
     1  0    0    0    0    0    0                 1  0    0    0    0    0    0

L = 17  H = 23                             L=7  H = 12
Mean = 20                    Band 1        Mean = 9.1                   Band 2
SD = 1.89                                  SD = 2.1
1 ---------***--------------------- Class 1  1 ---------*****--------------------- Class 1
L = 28  H = 34                             L = 28  H = 35
Mean = 31.23                 Band 1        Mean = 30.23                 Band 2
SD = 1.47                                  SD = 1.94
2 ------------------***------------- 2      2 --------------------****------------ 2
L = 28  H = 34                             L = 22  H = 28
Mean = 30.26                 Band 1        Mean = 24.73                 Band 2
SD = 1.73                                  SD = 1.76
3 ------------------***------------- 3      3 --------------****----------------- 3
L = 36  H = 42                             L = 32  H = 43
Mean = 38.93                 Band 1        Mean = 37.06                 Band 2
SD = 1.91                                  SD = 2.77
4 ------------------***------------- 4      4 ------------------*****------------ 4
L = 36  H = 53                             L = 36  H = 55
Mean = 44.38                 Band 1        Mean =45.19                  Band 2
SD = 4.29                                  SD = 4.75
5 ------------------**********---------- 5  5 ------------------**********---------- 5
L = 24  H = 64                             L = 39  H = 70
Mean = 49.4                  Band 1        Mean = 54.12                 Band 2
SD = 7.44                                  SD = 7.17
6 ------------------**********---- 6        6 ------------------**************** 6
```

```
L=4  H = 12
Mean = 8.5                   Band 3
SD = 3.5
1 ----------*******--------------------- Class 1
L = 42  H = 54
Mean = 46.85                 Band 3
SD = 3.58
2 --------------------*******---------- 2
L = 44  H = 59
Mean = 50.66                 Band 3
SD = 5.05
3 --------------------**********---- 3
L = 48  H = 55
Mean = 51.09                 Band 3
SD = 2.23
4 --------------------*****------- 4
L = 36  H = 57
Mean =48.80                  Band 3
SD = 5.51
5 --------------------**********----- 5
L = 35  H = 70
Mean = 53.28                 Band 3
SD = 8.13
6 --------------------************** 6
```

Class 1 = water
Class 2 = natural vegetation
Class 3 = agriculture
Class 4 = single-family residential
Class 5 = multiple-family residential
Class 6 = commercial complex/barren land

Figure 9-8   Bar graph spectral plots of data. Training statistics (the mean ±1σ) for six land-cover classes are displayed for three Landsat MSS bands. The simple display can be used to identify between-class separability for each class and single band.

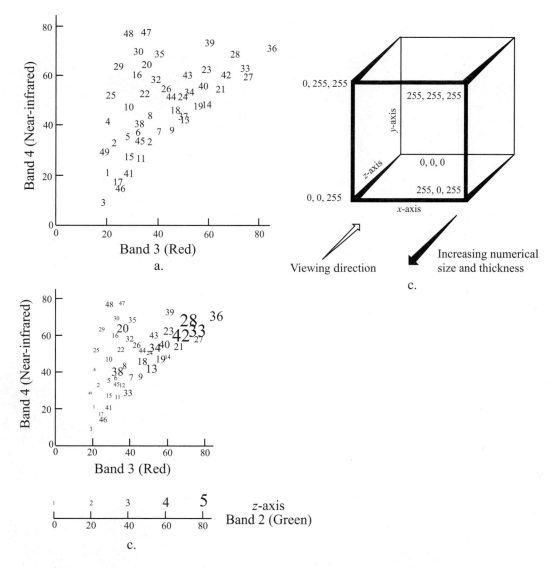

Figure 9-9   a) Cospectral mean vector plots of 49 clusters extracted from Charleston Landsat TM data bands 3 and 4. b) The logic for increasing numeral size and thickness along the *z*-axis. c) The introduction of band 2 information scaled according to size and thickness along the *z*-axis (Hodgson and Plews, 1989).

Feature space plots provide great insight into the actual information content of the image and the degree of between-band correlation. For example, in Figure 9-10a it is obvious that bands 1 and 3 are highly correlated and that atmospheric scattering in band 1 (blue) results in a significant shift of the brightness values down the *x*-axis. Conversely, plots of bands 2 (green) and 4 (near-infrared) and bands 3 (red) and 4 have a much greater distribution of pixels within the spectral space and some very interesting bright locations, which correspond with important land-cover types (Figures 9-10b and c). Finally, the plot of bands 4 (near-infrared) and 5 (middle-infrared) shows exceptional dispersion throughout

the spectral space and some very interesting bright locations (Figure 9-10d). For this reason, a spectral feature space plot of bands 4 and 5 will be used as a backdrop for the next graphic feature selection method.

Cospectral parallelepiped or *ellipse plots* in two-dimensional feature space provide useful visual between-class separability information (Jensen and Toll, 1982; Jain, 1989). They are created using the mean, μ_{ck}, and standard deviation, σ_{ck}, of training class statistics for each class c and band k. For example, the training statistics for five Charleston land-cover classes are portrayed in this manner and draped

Two-dimensional Feature Space Plots

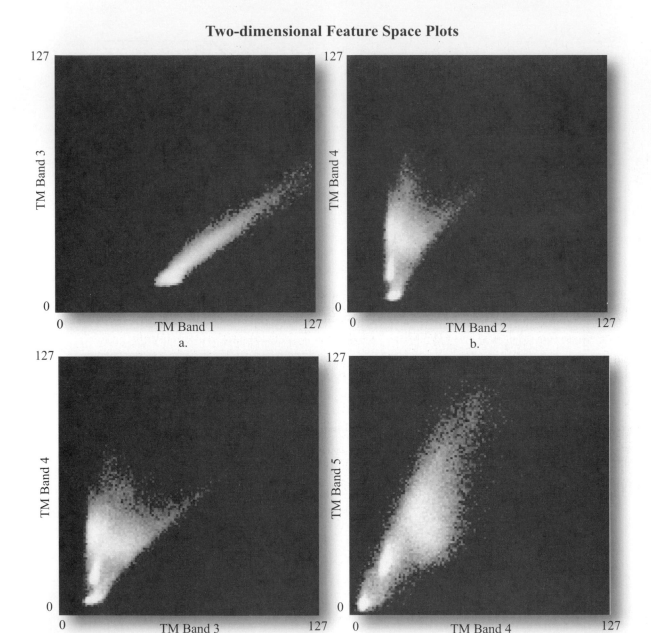

Figure 9-10   Two-dimensional feature space plots of four pairs of Landsat TM data of Charleston, SC. a) TM bands 1 and 3, b) TM bands 2 and 4, c) TM bands 3 and 4, and d) TM bands 4 and 5. The brighter a particular pixel is in the display, the more pixels within the scene having that unique combination of band values.

over the feature space plot of TM bands 4 and 5 in Figure 9-11. The lower and upper limits of the two-dimensional parallelepipeds (rectangles) were obtained using the mean $\pm 1\sigma$ of each band for each class. If only band 4 data were used to classify the scene, there would be confusion between classes 1 and 4, and if only band 5 data were used, there would be confusion between classes 3 and 4. However, when band 4

and 5 data are used at the same time to classify the scene there appears to be good between-class separability among the five classes (at least at $\pm 1\sigma$).

An evaluation of Figure 9-11 reveals that there are numerous water pixels in the scene found near the origin in bands 4 and 5. The water training class is located in this region. Simi-

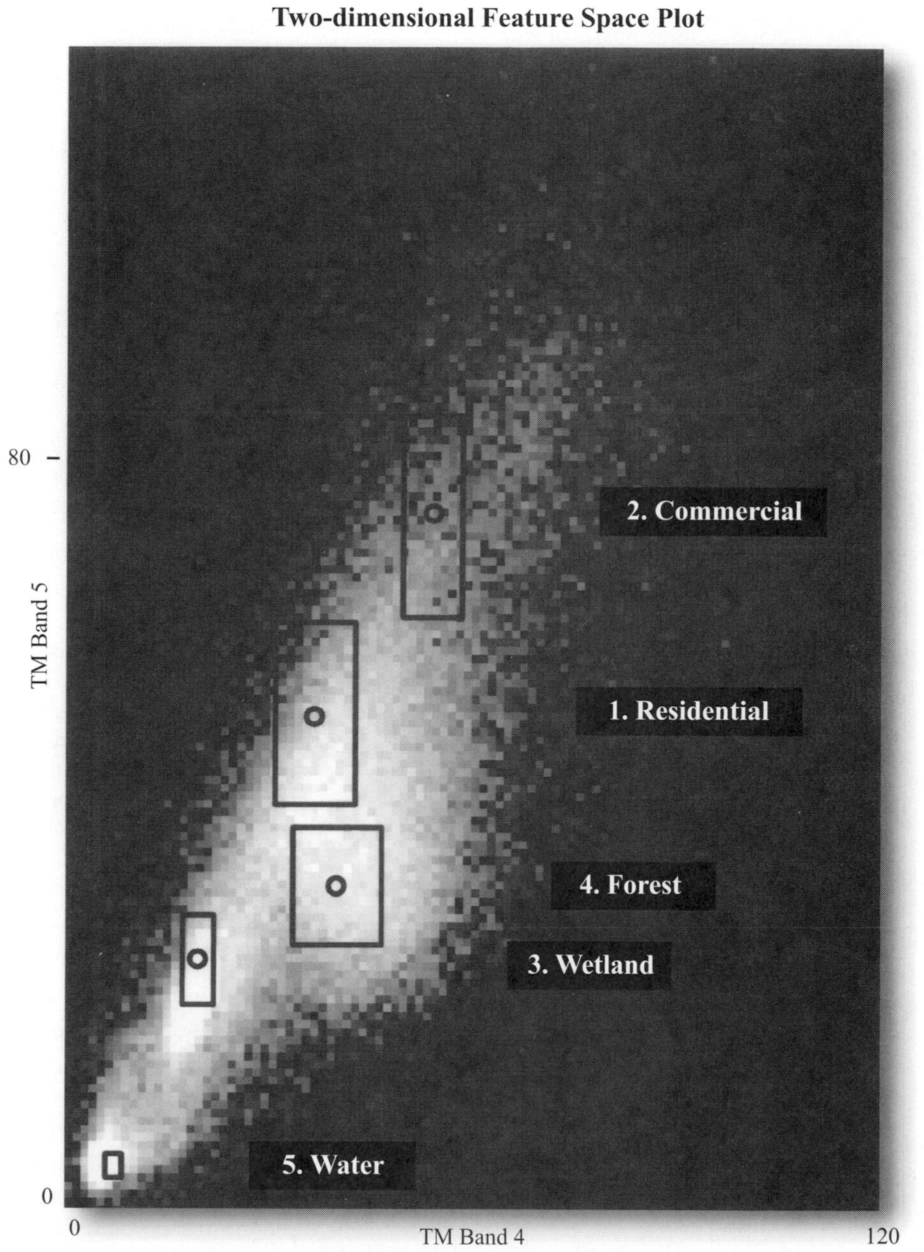

Two-dimensional Feature Space Plot

Figure 9-11 Plot of the Charleston, SC, Landsat TM training statistics for five classes measured in bands 4 and 5 displayed as cospectral parallelepipeds. The upper and lower limit of each parallelepiped is ±1σ. The parallelepipeds are superimposed on a feature space plot of bands 4 and 5.

larly, the wetland training class is situated within the bright wetland region of band 4 and 5 spectral space. However, it appears that training data were not collected in the heart of the wetland region of spectral space. Such information is valuable because we may want to collect additional training data in the wetland region to see if we can capture more of the essence of the feature space. In fact, there may be two or more wetland classes residing in this portion of spectral space. Sophisticated digital image processing systems allow the analyst to select training data directly from this type of display, which contains 1) the training class parallelepipeds and 2) the feature space plot. The analyst uses the cursor to interactively select training locations (they may be polygonal areas, not just parallelepipeds) within the feature space.

If desired, these feature space partitions can be used as the actual decision logic during the classification phase of the project. This type of interactive feature space partitioning is very useful.

It is possible to display three bands of training data at once using *trispectral parallelepipeds* or *ellipses* in synthetic three-dimensional feature space (Figure 9-12). Jensen and Toll (1982) present a method of displaying parallelepipeds in synthetic three-dimensional space and of interactively varying the viewpoint azimuth and elevation angles to enhance feature analysis and selection. Again, the mean, μ_{ck}, and standard deviation, σ_{ck}, of training class statistics for each class c and band k are used to identify the lower and upper threshold values for each class and band. The analyst then selects a combination of three bands to portray because it is not possible to use all six bands at once in a three-dimensional display. Landsat TM bands 4, 5, and 7 are used in the following example; however, the method is applicable to any three-band subset. Each corner of a parallelepiped is identifiable by a unique set of x, y, z coordinates corresponding to either the lower or upper threshold value for the three bands under investigation (Figure 9-12).

The corners of the parallelepipeds may be viewed from a vantage point other than a simple frontal view of the x, y axes using three-dimensional coordinate transformation equations. The feature space may be rotated about any of the axes, although rotation around the x- and y-axes normally provides a sufficient number of viewpoints. Rotation about the x-axis ϕ radians and the y-axis θ radians is implemented using the following equations (Hodgson and Plews, 1989):

$$\overset{P^{T'}}{[X, Y, Z, 1]} = \overset{P^{T}}{[BVx, BVy, BVz, 1]} \bullet \quad (9-5)$$

$$\begin{bmatrix} 1 & 0 & 0 & 0 \\ 0 & \cos\phi & -\sin\phi & 0 \\ 0 & \sin\phi & \cos\phi & 0 \\ 0 & 0 & 0 & 1 \end{bmatrix} \bullet \begin{bmatrix} \cos\theta & 0 & \sin\theta & 0 \\ 0 & 1 & 0 & 0 \\ -\sin\theta & 0 & \cos\theta & 0 \\ 0 & 0 & 0 & 1 \end{bmatrix}.$$

Negative signs of ϕ or θ are used for counterclockwise rotation and positive signs for clockwise rotation. This transformation causes the original brightness value coordinates, P^T, to be shifted about and contain depth information as vector $P^{T'}$. Display devices are two-dimensional (e.g., plotter surfaces or cathode-ray-tube screens); only the x and y elements of the transformed matrix $P^{T'}$ are used to draw the parallelepipeds.

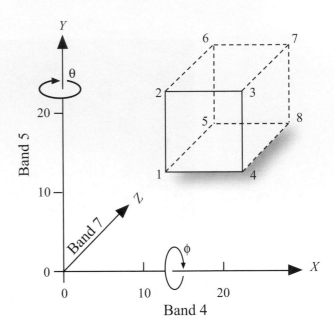

Figure 9-12 Simple parallelepiped displayed in pseudo three-dimensional space. Each of the eight corners represents a unique x, y, z coordinate corresponding to a lower or upper threshold value of the training data. For example, the original coordinates of point 4 are associated with 1) the upper threshold value of band 4, 2) the lower threshold value of band 5, and 3) the lower threshold value of band 7. The rotation matrix transformations cause the original coordinates to be rotated about the y-axis some θ radians, and the x axis some ϕ radians.

Manipulation of the transformed coordinates of the Charleston training statistics is shown in Figure 9-13. All three bands (4, 5, and 7) are displayed in Figure 9-13a, except that the band 7 statistics are perpendicular (orthogonal) to the sheet of paper. By rotating the display 45°, the contribution of band 7 becomes apparent (Figure 9-13b). This represents a synthetic three-dimensional display of the parallelepipeds. As the display is rotated another 45° to 90°, band 7 data collapse onto what was the band 4 axis (Figure 9-13c). The band 4 axis is now perpendicular to the page, just as band 7 was originally. The band 7, band 5 plot (Figure 9-13c) displays some overlap between wetland (3) and forest (4). By systematically specifying various azimuth and elevation angles, it is possible to display the parallelepipeds for optimal visual examination. This allows the analyst to obtain insight as to the consistent location of the training data in three-dimensional feature space.

In this example it is evident that just two bands, 4 and 5, provide as good if not better separation than all three bands used

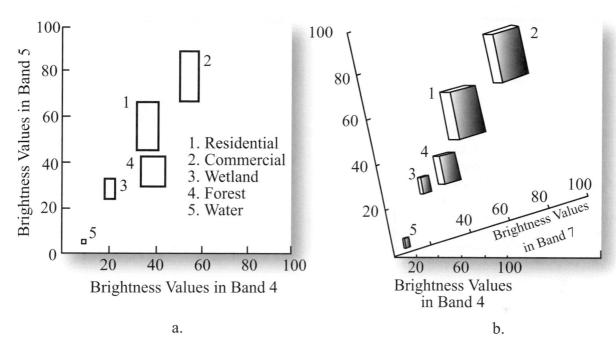

1. Residential
2. Commercial
3. Wetland
4. Forest
5. Water

a.

b.

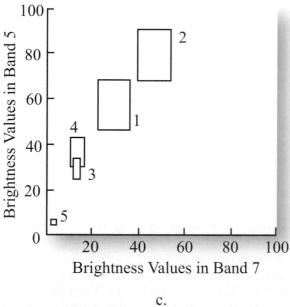

c.

Figure 9-13   Development of the three-dimensional parallelepipeds of the five Charleston, SC, training classes derived from Landsat Thematic Mapper data. Only bands 4, 5, and 7 are used in this investigation. The data are rotated about the y-axis, 0°, 45°, 90°. At 0° and 90° [parts (a) and (c), respectively], we are actually looking at only two bands, analogous to the two-dimensional parallelepiped boxes shown in Figure 9-11. The third band lies perpendicular to the page we are viewing. Between such extremes, however, it is possible to obtain optimum viewing angles for visual analysis of training class statistics using three bands at once. Part (b) displays the five classes at a rotation of 45°, demonstrating that the classes are entirely separable using this three-band combination. However, it probably is not necessary to use all three bands since bands 4 and 5 alone will discriminate satisfactorily between the five classes, as shown in part (a). There would be a substantial amount of overlap between classes if bands 5 and 7 were used.

together. However, this may not be the very best set of two bands to use. It might be useful to evaluate other two- or three-band combinations. In fact, a certain combination of perhaps four or five bands used all at one time might be superior. The only way to determine this is through statistical feature selection.

Statistical Methods of Feature Selection

Statistical methods of *feature selection* are used to quantitatively select which subset of bands (also referred to as features) provides the greatest degree of statistical separability between any two classes c and d. The basic problem of spectral pattern recognition is that given a spectral distribution of data in n bands of remotely sensed data, we must find a discrimination technique that will allow separation of the major land-cover categories with a minimum of error and a minimum number of bands. This problem is demonstrated diagrammatically using just one band and two classes in Figure 9-14. Generally, the more bands we analyze in a classification, the greater the cost and perhaps the greater the amount of redundant spectral information being used. When there is overlap, any decision rule that one could use to separate or distinguish between two classes must be concerned with two types of error (Figure 9-14):

1. A pixel may be assigned to a class to which it does not belong (an error of commission).

2. A pixel is not assigned to its appropriate class (an error of omission).

The goal is to select the optimum subset of bands and apply appropriate classification techniques to minimize both types of error in the classification process. If the training data for each class from each band are normally distributed, as suggested in Figure 9-14, it is possible to use either a transformed divergence or Jeffreys–Matusita distance equation to identify the optimum subset of bands to use in the classification procedure.

Divergence was one of the first measures of statistical separability used in the machine processing of remote sensor data, and it is still widely used as a method of feature selection (Mausel et al., 1990; Konecny, 2003). It addresses the basic problem of deciding what is the best q-band subset of n bands for use in the supervised classification process. The number of combinations C of n bands taken q at a time is:

$$C\left(\frac{n}{q}\right) = \frac{n!}{q!(n-q)!}.$$

(9-6)

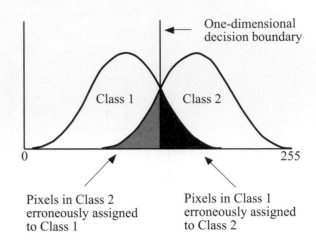

Figure 9-14   The basic problem in remote sensing pattern recognition classification is, given a spectral distribution of data in n bands (here just 1 band), to find an n-dimensional decision boundary that will allow the separation of the major classes (just 2 in this example) with a minimum of error and a minimum number of bands being evaluated. The dark areas of both distributions identify potential classification error.

Thus, if there are six TM bands and we are interested in the three best bands to use in the classification of the Charleston scene, this results in 20 combinations that must be evaluated:

$$C\left(\frac{6}{3}\right) = \frac{6!}{3!(6-3)!} = \frac{720}{6(6)}$$

(9-7)

$$= 20 \text{ combinations.}$$

If the best two-band combinations were desired, it would be necessary to evaluate 15 possible combinations.

Divergence is computed using the mean and covariance matrices of the class statistics collected in the training phase of the supervised classification. We will initiate the discussion by concerning ourselves with the statistical separability between just two classes, c and d. The degree of divergence or separability between c and d, $Diver_{cd}$, is computed according to the formula (Mausel et al., 1990):

$$Diver_{cd} = \frac{1}{2}tr[(V_c - V_d)(V_d^{-1} - V_c^{-1})]$$

(9-8)

$$+ \frac{1}{2}tr[(V_c^{-1} + V_d^{-1})(M_c - M_d)(M_c - M_d)^T]$$

where $tr[\ ]$ is the trace of a matrix (i.e., the sum of the diagonal elements), V_c and V_d are the covariance matrices for the two classes under investigation, c and d, and M_c and M_d are the mean vectors for classes c and d (Konecny, 2003). It should be remembered that the sizes of the covariance matrices V_c and V_d are a function of the number of bands used in the training process (i.e., if six bands were trained upon, then both V_c and V_d would be matrices 6×6 in dimension). Divergence in this case would be used to identify the statistical separability of the two training classes using six bands of training data. However, this is not the usual goal of applying divergence. What we actually want to know is the optimum subset of q bands. For example, if $q = 3$, what subset of three bands provides the best separation between these two classes? Therefore, in our example, we would proceed to systematically apply the algorithm to the 20 three-band combinations, computing the divergence for our two classes of interest and eventually identifying the subset of bands, perhaps bands 2, 3, and 6, that results in the largest divergence value.

But what about the case where there are more than two classes? In this instance, the most common solution is to compute the *average divergence, Diver*$_{avg}$. This involves computing the average over all possible pairs of classes c and d, while holding the subset of bands q constant. Then, another subset of bands q is selected for the m classes and analyzed. The subset of features (bands) having the maximum average divergence may be the superior set of bands to use in the classification algorithm. This can be expressed:

$$Diver_{avg} = \frac{\displaystyle\sum_{c=1}^{m-1} \sum_{d=c+1}^{m} Diver_{cd}}{C}. \qquad (9\text{-}9)$$

Using this, the band subset q with the highest average divergence would be selected as the most appropriate set of bands for classifying the m classes.

Unfortunately, outlying easily separable classes will weight average divergence upward in a misleading fashion to the extent that suboptimal feature subsets might be indicated as best (Richards and Jia, 1999; Ozkan and Erbek, 2003). Therefore, it is necessary to compute *transformed divergence, TDiver*$_{cd}$, expressed as:

$$TDiver_{cd} = 2000\left[1 - \exp\left(\frac{-Diver_{cd}}{8}\right)\right]. \qquad (9\text{-}10)$$

This statistic gives an exponentially decreasing weight to increasing distances between the classes. It also scales the divergence values to lie between 0 and 2000. For example, Table 9-7 demonstrates which bands are most useful when

taken 1, 2, 3, 4, or 5 at a time. There is no need to compute the divergence using all six bands since this represents the totality of the dataset. It is useful, however, to calculate divergence with individual channels ($q = 1$), since a single channel might adequately discriminate among all classes of interest.

A transformed divergence value of 2000 suggests excellent between-class separation. Above 1900 provides good separation, while below 1700 is poor. It can be seen that for the Charleston study, using any single band (Table 9-7a) would not produce results as acceptable as using bands 3 and 4 together (Table 9-7b). Several three-band combinations should yield good between-class separation for all classes. Most of them understandably include bands 3 and 4. But why should we use three, four, five, or six bands in the classification when divergence statistics suggest that very good between-class separation is possible using just two bands? We probably should not if the dimensionality of the dataset can be reduced by a factor of 3 (from 6 to 2) and classification results appear promising using just the two bands.

Other methods of feature selection are also based on determining the separability between two classes at a time. For example, the *Bhattacharyya distance* assumes that the two classes c and d are Gaussian and that the means M_c and M_d and covariance matrices V_c and V_d are available. It is computed (Duda et al., 2001):

$$Bhat_{cd} = \frac{1}{8}(M_c - M_d)^T\left(\frac{V_c + V_d}{2}\right)^{-1}(M_c - M_d) \qquad (9\text{-}11)$$

$$+ \frac{1}{2}\log_e\left[\frac{\left|\dfrac{V_c + V_d}{2}\right|}{\sqrt{(|V_c| \cdot |V_d|)}}\right].$$

To select the best q features (i.e., combination of bands) from the original n bands in an m-class problem, the Bhattacharyya distance is calculated between each $m(m - 1)/2$ pair of classes for each possible way of choosing q features from n dimensions. The best q features are those dimensions whose sum of the Bhattacharyya distance between the $m(m - 1)/2$ classes is highest (Haralick and Fu, 1983; Konecny, 2003).

A saturating transform applied to the Bhattacharyya distance measure yields the *Jeffreys–Matusita distance* (often referred to as the *JM* distance) (Ferro and Warner, 2002):

$$JM_{cd} = \sqrt{2(1 - e^{-Bhat_{cd}})}. \qquad (9\text{-}12)$$

Table 9-7.   Divergence statistics for the five Charleston, SC, land-cover classes evaluated using 1, 2, 3, 4, and 5 Thematic Mapper band combinations at one time.

Band Combinations	Average Divergence	Divergence (upper number) and Transformed Divergence (lower number) Class Combinations (1 = residential; 2 = commercial; 3 = wetland; 4 = forest; 5 = water)									
		1 2	1 3	1 4	1 5	2 3	2 4	2 5	3 4	3 5	4 5
a. One Band at a Time											
1		45	36	23	38	600	356	803	1	3	7
	1583	1993	1977	1889	1982	2000	2000	2000	198	651	1145
2		34	67	15	54	1036	286	1090	1	5	5
	1588	1970	2000	1786	1998	2000	2000	2000	246	988	890
3		54	107	39	160	1591	576	2071	1	3	1
	1525	1998	2000	1985	2000	2000	2000	2000	286	642	339
4		19	47	0	1238	209	13	3357	60	210	1466
	1748	1809	1994	70	2000	2000	1603	2000	1999	2000	2000
5		4	26	7	2645	77	29	5300	2	556	961
	1636	779	1920	1194	2000	2000	1947	2000	523	2000	2000
7		6	61	18	345	238	74	940	1	63	56
	1707	1061	1999	1795	2000	2000	2000	2000	213	1999	1998
b. Two Bands at a Time											
1 2		51	92	26	85	1460	410	1752	2	8	10
	1709	1997	2000	1919	2000	2000	2000	2000	463	1256	1457
1 3		56	125	40	182	1888	589	2564	2	7	11
	1709	1998	2000	1987	2000	2000	2000	2000	418	1196	1490
1 4		55	100	32	1251	941	446	3799	66	219	1525
	1996	1998	2000	1962	2000	2000	2000	2000	1999	2000	2000
1 5		54	71	28	3072	778	497	7838	6	585	1038
	1896	1998	2000	1939	2000	2000	2000	2000	1029	2000	2000
1 7		52	107	28	426	944	421	2065	3	63	76
	1852	1997	2000	1939	2000	2000	2000	2000	586	1999	2000
2 3		57	140	42	170	2099	593	2345	2	13	9
	1749	1998	2000	1990	2000	2000	2000	2000	524	1599	1382
2 4		35	103	28	1256	1136	356	3985	65	228	1529
	1992	1976	2000	1941	2000	2000	2000	2000	1999	2000	2000
2 5		35	86	20	2795	1068	328	6932	4	560	979
	1856	1976	2000	1826	2000	2000	2000	2000	760	2000	2000
2 7		37	111	24	423	1148	292	2192	2	69	66
	1829	1980	2000	1902	2000	2000	2000	2000	405	2000	1999

Table 9-7. Divergence statistics for the five Charleston, SC, land-cover classes evaluated using 1, 2, 3, 4, and 5 Thematic Mapper band combinations at one time. (Continued)

| | | Divergence (upper number) and Transformed Divergence (lower number) | | | | | | | | | |
| | | Class Combinations (1 = residential; 2 = commercial; 3 = wetland; 4 = forest; 5 = water) | | | | | | | | | |
Band Combinations	Average Divergence	1 2	1 3	1 4	1 5	2 3	2 4	2 5	3 4	3 5	4 5
3 4		101	124	61	1321	1606	905	4837	80	210	1487
	2000	2000	2000	1999	2000	2000	2000	2000	2000	2000	2000
3 5		59	114	45	3206	1609	740	9142	5	597	1024
	1895	1999	2000	1992	2000	2000	2000	2000	964	2000	2000
3 7		63	131	41	525	1610	606	3122	2	65	59
	1845	1999	2000	1989	2000	2000	2000	2000	469	1999	1999
4 5		21	52	11	4616	231	37	10376	98	889	2902
	1930	1851	1997	1468	2000	2000	1981	2000	2000	2000	2000
4 7		20	76	21	1742	309	79	4740	86	285	1599
	1970	1844	2000	1857	2000	2000	2000	2000	2000	2000	2000
5 7		6	62	24	2870	246	97	5956	5	598	989
	1795	1074	1999	1900	2000	2000	2000	2000	978	2000	2000

c. Three Bands at a Time

Band Combinations	Average Divergence	1 2	1 3	1 4	1 5	2 3	2 4	2 5	3 4	3 5	4 5
1 2 3		59	154	44	191	2340	613	2821	3	16	17
	1815	1999	2000	1992	2000	2000	2000	2000	643	1745	1774
1 2 4		95	142	40	1266	1662	675	4381	68	236	1573
	1999	2000	2000	1986	2000	2000	2000	2000	2000	2000	2000
1 2 5		58	118	32	3201	1564	604	9281	7	589	1045
	1909	1999	2000	1964	2000	2000	2000	2000	1129	2000	2000
1 2 7		57	146	30	493	1653	494	3176	4	69	80
	1868	1998	2000	1953	2000	2000	2000	2000	732	2000	2000
1 3 4		117	150	64	1329	1905	985	5120	86	219	1534
	2000	2000	2000	1999	2000	2000	2000	2000	2000	2000	2000
1 3 5		60	137	51	3569	1902	863	11221	7	622	1088
	1920	1999	2000	1997	2000	2000	2000	2000	1202	2000	2000
1 3 7		63	157	45	580	1935	669	3879	4	66	79
	1872	1999	2000	1993	2000	2000	2000	2000	731	1999	2000
1 4 5		82	105	36	4923	978	635	12361	104	906	2955
	1998	2000	2000	1979	2000	2000	2000	2000	2000	2000	2000
1 4 7		82	129	37	1777	1055	610	5452	93	288	1669
	1998	2000	2000	1980	2000	2000	2000	2000	2000	2000	2000
1 5 7		56	109	37	3405	956	508	8948	8	627	1077
	1924	1998	2000	1982	2000	2000	2000	2000	1261	2000	2000

Table 9-7.   Divergence statistics for the five Charleston, SC, land-cover classes evaluated using 1, 2, 3, 4, and 5 Thematic Mapper band combinations at one time. (Continued)

		Divergence (upper number) and Transformed Divergence (lower number)									
		Class Combinations (1 = residential; 2 = commercial; 3 = wetland; 4 = forest; 5 = water)									
Band Combinations	Average Divergence	1 2	1 3	1 4	1 5	2 3	2 4	2 5	3 4	3 5	4 5
2 3 4	2000	117 2000	156 2000	63 1999	1331 2000	2119 2000	956 2000	4971 2000	81 2000	229 2000	1530 2000
2 3 5	1908	62 1999	147 2000	47 1994	3221 2000	2120 2000	749 2000	9480 2000	6 1082	605 2000	1034 2000
2 3 7	1865	66 1999	160 2000	46 1994	541 2000	2113 2000	617 2000	3480 2000	3 661	74 2000	69 2000
2 4 5	1994	38 1984	108 2000	31 1956	4674 2000	1158 2000	385 2000	11402 2000	103 2000	896 2000	2946 2000
2 4 7	1996	40 1986	125 2000	34 1970	1771 2000	1191 2000	367 2000	5511 2000	90 2000	300 2000	1668 2000
2 5 7	1906	38 1982	113 2000	33 1968	3050 2000	1157 2000	365 2000	7757 2000	7 1113	594 2000	1006 2000
3 4 5	2000	106 2000	129 2000	65 1999	5031 2000	1622 2000	1037 2000	13505 2000	120 2000	914 2000	2935 2000
3 4 7	2000	111 2000	144 2000	63 1999	1841 2000	1644 2000	955 2000	6309 2000	102 2000	285 2000	1626 2000
3 5 7	1927	66 1999	134 2000	63 1999	3453 2000	1648 2000	823 2000	9900 2000	8 1268	631 2000	1054 2000
4 5 7	1979	22 1870	83 2000	26 1923	5003 2000	362 2000	114 2000	11477 2000	105 2000	944 2000	2994 2000
d. Four Bands at a Time											
1 2 3 4	2000	167 2000	177 2000	65 1999	1339 2000	2361 2000	1151 2000	5259 2000	87 2000	238 2000	1575 2000
1 2 3 5	1929	63 1999	165 2000	54 1998	3582 2000	2355 2000	876 2000	11525 2000	8 1294	630 2000	1095 2000
1 2 3 7	1888	67 2000	182 2000	49 1996	595 2000	2369 2000	683 2000	4222 2000	5 885	75 2000	87 2000
1 2 4 5	1999	115 2000	147 2000	46 1994	4971 2000	1696 2000	901 2000	13287 2000	108 2000	913 2000	2987 2000
1 2 4 7	1999	110 2000	165 2000	45 1993	1801 2000	1731 2000	868 2000	6161 2000	96 2000	303 2000	1725 2000
1 2 5 7	1932	61 1999	148 2000	41 1989	3564 2000	1665 2000	614 2000	10579 2000	9 1331	633 2000	1085 2000

Table 9-7. Divergence statistics for the five Charleston, SC, land-cover classes evaluated using 1, 2, 3, 4, and 5 Thematic Mapper band combinations at one time. (Continued)

Band Combinations	Average Divergence	Divergence (upper number) and Transformed Divergence (lower number) Class Combinations (1 = residential; 2 = commercial; 3 = wetland; 4 = forest; 5 = water)									
		1 / 2	1 / 3	1 / 4	1 / 5	2 / 3	2 / 4	2 / 5	3 / 4	3 / 5	4 / 5
1 3 4 5	2000	133 / 2000	156 / 2000	74 / 2000	5293 / 2000	1931 / 2000	1283 / 2000	15187 / 2000	127 / 2000	928 / 2000	2976 / 2000
1 3 4 7	2000	134 / 2000	172 / 2000	69 / 2000	1863 / 2000	1955 / 2000	1184 / 2000	6814 / 2000	110 / 2000	289 / 2000	1682 / 2000
1 3 5 7	1940	66 / 2000	159 / 2000	66 / 2000	3919 / 2000	1954 / 2000	901 / 2000	12411 / 2000	10 / 1397	665 / 2000	1129 / 2000
1 4 5 7	1999	88 / 2000	135 / 2000	42 / 1990	5422 / 2000	1105 / 2000	659 / 2000	13950 / 2000	112 / 2000	970 / 2000	3068 / 2000
2 3 4 5	2000	122 / 2000	161 / 2000	67 / 2000	5040 / 2000	2133 / 2000	1093 / 2000	13663 / 2000	121 / 2000	933 / 2000	2981 / 2000
2 3 4 7	2000	132 / 2000	173 / 2000	65 / 1999	1848 / 2000	2143 / 2000	1023 / 2000	6509 / 2000	103 / 2000	302 / 2000	1670 / 2000
2 3 5 7	1937	69 / 2000	163 / 2000	68 / 2000	3476 / 2000	2144 / 2000	837 / 2000	10308 / 2000	9 / 1370	639 / 2000	1062 / 2000
2 4 5 7	1997	41 / 1987	131 / 2000	38 / 1983	5079 / 2000	1229 / 2000	397 / 2000	12641 / 2000	110 / 2000	951 / 2000	3037 / 2000
3 4 5 7	2000	112 / 2000	148 / 2000	74 / 2000	5436 / 2000	1665 / 2000	1066 / 2000	14688 / 2000	125 / 2000	971 / 2000	3030 / 2000
e. Five Bands at a Time											
1 2 3 4 5	2000	176 / 2000	183 / 2000	75 / 2000	5302 / 2000	2384 / 2000	1422 / 2000	15334 / 2000	128 / 2000	947 / 2000	3019 / 2000
1 2 3 4 7	2000	176 / 2000	196 / 2000	71 / 2000	1871 / 2000	2393 / 2000	1316 / 2000	7015 / 2000	111 / 2000	305 / 2000	1726 / 2000
1 2 3 5 7	1948	70 / 2000	184 / 2000	72 / 2000	3940 / 2000	2386 / 2000	919 / 2000	12798 / 2000	11 / 1479	673 / 2000	1135 / 2000
1 2 4 5 7	2000	117 / 2000	171 / 2000	50 / 1996	5487 / 2000	1770 / 2000	920 / 2000	15021 / 2000	115 / 2000	977 / 2000	3101 / 2000
1 3 4 5 7	2000	138 / 2000	176 / 2000	80 / 2000	5803 / 2000	1979 / 2000	1294 / 2000	16829 / 2000	132 / 2000	994 / 2000	3089 / 2000
2 3 4 5 7	2000	134 / 2000	177 / 2000	77 / 2000	5443 / 2000	2161 / 2000	1130 / 2000	14893 / 2000	126 / 2000	987 / 2000	3072 / 2000

The *JM* distance has a saturating behavior with increasing class separation like transformed divergence. However, it is not as computationally efficient as transformed divergence.

Mausel et al. (1990) evaluated four statistical separability measures to determine which would most accurately identify the best subset of four bands from an eight-band (two date) set of multispectral data for a computer classification of six agricultural features. Supervised maximum likelihood classification was applied to all 70 possible four-band combinations. Transformed divergence and the Jeffreys–Matusita distance both selected the four-band subset (bands 3, 4, 7, and 8 in their example), which yielded the highest overall classification accuracy of all the band combinations tested. In fact, the transformed divergence and *JM*-distance measures were highly correlated with classification accuracy (0.96 and 0.97, respectively) when all 70 classifications were considered. The Bhattacharyya distance and simple divergence selected the eleventh and twenty-sixth ranked four-channel subsets, respectively.

Select the Appropriate Classification Algorithm

Various supervised classification algorithms may be used to assign an unknown pixel to one of m possible classes. The choice of a particular classifier or decision rule depends on the nature of the input data and the desired output. *Parametric* classification algorithms assumes that the observed measurement vectors X_c obtained for each class in each spectral band during the training phase of the supervised classification are Gaussian; that is, they are normally distributed (Schowengerdt, 1997). *Nonparametric* classification algorithms make no such assumption.

Several widely adopted nonparametric classification algorithms include:

- one-dimensional *density slicing* (e.g., using a single band of imagery; please refer to Chapter 8),

- parallelepiped,

- minimum distance,

- nearest-neighbor, and

- neural network and expert system analysis (Chapter 10).

The most widely adopted parametric classification algorithms is:

- the maximum likelihood algorithm.

It is instructive to review the logic of several classification algorithms in this and subsequent chapters.

Parallelepiped Classification Algorithm

This is a widely used digital image classification decision rule based on simple Boolean "and/or" logic. Training data in n spectral bands are used to perform the classification (Gibson and Power, 2000). Brightness values from each pixel of the multispectral imagery are used to produce an n-dimensional mean vector, $M_c = (\mu_{ck}, \mu_{c2}, \mu_{c3}, ..., \mu_{cn})$ with μ_{ck} being the mean value of the training data obtained for class c in band k out of m possible classes, as previously defined. σ_{ck} is the standard deviation of the training data class c of band k out of m possible classes. In this discussion we will evaluate all five Charleston classes using just bands 4 and 5 of the training data.

Using a one-standard deviation threshold (as shown in Figure 9-15), a parallelepiped algorithm decides BV_{ijk} is in class c if, and only if:

$$\mu_{ck} - \sigma_{ck} \leq BV_{ijk} \leq \mu_{ck} + \sigma_{ck} \qquad (9\text{-}13)$$

where

$$c = 1, 2, 3, ..., m, \quad \text{number of classes, and}$$
$$k = 1, 2, 3, ..., n, \quad \text{number of bands.}$$

Therefore, if the low and high decision boundaries are defined as

$$L_{ck} = \mu_{ck} - \sigma_{ck} \qquad (9\text{-}14)$$

and

$$H_{ck} = \mu_{ck} + \sigma_{ck}, \qquad (9\text{-}15)$$

the parallelepiped algorithm becomes

$$L_{ck} \leq BV_{ijk} \leq H_{ck}. \qquad (9\text{-}16)$$

These decision boundaries form an n-dimensional parallelepiped in feature space. If the pixel value lies above the low threshold and below the high threshold for all n bands evaluated, it is assigned to that class (see point a in Figure 9-15). When an unknown pixel does not satisfy any of the Boolean logic criteria (point b in Figure 9-15), it is assigned to an unclassified category. Although it is only possible to analyze visually up to three dimensions, as described in the section on computer graphic feature analysis, it is possible to create an n-dimensional parallelepiped for classification purposes.

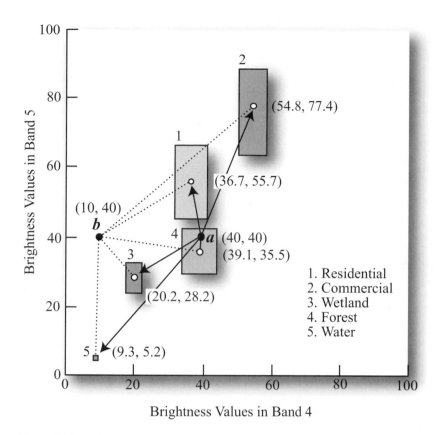

Figure 9-15 Points *a* and *b* are pixels in the image to be classified. Pixel *a* has a brightness value of 40 in band 4 and 40 in band 5. Pixel *b* has a brightness value of 10 in band 4 and 40 in band 5. The boxes represent the *parallelepiped* decision rule associated with a $\pm 1\sigma$ classification. The vectors (*arrows*) represent the distance from *a* and *b* to the mean of all classes in a *minimum distance to means* classification algorithm. Refer to Tables 9-8 and 9-9 for the results of classifying points *a* and *b* using both classification techniques.

We will review how unknown pixels *a* and *b* are assigned to the forest and unclassified categories in Figure 9-15. The computations are summarized in Table 9-8. First, the standard deviation is subtracted and added to the mean of each class and for each band to identify the lower (L_{ck}) and upper (H_{ck}) edges of the parallelepiped. In this case only two bands are used, 4 and 5, resulting in a two-dimensional box. This could be extended to *n* dimensions or bands. With the lower and upper thresholds for each box identified, it is possible to determine if the brightness value of an input pixel in each band, *k*, satisfies the criteria of any of the five parallelepipeds. For example, pixel *a* has a value of 40 in both bands 4 and 5. It satisfies the band 4 criteria of class 1 (i.e., $31.27 \leq 40 \leq 41.23$), but it does not satisfy the band 5 criteria. Therefore, the process continues by evaluating the parallelepiped criteria of classes 2 and 3, which are also not satisfied. However, when the brightness values of *a* are compared with class 4 thresholds, we find it satisfies the criteria for band 4 (i.e., $33.99 \leq 40 \leq 44.21$) and band 5 ($29.09 \leq 40 \leq 41.91$). Thus, the pixel is assigned to class 4, forest.

This same logic is applied to classify unknown pixel *b*. Unfortunately, its brightness values of 10 in band 4 and 40 in band 5 never fall within the thresholds of any of the parallelepipeds. Therefore, it is assigned to an unclassified category. Increasing the size of the thresholds to ± 2 or 3 standard deviations would increase the size of the parallelepipeds. This might result in point *b* being assigned to one of the classes. However, this same action might also introduce a significant amount of overlap among many of the parallelepipeds resulting in classification error. Perhaps point *b* really belongs to a class that was not trained upon (e.g., dredge spoil).

The parallelepiped algorithm is a computationally efficient method of classifying remote sensor data. Unfortunately, because some parallelepipeds overlap, it is possible that an unknown candidate pixel might satisfy the criteria of more than one class. In such cases it is usually assigned to the first class for which it meets all criteria. A more elegant solution is to take this pixel that can be assigned to more than one

Table 9-8.   Example of parallelepiped classification logic for pixels *a* and *b* in Figure 9-15.

	Class	Lower Threshold, L_{ck}	Upper Threshold, H_{ck}	Does pixel *a* (40, 40) satisfy criteria for this class in this band? $L_{ck} \le a \le H_{ck}$	Does pixel *b* (10, 40) satisfy criteria for this class in this band? $L_{ck} \le b \le H_{ck}$
1.	**Residential**				
	Band 4	$36.7 - 4.53 = 31.27$	$36.7 + 4.53 = 41.23$	Yes	No
	Band 5	$55.7 - 10.72 = 44.98$	$55.7 + 10.72 = 66.42$	No	No
2.	**Commercial**				
	Band 4	$54.8 - 3.88 = 50.92$	$54.8 + 3.88 = 58.68$	No	No
	Band 5	$77.4 - 11.16 = 66.24$	$77.4 + 11.16 = 88.56$	No	No
3.	**Wetland**				
	Band 4	$20.2 - 1.88 = 18.32$	$20.2 + 1.88 = 22.08$	No	No
	Band 5	$28.2 - 4.31 = 23.89$	$28.2 + 4.31 = 32.51$	No	No
4.	**Forest**				
	Band 4	$39.1 - 5.11 = 33.99$	$39.1 + 5.11 = 44.21$	Yes	No
	Band 5	$35.5 - 6.41 = 29.09$	$35.5 + 6.41 = 41.91$	Yes, assign pixel to class 4, forest. STOP.	No
5.	**Water**				
	Band 4	$9.3 - 0.56 = 8.74$	$9.3 + 0.56 = 9.86$	—	No
	Band 5	$5.2 - 0.71 = 4.49$	$5.2 + 0.71 = 5.91$	—	No, assign pixel to unclassified category. STOP.

class and use a *minimum distance to means* decision rule to assign it to just one class.

Minimum Distance to Means Classification Algorithm

The *minimum distance to means* decision rule is computationally simple and commonly used. When used properly it can result in classification accuracy comparable to other more computationally intensive algorithms such as the maximum likelihood algorithm. Like the parallelepiped algorithm, it requires that the user provide the mean vectors for each class in each band μ_{ck} from the training data. To perform a minimum distance classification, a program must calculate the distance to each mean vector μ_{ck} from each unknown pixel (BV_{ijk}) (Jahne, 1991; Lo and Yeung, 2002). It is possible to calculate this distance using Euclidean distance based on the Pythagorean theorem or "round the block" distance measures (Figure 9-16). In this discussion we demonstrate the method of minimum distance classification using Euclidean distance measurements applied to the two unknown points (*a* and *b*) shown in Figure 9-15.

The computation of the Euclidean distance from point *a* (40, 40) to the mean of class 1 (36.7, 55.7) measured in bands 4 and 5 relies on the equation:

$$Dist = \sqrt{(BV_{ijk} - \mu_{ck})^2 + (BV_{ijl} - \mu_{cl})^2} \qquad (9\text{-}17)$$

where μ_{ck} and μ_{cl} represent the mean vectors for class *c* measured in bands *k* and *l*. In our example this would be:

$$Dist_{a \text{ to class 1}} = \sqrt{(BV_{ij4} - \mu_{1,4})^2 + (BV_{ij5} - \mu_{1,5})^2}. \qquad (9\text{-}18)$$

The distance from point *a* to the mean of class 2 in these same two bands would be:

$$Dist_{a \text{ to class 2}} = \sqrt{(BV_{ij4} - \mu_{2,4})^2 + (BV_{ij5} - \mu_{2,5})^2}. \qquad (9\text{-}19)$$

Notice that the subscript that stands for class *c* is incremented from 1 to 2. By calculating the Euclidean distance from point *a* to the mean of all five classes, it is possible to determine which distance is shortest. Table 9-9 is a list of the mathematics associated with the computation of distances for the five land-cover classes. It reveals that pixel *a* should

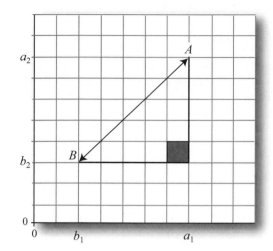

Euclidean distance

$$D_{AB} = \sqrt{\sum_{i=1}^{2} (a_i - b_i)^2}$$

Round the block distance

$$D_{AB} = \sum_{i=1}^{2} |(a_i - b_i)|$$

Figure 9-16   The distance used in a *minimum distance to means* classification algorithm can take two forms: the Euclidean distance based on the Pythagorean theorem and the "round the block" distance. The Euclidean distance is more computationally intensive.

be assigned to class 4 (forest) because it obtained the minimum distance of 4.59. The same logic can be applied to evaluate the unknown pixel *b*. It is assigned to class 3 (wetland) because it obtained the minimum distance of 15.75. It should be obvious that any unknown pixel will definitely be assigned to one of the five training classes using this algorithm. There will be no unclassified pixels.

Many minimum distance algorithms let the analyst specify a distance or threshold from the class means beyond which a pixel will not be assigned to a category even though it is nearest to the mean of that category. For example, if a threshold of 10.0 were specified, point *a* would still be classified as class 4 (forest) because it had a minimum distance of 4.59, which was below the threshold. Conversely, point *b* would not be assigned to class 3 (wetland) because its minimum distance of 15.75 was greater than the 10.0 threshold. Instead, point *b* would be assigned to an unclassified category.

When more than two bands are evaluated in a classification, it is possible to extend the logic of computing the distance between just two points in *n* space using the equation (Schalkoff, 1992; Duda et al., 2001):

$$D_{AB} = \sqrt{\sum_{i=1}^{n} (a_i - b_i)^2} .$$

(9-20)

Figure 9-16 demonstrates how this algorithm is implemented.

Hodgson (1988) identified six additional Euclidean-based minimum distance algorithms that decrease computation time by exploiting two areas: 1) the computation of the distance estimate from the unclassified pixel to each candidate class, and 2) the criteria for eliminating classes from the search process, thus avoiding unnecessary distance computations. Algorithms implementing these improvements were tested using up to 2, 4, and 6 bands of TM data and 5, 20, 50, and 100 classes. All algorithms were more efficient than the traditional Euclidean minimum distance algorithm.

A traditional minimum distance to means classification algorithm was run on the Charleston, SC, Landsat Thematic Mapper dataset using the training data previously described. The results are displayed as a color-coded thematic map in **Color Plate 9-2a**. The total number of pixels in each class are summarized in Table 9-10. Error associated with the classification is discussed in Chapter 13.

Nearest-Neighbor Classifiers

The parallelepiped classifier made use of the training class mean and standard deviation statistics for each band. The minimum distance classifier required the mean for each training class for each band. It is also possible to classify an unknown pixel measurement vector into *m* classes using just the training data brightness values in each band (i.e., the mean and standard deviation statistics are not used) and nearest-neighbor distance measurements (Duda et al., 2001). The most common nonparametric nearest-neighbor classifiers are:

• nearest-neighbor,

• *k*-nearest neighbor, and

• *k*-nearest-neighbor distance-weighted.

The simple *nearest-neighbor classifier* computes the Euclidean distance from the pixel to be classified to the nearest training data pixel in *n*-dimensional feature space and assigns it to that class (Schowengerdt, 1997). For example, in the hypothetical example in Figure 9-17, the nearest-

Table 9-9.   Example of minimum distance to means classification logic for pixels *a* and *b* in Figure 9-15.

Class	Distance from Pixel *a* (40, 40) to the Mean of Each Class	Distance from Pixel *b* (10, 40) to the Mean of Each Class
1. Residential	$\sqrt{(40-36.7)^2 + (40-55.7)^2} = 16.04$	$\sqrt{(10-36.7)^2 + (40-55.7)^2} = 30.97$
2. Commercial	$\sqrt{(40-54.8)^2 + (40-77.4)^2} = 40.22$	$\sqrt{(10-54.8)^2 + (40-77.4)^2} = 58.35$
3. Wetland	$\sqrt{(40-20.2)^2 + (40-28.2)^2} = 23.04$	$\sqrt{(10-20.2)^2 + (40-28.2)^2} = 15.75$ Assign pixel *b* to this class; it has the minimum distance.
4. Forest	$\sqrt{(40-39.1)^2 + (40-35.5)^2} = 4.59$ Assign pixel *a* to this class; it has the minimum distance.	$\sqrt{(10-39.1)^2 + (40-35.5)^2} = 29.45$
5. Water	$\sqrt{(40-9.3)^2 + (40-5.2)^2} = 46.4$	$\sqrt{(10-9.3)^2 + (40-5.2)^2} = 34.8$

Table 9-10.   Total number of pixels classified into each of the five Charleston, SC, land-cover classes shown in **Color Plate 9-2a**.

Class	Total Number of Pixels
1. Residential	14,398
2. Commercial	4,088
3. Wetland	10,772
4. Forest	11,673
5. Water	20,509

neighbor logic would assign the pixel to the commercial class.

The *k-nearest-neighbor classifier* searches away from the pixel to be classified in all directions until it encounters *k* user-specified training pixels (e.g., *k* = 5). It then assigns the pixel to the class with the majority of pixels encountered. For example, the pixel under investigation in Figure 9-17 would be assigned to the residential class if a *k*-nearest-neighbor classifier were applied because three of the five training class pixels encountered within the circle are residential.

The *k-nearest-neighbor distance-weighted classifier* uses the same *k*-nearest neighbors but weights them according to the distance-weighting logic discussed in Chapter 7. The pixel under investigation is assigned to the training class

pixel with the highest total weight. For example, the pixel under investigation in Figure 9-17 would be assigned to the commercial class once again if the *k*-nearest-neighbor distance-weighted algorithm were applied.

Nearest-neighbor classifiers can be relatively slow because of the number of distance calculations required between the unknown pixel measurement vector and all of the training pixels in the various bands (Hardin, 1994). Nearest-neighbor classifiers can yield useful results if the training data are well separated in *n*-dimensional feature space. Otherwise, it will probably be necessary to use a different algorithm (e.g., the minimum distance to means, maximum likelihood, neural network).

Maximum Likelihood Classification Algorithm

The aforementioned classifiers were based primarily on identifying decision boundaries in feature space based on training class multispectral distance measurements. The *maximum likelihood* decision rule is based on probability. It assigns each pixel having pattern measurements or features *X* to the class *i* whose units are most probable or likely to have given rise to feature vector *X* (Atkinson and Lewis, 2000; Lo and Yeung, 2002). In other words, the probability of a pixel belonging to each of a predefined set of *m* classes is calculated, and the pixel is then assigned to the class for which the probability is the highest. The maximum likelihood decision rule is still one of the most widely used supervised classification algorithms (Wu and Shao, 2002; McIver and Friedl, 2002).

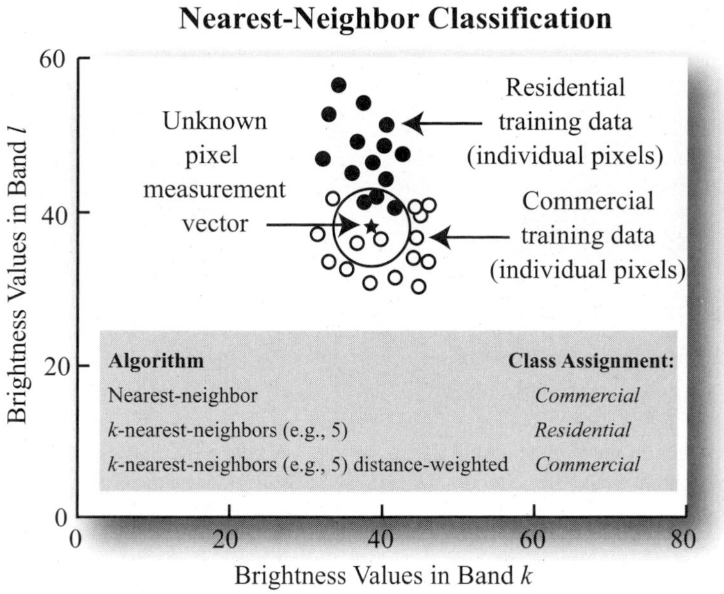

Nearest-Neighbor Classification

Figure 9-17   Hypothetical example of *nearest-neighbor classification*. Simple nearest-neighbor classification locates the nearest training class pixel in *n*-dimensional feature space from the unknown pixel measurement vector under consideration and assigns the unknown pixel the class of the training data. The *k*-nearest-neighbor classifier locates the nearest *k* (e.g., 5) training pixels (no matter what their class) in feature space and assigns the pixel under investigation to the majority class. The *k*-nearest-neighbor distance-weighted algorithm measures the distance to the same *k* training class pixels and weights each of them according to their distance to the pixel under investigation. The pixel is assigned the class with the highest total weight.

The maximum likelihood procedure assumes that the training data statistics for each class in each band are normally distributed (Gaussian). Training data with bi- or *n*-modal histograms in a single band are not ideal. In such cases the individual modes probably represent unique classes that should be trained upon individually and labeled as separate training classes. This should then produce unimodal, Gaussian training class statistics that fulfill the normal distribution requirement.

But how do we obtain the probability information we will need from the remote sensing training data we have collected? The answer lies first in the computation of *probability density functions*. We will demonstrate using a single class of training data based on a single band of imagery. For example, consider the hypothetical histogram (data frequency distribution) of forest training data obtained in band *k* shown in Figure 9-18a. We could choose to store the values contained in this histogram in the computer, but a more elegant solution is to approximate the distribution by a normal probability density function (curve), as shown superimposed on the histogram in Figure 9-18b. The estimated probability function for class w_i (e.g., forest) is computed using the equation:

$$\hat{p}(x|w_i) = \frac{1}{(2\pi)^{\frac{1}{2}}\hat{\sigma}_i} \exp\left[-\frac{1}{2}\frac{(x-\hat{\mu}_i)^2}{\hat{\sigma}_i^2}\right] \qquad (9\text{-}21)$$

where exp [ ] is *e* (the base of the natural logarithms) raised to the computed power, *x* is one of the brightness values on the *x*-axis, $\hat{\mu}_i$ is the estimated mean of all the values in the forest training class, and $\hat{\sigma}_i^2$ is the estimated variance of all the measurements in this class. Therefore, we need to store only the mean and variance of each training class (e.g., forest) to compute the probability function associated with any of the individual brightness values in it.

But what if our training data consists of multiple bands of remote sensor data for the classes of interest? In this case we compute an *n*-dimensional multivariate normal density function using (Swain and Davis, 1978):

$$p(X|w_i) = \frac{1}{(2\pi)^{\frac{n}{2}}|V_i|^{\frac{1}{2}}} \exp\left[-\frac{1}{2}(X-M_i)^T V_i^{-1}(X-M_i)\right] \qquad (9\text{-}22)$$

where $|V_i|$ is the determinant of the covariance matrix, V_i^{-1} is the inverse of the covariance matrix, and $(X-M_i)^T$ is the

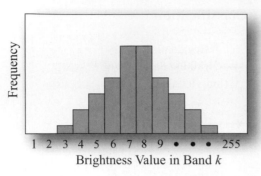

a. Histogram (data frequency distribution) of forest training data in a single band k.

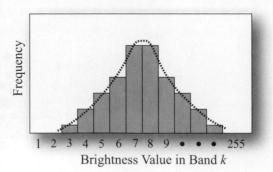

b. Data distribution approximated by a normal probability density function.

Figure 9-18   a) Hypothetical histogram of the forest training data in band k. b) Probability density function computed for the hypothetical training data using Equation 9-21. This function can be used to approximate the frequency of occurrence for every value on the x-axis.

transpose of the vector $(X - M_i)$. The mean vectors (M_i) and covariance matrix (V_i) for each class are estimated from the training data. For example, consider Figure 9-19 where the bivariate probability density functions of six hypothetical classes are arrayed in red and near-infrared feature space. It is bivariate because two bands are used. Note how the probability density function values appear to be normally distributed (i.e., bell-shaped). The vertical axis is associated with the probability of an unknown pixel measurement vector X being a member of one of the classes. In other words, if an unknown measurement vector has brightness values such that it lies within the wetland region, it has a high probability of being wetland.

If we assume that there are m classes, then $p(X/w_i)$ is the probability density function associated with the unknown measurement vector X, given that X is from a pattern in class w_i (Swain and Davis, 1978). In this case the *maximum likelihood decision rule* becomes (Richards and Jia, 1999):

Decide $X \in w_i$ if, and only if,

$$p(X|w_i) \cdot p(w_i) \geq p(X|w_j) \cdot p(w_j) \qquad (9\text{-}23)$$

for all i and j out of 1, 2, ... m possible classes.

Therefore, to classify a pixel in the multispectral remote sensing dataset with an unknown measurement vector X, a maximum likelihood decision rule computes the product $p(X|w_i) \cdot p(w_i)$ for each class and assigns the pattern to the class having the largest product (Swain and Davis, 1978). This assumes that we have some useful information about the prior probabilities of each class i (i.e., $p(w_i)$).

Maximum Likelihood Classification Without Prior Probability Information: In practice, we rarely have prior information about whether one class (e.g., forest) is expected to occur more frequently in a scene than any other class (e.g., 60% of the scene should be forest). This is called class a priori probability information (i.e., $p(w_i)$). Therefore, most applications of the maximum likelihood decision rule assume that each class has an equal probability of occurring in the landscape. This makes it possible to remove the prior probability term ($p(w_i)$) in Equation 9-23 and develop a simplified decision rule that can be applied to the unknown measurement vector X for each pixel in the scene (Swain and Davis, 1978; Schalkoff, 1992; Richards and Jia, 1999):

Decide unknown measurement vector X is in class i if, and only if,

$$p_i \geq p_j \qquad (9\text{-}24)$$

for all i and j out of 1, 2, ... m possible classes

and

$$p_i = -\frac{1}{2}\log_e|V_i| - \left[\frac{1}{2}(X - M_i)^T V_i^{-1}(X - M_i)\right] \qquad (9\text{-}25)$$

where M_i is the mean measurement vector for class i and V_i is the covariance matrix of class i for bands k through l. Therefore, to assign the measurement vector X of an unknown pixel to a class, the maximum likelihood decision rule computes the value p_i for each class. Then it assigns the pixel to the class that has the largest (or maximum) value. This assumes that we have no useful information about the prior probabilities of each class, that is, every class has an

Probability Density Functions Derived from Multispectral Training Data

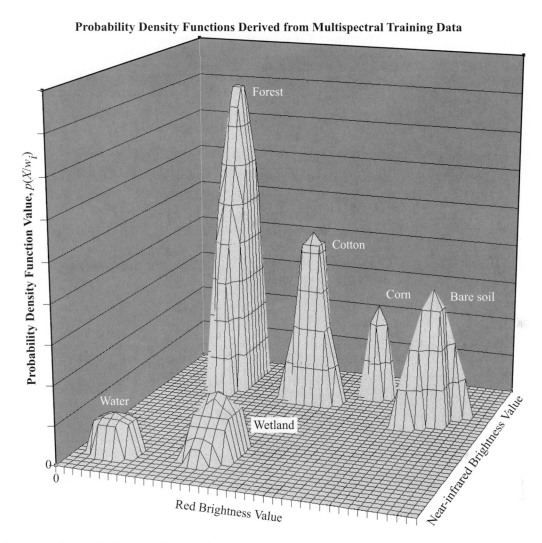

Figure 9-19   Example of normally distributed probability density functions associated with six hypothetical land-cover classes using two bands of multispectral data (red and near-infrared). The vertical axis is the probability of an unknown pixel measurement vector X being a member of one of the classes.

equal probability of occurring in the landscape (Richards and Jia, 1999).

Now let us consider the computations required. In the first pass, p_1 is computed with V_1 and M_1 being the covariance matrix and mean vectors for class 1. Next, p_2 is computed with V_2 and M_2 being the covariance matrix and mean vectors for class 2. This continues for all m classes. The pixel or measurement vector X is assigned to the class that produces the largest or maximum p_i. The measurement vector X used in each step of the calculation consists of n elements (the number of bands being analyzed). For example, if six Landsat TM bands (i.e., no thermal band) were being analyzed, each unknown pixel would have a measurement vector X of:

$$X = \begin{bmatrix} BV_{i,j,1} \\ BV_{i,j,2} \\ BV_{i,j,3} \\ BV_{i,j,4} \\ BV_{i,j,5} \\ BV_{i,j,7} \end{bmatrix}. \tag{9-26}$$

But what happens when the probability density functions of two or more training classes overlap in feature space? For example, consider two hypothetical normally distributed probability density functions associated with forest and agriculture training data measured in bands 1 and 2 (Figure 9-

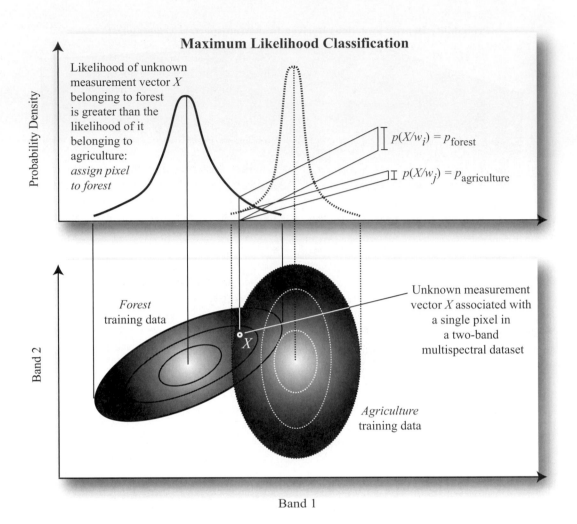

Figure 9-20   Hypothetical example of how the maximum likelihood decision rule functions when two training class probability density functions (forest and agriculture) measured in two bands overlap. The unknown measurement vector X associated with a single pixel in a two-band dataset would be assigned to forest because the probability density of its measurement vector X is greater for forest than for agriculture.

20). In this case, pixel X would be assigned to forest because the probability density of unknown measurement vector X is greater for forest than for agriculture.

Maximum Likelihood Classification with Prior Probability Information: Equation 9-24 assumes that each class has an equal probability of occurring in the terrain. Common sense reminds us that in most remote sensing applications there is a high probability of encountering some classes more often than others. For example, in the Charleston scene the probability of encountering various land covers is:

$$p(w_1) = p(\text{residential}) = 0.2,$$

$$p(w_2) = p(\text{commercial}) = 0.1,$$

$$p(w_3) = p(\text{wetland}) = 0.25,$$

$$p(w_4) = p(\text{forest}) = 0.1, \text{ and}$$

$$p(w_5) = p(\text{water}) = 0.35.$$

Thus, we would expect more pixels to be classified as water simply because it is more prevalent in the terrain. If we have such information, it is possible to include this valuable prior knowledge in the classification decision. We can do this by weighting each class i by its appropriate a priori probability, $p(w_i)$. The equation then becomes:

Decide unknown measurement vector X is in class i if, and only if,

$$p_i \cdot p(w_i) \ge p_j \cdot p(w_j) \qquad (9\text{-}27)$$

for all i and j out of 1, 2, ... m possible classes

and

$$p_i \cdot p(w_i) = \log_e p(w_i) - \frac{1}{2}\log_e |V_i|$$
$$- \left[\frac{1}{2}(X-M_i)^T V_i^{-1}(X-M_i)\right].$$

This Bayes's decision rule is identical to the maximum likelihood decision rule except that it does not assume that each class has equal probabilities (Hord, 1982; Richards and Jia, 1999). A priori probabilities have been used successfully as a way of incorporating the effects of relief and other terrain characteristics in improving classification accuracy (Strahler, 1980). Haralick and Fu (1983) provide an in-depth discussion of the probabilities and mathematics of the maximum likelihood and Bayes's decision rules. The maximum likelihood and Bayes classifications require many more computations per pixel than either the parallelepiped or minimum distance classification algorithms. They do not always produce superior results.

Unsupervised Classification

Unsupervised classification (also commonly referred to as *clustering*) is an effective method of partitioning remote sensor image data in multispectral feature space and extracting land-cover information (Loveland et al., 1999; Huang, 2002). Compared to supervised classification, unsupervised classification normally requires only a minimal amount of initial input from the analyst. This is because clustering does not normally require training data.

Unsupervised classification is the process whereby numerical operations are performed that search for natural groupings of the spectral properties of pixels, as examined in multispectral feature space. The clustering process results in a classification map consisting of m spectral classes (Lo and Yeung, 2002). The analyst then attempts *a posteriori* (after the fact) to assign or transform the *spectral* classes into thematic *information* classes of interest (e.g., forest, agriculture, urban). This may not be easy. Some spectral clusters may be meaningless because they represent mixed classes of Earth surface materials. It takes careful thinking by the analyst to unravel such mysteries. The analyst must understand the spectral characteristics of the terrain well enough to be able to label certain clusters as specific information classes.

Hundreds of clustering algorithms have been developed (Schowengerdt, 1997; Duda et al., 2001). Two examples of conceptually simple but not necessarily efficient clustering algorithms will be used to demonstrate the fundamental logic of unsupervised classification of remote sensor data.

Unsupervised Classification Using the Chain Method

The first clustering algorithm that will be discussed operates in a two-pass mode (i.e., it passes through the multispectral dataset two times). In the first pass, the program reads through the dataset and sequentially builds clusters (groups of points in spectral space). A mean vector is then associated with each cluster (Jain, 1989). In the second pass, a minimum distance to means classification algorithm similar to the one previously described is applied to the whole dataset on a pixel-by-pixel basis whereby each pixel is assigned to one of the mean vectors created in pass 1. The first pass, therefore, automatically creates the cluster signatures to be used by the minimum distance to means classifier.

Pass 1: Cluster Building

During the first pass, the analyst is required to supply four types of information:

1. R, a radius distance in spectral space used to determine when a new cluster should be formed (e.g., when raw remote sensor data are used, it might be set at 15 brightness value units),

2. C, a spectral space distance parameter used when merging clusters (e.g., 30 units) when N is reached,

3. N, the number of pixels to be evaluated between each major merging of the clusters (e.g., 2000 pixels), and

4. C_{max}, the maximum number of clusters to be identified by the algorithm (e.g., 20 clusters).

These can be set to default values if no initial human interaction is desired.

Starting at the origin of the multispectral dataset (i.e., line 1, column 1), pixels are evaluated sequentially from left to right as if in a chain. After one line is processed, the next line of data is considered. We will analyze the clustering of only the first three pixels in a hypothetical image and label them pixels 1, 2, and 3. The pixels have brightness values in just

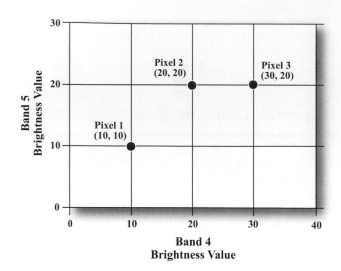

Figure 9-21 Original values of pixels 1, 2, and 3 as measured in bands 4 and 5 of the hypothetical remotely sensed data.

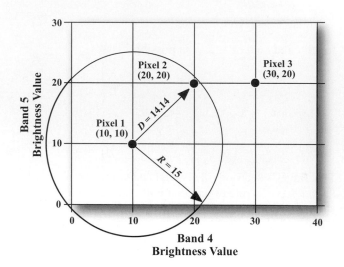

Figure 9-22 The distance (D) in two-dimensional spectral space between pixel 1 (cluster 1) and pixel 2 (cluster 2) in the first iteration is computed and tested against the value of R, the minimum acceptable radius. In this case, D does not exceed R; therefore, we merge clusters 1 and 2, as shown in Figure 9-23.

two bands, 4 and 5. Their spatial relationships in two-dimensional feature space are shown in Figure 9-21.

First, we let the brightness values associated with pixel 1 in the image represent the mean data vector of cluster 1 (i.e., $M_1 = \{10, 10\}$). Remember, it is an n-dimensional mean data vector with n being the number of bands used in the unsupervised classification. In our example, just two bands are being used, so $n = 2$. Because we have not identified all 20 spectral clusters (C_{max}) yet, pixel 2 will be considered as the mean data vector of cluster 2 (i.e., $M_2 = \{20, 20\}$). If the spectral distance D between cluster 2 and cluster 1 is greater than R, then cluster 2 will remain cluster 2. However, if the spectral distance D is less than R, then the mean data vector of cluster 1 becomes the average of the first and second pixel brightness values and the weight (or count) of cluster 1 becomes 2 (Figure 9-22). In our example, the distance D between cluster 1 (actually pixel 1) and pixel 2 is 14.14. Because the radius R was initially set at 15.0, pixel 2 does not satisfy the criteria for being cluster 2 because its distance from cluster 1 is <15. Therefore, the mean data vectors of cluster 1 and pixel 2 are averaged, yielding the new location of cluster 1 at $M_1 = \{15, 15\}$, as shown in Figure 9-23. The spectral distance D is computed using the Pythagorean theorem, as discussed previously.

Next, pixel 3 is considered as the mean data vector of cluster 2 (i.e., $M_2 = \{30, 20\}$). The distance from pixel 3 to the revised location of cluster 1, $M_1 = \{15, 15\}$, is 15.81 (Figure

9-23). Because it is >15, the mean data vector of pixel 3 becomes the mean data vector of cluster 2.

This cluster accumulation continues until the number of pixels evaluated is greater than N. At that point, the program stops evaluating individual pixels and looks closely at the nature of the clusters obtained thus far. It calculates the distance between each cluster and every other cluster. Any two clusters separated by a spectral distance less than C are merged. Such a new cluster mean vector is the weighted average of the two original clusters, and the weight is the sum of the two individual weights. This proceeds until there are no clusters with a separation distance less than C. Then the next pixel is considered. This process continues to iterate until the entire multispectral dataset is examined.

Schowengerdt (1997) suggests that virtually all the commonly used clustering algorithms use iterative calculations to find an optimum set of decision boundaries for the data set. It should be noted that some clustering algorithms allow the analyst to initially seed the mean vector for several of the important classes. The seed data are usually obtained in a supervised fashion, as discussed previously. Others allow the analyst to use a priori information to direct the clustering process.

Some programs do not evaluate every line and every column of the data when computing the mean vectors for the clusters.

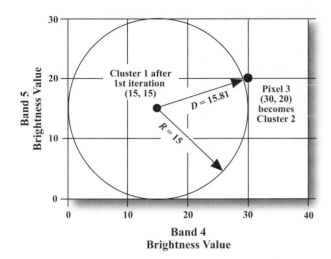

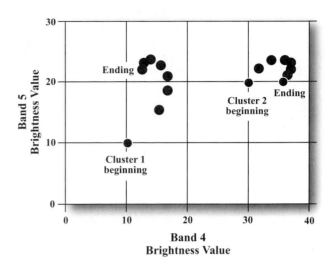

Figure 9-23   Pixels 1 and 2 now represent cluster 1. Note that the location of cluster 1 has migrated from (10, 10) to (15, 15), after the first iteration. Now, pixel 3 distance (*D*) is computed to see if it is greater than the minimum threshold, *R*. It is, so pixel location 3 becomes cluster 2. This process continues until all 20 clusters are identified. Then the 20 clusters are evaluated using a distance measure, *C* (not shown), to merge the clusters that are closest to one another.

Figure 9-24   How clusters migrate during the several iterations of a clustering algorithm. The final ending point represents the mean vector that would be used in phase 2 of the clustering process when the minimum distance classification is performed.

Instead, they may sample every *i*th row and *j*th column to identify the C_{max} clusters. If computer resources are abundant, then every pixel may be sampled. If resources are scarce, then acceptable results may usually be obtained by sampling the data. Obviously, a tremendous number of computations are performed during this initial pass through the dataset.

A hypothetical diagram showing the cluster migration for our two-band dataset is shown in Figure 9-24. Notice that as more points are added to a cluster, the mean shifts less dramatically since the new computed mean is weighted by the number of pixels currently in a cluster. The ending point is the spectral location of the final mean vector that is used as a signature in the minimum distance classifier applied in pass 2.

Pass 2: Assignment of Pixels to One of the C_{max} Clusters Using Minimum Distance Classification Logic

The final cluster mean data vectors are used in a minimum distance to means classification algorithm to classify all the pixels in the image into one of the C_{max} clusters. The analyst usually produces a cospectral plot display to document where the clusters reside in three-dimensional feature space.

It is then necessary to evaluate the location of the clusters in the image, label them if possible, and see if any should be combined. It is usually necessary to combine some clusters. This is where an intimate knowledge of the terrain is critical.

An unsupervised classification of the Charleston, SC, Landsat TM scene is displayed in **Color Plate 9-2b**. It was created using TM bands 2, 3, and 4. The analyst stipulated that a total of 20 clusters (C_{max}) be extracted from the data. The mean data vectors for each of the final 20 clusters are summarized in Table 9-11. These mean vectors represented the data used in the minimum distance classification of every pixel in the scene into one of the 20 cluster categories.

Cospectral plots of the mean data vectors for each of the 20 clusters using bands 2 and 3 and bands 3 and 4 are displayed in Figures 9-25 and 9-26, respectively. The 20 clusters lie on a diagonal extending from the origin in the band 2 versus band 3 plot. Unfortunately, the water cluster was located in the same spectral space as forest and wetland when viewed using just bands 2 and 3 (Figure 9-25). Therefore, this scatterplot was not used to *label* or assign the clusters to *information* classes. Conversely, a cospectral plot of bands 3 and 4 mean data vectors was relatively easy to interpret (Figure 9-26).

Cluster labeling is usually performed by interactively displaying all the pixels assigned to an individual cluster on the screen with a color composite of the study area in the back-

Table 9-11.   Results of clustering on Thematic Mapper Bands 2, 3, and 4 of the Charleston, SC, Landsat TM scene.

Cluster	Percent of Scene	Mean Vector			Class Description	Color Assignment
		Band 2	Band 3	Band 4		
1	24.15	23.14	18.75	9.35	Water	Dark blue
2	7.14	21.89	18.99	44.85	Forest 1	Dark green
3	7.00	22.13	19.72	38.17	Forest 2	Dark green
4	11.61	21.79	19.87	19.46	Wetland 1	Bright green
5	5.83	22.16	20.51	23.90	Wetland 2	Green
6	2.18	28.35	28.48	40.67	Residential 1	Bright yellow
7	3.34	36.30	25.58	35.00	Residential 2	Bright yellow
8	2.60	29.44	29.87	49.49	Parks, golf	Gray
9	1.72	32.69	34.70	41.38	Residential 3	Yellow
10	1.85	26.92	26.31	28.18	Commercial 1	Dark red
11	1.27	36.62	39.83	41.76	Commercial 2	Bright red
12	0.53	44.20	49.68	46.28	Commercial 3	Bright red
13	1.03	33.00	34.55	28.21	Commercial 4	Red
14	1.92	30.42	31.36	36.81	Residential 4	Yellow
15	1.00	40.55	44.30	39.99	Commercial 5	Bright red
16	2.13	35.84	38.80	35.09	Commercial 6	Red
17	4.83	25.54	24.14	43.25	Residential 5	Bright yellow
18	1.86	31.03	32.57	32.62	Residential 6	Yellow
19	3.26	22.36	20.22	31.21	Commercial 7	Dark red
20	0.02	34.00	43.00	48.00	Commercial 8	Bright red

ground. In this manner it is possible to identify the location and spatial association among clusters. This interactive visual analysis in conjunction with the information provided in the cospectral plot allows the analyst to group the clusters into information classes, as shown in Figure 9-27 and Table 9-11. It is instructive to review some of the logic that resulted in the final unsupervised classification in **Color Plate 9-2b**.

Cluster 1 occupied a distinct region of spectral space (Figure 9-27). It was not difficult to assign it to the information class

water. Clusters 2 and 3 had high reflectance in the near-infrared (band 4) with low reflectance in the red (band 3) due to chlorophyll absorption. These two clusters were assigned to the forest class and color-coded dark green (refer to Table 9-11). Clusters 4 and 5 were situated alone in spectral space between the forest (2 and 3) and water (1) and were comprised of a mixture of moist soil and abundant vegetation. Therefore, it was not difficult to assign both these clusters to a wetland class. They were given different color codes to demonstrate that, indeed, two separate classes of wetland were identified.

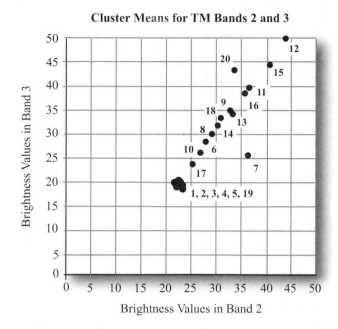

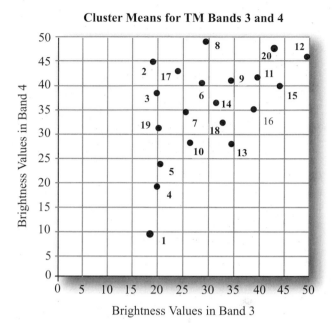

Figure 9-25 The mean vectors of the 20 clusters displayed in **Color Plate 9-2b** are shown here using only bands 2 and 3. The mean vector values are summarized in Table 9-11. Notice the substantial amount of overlap among clusters 1 through 5 and 19.

Figure 9-26 The mean vectors of the 20 clusters displayed in **Color Plate 9-2b** are shown here using only bands 3 and 4 data. The mean vector values are summarized in Table 9-11.

Six clusters were associated with residential housing. These clusters were situated between the forest and commercial clusters. This is not unusual since residential housing is composed of a mixture of vegetated and nonvegetated (asphalt and concrete) surfaces, especially at TM spatial resolutions of 30×30 meters. Based on where they were located in feature space, the six clusters were collapsed into just two: bright yellow (6, 7, 17) and yellow (9, 14, 18).

Eight clusters were associated with commercial land use. Four of the clusters (11, 12, 15, 20) reflected high amounts of both red and near-infrared energy as commercial land composed of concrete and bare soil often does. Two other clusters (13 and 16) were associated with commercial strip areas, particularly the downtown areas. Finally, there were two clusters (10 and 19) that were definitely commercial but that had a substantial amount of associated vegetation. They were found mainly along major thoroughfares in the residential areas where vegetation is more plentiful. These three subgroups of commercial land use were assigned bright red, red, and dark red, respectively (Table 9-11). Cluster 8 did not fall nicely into any group. It experienced very high near-infrared reflectance and chlorophyll absorption often associated with

very well kept lawns or parks. In fact, this is precisely what it was labeled, "parks and golf."

The 20 clusters and their color assignments are shown in Table 9-11. There is more information present in this unsupervised classification than in the supervised classification. Except for water, there are at least two classes in each land-use category that could be identified successfully using the unsupervised technique. The supervised classification simply did not sample many of these classes during the training process.

Unsupervised Classification Using the ISODATA Method

Another widely used clustering algorithm is the Iterative Self-Organizing Data Analysis Technique (ISODATA). ISODATA represents a comprehensive set of heuristic (rule of thumb) procedures that have been incorporated into an iterative classification algorithm (ERDAS, 2003; Stow et al., 2003; Rees et al., 2003). Many of the steps incorporated into the algorithm are a result of experience gained through experimentation. The ISODATA algorithm is a modification

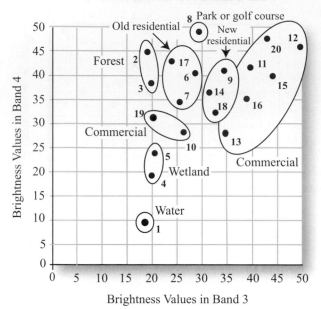

Cluster Means for TM Bands 3 and 4

Figure 9-27   Grouping (labeling) of the original 20 spectral clusters into information classes. The labeling was performed by analyzing the mean vector locations in bands 3 and 4.

of the *k*-means clustering algorithm, which includes a) merging clusters if their separation distance in multispectral feature space is below a user-specified threshold and b) rules for splitting a single cluster into two clusters (Schowengerdt, 1997).

ISODATA is self-organizing because it requires relatively little human input. A sophisticated ISODATA algorithm normally requires the analyst to specify the following criteria:

- C_{max}: the maximum number of clusters to be identified by the algorithm (e.g., 20 clusters). However, it is not uncommon for fewer to be found in the final classification map after splitting and merging take place.

- T: the maximum percentage of pixels whose class values are allowed to be *unchanged* between iterations. When this number is reached, the ISODATA algorithm terminates. Some datasets may never reach the desired percentage unchanged. If this happens, it is necessary to interrupt processing and edit the parameter.

- M: the maximum number of times ISODATA is to classify pixels and recalculate cluster mean vectors. The

ISODATA algorithm terminates when this number is reached.

- *Minimum members in a cluster (%):* If a cluster contains less than the minimum percentage of members, it is deleted and the members are assigned to an alternative cluster. This also affects whether a class is going to be split (see maximum standard deviation). The default minimum percentage of members is often set to 0.01.

- *Maximum standard deviation* (σ_{max}): When the standard deviation for a cluster exceeds the specified maximum standard deviation and the number of members in the class is greater than twice the specified minimum members in a class, the cluster is split into two clusters. The mean vectors for the two new clusters are the old class centers $\pm 1\sigma$. Maximum standard deviation values between 4.5 and 7 are typical.

- *Split separation value:* If this value is changed from 0.0, it takes the place of the standard deviation in determining the locations of the new mean vectors plus and minus the split separation value.

- *Minimum distance between cluster means (C):* Clusters with a weighted distance less than this value are merged. A default of 3.0 is often used.

ISODATA Initial Arbitrary Cluster Allocation

ISODATA is iterative because it makes a large number of passes through the remote sensing dataset until specified results are obtained, instead of just two passes. Also, ISODATA does not allocate its initial mean vectors based on the analysis of pixels in the first line of data the way the two-pass algorithm does. Rather, an initial arbitrary assignment of all C_{max} clusters takes place along an *n*-dimensional vector that runs between very specific points in feature space. The region in feature space is defined using the mean, μ_k, and standard deviation, σ_k, of each band in the analysis. A hypothetical two-dimensional example using bands 3 and 4 is presented in Figure 9-28a, in which five mean vectors are distributed along the vector beginning at the location $\mu_3 - \sigma_3$, $\mu_4 - \sigma_4$ and ending at $\mu_3 + \sigma_3$, $\mu_4 + \sigma_4$. This method of automatically seeding the original C_{max} vectors makes sure that the first few lines of data do not bias the creation of clusters. Note that the two-dimensional parallelepiped (box) does not capture all the possible band 3 and 4 brightness value combinations present in the scene. The location of the initial C_{max} mean vectors (Figure 9-28a) should move about somewhat to partition the feature space better. This takes place in the first and subsequent iterations. Huang (2002) developed an efficient version of ISODATA that provides improved automatic location of the initial mean vectors (i.e., seed clusters).

ISODATA First Iteration

With the initial C_{max} mean vectors in place, a pass is made through the database beginning in the upper left corner of the matrix. Each candidate pixel is compared to each cluster mean and assigned to the cluster whose mean is closest in Euclidean distance (Figure 9-28b). This pass creates an actual classification map consisting of C_{max} classes. It should be noted that some image processing systems process data line by line, and others process the data in a block or tiled data structure. The way that ISODATA is instructed to process the data (e.g., line by line or block by block) will have an impact on the creation of the mean vectors.

ISODATA Second to *M*th Iteration

After the first iteration, a new mean for each cluster is calculated based on the actual spectral locations of the pixels assigned to each cluster, instead of on the initial arbitrary calculation. This involves analysis of the following parameters: minimum members in a cluster (%), maximum standard deviation (σ_{max}), split separation, and minimum distance between cluster means (C). Then the entire process is repeated with each candidate pixel once again compared to the new cluster means and assigned to the nearest cluster mean (Figure 9-28c). Sometimes individual pixels do not change cluster assignment. This iterative process continues (Figure 9-28d) until there is (1) little change in class assignment between iterations (i.e., the T threshold is reached) or (2) the maximum number of iterations is reached (M). The final file is a matrix with C_{max} clusters in it, which must be labeled and recoded to become useful land-cover information. The fact that the initial mean vectors are situated throughout the heart of the existing data is superior to initiating clusters based on finding them in the first line of the data.

The iterative ISODATA algorithm is relatively slow, and image analysts are notoriously impatient. Analysts must allow the ISODATA algorithm to iterate enough times to generate meaningful mean vectors.

ISODATA Example 1: An ISODATA classification was performed using the Charleston, SC, Landsat TM band 3 and 4 data. The locations of the clusters (mean $\pm 2\sigma$) after just one iteration are shown in Figure 9-29a. The clusters are superimposed on the distribution of all brightness values found in TM bands 3 and 4. The location of the final mean

vectors after 20 iterations is shown in Figure 9-29b. The ISODATA algorithm has partitioned the feature space effectively. Requesting more clusters (e.g., 100) and allowing more iterations (e.g., 500) would partition the feature space even better. A classification map example is not provided because it would not be dramatically different from the results of the two-pass clustering algorithm since so few clusters were requested.

ISODATA Example 2: An ISODATA algorithm was applied to two bands of HyMap hyperspectral data (red and near-infrared) of an area near North Inlet, SC (**Color Plate 9-3**). The goal was to extract six land-cover classes from the data: water, wetland, roof/asphalt, forest, bare soil, and golf fairway. Only 10 clusters were extracted using ISODATA for demonstration purposes. The location of the 10 clusters (actually mean vectors) after 1–5 and 10 iterations are displayed in Figure 9-30. The red and near-infrared spectral characteristics of the final 10 mean vectors after 10 iterations are found in Figure 9-30f and **Color Plate 9-3b**. Typically, twenty or more clusters are requested. Therefore, Figure 9-30g shows where the 20 clusters reside after 20 iterations.

The 10 mean vectors shown in Figure 9-30f were used in the final pass of the ISODATA algorithm to create a thematic map consisting of 10 spectral classes. These spectral classes were investigated by an analyst and labeled as the 10 information classes shown in **Color Plate 9-3c**. The final step in the process would be to recode the 10 information classes into just 6 classes to create a normal 6-class legend and to compute the hectares in each land-cover class.

Unsupervised Cluster Busting

It is common when performing unsupervised classification using the chain algorithm or ISODATA to generate n clusters (e.g., 100) and have no confidence in labeling q of them to an appropriate information class (let us say 30 in this example). This is because (1) the terrain within the IFOV of the sensor system contained at least two types of terrain, causing the pixel to exhibit spectral characteristics unlike either of the two terrain components, or (2) the distribution of the mean vectors generated during the unsupervised classification process was not good enough to partition certain important portions of feature space. When this occurs, it may be possible to perform *cluster busting* if in fact there is still some unextracted information of value in the dataset.

First, all the pixels associated with the q clusters (30 in a hypothetical example) that are difficult to label (e.g., mixed

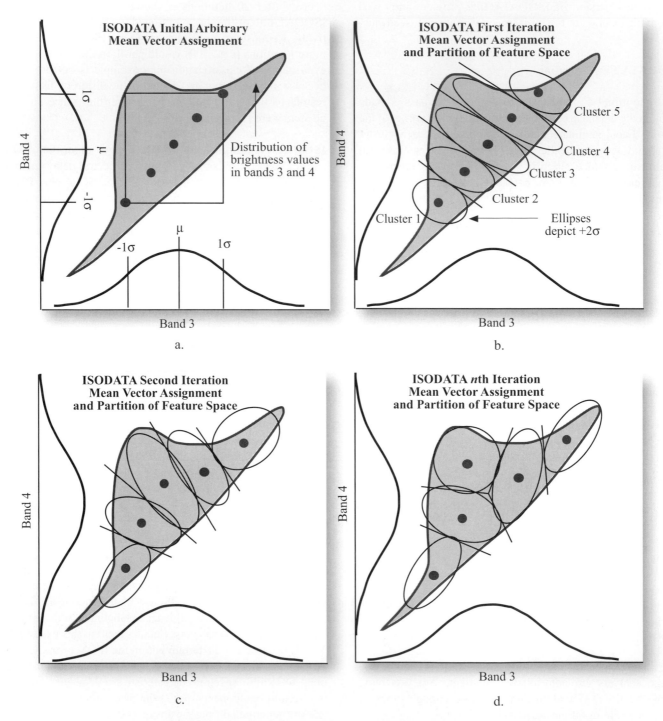

Figure 9-28   a) ISODATA initial distribution of five hypothetical mean vectors using $\pm 1\sigma$ standard deviations in both bands as beginning and ending points. b) In the first iteration, each candidate pixel is compared to each cluster mean and assigned to the cluster whose mean is closest in Euclidean distance. c) During the second iteration, a new mean is calculated for each cluster based on the actual spectral locations of the pixels assigned to each cluster, instead of the initial arbitrary calculation. This involves analysis of several parameters to merge or split clusters. After the new cluster mean vectors are selected, every pixel in the scene is assigned to one of the new clusters. d) This split–merge–assign process continues until there is little change in class assignment between iterations (the T threshold is reached) or the maximum number of iterations is reached (M).

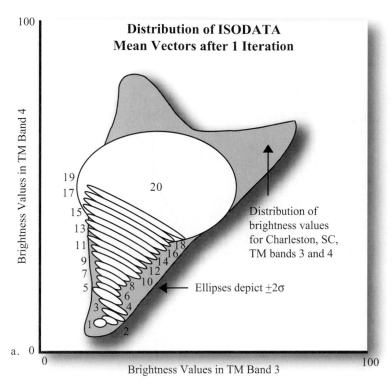

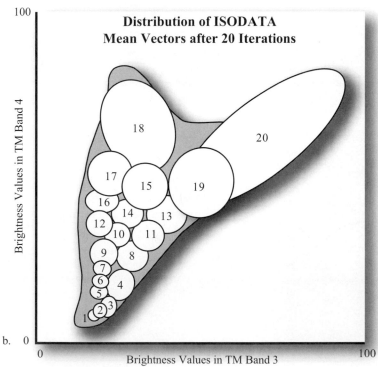

Figure 9-29 a) Distribution of 20 ISODATA mean vectors after just one iteration using Landsat TM band 3 and 4 data of Charleston, SC. Notice that the initial mean vectors are distributed along a diagonal in two-dimensional feature space according to the $\pm 2\sigma$ standard deviation logic discussed. b) Distribution of 20 ISODATA mean vectors after 20 iterations. The bulk of the important feature space (the gray background) is partitioned rather well after just 20 iterations.

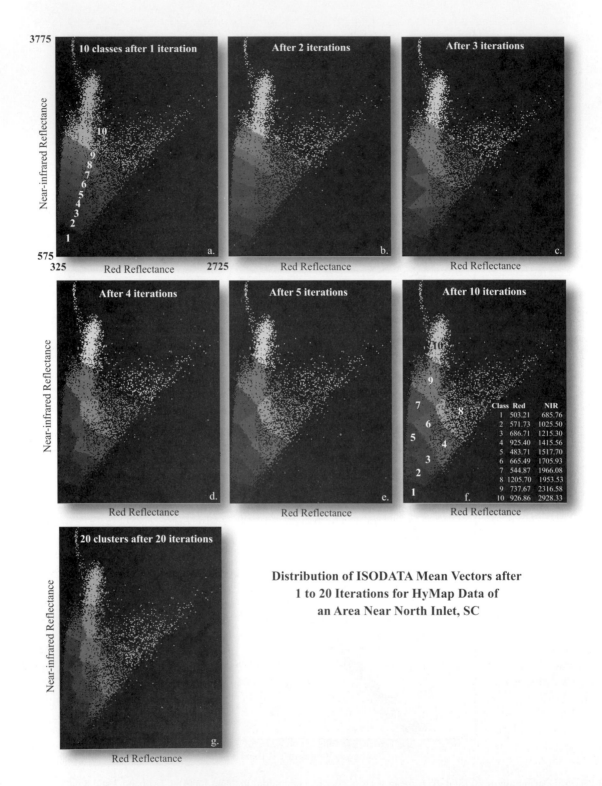

The following table appears within panel f. (After 10 iterations):

Class	Red	NIR
1	503.21	685.76
2	571.73	1025.50
3	686.71	1215.30
4	925.40	1415.56
5	483.71	1517.70
6	665.49	1705.93
7	544.87	1966.08
8	1205.70	1953.53
9	737.67	2316.58
10	926.86	2928.33

**Distribution of ISODATA Mean Vectors after
1 to 20 Iterations for HyMap Data of
an Area Near North Inlet, SC**

Figure 9-30   a–f) Distribution of 10 ISODATA mean vectors after 1–5 and 10 iterations using just two bands of HyMap hyperspectral data of an area near North Inlet, SC. The 10 mean vectors after 10 iterations were used to produce the thematic map shown in **Color Plate 9-3c**. g) Typically, more than 20 clusters are requested when using ISODATA. This graphic shows the distribution of the two-dimensional feature space for 20 clusters after 20 iterations.

clusters 13, 22, 45, 92, etc.) are all recoded to a value of 1 and a binary mask file is created. A mask program is then run using (1) the binary mask file and (2) the original remote sensor data file. The output of the mask program is a new multiband image file consisting of only the pixels that could not be adequately labeled during the initial unsupervised classification. The analyst then performs a new unsupervised classification on this file, perhaps requesting an additional 25 clusters. The analyst displays these clusters using standard techniques and keeps as many of these new clusters as possible (e.g., 15). Usually, there are still some clusters that contain mixed pixels, but the proportion definitely goes down. The analyst may want to iterate the process one more time to see if an additional unsupervised classification breaks out additional clusters. Perhaps five good clusters are extracted during the final iteration.

In this hypothetical example, the final cluster map would be composed of the 70 good clusters from the initial classification, 15 good clusters from the first cluster-busting pass (recoded as values 71 to 85), and 5 from the second cluster-busting pass (recoded as values 86 to 90). The final cluster map file may be put together using a simple GIS maximum dominate function. The final cluster map is then recoded to create the final classification map.

Fuzzy Classification

Geographical information (including remotely sensed data) is imprecise, meaning that the boundaries between different phenomena are fuzzy, or there is heterogeneity within a class, perhaps due to differences in species, health, age, and so forth.

For example, terrain in the southeastern United States often exhibits a gradual transition from water to forested wetland to deciduous upland forest (Figure 9-31a.) Normally, the greater the canopy closure, the greater the amount of near-infrared energy reflected from within the IFOV of a pixel along this continuum. Also, the greater the proportion of water in a pixel, the more near-infrared radiant flux absorbed. A hard classification algorithm applied to these remotely sensed data collected along this continuum would be based on classical set theory, which requires precisely defined set boundaries for which an element (e.g., a pixel) is either a member (true = 1) or not a member (false = 0) of a given set. For example, if we made a classification map using just a single near-infrared band (i.e., one-dimensional density slicing), the decision rules might be as shown in Figure 9-31a: 0 to 30 = water, 31 to 60 = forested wetland, and

61 to 90 = upland forest. The classic approach creates three discrete classes with specific class ranges, and no intermediate situations are allowed. Thus, using classical set theory, an unknown measurement vector may be assigned to one and only one class (Figure 9-31a). But everyone knows that the phenomena grade into one another and that mixed pixels are present, especially around the values of 24 to 36 and 55 to 70, as shown in the figure. Clearly, there needs to be a way to make the classification algorithms more sensitive to the imprecise (fuzzy) nature of the real world.

Fuzzy set theory provides some useful tools for working with imprecise data (Zadeh, 1965; Wang, 1990a,b). Fuzzy set theory is better suited for dealing with real-world problems than traditional logic because most human reasoning is imprecise and is based on the following logic. First, let X be a universe whose elements are denoted x. That is, $X = \{x\}$. As previously mentioned, membership in a classical set A of X is often viewed as a binary characteristic function x_A from X $\{0$ or $1\}$ such that $x_A(x) = 1$ if and only if $x \in A$. Conversely, a *fuzzy set B* in X is characterized by a *membership function f_B* that associates with each x a real number from 0 to 1. The closer the value of $f_B(x)$ is to 1, the more x belongs to B. Thus, a fuzzy set does not have sharply defined boundaries, and a set element (a pixel in our case) may have partial membership in several classes.

So how is fuzzy logic used to perform image classification? Figure 9-31b illustrates the use of fuzzy classification logic to discriminate among the three hypothetical land covers. The vertical boundary for water at brightness value 30 (Figure 9-31a) is replaced by a graded boundary that represents a gradual transition from water to forested wetland (Figure 9-31b). In the language of fuzzy set theory, BVs of less than 24 have a *membership grade* of 1 for water, and those greater than about 70 have a membership grade of 1 for upland forest. At several other locations a BV may have a membership grade in two classes. For example, at BV 30 we have membership grades of 0.5 water and 0.5 of forested wetland. At BV 60 the membership grades are 0.7 for forested wetland and 0.3 for upland forest. This membership grade information may be used by the analyst to create a variety of classification maps.

All that has been learned before about traditional hard classification is pertinent for fuzzy classification because training still takes place, feature space is partitioned, and it is possible to assign a pixel to a single class, if desired. However, the major difference is that it is also possible *to obtain information on the various constituent classes found in a mixed pixel*, if desired (Foody, 2000). It is instructive to review how this is done.

Conventional Classification

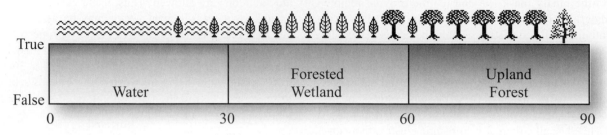

a.

Fuzzy Classification

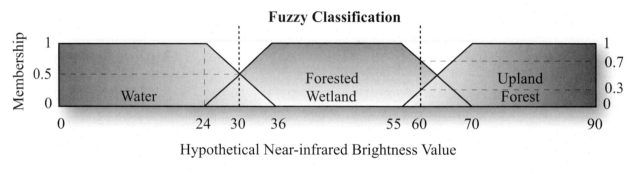

b.

Figure 9-31   (a) Conventional hard classification rules applied to discriminate among three land-cover classes. The terrain icons suggest that there is a gradual transition in near-infrared brightness value as one progresses from water to forested wetland to upland forest. A remote sensing system would be expected to record radiant flux from mixed pixels at the interface between the major land-cover types. Mixed pixels may also be encountered within a land-cover type as a result of differences in species, age, or functional health of vegetation. Despite these fuzzy conditions, a hard classification would simply assign a pixel to one and only one class. (b) The logic of a fuzzy classification. In this hypothetical example, a pixel having a near-infrared brightness value of <24 would have a *membership grade* value of 1.0 in water and 0 in both forested wetland and upland forest. Similarly, a brightness value of 60 would have a graded value of 0.70 for forested wetland, 0.30 for upland forest, and 0 for water. The membership grade values provide information on mixed pixels and may be used to classify the image using various types of logic.

First, the process of collecting training data as input to a fuzzy classification is somewhat different. Instead of selecting training sites that are as homogeneous as possible, the analyst may desire to select training areas that contain heterogeneous mixtures of biophysical materials in order to understand them better and to create a more accurate representation of the real world in the final classification map. Thus, a combination of pure (homogeneous) and mixed (heterogeneous) training sites may be selected.

Feature space partitioning may be dramatically different. For example, Figure 9-32a depicts a hypothetical hard partition of feature space based on classical set theory. In this

example, an unknown measurement vector (pixel) is assigned to the forested wetland class based on its location in a feature space that is partitioned using minimum distance to means criteria. *All* pixels inside a partitioned feature space are assigned to a single class, no matter how close they may be to a partition line. The assignment implies full membership in that class and no membership in other classes. In this example, it is likely that the pixel under investigation probably has a lot of forested wetland, considerable water, and perhaps a small amount of forest based on its location in feature space. Such information is completely lost when the pixel is assigned to a single class using hard feature space partitioning.

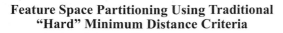

Feature Space Partitioning Using Traditional "Hard" Minimum Distance Criteria

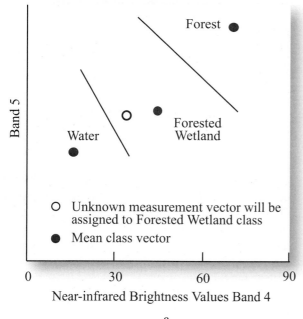

Unknown measurement vector will be assigned to Forested Wetland class

Mean class vector

Near-infrared Brightness Values Band 4

a.

Fuzzy Membership Grades Associated with an Unknown Measurement Vector

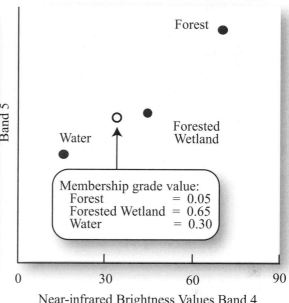

Membership grade value:
Forest   = 0.05
Forested Wetland = 0.65
Water   = 0.30

Near-infrared Brightness Values Band 4

b.

Figure 9-32   (a) Hypothetical hard partition of feature space based on classical set theory and a minimum distance decision rule. (b) Each unknown measurement vector (pixel) has membership grade values describing how close the pixel is to the *m* training class mean vectors. In this example, we are most likely working with a mixed pixel predominantly composed of forested wetland (0.65) and water (0.30) based on the membership grade values. This membership grade information may be used to classify the image using various types of logic.

Contrast this situation with a fuzzy partition of the feature space (Figure 9-32b). Rather than being assigned to a single class, the unknown measurement vector (pixel) now has *membership grade* values describing how close the pixel is to the *m* training class mean vectors (Wang, 1990a). In this example, the pixel being evaluated has values of forested wetland = 0.65, water = 0.30, and forest = 0.05. The values for all classes for each pixel must total 1.0.

When working with real remote sensor data, the actual fuzzy partition of spectral space is a family of fuzzy sets, F_1, F_2, ..., F_m on the universe X such that for every x that is an element of X (Wang, 1990b):

$$0 \le f_{F_i}(x) \le 1 \qquad (9\text{-}28)$$

$$\sum_{x \in X} f_{F_i}(x) > 0 \qquad (9\text{-}29)$$

$$\sum_{i=1}^{m} f_{F_i}(x) = 1 \qquad (9\text{-}30)$$

where F_1, F_2, ..., F_m represent the spectral classes, X represents all pixels in the dataset, m is the number of classes trained upon, x is a pixel measurement vector, and f_{F_i} is the membership function of the fuzzy set F_i ($1 \le i \le m$). The fuzzy partition may be recorded in a fuzzy partition matrix:

$$\begin{bmatrix} f_{F_1}(x_1) \; f_{F_1}(x_2) \; .... \; f_{F_1}(x_n) \\ f_{F_2}(x_1) \; f_{F_2}(x_2) \; .... \; f_{F_2}(x_n) \\ \vdots \qquad \vdots \qquad \vdots \qquad \vdots \\ f_{F_m}(x_1) \; f_{F_m}(x_2) \; .... \; f_{F_m}(x_n) \end{bmatrix}$$

where n is the number of pixels and x_i is the ith pixel's measurement vector ($1 \le i \le n$).

Fuzzy logic may be used to compute fuzzy mean and covariance matrices. For example, the fuzzy mean may be expressed (Wang, 1990a):

$$\mu_c^* = \frac{\sum_{i=1}^{n} f_c(x_i)x_i}{\sum_{i=1}^{n} f_c(x_i)} \tag{9-31}$$

where n is the total number of sample pixel measurement vectors, f_c is the membership function of class c, and x_i is a sample pixel measurement vector ($1 \le i \le n$). The fuzzy covariance matrix V_c^* is computed:

$$V_c^* = \frac{\sum_{i=1}^{n} f_c(x_i)(x_i - \mu_c^*)(x_i - \mu_c^*)^T}{\sum_{i=1}^{n} f_c(x_i)}. \tag{9-32}$$

When calculating a fuzzy mean for class c, a sample pixel measurement vector x is multiplied by its membership grade in c, $f_c(x)$ before being added to the sum. Similarly, in calculating a fuzzy covariance matrix for class c, $(x_i - \mu_c^*)(x_i - \mu_c^*)^T$ is multiplied by $f_c(x)$ before being added.

To perform a fuzzy feature space partition, a membership function must be defined for each class. The following example is based on the maximum likelihood classification algorithm with fuzzy mean μ^* and fuzzy covariance matrix V^* replacing the conventional mean and covariance matrix (Wang, 1990b). The following is the definition of the membership function for class c:

$$f_c(x) = \frac{P_c^*(x)}{\sum_{i=1}^{m} P_i^*(x)} \tag{9-33}$$

where

$$P_i^*(x) = \frac{1}{(2\pi)^{N/2}|V_i^*|^{1/2}} \times \tag{9-34}$$

$$\exp[-0.5(x - \mu_i^*)^T (V_i^*)^{-1}(x - \mu_i^*)]$$

and N is the dimension of the pixel vectors, m is the number of classes, and $1 \le i \le m$.

The membership grades of a pixel vector x depend on x's position in the spectral space (Wang, 1990a). $f_c(x)$ increases exponentially with the decrease of $(x - \mu_c^*)^T V_i^{*-1}(x - \mu_c^*)$, that is, the Mahalanobis distance between x and class c. The value

$$\sum_{i=1}^{m} P_i^*(x) \tag{9-35}$$

is a normalizing factor.

Applying this type of fuzzy logic creates a membership grade matrix for each pixel. An example based on the work of Wang (1990a) using Landsat MSS data of Hamilton City, Ontario, Canada, is shown in Table 9-12. Eight pixels (labeled A through H) in the scene are arrayed according to their membership grades. Homogeneous and mixed pixels may be differentiated easily by analyzing the membership grades; for example, pixels C, F, G, and H are mixed pixels, while A, B, D, and E are relatively homogeneous. Proportions of component cover classes in a pixel can be estimated from the membership grades. For example, it can be estimated that pixel A in the dataset contained 99% forest, pixel B contained 67% forest and 33% grass, pixel C contained 54% forest and 45% grass, and pixel D contained 87% grass and 12% bare soil (Table 9-12). This is very useful information. It may be used to produce one map or a series of maps that contain(s) robust ecological information because the map(s) may more closely resemble the real-world situation. For example, an analyst could apply simple Boolean logic to the membership-grade dataset to make a map showing only the pixels that had a forest grade value of >70% and a grass value of >20% (pixel B in the example would meet this criteria). Conversely, a hard partition can be derived from the fuzzy partition matrix by changing the maximum value in each column into a 1 and others into 0. A hardened classification map can then be generated by assigning the label of the row with the value 1 of each column to the corresponding pixel.

Scientists have also applied fuzzy logic to perform unsupervised classification and change detection (Bezdek et al., 1984; Fisher and Pathirana, 1993; Rignot, 1994). Fuzzy set theory is not a panacea, but it does offer significant potential for extracting information on the makeup of the biophysical materials within a mixed pixel, a problem that will always be with us (Ji and Jensen, 1996).

Table 9-12. Fuzzy classification membership grades for eight selected pixels (A through H) from Landsat MSS Data of Hamilton City, Ontario, Canada (Wang, 1990a).

Pixel	A	B	C	D	E	F	G	H
Water	0.00	0.00	0.00	0.00	0.00	0.00	0.00	0.00
Industrial	0.00	0.00	0.00	0.00	0.00	0.00	0.00	0.00
Residential	0.00	0.00	0.00	0.00	0.99	0.64	0.48	0.24
Forest	0.99	0.67	0.54	0.00	0.00	0.13	0.00	0.00
Grass	0.00	0.33	0.45	0.87	0.00	0.22	0.17	0.00
Pasture	0.00	0.00	0.00	0.00	0.00	0.00	0.00	0.14
Bare soil	0.00	0.00	0.00	0.12	0.00	0.00	0.35	0.62

Classification Based on Object-oriented Image Segmentation

The new millennium has seen the development of remote sensing systems such as IKONOS and QuickBird that produce remote sensor data with spatial resolutions of $\leq 1 \times 1$ m. Unfortunately, classification algorithms based on single-pixel analysis often are not capable of extracting the information we desire from high-spatial-resolution remote sensor data (Visual Learning Systems, 2002). For example, the spectral complexity of urban land-cover materials results in specific limitations using per-pixel analysis for the separation of human-made materials such as roads and roofs and natural materials such as vegetation, soil, and water (Herold et al., 2003). Furthermore, a significant but usually ignored problem with per-pixel characterization of land cover is that a substantial proportion of the signal apparently coming from the land area represented by a pixel comes from the surrounding terrain (Townshend et al., 2000). Improved algorithms are needed that take into account not only the spectral characteristics of a single pixel but those of the surrounding (contextual) pixels. In addition, we need information about the spatial characteristics of the surrounding pixels so that we can identify areas (or segments) of pixels that are homogeneous.

Object-oriented Image Segmentation and Classification

This need has given rise to the creation of image classification algorithms based on *object-oriented image segmentation*. The algorithms typically incorporate both spectral and spatial information in the image segmentation phase. The result is the creation of *image objects* defined as individual areas with shape and spectral homogeneity (Benz, 2001), which one may recognize as segments or patches in the landscape ecology literature (Forman, 1995; Turner et al., 2001; Herold et al., 2002; Batistella et al., 2003). In many instances, carefully extracted image objects can provide a greater number of meaningful features for image classification (Visual Learning Systems, 2002). In addition, objects don't have to be derived from just image data but can also be developed from any spatially distributed variable (e.g., elevation, slope, aspect, population density). Homogeneous image objects are then analyzed using traditional classification algorithms (e.g., nearest-neighbor, minimum distance, maximum likelihood) or knowledge-based approaches and fuzzy classification logic (Civco et al., 2002).

There are many algorithms that can be used to segment an image into relatively homogeneous image objects. Most can be grouped into two classes: edge-based algorithms and area-based algorithms. Unfortunately, the majority do not incorporate both spectral and spatial information, and very few have been used for remote sensing digital image classification.

Blaschke and Strobl (2001) suggest that one of the most promising approaches to remote sensing image segmentation was developed by Baatz and Schape (2000). The image segmentation involves looking at individual pixel values and their neighbors to compute a (Baatz et. al., 2001):

- color criterion (h_{color}), and

- a shape or spatial criterion (h_{shape}).

These two criteria are then used to create image objects (patches) of relatively homogeneous pixels in the remote sensing dataset using the general segmentation function (S_f) (Baatz et al., 2001; Definiens, 2003):

$$S_f = w_{color} \cdot h_{color} + (1 - w_{color}) \cdot h_{shape} \qquad (9\text{-}36)$$

where the user-defined weight for spectral color versus shape is $0 \le w_{color} \le 1$. If the user wants to place greater emphasis on the spectral (color) characteristics in the creation of homogeneous objects (patches) in the dataset, then w_{color} is weighted more heavily (e.g., $w_{color} = 0.8$). Conversely, if the spatial characteristics of the dataset are believed to be more important in the creation of the homogeneous patches, then shape should be weighted more heavily.

Spectral (i.e., color) heterogeneity (h) of an *image object* is computed as the sum of the standard deviations of spectral values of each layer (σ_k) (i.e., band) multiplied by the weights for each layer (w_k) (Kuehn, et al., 2002; Definiens, 2003):

$$h = \sum_{k=1}^{m} w_k \cdot \sigma_k. \qquad (9\text{-}37)$$

The *color criterion* is computed as the weighted mean of all changes in standard deviation for each channel k of the m band remote sensing dataset. The standard deviation σ_k are weighted by the object sizes n_{ob} (Definiens, 2003):

$$h = \sum_{k=1}^{m} w_k [n_{mg} \cdot \sigma_k{}^{mg} - (n_{ob1} \cdot \sigma_k{}^{ob1} + n_{ob2} \cdot \sigma_k{}^{ob2})] \quad (9\text{-}38)$$

where *mg* means merge.

The *shape criterion* is computed using two landscape ecology metrics: compactness and smoothness. Heterogeneity as deviation from a *compact* shape (*cpt*) is described by the ratio of the pixel perimeter length l and the square root of the number of pixels n forming an image object (i.e., a patch):

$$cpt = \frac{l}{\sqrt{n}}. \qquad (9\text{-}39)$$

Shape heterogeneity may also be described as *smoothness*, which is the ratio of the pixel perimeter length l and the shortest possible border length b of a box bounding the image object (i.e., a patch) parallel to the raster:

$$smooth = \frac{l}{b}. \qquad (9\text{-}40)$$

The shape criterion incorporates these two measurements using the equation (Definiens, 2003):

$$h_{shape} = w_{cpt} \cdot h_{cpt} + (1 - w_{cpt}) \cdot h_{smooth} \qquad (9\text{-}41)$$

where $0 \le w_{cpt} \le 1$ is the user-defined weight for the compactness criterion. The change in shape heterogeneity caused by each merge is evaluated by calculating the difference between the situation after and before image objects (*ob*) are merged. This results in the following algorithms for computing roughness and smoothness (Definiens, 2003):

$$h_{cpt} = n_{mg} \cdot \frac{l_{mg}}{\sqrt{n_{mg}}} - \left(n_{ob1} \cdot \frac{l_{ob1}}{\sqrt{n_{ob1}}} + n_{ob2} \cdot \frac{l_{ob2}}{\sqrt{n_{ob2}}} \right) \qquad (9\text{-}42)$$

$$h_{smooth} = n_{mg} \cdot \frac{l_{mg}}{b_{mg}} - \left(n_{ob1} \cdot \frac{l_{ob1}}{b_{ob1}} + n_{ob2} \cdot \frac{l_{ob2}}{b_{ob2}} \right) \qquad (9\text{-}43)$$

where n is the object size in pixels.

A *pixel neighborhood* function may be used to determine whether an image object should be grown or whether a new image object should be created (Definiens, 2003). The logic associated with plane 4 and diagonal 8 neighborhood functions are shown in Figure 9-33. In this example, the plane 4 neighborhood function results in the creation of two distinct image objects. The diagonal 8 neighborhood function results in a larger single-image object.

The user specifies the spectral (color) and spatial shape parameters (compactness and smoothness) criteria and the neighborhood function logic. A specially designed heuristic algorithm then applies these criteria to individual pixels in the scene and, in effect, grows homogeneous regions (or, if you like, regions with specified amounts of heterogeneity). Once a segment patch exceeds the user-specified parameters, it stops growing. The final result is a new segmented image consisting of image objects (patches) that contain relatively homogeneous spectral and spatial characteristics.

Image Segmentation Classification Example. To appreciate the creation of such image objects (i.e., segments or patches) and how they might be used to classify a remotely sensed image, consider the high-spatial-resolution ADAR 5000 image of a yacht harbor on Pritchard's Island, SC in the Ace Basin (Figure 9-34). The data were obtained on September 23, 1999, at a spatial resolution of 0.7×0.7 m in four bands (blue, green, red, and near-infrared). Note the significant amount of spatial information in the dataset including building rooftops oriented toward and away from the Sun,

Pixel Neighborhood Functions Used to Determine if Two Image Objects Should Be Merged

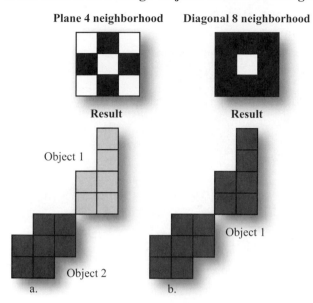

Figure 9-33 One of the criteria used to segment a remotely sensed image into image objects is a pixel neighborhood function, which compares an image object being grown with adjacent pixels. The information is used to determine if the adjacent pixel should be merged with the existing image object or be part of a new image object. a) In this example, if a plane 4 neighborhood function is selected, then two image objects would be created because the pixels under investigation are not connected at their plane borders. b) Pixels and objects are defined as neighbors in a diagonal 8 neighborhood if they are connected at a plane border *or* a corner point (Definiens, 2003). In this example, image object 1 can be expanded because it connects at a diagonal corner point. This resulted is a larger image object 1. Other types of neighborhood functions could be used.

automobiles on the concrete/asphalt pavement, upland forest texture, shadows, and smooth cordgrass (*Spartina alterniflora*) wetland along the major tributaries.

The *object-oriented image segmentation process* was weighted so that spectral information (color) was more important than spatial (shape) information (i.e., $w_{color} = 0.8$). Three spectral bands of ADAR 5000 data (green, red, and near-infrared) were used in the segmentation process. The spatial (shape) parameter was more heavily weighted to smoothness (0.9) than compactness (0.1). The user-specified inputs were then used to identify various "scales" of image segmentation wherein each successive scale file contained

larger segmented image objects. This process is commonly referred to as *multiresolution image segmentation* (Baatz and Schape, 2000). This example shows multiresolution image objects associated with four levels of aggregation or scale: 10, 20, 30, and 40 (Figure 9-34). A color-infrared color composite version is shown in **Color Plate 9-4**. Each image object (i.e., patch, segment, or polygon, if you prefer) in each segmentation scale is hierarchically related to all other files. In other words, the boundaries of one large polygon at segmentation scale 40 may be topologically linked to perhaps four smaller polygons at segmentation scale 10. The four smaller polygons at scale 10 will have exactly the same outer-perimeter coordinates as the larger polygon at scale 40.

One of the most important aspects of performing this type of image segmentation using remote sensor data is that each image object in the dataset (i.e., each polygon) contains not only the mean spectral values of all pixels found within each band but also various spatial measurements that characterize the shape of the polygon. Table 9-13 summarizes several image object metrics that may be computed for each polygon.

The *object-oriented classification* of a segmented image is substantially different from performing a per-pixel classification. First, the analyst is not constrained to using just spectral information. He or she may choose to use a) the mean spectral information in conjunction with b) various shape measures associated with each image object (polygon) in the dataset. This introduces flexibility and robustness. Once selected, the spectral and spatial attributes of each polygon are input to a variety of classification algorithms for analysis (e.g., nearest-neighbor, minimum distance, maximum likelihood).

The classification process is usually very fast because individual image objects (i.e., polygons) rather than individual pixels are assigned to specific classes. Consider the Pritchard's Island example where the dataset consists of 543 rows and 460 columns. A per-pixel classification would process 249,780 pixels. **Color Plate 9-5a** depicts the object-oriented classification of just 2391 polygons into nine classes using segmentation scale 10 data. Only 753 image objects (polygons) were classified in **Color Plate 9-5b** (segmentation scale 20 data); 414 in **Color Plate 9-5c** (segmentation scale 30 data), and 275 in **Color Plate 9-5d** (segmentation scale 40 data).

Examination of **Color Plate 9-5** reveals some interesting aspects of object-oriented classification. First, the user must decide what level of polygon aggregation is acceptable for

**Image Segmentation Based on Spectral (Green, Red, Near-infrared)
and Spatial (Smoothness and Compactness) Criteria**

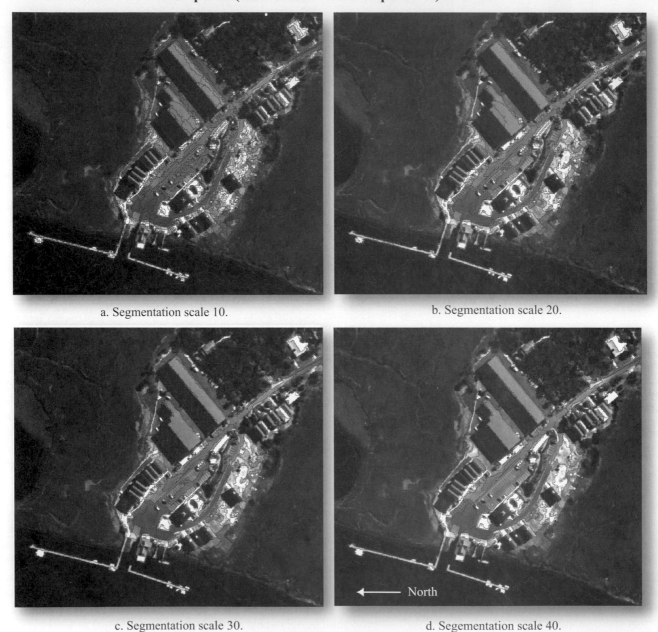

a. Segmentation scale 10.

b. Segmentation scale 20.

c. Segmentation scale 30.

d. Segementation scale 40.

Figure 9-34   A black and white near-infrared ADAR 5000 image of a yacht harbor on Pritchard's Island, SC in the Ace Basin obtained on
September 23, 1999 at a spatial resolution of 0.7 × 0.7 m. Multi-resolution image segmentation was performed using three
bands (green, red, and near-infrared) at four segmentation scales: 10, 20, 30, and 40. The segmentation process was weighted
so that spectral information (color) was more important than spatial (shape) information (weighted 0.8 to 0.2, respectively).
The spatial (shape) parameter was more heavily weighted to smoothness (0.9) than compactness (0.1). A color-infrared color
composite version is found in **Color Plate 9-3**. Classification of the four segmentation scale images is shown in **Color Plate
9-4**.

Table 9-13. A selection of various metrics that can be computed from individual image objects that have been segmented from a spatial dataset (adapted from Definiens, 2003) and landscape pattern and structure metrics that can be computed by analyzing all of the image objects (patches) associated with an image or classified thematic map (O'Neill et al, 1997; Jensen, 2000).

Image Object Metrics	Algorithm	Description
Mean	$$\mu_L = \frac{\sum\limits_{i=1}^{n} v_i}{n}$$	Image object (i.e., polygon, patch) mean value (μ_L) in a single layer is calculated by summing all the pixel values within the image object (v_i) and dividing by the total number of pixels in the object, n. If the image object is composed of spectral brightness values, then $v_i = BV_i$. Otherwise, v_i may be any type of data (e.g., radar backscatter, LIDAR elevation value, DEM elevation value).
Spectral Mean	$$b = \frac{\sum\limits_{i=1}^{n_L} \mu_i}{n_L}$$	For a single image object (i.e., polygon, patch), the sum of all mean layer values (μ_i) divided by the total number of layers (n_L) (i.e., bands).
Spectral Ratio of an Image Object	$$ratio_L = \frac{\mu_L}{\sum\limits_{i=1}^{n_L} \mu_i}$$	The mean of a single image object (μ_L) divided by the sum of all spectral layers (bands) associated with this image object (μ_i).
Standard Deviation of an Image Object	$$\sigma_L = \frac{\sum\limits_{i=1}^{n_L} (v_i - \mu_L)^2}{n-1}$$	For a single image object (i.e., polygon, patch) in a single layer, the sum of all pixel layer values (v_i) subtracted from the mean of the image object (μ_L) squared, divided by the total number of pixels in the polygon (n) minus one.
Mean Difference to Neighboring Image Object	$$\Delta c_L = \frac{1}{l} \sum\limits_{i=1}^{nn} l_{si}(\mu_i - \mu_{Li})$$	For a single image object (i.e., polygon, patch), the mean difference to a direct neighbor is calculated using l = border length in pixels of the image object, l_{si} = border length shared with a direct (attached neighbor), (μ_i) = image object mean value of layer i, (μ_{Li}) = neighboring image object mean value of layer i, and nn = number of neighbors.
Length-to-Width Ratio	$$\gamma = \frac{l}{w}$$	The length-to-width ratio (γ) is computed by dividing the length (l) by the width (w) of an image object. This is approximated by determining the dimensions of the smallest box that bounds the polygon.
Area	$$A = \sum\limits_{i=1}^{n} a_i$$	For georeferenced data, the area of an image object (A) equals the summation of the true area (a_i) of each of the n pixels in the image object. An image object consisting of six Landsat Thematic Mapper 30×30 m pixels would have an area of 6 pixels $\times 90m^2 = 540$ m^2.
Length	$$l = \sqrt{A \cdot \gamma}$$	Length of an image object is approximated using the length-to-width ratio. If the image object has a curved shape, it is better to break it into subjects.
Width	$$w = \sqrt{\frac{A}{\gamma}}$$	Width of an image object is approximated using the length-to-width ratio. If the image object has a curved shape, it is better to break it into sub-objects.

Table 9-13.   A selection of various metrics that can be computed from individual image objects that have been segmented from a spatial dataset (adapted from Definiens, 2003) and landscape pattern and structure metrics that can be computed by analyzing all of the image objects (patches) associated with an image or classified thematic map (O'Neill et al, 1997; Jensen, 2000).

Image Object Metrics	Algorithm	Description
Border Length	$bl = \sum_{i=1}^{n} e_i$	The border length of an image object is the sum of the number of edges (e_i) that touch *all* neighboring image objects (or that touch the outside edge of a scene).
Shape Index	$si = \dfrac{bl}{4 \times \sqrt{A}}$	The shape index of an image object is the border length (bl) divided by four times the square root of the area (A). The smoother the shape, the lower the value. The greater the value, the more fractal the shape.
Density	$d = \dfrac{\sqrt{n}}{1 + \sqrt{Var(X) + Var(Y)}}$	The density of an image object is the area of the object divided by its radius. It is approximated by computing the variance of all x- and y-coordinates of all n pixels forming the image object. Density is a surrogate for compactness. The more compact the object, the higher its density and the more the shape is like a square.
Asymmetry	$k = 1 - \dfrac{n}{m}.$	Image object asymmetry is one minus the ratio of the length of the minor (n) and major (m) axes of an ellipse enclosing the image object. The greater the asymmetry, the greater the value.
Classified Map Summary Metrics		
Dominance	$D = 1 - \left[\sum_{k=1}^{n} \dfrac{(-P_k \cdot \ln P_k)}{\ln(n)} \right]$	Dominance ($0 < D < 1$) identifies the extent to which the landscape is dominated by a single land-cover type where $0 < P_k < 1$ is the proportion of land-cover type k and n is the total number of land-cover types in the land-cover map.
Contagion	$C = 1 - \left[\sum_i \sum_j \dfrac{(-P_{ij} \cdot \ln P_{ij})}{2 \ln(n)} \right]$	Contagion ($0 < C < 1$) expresses the probability that land cover is more clumped than the random expectation where P_{ij} is the probability that a pixel of cover type i is adjacent to type j.
Fractal Dimension	F	The fractal dimension, F, of patches (image objects) indicates the extent of human reshaping of the landscape. It is calculated by regressing the log of the patch perimeter against the log of the patch area for each patch on the landscape. The index equals twice the slope of the regression line. Patches < 4 pixels are excluded.

the task at hand. In this example, a classification based on the use of segmentation scale 20 data (**Color Plate 9-5b**) is probably the best representation. It provides the greatest amount of accurate information for both very small objects and larger, geographically-extensive surface materials such as forest and wetland. Conversely, if only segmentation scale 40 data (**Color Plate 9-5d**) were used, then the classification of the wetland would contain serious error. Perhaps

the information that the user wants is best extracted using a combination of information present in image segmentation scale 10 for relatively small rooftop materials and segmentation scale 40 for identifying homogeneous forest areas. The user must decide which of the various segmented files are of most value. He or she can then enter very specific rules into the classification process if desired.

Object-oriented Considerations

Once the object-oriented image segmentation and classification is complete, it may be desirable to evaluate the characteristics of all the patches (image objects) in the map to address important applications. O'Neill (1997) suggests that the health of an ecosystem can be monitored if the following three landscape pattern and structure metrics (indices) are monitored through time: dominance, contagion, and fractal dimension (Table 9-13). Other studies use different landscape pattern and structure metrics (Batistella et al., 2003).

Per-pixel classifications often appear pixelated. Conversely, object-oriented classification can appear fractal. The analyst conducting an object-oriented classification should decide what level (scale) of segmentation is most acceptable for the task at hand and most easily understood by lay viewers when the final thematic map(s) are produced.

Segmentation into image objects allows the use of geographical and landscape ecology concepts involving neighborhood, distance, and location for analyzing remotely sensed data. It also facilitates the merging or fusion of multiple types of remote sensor data for a particular application (Kuehn et al., 2002). Object-oriented image segmentation and classification is a major paradigm shift when compared with per-pixel classification. It will become increasingly important not only for classification but also for use in change detection.

Incorporating Ancillary Data in the Classification Process

An analyst photo-interpreting a color aerial photograph of the terrain often has at his or her disposal (1) systematic knowledge about the soils, geology, vegetation, hydrology, and geography of the area, (2) the ability to visualize and comprehend the landscape's color, texture, height, and shadows, (3) the ability to place much of this diverse information in context to understand site conditions and associations among phenomena, and (4) historical knowledge about the area. Conversely, 95% of all remote sensing digital image classifications attempt to accomplish the same task using a single variable, an object's spectral reflectance characteristics (color) or black-and-white tone. Therefore, it is not surprising that there is error in remote-sensing-derived classification maps. Why should we expect the maps to be extremely accurate when the information provided to the classification algorithm is so rudimentary?

Numerous scientists recognize this condition and have attempted to improve the accuracy and quality of remote sensing–derived land-cover classification by incorporating ancillary data in the classification process (e.g., Strahler et al., 1978; Hutchinson, 1982; McIver and Friedl, 2002). *Ancillary data* are any type of spatial or nonspatial information that may be of value in the image classification process, including elevation, slope, aspect, geology, soils, hydrology, transportation networks, political boundaries, and vegetation maps. Ancillary data are not without error. Analysts who desire to incorporate ancillary data into the remote sensing classification process must be aware of several shortcomings.

Problems Associated with Ancillary Data

First, ancillary data were usually produced for a specific purpose and it was not to improve remote sensing classification accuracy. Second, the nominal, ordinal, or interval thematic attributes on the collateral maps may be inaccurate or incomplete. For example, Kuchler (1967) pointed out that polygonal boundaries between vegetation types on his respected regional maps may or may not actually exist on the ground! Great care must be exercised when generalizing the classes found on the ancillary map source materials as we try to make them compatible with the remote sensing investigation classes of interest.

Third, considerable ancillary information is stored in analog map format. The maps must be digitized, translated, rotated, rescaled, and often resampled to bring the dataset into congruence with the remote sensing map projection. During this process the locational attributes of the phenomena may be moved from their true planimetric positions. This assumes that the ancillary data were planimetrically accurate to begin with. Unfortunately, considerable ancillary data were never recorded in their proper planimetric positions. For example, old soil surveys published by the U.S. Soil Conservation Service were compiled onto uncontrolled photomosaics. Analysts trying to use such data must be careful that they do not introduce more error into the classification process than they are attempting to remove.

Approaches to Incorporating Ancillary Data to Improve Remote Sensing Classification Maps

Several approaches may be used to incorporate ancillary data in the image classification process that should improve results. These include incorporating the data before, during, or after classification through geographical stratification, classifier operations, and/or post-classification sorting. Ancillary data may also be incorporated using object-oriented image segmentation, neural networks, expert systems, and decision-tree classifiers.

Geographical Stratification

Ancillary data may be used *prior* to classification to subdivide the regional image into strata, which may then be processed independently. The goal is to increase the homogeneity of the individual stratified image datasets to be classified. For example, what if we wanted to locate spruce fir in the Colorado Rockies but often encountered misclassification up and down the mountainside? One approach would be to stratify the scene into just two files: one with elevations from 0 to 2600 ft above sea level (dataset 1) and another with elevation > 2600 ft ASL (dataset 2). We would then classify the two datasets independently. Spruce fir do not grow below 2600 ft ASL; therefore, during the classification process we would not label *any* of the pixels in dataset 1 spruce fir. This would keep spruce fir pixels from being assigned to forested areas that cannot ecologically support them. Errors of commission for spruce fir should be reduced when datasets 1 and 2 are put back together to compile the final map and are compared to a traditional classification. If specific ecological principles are known, the analyst could stratify the area further using slope and aspect criteria to refine the classification (Franklin and Wilson, 1992).

Stratification is a conceptually simple tool and, carefully used, can be effective in improving classification accuracy. Illogical stratification can have severe implications. For example, differences in training set selection for individual strata and/or the vagaries of clustering algorithms, if used, may produce different spectral classes on either side of strata boundaries (Hutchinson, 1982). Edge-matching problems become apparent when the final classification map is put together from the maps derived from the individual strata.

Classifier Operations

Several methods may be used to incorporate ancillary data during the image classification process. One of the most useful is the logical channel method. A *per-pixel logical channel* classification includes ancillary data as one of the channels (features) used by the classification algorithm. For example, a dataset might consist of three IKONOS bands of spectral data plus two additional bands (percent slope and aspect) derived from a digital elevation model. The entire five-band dataset is acted on by the classification algorithm. When ancillary data are incorporated into traditional classi-

fication algorithms as logical channels, the full range of information available in the ancillary data is used (Franklin and Wilson, 1992; Ricchetti, 2000). With logical channel addition, ancillary data are given equal weight to single spectral bands unless weights are assigned in a maximum likelihood classifier (Lawrence and Wright, 2001). Chen and Stow (2003) used the logical channel approach to classify land cover using multiple types of imagery obtained at different spatial resolutions.

The context of a pixel refers to its spatial relationship with any other pixel or group of pixels throughout the scene (Gurney and Townshend, 1983). *Contextual logical channel* classification occurs when information about the neighboring (surrounding) pixels is used as one of the features in the classification. *Texture* is one simple contextual measure that may be extracted from an $n \times n$ window (see Chapter 8) and then added to the original image dataset prior to classification (Jensen and Toll, 1982; Stow et al., 2003). It is important to remember that contextual information may also be derived from *non*image ancillary sources, such as maps showing proximity to roads, streams, and so on.

A second approach involves the use of a priori probabilities in the classification algorithm (Strahler, 1980; McIver and Friedl, 2002). The analyst gets the a priori probabilities by evaluating historical summaries of the region (e.g., last year cotton accounted for 80% of the acreage, hay 15%, and barley 5%). These statistics are incorporated directly into the maximum likelihood classification algorithm as weights $p(w_i)$ to the classes, as previously discussed. Prior probabilities can improve classification results by helping to resolve confusion among classes that are difficult to separate and by reducing bias when the training sample is not representative of the population being classified. McIver and Friedl (2002) point out that the use of prior probabilities in maximum likelihood classification are often problematic in practice. They developed a useful method for incorporating prior probabilities into a nonparametric decision-tree classifier.

Another approach involves image segmentation, as discussed in the previous section. This method can incorporate both spectral and nonspectral ancillary data, which are subjected to multiresolution segmentation to produce polygons (patches) that contain relatively uniform spectral and spatial characteristics. This is a straightforward way of merging both spectral and nonspectral information.

Ancillary data have been incorporated into modern classification methods such as expert systems and neural networks (Skidmore et al., 1997; Qiu and Jensen, 2003; Stow et al., 2003). These approaches incorporate the ancillary data directly into the classification algorithms and are usually not dependent on a priori weights. Chapter 10 describes how such systems work and the ease with which ancillary data may be introduced, including the fact that they are not confined to normally distributed data.

Machine-learning approaches have been used to establish rule-based classification systems where expert knowledge was inadequate (Huang and Jensen, 1997). Lawrence and Wright (2001) used rule-based classification systems based on classification and regression tree (CART) analysis to incorporate ancillary data into the classification process.

Post-classification Sorting

This method involves the application of very specific rules to (1) initial remote sensing classification results and (2) spatially distributed ancillary information. For example, Hutchinson (1982) classified Landsat MSS data of a desert area in California into nine initial classes. He then registered slope and aspect maps derived from a digital elevation model with the classification map and applied 20 if–then rules to the datasets (e.g., if the pixel was initially classified as an active sand dune and if the slope <1%, then the pixel is a dry lake bed). This eliminated confusion between several of the more prominent classes in this region [e.g., between the bright surfaces of a dry lake bed (playa) and the steep sunny slopes of large sand dunes]. Similarly, Cibula and Nyquist (1987) used post-classification sorting to improve the classification of Landsat MSS data for Olympic National Park. Topographic (elevation, slope, and aspect) and watershed boundary data (precipitation and temperature) were analyzed in conjunction with the initial land-cover classification using Boolean logic. The result was a 21-class forest map that was just as accurate as the initial map but contained much more information. Janssen et. al (1990) used an initial per-pixel classification of TM data and digital terrain information to improve classification accuracy for areas in the Netherlands by 12% to 20%. Post-classification sorting is only as good as the quality of the rules and ancillary data used.

The incorporation of ancillary data in the remote sensing classification process is important. However, the choice of variables to include is critical. Common sense suggests that the analyst thoughtfully select only variables with conceptual and practical significance to the classification problem at hand. Incorporating illogical or suspect ancillary information can rapidly consume limited data analysis resources and lead to inaccurate results.

References

American Planning Association, 2004a, *Land-Based Classification System*, Washington: American Planning Association, www.planning.org/lbcs.

American Planning Association, 2004b, *Land-Based Classification System Annotated Bibliography*, Washington: American Planning Association, www.planning.org/publications.

American Planning Association, 2004c, *Pictures of Common Land Uses*, Washington: American Planning Association, www.planning.org/lbcs/standards/pictureview.htm.

Anderson, J. R., Hardy, E., Roach, J. and R. Witmer, 1976, *A Land-Use and Land-Cover Classification System for Use with Remote Sensor Data*, Washington: U.S. Geological Survey, Professional Paper #964, 28 p.

Atkinson, P. M. and P. Lewis, 2000, "Geostatistical Classification for Remote Sensing: An Introduction," *Computers & Geosciences*, 26:361–371.

Baatz, M. and A. Schape, 2000, "Multiresolution Segmentation: An Optimization Approach for High Quality Multi-scale Image Segmentation," in Strobl, J., Blaschke, T., and G. Griesebner (Eds.), *Angewandte Geographische Informationsverarbeitung XII*, Heidelberg: Wichmann-Verlag, 12–23.

Baatz, M., Benz, U., Dehghani, S., Heymen, M., Holtje, A., Hofmann, P., Ligenfelder, I., Mimler, M., Sohlbach, M., Weber, M. and G. Willhauck, 2001, *eCognition User Guide*, Munich: Definiens Imaging GmbH, 310 p.

Batistella, M., Robeson, S. and E. F. Moran, 2003, "Settlement Design, Forest Fragmentation, and Landscape Change in Rondonia, Amazonia," *Photogrammetric Engineering & Remote Sensing*, 69(7):805–812.

Beauchemin, M. and K. B. Fung, 2001, "On Statistical Band Selection for Image Visualization," *Photogrammetric Engineering & Remote Sensing*, 67(5):571–574.

Benz, U., 2001, "Definiens Imaging GmbH: Object-Oriented Classification and Feature Detection," *IEEE Geoscience and Remote Sensing Society Newsletter*, (September), 16–20.

Bezdek, J. C., Ehrlich, R. and W. Full, 1984, "FCM: The Fuzzy c–Means Clustering Algorithm," *Computers & Geosciences*, 10(2):191–203.

Blaschke, T. and J. Strobl, 2001, "What's Wrong with Pixels? Some Recent Developments Interfacing Remote Sensing and GIS," *GIS*, Heidelberg: Huthig GmbH & Co., 6:12–17.

Botkin, D. B., Estes, J. E., MacDonald, R. B. and M. V. Wilson, 1984, "Studying the Earth's Vegetation from Space," *Bioscience*, 34(8):508–514.

Chen, D. and D. Stow, 2003, "Strategies for Integrating Information from Multiple Spatial Resolutions into Land-use/Land-cover Classification Routines," *Photogrammetric Engineering & Remote Sensing*, 69(11):1279–1287.

Cibula, W. G. and M. O. Nyquist, 1987, "Use of Topographic and Climatological Models in a Geographical Data Base to Improve Landsat MSS Classification for Olympic National Park," *Photogrammetric Engineering & Remote Sensing*, 53:67–75.

Civco, D. L., Hurd, J. D., Wilson, E. H., Song, M. and Z. Zhang, 2002, "A Comparison of Land Use and Land Cover Change Detection Algorithms," *Proceedings, ASPRS-ACSM Annual Conference and FIG XXII Congress*, Bethesda: American Society for Photogrammetry & Remote Sensing, 12 p.

Congalton, R. G. and K. Green, 1999, *Assessing the Accuracy of Remotely Sensed Data: Principles and Practices*, Boca Raton, FL: Lewis Publishers, 137 p.

Cowardin, L. M., Carter, V., Golet, F. C. and E. T. LaRoe, 1979, *Classification of Wetlands and Deepwater Habitats of the United States*, Washington: U.S. Fish & Wildlife Service, FWS/ OBS-79/31, 103 p.

Definiens, 2003, *eCognition Professional*, Version 3.0, Munich: Definiens Imaging GmbH, www.definiens-imaging.com.

DeFries, R. S. and J. R. G. Townshend, 1999, "Global Land Cover Characterization from Satellite Data: From Research to Operational Implementation?" *Global Ecology & Biogeography*, 8:367–379.

Dobson, J. R., Bright, E. A., Ferguson, R. L., Field, D. W., Wood, L. L, Haddad, K. D., Iredale, H., Jensen, J. R., Klemas, V. V., Orth, R. J. and J. P. Thomas, 1995, *NOAA Coastal Change Analysis Program (C-CAP): Guidance for Regional Implementation*, Washington: National Oceanic & Atmospheric Administration, NMFS 123, 92 p.

Duda, R. O., Hart, P. E. and D. G. Stork, 2001, *Pattern Classification*, New York: John Wiley & Sons, 654 p.

ESA, 2004a, *Review of Physiognomic Levels of a National Vegetation Classification*, Washington: Ecological Society of America, http://esa.sdsc.edu/vegwebpg.htm.

ESA, 2004b, *Development of Floristic Levels of a National Vegetation Classification*, Washington: Ecological Society of America, http://esa.sdsc.edu/vegwebpg.htm.

ERDAS, 2003, *ERDAS Field Guide*, Atlanta: Leica GeoSystems, 672 p.

EROS, 2003, "MODIS Land Cover Type (MOD 12)" in *Data Products Handbook*, Vol. 2, Sioux Falls, SD: EROS Data Center, 171–172.

Ferro, C. J. and T. A. Warner, 2002, "Scale and Texture in Digital Image Classification," *Photogrammetric Engineering & Remote Sensing*, 68(1):51–63.

FGDC, 1996, *Federal Register - 61 FR 39465*, Washington: Federal Geographic Data Committee, July 29, 1996.

FGDC, 2004, *Natural Vegetation Classification System*, Washington: Federal Geographic Data Committee, www.fgdc.gov/standards/status.

FGDC Vegetation Subcommittee, 1997, *Vegetation Classification Standard*, Washington: Federal Geographic Data Committee, FGDC-STD-005, 21 p.

Fisher, P. F. and S. Pathirana, 1993, "The Ordering of Multitemporal Fuzzy Land-cover Information Derived from Landsat MSS Data," *Geocarto International*, 8(3):5–14.

Foody, G. M., 1996, "Approaches for the Production and Evaluation of Fuzzy Land Cover Classifications from Remotely Sensed Data," *International Journal of Remote Sensing*, 17(7):1317–1340.

Foody, G. M., 2000, "Estimation of Sub-pixel Land Cover Composition in the Presence of Untrained Classes," *Computers & Geosciences*, 26:469–478.

Foody, G. M., 2002, "Status of Land Cover Classification Accuracy Assessment," *Remote Sensing of Environment*, 80:185–201.

Forman, R. T. T., 1995, *Land Mosaics: The Ecology of Landscapes and Regions*, Cambridge: Cambridge University Press, 652 p.

Franklin, S. E. and B. A. Wilson, 1992, "A Three-stage Classifier for Remote Sensing of Mountain Environments," *Photogrammetric Engineering & Remote Sensing*, 58(4):449–454.

Friedl, M. A., McIver, D. K., Hodges, J. C. F., Zhang, X. Y., Muchoney, D., Strahler, A. H., Woodcock, C. E., Gopal, S., Schneider, A., Cooper, A., Baccini, A., Gao, F. and C. Schaaf, 2002, "Global Land Cover Mapping from MODIS: Algorithms and Early Results," *Remote Sensing of Environment*, 83:287–302.

Gibson, P. J. and C. H. Power, 2000, *Introductory Remote Sensing*, New York: Routledge, 248 p.

Gong, P. and P. J. Howarth, 1992, "Frequency-based Contextual Classification and Gray-level Vector Reduction for Land-use Identification," *Photogrammetric Engineering & Remote Sensing*, 58(4):423–437.

Green, K. and R. G. Congalton, 2003, "An Error Matrix Approach to Fuzzy Accuracy Assessment: The NIMA GeoCover Project," in Lunetta, R. S. and J. G. Lyons (Eds.), *Geospatial Data Accuracy Assessment,* Las Vegas: U.S. Environmental Protection Agency, Report No. EPA/600/R-03/064.

Gurney, C. M. and J. R. G. Townshend, 1983, "The Use of Contextual Information in the Classification of Remotely Sensed Data," *Photogrammetric Engineering & Remote Sensing*, 49:55–64.

Haralick, R. M. and K. Fu, 1983, "Pattern Recognition and Classification," Chapter 18 in R. N. Colwell (Ed.), *Manual of Remote Sensing*, Falls Church, VA: American Society of Photogrammetry, 1:793–805.

Hardin, P. J., 1994, "Parametric and Nearest-neighbor Methods for Hybrid Classification: A Comparison of Pixel Assignment Accuracy," *Photogrammetric Engineering & Remote Sensing*, 60(12):1439–1448.

Herold, M., Guenther, S. and K. C. Clarke, 2003, "Mapping Urban Areas in the Santa Barbara South Coast using IKONOS and *eC*ognition," *eCognition Application Note*, Munchen: Definiens ImgbH, 4(1):2 p.

Herold, M., Scepan, J. and K. C. Clarke, 2002, "The Use of Remote Sensing and Landscape Metrics to Describe Structures and Changes in Urban Land Uses," *Environment and Planning A*, 34:1443–1458.

Hodgson, M. E., 1988, "Reducing the Computational Requirements of the Minimum-distance Classifier," *Remote Sensing of Environment*, 25:117–128.

Hodgson, M. E. and R. W. Plews, 1989, "*N*-dimensional Display of Cluster Means in Feature Space," *Photogrammetric Engineering & Remote Sensing*, 55(5):613–619.

Hodgson, M. E., Jensen, J. R., Tullis, J. A., Riordan, K. D. and C. M. Archer, 2003, "Synergistic Use of LIDAR and Color Aerial Photography for Mapping Urban Parcel Imperviousness," *Photogrammetric Engineering & Remote Sensing*, 69(9):973-980.

Hord, R. M., 1982, *Digital Image Processing of Remotely Sensed Data*, New York: Academic Press, 256 p.

Huang, K., 2002, "A Synergistic Automatic Clustering Technique for Multispectral Image Analysis," *Photogrammetric Engineering & Remote Sensing*, 68(1):33–40.

Huang, X. and J. R. Jensen, 1997, "A Machine Learning Approach to Automated Construction of Knowledge Bases for Image Analysis Expert Systems That Incorporate GIS Data," *Photogrammetric Engineering & Remote Sensing*, 63(10):1185–1194.

Hutchinson, C. F., 1982, "Techniques for Combining Landsat and Ancillary Data for Digital Classification Improvement," *Photogrammetric Engineering & Remote Sensing*, 48(1):123–130.

Jahne, B., 1991, *Digital Image Processing*, New York: Springer-Verlag, 219–230.

Jain, A. K., 1989, *Fundamentals of Digital Image Processing*, Englewood Cliffs, NJ: Prentice-Hall, 418–421.

Janssen, L. F., Jaarsma, J. and E. Van der Linden, 1990, "Integrating Topographic Data with Remote Sensing for Land-cover Classification," *Photogrammetric Engineering & Remote Sensing*, 56(11):1503–1506.

Jensen, J. R., 1983, "Urban–Suburban Land Use Analysis," Chapter 30 in R. N. Colwell (Ed.), *Manual of Remote Sensing*, Falls Church, VA: American Society of Photogrammetry, 2:1571–1666.

Jensen J. R., 2000, *Remote Sensing of The Environment: An Earth Resource Perspective*, Upper Saddle River, NJ: Prentice-Hall, 544 p.

Jensen, J. R. and D. C. Cowen, 1999, "Remote Sensing of Urban/Suburban Infrastructure and Socioeconomic Attributes," *Photogrammetric Engineering & Remote Sensing*, 65:611–622.

Jensen, J. R. and D. L. Toll, 1982, "Detecting Residential Land Use Development at the Urban Fringe," *Photogrammetric Engineering & Remote Sensing*, 48:629–643.

Jensen, J. R., D. J. Cowen, S. Narumalani, J. D. Althausen, and O. Weatherbee, 1993, "An Evaluation of CoastWatch Change Detection Protocol in South Carolina," *Photogrammetric Engineering & Remote Sensing*, 59(6):1039–1046.

Jensen, J. R., Qiu, F. and M. Ji, 2000, "Predictive Modelling of Coniferous Forest Age Using Statistical and Artificial Neural Network Approaches Applied to Remote Sensor Data," *International Journal of Remote Sensing*, 20(14):2805-2822.

Jensen, J. R., Qiu, F. and K. Patterson, 2001, "A Neural Network Image Interpretation System to Extract Rural and Urban Land Use and Land Cover Information from Remote Sensor Data," *Geocarto International: A Multidisciplinary Journal of Remote Sensing and GIS*, 16(1):19–28.

Jensen, J. R., Botchway, K., Brennan-Galvin, E., Johannsen, C., Juma, C., Mabogunje, A., Miller, R., Price, K., Reining, P., Skole, D., Stancioff, A. and D. R. F. Taylor, 2002, *Down to Earth: Geographic Information for Sustainable Development in Africa*, Washington: National Research Council, 155 p.

Ji, M. and J. R. Jensen, 1996, "Fuzzy Training in Supervised Digital Image Processing," *Geographic Information Science*, 2(1):1–11.

Ji, M. and J. R. Jensen, 1999, "Effectiveness of Subpixel Analysis in Detecting and Quantifying Urban Imperviousness from Landsat Thematic Mapper Imagery," *Geocarto International: A Multidisciplinary Journal of Remote Sensing and GIS*, 14(4):39-49.

Karaska, M. A., Huguenin, R. L., Van Blaricom, D., Savitsky, B. and J. R. Jensen, 1997, "Subpixel Classification of Bald Cypress and Tupelo Gum Trees in Thematic Mapper Imagery," *Photogrammetric Engineering & Remote Sensing*, 63(6):717–725.

Kerr, J. and M. Ostrovsky, 2003, "From Space to Species: Ecological Applications for Remote Sensing," *Trends in Ecology and Evolution*, 18(6):299–305.

Konecny, G., 2003, *Geoinformation: Remote Sensing, Photogrammetry and Geographic Information Systems*, New York: Taylor & Francis, 248 p.

Kuchler, A. W., 1967, *Vegetation Mapping*, New York: Ronald Press, 472 p.

Kuehn, S., Benz, U. and J. Hurley, 2002, "Efficient Flood Monitoring Based on RADARSAT-1 Images Data and Information Fusion with Object-Oriented Technology," *Proceedings, IGARSS*, 3 p.

Lawrence, R. L. and A. Wright, 2001, "Rule-Based Classification Systems Using Classification and Regression Tree (CART)

Analysis," *Photogrammetric Engineering & Remote Sensing*, 67(10):1137–1142.

Liu, X. H., Skidmore, A. K. and H. V. Oosten, 2002, "Integration of Classification Methods for Improvement of Land-cover Map Accuracy," *ISPRS Journal of Photogrammetry & Remote Sensing*, 257–268.

Lo, C. P. and A. K. Yeung, 2002, *Concepts and Techniques of Geographic Information Systems*, Upper Saddle River, NJ: Prentice-Hall, 492 p.

Loveland, T. R., Zhiliang, Z., Ohlen, D. O., Brown, J. F., Reed, B. C. and L. Yang, 1999, "An Analysis of the IGBP Global Land-Cover Characterization Process," *Photogrammetric Engineering & Remote Sensing*, 65(9):1021–1032.

Lunetta, R. S. and C. D. Elvidge, 1998, *Remote Sensing Change Detection: Environmental Monitoring Methods and Applications*, Chelsea, MI: Ann Arbor Press, 318 p.

Lunetta, R. S. and J. G. Lyons (Eds.), 2003, *Geospatial Data Accuracy Assessment*, Las Vegas: Environmental Protection Agency, Report No. EPA/600/R-03/064.

Lunetta, R. S., Congalton, R. G., Fenstermaker, L. K, Jensen, J. R., McGwire, K. C. and L. R. Tinney, 1991, "Remote Sensing and Geographic Information System Data Integration: Error Sources and Research Issues," *Photogrammetric Engineering & Remote Sensing*, 57(6):677–687.

Mausel, P. W., Kamber, W. J. and J. K. Lee, 1990, "Optimum Band Selection for Supervised Classification of Multispectral Data," *Photogrammetric Engineering & Remote Sensing*, 56(1):55–60.

McIver D. K. and M. A. Friedl, 2002, "Using Prior Probabilities in Decision-tree Classification of Remotely Sensed Data," *Remote Sensing of Environment*, 81:253-261.

Narumalani, S., Hlady, J. T. and J. R. Jensen, 2002, "Information Extraction from Remotely Sensed Data," in Bossler, J. D., Jensen, J. R., McMaster, R. B. and C. Rizos (Eds.), *Manual of Geospatial Science and Technology*, New York: Taylor & Francis, 298–324.

Narumalani, S., Zhou, Y. and J. R. Jensen, 1997, "Application of Remote Sensing and Geographic Information Systems to the Delineation and Analysis of Riparian Buffer Zones," *Aquatic Botany*, 58(1997):393–409.

NASA Earth Observatory, 2002, "NASA's *Terra* Satellite Refines Map of Global Land Cover," *Earth Observatory News*, 02-126,

August 13, 2002, http://earthobservatory.nasa.gov/Newsroom/LCC/.

NASA GSFC, 2004, *MODIS Land Cover and Land Cover Change Data Product #12*, Greenbelt, MD: NASA Goddard Space Flight Center, http://modis.gsfc.nasa.gov/data/dataproucts.php?MOD_NUMBER=12.

NOAA, 2004, *Coastal Change Analysis Program (C-CAP)*, Charleston: NOAA Coastal Services Center, http://www. csc. noaa.gov/crs/lca/ccap_program.html.

O'Neill, R. V., Hunsaker, C. T., Jones, K. B., Ritters, K. H., Wickham, J. D., Schwarz, P., Goodman, I. A., Jackson, B. and W. S. Bailargeon, 1997, "Monitoring Environmental Quality at the Landscape Scale," *BioScience*, 47(8):513–519.

Ozkan, C. and F. S. Erbek, 2003, "The Comparison of Activation Functions for Multispectral Landsat TM Image Classification," *Photogrammetric Engineering & Remote Sensing*, 69(11):1225–1234.

Qui, F. and J. R. Jensen, 2004, "Opening the Neural Network Black Box and Breaking the Knowledge Acquisition Bottleneck of Fuzzy Systems for Remote Sensing Image Classification," *International Journal of Remote Sensing*, in press.

Rees, W. G., Williams, M. and P. Vitebsky, 2003, "Mapping Land Cover Change in a Reindeer Herding Area of the Russian Arctic Using Landsat TM and ETM+ Imagery and Indigenous Knowledge," *Remote Sensing of Environment*, 85:441–452.

Ricchetti, E., 2000, "Multispectral Satellite Image and Ancillary Data Integration for Geological Classification," *Photogrammetric Engineering & Remote Sensing*, 66(4):429–435.

Richards, J. A., Landgrebe, D. A. and P. H. Swain, 1982, "A Means for Utilizing Ancillary Information in Multispectral Classification," *Remote Sensing of Environment*, 12:463–477.

Richards, J. A. and X. Jia, 1999, *Remote Sensing Digital Image Analysis: An Introduction*, Berlin: Springer, 363 p.

Rignot, E. J., 1994, "Unsupervised Segmentation of Polarimetric SAR Data," *NASA Tech Briefs*, 18(7):46–47.

Schalkoff, R., 1992, *Pattern Recognition: Statistical, Structural and Neural Approaches*, New York: John Wiley, 364 p.

Schowengerdt, R. A., 1997, *Remote Sensing: Models and Methods for Image Processing*, 2nd ed., San Diego, CA: Academic Press, 522 p.

Skidmore, A. K., Turner, B. J., Brinkhof, W. and E. Knowles, 1997, "Performance of a Neural Network Mapping Forests Using GIS and Remotely Sensed Data," *Photogrammetric Engineering & Remote Sensing*, 63(5):501–514.

Stow, D., Coulter, L., Kaiser, J., Hope, A., Service, D., Schutte, K. and A. Walters, 2003, "Irrigated Vegetating Assessments for Urban Environments," *Photogrammetric Engineering & Remote Sensing*, 69(4):381–390.

Strahler, A. H., 1980, "The Use of Prior Probabilities in Maximum Likelihood Classification of Remotely Sensed Data," *Remote Sensing of Environment*, 10:135–163.

Strahler, A. H., 1999, *MODIS Land Cover Product Algorithm Theoretical Basis Document (ATBD)*, Version 5.0, Washington: NASA, 72 p.

Strahler, A. H., Logan, T. L. and N. A. Bryant, 1978, "Improving Forest Cover Classification Accuracy from Landsat by Incorporating Topographic Information," *Proceedings, 12th International Symposium on Remote Sensing of the Environment*, 927–942.

Swain, P. H. and S. M. Davis, 1978, *Remote Sensing: The Quantitative Approach*, New York: McGraw-Hill, 166–174.

Townshend, J. R. G., 1999, *MODIS Enhanced Land Cover and Land Cover Change Product*, Washington: NASA, 93 p.

Townshend, J. R. G. and C. O. Justice, 2002, "Towards Operational Monitoring of Terrestrial Systems by Moderate-resolution Remote Sensing," *Remote Sensing of Environment*, 83:351–359.

Townshend, J. R. G., Huang, C., Kalluri, S., DeFries, R., Liang, S. and K. Yang, 2000, "Beware of Per-pixel Characterization of Land Cover," *International Journal of Remote Sensing*, 21(4):839–843.

Tullis, J. A. and J. R. Jensen, 2003, "Expert System House Detection in High Spatial Resolution Imagery Using Size, Shape, and Context," *Geocarto International*, 18(1):5–15.

Turner, M. G., Gardner, R. H. and R. V. O'Neill, 2001, *Landscape Ecology in Theory and Practice: Pattern and Process*, New York: Springer-Verlag, 352 p.

USGS, 2004, *USGS National Land Cover Data*, Sioux Falls: EROS Data Center, http://landcover.usgs.gov/prodescription.html.

Visual Learning Systems, 2002, *User Manual: Feature Analyst Extension for ArcView/ArcGIS*, Missoula, MT: Visual Learning Systems.

Wang, F., 1990a, "Improving Remote Sensing Image Analysis through Fuzzy Information Representation," *Photogrammetric Engineering & Remote Sensing*, 56(8):1163–1169.

Wang, F., 1990b, "Fuzzy Supervised Classification of Remote Sensing Images," *IEEE Transactions on Geoscience and Remote Sensing*, 28(2):194–201.

Welch, R. A., 1982, "Spatial Resolution Requirements for Urban Studies," *International Journal of Remote Sensing*, 3:139–146.

Wu, W. and G. Shao, 2002, "Optimal Combinations of Data, Classifiers, and Sampling Methods for Accurate Characterizations of Deforestation," *Canadian Journal of Remote Sensing*, 28(4):601–609.

Zadeh, L. A., 1965, "Fuzzy Sets," *Information and Control*, 8:338–353.

Zhang, X., Friedl, M. A., Schaff, C. B., Strahler, A. H., Hodges, J. C. F, Gao, F., Reed, C. and A. Huete, 2003, "Monitoring Vegetation Phenology Using MODIS," *Remote Sensing of Environment*, 84:471–475.

Information Extraction Using Artificial Intelligence 10

A *rtificial intelligence (AI) is:*

> the study of how to make computers do things which, at the
> moment, people do better (Rich and Knight, 1991).

But how do we know when an artificially intelligent system has been created? Ideally we could use the *Turing test,* which suggests that if we are unable to distinguish a computer's response to a problem of interest from a human's response to the same problem, then the computer system is said to have *intelligence* (Turing, 1950). The test is for an artificial intelligence program to have a blind conversation with an interrogator for 5 minutes. The interrogator has to guess if the conversation is with an artificial intelligence program or with a real person. The AI program passes the test if it fools the interrogator 30% of the time. Unfortunately, it is very difficult for most artificial intelligence systems to pass the Turing test. For this reason, "the field of AI as a whole has paid little attention to Turing tests," preferring instead to forge ahead developing artificial intelligence applications that simply work (Russell and Norvig, 2003).

Artificial intelligence research was initiated in 1955 when Allen Newell and Herbert Simon at the RAND Corporation proved that computers could do more than calculate.

> They demonstrated that computers were physical symbol systems
> whose symbols could be made to stand for anything, including
> features of the real world, and whose programs could be used as
> rules for relating these features. In this way computers could be
> used to simulate certain important aspects of intelligence. Thus,
> the information-processing model of the mind was born (Dreyfus
> and Dreyfus, 2001).

Unfortunately, artificial intelligence was oversold in the 1960s much like remote sensing was oversold in the 1970s. General artificial intelligence problem solving was found to be much more difficult than originally anticipated. Scientists could not get computers to solve problems that were routinely solved by human experts. Therefore, scientists instead started to investigate the development of artificial intelligence applications in "microworlds," or very narrow topical areas. This led to the creation of the first useful artificial intelligence systems for select applications, e.g., games, disease diagnosis (MYCIN), spectrograph analysis (DENDRAL). MYCIN was developed at Stanford University in 1976 to aid physicians in diagnosing and treating patients with infectious blood diseases caused by bacteria in the blood and meningitis. These diseases can be fatal if not recognized and treated quickly. The DENDRAL program solved the problem of inferring molecular structure from the information provided by a mass spectrometer (Buchanan and Lederberg, 1971). DENDRAL was the first successful knowledge-intensive system deriving its expertise from large numbers of special-purpose rules. The field has advanced at a tremendous rate. NASA's REMOTE AGENT program became the first on-board autonomous planning program to control the scheduling of operations for a spacecraft travelling a hundred million miles from Earth (Jonsson et al., 2000). Such expert systems are based on the use of knowledge or rules derived from human experts. The knowledge is extracted from the expert by a knowledge engineer and turned into rule-based reasoning that can, we hope, be performed by a computer.

 Expert Systems

A knowledge-based *expert system* may be defined as "a system that uses human knowledge to solve problems that normally would require human intelligence" (PC AI, 2002). It is the ability to "solve problems efficiently and effectively in a narrow problem area" (Waterman, 1986) and "to perform at the level of an expert" (Liebowitz, 1988). Expert systems represent the expert's domain (i.e., subject matter) knowledge base as data and rules within the computer. The rules and data can be called upon when needed to solve problems. A different problem within the domain of the knowledge base can be solved using the same program without reprogramming.

Knowledge-based expert systems continue to be used extensively in remote sensing research. An early overview of expert systems related to image processing is found in Matsuyama (1989). Moller-Jensen (1997) used texture and an expert system to inventory urban land cover. Muchoney et al. (2000) used a decision-tree classifier to extract land-cover information from MODIS data of Central America. Tso and Mather (2001) summarized numerous applications of *hierarchical decision-tree classifiers*—the most general type of knowledge-based classifier (Hansen et al., 1996; 2000). Pal and Mather (2003) assessed the effectiveness of decision-tree methods for land-cover classification. Krapivin and Phillips (2001) used an expert system to model what it would take to stabilize the Aral and Caspian Sea water levels to the 1960 level within 10 to 12 years. Expert systems may also be used to detect change in complex heterogeneous environments (e.g., Stefanov et al., 2001; Yang and Chung, 2002; Stow et al., 2003).

A knowledge-based expert system consists of the components shown in Figure 10-1, including:

- human expert,

- user interface,

- knowledge base (rule-based domain),

- inference engine,

- on-line databases, and

- user.

It is useful to review the characteristics of each of these components.

Expert System User Interface

The expert system user interface should be easy to use, interactive, and interesting. It should also be intelligent and accumulate user preferences in an attempt to provide the most pleasing communication environment possible. Figure 10-2 depicts a commercially available Knowledge Engineer interface that can be used to develop remote sensing–assisted expert systems. This expert system *shell* was built using object-oriented programming and is easy to use. All of the hypotheses, rules, and conditions for an entire expert system may be viewed and queried from the single user interface.

Creating the Knowledge Base

Images, books, articles, manuals, and periodicals have a tremendous amount of information in them. Practical experience in the field with vegetation, soils, rocks, water, atmosphere, and urban infrastructure is also tremendously important. However, a human must comprehend the information and experiences and turn it into knowledge for it to be useful. Many human beings have trouble interpreting and understanding the information in images, books, articles, manuals, and periodicals. Similarly, some do not obtain much knowledge from field work. Fortunately, some laypersons and scientists are particularly adept at processing their knowledge using three different problem-solving approaches:

1. algorithms using conventional computer programs,

2. heuristic knowledge-based expert systems, and

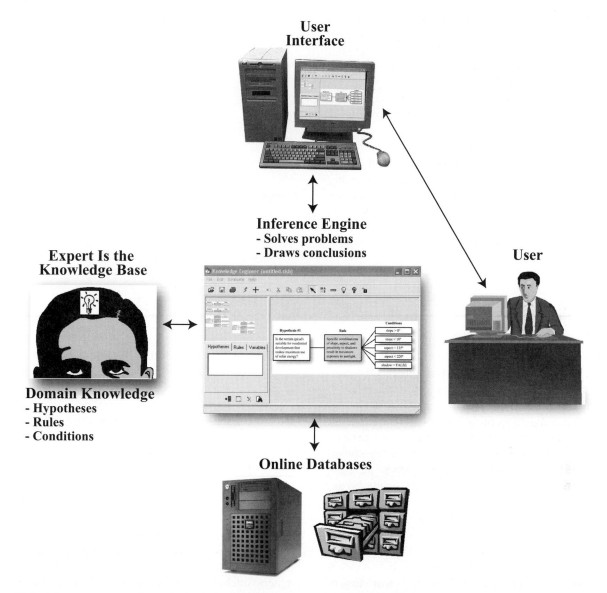

Figure 10-1   The components of a typical rule-based expert system. The domain (thematic) knowledge contained in an expert's mind is extracted in the form of a *knowledge base* that consists of hypotheses (problems), rules, and conditions that satisfy the rules. A *user interface* and an *inference engine* are used to encode the knowledge base rules, extract the required information from *online databases*, and solve problems. Hopefully, the information is of value to the *user* who queries the expert system.

3. artificial neural networks.

Neural networks are discussed later in the chapter.

Algorithmic Approaches to Problem Solving

Conventional algorithmic computer programs contain little knowledge other than the basic algorithm for solving a specific problem, the necessary boundary conditions, and data. The knowledge is usually embedded in the programming code. As new knowledge becomes available, the program has to be changed and recompiled (Table 10-1).

Heuristic Knowledge-based Expert System Approaches to Problem Solving

Knowledge-based expert systems, on the other hand, collect many small fragments of human know-how for a specific application area (domain) and place them in a *knowledge base* that is used to reason through a problem, using the

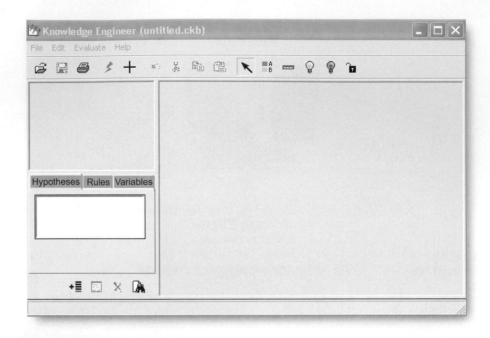

Figure 10-2   The Knowledge Engineer interface used in ERDAS Imagine's rule-based expert system shell (courtesy ERDAS/Leica Geosystems, 2002a and b).

knowledge that is most appropriate. Characteristics that distinguish knowledge-based expert systems from conventional algorithmic systems are summarized in Table 10-1. *Heuristic knowledge* is defined as "involving or serving as an aid to learning, discovery, or problem solving by experimental and especially by trial-and-error methods. Heuristic computer programs often utilize exploratory problem-solving and self-educating techniques (as the evaluation of feedback) to improve performance" (Merriam-Webster, 2003).

The Problem with Experts: Unfortunately, most experts really do not know *exactly* how they perform their expert work (Dreyfus and Dreyfus, 2001). Much of their expertise is derived from experiencing life and observing hundreds or even thousands of case studies. It is difficult for the experts to understand the intricate workings of complex systems much less be able to break them down into their constituent parts and then mimic the decision-making process of the human mind. Therefore, how does one get the knowledge embedded in the mind of an expert into formal rules and conditions necessary to create an expert system to solve relatively narrowly defined hypotheses (problems)? This is the responsibility of the *knowledge engineer*.

The knowledge engineer interrogates the domain expert and extracts as many rules and conditions as possible that are relevant to the hypotheses (problems) being examined. Ideally,

the knowledge engineer has unique capabilities that allow him or her to help build the most appropriate rules. This is not easy. The knowledge engineering process can be costly and time-consuming.

Recently, it has become acceptable for a domain expert (e.g., biologist, geographer, agronomist, forester) to create his or her own knowledge-based expert system by querying oneself and hopefully accurately specifying the rules associated with the problem at hand, for example, using ERDAS Imagine's expert system Knowledge Engineer (ERDAS/Leica Geosystems, 2002a and b). When this activity takes place, the expert must have a wealth of knowledge in a certain domain and the ability to formulate a hypothesis and parse the rules and conditions into understandable elements that are amenable to the "knowledge representation process."

The Knowledge Representation Process

The *knowledge representation process* normally involves encoding information from verbal descriptions, rules of thumb, images, books, maps, charts, tables, graphs, equations, etc. Hopefully, the knowledge base contains sufficient high-quality rules to solve the problem under investigation. *Rules* are normally expressed in the form of one or more "IF condition THEN action" statements. The *condition* portion of a rule statement is usually a fact, e.g., the pixel under

Table 10-1. Characteristics that distinguish knowledge-based expert systems from conventional algorithmic problem-solving systems (adapted from Darlington, 1996).

Characteristics	Knowledge-based Expert System	Conventional Algorithmic System
Paradigm	**Heuristic**. Based on rules of thumb. Solution steps are implicit (not determined by the programmer). The "solution" is not always correct. Declarative problem-solving paradigm.	**Algorithmic**. Solution steps explicitly written by programmer. Correct answer usually given. Procedural problem-solving paradigm.
Method of Operation	**Reasons with symbols**. Infers conclusions from known premises. Inference engine decides the order in which premises are evaluated.	Manipulates numeric data by sorting, calculating, and processing data to produce information.
Processing Unit	**Knowledge**. Usually represented in the form of rules and conditions. Knowledge is active in that an expert system can reason with knowledge to infer new knowledge from given data.	**Data**. Typically represented in the form of arrays or records in languages like C^{++}. Data is typically passive in that it does not give rise to further generations of data.
Control Mechanism	**Inference engine** is usually separate from domain knowledge.	Data or information and control are usually integrated.
Fundamental Components	**Expert system** = inference + knowledge.	**Conventional algorithmic system** = algorithm + data.
Explanation Capability	**Yes.** An explicit trace of the chain of steps underlying the reasoning processes allows a user to find out how the expert system arrived at its conclusion or why the system is asking for an answer to a particular question.	**No**.

investigation must reflect > 45% of the incident near-infrared energy. When certain rules are applied, various operations may take place such as adding a newly derived derivative fact to the database or firing another rule. Rules can be implicit (slope is high) or explicit (e.g., slope > 70%). It is possible to chain together rules, e.g., IF *c* THEN *d*; IF *d* THEN *e*; therefore IF *c* THEN *e*. It is also possible to attach confidences (e.g., 80% confident) to facts and rules. For example, a typical rule used by the MYCIN expert system is (Darlington, 1996):

IF the stain of the organism is gram-negative
   AND the morphology of the organism is rod
      AND the aerobicity of the organism is anaerobic
         THEN there is strong suggestive evidence (0.8) that the class of the organism is *Enterobacter iaceae.*

Following the same format, a typical remote sensing rule might be:

IF blue reflectance is (Condition) < 15%
   AND green reflectance is (Condition) < 25%
      AND red reflectance is (Condition) < 15%
         AND near-infrared reflectance is (Condition) > 45%

THEN there is strong suggestive evidence (0.8) that the pixel is vegetated.

Decision Trees: The best way to conceptualize an expert system is to use a *decision-tree structure* where rules and conditions are evaluated in order to test hypotheses (Figure 10-3). When decision trees are organized with hypotheses, rules, and conditions, each hypothesis may be thought of as the trunk of a tree, each rule a limb of a tree, and each condition a leaf. This is commonly referred to as a *hierarchical decision-tree classifier* (e.g., Swain and Hauska, 1977; Jensen, 1978; Kim and Landgrebe, 1991; DeFries and Chan, 2000; Stow et al., 2003; Zhang and Wang, 2003). The purpose of using a hierarchical structure for labeling objects is to gain a more comprehensive understanding of relationships among objects at different scales of observation or at different levels of detail (Tso and Mather, 2001). A decision tree takes as input an object or situation described by a set of attributes and returns a decision. The input attributes can be discrete or continuous. The output value can also be discrete or continuous. Learning a discrete-valued function is called *classification learning*. Learning a continuous function is called *regression* (Lawrence and Wright, 2001). We will concentrate on Boolean classification wherein each example is

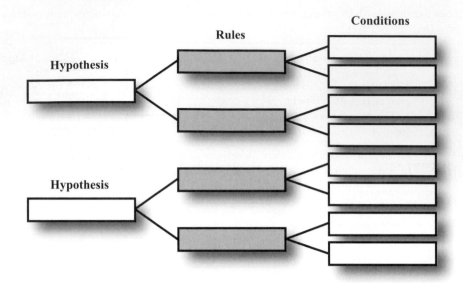

Figure 10-3   The fundamental building blocks of an expert system include hypotheses (problems), rules, and conditions. The rules and conditions operate on data (information). It is possible to address more than one hypothesis in an expert system.

classified as true (positive) or false (negative). A decision tree reaches its decision by performing a sequence of tests (Russell and Norvig, 2003).

Decision trees predict class membership by recursively partitioning a dataset into more homogeneous subsets (DeFries and Chan, 2000). We will first use a simple hypothetical GIS-related example to demonstrate the logic that should be adopted when building the knowledge domain.

Identify the hypothesis: The expert in charge of creating the knowledge domain identifies a hypothesis (problem) to be addressed. This may be a formal hypothesis to be tested using inductive logic and confidence levels or an informal hypothesis that is in search of a logical conclusion:

- ***Hypothesis 1:*** the terrain (pixel) is suitable for residential development that makes maximum use of solar energy (i.e., I will be able to put solar panels on my roof).

Specify the expert system rules: Heuristic rules that the expert has learned over time are the heart and soul of an expert system. If the expert's heuristic rules of thumb are indeed based on correct principles, then the expert system will most likely function properly. If the expert does not understand all the subtle nuances of the problem, has left out important variables or interaction among variables, or applied too much significance (weight) to certain variables, the expert system outcome may not be accurate. Therefore, the creation of accurate, definitive rules is extremely impor-

tant. Each rule provides the specific conditions to accept the hypothesis to which it belongs. A single rule that might be associated with hypothesis 1 is:

- specific combinations of terrain slope, aspect, and proximity to shadows result in maximum exposure to sunlight.

Specify the rule conditions: The expert would then specify one or more conditions that must be met for each rule. For example, conditions for the rule stated above might include:

- slope > 0 degrees, AND

- slope < 10 degrees (i.e., the terrain should ideally lie on terrain with 1 to 9 degrees slope), AND

- aspect > 135 degrees, AND

- aspect < 220 degrees (i.e., in the Northern Hemisphere the terrain should ideally face south between 136 and 219 degrees to obtain maximum exposure to sunlight), AND

- the terrain is not intersected by shadows cast by neighboring terrain, trees, or other buildings (derived from a viewshed model).

In this case, the hierarchical decision-tree diagram would look like Figure 10-4.

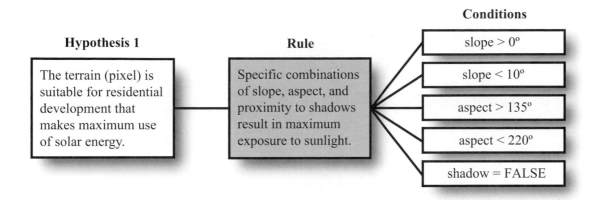

Figure 10-4 A human-derived decision-tree expert system with a rule and conditions to be investigated by an inference engine to test hypothesis 1.

To function properly, the expert system must have access to the data (variables) required by the rules. In this case three types of spatial information would be required for one rule: 1) slope, 2) aspect, and 3) a shadow viewshed file derived from a detailed digital elevation model.

Inference Engine

The terms *reasoning* and *inference* are generally used to describe any process by which conclusions are reached (Russell and Norvig, 2003). Thus, the hypotheses, rules, and conditions are passed to the *inference engine* where the expert system is implemented. One or more conditional statements within each rule are evaluated using the spatial data (e.g., 135 < aspect < 220). Multiple conditions within a rule are evaluated based on Boolean AND logic. While all of the conditions within a rule must be met to satisfy the rule, any single rule within a hypothesis can cause that hypothesis to be accepted or rejected. In some cases, rules within a hypothesis disagree on the outcome and a decision must be made using rule confidences (e.g., a confidence of 0.8 in a preferred rule and a confidence of 0.7 in another) or the order of the rules (e.g., preference given to the first) as the factor. The confidences and order associated with the rules are normally stipulated by the expert.

The inference engine interprets the rules in the knowledge base to draw conclusions. The inference engine may use backward- or forward-chaining strategies or both. Both backward and forward inference processes consist of a chain of steps that can be traced by the expert system. This enables expert systems to explain their reasoning processes, which is an important and positive characteristic of expert systems (Darlington, 1996). You would expect a doctor to explain how he or she came to a certain diagnosis regarding your health. An expert system can provide explicit information about how a particular conclusion (diagnosis) was reached.

An expert system *shell* provides a customizable inference engine. Expert system shells come equipped with an inference mechanism (backward chaining, forward chaining, or both) and require knowledge to be entered according to a specified format. Expert system shells qualify as languages, although certainly with a narrower range of application than most programming languages (PC AI, 2002). Typical artificial intelligence programming languages include LISP, developed in the 1950s, PROLOG, developed in the 1970s, and now object-oriented languages such as C^{++}.

On-line Databases

The rules and conditions may be applied and evaluated using data and/or information stored in on-line databases. The databases can take a variety of forms. It can be spatial and consist of remotely sensed images and thematic maps in raster and vector format. However, the database may also consist of charts, graphs, algorithms, pictures, and text that are considered important by the expert. The database should contain detailed, standardized metadata.

Expert Systems Applied to Remote Sensor Data

The use of expert systems in remote sensing research will be demonstrated using two different methodologies used to create the rules and conditions in the knowledge base. The first expert system classification is based on the use of formal rules developed by a human expert. The second example

Table 10-2. A hypothesis (class), variables, and conditions necessary to extract white fir (*Abies concolor*) forest cover information from Maple Mountain, Utah, using remote sensing and digital elevation model data. The Boolean logic with which these variables and conditions are organized within a chain of inference may be controlled by the use of rules and sub-hypotheses.

Hypothesis	Variables	Conditions
White Fir (*Abies concolor*)	Aspect Elevation Slope Multispectral	Aspect = 300 to 45 degrees Elevation >1200 m Slope = 25 to 50 degrees Remote sensing reflectance 　　TM band 1 Blue = 44 to 52 　　TM band 2 Green = 31 to 40 　　TM band 3 Red = 22 to 32 　　TM band 4 Near-infrared = 30 to 86 　　TM band 5 Mid-infrared = 19 to 47 NDVI = 0.2 to 0.7

involves expert system rules not generated by humans, but derived automatically by an inductive machine-learning algorithm based on training data that is input by humans into the system. Both methods will be used to identify white fir forest stands on Maple Mountain in Utah County, Utah, using Landsat Enhanced Thematic Mapper Plus (ETM⁺) imagery and several topographic variables extracted from a digital elevation model of the area. The comparison will demonstrate important characteristics of human expert versus machine-learning rule development strategies.

Hypotheses, Rules, and Conditions Specified by an Expert Engineer

This example is based on rules specified by a human expert to map white fir (*Abies concolor*) forest on Maple Mountain located in Utah County, Utah. Maple Mountain rises from the 5000 ft valley floor to 10,200 ft above sea level (ASL). A Landsat Enhanced Thematic Mapper Plus panchromatic image of the mountain and valley obtained on August 10, 1999, is shown in Figure 10-5a. A color composite is shown in **Color Plate 10-1a**.

The goal of this exercise is to accurately extract the spatial distribution of forest land-cover information using the remotely sensed data, a digital elevation model (DEM), and DEM-derived products. A U.S. Geological Survey 30 × 30 m digital elevation model is shown in Figure 10-5c. Contour, shaded relief, slope, and aspect were extracted from the digital elevation model (Figure 10-5d through f).

Hypotheses to Be Tested: The hypothesis (class) to be tested (extracted) from the spatial data was white fir (*Abies concolor*). Many other types of land cover are present in the scene but we will focus our attention on white fir. The overall structure of the expert system logic is displayed in decision-

tree format in the expert system interface shown in Figure 10-6. Note that the hypothesis represents the base of the decision tree (laid on its side).

Rules (variables): A human expert who worked for the Uinta National Forest developed the knowledge base (hypotheses, rules, and conditions) to identify white fir habitat from all other land cover in this area. The rules and conditions were based on remote sensing multispectral reflectance characteristics and derivatives (e.g., NDVI), elevation above sea level, and microclimate and soil moisture conditions that are controlled primarily by terrain slope and aspect (Table 10-2).

Conditions: The expert identified very specific conditions that are associated with the remote sensing reflectance data, elevation, slope and aspect. This part of Utah is in a semiarid mountain and range province and receives relatively little moisture in the summer months. Therefore, south-facing slopes have reduced soil moisture conditions throughout much of the summer. The expert knows that white fir requires substantial year-round soil moisture. Therefore, it thrives best on north-facing slopes ($300 \leq$ aspect ≤ 45 degrees). White fir favors elevations > 1900 m ASL. It outcompetes other forest types when the slope is between 25 and 50 degrees.

White fir forest physiology causes it to have relatively distinct red and near-infrared reflectance characteristics when compared with other vegetation types on the mountain. This translates into useful normalized difference vegetation index (NDVI) values,

$$\text{NDVI} = \frac{\rho_{TM4} - \rho_{TM3}}{\rho_{TM4} + \rho_{TM3}}, \tag{10-1}$$

Maple Mountain in Utah County, UT

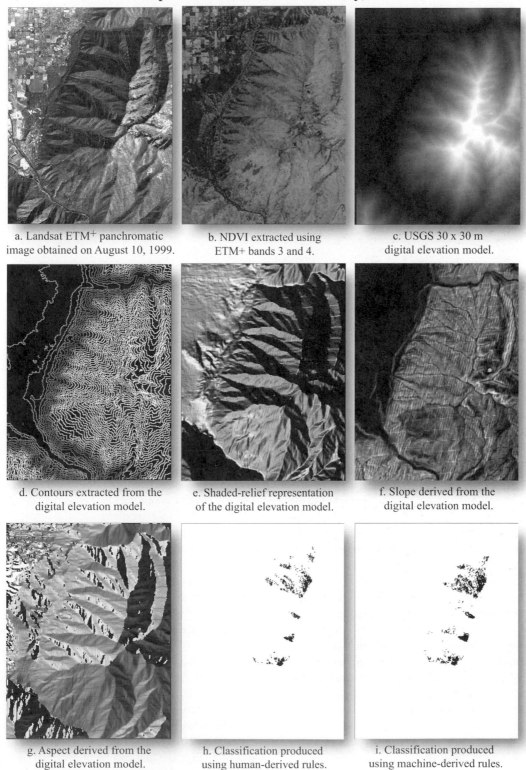

a. Landsat ETM+ panchromatic image obtained on August 10, 1999.

b. NDVI extracted using ETM+ bands 3 and 4.

c. USGS 30 x 30 m digital elevation model.

d. Contours extracted from the digital elevation model.

e. Shaded-relief representation of the digital elevation model.

f. Slope derived from the digital elevation model.

g. Aspect derived from the digital elevation model.

h. Classification produced using human-derived rules.

i. Classification produced using machine-derived rules.

Figure 10-5   a–g) Datasets used to identify white fir (*Abies concolor*) on Maple Mountain in Utah County, Utah. h–i) Classification maps derived using an expert system based on human (h) and machine-derived (i) rules.

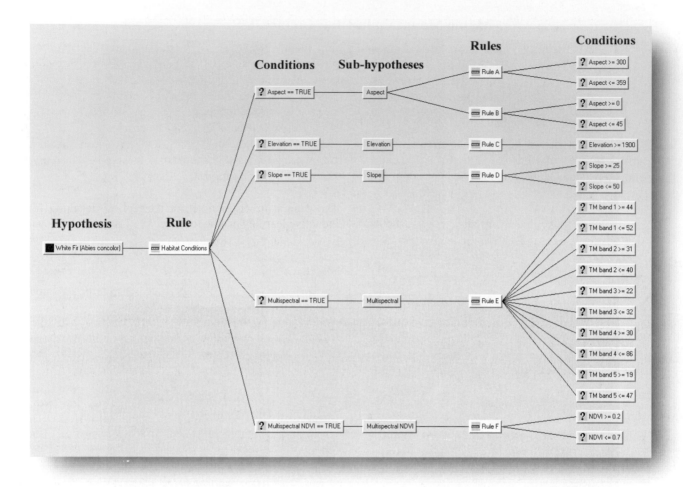

Figure 10-6   Hierarchical decision tree associated with a knowledge-based expert system classification of white fir (*Abies concolor*) on Maple Mountain, Utah. A human expert identified the hypotheses, rules, and conditions.

based on Landsat 7 ETM$^+$ band 3 (red) and 4 (near-infrared) reflectance data. The specific multispectral remote sensing conditions for each band and for the NDVI are summarized in Table 10-2. The hypothesis, rules, and conditions specified in the expert system interface are summarized in Figure 10-6.

Inference Engine: The hypothesis, rules, and conditions created by the expert were passed to the inference engine. The inference engine processed the rules and conditions in conjunction with the required spatial data. This produced the classification map depicting the spatial distribution of white fir displayed in planimetric map format in Figure 10-5h and draped over the Landsat ETM$^+$ color composite and digital elevation model in **Color Plate 10-1e**.

Unfortunately, this classification took quite a bit of research on the part of the human expert to come up with the expert

system rules and conditions. It would be much better if there were some way for the expert system to generate its own rules based on training data (DeFries and Chan, 2000). This leads us to a discussion of machine learning.

Rules and Conditions Based on Machine Learning

The heart of an expert system is its knowledge base (Luger and Stubblefield, 1993). The usual method of acquiring knowledge in a computer-usable format to build a knowledge base involves human domain experts and knowledge engineers, as previously discussed. The human domain expert explicitly expresses his or her knowledge about a subject in a language that can be understood by the knowledge engineer. The knowledge engineer translates the domain knowledge into a computer-usable format and stores it in the knowledge base.

This process presents a well-known problem in creating expert systems that is often referred to as the *knowledge acquisition bottleneck*. The reasons are (Bratko, et al., 1989):

- the process requires the engagement of the domain expert and/or knowledge engineer over a long period of time, and

- although experts are capable of using their knowledge for decisionmaking, they are often incapable of articulating their knowledge explicitly in a format that is sufficiently systematic, correct, and complete to be used in a computer application.

Remote sensing scientists have acknowledged these difficulties when building knowledge bases for image analysis (Argialas and Harlow, 1990). For example, operational monitoring of land cover from satellite requires automated procedures for analyzing large volumes of data (DeFries and Chan, 2000). To solve such problems, much effort has been exerted in the artificial intelligence community to automate the building of expert system knowledge bases (e.g., Huang and Jensen, 1997; Kubat et al., 1998; Tso and Mather, 2001; Gahegan, 2003; Tullis and Jensen, 2003).

Machine learning is defined as the science of computer modeling of learning processes. It enables a computer to acquire knowledge from existing data or theories using certain inference strategies such as induction or deduction. We will focus only on inductive learning and its application in building knowledge bases for image analysis expert systems.

A human being has the ability to make accurate generalizations from a few scattered facts provided by a teacher or the environment using inductive inferences. This is called *inductive learning* (Huang and Jensen, 1997). In machine learning, the process of inductive learning can be viewed as a heuristic search through a space of symbolic descriptions for plausible general descriptions, or concepts, that explain the input training data and are useful for predicting new data. Inductive learning can be formulated using the following symbolic formulas (Michalski, 1983):

$$\forall i \in I \qquad (E_i \Rightarrow D_i) \qquad\qquad (10\text{-}2)$$

$$\forall\, i, j \in I \qquad (E_i \Rightarrow \sim D_i),\ \text{if } (\, j \neq i) \qquad (10\text{-}3)$$

where D_i is a symbolic description of class i, E_i is a predicate that is true only for the training events of class i, I is a set of class names, $\sim$ stands for "negation," and $\Rightarrow$ stands for "implication." Expression 10-2 is called the *completeness condition* and states that every training event of some class must satisfy the induced description D_i of the same class.

However, the opposite does not have to hold because D_i is equivalent to or more general than E_i. This means that D_i may include some features that do not exist in some examples in E_i. Expression 10-3 is called the *consistency condition* and states that if an event satisfies a description of some class, it cannot be a member of a training set of any other class. The task of inductive learning is to find through the space of descriptions the general description set $D = \{D_1, D_2, ..., D_i\}$ for the class set $K = \{K_1, K_2, ..., K_i)$ that satisfies the completeness condition and also, in some cases, the consistency condition (Huang and Jensen, 1997).

The general description set, or concept, D, resulting from inductive learning can be represented by a variety of formalisms, including *production rules* (Quinlan, 1993). This means that inductive learning can be used to build knowledge bases for expert systems because production rules are the most popular form of knowledge representation in expert systems (Giarratano and Riley, 1994). A motivation for the use of this approach to build a knowledge base is that it requires only a few good examples to function as *training data*. This is often much easier than explicitly extracting complete general theories from the domain expert.

There are a number of inductive learning algorithms such as Quinlan's (1993; 2003) C5.0 machine-learning program, which will be demonstrated here. It has the following advantages:

- The knowledge learned using C5.0 can be stored in a production rule format that can be used to create a knowledge base for a rule-based expert system.

- C5.0 is flexible. Unlike many statistical approaches, it does not depend on assumptions about the distribution of attribute values or the independence of the attributes themselves (Quinlan, 2003). This is very important when incorporating ancillary GIS data with remotely sensed data because they usually have different attribute value distributions, and some of the attributes may be correlated.

- C5.0 is based on a decision-tree learning algorithm that is one of the most efficient forms of inductive learning. The time taken to build a decision tree increases only linearly with the size of the problem (Quinlan, 2003).

The procedure of applying the inductive learning technique to automatically build a knowledge base for a remote sensing image analysis expert system that incorporates GIS data involves *training*, *decision-tree generation*, and the creation of *production rules*. The resultant production rules compose

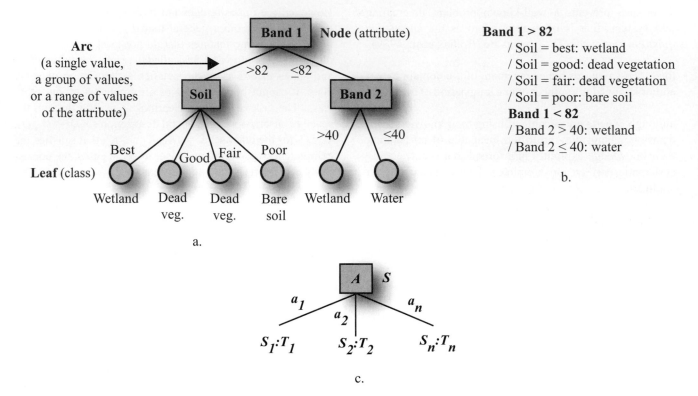

Figure 10-7   a) Example of a decision tree. A dataset consisting of three attributes (band 1, band 2, and soils) is classified into one of four classes: water, wetland, dead vegetation, and bare soil. b) Another method of presenting the decision tree shown in (a). c) A decision tree generated by dividing dataset S into subsets (adapted from Huang and Jensen, 1997).

the knowledge base and can be used by an expert system inference engine to perform the final image classification.

Training: The objective of training is to provide examples of the concepts to be learned. When building a knowledge base for image classification, the examples should be a set of training objects, each of which is represented by an attribute value class vector such as:

[attribute_1, ..., attribute_n, class_i]

The classification scheme must be developed at this stage along with the attributes to be used in learning and classification. Attributes (e.g., spectral reflectance values in various bands) are then collected for each class of information. Stratified random sampling may be the most appropriate way to locate the candidate training class pixels because it guarantees that a minimum number of samples are selected from each class (strata) (Congalton, 1988). The learning algorithm attempts to extract from this attribute training dataset some generalized concepts, i.e., rules that can be used to classify the remaining data.

Decision Tree Generation: The C5.0 learning algorithm first generates decision trees from the training data. The decision trees are then transformed into production rules. A raw decision tree (prior to the formation of production rules) can be viewed as a classifier composed of leaves that correspond to classes, decision nodes that correspond to attributes of the data being classified, and arcs that correspond to alternative values for these attributes. A hypothetical example of a decision tree is shown in Figure 10-7a and b.

A recursive "divide and conquer" strategy is used by C5.0 to generate a decision tree from a set of training data. The training dataset S is divided into subsets $S_1, ..., S_n$ according to $a_1, ..., a_n$, which are the possible values of a single attribute A. This generates a decision tree with A being the root and $S_1, ..., S_n$ corresponding to subtrees $T_1, ..., T_n$ (Figure 10-7c).

The stop condition for such a procedure will eventually be satisfied, resulting in a final decision tree. The goal is to build a decision tree that is as small as possible. This ensures that the decisionmaking by the tree is efficient. The goal is realized best by selecting the most informative attribute at each node so that it has the power to divide the dataset corresponding to the node into subsets as pure (homogeneous) as possible (DeFries and Chan, 2000). C5.0's attribute selection criteria are based on the entropy measure from communication theory where, at each node, the attribute with the minimum entropy is selected to divide the dataset.

From Decision Trees to Production Rules: Decision trees are often too complex to be understood, especially when they are large. A decision tree is also difficult to maintain and update. Therefore, it is desirable to transform a decision tree to another representation of knowledge that is adopted commonly in expert systems, such as production rules.

A *production rule* can be expressed in the following general form (Jackson, 1990):P

$$P_1, ..., P_m \rightarrow Q_1, ..., Q_m \qquad (10\text{-}4)$$

with the meaning

- if *premises* (or conditions) P_1 and ... P_m are true,

- then perform *actions* Q_1 and ... Q_m.

In fact, each path from the root node to a leaf in a decision tree can be translated into a production rule. For example, the path from the root node to the leftmost leaf in the decision tree in Figure 10-7a can be represented by a production rule such as

$$(\text{band } 1 > 82), (\text{soil} = \text{poor}) \rightarrow (\text{class} = \text{bare soil}).$$

Several problems must be solved when transforming a decision tree into production rules. First, individual rules transformed from the decision tree may contain irrelevant conditions. C5.0 uses a pessimistic estimate of the accuracy of a rule to decide whether a condition is irrelevant and should be deleted. Second, the rules may cease to be mutually exclusive and collectively exhaustive. Some rules may be duplicative or may conflict. This is a common problem for rule-base building using either a manual or automated approach. Usually, a rule-based system has some conflict resolution mechanism to deal with this problem. The approach adopted by C5.0 includes the use of rule confidence voting and a default class. In other words, each rule is assigned a confidence level based on false-positive errors (the number of training objects that were incorrectly classified as class C by a rule). If an object can be assigned to more than one class by two or more rules, a summary vote of all rule confidences concerned is tabulated and the class with the highest number of votes wins (Quinlan, 2003).

Unfortunately, some objects to be classified may satisfy no rules or they may satisfy conflicting rules with equal confidence levels. When this occurs, a default class may be used. There are a number of ways a default class can be identified. One approach is to have the default class be the one that contains the most training objects not satisfying any rule.

The quality of the resultant rules can be evaluated by predicting error rates derived by applying the rules on a test dataset. Because the rules are easy to understand, they can also be examined by human experts. With caution, they may be edited directly.

Case Study: To demonstrate the utility of machine learning for the creation of production rules, the C5.0 algorithm was trained with the Maple Mountain data previously discussed. This resulted in the creation of the rules summarized in Figure 10-8a. The classification map produced from the machine-learning rules is shown in Figure 10-5i and Figure 10-8b.

Thus, *machine-learning* algorithms may be used to train expert systems by creating rules and conditions directly from the training data without human intervention (e.g., Huang and Jensen, 1997; Friedl et al., 1999). This is a very important advancement because the expert system can adapt when new learning data are provided. Otherwise, an expert system cannot learn by example.

Advantages of Expert Systems

Expert system decision-tree classifiers have several characteristics that make them attractive when compared with other classification methods (Hansen et al., 1996; DeFries and Chan, 2000):

- You can evaluate the output of the expert system (e.g., whether a certain pixel is ideal or not for residential development) and work backward to identify how a conclusion was reached. This is in contrast to neural networks where the exact nature of the decision process is lost in the weights used in the hidden layers (Qiu and Jensen, 2004).

- Until recently, maximum likelihood classification was the most common method used for supervised classification of remote sensor data (McIver and Friedl, 2002). This method assumes that the probability distributions for the input classes possess a multivariate normal distribution. Increasingly, nonparametric classification algorithms are

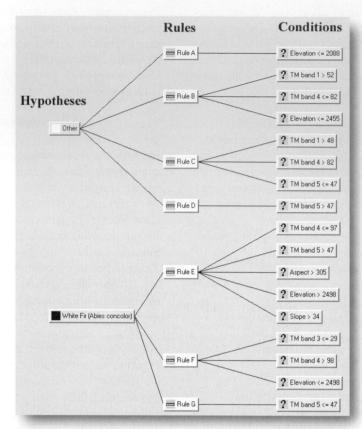

a. Expert system rules for classifying white fir (*Abies concolor*) derived using machine learning.

b. Classification of white fir (*Abies concolor*) on Maple Mountain based on rules derived using machine learning, draped over a three-dimensional view of Landsat ETM+ imagery.

Figure 10-8   a) Hierarchical decision tree associated with an expert system classification of white fir (*Abies concolor*) on Maple Mountain in Utah County, Utah. A machine-learning algorithm stipulated the hypotheses, rules, and conditions. b) Results of applying the machine-derived rules.

being used such as decision trees, which make no assumptions regarding the distribution of the data.

- The decision tree can reveal nonlinear and hierarchical relationships among the input variables and use them to predict class membership.

- A large body of evidence demonstrates the ability of machine-learning techniques (particularly decision trees and neural networks) to deal effectively with tasks that involve highly dimensional data (e.g., hyperspectral) (Lawrence and Wright, 2001). Reducing a dataset to just two or three variables (e.g., bands) is becoming an outdated notion (Gahegan, 2003). The new thinking is to let the geographical data itself "have a stronger voice" rather than let statistics derived from the dataset dictate the analysis (e.g., the means and covariance matrices used in maximum likelihood classification).

Neural Networks

A neuron is a cell in the brain whose principal function is the collection, processing, and dissemination of electrical signals (Russell and Norvig, 2003). Neural networks simulate the thinking process of human beings, whose brains use interconnected neurons to process incoming information (Jensen et al., 1999; Hengl, 2002). A neural network reaches a solution not via a step-by-step algorithm or a complex logical program, but in a nonalgorithmic, unstructured fashion based on the adjustment of the weights connecting the neurons in the network (Rao and Rao, 1993). Neural networks have been used to classify various types of remote sensor data and have in certain instances produced results superior to those of traditional statistical methods (e.g., Benediktsson et al., 1990; Foody et al., 1995, 1997; Foody, 1996; Atkinson and Tatnall, 1997; Jensen et al., 1999; Ji, 2000). This success can be attributed to two of the important advantages of neural networks: 1) freedom from normal distribution requirements and 2) the ability to adaptively simulate complex and nonlinear patterns given proper topological structures (Atkinson and Tatnall, 1997; Jensen et al., 1999).

Components and Characteristics of a Typical Artificial Neural Network Used to Extract Information from Remotely Sensed Data

The topological structure of a typical back-propagation neural network is shown in Figure 10-9 (Civco, 1993; Bene-

diktsson and Sveinsson, 1997). The artificial neural network normally contains neurons arranged in three types of layer:

- an input layer,

- a hidden layer(s), and

- an output layer.

The neurons in the input layer might be the multispectral reflectance values for individual pixels plus their texture, surface roughness, terrain elevation, slope, aspect, etc. The use of neurons in the hidden layer(s) enables the simulation of nonlinear patterns in the input data. A neuron in the output layer might represent a single thematic map land-cover class, e.g., agriculture.

Much like a supervised classification, an artificial neural network requires *training* and *testing (classification)* to extract useful information from the remotely sensed and ancillary data (Foody and Arora, 1997; Qiu and Jensen, 2004).

Training an Artificial Neural Network

In the *training* phase, the analyst selects specific x, y locations in the input image with known attributes (e.g., agriculture, upland pine) as training sites (Jensen et al., 2001). The per-pixel spectral information (e.g., spectral reflectance in red and near-infrared bands) and ancillary information (e.g., elevation, slope, aspect) for each training site is then collected and passed to the input layer of the neural network. At the same time, the true target (class) value of this exact location (e.g., agriculture) is sent to the output layer by assigning the neuron representing this class a membership value of 1, while all the other output neurons are assigned a value of 0.

It is important to remember that neural network training based on the examples obtained from an image and other ancillary data acquired at a specific time and location may be applicable only in the immediate geographic area and perhaps only for a given season. Thus, they may not be extendable through space or time.

Learning is usually accomplished by adjusting the weights using a back-propagation algorithm. For each training example, the output of the network is compared with the true target (class) value. The difference between the target and output value is regarded as an error and is passed back to previous layers in the network to update the connection weights. The magnitude of the adjustment is proportional to the absolute values of the error (discussed shortly in the mathematics section). After many iterations, the root mean square (RMS)

Components of a Typical Artificial Neural Network

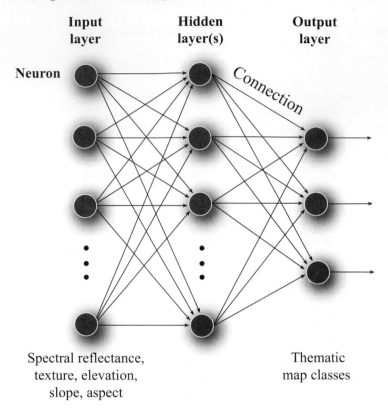

Figure 10-9   Components of a typical back-propagation artificial neural network (ANN). The network consists of input, hidden, and output layers. The input layer may contain information about individual training pixels including percent spectral reflectance in various bands and ancillary data such as elevation, slope, etc. Each layer consists of nodes that are interconnected. This interconnectedness allows information to flow in multiple directions (i.e., back-propagation can take place) as the network is trained. The strength (or weight) of these interconnections is eventually *learned* by the neural network and stored. These weights are used during the *testing* (classification) procedure. The more representative the training data, the more likely that the neural network will develop weights in the hidden layer that mirror reality and result in an accurate classification. The output layer may represent individual thematic map classes such as water or forest (adapted from Jensen et al., 2001).

error diminishes to a small value less than a predefined tolerance, and further iterations will not improve the performance of the network (Jensen et al., 1999). At this time, the system achieves convergence and the training process is completed. The rules inherent in the examples are stored in the hidden weights in the hidden layer(s) for use in the testing (classification) phase.

Testing (Classification)

During the *test* or *classification* phase, the spectral and ancillary characteristics of every pixel in the scene are passed to input neurons of the neural network. The neural network evaluates each pixel using the weights stored in the hidden layer neurons to produce a predicted value for every neuron

of the output layers. The value obtained for every output neuron is a number between 0 and 1 that gives the fuzzy membership grade of the pixel belonging to the class represented by that neuron. Defuzzification of these maps using a local maximum function leads to a hard classification map where each pixel is assigned to a unique class that has the highest fuzzy membership grade (Jensen et al., 2001).

Mathematics of the Artificial Neural Network

An artificial neural network (ANN) is defined by neurons, topological structure, and learning rules. The neuron is the fundamental processing unit of an ANN for computation (Jensen et al., 1999). Analogous to the human brain's biological neuron, an artificial neuron is composed of inputs (den-

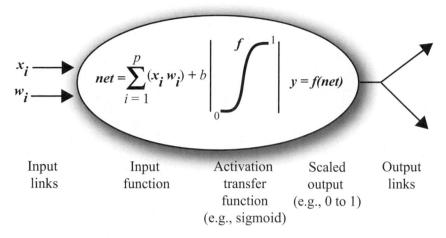

Input   Input   Activation   Scaled   Output
links   function   transfer   output   links
                              function   (e.g., 0 to 1)
                            (e.g., sigmoid)

Figure 10-10   Mathematical model of a neuron. The output of the neuron is y, which is a function of the summation of the input values (x_i) multiplied times their individual weights (w_i) in conjunction with an activation function (f). A sigmoid activation function is used in this example.

drites), weights (synapses), processing units (cell bodies), and outputs (axons) (Hagan, et al., 1996). Each input x_i is multiplied by the scalar weight w_i to form $x_i w_i$, a term that is sent to the "summing unit" of the processing unit (Figure 10-10). An offset, b, may be added to the total. The summation output:

$$net = \sum_{i=1}^{p} (x_i w_i) + b, \qquad (10\text{-}5)$$

referred to as net input, goes into an activation transfer function f that produces scaled neuron output y (between 0 and 1, or −1 to 1) through a transform algorithm. Thus, y can be calculated as (Jensen et al., 1999):

$$y = f(net) = f\left[\sum_{i=1}^{p} (x_i w_i) + b\right]. \qquad (10\text{-}6)$$

The topological structure of an ANN defines the overall architecture of the network, including the framework and interconnection of neurons that are organized into layers. As previously discussed, a typical ANN consists of three or more layers (Figure 10-9): one input layer, one output layer, and one or more hidden layers (Jensen et al., 2001). Neurons within and between layers can be connected to form a user-defined, task-oriented network. Decision processes can be duplicated or approximated through an ANN's ability to learn and recall. Learning is made possible by feeding the input layer with training data. By comparing the current activation of neurons in the output layer to a desired output

response, the difference can be obtained and used to adjust weights connecting neurons in the network. It is the weights that primarily determine the behavior of the network, so the goal of learning is to achieve a set of weights that will produce an output most resembling the target (Jensen et al., 1999). This adaptive learning process repeats until changes in network weights drop below a preset threshold, which indicates a user-defined accuracy. Once learned, the network can recall stored knowledge to perform either classification or prediction on new input data (Lloyd, 1996).

The desirable properties of non-normality and nonlinearity of an ANN can be attributed not only to the network's massively parallel distributed structure, but also to the transfer function for each neuron (Haykin, 1994). A neuron is basically a nonlinear device that may take continuous and differential functions as the transfer function. Therefore, ANNs are capable of modeling any complex system, be it physical or human. Image classification using ANNs takes advantage of this relationship to incorporate non-normally distributed numerical and categorical GIS data and image spatial information (e.g., texture and context) into the process (Bischof et al., 1992; Civco, 1993; Gong et al., 1996; Qiu and Jensen, 2004).

There are two commonly used artificial neural network learning paradigms:

1) a pattern associator based on a least-mean-square algorithm, and

2) an error back-propagation algorithm.

The Pattern Associator ANN Algorithm

A pattern associator (PA) model is a primitive neural network that consists of only one input layer and one output layer and operates under the assumption of linearity (Haykin, 1994). It is based on the least-mean-square (LMS) algorithm. The LMS algorithm is driven by the difference between the target activation and the obtained activation. The idea is to adjust the weights of the connection to reduce the mean-square error. First, the network is initialized by setting all the weights to zero (Haykin, 1994):

$$w_k(1) = 0 \text{ for } k = 1, 2, 3, ..., p \tag{10-7}$$

where p is the number of inputs and $w_k(1)$ is the weight associated with input k in the first iteration. For every iteration t, $t = 1, 2,...$, the obtained activation y can be calculated:

$$y(t) = \sum_{j=1}^{p} [x_j(t)\hat{w}_j(t)] \tag{10-8}$$

where $x_j(t)$ is the value of input j at iteration t.

Thus, the difference between the desired activation $d(t)$ and obtained activation $y(t)$ at iteration t can be computed:

$$e(t) = d(t) - y(t). \tag{10-9}$$

Then, in the next iteration $t + 1$, the weight of input k is adjusted as

$$\hat{w}_k(t+1) = w_k(t) + \eta \cdot e(t) \cdot x_k(t) \tag{10-10}$$

where η is a positive constant known as the learning rate.

If $y(t)$ is greater than $d(t)$, $e(t)$ computed from Equation (10-9) will be a negative value. When applied to Equation (10-10), $w_k(t + 1)$ will be adjusted (penalized) to a smaller number. If $d(t)$ is greater than $y(t)$, $w_k(t + 1)$ will be adjusted (rewarded) to a larger number. The difference between the desired and obtained activation is thus minimized at the learning rate η. The ideal choice for η is the largest value that does not lead to oscillation (Benediktsson, 1990; Jensen et al., 1999).

The traditional PA usually uses a linear transfer function in its output neuron, so that the activation of the output neuron is simply equal to the net input (Jensen et al., 1999). In this sense, the LMS learning process emulates a linear regression analysis, except that it is free from any assumption of data distribution. Like regression analysis, the traditional PA fails to achieve a good result when the data being modeled are not linear. Variants to the linear PA adopt nonlinear transform functions such as a continuous sigmoidal function that resembles regression with a logarithmic transformation. These variants have the ability of modeling simple nonlinearity without distribution constraints.

The Error Back-propagation ANN Algorithm

Back-propagation (BP) neural networks were first proposed by Werbos (1974) to improve the efficiency of multilayer network training (Jensen et al., 1999). This network has a hierarchical architecture with fully interconnected layers. Typically, one or more hidden layers are used to enable the network to learn complex tasks. The training is based on an error-correction learning rule. Each neuron in the network may include a nonlinear transfer function at the output end, producing smooth signals to other neurons. One of the common forms of nonlinearity is a sigmoidal transfer function defined by the logistic function (Haykin, 1994; Hagan et al., 1996):

$$f(net) = \frac{1}{1 + e^{(-net)}}. \tag{10-11}$$

Initialized with all the synaptic weights and thresholds set to small random numbers, the network is fed with training patterns. Each learning iteration of the network consists of two passes: a forward pass and a backward pass. In the forward pass, the net input of the jth neuron of layer l is calculated:

$$net_j^l(t) = \sum_{i=0}^{p} (y_i^{(l-1)}(t)w_{ji}^l(t)) \tag{10-12}$$

and the output activation is computed using the transfer function (Equation 10-11):

$$y_j^l(t) = f[net_j^l(t)]. \tag{10-13}$$

The difference between the desired response and output activation of neuron j at the output layer is obtained by:

$$e_j(t) = d_j(t) - o_j(t). \tag{10-14}$$

In the backward pass, a local error gradient, δ, is computed layer by layer:

$$\delta_j^L(t) = e_j^L(t)o_j(t)[1 - o_j x(t)] \tag{10-15}$$

for neuron j in output layer L, and

$$\delta_j^l(t) = y_j^l(t)[1 - y_j^l(t)] \sum_k \delta_j^{(l+1)}(t) w_{kj}^{(l+1)}(t) \quad (10\text{-}16)$$

for neuron j in hidden layer l.

With this information, the synaptic weights in layer l can be adjusted according to a generalized LSM rule (Haykin, 1994; Hagan et al., 1996):

$$w_{ji}^l(t+1) = w_{ji}^l(t) + \alpha \left[w_{ji}^l(t) - w_{ji}^l(t-1) \right] + \eta \delta_j^l(t) y_i^{(l-1)}(t) \quad (10\text{-}17)$$

where η is the learning rate and α is the momentum constant for speeding up learning without running into the risk of oscillation. In a special case when α equals zero, the generalized LSM rule becomes the standard LSM rule. In the backward pass, weights are adjusted in a manner that changes are proportional to the error gradient. Because this pass is started from the output layer and error correction is propagated backward to the previous layer, this process is called *error back-propagation*. The knowledge inherent in the training data is obtained through the iteration of forward and backward passes until all the parameters in the network are stabilized (Jensen et al., 2001). This is signified by the decrease of average squared error to a minimum or acceptable level.

Advantages of Artificial Neural Networks

Artificial neural networks have been employed to process multispectral remote sensing images and often achieve improved accuracy compared to those obtained using traditional statistical methods (Heermann and Khazenie, 1992; Civco, 1993; Benediktsson and Sveinsson, 1997; Foody et al. 1995; Gong, 1996). These successes are underpinned by many salient characteristics of neural networks, including:

- A single neuron simulates the computation of a multivariate linear regression model (Hewitson and Crane, 1994; Jensen et al., 1999).

- A neural network makes no a priori assumptions of normal and linear data distribution due to its operation in a nonparametric fashion (Lloyd, 1996; Foody el al., 1995).

- Neural networks are able to learn from existing examples adaptively, which makes the classification objective.

- The nonlinear patterns are "learned" from the empirical examples instead of "prespecified" by an analyst based on prior knowledge of the datasets (Hagan, et al., 1996).

- The noisy information inevitably included in the examples supplies a trained neural network with the ability to generalize, which makes neural networks robust solutions in the presence of previously unseen, incomplete, or imprecise data (Fausett, 1994). The knowledge in a traditional expert system that must be extracted from knowledgeable experts of a domain area may be subjective and incomplete. This is because the experts may have a biased or even incorrect understanding of reality, they may not be aware of underlying rules they have used, and they may have difficulty articulating these rules (NeuralWare, 1996). Conversely, knowledge in a neural network is acquired through learning by empirical (real-world) training examples. Although experts are essential in selecting and preparing sample training data, generally their personal biases are excluded from the knowledge acquisition process.

- Knowledge in an expert system is represented by logical rules made up of binary predicates. Numerical attributes have to be converted to binary true/false statements, which may cause a large amount of information to be lost in the simplification process (Gahegan, 2003). On the other hand, a neural network can embrace data in all formats as long as the data are converted to a numeric representation (Roli et al., 1996; Brand and Gerritsen, 2004b).

- Most rule-based expert systems fail to generalize a predictable inference if an appropriate match with the perfect rules that must be articulated by experts cannot be obtained. Conversely, the knowledge in a neural network derived from real-world training examples inevitably contains noise no matter how careful the examples are selected. Therefore, neural networks are good at generalizing both discrete and continuous data and have a capability to interpolate or adapt to the patterns never seen in the training process. Thus, neural networks are tolerant of noise and missing data and attempt to find the best fit for input patterns (Russell and Norvig, 2003; Kasabov, 1996).

- Finally, neural networks continuously adjust the weights as more training data are provided in a changing environment. Thus, they continuously *learn*.

Foody and Arora (1997) evaluated four factors that affect image classification using neural networks. They found that variations in the dimensionality of the remote sensing dataset, as well as the training and testing set characteristics, had a significant effect on classification accuracy. The network architecture, specifically the number of hidden units and layers, did not, however, have a significant effect on

Table 10-3.   Summary of differences and similarities between using a machine learning–based decision-tree expert system and a neural network (adapted from Brand and Gerritsen, 2004ab).

	Decision Tree Expert System Based on Machine Learning	Neural Network
Classification	Yes	Yes
Regression	Yes (classification and regression tree - CART)	Yes
Directly handles categorical data	Yes	No (categorical data must be converted to numeric values)
Number of passes	Few (no more than the number of layers in the decision tree, generally related to the number of input variables)	Many (called epochs); often > 1000
Ability to understand the decision making process	Straightforward	Difficult
Ability to model a relationship between the dependent variable and a ratio of two of the independent variables	Can only be approximated in a stepwise fashion	Easy
Can model trigonometric and logarithmic relationships	Difficult	Easy

classification accuracy. Therefore, it is important to consider a broader set of issues in addition to network architecture when using an artificial neural network for image classification.

Limitations of Artificial Neural Networks

Despite the excellent performance of neural networks in image classification, it is usually difficult to explain in a comprehensive fashion the process through which a given decision or output has been obtained from a neural network (Andrews et al., 1995; Qiu and Jensen, 2004). There exists an inherent inability to represent knowledge acquired by the network in an explicit form with simple "if–then" rules. The rules of image classification and interpretation learned by the neural network are buried in the weights of the neurons of the hidden layers. It is difficult to interpret these weights due to their complex nature (Lein, 1997). Therefore, a neural network is often accused of being a *black box* (Benitez et. al., 1997; Qiu and Jensen, 2004).

For these reasons, neural networks may not be accepted as trusted solutions in critical real-world applications. Using neural networks, an analyst might find it difficult to gain an understanding of the problem at hand because of the lack of an explanatory capability to provide insight into the charac-

teristics of the dataset. For the same reason, it is difficult to incorporate human expertise to simplify, accelerate, or improve the performance of image classification; a neural network always has to learn from scratch (Nauck et al., 1997). For neural networks to be widely applied in complex remote sensing image classification tasks, an explanation capability should eventually be an integral part of the functionality of a trained neural network (Andrews et al., 1995). A possible solution is to combine neural networks with fuzzy logic so that the knowledge residing in the hidden neurons of the network can be extracted in the form of fuzzy "if–then rules" (Qiu, 1999; Qiu and Jensen, 2004).

Neural Networks versus Expert Systems Developed Using Machine Learning

A summary of differences and similarities associated with a decision-tree expert system developed using machine learning and a neural network are summarized in Table 10-3 (Brand and Gerritsen, 2004a and b). Below are some additional considerations:

- A remote sensing project should not assume that one machine-learning algorithm is superior to another. There are simply too many parameters to be considered.

- Decision-tree expert systems may gain in popularity because they have fewer parameters to adjust than neural networks.

- Adjusting the parameters can have a dramatic affect (e.g., architecture of the ANN or decision tree). Don't expect a single method to work best for all cases.

- A hybrid system that incorporates both expert systems and neural networks may provide the best solution. For example, an expert system may be used to isolate areas where an ANN should focus (Liljeblad, 2003). It could also be used to identify optimum ANN architecture.

 References

Andrews, R., Diederich, J. and A. B. Tickle, 1995, "Survey and Critique of Techniques for Extracting Rules from Trained Artificial Neural Networks," *Knowledge-based Systems*, 8(6):379–389.

Argialas, D. and C. Harlow, 1990, "Computational Image Interpretation Models: An Overview and Perspective," *Photogrammetric Engineering & Remote Sensing*, 56(6):871–886.

Atkinson, P. M. and A. R. L. Tatnall, 1997, "Neural Networks in Remote Sensing," *International Journal of Remote Sensing*, 18(4):699–709.

Benediktsson, J. A. and J. R. Sveinsson, 1997, "Feature Extraction for Multisource Data Classification with Artificial Neural Networks," *International Journal of Remote Sensing*, 18:727–740.

Benediktsson, J. A., Swain, P. H. and O. K. Ersoy, 1990, "Neural Network Approaches Versus Statistical Methods in Classification of Multisource Remote Sensing Data," *IEEE Transactions on Geoscience and Remote Sensing*, 28:540–552.

Benitez, J. M., Castro, J. L. and I. Requena, 1997, "Are Artificial Neural Networks Black Boxes?" *IEEE Transactions on Neural Networks*, 8(5):1156–1164.

Bischof, H., Schneider, W. and A. Pi, 1992, "Multispectral Classification of Landsat Images Using Neural Networks," *IEEE Transactions on Geoscience and Remote Sensing*, 30:482–490.

Brand, E. and R. Gerritsen, 2004a, "Decision Trees," *Data Base Mining Solutions*, www.dbmasmag.com/9807m05. html.

Brand, E. and R. Gerritsen, 2004b, "Neural Networks," *Data Base Mining Solutions*, www.dbmasmag.com/9807m06. html.

Bratko, I., Kononenko, I., Lavrac, N., Mozetic, I. and E. Roskar, 1989, "Automatic Synthesis of Knowledge: Ljubljana Research," Y. Kodratoff and A. Hutchinson (Eds.), *Machine and Human Learning*, Columbia, MD: GP Publishing, 25–33.

Buchanan, B. G. and J. Lederberg, 1971, "The Heuristic DENDRAL Program for Explaining Empirical Data," *IFIP Congress* (1):179–188.

Civco, D. L., 1993, "Artificial Neural Network for Land Cover Classification and Mapping," *International Journal of Geographical Information System*, 7:179–186.

Congalton, R. G., 1988, "A Comparison of Sampling Schemes Used in Generating Error Matrices for Assessing the Accuracy of Maps Generated from Remotely Sensed Data," *Photogrammetric Engineering & Remote Sensing*, 54(5):593–600.

Darlington, K., 1996, "Basic Expert Systems," *Information Technology in Nursing*, London: British Computer Society, http://www.scism.sbu.ac.uk/~darlink.

DeFries, R. S. and J. C. Chan, 2000, "Multiple Criteria for Evaluating Machine Learning Algorithms for Land Cover Classification," *Remote Sensing of Environment*, 74:503–515.

Dreyfus, H. L. and S. E. Dreyfus, 2001, "From Socrates to Expert Systems: The Limits and Dangers of Calculative Rationality," *Selected Papers of Hubert Dreyfus*, Berkeley: Dept. of Philosophy, Regents of the University of California, March, http://ist-socrates.berkeley.edu/~hdreyfus/html/paper_socrates.html.

ERDAS/Leica Geosystems, 2002a, *ERDAS Field Guide*, 6th ed., Atlanta: ERDAS/Leica Geosystems, LLC, 237–239.

ERDAS/Leica Geosystems, 2002b, "Image Expert Classifier" in *ERDAS Imagine 8.6 Tour Guides*, Atlanta: ERDAS/Leica Geosystems, LLC, 559–594.

Fausett, L. V., 1994, *Fundamentals of Neural Networks*, Upper Saddle River, NJ: Prentice-Hall, 461 p.

Foody, G. M., 1996, "Fuzzy Modelling of Vegetation from Remotely Sensed Imagery," *Ecological Modeling*, 85:2–12.

Foody, G. M. and M. K. Arora, 1997, "An Evaluation of Some Factors Affecting the Accuracy of Classification by an Artificial Neural Network," *International Journal of Remote Sensing*, 18(4):799–810.

Foody, G. M., Lucas, R. M., Curran, P. J. and M. Honzak, 1997, "Non-linear Mixture Modelling without End-members Using an

Artificial Neural Network," *International Journal of Remote Sensing*, 18:937–953.

Foody, G. M., McCulloch, M. B. and W. B. Yates, 1995, "Classification of Remotely Sensed Data by an Artificial Neural Network: Issues Related to Training Data Characteristics," *Photogrammetric Engineering & Remote Sensing*, 61:391–401.

Friedl, M. A., Brodley, C. E. and A. H. Strahler, 1999, "Maximizing Land Cover Classification Accuracies Produced by Decision Trees at Continental to Global Scales," *IEEE Transactions on Geoscience and Remote Sensing*, 37:969–977.

Gahegan, M., 2003, "Is Inductive Machine Learning Just Another Wild Goose (or Might It Lay the Golden Egg)?" *International Journal of Geographical Information Science*, 17(1):69–92.

Giarratano, J. and G. Riley, 1994, *Expert Systems: Principles and Programming*, 2nd ed., Boston: PWS Publishing.

Gong, P., 1996, "Integrated Analysis of Spatial Data from Multiple Sources: Using Evidential Reasoning and Artificial Neural Network Techniques for Geological Mapping," *Photogrammetric Engineering and Remote Sensing*, 62:519–523.

Gong, P., Pu, R. and J. Chen, 1996, "Mapping Ecological Land Systems and Classification Uncertainties from Digital Elevation and Forest-cover Data Using Neural Networks," *Photogrammetric Engineering & Remote Sensing*, 62:1249–1260.

Hansen, M. C., DeFries, R. S., Townshend, J. R. G. and R. Sohlbert, 2000, "Global Land Cover Classification at 1 km Resolution Using a Classification Tree Approach," *International Journal of Remote Sensing*, 21:1331–1364.

Hansen, M. C., Dubayah, R. and R. S. DeFries, 1996, "Classification Trees: An Alternative to Traditional Land Cover Classifiers," *International Journal of Remote Sensing*, 17:1075–1081.

Hagan, M. T., Demuth, H. B., and Beale, M., 1996, *Neural Network Design*, Boston: PWS Publishing.

Haykin, S., 1994, *Neural Networks: A Comprehensive Foundation*, New York: Macmillan College Publishing, 696 p.

Hengl, T., 2002, "Neural Network Fundamentals: A Neural Computing Primer," *Personal Computing Artificial Intelligence*, 16(3):32–43, www.pcai.com.

Heermann, P. D. and N. Khazenie, 1992, "Classification of Multispectral Remote Sensing Data Using a Back-propagation Neural

Network," *IEEE Transactions on Geoscience and Remote Sensing*, 30(1):81–88.

Hewitson, B. C. and R. G. Crane, 1994, *Looks and Use: Neural Nets: Applications in Geography*, Hewitson, B. C. and R. G. Crane (Eds.), Boston: Kluwer Academic Publishers.

Hornik, K., Stingchcombe, M. and H. Whitee, 1989, "Multilayer Feedforward Networks Are Universal Approximators," *Neural Networks*, 2: 359–366.

Huang, X. and J. R. Jensen, 1997, "A Machine-Learning Approach to Automated Knowledge-base Building for Remote Sensing Image Analysis with GIS Data," *Photogrammetric Engineering & Remote Sensing*, 63(10):1185–1194.

Jackson, P., 1990, *Introduction to Expert Systems*, 2nd ed., Wokingham, England: Addison-Wesley.

Jensen, J. R., 1978, "Digital Land Cover Mapping Using Layered Classification Logic and Physical Composition Attributes," *American Cartographer*, 5(2):121–132.

Jensen, J. R., 2000, *Remote Sensing of the Environment: An Earth Resource Perspective*, Upper Saddle River, NJ: Prentice-Hall, 544 p.

Jensen, J. R., Qiu, F. and M. Ji, 1999, "Predictive Modeling of Coniferous Forest Age Using Statistical and Artificial Neural Network Approaches Applied to Remote Sensing Data," *International Journal of Remote Sensing*, 20(14):2805–2822.

Jensen, J. R., Qiu, F. and K. Patterson, 2001, "A Neural Network Image Interpretation System to Extract Rural and Urban Land Use and Land Cover Information from Remote Sensor Data," *Geocarto International*, 16(1):19-28.

Ji, C. Y., 2000, "Land-Use Classification of Remotely Sensed Data Using Kohonen Self-Organizing Feature Map Neural Networks," *Photogrammetric Engineering & Remote Sensing*, 66(12):1451–1460.

Jonsson, A., Morris, P., Muscettola, N., Rajan, K. and B. Smith, 2000, "Planning in Interplanetary Space: Theory and Practice," in *Proceedings, 5th International Conference on Artificial Intelligence Planning Systems (AIPS-00)*, Breckenridge, CO: AAAI Press, 177–186.

Kasabov, K. N., 1996, *Foundations of Neural Networks, Fuzzy Systems and Knowledge Engineering*, Cambridge, MA: MIT Press, 550 p.

Kim, B. and D. A. Landgrebe, 1991, "Hierarchical Classifier Design in High-dimensional, Numerous Class Cases," *IEEE Transactions on Geoscience and Remote Sensing*, 29:518–528.

Krapivin, V. F. and G. W. Phillips, 2001, "A Remote Sensing-Based Expert System to Study the Aral-Caspian Aquageosystem Water Regime," *Remote Sensing of Environment*, 75:201–215.

Kubat, M., Holte, R. C. and S. Matwin, 1998, "Machine Learning for the Detection of Oil Spills in Satellite Radar Images," *Machine Learning*, 30:195–215.

Lawrence, R. L. and A. Wright, 2001, "Rule-based Classification Systems Using Classification and Regression Tree (CART) Analysis," *Photogrammetric Engineering & Remote Sensing*, 7(10):1137–1142.

Lein, J. K., 1997, *Environment Decision Making: An Information Technology Approach*, Malden, MA: Blackwell Science, 213 p.

Liebowitz, J., 1988, *Introduction to Expert Systems*, Santa Cruz, CA: Mitchell Publishing, p. 3.

Liljeblad, D., 2003, *Enhancing Trading with Technology: A Neural Network – Expert System Hybrid Approach*, Unpublished Masters Thesis, Department of Informatics, Gothenburg University, Sweden, 45 p.

Lloyd, R. 1996, *Spatial Cognition, Geographic Environments*, Dordrecht, Netherlands: Kluwer Academic Publishers, 287 p.

Luger, G. F. and W. A. Stubblefield, 1993, *Artificial Intelligence: Structures and Strategies for Complex Problem Solving*, 2nd ed., Redwood City, CA: Benjamin/Cummings Publishing, 740 p.

Matsuyama, T., 1989, "Expert Systems for Image Processing: Knowledge-based Composition of Image Analysis Processes," *Computer Vision, Graphics and Image Processing*, 48(1):22–49.

McIver, D. K. and M. A. Friedl, 2002, "Using Prior Probabilities in Decision-Tree Classification of Remotely Sensed Data," *Remote Sensing of Environment*, 81:253–261.

Merriam-Webster, 2003, *Merriam-Webster Dictionary*, Springfield, MA: Merriam-Webster; http://www.m-w.com/dictionary.htm.

Michalski, R. S., 1983, "A Theory and Methodology of Inductive Learning," in Michalski, R. S., Carbonell, S. and T. M. Mitchell (Eds.), *Machine Learning*, Vol. 1, San Mateo, CA: Morgan Kaufmann Publishers.

Moller-Jensen, L., 1997, "Classification of Urban Land Cover Based on Expert Systems, Object Models and Texture," *Computers, Environment and Urban Systems*, 21(3/4):291–302.

Muchoney, D., Borak, J., Chi, H., Friedl, M., Gopal, S., Hodges, J., Morrow, N. and A. H. Strahler, 2000, "Application of MODIS Global Supervised Classification Model to Vegetation and Land Cover Mapping in Central America," *International Journal of Remote Sensing*, 21:1115–1138.

Nauck, D., Klawoon, F. and R. Kruse, 1997, *Foundation of Neuro-Fuzzy Systems*, Chichester, England: John Wiley & Sons, 320 p.

NeuralWare, 1996, *Using NeuralWorks*, Pittsburgh: NeuralWare, 157 p.

Pal, M. and P. M. Mather, 2003, "An Assessment of the Effectiveness of Decision Tree Methods for Land Cover Classification," *Remote Sensing of Environment*, 86:554–565.

PC AI, 2002, "Expert Systems," *Personal Computing Artificial Intelligence Electronic Magazine*, www.PCAI.com, Feb. 14, 2002.

Qiu, F., 1999, "Remote Sensing Image Classification Based on Automatic Fuzzy Rule Extraction by Integrating Fuzzy Logic and Neural Network Models," *Proceedings of the American Society for Photogrammetry & Remote Sensing Annual Conference* (CD-ROM), May 17–21, Portland, Oregon.

Qiu, F. and J. R. Jensen, 2004, "Opening the Black Box of Neural Networks for Remote Sensing Image Classification," *International Journal of Remote Sensing*, in press.

Quinlan, J. R., 1993, *C4.5: Programs for Machine Learning*, San Mateo, CA: Morgan Kaufmann Publishers.

Quinlan, J. R., 2003, *Data Mining Tools See5 and C5.0*, St. Ives NSW, Australia: RuleQuest Research, http://www.rulequest.com/see5-info.html.

Rao, V. B. and H. V. Rao. 1993, *C^{++} Neural Network and Fuzzy Logic*, New York: Management Information, 408 p.

Rich, E. and K. Knight, 1991, *Artificial Intelligence,* 2nd ed., New York: McGraw-Hill.

Roli, F., Serpico, S. B. and G. Vernazza, 1996, "Neural Networks for Classification of Remotely Sensed Images," in C. H. Chen (Ed.), *Fuzzy Logic and Neural Network Handbook*, New York: McGraw-Hill, 15.1–15.28.

Russell, S. J. and P. Norvig, 2003, *Artificial Intelligence: A Modern Approach*, 2nd ed., Upper Saddle River, NJ: Prentice-Hall, 1080 p.

Stefanov, W. L., Ramsey, M. S. and P. R. Christensen, 2001, "Monitoring Urban Land Cover Change: An Expert System Approach to Land Cover Classification of Semiarid to Arid Urban Centers," *Remote Sensing of Environment*, 77:173–185.

Stow, D., Coulter, L., Kaiser, J., Hope, A., Service, D., Schutte, K. and A. Walters, 2003, "Irrigated Vegetating Assessments for Urban Environments," *Photogrammetric Engineering and Remote Sensing*, 69(4):381–390.

Swain, P. H. and H. Hauska, 1977, "The Decision Tree Classifier: Design and Potential," *IEEE Transactions on Geoscience and Remote Sensing*, 15:142–147.

Tso, B. and P. M. Mather, 2001, *Classification Methods for Remotely Sensed Data*, New York: Taylor & Francis, 332 p.

Tullis, J. A. and J. R. Jensen, 2003, "Expert System House Detection in High Spatial Resolution Imagery Using Size, Shape, and Context," *Geocarto International*, 18(1):5–15.

Turing, A. M., 1950, "Computing Machinery and Intelligence," *Mind*, 59:439–460.

Werbos, P. J., 1974, *Beyond Regression: New Tools for Prediction and Analysis in Behavioral Sciences*, Ph.D. Thesis, Cambridge, MA: Harvard University.

Waterman, D., 1986, *A Guide to Expert Systems*, Reading, MA: Addison-Wesley, p. xvii.

Yang, C. and P. Chung, 2002, "Knowledge-based Automatic Change Detection Positioning System for Complex Heterogeneous Environments," *Journal of Intelligent and Robotic Systems*, 33:85–98.

Zhang, Q. and J. Wang, 2003, "A Rule-based Urban Land Use Inferring Method for Fine-resolution Multispectral Imagery," *Canadian Journal of Remote Sensing*, 29(1):1–13.

Thematic Information Extraction: Hyperspectral Image Analysis

11

I*maging spectrometry* is defined as:

> the simultaneous acquisition of images in many relatively narrow, contiguous and/or noncontiguous spectral bands throughout the ultraviolet, visible, and infrared portions of the electromagnetic spectrum.

Most multispectral remotely sensed data obtained to date have been acquired in 3 to 10 spectral bands with relatively broad bandwidths [e.g., aerial photography, Landsat Multispectral Scanner (MSS), Landsat Thematic Mapper (TM, ETM$^+$), SPOT (HRV), IKONOS, QuickBird]. Conversely, *hyperspectral* remote sensing systems typically collect data in at least 10 spectral bands with relatively narrow bandwidths (Baltsavias, 2002). Examples include the Moderate Resolution Imaging Spectrometer (MODIS) onboard NASA's *Terra* and *Aqua* satellites with 36 bands and NASA's Airborne Visible/Infrared Imaging Spectrometer (AVIRIS) with 224 bands. However, there is no universally agreed upon minimum number of bands or bandwidth dimension required for a dataset to be considered hyperspectral.

The value of using a hyperspectral imaging spectrometer lies in its ability to provide a high-resolution reflectance spectrum for each picture element in the image (Goetz et al., 1985; Karaska et al., 2004). Many, although not all, surface materials have diagnostic absorption features that are only 10 to 20 nm wide. Therefore, hyperspectral sensors that acquire data in many contiguous or noncontiguous 10-nm bands throughout the 400 to 2500-nm region of the electromagnetic spectrum sometimes capture spectral data with sufficient resolution for the direct identification of those materials (Curran et al., 1998; Jensen, 2000).

 Multispectral versus Hyperspectral Data Collection

Many multispectral remote sensing systems use a scanning mirror and focus the reflected or emitted radiant flux onto just a few detectors that are sensitive to relatively broad portions of the electromagnetic spectrum. To improve image geometry, some multispectral remote sensing systems use linear array technology to obtain data [e.g., the SPOT Image High Resolution Visible (HRV) sensor system and the Indian LISS]. Multispectral remote sensing systems are reviewed in Chapter 2.

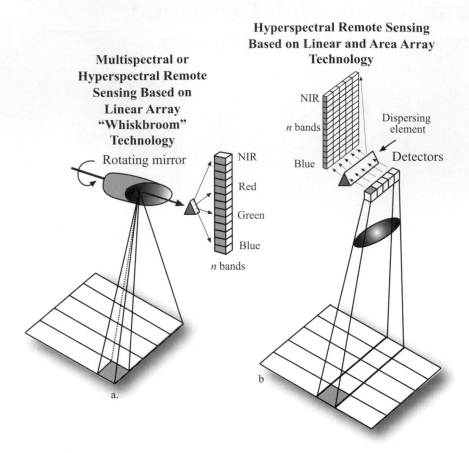

Figure 11-1 a) Multispectral or hyperspectral "whiskbroom" remote sensing system based on linear array detector technology and a scanning mirror (e.g., AVIRIS). b) A hyperspectral instrument based on linear and area array technology.

The collection of spectral reflectance and emittance information in hundreds of spectral bands requires different approaches to sensor system design. Two approaches to imaging spectrometry are shown in Figure 11-1. The "whiskbroom" scanner linear array approach is analogous to the scanner approach used for Landsat MSS or Landsat ETM+, except that radiant flux from within the IFOV is passed onto a spectrometer, where it is dispersed and focused onto a linear array of detectors (Figure 11-1a). The terrain within each IFOV (i.e., pixel) is sensed in as many spectral bands as there are detector elements in the linear array. This is basically how NASA's AVIRIS collects data. Other imaging spectrometers make use of linear and two-dimensional area arrays of detectors. In this situation, there is a dedicated column of spectral detector elements for each cross-track pixel in the scene (Figure 11-1b). This configuration improves image geometry and radiometry because a scanning mirror is not used, allowing each detector to dwell longer on an IFOV, thereby resulting in a more accurate recording of the radiant flux as it leaves the landscape.

Selected *suborbital airborne* hyperspectral remote sensing systems include AVIRIS (JPL, NASA), DAIS 7915 (DLR), CASI and CASI 2 (Itres, Inc., Canada), HyMap (Integrated Spectronics, Inc., Australia), ROSOS (DLR), AISA (Specim, Finland), HYDICE (NRL, USA), EPS A and H Series (GER, Inc., USA), ASAS (GSFC, NASA), SMOFTS (Hawaii Institute of Geophysics and Planetology). Selected *spaceborne* hyperspectral systems include MERIS on Envisat (European Space Agency), MODIS on EOS *Terra* and *Aqua* (NASA), MOS on IRS-3P (India), Hyperion on EO-1 (NASA), CHRIS on PROBA (European Space Agency), and COIS on NEMO (US Navy) (Baltsavias, 2002).

Traditional broadband remote sensing systems such as Landsat MSS and SPOT HRV *undersample* the information available from a reflectance spectrum by making only a few measurements in spectral bands sometimes several hundred nanometers wide. Conversely, imaging spectrometers sample at close intervals (bands typically on the order of tens of nanometers wide) and have a sufficient number of spectral

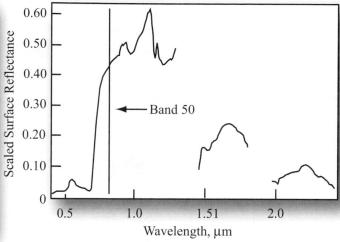

a. Subset of Run 08 Scene 05 AVIRIS
band 50 data (3.4 x 3.4 m).

b. Scaled surface reflectance curve for a
pixel of loblolly pine (*Pinus taeda*).

Figure 11-2 a) Band 50 subset of Airborne Visible/Infrared Imaging Spectrometer (AVIRIS) imagery obtained on July 26, 1999 at the Savannah River Site near Aiken, SC. b) Scaled surface reflectance spectrum for a pixel of loblolly pine (*Pinus taeda*) forest at the location shown in (a).

bands to allow construction of *spectra* that closely resemble those acquired by laboratory spectroradiometers. Analysis of imaging spectrometer data allows extraction of a detailed spectrum for each picture element in the image. Such spectra may allow direct identification of specific materials within the IFOV of the sensor based upon their reflectance/absorptance characteristics, including minerals, atmospheric gases, vegetation, snow and ice, and dissolved matter in water bodies (e.g., Kruse, 1994; Nolin and Dozier, 2000; Haboudane et al., 2002). For example, Figure 11-2 depicts the scaled surface reflectance characteristics of hyperspectral data obtained over a loblolly pine (*Pinus taeda*) forest on the Savannah River Site near Aiken, SC, within the wavelength interval of 400 to 2500 nm.

Steps to Extract Information from Hyperspectral Data

Scientists have identified digital image processing methods that may be applied in a relatively straightforward sequence to extract useful information from hyperspectral remote sensor data. Many of the operations identified in Figure 11-3 are discussed in this chapter.

The analysis of hyperspectral data usually requires sophisticated digital image processing software (e.g., ENVI, the Environment for Visualizing Images; ERDAS Imagine; IDRISI, PCI Geomatica). This software must be able to:

1. calibrate (convert) the raw hyperspectral at-sensor radiance to apparent surface reflectance and ultimately to surface reflectance. This requires the removal of atmospheric attenuation, topographic effects due to slope and aspect, and any sensor system–induced electronic anomalies.

2. analyze the calibrated remote sensing–derived scaled surface reflectance data to determine its constituent materials by a) deriving endmembers from the hyperspectral data itself, b) comparing *in situ* spectroradiometric data obtained in the field at the time of the overflight with hyperspectral image characteristics, and/or c) comparing hyperspectral image characteristics with a library of laboratory-based spectra such as that provided by Johns Hopkins University, the U.S. Geological Survey, and NASA's Jet Propulsion Laboratory (ASTER Spectral Library, 2004).

To understand how to analyze hyperspectral data, it is useful to examine the image processing procedures in the context of an actual application. This chapter summarizes how high spatial (3.4 × 3.4 m) and spectral (224 bands) resolution AVIRIS remote sensor data were analyzed to detect possible

> **General Steps Used to Extract**
> **Information from Hyperspectral Data**

State the nature of the information extraction problem.
    * Specify the geographic region of interest.
    * Define the classes or biophysical materials of interest.
Acquire appropriate remote sensing and initial ground reference data.
    * Select remotely sensed data based on the following criteria:
        - Remote sensing system considerations:
            - Spatial, spectral, temporal, and radiometric resolution
        - Environmental considerations:
            - Atmospheric, soil moisture, phenological cycle, etc.
    * Obtain initial ground reference data based on:
        - a priori knowledge of the study area
Process hyperspectral data to extract thematic information.
    * Subset the study area from the hyperspectral data flight line(s).
    * Conduct initial image quality assessment:
        - Visual individual band examination
        - Visual examination of color composite images
        - Animation
        - Statistical individual band examination; signal-to-noise ratio
    * Radiometric correction:
        - Collect necessary *in situ* spectroradiometer data (if possible)
        - Collect *in situ* or environmental data (e.g., using radiosondes)
        - Perform pixel-by-pixel correction (e.g., ACORN)
        - Perform band-by-band spectral polishing
        - Empirical line calibration
    * Geometric correction/rectification:
        - Use onboard navigation and engineering data (GPS
          and Inertial Navigation System information)
        - Nearest-neighbor resampling
    * Reduce the dimensionality of hyperspectral dataset:
        - Minimum Noise Fraction (MNF) transformation
    * End-member determination - locate pixels with relatively pure
      spectral characteristics:
        - Pixel Purity Index (PPI) mapping
        - N-dimensional end-member visualization
    * Methods of mapping and matching using hyperspectral data:
        - Spectral Angle Mapper (SAM)
        - Subpixel classification (linear spectral unmixing)
        - Spectroscopic library matching techniques
        - Matched filter or mixture-tuned matched filter
        - Indices developed for use with hyperspectral data
        - Derivative spectroscopy
Perform accuracy assessment (Chapter 13).
    * Select method:
        - Qualitative confidence-building
        - Statistical measurement
    * Determine number of observations required by class.
    * Select sampling scheme.
    * Obtain ground reference test information.
    * Create and analyze error matrix:
        - Univariate and multivariate statistical analysis
Accept or reject previously stated hypothesis.
Distribute results if accuracy is acceptable.

Figure 11-3 A generalized flow chart useful for the extraction of quantitative biophysical and thematic information from hyperspectral remote sensor data.

Table 11-1. Characteristics of the NASA Jet Propulsion Laboratory Airborne Visible/Infrared Imaging Spectrometer remote sensing system.

Sensor	Technology	Spectral Range (nm)	Nominal Channel Bandwidth (nm)	Data-collection Mode	Dynamic Range (bits)	IFOV (mrad)	Total Field of View
AVIRIS	Whiskbroom linear array	400 – 2500	10	224 bands	16	1.0	30°

stress associated with Bahia grass–covered clay caps on waste sites on the U.S. Department of Energy Savannah River Site (SRS) near Aiken, SC.

NASA's Airborne Visible/Infrared Imaging Spectrometer

Characteristics of the AVIRIS sensor system are summarized in Table 11-1. AVIRIS acquires images in 224 bands, each 10 nm wide in the 400- to 2500-nm region of the electromagnetic spectrum, using a whiskbroom scanning mirror and linear arrays of silicon (Si) and indium-antimonide (InSb), as configured in Figure 11-1a (Green et al., 1998; Analytical Imaging and Geophysics, 1999). The sensor is typically flown onboard the NASA ER-2 aircraft at 20 km above ground level (AGL) and has a 30° total field of view and an instantaneous field of view (IFOV) of 1.0 mrad, which yields 20 × 20 m pixels. The data are recorded in 16 bits.

In 1999, the AVIRIS was also flown in low-altitude mode (AVIRIS-LA) on a NOAA De Havilland DH-6 Twin Otter aircraft that acquired data at an altitude of 12,500 ft AGL. This resulted in hyperspectral data with a spatial resolution of approximately 3.4 × 3.4 m (Green et al., 1998; JPL, 1999). Figure 11-4 depicts a single band of AVIRIS 3.4 × 3.4 m hyperspectral imagery (band 30; 655.56 nm) obtained over the Savannah River Site near Aiken, SC, and radiance data extracted for a single pixel.

Both public (e.g., NASA Jet Propulsion Laboratory) and commercial (e.g., HyMap, CASI 2) data providers often preprocess the hyperspectral data before transferring it to the user for image analysis. For example, personnel at NASA's Jet Propulsion Laboratory Data Facility preprocess the raw digital AVIRIS data to remove fundamental geometric and radiometric errors associated with the AVIRIS off-nadir data collection and the roll, pitch, and yaw of the aircraft platform at the time of data acquisition (Clark et al., 1998; Boardman, 1999).

Subset Study Area from Flight Lines

The aforementioned geometric corrections are applied to each flight line (often referred to as a *run*) of AVIRIS data (Boardman, 1999). Each AVIRIS flight line is then subset into individual scenes each 512 pixels long by 746 pixels wide. The individual scenes may be viewed via the Internet as quick-look image files. Scenes 04, 05, and 06 extracted from run 08 obtained over the Savannah River Site on July 26, 1999, are displayed in **Color Plate 11-1**. There is no overlap of the successive scenes. Therefore, it is a relatively straightforward task to mosaic the scenes into a single flight line. A datacube of run 8 scene 05 is shown in **Color Plate 11-1d**. The dark regions of the 224-band datacube correspond to atmospheric absorption bands.

Initial Image Quality Assessment

It is important to perform an initial assessment of the quality of the hyperspectral data. A number of methods can be used to assess image quality including visual examination of individual bands of imagery, examination of multiple-band color composites, viewing of an animated movie of the individual bands, and statistical evaluation of individual bands including a quantitative evaluation of individual band signal-to-noise ratios (SNRs).

Visual Individual Band Examination

Although it is a tedious process, sometimes there is simply no substitute for painstakingly viewing each band of data to judge its quality. Many bands appear crisp and sharp with good grayscale tonal variation. For example, Figure 11-5a depicts AVIRIS band 50 (centered at 0.8164 μm) from run 08 scene 05. Compare that with images of AVIRIS band 112 (1.405 μm) and band 163 (1.90 μm) shown in Figure 11-5b and c. These images were obtained in two of the prominent

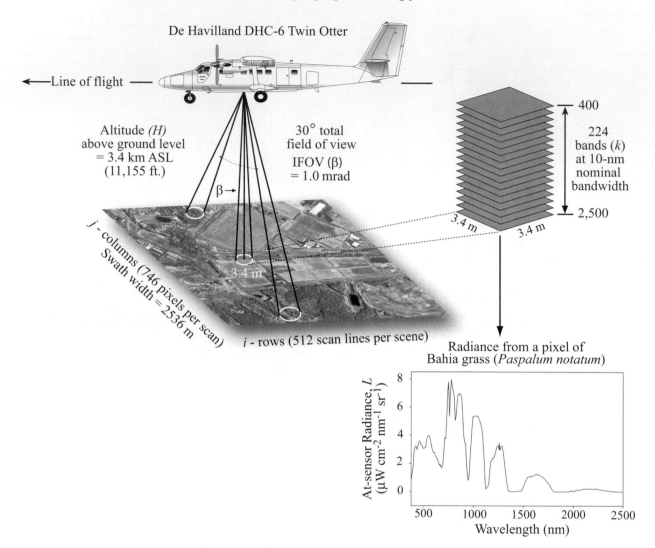

Figure 11-4 Conceptual diagram of low-altitude imaging spectroscopy performed using the NASA Jet Propulsion Laboratory's Airborne Visible/Infrared Imaging Spectrometer (AVIRIS). The scanner mirror focuses radiant flux onto linear arrays that contain 224 detector elements with a spectral sensitivity ranging from 400 to 2,500 nm. A spectrum of radiance (L) is obtained for each picture element. The radiance data can be processed to yield percent reflectance information. Most AVIRIS overflights obtain data at 20×20 m pixels. Sometimes the AVIRIS sensor is flown on other aircraft and at different altitudes above ground level. This particular AVIRIS scene of waste sites on the Savannah River Site near Aiken, SC, was acquired at 11,155 ft above sea level in a De Havilland DHC-6 Twin Otter on July 26, 1999. This resulted in 3.4×3.4 m spatial resolution data. Only the band 30 (centered on 655.36 nm) image is displayed (adapted from Filippi, 1998).

atmospheric absorption windows. Therefore, it is not surprising that they are of poor quality. The analyst normally keeps a list of bands that exhibit considerable atmospheric noise. Such bands are often deleted from further analysis, as described in subsequent sections. However, the information in the atmospheric absorption bands may be used by certain atmospheric correction algorithms.

It is very important to determine early in a project if certain bands or particular regions within a given band contain null data values (e.g., values of –9999) or if there are serious line dropouts (i.e., an entire line has a value of –9999). Also, an examination of the individual bands and color composites provides information on whether there are serious geometric problems with the data such as extreme image warping.

Assessing the Quality of Individual AVIRIS Scenes Band by Band

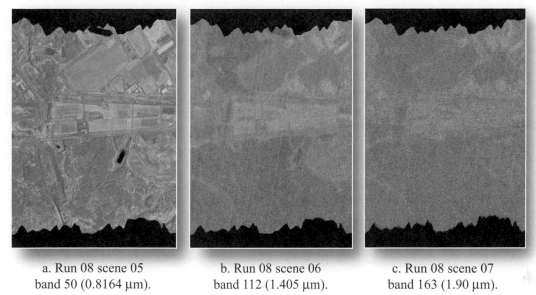

a. Run 08 scene 05	b. Run 08 scene 06	c. Run 08 scene 07
band 50 (0.8164 µm).	band 112 (1.405 µm).	band 163 (1.90 µm).

Figure 11-5 Individual bands of AVIRIS data from run 08 scenes 05, 06, and 07. a) An image of band 50 after contrast stretching. b) A contrast-stretched image of band 112. c) A contrast-stretched image of band 163. The near-infrared band 50 image is relatively free of attenuation, while bands 112 and 163 exhibit the effects of considerable atmospheric attenuation.

Deep notches along the edges of a scene often indicate spatial image distortion which must be corrected during the geometric rectification phase of the project.

Visual Examination of Color Composite Images Consisting of Three Bands

One of the most useful ways to determine if hyperspectral imagery has information of value is to select representative bands from the near-infrared (e.g., AVIRIS band 50), red (e.g., band 30), and green (e.g., band 20) portions of the spectrum and create a false color composite of the scene (**Color Plate 11-1**). Hopefully, the individual bands are co-registered and contain spectral information of value. The analyst can select any color composite of three bands to evaluate. Looking at numerous three-band color composites from the 224 possible bands can be a tedious process. However, the exercise does provide valuable qualitative information about the individual scenes and bands in the hyperspectral dataset.

Animation

Hyperspectral datasets usually contain hundreds of bands. Therefore, most image processing systems designed for

hyperspectral analysis have an image animation function whereby the analyst selects a certain time rate at which individual bands are displayed on the screen, e.g., every 5 seconds. Examination of hyperspectral bands in this manner allows the analyst to 1) identify individual bands that have serious atmospheric attenuation or electronic noise problems, as previously discussed, and 2) determine if any misregistration (offset) of bands exists.

Occasionally, linear features in the scene such as roads or the edges of water bodies may appear to shift slightly when you are viewing the animated bands. This does not necessarily indicate that the bands are misregistered, but rather that the detectors in one band are more or less sensitive than those in an adjacent band. Sometimes with AVIRIS data there is a slew effect in the detector array readout, which translates to a shift of the IFOV in areas of high spatial frequency. If the displacement appears to be severe, then a more detailed examination should be conducted. This might entail creating color composites of multiple bands to determine if there is actually band-to-band misregistration.

Statistical Individual Band Examination

Examination of the univariate statistics (e.g., mean, median, mode, standard deviation, range) of the individual bands can be helpful when assessing image quality. For example, if a band's 16-bit brightness values are constrained to a very small range with a very small standard deviation, this may indicate a serious problem. When a band is believed to have problems, there is no substitute for analyzing its univariate statistics and histogram. In order to detect many absorption features associated with materials found in nature, a noise level that is approximately an order of magnitude smaller than the absorption depth is required. Therefore, evaluation of each band's signal-to-noise ratio is of value (Goetz and Calvin, 1987; Curran and Dungan, 1989; Wolf, 1997).

 Radiometric Calibration

As discussed in Chapter 9, many remote sensing investigations do not concern themselves with radiometric correction of the data to remove the deleterious effects of atmospheric absorption and scattering. However, to use hyperspectral remote sensor data properly, it is generally accepted that the data must be radiometrically corrected. In addition to removing atmospheric effects, this process normally involves transforming the hyperspectral data from at-sensor radiance, L_s *(μW cm-2 nm-1 sr-1)* to scaled surface reflectance (Mustard et al., 2001). This allows the remote sensor–derived spectral reflectance data (often referred to as *spectra*) to be quantitatively compared with *in situ* spectral reflectance data obtained on the ground using a handheld spectroradiometer or with laboratory-derived spectra (Figure 11-6a). Such spectroradiometric data are often stored in spectral libraries.

In Situ Data Collection

Whenever possible, it is desirable to obtain handheld *in situ* spectroradiometer measurements on the ground at approximately the same time as the remote sensing overflight (Clark et al., 2002). These measurements should be made with well-respected spectroradiometers (e.g., Spectron Engineering, Analytical Spectral Devices) that cover the same spectral range as the hyperspectral remote sensing system (e.g., 400 to 2500 nm). Ideally, several calibrated spectroradiometers are used to collect data of the most important materials in the study area at approximately the same time of day and under the same atmospheric conditions as the remote sensing data collection. If this is not possible, then data should be collected at the same time of day on days before and after the data collection. Of coarse, each spectroradiometer measurement taken in the field should be calibrated using a standard reference panel (e.g., Spectralon). Spectroradiometer measurements made in the laboratory under controlled illumina-

tion conditions are also valuable as long as the materials (especially vegetation samples) have been properly maintained and analyzed as soon as possible. *In situ* spectroradiometer spectral reflectance measurements can be useful in the successful transformation of AVIRIS-type hyperspectral data from radiance to scaled surface reflectance.

Radiosondes

A radiosonde can provide valuable information about atmospheric temperature, pressure, relative humidity, wind speed, ozone, and wind direction aloft. A radiosonde launched at the approximate time of remote sensing data collection is very desirable (Figure 11-6c).

Radiative Transfer-based Atmospheric Correction

Ideally, the analyst knows the exact nature of the atmospheric characteristics above each picture element at the time of hyperspectral data collection, e.g., barometric pressure, water vapor, amount of atmospheric molecular (Rayleigh) scattering, etc. Unfortunately, the atmosphere is variable even over relatively short geographic distances and time periods. One method of atmospheric correction uses the remote sensing–derived radiance data in very selective narrow bands to infer information about the atmospheric conditions above each pixel. This information is then used to remove atmospheric effects from each pixel. Three robust algorithms used to remove the effects of atmospheric attenuation from individual pixels of hyperspectral data are:

• ATmosphere REMoval (ATREM),

• Atmospheric CORrection Now (ACORN), and

• ATmospheric CORrection (ATCOR).

ATREM [Center for the Study of Earth from Space (CSES), University of Colorado] was based on the radiative transfer code *Second Simulation of the Satellite Signal in the Solar Spectrum (6S)* (Tanre et al., 1986). It is no longer supported but can be used to introduce radiometric correction principles. Most commercially available radiative transfer-based atmospheric correction programs (e.g., ACORN and ATCOR) are based on the MODTRAN radiative transfer code (Berk et al., 1989; 1998). An example of ACORN atmospheric correction of hyperspectral data is provided in Chapter 6. This section demonstrates the use of ATREM to correct the 1999 Savannah River Site hyperspectral AVIRIS data.

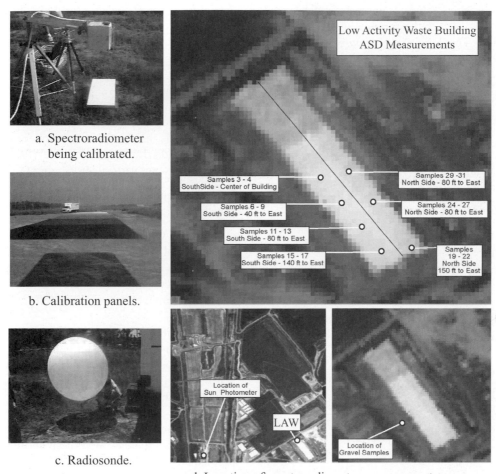

a. Spectroradiometer being calibrated.

b. Calibration panels.

c. Radiosonde.

d. Location of spectroradiometer measurements on a rooftop, a gravel site next to the building, and the location of a nearby Sun photometer.

Figure 11-6 *In situ* spectroradiometer measurements can be very valuable when used to modify radiative transfer modeling results. They are especially useful for empirical line calibration and as *boost spectra*. a) A spectroradiometer being calibrated prior to a Bahia grass measurement. b). Calibration targets. c) A radiosonde launched at approximately the same time as the hyperspectral data collection. Atmospheric visibility, temperature, pressure, relative humidity, wind speed, and ozone are valuable for use in per-pixel atmospheric correction programs such as ACORN, ATCOR, and ATREM. d) The location of numerous *in situ* spectroradiometer measurements obtained on top of the Low Activity Waste (LAW) building on the Savannah River Site and adjacent to the building (Jensen et al., 2003).

There are approximately 30 gases in the atmosphere. Most of them do not produce observable absorption features in the 400- to 2500-nm region at AVIRIS 10-nm spectral resolution. However, seven gases (water vapor, H_2O; carbon dioxide, CO_2; ozone, O_3; nitrous oxide, N_2O; methane, CH_4; carbon monoxide, CO; and oxygen, O_2) do produce observable features in remotely sensed images obtained in this region (Gao et al., 1999). The impact of these gases on the atmospheric transmission of energy in the wavelength interval from 400 to 2500 nm is summarized in Figure 6-14.

The scattering and absorption processes in the atmosphere are commonly referred to as the atmospheric radiative transfer processes. ATREM calculates the amount of molecular (Rayleigh) scattering and absorption present using the radiative transfer code *6S* and user-specified aerosol and atmospheric models (Research Systems, Inc., 2002). The following paragraphs describe how at-sensor radiance (L_s) is transformed into *apparent reflectance* ($\rho*$) and then into *scaled surface reflectance* $\rho(\lambda)$, which is sufficient for the topographically level Savannah River Site study area.

Using the *6S* code, the apparent reflectance, $\rho^*(\theta_o, \phi_o, \theta_v, \phi_v, \lambda)$, is used in the formulation of the radiative transfer problem. The definition of *apparent reflectance* is (Gao et al., 1999):

$$\rho^*(\theta_o, \phi_o, \theta_v, \phi_v, \lambda) = \frac{\pi L_s(\theta_o, \phi_o, \theta_v, \phi_v, \lambda)}{\mu_o E_o(\lambda)} \quad (11\text{-}1)$$

where

θ_o = the solar zenith angle
ϕ_o = the solar azimuth angle
θ_v = the sensor zenith angle
ϕ_v = the sensor azimuth angle
λ = wavelength
L_s = total radiance measured by the sensor
E_o = the solar irradiance at the top of the atmosphere
$\mu_o = \cos\theta_o$.

For a horizontal surface with uniform Lambertian reflectance, apparent reflectance can also be expressed as (Tanre et al., 1986):

$$\rho^*(\theta_o, \phi_o, \theta_v, \phi_v, \lambda) = T_g(\theta_o, \theta_v, \lambda) \times \quad (11\text{-}2)$$

$$\left[\rho_a(\theta_o, \phi_o, \theta_v, \phi_v, \lambda) + \frac{T(\theta_o, \lambda)T(\theta_v, \lambda)\rho(\lambda)}{1 - \rho(\lambda)S(\lambda)} \right]$$

where

$$T_g(\theta_o, \theta_v, \lambda) = \prod_{i=1}^{n} T_{g_i}(\theta_o, \theta_v, \lambda)$$

and

T_g = the gaseous transmittance in the Sun-surface-sensor path
ρ_a = the atmospheric reflectance
$T(\theta_o)$ = the downward scattering transmittance
$T(\theta_v)$ = the upward scattering transmittance
S = the spherical albedo of the atmosphere
$\rho(\lambda)$ = the surface reflectance
ρ^* = the apparent reflectance
T_{g_i} = the transmittance of the ith gas in the Sun-surface-sensor path
n = the number of gases.

Equation (11-2) can be solved for *surface reflectance* using the equation (Teillet, 1989; Gao et al., 1999):

$$\rho(\lambda) = \frac{\left(\dfrac{\rho^*(\theta_o, \phi_o, \theta_v, \phi_v, \lambda)}{T_g(\theta_s, \theta_v, \lambda)} - \rho_a(\theta_o, \phi_o, \theta_v, \phi_v, \lambda) \right)}{T(\theta_s, \lambda)T(\theta_v, \lambda) + S(\lambda)\left(\dfrac{\rho^*(\theta_o, \phi_o, \theta_v, \phi_v, \lambda)}{T_g(\theta_o, \theta_v, \lambda)} - \rho_a(\theta_o, \phi_o, \theta_v, \phi_v, \lambda) \right)} . \quad (11\text{-}3)$$

Therefore, given at-sensor measured radiance, L_s, the apparent surface reflectance (ρ^*) is calculated using Equation 11-1. Scaled surface reflectance $\rho(\lambda)$ is then computed using Equation 11-3.

The scattering terms, ρ_a, $T(\theta_o)$, $T(\theta_v)$, and S, are calculated using the *6S* code (Vermote et al., 1996) with a user-selected aerosol model. The absorption term T_g is calculated using the Malkmus narrow-band spectral model with a user-selected standard atmospheric model (temperature, pressure, and water vapor vertical distributions) or a user-supplied atmospheric model (Gao et al., 1999). The algorithm assumes that the amounts of six gases (CO_2, O_3, N_2O, CO, CH_4, and O_2) are uniform across the scene. The amount of atmospheric water vapor is derived pixel by pixel from the hyperspectral data using the 0.94-μm and the 1.14-μm water vapor bands and a three-channel ratioing technique. The derived water vapor values are then used to model water vapor absorption effects over the 400- to 2500-nm region.

The output of ATREM consists of scaled surface reflectance data stored in a hyperspectral image cube that has the same dimensions as the input image cube (**Color Plate 11-1d**). In addition, it provides a separate single-band water vapor image with the same spatial dimensions as the input data.

ATREM allows the use of only a single mean elevation value to characterize the elevation of an entire scene. This is not problematic in areas like the Savannah River Site where the variation in local relief is only about 100 m. However, in mountainous terrain the significant differences in local relief, slope, and aspect should be accounted for. In such cases the scaled surface reflectance data can be converted to absolute surface reflectance if surface topography (a digital elevation model) is known.

ATREM Input Parameters and Radiative Transfer Modeling Results

To run ATREM, the analyst provides several types of information. The following parameters were used to atmospherically correct AVIRIS data from the Savannah River Site:

1) altitude of the sensor at the time of data collection (e.g., 3.4 km above sea level);

2) date and time of data collection (GMT);

3) the geographic center of each scene (latitude and longitude);

4) wavelength for the channel center of each band;

5) the full-width-half-maximum (FWHM) value for each band used in the analysis;

6) the channel ratio input parameters used to assess the atmospheric constituents on a pixel-by-pixel basis. If the site contains a large amount of iron-rich soils and minerals, such as in the southeastern United States, then it is good practice to use only the 1.14-μm water vapor absorption band. If iron-dominated soils do not exist in a scene, both the 0.94 and 1.14 μm bands may be used;

7) a standard atmospheric model (e.g., mid-latitude summer);

8) the selection of atmospheric gases to be modeled;

9) total column ozone amount;

10) an aerosol model (e.g., continental) and the average atmospheric visibility (e.g., 8 km);

11) average surface elevation (e.g., 50 m ASL) and

12) a scaled, calibrated radiance hyperspectral dataset (e.g., AVIRIS, HyMap, HYDICE, CASI 2).

To appreciate the importance of atmospherically correcting the hyperspectral data prior to data analysis, consider the application of the ATREM program to run 8 scene 05 of the Savannah River Site. An example of a raw AVIRIS loblolly pine (*Pinus taeda*) spectrum for a pixel located at 3,683,437 Northing and 437,907 Easting UTM (sample 198, line 123) is shown in Figure 11-7a. The *y*-axis measures at-sensor radiance, L (μW cm^{-2} nm^{-1} sr^{-1}). For all practical purposes, the raw radiance spectra is not very informative for Earth surface inquiries. Notice the dramatic increase in radiance in the shorter wavelength (blue) portion of the spectrum caused by increased atmospheric Rayleigh scattering. Basically, the radiance spectra do not resemble the spectra that should be associated with a healthy, photosynthesizing pine canopy.

Application of the ATREM program to the raw AVIRIS data resulted in the scaled surface reflectance spectrum shown in Figure 11-7b. The ATREM-corrected spectra appears as it should, with chlorophyll absorption bands in the blue and red, a peak in the green portion of the spectrum, increased reflectance throughout the near-infrared, and depressions surrounding the atmospheric absorption bands. Note that discontinuities in the spectra represent atmospheric absorption bands that are not plotted.

The quality of the atmospheric correction and radiometric calibration to scaled surface reflectance is very important. If the atmospheric correction and radiometric calibration are poor, it will be difficult for the analyst to compare remote sensing–derived spectra with *in situ*–derived spectra stored in spectral libraries.

Band-by-Band Spectral Polishing

Notice in Figure 11-7b that there is noise in the loblolly pine (*Pinus taeda*) spectrum even after the data were atmospherically corrected on a pixel-by-pixel basis. This is because there is still cumulative error in the hyperspectral dataset due to sensor system anomalies and the limited accuracy of the standards, measurements, and models used and calibrations performed along the signal processing chain. This cumulative error can be as much as a few percent in each band of data.

There are additional techniques that can be used to remove some of this cumulative error. Normally this involves deriving a mild linear correction that is applied to the entire band and the entire scene at one time (i.e., it is not a pixel-by-pixel correction). The methods "polish" out some of this error in an attempt to improve the accuracy of the scaled surface reflectance data (Boardman, 1997).

One useful algorithm is the Empirical Flat Field Optimal Reflectance Transformation, or EFFORT (Boardman, 1998). It requires ATREM-corrected data as input. In addition, it can incorporate *in situ* spectroradiometer spectra for those materials most commonly in the scene. For example, EFFORT was run using the ATREM-corrected data previously discussed. In addition, two *in situ* spectroradiometer sweet gum (*Liquidambar styraciflua*) spectra (one representing an average of multiple individual spectra, and one unaveraged spectrum) and two loblolly pine spectra were also input. This set of four (4) spectra was applied to the ATREM-corrected bands. An EFFORT treatment was also applied using a laboratory-derived Bahia grass spectrum, but it was applied to only a limited number of visible and near-infrared channels due to its limited spectral range (400 – 800 nm). Bahia grass was judged to be important because it is grown on many of the clay-caps. Loblolly pine and sweet gum are important because they are the vegetation types often impacted downslope of clay-capped SRS waste sites.

These *in situ* spectral reflectance measurements are sometimes referred to as "reality boost spectra." Reality boost spectra with sharp edges such as the vegetation red edge tend to yield the best EFFORT-corrected hyperspectral data. Fig-

Pixel-by-Pixel Atmospheric Correction of AVIRIS Data

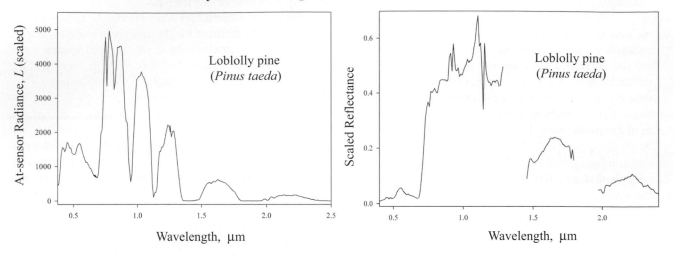

a. Run 08 scene 05,
original AVIRIS data.

b. Run 08 scene 05,
atmospherically corrected using ATREM.

Band-by-Band Atmospheric Correction of AVIRIS Data

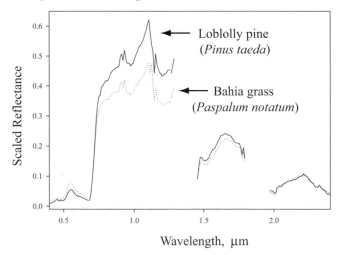

c. Run 08 scene 05,
ATREM- and EFFORT-corrected AVIRIS data.

Figure 11-7 a) An example of an original AVIRIS vegetation spectral profile for a single pixel of loblolly pine located at 3,683,437 Northing and 437,907 Easting in the Universal Transverse Mercator (UTM) projection. The y-axis is in units of radiance (L). b) The results of applying the ATmosphere REMoval program (ATREM) to the original AVIRIS data. This is the scaled surface reflectance data for the loblolly pine pixel. c) The results of applying the Empirical Flat Field Optimal Reflectance Transformation (EFFORT) to the ATREM-corrected loblolly pine data. This resulted in a mild band-by-band, entire-scene correction. The scaled surface reflectance for a pixel of Bahia grass (*Paspalum notatum*) in the same scene is also plotted. Some of the artifacts of the ATREM correction are retained in all resultant spectral curves, but some artifacts have been partially removed in the EFFORT-corrected spectra.

ure 11-7c depicts the same loblolly pine pixel in the hyperspectral dataset after it was spectrally polished using EFFORT. Notice that some artifacts in the atmospheric correction are removed and that the vegetation spectrum is much improved. The spectral reflectance curve for a nearby pixel of Bahia grass (*Paspalum notatum*) is also shown in Figure 11-7c for comparison. Note the dramatic differences in the reflectance characteristics between the loblolly pine and Bahia grass, especially in the visible and near-infrared portion of the spectrum.

When running programs like EFFORT, the user should first mask any invalid data from each band in the dataset. This involves applying a mask that effectively removes all the pixels with null data values from the dataset in all 224 bands. The analyst can also provide a *bad bands list* based on the visual and quantitative analysis of all 224 hyperspectral bands, as previously discussed.

Before running programs like EFFORT, it is also important that the user avoid wavelength ranges that contain noise such as the 1.4 μm and 1.9 μm water vapor absorption bands. Also, the user should experiment with the order of the polynomial that is fit to the data. The lower the order of the polynomial, the more error suppression but possible loss of valuable spectral information. A higher-order polynomial will produce a spectrum that fits the data better but also may fit some error features. This example was based on a 10th-order polynomial. Programs like EFFORT perform a relatively mild adjustment to the ATREM-corrected data. An additional method of improving the EFFORT spectral polishing is to carefully subset out only those areas of greatest importance and use 'reality boost' spectra specific to that area of interest. Adler-Golden et al. (2002) provide an algorithm for removing shadows in spectral imagery.

Empirical Line Calibration Atmospheric Correction

As discussed in Chapter 6, it is also possible to use *in situ* spectroradiometric data to atmospherically correct hyperspectral data using empirical line calibration. This method applies a single adjustment to all the pixels in a single band based on a model that is derived from the regression of *in situ* spectroradiometer measurements at specific locations (e.g., deep nonturbid water bodies, asphalt parking lots, bare soil, concrete) with radiance measurements extracted from the hyperspectral data at the same geographic locations. A empirical line calibration example was provided in Chapter 6.

 Geometric Correction of Hyperspectral Remote Sensor Data

It is important to geometrically correct the remote sensor data to a known datum and map projection at some point in the hyperspectral data analysis chain. Some image analysts prefer to perform all the information extraction on unrectified, atmospherically corrected hyperspectral data, and then at the very end geometrically correct the derived information for distribution to the public. Other scientists geometrically correct the hyperspectral data early in the process because they want to relate *in situ* spectroradiometer spectral reflectance measurements at known *x, y* locations in the field with the same locations in the hyperspectral data. It is also useful to geometrically rectify the hyperspectral data early in the processing chain if field, collected data are to be used to train a classifier. For example, the GPS *x, y* coordinates of a lake polygon inventoried on the ground might be used to extract training statistics or identify candidate water end-member pixels in a rectified hyperspectral dataset. This cannot be done if the hyperspectral data remain unrectified.

The geometric rectification techniques described in Chapter 7 may be applied to hyperspectral data. The process just takes longer because there are more bands to process. Most scientists use nearest-neighbor resampling when geometrically correcting hyperspectral data. However, *n*-term polynomial and rubber-sheet image warping techniques often do not yield satisfactory results when applied to aircraft-derived imaging spectrometer data. In such cases, on-board engineering and navigation data are also used (Clark et al., 1998). Properly configured aerial platforms are usually equipped with onboard GPS and inertial navigation system (INS) technology, which provide data that can be applied to the geometric correction problem.

 Reducing the Dimensionality of Hyperspectral Data

The number of spectral bands associated with a remote sensing system is referred to as its *data dimensionality*. Many remote sensing systems have relatively low data dimensionality because they collect information in relatively few regions of the electromagnetic spectrum. For example, the SPOT 1–3 HRV sensors recorded spectral data in three coarse bands. The Landsat 7 Enhanced Thematic Mapper Plus records data in seven bands. Hyperspectral remote sensing systems such as AVIRIS and MODIS obtain data in 224 and 36 bands, respectively. Ultraspectral remote sensing systems collect data in many hundreds of bands (Belokon, 1997).

As one might expect, data dimensionality is one of the most important issues when analyzing remote sensor data. The greater the number of bands in a dataset (i.e., its dimensionality), the more pixels that must be stored and processed by the digital image processing system. Storage and processing consume valuable resources. Therefore, significant attention has been given to developing methods to reduce the dimensionality of hyperspectral data while retaining the information content inherent in the imagery.

Statistical measures such as the optimum index factor (OIF), transformed divergence, and principal components analysis have been used for decades to reduce the dimensionality of multispectral data. These methods are described in Chapters 7, 8, and 9. Unfortunately, these methods generally are not sufficient for reducing the dimensionality of hyperspectral data.

Hyperspectral data contain a tremendous amount of redundant spectral information. This is not surprising when one considers that the individual channels often have a nominal bandwidth of only 10 nm. Thus, in the spectral region from 820 to 920 nm we can expect to find 10 bands measuring the amount of near-infrared radiant flux exiting the Earth's surface. While there are certainly subtle differences in the amount of radiant flux recorded in each of these bands, there will probably be a significant amount of redundant spectral information. Statistical analysis usually reveals that many of these 10 bands are highly correlated. Therefore, one can use statistical methods to a) delete some of the unwanted redundant bands or b) transform the data so that the information content is preserved while reducing the dimensionality of the dataset. In addition, it is hoped that the selected data dimensionality reduction method will also remove some of the noise in the hyperspectral dataset.

Minimum Noise Fraction Transformation

A useful algorithm for reducing the dimensionality of hyperspectral data and minimizing the noise in the imagery is the minimum noise fraction (MNF) transformation (Green et al., 1988). The MNF is used to determine the true or inherent dimensionality of the hyperspectral data, to identify and segregate noise in the data, and to reduce the computation requirements of further hyperspectral processing by collapsing the useful information into a much smaller set of MNF images (Boardman and Kruse, 1994). The MNF transformation applies two cascaded principal components analyses (Chen et al., 2003). The first transformation decorrelates and rescales noise in the data. This results in transformed data in which the noise has unit variance and no band-to-band correlation. A second principal components analysis results in the creation of a) coherent MNF eigenimages that contain useful information and b) noise-dominated MNF eigenimages. Basically, the noise in the hyperspectral dataset is separated from the useful information. This is very important because subsequent hyperspectral data analysis procedures function best when the hyperspectral data contain coherent, useful information and very little noise.

Both the eigenvalues and the output MNF eigenimages are used to determine the *true* dimensionality of the data. Application of the MNF transformation often results in a significant decrease in the number of bands that must be processed during subsequent endmember analysis. In the case of AVIRIS data, the process can result in a reduction from 224 bands to < 20 useful MNF bands. This number of bands is actually quite manageable and exhibits the dimensionality of typical multispectral datasets. Therefore, some scientists have input the MNF bands that describe most of the variance directly into traditional classification algorithms (e.g., maximum likelihood) (Underwood et al., 2003). In the SRS example, the bands dominated by atmospheric noise were discarded prior to running the MNF transformation.

MNF output bands that contain useful image information typically have an eigenvalue an order of magnitude greater than those that contain mostly noise. An analyst can determine the information content of individual MNF eigenimages by a) displaying them on the CRT screen and visually analyzing them or b) plotting their eigenvalues or the amount of variance explained by each eigenvalue. For example, an MNF transformation was applied to AVIRIS run 8 scene 05. In this example, the first 19 MNF bands could be selected for further analysis based on a dual inspection of the spatial coherency of the eigenimages and the corresponding eigenvalue plot. Figure 11-8 depicts the first 10 MNF eigenimages. Generally, the more spatially coherent the image, the less noise and greater the information content. A graph of the eigenvalues by band reveals that the first 10 eigenimages contain most of the valuable information. MNF eigenimages with values close to 1 contain mostly noise. MNF bands with eigenvalues > 1 account for most of the variance in the hyperspectral dataset. However, the MNF eigenimages maintain a relatively high degree of spatial coherency through component 19 (not shown).

Care should be exercised in the selection of the MNF bands to keep for subsequent hyperspectral image processing. For example, only the first 19 MNF bands associated with Run 8 Scene 5 were deemed useful for subsequent hyperspectral endmember analysis of the Savannah River Site study area, though a smaller subset could be justified. The number of MNF transform eigenimages passed to subsequent analytical steps (and hence, the inherent dimensionality of the dataset) will determine the number of endmembers, or unique materials that can be defined (see next section).

The aforementioned example was based on an entire scene of AVIRIS data. It is often useful to mask and/or subset the scene according to the most important features of interest. For example, consider **Color Plate 11-2a and b** where only the clay-capped hazardous waste sites associated with the Mixed Waste Management Facility and the H-area seepage basins are retained after masking. **Color Plate 11-2a** is a masked false-color composite of three AVIRIS bands (RGB = 50, 30, 20). **Color Plate 11-2b** is a masked false-color composite of MNF bands (RGB = 3, 2, 1). A binary mask was applied to the image such that features not of interest, including null data values, were excluded from further analysis. The MNF transform was then applied to the masked dataset, and the result was propagated to the next step in the hyperspectral image processing chain.

Endmember Determination: Locating the Spectrally Purest Pixels

The primary goal of most hyperspectral analyses was to remotely identify the physical or chemical properties of materials found within the instantaneous-field-of-view of the sensor system. In the Savannah River Site application, the goal was to determine if there was any vegetation stress or unusual characteristics associated with a) grass covered, relatively flat clay-capped hazardous waste sites on the Savannah River Site, or b) areas downslope of the clay-capped hazardous waste sites. To identify such conditions, it was useful to first identify the major materials found within the AVIRIS hyperspectral scene. These are typically called *endmembers* and represent relatively *pure* materials such as water, asphalt, concrete, healthy full-canopy grass cover (e.g., Bahia grass or Centipede grass), healthy full-canopy forest cover (e.g., loblolly pine or sweet gum), and perhaps even stressed full-canopy grass cover or stressed full-canopy forest cover.

Most of the time the spectral reflectance entering the IFOV of the sensor system at the instant of data collection is a function of the radiant flux from a variety of endmember materials. If we can identify the spectral characteristics of

the endmember materials, then it may be possible to identify pixels that contain varying *proportions* of these materials. There are several methods of identifying the most spectrally pure pixels in a multispectral or hyperspectral scene, including:

- pixel purity index, and

- *n*-dimensional visualization of endmembers in feature space.

Pixel Purity Index Mapping

It is usually difficult to identify pure endmembers in a remote sensing dataset because very few pixels contain just one type of biophysical material. Therefore, a rigorous mathematical method of determining the most spectrally pure pixels is to repeatedly project *n*-dimensional scatterplots of the clean minimum noise fraction images (e.g., not just 3 bands as in the previous example but *all* useful MNF bands) onto a random unit vector. Each time the spectral data are projected, the most extreme pixels in each projection are noted (Research Systems, Inc., 2003). By carefully keeping track of which pixels in the scene are repeatedly identified as "extreme" pixels, it is possible to create a *pixel purity index* (PPI) image. Basically, the higher the pixel value in the pixel purity index image, the greater the number of times it was judged to be a spectrally extreme pixel (e.g., a pure water or pure vegetation endmember). It is important to include as input to the PPI calculation only MNF images that contain spectrally valuable information. MNF components that contain significant noise levels should not be included in the analysis.

A pixel purity index image derived from an analysis of masked minimum-noise-fraction images is shown in Figure 11-9a. Note that there are not a tremendous number of relatively pure pixels in the scene. In fact, only 4155 pixels in the masked clay-cap image were identified as being pure to some degree. Furthermore, a threshold is typically used to create a subset of the most pure PPI pixels to be used in subsequent analyses (Figure 11-9b). In this case, only 1207 of the 4155 pixels were selected for further processing after 70,000 PPI iterations using a conservative threshold value of 10 "hits."

A PPI image simply identifies the location of the most pure pixels (i.e., endmember candidates). Unfortunately, it is difficult to label the types of endmembers by just viewing the two-dimensional pixel purity index image. Therefore, it is common to use an *n*-dimensional visualization technique.

Hyperspectral Data Dimensionality and Noise Reduction

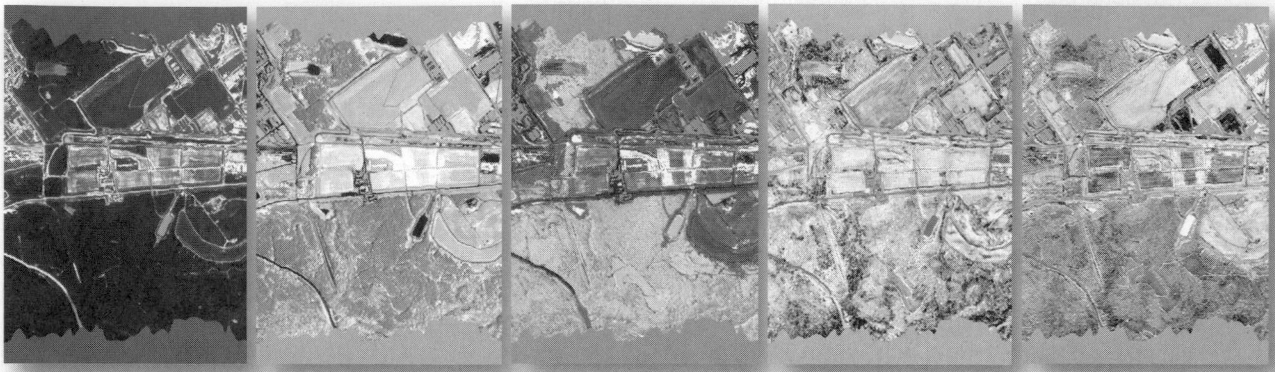

a. MNF component 1. b. MNF component 2. c. MNF component 3. d. MNF component 4. e. MNF component 5.

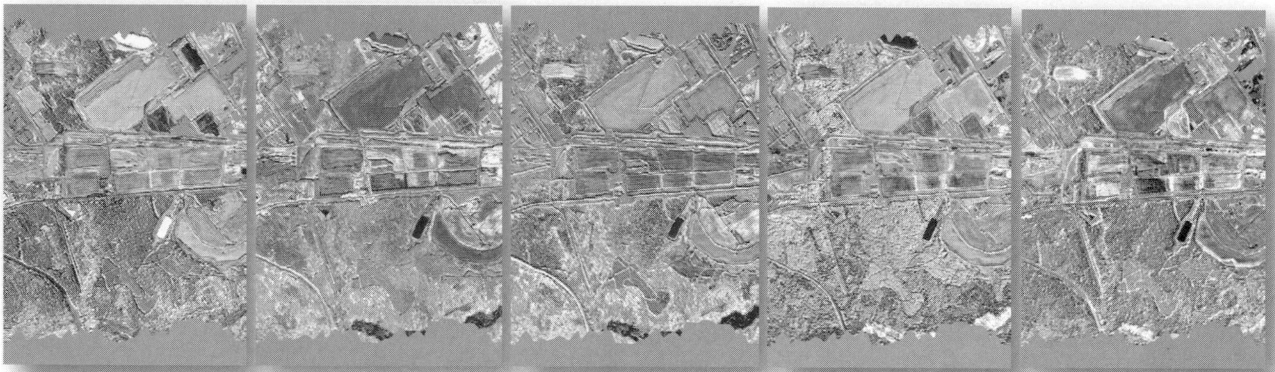

f. MNF component 6. g. MNF component 7. h. MNF component 8. i. MNF component 9. j. MNF component 10.

Minimum Noise Fraction

6.0×10^4
5.0×10^4
4.0×10^4
3.0×10^4
2.0×10^4
1.0×10^4
Eigenvalue
Percent
1 79.16
2 88.63
3 92.74
4 93.98
5 95.17
6 95.96
7 96.96
8 97.22
9 97.65
10 98.01
5
10
15
20
Eigenvalue Number

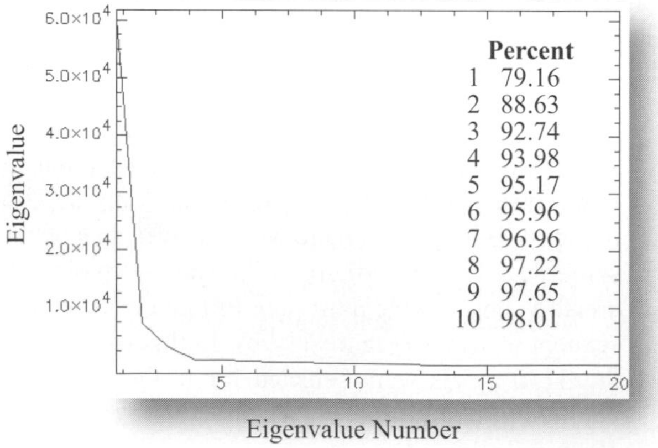

k. Plot of eigenvalues associated with 19 minimum noise fraction (MNF) bands and the cumulative amount of variance accounted for by the first 10 eigenimages.

Figure 11-8 a–j) The first 10 eigenimages (MNF bands) extracted from run 8 scene 05. k) A plot of the eigenvalues associated with the first 19 MNF eigenimages. Note that as the MNF eigenimage number increments, the eigenvalue and the coherent information in the eigenimages decrease. However, there is still a degree of spatial coherency in MNF band 10.

Pixel Purity Index Images

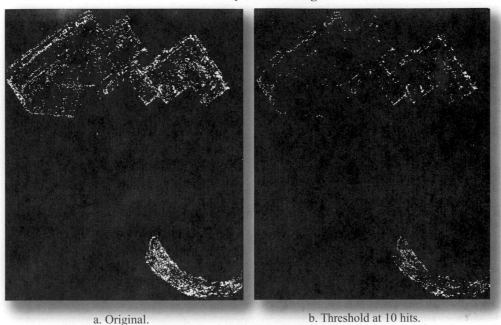

| a. Original. | b. Threshold at 10 hits. |

Figure 11-9 Pixel purity index (PPI) image of a portion of run 8, scene 05 derived from masked minimum noise fraction AVIRIS data. The pixels in (a) denote the original output from the PPI analysis. The pixels highlighted in (b) represent a subset of those pixels selected using a threshold of 10 hits.

n-dimensional Endmember Visualization

To identify and label the most spectrally pure endmembers in the two-dimensional pixel purity index image, it is necessary to systematically view where the relatively few pure vegetation pixels reside in *n*-dimensional feature space. For example, Figure 11-10 depicts a display of the most spectrally pure pixels found in the PPI image shown in Figure 11-9b. By using more than two of the MNF bands at a time it is possible to interactively view and rotate the endmembers in *n*-dimensional spectral space on the CRT or LCD screen (Boardman, 1993). The analyst should attempt to locate the corners of the data cloud in *n*-dimensional space as the cloud rotates. The purest pixels lie at the convex corners of the data cloud. Clusters can be defined manually, or an algorithm can be used to precluster the data, and the endmembers can be subsequently refined manually (Research Systems, Inc., 2003). In this context, *n* equals the number of MNF bands, and given the dimensionality of the simplex formed by the cluster means, the number of endmembers that can be defined is equal to $n + 1$. Thus, there are theoretically $n + 1$ endmembers if the data are truly *n*-dimensional (Boardman, 1993).

By comparing the actual spectra of an endmember found within the *n*-dimensional CRT display with where it is actually located in *x, y* space in the MNF or surface reflectance image, it is possible to label the pure pixel as a specific type of endmember, e.g., water, vegetation, or bare soil. Endmembers that are derived in this manner may then be used to perform spectral matching or classification. More automated methods of endmember determination are available (e.g., Winter et al., 1999).

To demonstrate these methods, let us examine the clay-capped hazardous waste sites in the run 8 scene 05 dataset to identify and label the following spectra:

• healthy Bahia grass, and

• potentially stressed Bahia grass.

Masked MNF images were analyzed to produce the pixel purity index image (and spectral profiles of the most pure pixels) previously discussed (Figure 11-9). These data were then displayed as data clouds in *n*-dimensional feature space (Figure 11-10). Spectra (scaled reflectance values) for one healthy vegetation endmember (37) and three potentially stressed vegetation endmembers (15, 25, and 36) over the

Three-dimensional Visualization of Five Endmembers

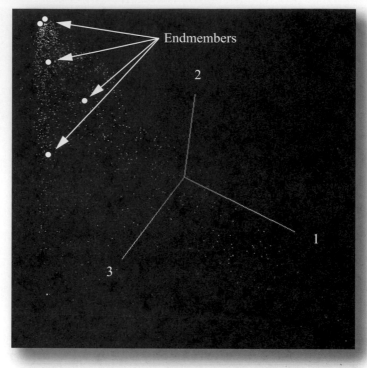

Figure 11-10 *n*-dimensional visualization of four potential image-derived vegetation endmembers and one shade endmember derived from the pixel purity index analysis. Endmembers associated with other materials (e.g., asphalt, concrete) are not highlighted. This particular example is only a three-dimensional visualization. Other dimensions were viewed during the analysis. Note that some of the highlighted pixels are not immediately recognizable as endmembers. Other data cloud rotations were necessary to discern all the endmember regions of interest.

wavelength interval from 0.4 to 2.43 μm are superimposed in Figure 11-11a. The same spectra are replotted in the visible through near-infrared region (0.4 – 0.9 μm) in Figure 11-11b to highlight certain reflectance characteristics. The numerical names assigned to these endmember spectra (e.g., 37) are an artifact of the interative endmember selection process.

The clay-caps on both the Mixed Waste Management Facility and the H-area seepage basins contain a near-monoculture of Bahia grass (there are a few patches of Centipede). Bahia grass endmember 37 exhibited all the characteristics of healthy vegetation, including a) strong chlorophyll *a* and *b* absorption bands in the blue (0.43 – 0.45 μm) and red (0.65 – 0.66 μm) portions of the spectrum, b) a relative lack of absorption in the wavelength interval between the two chlorophyll absorption bands, producing an increase in reflectance at approximately 0.54 μm in the green portion of the

spectrum, c) a substantial increase in reflectance throughout the near-infrared portion of the spectrum (0.7 – 1.2 μm) due to scattering and reflectance in the spongy mesophyll portion of the plant, and d) typical absorption/reflectance characteristics in the middle-infrared (1.45 – 1.79 μm and 2.08 – 2.35 μm) wavelengths (Jensen, 2000).

Conversely, each potentially stressed Bahia grass endmember (15, 25, and 36) exhibited spectral reflectance characteristics that were significantly different from the healthy Bahia grass spectra. Relative to the healthy turfgrass spectrum, each exhibited a) a slight to moderate increase in red reflectance (15 most; 25 intermediate; 36 least); b) a slight to moderate red-edge shift toward shorter wavelengths in the transition region from red to near-infrared wavelengths (15 most; 25 intermediate; 36 least), and c) a decrease in near-infrared reflectance.

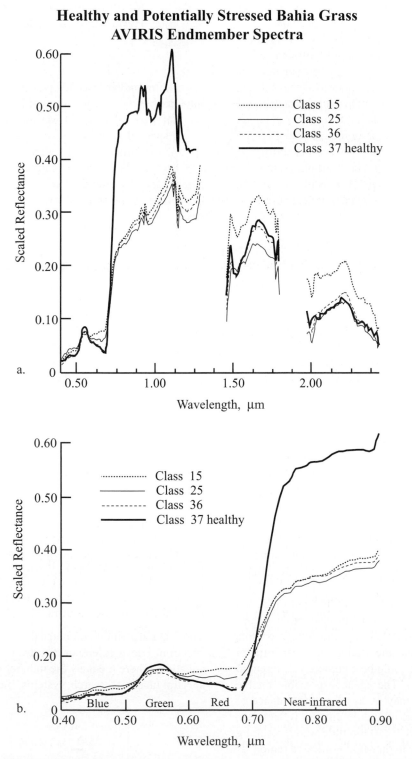

Figure 11-11 a) One healthy (37) and three potentially stressed Bahia grass image-derived endmember spectral profiles derived using a pixel purity index image and *n*-dimensional visualization endmember analysis. b) Only the visible through near-infrared region (0.4 – 0.90 μm) is plotted to highlight the increase in red reflectance, the apparent red-edge shift, and decrease in near-infrared reflectance of the potentially stressed vegetation endmembers.

Endmember 15 consistently reflected more energy in both middle-infrared regions (1.45 – 1.79 μm and 2.08 – 2.35 μm) than the healthy Bahia grass endmember (37). Endmember 25 exhibited a decrease in middle-infrared reflectance in the 1.45- to 1.79-μm region and approximately the same reflectance as the healthy vegetation in the 2.08- to 2.35-μm region. Finally, endmember 36 exhibited a decrease in middle-infrared reflectance in the 1.45- to 1.79-μm region and a slight increase in reflectance in the 2.08- to 2.35-μm region. While the identified endmembers are spectrally pure in theory, it is possible that albedo contributions from other materials are present in the pixels (e.g., soil background reflectance). However, the extracted pixels are the purest relative to all others in the image.

To illustrate the utility of the endmember concept further, consider the reflectance profiles obtained at two randomly selected healthy and potentially stressed areas on the Bahia grass clay-cap (Figure 11-12ab). Note that these non-endmember profiles retain the same general relationships as the image-derived endmember profiles (Figure 11-11ab).

Before proceeding, one should ask, "How can we be sure that the Bahia grass is actually undergoing some stress?" Fortunately, we have some laboratory Bahia grass spectra that depict how healthy and stressed Bahia grass reflect electromagnetic energy in a controlled environment. For example, Figure 11-13 depicts the spectra of healthy Bahia grass and Bahia grass subjected to various concentrations of copper (mg/ml) (Schuerger, unpublished data). Note the increase in reflectance in the red region, the red-edge shift, and the decrease in reflectance throughout the near-infrared region as the copper concentration is increased. Unfortunately, the laboratory spectroradiometer did not record Bahia grass spectral information in the middle-infrared region. Thus, the AVIRIS hyperspectral results only suggest that the spectra labeled "potentially stressed Bahia grass" are indicative of some type of stressing agent. It is not suggested that the stress on the MWMF is due to the absorption of copper, but rather that the stressed Bahia spectra simply exhibits stressed behavior throughout the visible and near-infrared portions of the spectrum (it could be completely different in the middle-infrared region). The agent responsible for the potential vegetation stress may be a) an overabundance of water, resulting in a decrease in middle-infrared reflectance, as with endmembers 25 and 36; b) too little water, resulting in an increase in reflectance in the middle-infrared region, as with endmember 15; or c) some other stressing agent related to clay-capped constituents.

Mapping and Matching Using Hyperspectral Data

Algorithms may be used to convert hyperspectral reflectance into biophysical information or thematic maps, including the spectral angle mapper (SAM), subpixel classification (e.g., linear spectral unmixing), spectroscopic library matching, specialized hyperspectral indices, and derivative spectroscopy. Scientists have also applied traditional classification algorithms, decision-tree classifiers, and artificial neural networks to hyperspectral data with varying degrees of success (e.g., Goel et al., 2003; Underwood et al., 2003).

Spectral Angle Mapper

Spectral angle mapping (SAM) algorithms take an atmospherically corrected unlabeled pixel (e.g., composed of an AVIRIS measurement vector of n brightness values) and compare it with reference spectra in the same n-dimensions. In this example, only 175 of the original 224 AVIRIS bands were used. The reference spectra may be obtained using 1) calibrated *in situ*- or laboratory-derived spectroradiometer measurements stored in an ASCII or binary spectral library, 2) theoretical calculations, and/or 3) multispectral or hyperspectral image endmember analysis procedures, as previously discussed. The algorithm compares the angle (α) between the reference spectrum (r) and the hyperspectral image pixel measurement vector (t) in n-dimensions and assigns it to the reference spectrum class that yields the smallest angle (Boardman and Kruse, 1994). For example, the simple two-dimensional (2 band) illustration in Figure 11-14 suggests that unknown material t has spectra that are more similar to reference spectrum r than the spectrum for material k, i.e., the angle (α) in radians between r and t is smaller.

The SAM algorithm computes the similarity of an unknown spectrum t to a reference spectrum r using the following equation, where n equals the number of bands in the hyperspectral image (Research Systems, Inc., 2003):

$$\alpha = \cos^{-1}\left(\frac{\sum\limits_{i=1}^{n} t_i r_i}{\left(\sum\limits_{i=1}^{n} t_i^2\right)^{\frac{1}{2}} \left(\sum\limits_{i=1}^{n} r_i^2\right)^{\frac{1}{2}}} \right). \qquad (11\text{-}4)$$

Selected Healthy and Stressed Bahia Grass AVIRIS Spectra

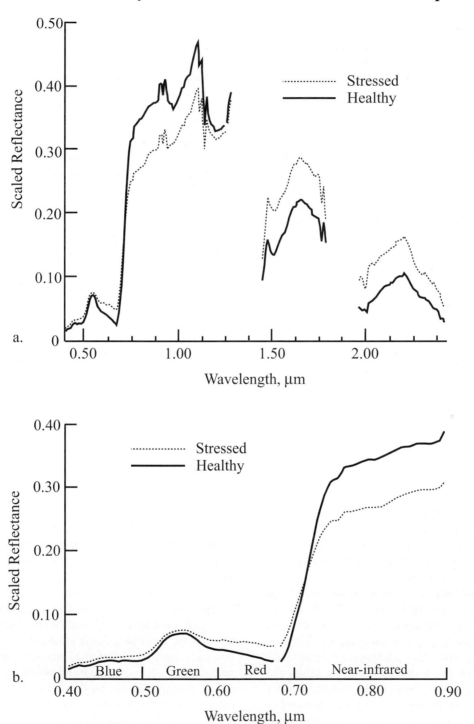

Figure 11-12 a) Spectra located within areas of homogeneous healthy and stressed Bahia grass. b) The visible through near-infrared region is rescaled to highlight the increase in red reflectance, the red-edge shift, and decrease in near-infrared reflectance of the potentially stressed vegetation endmembers.

Laboratory-derived Bahia Grass Spectra

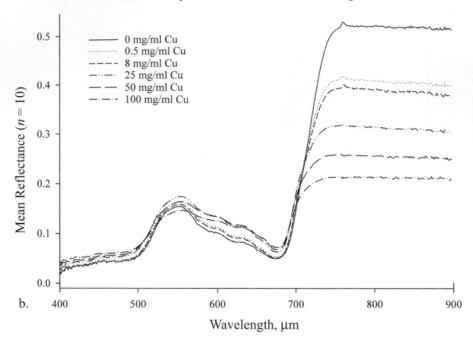

Figure 11-13 Laboratory spectral reflectance of healthy and copper-stressed Bahia grass. The spectra were not obtained throughout the middle-infrared portion of the spectrum (Schuerger, unpublished data).

Basically, for each reference spectrum r (this could be an *in situ*–derived endmember or a remote sensing–derived endmember, as with the four endmembers previously discussed) a spectral angle (α) is computed for each unknown image spectrum (pixel) in the dataset. This angle becomes the pixel value in the output SAM image with one output image for each reference spectrum. This creates an entirely new SAM datacube with the number of bands equal to the number of reference spectra used in the mapping.

The SAM algorithm was applied to the previously described endmembers. Thematic maps of the spatial distribution of healthy (derived from endmember 37) and potentially stressed Bahia grass (derived from endmembers 15, 25, and 36) are shown in **Color Plate 11-3**. The darker the pixel (i.e., black, blue) the smaller the SAM angle and the closer the match. Thus, pixels in **Color Plate 11-3a** that had spectra very similar to that of the healthy Bahia grass endmember spectra (37) yielded very small angles and were color-coded as being black or dark blue. Note that the spatial distribution of the healthy vegetation derived from endmember 37 is almost the inverse of the maps derived from the potential stress endmembers (15, 25, and 36) in **Color Plates 11-3 b** through **d**.

It is possible to convert the output from the spectral angle mapper subroutine into a hardened classification map. For example, **Color Plate 11-4** depicts a hardened classification map derived using just the four endmembers previously discussed, one class for each endmember. A threshold angle value of 0.1 radian was used. Brown areas represent healthy vegetation (based on endmember 37) and the other three classes represent potentially stressed vegetation.

The clay-capped hazardous waste sites ideally have a relatively homogenous cover of Bahia grass. Maps such as this suggest that on this date the Bahia grass may not have been uniformly distributed and that certain locations within the site may have stressed Bahia grass. This stress could be the result of a number of factors. The display provides valuable spatial information that can be of significant value to the personnel responsible for clay-cap maintenance. It can help them to focus their attention on certain locations on the clay-cap to ensure its continued integrity.

It is also possible to use the laboratory-derived Bahia grass reference spectra as endmember spectra to be input to SAM. A potentially stressed Bahia grass classification map derived from averaged laboratory spectra with a 0.5 mg/ml copper

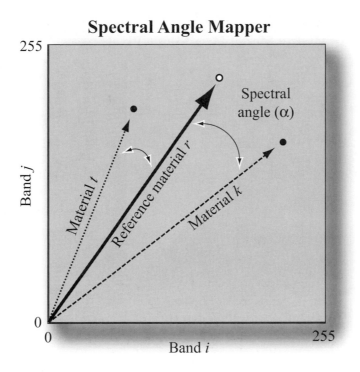

Spectral Angle Mapper

Figure 11-14 The logic associated with the Spectral Angle Mapper (SAM) algorithm (adapted from Kruse et al., 1994; Research Systems, Inc., 2003).

treatment is shown in **Color Plate 11-5**. The darker the pixel (blue, black, etc.), the smaller the angle and the closer the match. It visually correlates well with the results obtained from the remote sensing–derived spectra.

Subpixel Classification (Linear Spectral Unmixing)

The energy recorded by a remote sensing detector is a function of the amount of energy reflected or emitted by the materials within the IFOV. For example, consider the single pixel shown in Figure 11-15a, which is composed of 50% water, 25% bare soil, and 25% vegetation. Should this pixel be classified as water because its dominant constituent is water? Wouldn't it be more useful to determine and report the actual proportion (or abundance) of the pure endmember (class) materials within the pixel? This process is commonly referred to as *subpixel sampling, linear spectral unmixing,* or *spectral mixture analysis* (Okin et al., 2001; Segl et al., 2003).

To appreciate linear spectral unmixing, let us assume for a moment that we have a remotely sensed image consisting of just two bands ($k = 2$) and that there are literally only three types of pure endmember materials in the scene ($m = 3$):

water, vegetation, and bare soil — E_a, E_b, and E_c, respectively. The linear mixing associated with any pixel in the scene can be described using the matrix notation:

$$\left[BV_{ij} \right] = \left[E \right] \left[f_{ij} \right] + \varepsilon_{ij} \tag{11-5}$$

where $BV_{i,j}$ is the k-dimensional spectral vector of the pixel at location i,j under investigation, f_{ij} is the $m \times 1$ vector of m endmember fractions (e.g., 0.5, 0.25, 0.25) for the pixel at location i,j, and E is the $k \times m$ signature matrix, with each column containing one of the endmember spectral vectors (Schowengerdt, 1997). The relationship contains some noise, therefore ε_{ij} represents the residual error. This equation assumes that we have identified all of the theoretically pure classes (endmembers) in the scene such that their proportions (f_{ij}) will sum to 1 at each pixel and that all endmember fractions are positive.

If we assume that the data contain no noise (ε_{ij}), Equation 11-2 becomes:

$$\left[BV_{ij} \right] = \left[E \right] \left[f_{ij} \right] \tag{11-6}$$

and we can use the following matrix notation to solve for f_{ij} on a pixel-by-pixel basis:

Linear Mixing Model for a Single Pixel

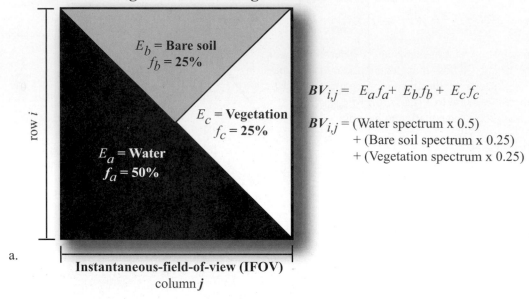

$$BV_{i,j} = E_a f_a + E_b f_b + E_c f_c$$

$BV_{i,j}$ = (Water spectrum x 0.5)
+ (Bare soil spectrum x 0.25)
+ (Vegetation spectrum x 0.25)

Endmembers in Two-dimensional Feature Space

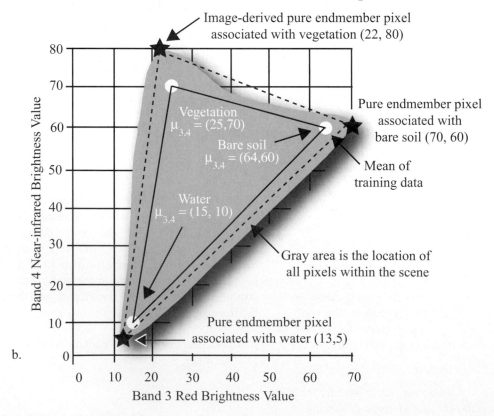

Figure 11-15 a) Linear mixture modeling associated with a single pixel consisting of water, vegetation, and bare soil. b) The location of hypothetical mean vectors and endmembers in two-dimensional feature space (red and near-infrared).

$$\begin{bmatrix} BV_3 \\ BV_4 \end{bmatrix} = \begin{bmatrix} E_{water3} & E_{veg3} & E_{bsoil3} \\ E_{water4} & E_{veg4} & E_{bsoil4} \end{bmatrix} \begin{bmatrix} f_{water} \\ f_{veg} \\ f_{bsoil} \end{bmatrix}. \quad (11\text{-}7)$$

Unfortunately, this relationship is under-determined because there are fewer bands than endmembers. Fortunately, because we know that the linear combination of all three endmembers must sum to 1 (i.e., $f_{water} + f_{vegetation} + f_{bare\,soil} = 1$), we can augment the equation to become:

$$\begin{bmatrix} BV_3 \\ BV_4 \\ 1 \end{bmatrix} = \begin{bmatrix} E_{water3} & E_{veg3} & E_{bsoil3} \\ E_{water4} & E_{veg4} & E_{bsoil4} \\ 1 & 1 & 1 \end{bmatrix} \begin{bmatrix} f_{water} \\ f_{veg} \\ f_{bsoil} \end{bmatrix}. \quad (11\text{-}8)$$

This equation can be inverted to solve for the exact fraction (proportion) of endmember materials found within each pixel:

$$\begin{bmatrix} f_{water} \\ f_{veg} \\ f_{bsoil} \end{bmatrix} = \begin{bmatrix} E_{water3} & E_{veg3} & E_{bsoil3} \\ E_{water4} & E_{veg4} & E_{bsoil4} \\ 1 & 1 & 1 \end{bmatrix}^{-1} \begin{bmatrix} BV_3 \\ BV_4 \\ 1 \end{bmatrix}. \quad (11\text{-}9)$$

Please note that E is now E^{-1}.

Now let us use Equation 11-6 to determine the proportion of endmember materials in pixels in a hypothetical scene. Consider the two-dimensional feature space plot shown in Figure 15b. This scene consists of just two bands of data (bands 3 and 4; red and near-infrared, respectively). The gray cloud represents the co-occurrence of all pixel values in the scene in bands 3 and 4. There are no pixel values in the scene outside the gray area convex hull. The circles are mean vectors associated with water, vegetation, and bare soil obtained through supervised training. They are present to help the reader appreciate the difference between mean vectors and endmembers and are not used in this computation. The three stars in the display represent pure endmember pixels associated with water, vegetation, and bare soil. These endmembers were derived from the image itself using pixel purity index assessment and/or n-dimensional visualization or by obtaining accurate spectroradiometer measurements on the ground of pure water, vegetation, and bare soil.

This graphic contains enough data for us to derive the information necessary to predict the proportion of each endmember in each pixel in the scene. The values found in Figure 11-15b are summarized in Table 11-2.

Table 11-2. Brightness values for the three image-derived endmembers shown in Figure 11-15b.

Band	Water	Vegetation	Bare Soil
3	13	22	70
4	5	80	60

Table 11-3. Matrices used to perform spectral unmixing. The inverse matrix E^{-1} was derived from the endmember brightness values in E.

E	E⁻¹
$\begin{bmatrix} 13 & 22 & 70 \\ 5 & 80 & 60 \\ 1 & 1 & 1 \end{bmatrix}$	$\begin{bmatrix} -0.0053 & -0.0127 & 1.1322 \\ -0.0145 & 0.0150 & 0.1137 \\ 0.0198 & -0.0024 & -0.2460 \end{bmatrix}$

It is possible to compute the fractions of each endmember class found in each pixel in the scene by a) placing the inverse matrix (E^{-1}) coefficients in Table 11-3 into Equation 11-6, and b) repeatedly placing new values of BV_3 and BV_4 in Equation 11-6 associated with each pixel in the scene. For example, if the BV_3 and BV_4 values for a single pixel were 25 and 57, respectively, then the proportion (abundance) of water, vegetation, and bare soil found within this single pixel would be 27%, 61%, and 11%, as shown here:

$$\begin{bmatrix} f_{water} \\ f_{veg} \\ f_{bsoil} \end{bmatrix} = \begin{bmatrix} -0.0053 & -0.0127 & 1.1322 \\ -0.0145 & 0.0150 & 0.1137 \\ 0.0198 & -0.0024 & -0.2460 \end{bmatrix} \begin{bmatrix} 25 \\ 57 \\ 1 \end{bmatrix}$$

$$\begin{bmatrix} 0.27 \\ 0.61 \\ 0.11 \end{bmatrix} = \begin{bmatrix} -0.0053 & -0.0127 & 1.1322 \\ -0.0145 & 0.0150 & 0.1137 \\ 0.0198 & -0.0024 & -0.2460 \end{bmatrix} \begin{bmatrix} 25 \\ 57 \\ 1 \end{bmatrix}$$

This logic may be extended to hyperspectral imagery and the overdetermined problem, where there are more bands than endmembers. Schowengerdt (1997) and Richards and Jia (1999) describe the more complex pseudo-matrix inversion that must be adopted when analyzing hyperspectral data.

Spectral mixture analysis (SMA) assumes a pixel's spectrum is a linear combination of a finite number of spectrally distinct endmembers. Spectral mixture analysis uses the dimensionality of hyperspectral data to produce a suite of abundance (fraction) images for each endmember. Each fraction image depicts a subpixel estimate of endmember relative abundance as well as the spatial distribution of the endmember (Adams et al., 1995). When the endmembers include vegetation, the endmember fraction is proportional to the areal abundance of projected canopy cover (Roberts et al., 1993; Williams and Hunt, 2002). Although this technique is intuitively very appealing, many scientists find it difficult to identify all of the pure endmembers in a scene.

McGwire et al. (2000) found that endmembers (average green leaf, soil, and shadow) derived from hyperspectral data in arid environments were more highly correlated with vegetation percent cover than when the hyperspectral data are processed using traditional narrow-band and broadband vegetation indices (e.g., NDVI, SAVI). Williams and Hunt (2002) used mixture-tuned matched filtering, a special case of subpixel analysis, to estimate leafy spurge canopy cover and map its distribution. Segl et al. (2003) used endmembers derived from hyperspectral data and linear spectral unmixing to identify urban surface cover types.

Some have suggested that if we obtain very-high-resolution imagery we will not have to be concerned about mixed pixels. This is not true. Higher-spatial-resolution imagery (e.g., 1×1 m) normally requires the identification of additional relevant endmembers that must be considered such as pure shadow between leaves in the vegetation, sunglint on water, or individual types of minerals in the bare soil. Thus, mixed pixels are with us at any spatial resolution.

Spectroscopic Library Matching Techniques

One of the great benefits of conducting hyperspectral remote sensing research is that a detailed spectral response calibrated to percent reflectance (spectra) can be obtained for a pixel (e.g., AVIRIS measures in 224 bands from 400 to 2500 nm). The remote sensing–derived spectra can be compared with *in situ*– or laboratory-derived spectra stored in a spectral library. The laboratory-derived spectra are usually considered more accurate because they have been made under controlled illumination conditions, and the atmosphere is not a factor. Laboratory-derived spectra may be found at the ASTER Spectral Library (2004), which contains the:

• Johns Hopkins Spectral Library,

• NASA Jet Propulsion Laboratory Spectral Library, and

• U.S. Geological Survey spectral library.

Several digital image processing software vendors provide library spectra as an appendix to their software (e.g., ERDAS, ENVI, PCI Geomatica).

Absorption features are present in the remote sensing– or laboratory-derived reflectance spectra (seen as localized dips) due to the existence of specific minerals, chlorophyll *a* and *b*, water, and/or other materials within the pixel IFOV. The absorption features in the spectra are characterized by their spectral locations (i.e., the bands affected), their depths, and widths.

Considerable attention has been given to developing library matching techniques that allow remote sensing–derived spectra to be compared with spectra that were previously collected in the field or in the laboratory. For example, we may have remote sensing–derived spectra from an agricultural field and desire to compare it with a library containing a significant variety of agricultural spectra. We would compare each pixel's spectrum with each of the agricultural spectra in the library and assign the pixel to the class its spectra most closely resembles.

Comparing the remote sensing–derived spectra with the stored laboratory-derived spectra is no small matter. First, if we are using AVIRIS data, then each pixel has the potential of being represented by 224 distinct spectral band measurements. Each of these distinct measurements would have to be compared with each of the related bands in the library reference spectra. The library may contain hundreds of spectra and the remote sensing scene often consists of millions of pixels. The computation required to conduct true band-to-band comparison between remote sensing–derived spectra and library spectra is daunting. Therefore, various coding techniques have been developed to represent an entire pixel spectrum in a simple yet effective manner so that it can be efficiently compared with library spectra.

One simple coding technique is *binary spectral encoding* (Jia and Richards, 1993). Binary spectral encoding can be used to transform a hyperspectral reflectance spectrum into simple binary information. The algorithm is (Richards and Jia, 1999):

$$b(k) = 0 \qquad \text{if } p(k) \le T_1, \qquad (11\text{-}10)$$

otherwise

$$b(k) = 1$$

where $p(k)$ is the reflectance value of a pixel in the kth spectral band, T_1 is a user-selected threshold for creating the binary code, and $b(k)$ is the output binary code symbol for the pixel in the kth spectral band. T_1 may be the average reflectance value of the entire spectrum for a pixel or a user-specified value corresponding to a unique absorption spectra of particular interest. For example, consider Figure 11-16a, which displays:

- an *in situ* vegetation reflectance spectrum measured in a laboratory using a spectroradiometer, and

- vegetation and water spectra obtained using an airborne hyperspectral sensor.

A threshold of 10% reflectance was chosen because it captured nicely the chlorophyll a and b absorption bands in the blue and red portions of the spectrum. With the threshold set at 10%, it was a straightforward task to binary encode each of the three spectra and then compare their binary codewords to see how similar or different they were. For example, the *in situ* vegetation spectra exceeded the threshold in the region from 540 to 580 nm and again in the region from 700 to 800 nm. Therefore, each of these wavelength regions (bands) received a value of 1 in their codewords in Figure 11-16b. The remote sensing–derived vegetation spectra had very similar characteristics, exceeding the threshold from 540 to 560 nm and from 720 to 800 nm. Once again, all these bands were coded to 1. The remote sensing–derived water spectra never exceeded the threshold and therefore received zeros in all bands.

Thus, each of the original spectra was turned into a 16-digit codeword, c_1, c_2, and c_3. But how do we use these three binary codewords (each of length $Z = 16$) to determine if the binary spectra of one class is identical, similar to, or completely different from the spectra of another class? Two spectra that have been binary encoded can be compared by computing the Hamming distance between their binary codewords using the formula:

$$\text{Dist}_{Ham}(c_i, c_j) = \sum_{k=1}^{N=\text{bands}} [c_i(k) \oplus c_j(k)] \qquad (11\text{-}11)$$

where c_i and c_j are two spectral codewords of length L and the $\oplus$ symbol represents exclusive OR Boolean logic. The algorithm compares each bit of the two codewords under examination and outputs a 0 where they agree and a 1 where

they are different. The result is a Hamming distance codeword of length Z.

Thus, the Hamming distance is computed by summing the number of times the binary digits are different. For example, consider the computation of the Hamming distance between the laboratory *in situ* vegetation codeword and the remote sensing–derived vegetation codeword. The two codewords differ at only two locations in the 16-bit lengths and therefore have a Hamming distance of 2. Conversely, when the laboratory *in situ* vegetation codeword is compared with the water codeword, it differs at nine locations, resulting in a Hamming distance of 9.

The more similar one spectra is to another, the lower its Hamming distance. Thus, one can a) identify the appropriate threshold for a material or land-cover class of interest in light of existing absorption bands, b) compute a binary codeword at each pixel, and c) compare this binary codeword at each pixel with m library spectra that have been binary coded in a similar manner. Each pixel could be assigned to the class that results in the smallest Hamming distance.

Richards and Jia (1999) provide examples of how more than one threshold can be used a) within a chosen absorption band and b) at numerous disjoint locations along the spectra. For example, Figure 11-17 depicts the location of three unique thresholds that could be used to perform binary encoding using just the information at T_1 (540 to 560 nm), T_2 (660 to 680 nm), and T_3(740 to 760 nm) set at 10%, 5%, and 40% scaled reflectance, respectively.

 ## Indices Developed for Hyperspectral Data

Traditional vegetation indices (discussed in Chapter 8) developed for use with broadband multispectral data can be used with hyperspectral data. Blonski et al. (2002) provide an algorithm that will synthesize multispectral bands (e.g., Landsat ETM$^+$ bands) from hyperspectral data (e.g., AVIRIS), if desired. In addition, there are indices developed especially for use with hyperspectral data.

Normalized Difference Vegetation Index — NDVI

As discussed in Chapter 8, the normalized difference vegetation index is based on the equation:

$$\text{NDVI} = \frac{\rho_{nir} - \rho_{red}}{\rho_{nir} + \rho_{red}}.$$

Library Matching Using Binary Encoding

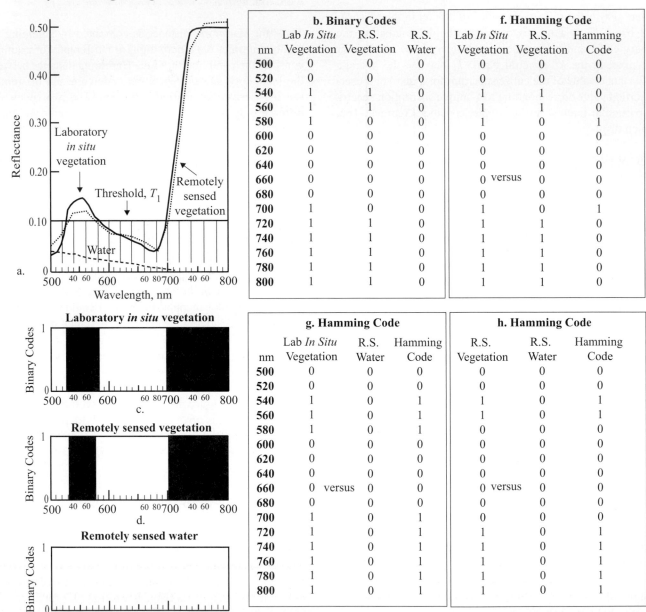

Figure 11-16 a) *In situ* vegetation spectrum obtained using a spectroradiometer in a controlled laboratory environment and vegetation and water spectra obtained from a remote sensing instrument. b) Binary encoding of the *in situ* vegetation, remotely sensed vegetation, and water data. c) Graph of the binary encoding of the *in situ* vegetation data. d) Graph of the binary encoding of the remote sensing–derived vegetation data. e) Graph of the binary encoding of the remote sensing–derived water data. f) Hamming code used in the computation of the Hamming distance between *in situ* vegetation and remote sensing–derived vegetation. g) Hamming code used in the computation of the Hamming distance between *in situ* vegetation and remote sensing–derived water. h) Hamming code used in the computation of the Hamming distance between remote sensing–derived vegetation and water.

Library Matching Using Binary Encoding at Selected Locations

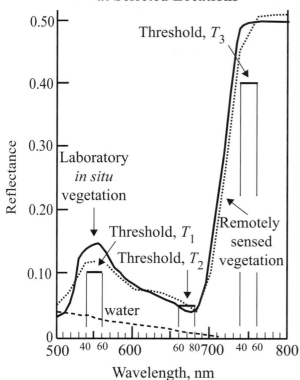

Figure 11-17 The location of three thresholds that could be used to perform binary encoding of the three spectra from 540 to 560 nm, 660 to 680 nm, and 740 to 760 nm set at 10%, 5%, and 40% reflectance, respectively.

Generally, reflectances from a red channel centered around 660 nm and a near-infrared channel centered on 860 nm are used to calculate the NDVI. The near-infrared band empirically corresponds to the long-wavelength shoulder of the chlorophyll red-edge, and the red band is associated with the maximum chlorophyll absorption. The following is one possible narrow-band implementation of the standard NDVI:

$$\text{NDVI}_{\text{narrow band}} = \frac{\rho_{(860nm)} - \rho_{(660nm)}}{\rho_{(860nm)} + \rho_{(660nm)}}. \qquad (11\text{-}12)$$

A narrow-band NDVI image derived from an analysis of AVIRIS bands 29 (red) and 51 (near-infrared) is shown in **Color Plate 11-6**. As expected, the lower biomass areas in the NDVI image correspond generally to stressed areas in the thematic map. Additional vegetation indices such as those proposed by Thenkabail et al. (2000) might yield superior results.

Narrow-band Derivative-based Vegetation Indices

A variety of narrow-band derivative-based vegetation indices may be computed. For instance, several derivative vegetation indices that measure the amplitude of the chlorophyll red-edge using hyperspectral data in the 626-nm to 795-nm spectral range can be implemented (Elvidge and Chen, 1995):

$$\text{1DL\_DGVI} = \sum_{\lambda_1}^{\lambda_n} |\rho'(\lambda_i) - R'(\lambda_1)| \Delta\lambda_i \qquad (11\text{-}13)$$

$$\text{1DZ\_DGVI} = \sum_{\lambda_1}^{\lambda_n} |\rho'(\lambda_i)| \Delta\lambda_i \qquad (11\text{-}14)$$

$$\text{2DZ\_DGVI} = \sum_{\lambda_1}^{\lambda_n} |\rho''(\lambda_i)| \Delta\lambda_i \qquad (11\text{-}15)$$

$$\text{3DZ\_DGVI} = \sum_{\lambda_1}^{\lambda_n} |\rho'''(\lambda_i)| \Delta\lambda_i \qquad (11\text{-}16)$$

where i is the band number; λ_i is the center wavelength for the ith band; $\lambda_1 = 626$ nm; $\lambda_2 = 795$ nm; ρ' is the first derivative reflectance; and ρ'' is the second derivative reflectance. Higher-order derivative-based indices are also possible. For instance, ρ''' is the third-order derivative reflectance.

Yellowness Index — YI

The yellowness index (YI) constitutes a descriptor of leaf chlorosis exhibited by stressed vegetation. It measures the change in shape of the reflectance spectra in the interval between the 0.55-μm reflectance maximum and the 0.65-μm minimum. The YI is a three-point gross measure of green–red spectral shape and is thus computed using only wavelengths in the visible spectrum. The justification is that the specified visible region tends to be relatively insensitive to changing leaf water content and structure (Philpot, 1991; Adams et al., 1999):

$$\text{YI} \propto \frac{\rho(\lambda_{-1}) - 2\rho(\lambda_0) + \rho(\lambda_{+1})}{\Delta\lambda^2} \cong \frac{d^2\rho}{d\lambda^2} \qquad (11\text{-}17)$$

where $\rho(\lambda_0)$ is the band center reflectance, $\rho(\lambda_{-1})$ and $\rho(\lambda_{+1})$ denote the reflectance of the lower and higher wavebands, respectively, and $\Delta\lambda$ is the spectral distance (in μm) between wavebands (i.e., $\Delta\lambda = \lambda_0 - \lambda_{-1} = \lambda_{+1} - \lambda_0$). The goal is to select wavelengths such that the band separation ($\Delta\lambda$) is as large as possible while constraining all three channels in the spectral range between approximately 0.55 μm and 0.68 μm. The YI can be computed as 0.1 multiplied by the negative of the finite approximation of the second derivative as a means of downscaling the range of output values; thus, the resultant relationship would be that as yellowness increases, so do positive YI values (Adams et al., 1999). The YI is in units of relative reflectance μm^{-2} (RRU μm^{-2}). The YI magnitude is sensitive to the λ_c and $\Delta\lambda$ values. Using 1999 AVIRIS data, each value in the YI equation can be the average of three adjacent bands; λ_0 centered at 0.61608 μm (band 26); λ_{-1} centered at 0.56696 μm (band 21); and λ_{+1} centered at 0.66518 μm (band 31) based on Adams et al. (1999).

Physiological Reflectance Index — PRI

The physiological reflectance index (PRI) is a narrow-band index that has been correlated with the epoxidation state of the xanthophyll pigments and with photosynthetic efficiency with respect to control (unstressed) and nitrogen-stressed canopies (Gamon et al., 1992). However, the PRI is generally not well correlated with water-stressed canopies experiencing midday wilting. The PRI employs the reflectance at 531 nm and a reference channel in order to minimize the effects of diurnal Sun angle changes. The PRI can track diurnal changes in photosynthetic efficiency. In addition, the PRI may be useful in situations where *in situ* spectroradiometer data are acquired at a different time or light regime geometry than the remote sensor data. The general PRI formula is:

$$PRI = \frac{\rho_{(ref)} - \rho_{(531nm)}}{\rho_{(ref)} + \rho_{(531nm)}} \qquad (11\text{-}18)$$

where ρ_{ref} is a reference wavelength and ρ_{531} is the reflectance at 531 nm. The best PRI noted in Gamon et al. (1992) was:

$$PRI = \frac{\rho_{(550nm)} - \rho_{(531nm)}}{\rho_{(550nm)} + \rho_{(531nm)}}. \qquad (11\text{-}19)$$

A reference wavelength of 550 nm seems to be appropriate at the canopy level. For leaf-scale spectra, a reference wavelength of 570 nm is likely better for a xanthophyll signal. Note that a single PRI is not likely to be applicable across all spatial and temporal scales as well as various canopy types

and diurnally dynamic canopy structures. AVIRIS implementation for the 1999 data might include bands 19 (centered at 547.32 nm) and 17 (centered at 527.67 nm), respectively.

In contrast with the PRI, the NDVI does not accurately indicate real-time photosynthetic fluxes. While NDVI is often sensitive at low leaf area index (LAI) values, it often saturates at high LAI values. The PRI may be able to indicate short-term changes in photosynthetic efficiency, especially in canopies with high LAI values where the NDVI is least effective (Gamon et al., 1992).

Normalized Difference Water Index — NDWI

The normalized difference water index (NDWI) may be used to remotely determine vegetation liquid water content. Two near-infrared channels are used in the computation of the NDWI; one is centered at approximately 860 nm and the other at 1240 nm (Gao, 1996):

$$NDWI = \frac{\rho_{(860nm)} - \rho_{(1240nm)}}{\rho_{(860nm)} + \rho_{(1240nm)}}. \qquad (11\text{-}20)$$

AVIRIS bands that might be used include bands 55 (864.12 nm) and 94 (1237.94 nm).

Red-edge Position Determination — REP

The red edge position (REP) is defined as the point of maximum slope on a vegetation reflectance spectrum between the red and near-IR wavelengths. The REP is useful because it is strongly correlated with foliar chlorophyll content and can be a sensitive indicator of vegetation stress. Although the AVIRIS sensor has a nominal band separation of 10 nm, subtle REP shifts may still not be discernible since the spectra are still sampled coarsely relative to data from laboratory spectrometers (Dawson and Curran, 1998).

A linear method proposed by Clevers (1994) can be implemented that makes use of four narrow bands and is computed as follows:

$$REP = 700 + 40\left[\frac{\rho_{(red\ edge)} - \rho_{(700nm)}}{\rho_{(740nm)} - \rho_{(700nm)}}\right] \qquad (11\text{-}21)$$

where

$$\rho_{(red\ edge)} = \frac{\rho_{(670nm)} + \rho_{(780nm)}}{2}. \qquad (11\text{-}22)$$

A derivative-based REP algorithm exists that is of low computational complexity and is appropriate for canopy-scale studies (Dawson and Curran, 1998).

Crop Chlorophyll Content Prediction

Haboudane et al. (2002) developed a narrow-band vegetation index that integrates the advantages of indices that minimize soil background effects and indices that are sensitive to chlorophyll concentration. The Transformed Chlorophyll Absorption in Reflectance Index (TCARI) (Daughtry et al., 2000) is:

$$TCARI = 3\left[(\rho_{700} - \rho_{670}) - 0.2(\rho_{700} - \rho_{550})\left(\frac{\rho_{700}}{\rho_{670}}\right)\right]. \quad (11\text{-}23)$$

The Optimized Soil-Adjusted Vegetation Index (Rondeaux et al., 1996) belongs to the Soil-Adjusted Vegetation Index (SAVI) family (Huete, 1988) and is:

$$OSAVI = \frac{(1 + 0.16)(\rho_{800} - \rho_{670})}{(\rho_{800} + \rho_{670} + 0.16)}. \quad (11\text{-}24)$$

The ratio

$$\frac{TCARI}{OSAVI} \quad (11\text{-}25)$$

is sensitive to chlorophyll content variations and resistant to variations of leaf area index and solar zenith angle. Evaluation with ground truth resulted in an $r^2 = 0.81$ between estimated and field-measured chlorophyll content data. Haboudane et al. (2002) suggest that the ratio index is insensitive to LAI variations for LAI values ranging from 0.5 to 8 and that it might be used operationally In precision agriculture because it allows an accurate estimation of crop photosynthetic pigments without a priori knowledge of crop canopy architecture.

Blonski et al. (2002) provide algorithms that can be used to synthesize multispectral bands (e.g., Landsat ETM$^+$) from hyperspectral data (e.g., AVIRIS) for use with indices typically based on broadband multispectral data.

Derivative Spectroscopy

Differentiation of a curve or its mathematical function estimates the slope over the entire interval. Deriving the slope of spectroradiometer-derived curves is called *derivative spectroscopy*. Derivative spectroscopy methods were originally developed in analytical chemistry to eliminate background signals and resolve overlapping spectral features (Demetriades-Shah et al., 1990). The concept has also been applied to differentiating remote sensing–derived spectra. Differentiation does not yield more information than exists in the original spectral channels. It can be used, however, to emphasize desired information while suppressing or eliminating information not of interest. For example, background absorption or reflectance signals caused by stray light can be eliminated (Talsky, 1994). Spectral features with sharp structures may be enhanced compared with broader-structured features (Chadburn, 1982).

Because of the inherent differences between laboratory- and remote sensor–based data, not all laboratory-based spectroscopic procedures translate well to remotely sensed inquiries. Laboratory analysis is characterized by controlled illumination sources and viewing geometries, as well as by the typical assumption of homogenous target samples and the use of a known standard. Conversely, imaging spectroscopic remote sensing entails a natural illumination source, mixed pixels, varying topography, generally coarser spectral resolution than that of laboratory spectrophotometers, and the general lack of useful reference standards (Tsai and Philpot, 1998). Given these differences, the application of laboratory spectroscopic techniques to remote sensor data must be carefully considered. Derivative techniques have been applied to remote sensor imagery, but such applications have been rather limited to date (e.g., Demetriades-Shah et al., 1990; Philpot, 1991; Li et al., 1993; Penuelas et al., 1993; Tsai and Philpot, 1998). Nevertheless, there are advantages to using derivative-based analytical techniques with remotely sensed data. Derivative spectra are sometimes more informative than zero-order reflectance spectra when attempting to glean certain information or relationships from the data. For example, Malthus and Madeira (1993) found first-order derivative spectra in the visible wavelengths to be more highly correlated with percent leaf surface area infected by the fungus *Botrytis fabae* than the original zero-order reflectance data.

First-, second-, and third-order derivative spectra may be computed on a pixel-by-pixel basis. The differentiation can be performed using three-point Lagrangian formulas (Hildebrand, 1956):

$$a'_{-1} = \frac{1}{2h}(-3a_{-1} + 4a_0 - a_1) + \frac{h^2}{3}a(\xi) \quad (11\text{-}26)$$

$$a'_0 = \frac{1}{2h}(-a_{-1} + a_1) - \frac{h^2}{6}a(\xi) \quad (11\text{-}27)$$

$$a'_1 = \frac{1}{2h}(a_{-1} - 4a_0 + 3a_1) + \frac{h}{3}a(\xi) \qquad (11\text{-}28)$$

where the subscripts 0, −1, and 1 denote the first derivative at the center point, and points to the left and right of the center point, respectively. h and ξ are distance and error terms, respectively. Higher-order derivative spectra are calculated in an analogous manner.

In remote sensing investigations, higher-order derivatives (second-order and higher) are relatively insensitive to illumination intensity variations, due to cloud cover, Sun angle variance, or topographic effects. In addition, derivatives are usually insensitive to changes in spectral solar flux and skylight (Tsai and Philpot, 1998). Derivative techniques can also be used to address interference from soil background reflectance in vegetation remote sensing studies (i.e., separating the vegetative signal from background noise). Second-order derivative spectra, which are insensitive to soil reflectance, specifically mitigate this problem, while first-order derivative spectra do not (Demetriades-Shah, 1990; Li et al., 1993).

Derivative spectroscopic techniques can also be disadvantageous. The signal-to-noise ratio (SNR) degrades at increasingly higher derivative orders (Talsky, 1994). Derivative spectra are also sensitive to noise. Thus, a prior smoothing step is often necessary (Dawson and Curran, 1998). Algorithms based on least-square fitting are commonly used to reduce random noise in the spectral data (Tsai and Philpot, 1998).

The 16-bit radiometric resolution of AVIRIS data facilitates the creation of quality higher-order derivative spectra. For example, a second-order derivative image of the MWMF on the Savannah River Site is displayed in **Color Plate 11-7**. Approximately the same areas of healthy vegetation and potential stressed vegetation are apparent in the derivative image, but they are more visually striking. In this example, only an initial smoothing was performed using EFFORT prior to the derivative calculation.

References

Adams, J. B., Sabot, D. E., Kapos, V., Almeida, F., Roberts, D. A., Smith, M. O. and A. R. Gillespie, 1995, "Classification of Multispectral Images Based on Fractions of Endmembers: Application to Land Cover Change in the Brazilian Amazon," *Remote Sensing of Environment*, 52:137–154.

Adler-Golden, S. M., Matthew, M. W., Anderson, G. P., Felde, G. W. and J. A. Gardner, 2002, "An Algorithm for De-shadowing Spectral Imagery," *Proceedings, Annual JPL AVIRIS Conference*, Pasadena: NASA Jet Propulsion Laboratory, http://eis.jpl.nasa.gov/aviris/html/aviris.biblios.html#2002, 8 p.

Adams, M. L., Philpot, W. D. and W. A. Norvell, 1999, "Yellowness Index: An Application of the Spectral Second Derivative to Estimate Chlorosis of Leaves in Stressed Vegetation," *International Journal of Remote Sensing*, 20(18):3663–3675.

Analytical Imaging and Geophysics, 1999, *Hyperspectral Data Analysis and Image Processing Workshop*, Boulder, CO: Analytical Imaging and Geophysics, March 15–19, 310 p.

ASTER Spectral Library, 2004, contains the *Johns Hopkins University Spectral Library, Jet Propulsion Laboratory Spectral Library,* and the *U.S. Geological Survey Spectral Library*, Pasadena: NASA Jet Propulsion Laboratory, http://speclab.jpl.nasa.gov.

Baltsavias, E. P., 2002, "Special Section on Image Spectroscopy and Hyperspectral Imaging," *ISPRS Journal of Photogrammetry & Remote Sensing*, 57:169–170.

Belokon, W. F., 1997, *Multispectral Imagery Reference Guide*, Fairfax, VA: Logicon Geodynamics, 101 p.

Berk, A., Bernstein, L. S., Anderson, G. P., Acharya, P. K., Robertson, D. C., Chetwynd, J. H. and S. M. Adler-Golden, 1998, "MODTRAN Cloud and Multiple Scattering Upgrades with Application to AVIRIS," *Remote Sensing of Environment,* 65:367–375.

Berk, A., Bernstein, L. S. and D. C. Robertson, 1989, *MODTRAN: A Moderate Resolution Model for LOWTRAN 7*, GL-TR-89-0122, Air Force Geophysics Lab, Hanscom AFB, MA, 38 p.

Blonski, S., Gasser, G., Russell, J., Ryan, R., Terrie, G. and V. Zanoni, 2002, "Synthesis of Multispectral Bands from Hyperspectral Data: Validation Based on Images Acquired by AVIRIS, Hyperion, ALI, and ETM⁺," *Proceedings, Annual AVIRIS Conference*, Pasadena: NASA Jet Propulsion Laboratory: Pasadena, http://eis.jpl.nasa. gov/aviris/html/aviris.biblios.html#2002, 9 p.

Boardman, J. W., 1993, "Automating Spectral Unmixing of AVIRIS Data Using Convex Geometry Concepts," *Summaries of the 4th Annual JPL Airborne Geoscience Workshop*, Pasadena: Jet Propulsion Laboratory, JPL Publication 93-26, 1:11–14.

Boardman, J. W., 1997, "Mineralogic and Geochemical Mapping at Virginia City, Nevada, Using 1995 AVIRIS Data," *Proceedings,*

12th Thematic Conference on Geological Remote Sensing, Ann Arbor: Environmental Research Institute of Michigan, 21–28.

Boardman, J. W., 1998, "Post-ATREM Polishing of AVIRIS Apparent Reflectance Data Using EFFORT: A Lesson in Accuracy Versus Precision," *Summaries of the 7th JPL Airborne Earth Science Workshop*, Pasadena: Jet Propulsion Laboratory, 1:53.

Boardman, J. W., 1999, "Precision Geocoding of Low Altitude AVIRIS Data: Lessons Learned from 1998," *Summaries of the 8th JPL Airborne Earth Science Workshop*, Pasadena: Jet Propulsion Laboratory, BOA-1 to BOA-6.

Boardman, J. W. and J. F. Huntington, 1997, "Mineralogic and Geochemical Mapping at Virginia City, Nevada Using 1995 AVIRIS Data," *12th International Conference and Workshops on Applied Geologic Remote Sensing*, Denver, CO: 191–196.

Boardman, J. W. and F. A. Kruse, 1994, "Automated Spectral Analysis: A Geological Example Using AVIRIS Data, North Grapevine Mountains, Nevada," *Proceedings, 10th Thematic Conference on Geologic Remote Sensing*, Ann Arbor: Environmental Research Institute of Michigan, Vol I: 407–418.

Chadburn, B. P., 1982, "Derivative Spectroscopy in the Laboratory: Advantages and Trading Rules," *Anal. Proc.*, 54:42–43.

Chen, C. M., Hepner, G. F. and R. R. Forster, 2003, "Fusion of Hyperspectral and Radar Data Using the IHS Transformation to Enhance Urban Surface Features," *ISPRS Journal of Photogrammetry & Remote Sensing*, 58:19–30.

Clark, R. N., Swayze, G. A., Livo, K. E., Kokaly, R. F., King, R. V., Dalton, J. B., Vance, J. S., Rockwell, B. W., Hoefen, T. and R. R. McDougal, 2002, "Surface Reflectance Calibration of Terrestrial Imaging Spectroscopy Data: A Tutorial Using AVIRIS," *Proceedings, Annual JPL AVIRIS Conference*, Pasadena: NASA Jet Propulsion Laboratory, http://eis.jpl.nasa.gov/aviris/html/aviris.biblios.html#2002, 21 p.

Clark, R. N., Livo, K. E. and R. F. Kokaly, 1998, "Geometric Correction of AVIRIS Imagery Using On-Board Navigation and Engineering Data," *Summaries of the 7th JPL Airborne Earth Science Workshop*, Pasadena: Jet Propulsion Laboratory, January 12-14, 1998.

Clevers, J. G., 1994, "Imaging Spectrometry in Agriculture: Plant Vitality and Yield Indicators," in Hill, J. and J. Megier (Eds.), *Imaging Spectrometry: A Tool for Environmental Observations*, Dordrecht, Netherlands: Kluwer Academic, 193–219.

Curran, P. J. and J. L. Dungan, 1989, "Estimation of Signal-to-Noise: A New Procedure Applied to AVIRIS data," *IEEE Transactions on Geoscience and Remote Sensing*, 27(5):620–628.

Curran, P. J., Milton, E. J., Atkinson, P. M. and G. M. Foody, 1998, "Remote Sensing: from Data to Understanding," Chapter 3 in Longley, P. A., Brooks, S. M., McDonnell, R. and B. Macmillan (Eds.), *Geocomputation: A Primer*, New York: John Wiley & Sons, 33–59.

Dawson, T. P. and P. J. Curran, 1998, "A New Technique for Interpolating the Reflectance Red Edge Position," *International Journal of Remote Sensing*, 19(11): 2133–2139.

Demetriades-Shah, T. H., Steven, M. D. and J. A. Clark, 1990, "High Resolution Derivative Spectra in Remote Sensing," *Remote Sensing of Environment*, 33: 55–64.

Daughtry, C. S. T., Walthall, C. L., Kim, M. S., Brown de Colstoun, E., and J. E. McMurtrey III, 2000, "Estimating Corn Leaf Chlorophyll Concentration from Leaf and Canopy Reflectance," *Remote Sensing of Environment*, 74:229–239.

Elvidge, C. D. and Z. Chen, 1995, "Comparison of Broad-band and Narrow-band Red and Near-infrared Vegetation Indices," *Remote Sensing of Environment*, 54(1):38–48.

Filippi, A. M., 1998, *Hyperspectral Image Classification Using a Batch Descending Fuzzy Learning Vector Quantization Artificial Neural Network: Vegetation Mapping at the John F. Kennedy Space Center*, unpublished Master's Thesis, Columbia: University of South Carolina, 276 p.

Gao, B. C., 1996, "NDWI: A Normalized Difference Water Index for Remote Sensing of Liquid Water from Space," *Remote Sensing of Environment,* 58:257–266.

Gao, B. C., Heidebrecht, K. B. and A. F. H. Goetz, 1999, *ATmosphere REMoval Program (ATREM) User's Guide*, Boulder, CO: Center for the Study of the Earth from Space, 31 p.

Goel, P. K., Prasher, S. O., Patel, R. M., Landry, J. A., Bonnell, R. B. and A. A. Viau, 2003, "Classification of Hyperspectral Data by Decision Trees and Artificial Neural Networks to Identify Weed Stress and Nitrogen Status of Corn," *Computers and Electronics in Agriculture*, 39:67–93.

Goetz, A. F. and W. M. Calvin, 1987, "Imaging Spectrometry: Spectral Resolution and Analytical Identification of Spectral Features," in G. Vane (Ed.), *Imaging Spectroscopy II*, Bellingham, WA: Photo-Optical Instrument Engineers, 158–165.

Goetz, A. F., Vane, G., Solomon, J. E. and B. N. Rock, 1985, "Imaging Spectrometry for Earth Remote Sensing," *Science*, 228(4704): 1147–1153.

Green, A. A., Berman, M., Switzer, P. and M. D. Craig, 1988, "A Transformation for Ordering Multispectral Data in Terms of Image Quality with Implications for Noise Removal," *IEEE Transactions on Geoscience and Remote Sensing*, 26(1):65–74.

Green, R. O., Eastwood, M. L., Sarture, C. M., Chrien, T. G., Aronsson, M., Chippendale, B. J., Faust, J. A., Pavri, B. E., Chovit, C. J., Solis, M., Olah, M. R. and O. Williams, 1998, "Imaging Spectroscopy and the Airborne Visible/Infrared Imaging Spectrometer (AVIRIS)," *Remote Sensing of the Environment*, 65:227–248.

Haboudane, D., Miller, J. R., Tremblay, N., Zarco-Tejada, P. J. and L. Dextraze, 2002, "Integrated Narrow-band Vegetation Indices for Prediction of Crop Chlorophyll Content for Application to Precision Agriculture," *Remote Sensing of Environment*, 81:416–426.

Hildebrand, F. B., 1956, *Introduction to Numerical Analysis*, New York: McGraw-Hill, 511 p.

Huete, A. R., 1988, "A Soil-adjusted Vegetation Index (SAVI)," *Remote Sensing of Environment*, 25:295–309.

Jensen, J. R., 2000, *Remote Sensing of the Environment: An Earth Resource Perspective*, Upper Saddle River, NJ: Prentice-Hall, 544 p.

Jia, X. and J. A. Richards, 1993, "Binary Coding of Imaging Spectrometry Data for Fast Spectral Matching and Classification," *Remote Sensing of Environment*, 43:47–53.

JPL, 1999, *Airborne Visible/Infrared Imaging Spectrometer*, Pasadena: Jet Propulsion Laboratory, http://aviris.jpl.nasa.gov/.

Karaska, M. A., Hugenin, R. L., Beacham, J. L., Wang, M., Jensen, J. R. and R. S. Kaufmann, 2004, "AVIRIS Measurements of Chlorophyll, Suspended Minerals, Dissolved Organic Carbon, and Turbidity in the Neuse River, North Carolina," *Photogrammetric Engineering & Remote Sensing*, 70(1):125–133.

Kruse, F. A., 1994, "Imaging Spectrometer Data Analysis — A Tutorial," *Proceedings, International Symposium on Spectral Sensing Research*, 1994, 11 pp.

Kruse, F. A., Kierein-Young, K. S. and J. W. Boardman, 1990, "Mineral Mapping at Cuprite, Nevada, with a 63-Channel Imaging Spectrometer," *Photogrammetric Engineering & Remote Sensing*, 56(1):83–92.

Li, Y., Demetriades-Shah, T. H., Kanemasu, E. T., Shultis, J. K. and K. B. Kirkham, 1993, "Use of Second Derivatives of Canopy Reflectance for Monitoring Prairie Vegetation over Different Soil Backgrounds," *Remote Sensing of Environment*, 44:81–87.

Malthus, T. J. and A. C. Madeira, 1993, "High Resolution Spectroradiometry: Spectral Reflectance of Field Bean Leaves Infected by *Botrytis fabae*," *Remote Sensing of Environment*, 45:107–116.

McGwire, K., Minor, T. and L. Fenstermaker, 2000, "Hyperspectral Mixture Modeling for Quantifying Sparse Vegetation Cover in Arid Environments," *Remote Sensing of Environment*, 72:360–374.

Mustard, J. F., Staid, M. I. and W. J. Fripp, 2001, "A Semianalytical Approach to the Calibration of AVIRIS Data to Reflectance over Water Application in a Temperate Estuary," *Remote Sensing of Environment*, 75:335–349.

Nolin, A. W. and J. Dozier, 2000, "A Hyperspectral Method for Remotely Sensing the Grain Size of Snow," *Remote Sensing of Environment*, 74:207–216.

Okin, G. S., Roberts, D. A., Muray, B. and W. J. Okin, 2001, "Practical Limits on Hyperspectral Vegetation Discrimination in Arid and Semiarid Environments," *Remote Sensing of Environment*, 77:212–225.

Penuelas, J., Gamon, J. A., Griffin, K. L. and C. B. Field, 1993, "Assessing Community Type, Plant Biomass, Pigment Composition, and Photosynthetic Efficiency of Aquatic Vegetation from Spectral Reflectance," *Remote Sensing of Environment*, 46:110–118.

Philpot, W. D., 1991, "The Derivative Ratio Algorithm: Avoiding Atmospheric Effects in Remote Sensing," *IEEE Transactions on Geoscience and Remote Sensing*, 29(3):350–357.

Research Systems, Inc., 2003, *ENVI User's Guide*, Boulder, CO: Research Systems, Inc.

Richards, J. A. and X. Jia, 1999, *Remote Sensing Digital Image Analysis*, New York: Springer-Verlag, 363 p.

Roberts, D. A., Smith, M. O. and J. B. Adams, 1993, "Green Vegetation, Nonphotosynthetic Vegetation and Soils in AVIRIS Data," *Remote Sensing of Environment*, 44:255–269.

Rondeaux, G., Steven, M. and F. Baret, 1996, "Optimization of Soil-adjusted Vegetation Indices," *Remote Sensing of Environment*, 55:95–107.

Schuerger, A. C., unpublished, "Use of Laser-induced Fluorescence Spectroscopy and Hyperspectral Imaging to Detect Plant Stress in Bahiagrass Grown under Different Concentrations of Zinc and Copper."

Schowengerdt, R. A., 1997, *Remote Sensing: Models and Methods for Image Processing*, New York: Academic Press, 522 p.

Segl, K., Roessner, S., Heiden, U. and H. Kaufmann, 2003, "Fusion of Spectral and Shape Features for Identification of Urban Surface Cover Types Using Reflective and Thermal Hyperspectral Data," *ISPRS Journal of Photogrammetry & Remote Sensing*, 58:99–112.

Talsky, G., 1994, *Derivative Spectrophotometry: Low and Higher Order*, New York: VCH Publishers, 228 p.

Tanre, D., Deroo, C., Duhaut, P., Herman, M., Morcrette, J. J., Perbos, J. and P. Y. Deschamps, 1986, *Simulation of the Satellite Signal in the Solar Spectrum (6S) User's Guide,* U.S.T. de Lille, 59655 Villeneuve d'Ascq, France: Laboratoire d'Optique Atmospherique.

Thenkabail, P. S., Smith, R. B. and E. De Paww, 2000, "Hyperspectral Vegetation Indices and Their Relationships with Agricultural Crop Characteristics," *Remote Sensing of Environment,* 71:158–182.

Teillet, P. M., 1986, "Image Correction for Radiometric Effects in Remote Sensing," *International Journal of Remote Sensing*, 7(12):1637–1651.

Tsai, F. and W. Philpot, 1998, "Derivative Analysis of Hyperspectral Data," *Remote Sensing of Environment*, 66:41–51.

Underwood, E., Ustin, S. and D. DiPietro, 2003, "Mapping Nonnative Plants Using Hyperspectral Imagery," *Remote Sensing of Environment*, 86:150–161.

Vermote, E., Tanre, D., Deuze, J. L., Herman, M. and J. J. Morcrette, 1996, *Second Simulation of the Satellite Signal in the Solar Spectrum (6S)*, Washington: NASA Goddard Space Flight Center Code 923, 54 p.

Wolfe, W. L., 1997, *Introduction to Imaging Spectrometers*, Vol. TT25, Bellingham, WA: SPIE Optical Engineering Press, 148 p.

Williams, A. P. and E. R. Hunt, 2002, "Estimation of Leafy Spurge Cover from Hyperspectral Imagery Using Mixture Tuned Matched Filtering," *Remote Sensing of Environment*, 82:446–456.

Winter, M. E., Schlangen, M. and E. M. Winter, 1999, "Comparison of Autonomous Hyperspectral Endmember Determination Methods," *Proceedings, ASPRS Annual Conference,* Portland, OR: American Society for Photogrammetry & Remote Sensing, 444–451.

Digital Change Detection 12

B iophysical materials and human-made features on the surface of Earth are inventoried using remote sensing and *in situ* techniques. Some of the data are fairly static; they do not change over time. Conversely, some biophysical materials and human-made features are dynamic, changing rapidly. It is important that such changes be inventoried accurately so that the physical and human processes at work can be more fully understood (Lunetta and Elvidge, 2000; Zhan et al., 2002). In fact, it is believed that land-use/land-cover change is a major component of global change with an impact perhaps greater than that of climate change (Skole, 1994; Foody, 2001). It is not surprising, therefore, that significant effort has gone into the development of change detection methods using remotely sensed data (e.g., Jensen et al., 1997; Maas, 1999; Song et al., 2001; Arzandeh and Wang, 2003; Lunetta and Lyons, 2003). This chapter reviews how change information is extracted from digital remotely sensed data. It summarizes the remote sensor system and environmental parameters that must be considered when change detection takes place. Several of the most widely used change detection algorithms are introduced and demonstrated.

 Steps Required to Perform Change Detection

The general steps required to perform digital change detection using remotely sensed data are summarized in Figure 12-1.

Change Detection Geographic Region of Interest

The dimensions of the change detection region of interest (ROI) must be carefully identified and held constant throughout a change detection project. The geographic ROI (e.g., a county, state, or watershed) is especially important in a change detection study because it must be completely covered by *n* dates of imagery. Failure to ensure that each of the multiple-date images covers the geographic area of interest results in change detection maps with data voids that are problematic when computing change statistics.

Change Detection Time Period

Sometimes change detection studies are overly ambitious in their attempt to monitor changes in the landscape. Sometimes the time period selected over which change is to be monitored is too short or too long to capture the information of interest. Therefore, the analyst must be careful to identify the optimal change detection time period(s). This selection, of course, is dictated by the nature of the problem. Traffic transportation studies might require a change detection period of just a few seconds or minutes. Conversely, images obtained monthly or seasonally might be sufficient to monitor the greening of a continent. Careful selection of the change detection time period can ensure that resource analysis funds are not wasted.

Select an Appropriate Land-use/Land-cover Classification System

As discussed in Chapter 9, it is wise to use an established, standardized land-cover/land-use classification system for change detection, such as the following:

- American Planning Association *Land-Based Classification Standard* (LBCS),

- U.S. Geological Survey *Land Use/Land Cover Classification System for Use with Remote Sensor Data*,

- U.S. *National Vegetation Classification System* (NVCS),

- U.S. Fish and Wildlife Service *Classification of Wetlands and Deepwater Habitats of the United States*, and

- International Geosphere-Biosphere Program *Land Cover Classification System*.

The use of standardized classification systems allows change information to be compared with other studies.

Hard and Fuzzy Change Detection Logic

Most change detection studies have been based on the comparison of multiple-date *hard* land-cover classifications of remotely sensed data. The result is the creation of a *hard* change detection map consisting of information about the change in discrete categories (e.g., change in forest, agriculture). This is still very important and practical in many instances, but we now recognize that it is ideal to capture both discrete and fuzzy changes in the landscape (refer to Chapter 9 for a discussion about fuzzy land-cover classification).

Land-cover changes may range from no landscape alteration whatsoever, through modifications of variable intensity, to a full transformation or conversion to an entirely new class (e.g., the Denver, CO, example later in this chapter). Scientists now believe that replacing the Date *n* and Date *n + 1* hard classification maps typically used in a change detection project with fuzzy classification maps will result in more informative and accurate land-cover change information (Foody, 2001; Woodcock et al., 2001).

Per-pixel or Object-oriented Change Detection

The majority of digital image change detection has been based on processing Date *n* and Date *n + 1* classification maps pixel by pixel. This is commonly referred to as *per pixel* change detection. *Object-oriented* change detection involves the comparison of two or more scenes consisting of many relatively homogenous image objects (patches or segments) that were identified using the techniques discussed in Chapter 9. The smaller number of relatively homogeneous image objects in the two scenes are then subjected to change detection techniques discussed in this chapter.

Remote Sensing System Considerations

Successful remote sensing change detection requires careful attention to:

- remote sensor system considerations, and

- environmental characteristics.

Failure to understand the impact of the various parameters on the change detection process can lead to inaccurate results (Dobson et al., 1995; Yuan and Elvidge, 1998). Ideally, the remotely sensed data used to perform change detection is acquired by a remote sensor system that holds the following resolutions constant: temporal, spatial (and look angle), spectral, and radiometric. It is instructive to review each of these parameters and identify why they can have a significant impact on the success of a remote sensing change detection project.

Temporal Resolution

Two important temporal resolutions should be held constant during change detection using multiple dates of remotely sensed data. First, the data should be obtained from a sensor system that acquires data at approximately the *same time of day*. For example, Landsat Thematic Mapper data are acquired before 9:45 a.m. for most of the conterminous United States. This eliminates diurnal Sun angle effects that

---

General Steps Used to Conduct Digital Change Detection Using Remote Sensor Data

State the nature of the change detection problem.
      * Specify change detection geographic region of interest.
      * Specify change detection time period (e.g., daily, seasonal, yearly).
      * Define the classes of interest in a classification system.
      * Select hard and/or fuzzy change detection logic.
      * Select per-pixel or object-oriented change detection.
Considerations of significance when performing change detection.
      * Remote sensing system considerations:
          - Spatial, spectral, temporal, and radiometric resolution
      * Environmental considerations:
          - Atmospheric conditions
          - Soil moisture conditions
          - Phenological cycle characteristics
          - Tidal stage, etc.
Process remote sensor data to extract change information.
      * Acquire appropriate change detection data:
          - *In situ* and collateral data
          - Remotely sensed data:
              - Base year (time n)
              - Subsequent year(s) (time $n - 1$ or $n + 1$)
      * Preprocess the multiple-date remote sensor data:
          - Geometric correction
          - Radiometric correction (or normalization)
      * Select change detection algorithm.
      * Apply appropriate image classification logic if necessary:
          - Supervised, unsupervised, hybrid
      * Perform change detection using GIS algorithms:
          - Highlight selected classes using change detection matrix
          - Generate change-map products
          - Compute change statistics
Perform accuracy assessment.
      * Select method:
          - Qualitative confidence building
          - Statistical measurement
      * Determine number of samples required by class.
      * Select sampling scheme.
      * Obtain ground reference test information.
      * Create and analyze change detection error matrix:
          - Univariate and multivariate statistical analysis
Accept or reject previously stated hypothesis.
Distribute results if accuracy is acceptable.

Figure 12-1 The general steps used to perform digital change detection of remotely sensed data.

can cause anomalous differences in the reflectance properties of the remotely sensed data. Second, whenever possible it is desirable to use remotely sensed data acquired on *anniversary dates*, for example, February 1, 2004, and February 1, 2005. Using anniversary date imagery minimizes the influence of seasonal Sun-angle and plant phenological differences that can negatively impact a change detection project (Jensen et al., 1993a).

Spatial Resolution and Look Angle

Accurate spatial registration of at least two images is essential for digital change detection. Ideally, the remotely sensed data are acquired by a sensor system that collects data with the same *instantaneous field of view* on each date. For example, Landsat Thematic Mapper data collected at 30×30 m spatial resolution on two dates are relatively easy to register to one another. It is possible to perform change detection using data collected from two different sensor systems with different IFOVs, for example, Landsat TM data (30×30 m) for Date 1 and SPOT HRV XS data (20×20 m) for Date 2. In such cases, it is usually necessary to decide on a representative minimum mapping unit (e.g., 20×20 m) and then resample both datasets to this uniform pixel size. This does not present a significant problem as long as the image analyst remembers that the information content of the resampled data can never be greater than the IFOV of the original sensor system (i.e., even though the Landsat TM data may be resampled to 20×20 m pixels, the information was still acquired at 30×30 m resolution and we should not expect to be able to extract additional spatial detail from the dataset).

Geometric rectification algorithms discussed in Chapter 6 are used to register the images to a standard map projection (Universal Transverse Mercator for most U.S. projects). Rectification should result in the two images having a root mean square error (RMSE) of ≤ 0.5 pixel. Misregistration of the two images may result in the identification of spurious areas of change between the datasets. For example, just one pixel misregistration may cause a stable road on the two dates to show up as a new road in the change image. Gong et al. (1992) suggest that adaptive grayscale mapping (a form of spatial filtering) be used in certain instances to remove change detection misregistration noise.

Some remote sensing systems like SPOT, IKONOS, and QuickBird collect data at off-nadir *look angles* as much as $\pm 20°$; that is, the sensors obtain data of an area on the ground from an *oblique* vantage point. Two images with significantly different look angles can cause problems when used for change detection purposes. For example, consider a

maple forest consisting of very large, randomly spaced trees. A SPOT image acquired at $0°$ off-nadir will look directly down on the top of the canopy. Conversely, a SPOT image acquired at $20°$ off-nadir will record reflectance information from the side of the canopy. Differences in reflectance from the two datasets may cause spurious change detection results. Therefore, the data used in a remote sensing digital change detection should be acquired with approximately the same look angle, if possible.

Spectral Resolution

A fundamental assumption of digital change detection is that a difference exists in the spectral response of a pixel on two dates if the biophysical materials within the IFOV have changed between dates. Ideally, the spectral resolution of the remote sensor system is sufficient to record reflected radiant flux in spectral regions that best capture the most descriptive spectral attributes of the object. Unfortunately, different sensor systems do not record energy in exactly the same portions of the electromagnetic spectrum (i.e., bandwidths). For example, the Landsat multispectral scanner (MSS) recorded energy in four relatively broad multispectral bands. SPOT 1, 2, and 3 HRV sensors collect data in three relatively broad multispectral bands and one panchromatic band. The Landsat 7 Enhanced Thematic Mapper Plus (ETM$^+$) collects data in six relatively broad optical bands, one thermal infrared band, and one broad panchromatic band (Chapter 2). Ideally, the same sensor system is used to acquire imagery on multiple dates. When this is not possible, the analyst should *select bands that approximate one another*. For example, Landsat MSS bands 4 (green), 5 (red), and 7 (near-infrared) and SPOT bands 1 (green), 2 (red), and 3 (near-infrared), can be used successfully with Landsat ETM$^+$ bands 2 (green), 3 (red), and 4 (near-infrared). Many of the change detection algorithms do not function well when bands from one sensor system do not match those of another sensor system (e.g., utilizing the Landsat TM band 1 (blue) with either SPOT or Landsat MSS data may not be wise).

Radiometric Resolution

An analog-to-digital conversion of the satellite remote sensor data usually results in 8-bit brightness values ranging from 0 to 255 (Table 2-2). Ideally, the sensor systems collect the data at the *same radiometric precision on both dates*. When the radiometric resolution of data acquired by one system (e.g., Landsat MSS 1 with 6-bit data) is compared with data acquired by a higher radiometric resolution instrument (e.g., Landsat TM with 8-bit data), the lower-resolution data (e.g., 6 bits) should be decompressed to 8 bits for change detection purposes. The precision of decompressed brightness values can never be better than the original, non-compressed data. Ideally, the brightness values associated with both dates of imagery are converted to apparent surface reflectance, which eliminates the problem.

Environmental Considerations of Importance When Performing Change Detection

Failure to understand the impact of various environmental characteristics on the remote sensing change detection process can lead to inaccurate results. When performing change detection, it is desirable to hold environmental variables as constant as possible.

Atmospheric Conditions

There should be no clouds (including stratus) or extreme humidity on the days that remote sensing data are collected. Even a thin haze can alter spectral signatures in satellite images enough to create the false impression of spectral change between two dates. Obviously, 0% cloud cover is preferred for satellite imagery and aerial photography. At the upper limit, cloud cover >20% is usually unacceptable. It should also be remembered that clouds not only obscure terrain, but the cloud shadow also causes major image classification problems. Areas obscured by clouds or affected by cloud shadow will filter through the entire change detection process, limiting the utility of the change detection product. Therefore, analysts must use good judgment in evaluating such factors as the specific locations affected by cloud cover and shadow and the availability of timely surrogate data for obscured areas. Substituting information derived from the interpretation of aerial photography for a cloud-shrouded area might be an option. Even when the stated cloud cover is 0%, it is advisable to browse the proposed image to confirm the cloud cover estimate.

Assuming no cloud cover, the use of anniversary dates helps to ensure general, seasonal agreement between the atmospheric conditions on the two dates. However, if dramatic differences exist in the atmospheric conditions on the *n* dates of imagery to be used in the change detection process, it may be necessary to *remove the atmospheric attenuation in the imagery.* Radiative transfer-based atmospheric correction algorithms may be used to radiometrically correct the remote sensor data (Kim and Elman; 1990; Song et al., 2001). For mountainous areas, topographic effects may also have to be removed (Civco, 1989). If it is not possible to perform an absolute radiometric correction, then image-to-image normalization might be a viable alternative as dis-

cussed in Chapter 6 (Eckhardt et al., 1990; Jensen et al., 1995; Du et al., 2002).

Soil Moisture Conditions

Ideally, *the soil moisture conditions should be identical* for the *n* dates of imagery used in a change detection project. Extremely wet or dry conditions on one of the dates can cause serious change detection problems. Therefore, when selecting the remotely sensed data to be used for change detection, it is very important not only to look for anniversary dates, but also to review precipitation records to determine how much rain or snow fell in the days and weeks prior to remote sensing data collection. When soil moisture differences between dates are significant for only certain parts of the study area (perhaps due to a local thunderstorm), it may be necessary to stratify (cut out) those affected areas and perform a separate change detection analysis, which can be added back in the final stages of the project.

Phenological Cycle Characteristics

Natural ecosystems go through repeatable, predictable cycles of development. Human beings often modify the landscape in repeatable, predictable stages. These cycles of predictable development are often referred to as phenomenological or *phenological cycles.* Image analysts use these cycles to identify when remotely sensed data should be collected to obtain the maximum amount of usable change information. Therefore, analysts must be intimately familiar with the *biophysical* characteristics of the vegetation, soils, and water constituents of ecosystems and their phenological cycles. Likewise, it imperative that they understand the phenological cycles associated with human-made development, such as residential expansion at the urban/rural fringe.

Vegetation Phenology: Vegetation grows according to relatively predictable diurnal, seasonal, and annual phenological cycles. Obtaining near-anniversary images greatly minimizes the effects of seasonal phenological differences that may cause spurious change to be detected in the imagery. When attempting to identify change in agricultural crops, the analyst must be aware of when the crops were planted. Ideally, monoculture crops (e.g., corn, wheat) *are planted at approximately the same time of year* on the two dates of imagery. A month lag in planting date between fields of the same crop can cause serious change detection error. Second, the monoculture crops should be *the same species.* Different species of a crop can cause the crop to reflect energy differently on the multiple dates of anniversary imagery. In addition, changes in row spacing and direction can have an impact. These observations suggest that the analyst must

Phenological Cycle of Cattails and Waterlilies in Par Pond in South Carolina

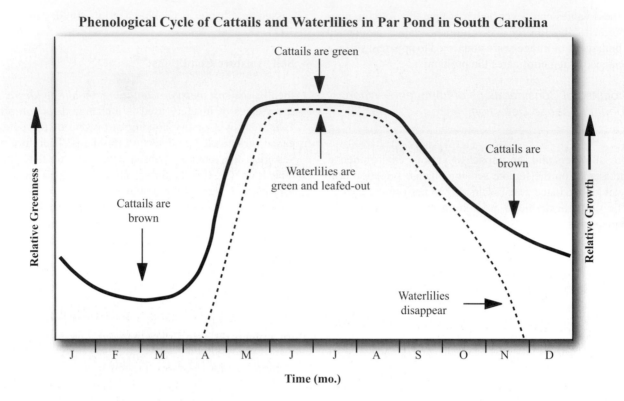

Figure 12-2   Yearly phenological cycle of cattails and waterlilies in Par Pond, SC.

know the crop's *biophysical* characteristics as well as the *cultural* land-tenure practices in the study area so that the most appropriate remotely sensed data can be selected for change detection.

Natural vegetation ecosystems such as wetland aquatic plants, forests, and rangeland have unique phenological cycles. For example, consider the phenological cycle of cattails and waterlilies found in lakes in the southeastern United States (Figure 12-2). Cattails persist year round in lakes and are generally found in shallow water adjacent to the shore (Jensen et al., 1993b). They begin greening up in early April and often have a full, green canopy by late May. Cattails senesce in late September to early October, yet they are physically present and appear brown through the winter months. Conversely, waterlilies and other nonpersistent species do not live through the winter. They appear at the outermost edge of the cattails in early May and reach full emergence 6 to 8 weeks later. The waterlily beds usually persist above water until early November, at which time they disappear. The phenological cycles of cattails and waterlilies dictate the most appropriate times for remote sensing data acquisition. The spatial distribution of cattails is best derived from remotely sensed data acquired in the early spring

(April or early May) when the waterlilies have not yet developed. Conversely, waterlilies do not reach their full development until the summer, thus dictating late summer or early fall as a better period for remote sensing data acquisition and measurement. It will be shown later in this chapter that SPOT panchromatic imagery collected in April and October of most years may be used to identify change in the spatial distribution of these species in southeastern lakes.

Urban–Suburban Phenological Cycles: Urban-suburban landscapes also have phenological cycles. For example, consider the residential development from 1976 to 1978 in the 6-mi^2 portion of the Fitzsimmons 7.5-minute quadrangle near Denver, CO. Aerial photographs obtained on October 8, 1976, and October 15, 1978, reveal dramatic changes in the landscape (Figures 12-3). Most novice image analysts assume that change detection in the urban–rural fringe will capture the residential development in the two most important stages: rural undeveloped land and completely developed residential. Jensen (1981) identified 10 stages of residential development taking place in this region based on evidence of clearing, subdivision, transportation, buildings, and landscaping (Figure 12-4). The remotely sensed data will most likely capture the development in all 10 stages of

Residential Development near Denver, CO, from October 8, 1976 to October 15, 1978

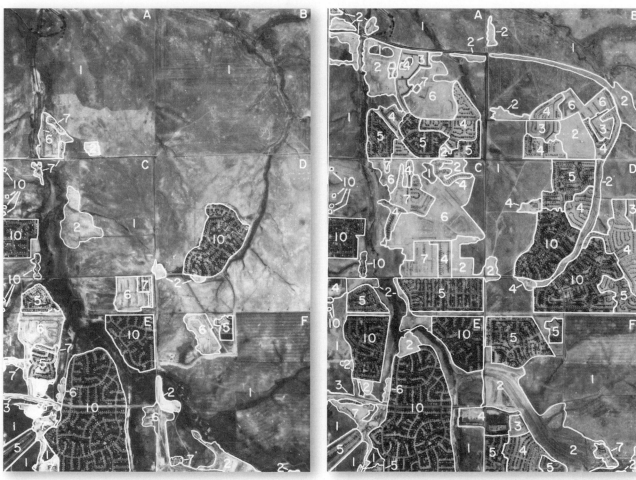

a. October 8, 1976.   b. October 15, 1978.

Figure 12-3   a) Panchromatic aerial photograph of a portion of the Fitzsimmons 7.5-minute quadrangle near Denver, CO, on October 8, 1976. The original scale was 1:52,800. The land cover was visually photo-interpreted and classified into 10 classes of residential development using the logic shown in Figure 12-4. b) Panchromatic aerial photograph of a portion of the Fitzsimmons 7.5-minute quadrangle on October 15, 1978. The original scale was 1:57,600. Comparison with the 1976 aerial photograph reveals substantial residential land development since October 8, 1976.

development. Many of these stages may appear spectrally similar to other phenomena. For example, it is possible that stage 10 pixels (subdivided, paved roads, building, and completely landscaped) may look exactly like stage 1 pixels (original land cover) in multispectral feature space if a relatively coarse spatial resolution sensor system such as the Landsat MSS (79×79 m) is used. This can cause serious change detection problems. Therefore, the analyst must be intimately aware of the phenological cycle of all urban phenomena being investigated, as well as the natural ecosystems.

Effects of Tidal Stage on Change Detection

Tidal stage is a crucial factor when conducting change detection in the coastal zone. Ideally, the tidal stage is identical on multiple-date images used for change detection. Sometimes this severe constraint can rule out the use of satellite remote sensing systems that cannot collect data off-nadir to meet the stringent tidal requirements. In such cases, the only way to obtain remote sensor data in the coastal zone that meets the stringent tidal requirements is to use suborbital sensors that can be flown at the exact time required. For most regions,

Dichotomous Key Used to Identify Progressive Stages of Residential Development

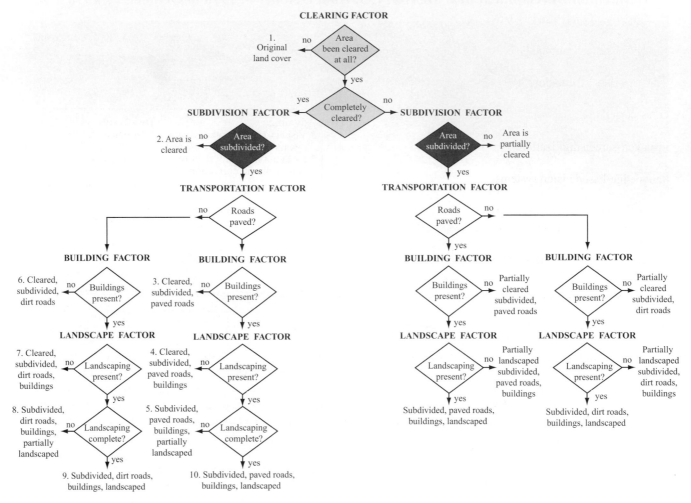

Figure 12-4   Dichotomous key used to identify progressive stages of residential development. Such development in Denver, CO, normally begins by clearing the terrain of vegetation prior to subdivision. In many geographic areas, such as the eastern and southeastern United States, however, some natural vegetation is usually left as landscaping. The absence or existence of natural vegetation dramatically affects the range of signatures that a parcel of land undergoes as it progresses from natural vegetation (1) to fully landscaped residential housing (10).

images to be used for change detection acquired at mean low tide (MLT) are preferred, 1 or 2 ft above MLT are acceptable, and 3 ft or more are generally unacceptable (Jensen et al., 1993a).

Selection of a Change Detection Algorithm

The selection of an appropriate change detection algorithm is very important (Jensen et al., 1995; 2002). First, it will have a direct impact on the type of image classification to be performed (if any). Second, it will dictate whether important

"from–to" change information can be extracted from the imagery. Many change detection projects require that "from–to" information be readily available in the form of maps and tabular summaries. Change detection algorithms commonly used include:

- write function memory insertion

- multi-date composite image

- image algebra (e.g., band differencing, band ratioing)

- post-classification comparison

- binary mask applied to date 2

- ancillary data source used as date 1

- spectral change vector analysis

- chi-square transformation

- cross-correlation

- visual on-screen digitization

- knowledge-based vision systems.

It is instructive to review these change detection alternatives and provide specific examples where appropriate.

Change Detection Using Write Function Memory Insertion

A simple yet powerful method of *visual* change detection involves the use of the three write function memory (WFM) banks found on the graphics card of every digital image processing system (see Chapters 3 and 5). Basically, individual bands (or derivative products) from multiple dates may be inserted into each of the three WFM banks (red, green, or blue) (Figure 12-5) to identify change in the imagery. To appreciate the technique, consider the change that has taken place on the shoreline of Lake Mead, NV. Lake Mead's watershed had experienced a severe drought for the past 5 years, resulting in a significant drawdown of the lake. Figure 12-6a depicts Landsat ETM$^+$ imagery (band 4; $0.75 - 0.90$ µm) obtained on May 3, 2000. Figure 12-6b is an ASTER image (band 3; $0.76 - 0.86$ µm) of the same geographic area obtained on April 19, 2003. Color versions of these two images are found in **Color Plate 12-1a,b**. Note that both datasets record approximately the same near-infrared radiant flux and that the images are within 14 days of being anniversary dates. The ETM$^+$ and ASTER images were resampled to 30×30 m (nearest-neighbor; RMSE $\pm$ 0.5 pixel).

It is possible using write function memory insertion to view multiple dates of registered imagery at one time on the display screen to highlight changes. For example, **Color Plate 12-1c** is a WFM insertion change detection that highlights the land exposed by the drawdown of Lake Mead. It was produced by placing the 2003 ASTER image (band 3; $0.76 - 0.86$ µm) in the red image plane and the 2000 ETM$^+$ image (band 4; $0.75 - 0.90$ µm) in both the green and blue memory planes, respectively. The result is a dramatic display of the area that has changed from 2000 to 2003 highlighted in red.

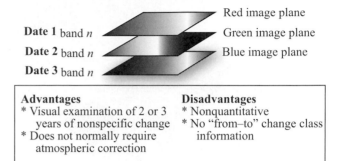

Write Function Memory Insertion Change Detection

Date 1 band *n* — Red image plane
Date 2 band *n* — Green image plane
Date 3 band *n* — Blue image plane

Advantages
* Visual examination of 2 or 3 years of nonspecific change
* Does not normally require atmospheric correction

Disadvantages
* Nonquantitative
* No "from–to" change class information

Figure 12-5 Advantages and disadvantages of write function memory insertion change detection.

Placing the 2003 ASTER image in the green function memory plane and the 2000 ETM$^+$ data in the red and blue image memory planes results in the change being displayed in shades of green (not shown).

WFM insertion may be used to visually examine virtually any type of registered, multiple-date information. For example, Franklin et al. (2002) placed Landsat TM Kauth-Thomas wetness index images from different years in the red, green, and blue function memories to visually highlight forest structure changes.

Advantages of write function memory insertion change detection include the possibility of looking at two and even three dates of remotely sensed imagery (or derivative products) at one time. Also, it is generally not necessary to atmospherically correct the remote sensor data used in write function memory insertion (unless the data are NDVI-related). Unfortunately, the technique does not provide quantitative information on the amount of hectares changing *from* one land-cover category *to* another. Nevertheless, it is an excellent analog method for qualitatively assessing the amount of change in a region, which might help with the selection of one of the more quantitative change detection techniques.

Multidate Composite Image Change Detection

Numerous researchers have rectified multiple dates of remotely sensed imagery (e.g., selected bands of two IKONOS scenes of the same region) and placed them in a single dataset (Figure 12-7). This composite dataset can then be analyzed in a number of ways to extract change information. First, a traditional classification using all *n* bands (six in

Lake Mead, Nevada

a. Landsat ETM+ band 4 (NIR) May 3, 2000.

b. ASTER band 3 (NIR) April 19, 2003.

Figure 12-6   a) Landsat ETM$^+$ imagery of a portion of Lake Mead in Nevada obtained on May 3, 2000. b) ASTER data of Lake Mead obtained on April 19, 2003 (images courtesy NASA Earth Observatory).

the example in Figure 12-7) may be performed. Unsupervised classification techniques will result in the creation of change and no-change clusters. The analyst must then label the clusters accordingly.

Other researchers have subjected the registered composite image dataset to principal component analysis (PCA) to detect change (Fung and LeDrew, 1988; Eastman and Fulk, 1993; Bauer et al., 1994; Yuan and Elvidge, 1998; Maas, 1999). A PCA based on variance–covariance matrices or a standardized PCA based on analysis of correlation matrices is then performed. This results in the computation of eigenvalues and factor loadings used to produce a new, uncorrelated PCA image dataset. The major components of the derived PCA dataset tend to account for variation in the image data that is *not* due to land-cover change, and they are termed *stable components*. Minor components tend to enhance spectral contrasts between the two dates, and they are termed *change components* (Collins and Woodcock, 1996). The difficulty arises when trying to interpret and

Multiple-date Composite Image Change Detection

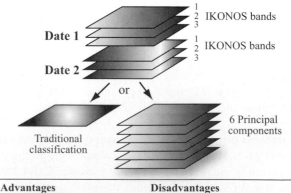

Figure 12-7   Diagram of multiple-date composite image change detection.

Multiple-Date Composite Image Change Detection
Based on Principal Components Analysis

Lake Mead, NV, dataset consists of
* Landsat ETM$^+$ data obtained on May 3, 2000 (bands 2, 3, 4)
* ASTER data obtained on April 19, 2003 (bands 1, 2, 3)

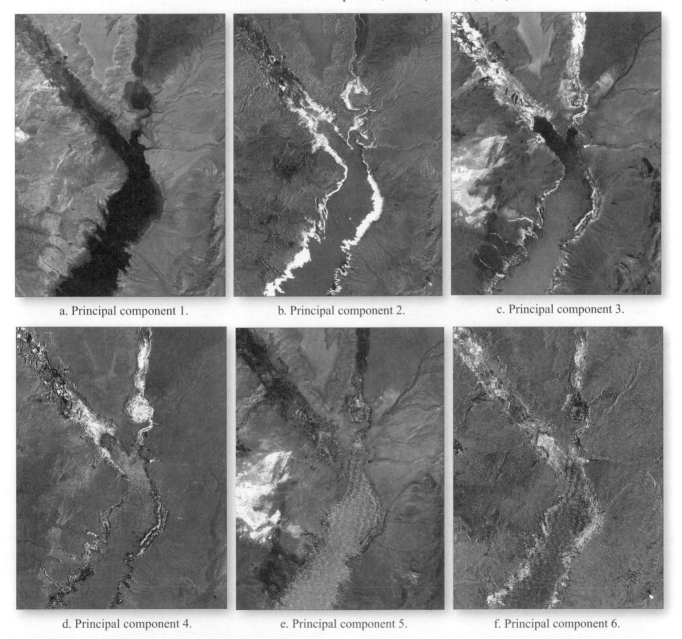

a. Principal component 1.　　　b. Principal component 2.　　　c. Principal component 3.

d. Principal component 4.　　　e. Principal component 5.　　　f. Principal component 6.

Figure 12-8 Principal components derived from a multiple-date dataset consisting of Landsat ETM$^+$ and ASTER imagery. Principal component 2 contains change information. The first three principal components were placed in various write function memory banks to highlight more subtle changes, as shown in **Color Plate 12-2**.

label each component image. Nevertheless, the method is of value and is used frequently. The advantage of this technique is that data do not have to be atmospherically corrected and only a single classification is required. Unfortunately, it is often difficult to label the change classes, and from–to change class information may not be available.

An example of a multiple-date composite image change detection is shown in Figure 12-8. The two three-band Lake Mead, NV, datasets (Landsat ETM[+] and ASTER) were merged into a single six-band dataset and subjected to a principal components analysis. This resulted in the creation of the six principal component images shown in Figure 12-8. Note that principal component 2 is a *change component image* (Figure 12-8b) containing detailed information about the area exposed by the lake drawdown. More subtle change information can be visually extracted from the multiple-date component dataset by placing the first three principal components in various write function memory banks, as shown in **Color Plate 12-2**.

Image Algebra Change Detection

It is possible to identify the amount of change between two rectified images by band ratioing or image differencing (Green et al., 1994; Maas, 1999; Song et al., 2001). Image differencing involves subtracting the imagery of one date from that of another (Figure 12-9). If the two images have almost identical radiometric characteristics (i.e., the data have been normalized or atmospherically corrected), the subtraction results in positive and negative values in areas of radiance change and zero values in areas of no change. The results are stored in a new change image. When 8-bit data are analyzed in this manner, the potential range of difference values found in the change image is –255 to 255 (Figure 12-10). The results can be transformed into positive values by adding a constant, c (e.g., 127). The operation is expressed as:

$$\Delta BV_{ijk} = BV_{ijk}(1) - BV_{ijk}(2) + c \qquad (12\text{-}1)$$

where

ΔBV_{ijk} = change pixel value
$BV_{ijk}(1)$ = brightness value on date 1
$BV_{ijk}(2)$ = brightness value on date 2
$\quad c$ = a constant (e.g., 127)
$\quad i$ = line number
$\quad j$ = column number
$\quad k$ = a single band (e.g., IKONOS band 3).

The change image produced using image differencing usually yields a *BV* distribution approximately Gaussian, where

Image Algebra Change Detection

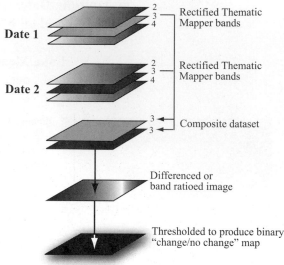

Advantages
* Normally does not require atmospheric correction
* Efficient method of identifying pixels that have changed in brightness value between dates

Disadvantages
* No "from–to" change information available
* Requires careful selection of the "change/no change" threshold

Figure 12-9   Diagram of image algebra change detection.

pixels of no *BV* change are distributed around the mean and pixels of change are found in the tails of the distribution (Song et al., 2001). It is not necessary to add the constant c in Equation 12-1 if the image differencing output file is allowed to be floating point, i.e., the differenced pixel values can range from –255 to 255 (Figure 12-10). Band ratioing involves exactly the same logic, except a ratio is computed with values ranging from $\frac{1}{255}$ to 255 and the pixels that did not change have a ratio value of 1 in the change image.

Image differencing change detection will be demonstrated using two datasets. The first example involves the Landsat ETM[+] imagery (band 4; 0.75 – 0.90 μm) obtained on May 3, 2000, and the ASTER imagery (band 3; 0.76 – 0.86 μm) of Lake Mead obtained on April 19, 2003 (Figure 12-11a and b). These two images were differenced, resulting in the change image histogram shown in Figure 12-11c. Note that the change image histogram is symmetrical, suggesting that one image was histogram-matched to the other. Note that most of the scene did not change between 2000 and 2003; therefore, the vast majority of the pixels are found around the value 0 in the histogram. However, where the lake was drawn down, exposing bedrock, and where new vegetation has grown on the exposed terrain has resulted in significant change documented in the change image in Figure 12-11d.

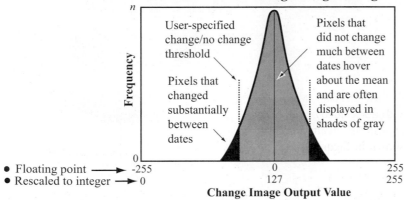

**Image Differencing Change Detection:
Scaling Alternatives and Placement of User-specified
Thresholds in the Change Image Histogram**

Figure 12-10   Image differencing change detection using two dates of 8-bit remote sensor data results in an output change image that can range from –255 to 255. The entire data range may be preserved if the data are stored in floating point format. The data may also be rescaled to 8-bit (0 to 255) data using Equation 12-1 and a constant. Pixels that had approximately the same brightness value (or reflectance if the data were radiometrically corrected) on both dates will produce change image pixel values that hover around 0 or 127, depending upon the scaling. Pixel values that changed dramatically between the two dates will show up in the tails of the change image histogram. Analysts can highlight certain types of change by identifying thresholds in one or both of the tails in the change image. The user-specified thresholds are usually not symmetrical about the mean.

There are actually two types of change in Figure 12-11d, bright white and black. All pixels in the change histogram below the first analyst-specified threshold were assigned black and all those above the second threshold were assigned white. The effect is even more dramatic when these two types of change are color-coded red and green, as shown in **Color Plate 12-1d**.

Figure 12-12 depicts the result of performing image differencing on April 26, 1989, and October 4, 1989, SPOT panchromatic imagery of Par Pond in South Carolina (Jensen et al., 1993b). The data were rectified, normalized, and masked using the methods previously described. The two files were then differenced and a change detection threshold was selected. The result was a change image showing the waterlilies that grew from April 26, 1989, to October 4, 1989 highlighted in gray (Figure 12-12c). The hectares of waterlily change are easily computed. Such information is used to evaluate the effect of various industrial activities on inland wetland habitat.

A critical element of both image differencing and band ratioing change detection is deciding where to place the threshold boundaries between "change" and "no-change" pixels displayed in the change image histogram. The threshold boundaries are rarely known a priori, but have to be found empirically. Sometimes a standard deviation from the mean is

selected and tested. Conversely, most analysts prefer to experiment empirically, placing the threshold at various locations in the tails of the distribution until a realistic amount of change is encountered (Figures 12-10 and 12-11). Thus, the amount of change selected and eventually recoded for display is often subjective and must be based on familiarity with the study area. Unfortunately, image differencing simply identifies the areas that may have changed and provides no "from–to" change information. Nevertheless, the technique is valuable when used in conjunction with other techniques such as the multiple-date change detection using a binary change mask.

Differencing Vegetation Index Images: Image differencing does not have to be based on just the individual bands of remote sensor data. It may also be extended to comparing vegetation index information derived from multiple dates of imagery. For example, scientists have computed a normalized difference vegetation index (NDVI) on two dates and then subtracted one from another to determine change (Yuan and Elvidge, 1998; Lyon et al., 1998; Song et al., 2001):

$$\Delta NDVI_{ij} = NDVI_{ij}(1) - NDVI_{ij}(2) + c, \qquad (12\text{-}2)$$

where

$$\Delta NDVI_{ij} = \text{change in NDVI value}$$
$$NDVI_{ij}(1) = \text{NDVI value on Date 1}$$
$$NDVI_{ij}(2) = \text{NDVI value on Date 2}$$

Image Differencing Change Detection

a. Landsat ETM$^+$ band 4 (NIR) May 3, 2000.

b. ASTER band 3 (NIR) April 19, 2003.

Lake Mead, Nevada
2003 ASTER − 2000 ETM$^+$

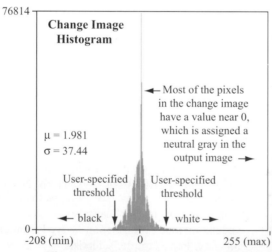

Change Image Histogram

$\mu = 1.981$
$\sigma = 37.44$

← Most of the pixels in the change image have a value near 0, which is assigned a neutral gray in the output image →

User-specified threshold   User-specified threshold

← black   white →

76814

0

-208 (min)   0   255 (max)

c. Histogram of the floating point image created by subtracting 2000 ETM$^+$ data from 2003 ASTER data. The symmetrical distribution confirms that the images were histogram-matched prior to processing.

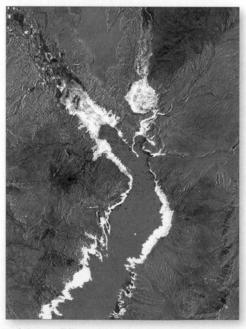

d. Image differencing change detection based on Landsat ETM$^+$ band 4 May 3, 2000 data and ASTER band 3 April 19, 2003 data.

Figure 12-11   a) Landsat ETM$^+$ imagery of a portion of Lake Mead in Nevada obtained on May 3, 2000. b) ASTER data of Lake Mead obtained on April 19, 2003. (c) The histogram of a change image produced by subtracting the ETM$^+$ 2000 data from the ASTER 2003 data. d) Map showing the change as a function of the two thresholds identified in the change image histogram. Values near 0 are shown in shades of gray. Values below the threshold are in black and those above the threshold are in white (images courtesy NASA Earth Observatory).

Image Differencing Change Detection of Waterlily Growth on Par Pond in South Carolina

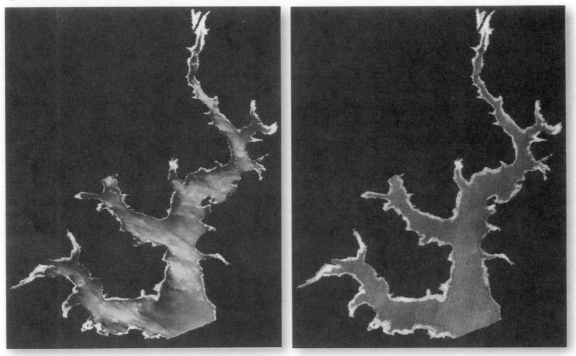

a. SPOT panchromatic image April 26, 1989.   b. SPOT panchromatic image October 4, 1989.

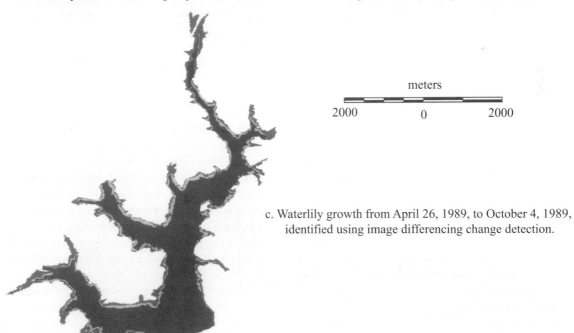

meters

2000   0   2000

c. Waterlily growth from April 26, 1989, to October 4, 1989, identified using image differencing change detection.

Figure 12-12   a) Rectified and masked SPOT panchromatic data of Par Pond located on the Savannah River Site in South Carolina obtained on April 26, 1989. b) SPOT panchromatic data of Par Pond located on the Savannah River Site in South Carolina
obtained on October 4, 1989. c) A map depicting the change in waterlilies from April 26, 1989, to October 4, 1989, using
image differencing logic (Jensen et al., 1993b).

i = line number
j = column number
c = a constant.

It is not necessary to use the constant c in Equation 12-2 if the image differencing output file is allowed to be floating point. The individual images used to perform NDVI change detection should be atmospherically corrected. Change detection based on differencing multiple-date Kauth-Thomas transformations (e.g., change in brightness, greenness, and/or wetness) have also been widely adopted (Ridd and Liu, 1998; Franklin et al., 2002).

Post-classification Comparison Change Detection

Post-classification comparison change detection is a heavily used quantitative change detection method (e.g., Jensen et al., 1995; 2002; Yuan and Elvidge, 1998; Maas, 1999; Song et al., 2001; Civco, 2002; Arzandeh and Wang, 2003). It requires rectification and classification of each remotely sensed image (Figure 12-13). The two maps are then compared on a pixel-by-pixel basis using a *change detection matrix,* to be discussed. Unfortunately, every error in the individual date classification map will also be present in the final change detection map (Rutchey and Velcheck, 1994). Therefore, it is imperative that the individual classification maps used in the post-classification change detection method be as accurate as possible (Arzandeh and Wang, 2003).

To demonstrate the post-classification comparison change detection method, consider the Kittredge (40 river miles inland from Charleston, SC) and Fort Moultrie, SC, study areas (Jensen et al., 1993a) (**Color Plate 12-3**). Nine land-cover classes were inventoried on each date (**Color Plate 12-4**). The 1982 and 1988 classification maps were then compared on a pixel-by-pixel basis using an $n \times n$ GIS matrix algorithm whose logic is shown in **Color Plate 12-5a**. This resulted in the creation of a change image map consisting of brightness values from 1 to 81. The analyst then selected specific "from–to" classes for emphasis. Only a select number of the 72 possible off-diagonal "from–to" land-cover change classes summarized in the change matrix (**Color Plate 12-5a**) were selected to produce the change detection maps (**Color Plate 12-6a** and **b**). For example, all pixels that changed from any land cover in 1982 to Developed Land in 1988 were color coded red (RGB = 255, 0, 0) by selecting the appropriate "from–to" cells in the change detection matrix (10, 19, 28, 37, 46, 55, 64, and 73). Note that the change classes are draped over a Landsat TM band 4 image of the study area to facilitate orientation. Similarly, all

Post-classification Comparison Change Detection

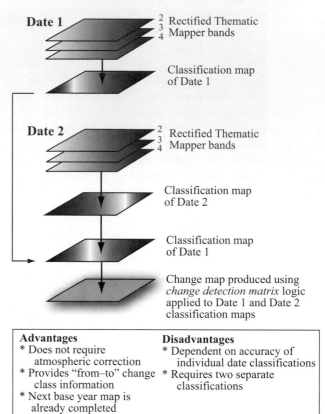

Figure 12-13 Diagram of post-classification comparison change detection.

pixels in 1982 that changed to Estuarine Unconsolidated Shore by December 19, 1988 (cells 9, 18, 27, 36, 45, 54, 63, and 72), were depicted in yellow (RGB = 255, 255, 0). If desired, the analyst could highlight very specific changes such as all pixels that changed from Developed Land to Estuarine Emergent Wetland (cell 5 in the matrix) by assigning a unique color look-up table value (not shown). A color-coded version of the change detection matrix can be used as an effective "from–to" change detection map legend.

Post-classification comparison change detection is widely used and easy to understand. When conducted by skilled image analysts, it represents a viable technique for the creation of change detection products. Advantages include the detailed "from–to" information that can be extracted and the fact that the classification map for the next base year is already complete (Arzandeh and Wang, 2003). However, the accuracy of the change detection depends on the accuracy of the two separate classification maps.

Change Detection Using a Binary Change Mask Applied to Date 2

This method of change detection is very effective. First, the analyst selects the base image referred to as Date 1 at time n. Date 2 may be an earlier image ($n-1$) or a later image ($n+1$). A traditional classification of Date 1 is performed using rectified remote sensor data. Next, one of the bands (e.g., band 3 in Figure 12-14) from both dates of imagery is placed in a new dataset. The two-band dataset is then analyzed using various image algebra change detection functions (e.g., band ratio, image differencing) to produce a new change image file. The analyst usually selects a threshold value to identify areas of "change" and "no change" in the new image, as discussed in the section on image algebra change detection. The change image is then recoded into a binary mask file consisting of areas that have changed between the two dates. Great care must be exercised when creating the "change/no change" binary mask (Jensen et al., 1993a). The change mask is then overlaid onto Date 2 of the analysis and only those pixels that were detected as having changed are classified in the Date 2 imagery. A traditional post-classification comparison can then be applied to yield "from–to" change information.

This method may reduce change detection errors (omission and commission) and provides detailed "from–to" change class information. The technique reduces effort by allowing analysts to focus on the small amount of area that has changed between dates. In most regional projects, the amount of actual change over a 1- to 5-year period is probably no greater than 10% of the total area. The method is complex, requiring a number of steps, and the final outcome is dependent on the quality of the "change/no change" binary mask used in the analysis. Nevertheless, this is a very useful change detection algorithm.

Change Detection Using an Ancillary Data Source as Date 1

Sometimes there exists a land-cover data source that may be used in place of a traditional remote sensing image in the change detection process. For example, the U.S. Fish and Wildlife Service conducted a National Wetland Inventory (NWI) of the United States at 1:24,000 scale. Some of these data have been digitized. Instead of using a remotely sensed image as Date 1 in a coastal change detection project, it is possible to substitute the digital NWI map of the region (Figure 12-15). In this case, the NWI map is recoded to be compatible with the classification scheme being used. Next, Date

**Change Detection Using a
Binary Change Mask Applied to Date 2**

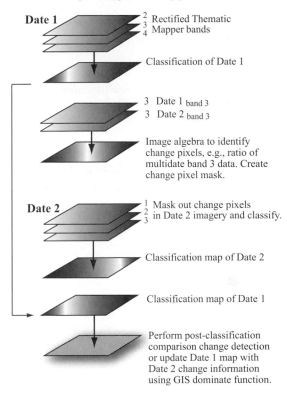

Figure 12-14   Diagram of change detection using a binary change mask applied to Date 2.

2 of the analysis is classified and then compared on a pixel-by-pixel basis with the Date 1 information using post-classification comparison methods. Traditional "from–to" information can then be derived.

Advantages of the method include the use of a well-known, trusted data source (e.g., NWI) and the possible reduction of errors of omission and commission. Detailed "from–to" information may be obtained using this method. Also, only a single classification of the Date 2 image is required. It may also be possible to update the NWI map (Date 1) with more current wetland information (this would be done using a GIS dominate function and the new wetland information found in the Date 2 classification). The disadvantage is that the NWI data must be digitized, generalized to be compatible with a

Change Detection Using an Ancillary Data Source as Date 1

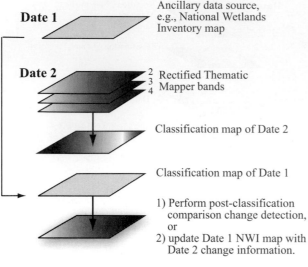

Date 1 — Ancillary data source, e.g., National Wetlands Inventory map

Date 2 — 2 3 4 Rectified Thematic Mapper bands

Classification map of Date 2

Classification map of Date 1

1) Perform post-classification comparison change detection, or
2) update Date 1 NWI map with Date 2 change information.

Advantages	Disadvantages
* May reduce change detection errors (omission and commission)	* Depends on quality of ancillary information
* Provides "from–to" change class information	
* Requires a single classification	

Figure 12-15   Diagram of change detection using ancillary data source as Date 1.

classification scheme, and then converted from vector to raster format to be compatible with the raster remote sensor data. Manual digitization and subsequent conversion introduce error into the database, which may not be acceptable (Lunetta et al., 1991).

Spectral Change Vector Analysis

When land undergoes a change or disturbance between two dates, its spectral appearance normally changes. For example, consider the red and near-infrared spectral characteristics of a single pixel displayed in two-dimensional feature space (Figure 12-16a). It appears that the land cover associated with this particular pixel has changed from Date 1 to Date 2 because the pixel resides at a substantially different location in the feature space on Date 2. The vector describing the direction and magnitude of change from Date 1 to Date 2 is a *spectral change vector* (Malila, 1980; Chen et al., 2003). The total change magnitude per pixel (CM_{pixel}) is computed by determining the Euclidean distance between

end points through n-dimensional change space (Michalek et al., 1993):

$$CM_{pixel} = \sum_{k=1}^{n} [BV_{ijk(date2)} - BV_{ijk(date1)}]^2 \qquad (12\text{-}3)$$

where $BV_{i,j,k(date2)}$ and $BV_{i,j,k(date1)}$ are the Date 1 and Date 2 pixel values in band k.

A scale factor (e.g., 5) can be applied to each band to magnify small changes in the data if desired. The change direction for each pixel is specified by whether the change is positive or negative in each band. Thus, 2^n possible types of changes can be determined per pixel (Virag and Colwell, 1987). For example, if three bands are used there are 2^3 or 8 types of changes or sector codes possible (Table 12-1). To demonstrate, let us consider a single registered pixel measured in three bands (1, 2, and 3) on two dates. If the change in band 1 was positive (e.g., $BV_{i,j,1(date\ 2)} = 45$; $BV_{i,j,1(date\ 1)} = 38$; $BV_{change} = 45 - 38 = 7$), and the change in band 2 was positive (e.g., $BV_{i,j,2(date2)} = 20$; $BV_{i,j,2(date1)} = 10$; $BV_{change} = 20 - 10 = 10$), and the change in band 3 was negative (e.g., $BV_{i,j,3(date2)} = 25$; $BV_{i,j,3(date1)} = 30$; $BV_{change} = 25 - 30 = -5$), then the change magnitude of the pixel would be $CM_{pixel} = 7^2 + 10^2 - 5^2 = 174$, and the change sector code for this pixel would be "+, +, –" and have a value of 7, as shown in Table 12-1 and Figure 12-17. For rare instances when pixel values do not change at all between the two dates, a default direction of + may be used to ensure that all pixels are assigned a direction (Michalek et al., 1993).

Change vector analysis outputs two geometrically registered files; one contains the sector code and the other contains the scaled vector magnitudes. The change information may be superimposed onto an image of the study area with the change pixels color-coded according to their sector code. This multispectral change magnitude image incorporates both the change magnitude and direction information (Figure 12-16a). The decision that a change has occurred is made if a threshold is exceeded (Virag and Colwell, 1987). The threshold may be selected by examining deep-water areas (if present), which should be unchanged, and recording their scaled magnitudes from the change vector file. Figure 12-16b illustrates a case in which no change would be detected because the threshold is not exceeded. Conversely, change would be detected in Figures 12-16c and d because the threshold was exceeded. The other half of the information contained in the change vector, that is, its direction, is also shown in Figure 12-16c and d. Direction contains information about the type of change. For example, the direction of

Spectral Change Vector Analysis

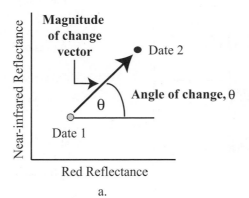

a.

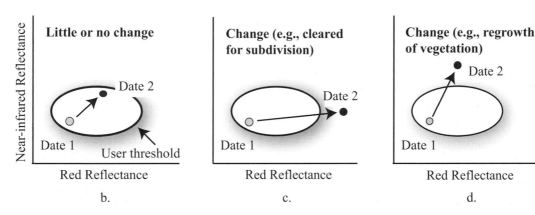

b.

c.

d.

Figure 12-16   Schematic diagram of the spectral change detection method.

change due to clearing should be different from change due to regrowth of vegetation.

Change vector analysis has been applied successfully to forest change detection in northern Idaho (Malila, 1980) and for monitoring changes in mangrove and reef ecosystems along the coast of the Dominican Republic (Michalek et al., 1993). It is the change detection algorithm of choice for producing the MODIS Vegetative Cover Conversion (VCC) product being compiled on a global basis using 250 m surface reflectance data (Zhan et al., 2002). The method is based on measuring the change in reflectance in just two bands, red ($\Delta\rho_{red}$) and near-infrared ($\Delta\rho_{nir}$), between two dates and using this information to compute the change magnitude per pixel,

$$CM_{pixel} = \sqrt{(\Delta\rho_{red})^2 + (\Delta\rho_{nir})^2} \qquad (12\text{-}4)$$

and change angle (θ):

Table 12-1.   Sector code definitions for change vector analysis using three bands (+ indicates pixel value increase from Date 1 to Date 2; – indicates pixel value decrease from Date 1 to Date 2) (Michalek et al., 1993).

Sector Code	Change Detection		
	Band 1	**Band 2**	**Band 3**
1	–	–	–
2	–	–	+
3	–	+	–
4	–	+	+
5	+	–	–
6	+	–	+
7	+	+	–
8	+	+	+

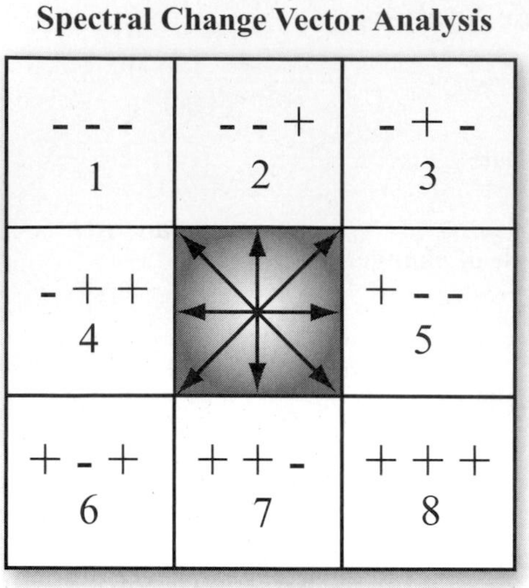

Spectral Change Vector Analysis

Figure 12-17   Possible change sector codes for a pixel measured in three bands on two dates.

$$\theta_{\text{pixel}} = \arctan\left(\frac{\Delta\rho_{\text{red}}}{\Delta\rho_{\text{nir}}}\right). \tag{12-5}$$

The change magnitude and angle information is then analyzed using decision-tree logic to identify important types of change in the multiple-date MODIS imagery (Zhan et al., 2002). Chen et al. (2003) developed an improved change vector analysis methodology that assists in the determination of the change magnitude and change direction thresholds when producing land-cover change maps.

Chi-square Transformation Change Detection

Ridd and Liu (1998) introduced a chi-square transformation change detection algorithm. It will work on any type of imagery but let us for the moment apply it to six bands of Landsat TM data obtained on two dates. The chi-square transformation is:

$$Y_{\text{pixel}} = (X - M)^T \Sigma^{-1} (X - M) \tag{12-6}$$

where Y_{pixel} is the digital value of the pixel in the output change image, X is the vector of the difference of the six digital values between the two dates for each pixel, M is the vector of the mean residuals of each band for the entire image, T is the transverse of the matrix, and Σ^{-1} is the inverse covariance matrix of the six bands between the two dates. The usefulness of the transformation in this context

rests on the fact that Y is distributed as a chi-square random variable with p degrees of freedom where p is the number of bands. $Y = 0$ represents a pixel of no change. The user creates the output change image and highlights pixels with varying amounts of Y.

Cross-correlation Change Detection

Cross-correlation change detection makes use of an existing Date 1 digital land-cover map and a Date 2 unclassified multispectral dataset (Koeln and Bissonnette, 2000). The Date 2 multispectral dataset does not need to be atmospherically corrected or converted to percent reflectance: the original brightness values are sufficient. Several passes through the datasets are required to implement cross-correlation change detection. First, every pixel in the Date 1 land-cover map associated with a particular class c (e.g., forest) out of m possible classes is located in the Date 2 multispectral dataset (Figure 12-18). The mean (μ_{ck}) and standard deviation (σ_{ck}) of all the brightness values in each band k in the Date 2 multispectral dataset associated with class c (e.g., forest) in the Date 1 land-cover map are computed. Next, every pixel (BV_{ijk}) in the Date 2 scene associated with class c is compared with the mean (μ_{ck}) and divided by the standard deviation (σ_{ck}). This value is summed and squared over all k bands. This is performed for each class. The result is a Z-score associated with each pixel in the scene:

$$Z_{ijc} = \sum_{k=1}^{n} \left(\frac{BV_{ijk} - \mu_{c_k}}{\sigma_{c_k}}\right)^2 \tag{12-7}$$

where

Z_{ijc} is the Z-score for a pixel at location i,j in the Date 2 multispectral dataset associated with a particular class c found in the Date 1 land-cover map;

c is the Date 1 land-cover class under investigation;

n is the number of bands in the Date 2 multispectral image;

k is the band number in the Date 2 multispectral image;

BV_{ijk} is the brightness value of a pixel (or reflectance) at location i,j in band k of the Date 2 multispectral dataset associated with a particular class found in the Date 1 land-cover map;

Cross-correlation Change Detection

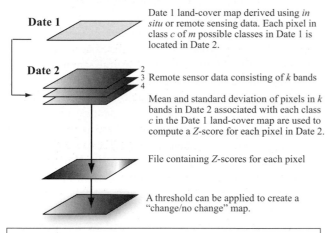

Date 1 land-cover map derived using *in situ* or remote sensing data. Each pixel in class *c* of *m* possible classes in Date 1 is located in Date 2.

Remote sensor data consisting of *k* bands

Mean and standard deviation of pixels in *k* bands in Date 2 associated with each class *c* in the Date 1 land-cover map are used to compute a Z-score for each pixel in Date 2.

File containing Z-scores for each pixel

A threshold can be applied to create a "change/no change" map.

Advantages	Disadvantages
* Not necessary to atmospherically correct or normalize Date 2 image data	* Depends on quality of Date 1 classification
* Requires a single classification	* Does not provide "from–to" change class information

Figure 12-18   Diagram of cross-correlation change detection.

μ_{ck} is the mean of all pixel brightness values found in band k of the Date 2 multispectral dataset associated with a particular class c in the Date 1 land-cover map;

σ_{ck} is the standard deviation of all pixel brightness values found in band k of the Date 2 multispectral dataset associated with a particular class c in the Date 1 land-cover map.

The mean (μ_{ck}) and standard deviation (σ_{ck}) values derived from the cross-correlation of a Date 1 four-class land-cover map with a Date 2 three-band multispectral dataset would be stored in a table like Table 12-2 and used in Equation 12-7.

The Z-statistic describes how close a pixel's response is to the expected spectral response of its corresponding class value in the land-cover map (Civco et al., 2002). In the output file, the greater the Z-score of an individual pixel, the greater the probability that its land cover has changed from Date 1 to Date 2. If desired, the image analyst can examine the Z-score image file and select a threshold that can be used to identify all pixels in the scene that have changed from Date 1 to Date 2. This information can be used to prepare a "change/no change" map of the area.

Advantages associated with cross-correlation change detection include the fact that it is not necessary to perform an atmospheric correction on any dataset. It also eliminates the

Table 12-2.   Hypothetical mean and standard deviations associated with a cross-correlation change detection analysis of a Date 1 land-cover map with four classes and a Date 2 multispectral image that contains three bands.

Land Cover Class	Band 1		Band 2		Band 3	
1	$\mu_{1,1}$	$\sigma_{1,1}$	$\mu_{1,2}$	$\sigma_{1,2}$	$\mu_{1,3}$	$\sigma_{1,3}$
2	$\mu_{2,1}$	$\sigma_{2,1}$	$\mu_{2,2}$	$\sigma_{2,2}$	$\mu_{2,3}$	$\sigma_{2,3}$
3	$\mu_{3,1}$	$\sigma_{3,1}$	$\mu_{3,2}$	$\sigma_{3,2}$	$\mu_{3,3}$	$\sigma_{3,3}$
4	$\mu_{4,1}$	$\sigma_{4,1}$	$\mu_{4,2}$	$\sigma_{4,2}$	$\mu_{4,3}$	$\sigma_{4,3}$

problems associated with phenological differences between dates of imagery. Unfortunately, this change detection method is heavily dependent upon the accuracy of the Date 1 land-cover classification. If it has serious error, then the cross-correlation between Date 1 and Date 2 will contain error. Every pixel in the Date 1 land-cover map must be assigned to a class. The method does not produce any "from–to" change detection information.

Knowledge-based Vision Systems for Detecting Change

The use of expert systems to detect change automatically in an image with very little human interaction is still in its infancy. In fact, most scientists attempting to develop such systems have significant human intervention and employ many of the aforementioned change detection algorithms in the creation of a knowledge-based change detection vision system. For example, Wang (1993) used a preprocessor to (1) perform image differencing, (2) create a change mask (using principal components analysis), (3) perform automated fuzzy supervised classification, and (4) extract attributes. Possible urban change areas were then passed to a rule-based interpreter, which produced a change image.

Visual On-screen Change Detection and Digitization

A considerable amount of high-resolution remote sensor data is now available (e.g., IKONOS and QuickBird 1 × 1 m, U.S.G.S. National Aerial Photography Program). Much of these data are being rectified and used as planimetric base maps or orthophotomaps. Often the aerial photography data are scanned (digitized) at high resolutions into digital image files (Light, 1993). These photographic datasets can then be registered to a common base map and compared to identify change. Digitized high-resolution aerial photography displayed on a CRT screen can be easily interpreted using standard photo interpretation techniques and the fundamental elements of image interpretation including size, shape, shadow, texture, etc. (Jensen, 2000). Therefore, it is becoming increasingly common for analysts to visually interpret both dates of aerial photography (or other type of remote sensor data) using heads-up on-screen digitizing and to compare the various images to detect change. The process is especially easy when 1) both digitized photographs (or images) are displayed on the CRT at the same time, side by side, and 2) they are topologically linked through object-oriented programming so that a polygon drawn around a feature on one photograph will also be drawn around the same object on the other photograph.

A good example of this methodology is shown in Figure 12-19. Hurricane Hugo with its 135-mph winds and 20-ft. storm surge struck the South Carolina coastline near Sullivan's Island on September 22, 1989. Vertical black-and-white aerial photographs obtained on July 1, 1988, were scanned at 500 dots per inch resolution using a Zeiss drum microdensitometer, rectified to the South Carolina State Plane Coordinate System, and resampled to 0.3×0.3 m pixels (Figure 12-19a). Aerial photographs acquired on October 5, 1989, were digitized in a similar manner and registered to the 1988 digital database (Figure 12-19b). Image analysts then performed on-screen digitization to identify the following features (Figure 12-20):

- buildings with no damage

- buildings partially damaged

- buildings completely damaged

- buildings that were moved

- buildings that might not be able to be rebuilt because they fell within certain SC Coastal Council beachfront management setback zones (base, 20-year, and 40-year)

- areas of beach erosion due to Hurricane Hugo

- areas of beach accretion due to Hurricane Hugo.

Digital classification of the digitized aerial photography on each date, performing image arithmetic (image differencing or band ratioing), or even displaying the two dates in different function memories did not work well for this type of data. The on-screen digitization procedure was the most useful for identifying housing and geomorphological change caused by Hurricane Hugo.

On-screen photo interpretation of digitized aerial photography, high-resolution aircraft multispectral scanner data, or high-resolution satellite data (e.g., SPOT panchromatic 10×10 m; QuickBird panchromatic 61×61 cm) is becoming very important for correcting or updating urban infrastructure databases. For example, the Bureau of the Census TIGER files represent a major resource for the development of GIS databases. For several reasons, the Bureau of the Census was forced to make a number of compromises during the construction of these nationwide digital cartographic files. As a result, the users of these files must develop their own procedures for dealing with some of the geometric inconsistencies in the files. One approach to solving these problems is to use remotely sensed image data as a source of current and potentially more accurate information (Cowen et al., 1991). For example, Figure 12-21a depicts U.S. Bureau of the Census TIGER road information draped over SPOT 10×10 m panchromatic data of an area near Irmo, SC. Note the serious geometric errors in the TIGER data. An analyst used heads-up, on-screen digitizing techniques to move roads to their proper planimetric positions and to add entirely new roads to the TIGER ARC-Info database (Figure 12-21b). All roads in South Carolina were updated using this type of logic and SPOT panchromatic data.

Finally, it is sometimes useful to simply visually examine multiple dates of remotely sensed imagery to appreciate processes at work. For example, consider the change in the Aral Sea in Kazakhstan from 1973 to 2000. The shoreline for a portion of the Aral Sea recorded by the Landsat MSS in 1973 and 1987 and the Landsat ETM$^+$ in 2000 is shown in **Color Plate 12-7**.

The Aral Sea is actually a lake, a body of fresh water. Unfortunately, more than 60 percent of the lake has disappeared in the last 30 years. Farmers and state offices in Uzbekistan, Kazakhstan, and Central Asian states began diverting river water to the lake in the 1960s to irrigate cotton fields and rice paddies. In 1965, the Aral Sea received about 50 cubic kilometers of fresh water per year—a number that fell to zero by the early 1980s. Concentrations of salts and minerals began to rise in the lake. The change in chemistry led to alterations in the lake's ecology, causing precipitous drops in the fish population. The commercial fishing industry employed 60,000 people in the early 1960s. By 1977, the fish harvest was reduced by 75 percent, and by the early 1980s the commercial fishing industry was gone.

Hurricane Hugo Impacts Sullivan's Island, SC, in 1989

a. Pre-Hurricane Hugo orthophotomap July 1, 1988.

b. Post-Hurricane Hugo orthophoto October 5, 1989.

Figure 12-19   a) Panchromatic orthophotomap of Sullivan's Island, SC, obtained on July 1, 1988, prior to Hurricane Hugo. The data were rectified to State Plane Coordinates and resampled to 0.3×0.3 m spatial resolution. b) Panchromatic aerial photograph of Sullivan's Island obtained on October 5, 1989, after Hurricane Hugo. The data were rectified to State Plane Coordinates and resampled to 0.3×0.3 m spatial resolution.

Impact of Hurricane Hugo Extracted Using Visual On-screen Change Detection

Figure 12-20   Change information overlaid on October 5, 1989, post-Hurricane Hugo aerial photograph, Sullivan's Island, SC. Completely destroyed houses are outlined in white. Partially destroyed houses are outlined in black. A white arrow indicates the direction of houses removed from their foundations. Three beachfront management setback lines are shown in white (base, 20 year, 40 year). Areas of beach erosion are depicted as black lines. Areas of beach accretion caused by Hurricane Hugo are shown as dashed black lines.

The shrinking Aral Sea has also had a noticeable effect on the region's climate. The growing season is now shorter, causing many farmers to switch from cotton to rice, which requires even more diverted water. A secondary effect of the reduction in the Aral Sea's overall size is the rapid exposure of the lake bed. Strong winds that blow across this part of Asia routinely pick up and deposit tens of thousands of tons of now-exposed soil every year. This process has caused a reduction in air quality for nearby residents and affected crop yields due to the heavily salt-laden particles falling on arable land.

a. Rectified SPOT 10 x 10 m panchromatic data of an area near Irmo, SC, overlaid with the TIGER road network.

b. Adjustment of the TIGER road network using visual on-screen digitization.

Figure 12-21   a) U.S. Bureau of the Census TIGER road network data overlaid on SPOT 10×10 m panchromatic data of an area near Irmo, SC. b) Correction of the TIGER data based on visual on-screen movement of roads in error and digitization of entirely new roads.

Environmental experts agree that the current situation cannot be sustained. Yet, driven by poverty and their dependence upon exports, officials in the region have failed to take any preventive action, and the Aral Sea continues to shrink (NASA Aral Sea, 2004).

 Atmospheric Correction for Change Detection

Now that many of the most widely adopted change detection algorithms have been identified, it is useful to provide some general guidelines about when it is necessary to atmospherically correct the individual dates of imagery used in the change detection process.

When Atmospheric Correction Is Necessary

Atmospheric correction of multiple-date remote sensor data is required when the individual date images used in the change detection algorithm are based on linear transformations of the data, e.g., a normalized difference vegetation index image is produced for Date 1 and Date 2. The additive effects of the atmosphere on each date contaminate the NDVI values and the modification is not linear (Song et al., 2001). Contributions from the atmosphere to NDVI values are significant and can amount to 50% or more over thin or broken vegetation cover (McDonald et al., 1998; Song et al., 2001). Similarly, the imagery should be atmospherically corrected if the change detection is based on multiple-date red/near-infrared ratioed images (e.g., Landsat TM 4/TM 3). This suggests that it may be necessary to normalize or atmospherically correct the multiple-date imagery used to compute the linearly transformed data (e.g., NDVI) when the goal is to identify *biophysical* change characteristics through time rather than just land-cover change through time (Yuan and Elvidge, 1998; Song et al., 2001; Du et al., 2002).

A change/no change map produced using image differencing logic and atmospherically corrected data normally looks different from a change/no change map produced using image differencing logic and non-atmospherically corrected data if the threshold boundaries are held constant in the change image histograms. However, if the analyst selects the appropriate thresholds in the two tails of the change detection image histogram, it doesn't really matter whether the change detection map was produced using atmospherically corrected or non-atmospherically corrected data. But, if the analyst desires that all stable classes in the change image have a value of 0 in the change histogram (refer to Figures 12-10 and 12-11), then it is useful to normalize one image to another or atmospherically correct both images to percent reflectance values prior to performing image differencing.

Obtaining quality training data is very expensive and time-consuming because it usually involves people and field-work. Therefore, it will become increasingly important to be able to extend training data through both time and space. In other words, training data extracted from a Date 1 image should be able to be extended to a Date 2 image of the same geographic area (signature extension through time) or perhaps even to a Date 1 or Date 2 image of a neighboring geographic area (signature extension through space). Extending training data through space and time will require that each image evaluated be atmospherically corrected to surface reflectance whenever possible using one of the techniques described in Chapter 6. Nevertheless, it is not always necessary to correct remote sensor data when classifying individual dates of imagery or performing change detection.

When Atmospheric Correction Is Unnecessary

A number of studies have documented that it is unnecessary to correct for atmospheric effects prior to image classification if the spectral signatures characterizing the desired classes are derived from the image to be classified (e.g., Kawata et al., 1990). This is because atmospherically correcting a single date of imagery is often equivalent to subtracting a constant from all pixels in a spectral band. This action simply translates the origins in multidimensional feature space. The class means may change, but the variance–covariance matrix remains the same irrespective of atmospheric correction. In other words, atmospheric correction is unnecessary as long as the training data and the data to be classified are in the same relative scale (corrected or uncorrected) (Song et al., 2001). This suggests that it is not necessary to atmospherically correct Landsat TM data obtained on Date 1 if it is going to be subjected to a maximum likelihood classification algorithm and all the training data are derived from the Date 1 imagery. The same holds true when a Date 2 image is classified using training data extracted from the Date 2 image. Change between the Date 1 and Date 2 classification maps derived from the individual dates of imagery (corrected or uncorrected) can easily be compared in a post-classification comparison.

Atmospheric correction is also unnecessary when change detection is based on classification of multiple-date composite imagery in which the multiple dates of remotely sensed images are rectified and placed in a single dataset and then

classified as if it were a single image (e.g., multiple-date principal components change detection). Only when training data from one time and/or place are applied in another time and/or place is atmospheric correction necessary for image classification and many change detection algorithms.

 ## Summary

A one-time inventory of natural resources is often of limited value. A time series of images and the detection of change provides significant information on *the resources at risk* and may be used in certain instances to identify *the agents of change*. Change information is becoming increasingly important in local, regional, and global environmental monitoring (Woodcock et al., 2001). This chapter identifies the remote sensor system and environmental variables that should be considered whenever a remote sensing change detection project is initiated. Several useful change detection algorithms are reviewed. Scientists are encouraged to carefully review and understand these principles so that accurate change detection can take place.

 ## References

Arzandeh, S. and J. Wang, 2003, "Monitoring the Change of Phragmites Distribution Using Satellite Data," *Canadian Journal of Remote Sensing*, 29(1):24–35.

Bauer, M. E., Burk, T. E., Ek, A. R., Coppin, P. R., Lime, S. D., Walsh, T. A., Walters, D. K., Befort, W. and D. F. Heinzen, 1994, "Satellite Inventory of Minnesota Forest Resources," *Photogrammetric Engineering & Remote Sensing* 60(3):287–298.

Chen, J., Gong, P., He, C., Pu, R. and P. Shi, 2003, "Land-Use/Land-Cover Change Detection Using Improved Change-Vector Analysis," *Photogrammetric Engineering & Remote Sensing* 69(4):369–379.

Civco, D. L., 1989, "Topographic Normalization of Landsat Thematic Mapper Digital Imagery," *Photogrammetric Engineering & Remote Sensing*, 55(9):1303–1309.

Civco, D. L., Hurd, J. D., Wilson, E. H., Song, M. and Z. Zhang, 2002, "A Comparison of Land Use and Land Cover Change Detection Methods," *Proceedings, ASPRS-ACSM Annual Conference and FIG XXII Congress,* Bethesda, MD: American Society for Photogrammetry & Remote Sensing, 10 p, CD.

Collins, J. B. and C. E. Woodcock, 1996, "An Assessment of Several Linear Change Detection Techniques for Mapping Forest Mortality Using Multitemporal Landsat TM Data," *Remote Sensing of Environment*, 56:66–77.

Cowen, D. J., J. R. Jensen, and J. Halls, 1991, "Maintenance of TIGER Files Using Remotely Sensed Data," *Proceedings, American Society for Photogrammetry & Remote Sensing*, (4):31–40.

Dobson, J. E., Ferguson, R. L., Field, D. W., Wood, L. L., Haddad, K. D., Iredale, H., Jensen, J. R., Klemas, V. V., Orth, R. J. and J. P. Thomas, 1995, *NOAA Coastal Change Analysis Project (C-CAP): Guidance for Regional Implementation,* Washington: NOAA, NMFS 123, 92 p.

Du, Y., Teillet, P. M. and J. Cihlar, 2002, "Radiometric Normalization of Multitemporal High-resolution Satellite Images with Quality Control for Land Cover Change Detection," *Remote Sensing of Environment*, 82:123–134.

Eastman, J. R. and M. Fulk, 1993, "Long Sequence Time Series Evaluation Using Standardized Principal Components," *Photogrammetric Engineering & Remote Sensing*, 59(6):991–996.

Eckhardt, D. W., Verdin, J. P. and G. R. Lyford, 1990, "Automated Update of Irrigated Lands GIS Using SPOT HRV Imagery," *Photogrammetric Engineering & Remote Sensing*, 56(11):1515–1522.

Foody, G. M., 2001, "Monitoring the Magnitude of Land-Cover Change Around the Southern Limits of the Sahara," *Photogrammetric Engineering & Remote Sensing*, 67(7):841–847.

Franklin, S. E., Lavigne, M. B., Wulder, M. A. and T. M. McCaffrey, 2002, "Large-area Forest Structure Change Detection: An Example," *Canadian Journal of Remote Sensing*, 28(4):588–592.

Fung, T. and E. LeDrew, 1988, "The Determination of Optimal Threshold Levels for Change Detection Using Various Accuracy Indices," *Photogrammetric Engineering & Remote Sensing,* 54(10):1449–1454.

Green, K., Kempka, D. and L. Lackey, 1994, "Using Remote Sensing to Detect and Monitor Land-cover and Land-use Change," *Photogrammetric Engineering & Remote Sensing* 60(3):331–337.

Gong, P., LeDrew, E. F. and J. R. Miller, 1992, "Registration-noise Reduction in Difference Images for Change Detection," *International Journal of Remote Sensing,* 13(4):773–779.

Jensen, J. R., 1981, "Urban Change Detection Mapping Using Landsat Digital Data," *The American Cartographer*, 8(2):127–147.

Jensen, J. R., 2000, *Remote Sensing of the Environment: An Earth Resource Perspective*, Upper Saddle River, NJ: Prentice-Hall, 544 p.

Jensen, J. R., Botchway, K., Brennan-Galvin, E., Johannsen, C., Juma, C., Mabogunje, A., Miller, R., Price, K., Reining, P., Skole, D., Stancioff, A. and D. R. F. Taylor, 2002, *Down to Earth: Geographic Information for Sustainable Development in Africa*, Washington: National Research Council, 155 p.

Jensen, J. R., Cowen, D. J., Narumalani, S., Althausen, J. D. and O. Weatherbee, 1993a, "An Evaluation of CoastWatch Change Detection Protocol in South Carolina," *Photogrammetric Engineering & Remote Sensing*, 59(6):1039–1046.

Jensen, J. R., Huang, X. and H. E. Mackey, 1997, "Remote Sensing of Successional Changes in Wetland Vegetation as Monitored During a Four-Year Drawdown of a Former Cooling Lake," *Applied Geographic Studies*, 1:31–44.

Jensen, J. R., S. Narumalani, O. Weatherbee and H. E. Mackey, 1993b, "Measurement of Seasonal and Yearly Cattail and Waterlily Changes Using Multidate SPOT Panchromatic Data," *Photogrammetric Engineering & Remote Sensing*, 59(4):519–525.

Jensen, J. R., Rutchey, K., Koch, M. and S. Narumalani, 1995, "Inland Wetland Change Detection in the Everglades Water Conservation Area 2A Using a Time Series of Normalized Remotely Sensed Data," *Photogrammetric Engineering & Remote Sensing*, 61(2):199–209.

Kawata, Y., Ohtani, A., Kusaka, T. and S. Ueno, 1990, "Classification Accuracy for the MOS-1 MESSR Data Before and After the Atmospheric Correction," *IEEE Transactions Geoscience Remote Sensing*, 35:1–13.

Kim, H. H. and G. C. Elman, 1990, "Normalization of Satellite Imagery," *International Journal of Remote Sensing*, 11(8):1331–1347.

Koeln, G. and J. Bissonnette, 2000, "Cross-correlation Analysis: Mapping Land Cover Change with a Historic Land Cover Database and a Recent, Single-date Multispectral Image," *Proceedings, American Society for Photogrammetry & Remote Sensing,* Bethesda, MD: ASPRS, 8 p., CD.

Light, D., 1993, "The National Aerial Photography Program as a Geographic Information System Resource," *Photogrammetric Engineering & Remote Sensing*, 59(1):61–65.

Lyon, J. G., Yuan, D., Lunetta, R. S. and C. D. Elvidge, 1998, "A Change Detection Experiment Using Vegetation Indices," *Photogrammetric Engineering & Remote Sensing*, 64:143–150.

Lunetta, R. S. and C. Elvidge, 2000, *Remote Sensing Change Detection: Environmental Monitoring Methods and Applications*, New York: Taylor & Francis, 340 p.

Lunetta, R. L. and J. G. Lyons (Eds.), 2003, *Geospatial Data Accuracy Assessment,* Las Vegas: US Environmental Protection Agency, Report No. EPA/600/R-03/064, 335 p.

Lunetta, R. S., Congalton, R. G., Fenstermaker, L. K., Jensen, J. R., McGwire, K. C. and L. R. Tinney, 1991, "Remote Sensing and Geographic Information System Data Integration: Error Sources and Research Issues," *Photogrammetric Engineering & Remote Sensing,* 57(6):677–687.

Maas, J. F., 1999, "Monitoring Land-cover Changes: A Comparison of Change Detection Techniques," *International Journal of Remote Sensing*, 20(1):139–152.

Malila, W. A., 1980, "Change Vector Analysis: An Approach for Detecting Forest Changes with Landsat," *Proceedings, LARS Machine Processing of Remotely Sensed Data Symposium,* W. Lafayette, IN: Laboratory for the Applications of Remote Sensing, pp. 326–336.

McDonald, A. J., Gemmell, F. M. and P. E. Lewis, 1998, "Investigation of the Utility of Spectral Vegetation Indices for Determining Information on Coniferous Forests," *Remote Sensing of Environment*, 66:250–272.

Michalek, J. L., Wagner, T. W., Luczkovich, J. J. and R. W. Stoffle, 1993, "Multispectral Change Vector Analysis for Monitoring Coastal Marine Environments," *Photogrammetric Engineering & Remote Sensing*, 59(3):381–384.

NASA Aral Sea, 2004, *The Shrinking Aral Sea*, Washington: NASA Earth Observatory, http://earthobservatory.nasa.gov/Newsroom/NewImages/images.php3?img_id=4819.

Ridd, M. K. and J. Liu, 1998, "A Comparison of Four Algorithms for Change Detection in an Urban Environment," *Remote Sensing of Environment*, 63:95–100.

Rutchey, K. and L. Velcheck, 1994, "Development of an Everglades Vegetation Map Using a SPOT Image and the Global Positioning System," *Photogrammetric Engineering & Remote Sensing*, 60(6):767–775.

Skole, D., 1994, "Data on Global Land-cover Change: Acquisition, Assessment and Analysis," in W. B. Meyer and B. L. Turner, (Eds.), *Changes in Land Use and Land Cover: A Global Perspective,* Cambridge: Cambridge University Press, 437–471.

Song, C., Woodcock, C. E., Seto, K. C., Lenney, M. P. and S. A. Macomber, 2001, "Classification and Change Detection Using Landsat TM Data: When and How to Correct Atmospheric Effects," *Remote Sensing of Environment*, 75:230–244.

Virag, L. A. and J. E. Colwell, 1987, "An Improved Procedure for Analysis of Change in Thematic Mapper Image-Pairs," *Proceedings of the 21st International Symposium on Remote Sensing of Environment*, Ann Arbor: Environmental Research Institute of Michigan, 1101–1110.

Wang, F., 1993, "A Knowledge-based Vision System for Detecting Land Changes at Urban Fringes," *IEEE Transactions on Geoscience & Remote Sensing*, 31:136–145.

Woodcock, C. E., Macomber, S. A., Pax-Lenney, M. and W. B. Cohen, 2001, "Monitoring Large Areas for Forest Change Using Landsat: Generalization Across Space, Time and Landsat Sensors," *Remote Sensing of Environment*, 78:194–203.

Yuan, D. and C. Elvidge, 1998, "NALC Land Cover Change Detection Pilot Study: Washington D.C. Area Experiments," *Remote Sensing of Environment*, 66:166–178.

Zhan, X., Sohlberg, R. A., Townshend, J. R. G., DiMiceli, C., Carroll, M. L., Eastman, J. C., Hansen, M. C. and R. S. DeFries, 2002, "Detection of Land Cover Changes Using MODIS 250 m Data," *Remote Sensing of Environment*, 83:336–350.

Thematic Map Accuracy Assessment 13

Information derived from remotely sensed data are becoming increasingly important for environmental models at local, regional, and global scales (Johannsen et al., 2003). The remote sensing–derived thematic information may be in the form of thematic maps or statistics derived from area-frame sampling techniques. The thematic information must be accurate because important decisions are made throughout the world using the information (Muchoney and Strahler, 2002; Kyriakidis et al., 2003).

Unfortunately, the thematic information contains error. Scientists who create remote sensing–derived thematic information should recognize the sources of the error, minimize it as much as possible, and inform the user how much confidence he or she should have in the thematic information. Remote sensing–derived thematic maps should normally be subjected to a thorough accuracy assessment before being used in scientific investigations and policy decisions (Stehman and Czaplewski, 1998; Paine and Kiser, 2003).

 Land-use and Land-cover Map Accuracy Assessment

The steps generally taken to assess the accuracy of thematic information derived from remotely sensed data are summarized in Figure 13-1. First, it is important to clearly state the nature of the thematic accuracy assessment problem at hand, including:

- what the accuracy assessment is expected to accomplish,

- the classes of interest (discrete or continuous), and

- the sampling design and sampling frame (consisting of area and list frames).

The objectives of the accuracy assessment should be clearly identified. Sometimes a simple qualitative examination may be appropriate if the remote sensing–derived data are to be used as general undocumented information. Conversely, if the remote sensing–derived thematic information will impact the lives of human beings, flora, fauna, etc., then it may be necessary to conduct a thorough probability design-based accuracy assessment.

The accuracy assessment sampling design is the protocol (i.e., instructions) by which the reference sample units are selected (Stehman and Czaplewski, 1998). Sampling designs generally require a *sampling frame* that consists of "the materials or devices which delimit, identify, and allow access to the elements of the target population" (Sarndal et al., 1992). There are two important elements associated with the sampling frame:

• the *area frame,* and

• the *list frame.*

In a remote sensing context, the area frame represents the exact geographic dimension of the entire study area. The list frame is simply a list of all the possible *sampling units* within the area frame. The actual sample is selected solely from this list of sampling units.

The accuracy assessment process is dependent upon two types of *sampling units* (Stehman and Czaplewski, 1998):

• *point,* and

• *areal.*

Point sampling units have no areal extent. When conducting remote sensing–related accuracy assessment, analysts typically select one of three types of *areal* sampling units: individual pixels (e.g. Landsat TM 30 × 30 m pixels), polygons, or fixed-area plots. There is no consensus on the ideal type of sampling unit.

Once the accuracy assessment problem has been stated, then there are two methods that may be used to validate the accuracy (or assess the error) of a remote sensing–derived thematic map:

• qualitative confidence-building assessment, and

• statistical measurements.

A *confidence-building assessment* involves the visual examination of the thematic map associated with the overall area frame by knowledgeable individuals to identify any gross errors (sometimes referred to as blunders) such as urban areas in water bodies or unrealistic classes on mountain tops. If major errors are identified and we have no qualitative confidence in the map, then the thematic information should be discarded and the classification process iterated using more appropriate logic. If the thematic map seems reasonable, then the analyst may proceed to statistical confidence-building measurements (Morisette et al., 2003).

Statistical measurements may be subdivided into two types: (Stehman, 2000, 2001):

• model-based inference, and

• design-based inference.

Model-based inference is not concerned with the accuracy of the thematic map. It is concerned with estimating the error of the remote sensing classification *process* (or model) that generated the thematic map. Model-based inference can provide the user with a quantitative assessment of each classification decision. For example, each pixel in a MODIS-derived land-cover product has a confidence value that measures how well the pixel fits the training examples presented to the classifier (Morisette et al., 2003).

Design-based inference is based on statistical principles that infer the statistical characteristics of a finite population based on the sampling frame. Some common statistical measurements include producer's error, consumer's error, overall accuracy, and Kappa coefficient of agreement. Design-based inference is expensive yet powerful because it provides unbiased map accuracy statistics using consistent estimators.

Before we describe how to extract the unbiased statistical measurements, however, it is instructive to briefly review the sources of error that may be introduced into a remote sensing–derived product (e.g., a land-use or land-cover map) (Figure 13-2). This will help us appreciate why an accuracy assessment is so important when the remote sensing–derived information is used to make decisions.

 Sources of Error in Remote Sensing–derived Thematic Products

First, error may be introduced during the remote sensing data acquisition process (Figure 13-2). It is quite common for remote sensing system detectors, cameras, RADARs, LIDARs, etc. to become miscalibrated. This can result in inaccurate remote sensing measurements (e.g., multispectral radiance, RADAR backscatter, LIDAR laser intensity). The aircraft or spacecraft platform may roll, tilt, or yaw during data acquisition. Erroneous commands may be put into the data-collection or inertial navigation system from ground control personnel. Finally, the scene being imaged may be randomly affected by unwanted haze, smog, fog, dust, high relative humidity, or sunglint that can dramatically affect the quality and accuracy of the information that is extracted.

Considerable effort is spent preprocessing the remotely sensed data to remove both the geometric and radiometric error inherent in the original imagery (see Chapters 6 and 7).

General Steps to Assess the Accuracy of Thematic Information Derived from Remotely Sensed Data

State the nature of the thematic accuracy assessment problem.
    * State what the accuracy assessment is expected to accomplish.
    * Identify classes of interest (discrete or continuous).
    * Specify the sampling frame within the sampling design:
        - *area* frame (the geographic region of interest)
        - *list* frame (consisting of points or areal sampling units)
Select method(s) of thematic accuracy assessment.
    * Confidence-building assessment:
        - Qualitative
    * Statistical measurement:
        - Model-based inference (concerned with image processing methodology)
        - Design-based statistical inference of thematic information
Compute total number of observations required in the sample.
    * Observations per class.
Select sampling design (scheme).
    * Random.
    * Systematic.
    * Stratified random.
    * Stratified systematic unaligned sample.
    * Cluster sampling.
Obtain ground reference data at observation locations using a response design.
    * Evaluation protocol.
    * Labeling protocol.
Error matrix creation and analysis.
    * Creation:
        - Ground reference test information (columns)
        - Remote sensing classification (rows)
    * Univariate statistical analysis:
        - Producer's accuracy
        - User's accruacy
        - Overall accuracy
    * Multivariate statistical analysis:
        - Kappa coefficient of agreement; conditional Kappa
        - Fuzzy
Accept or reject previously stated hypothesis.
Distribute results if accuracy is acceptable.
    * Accuracy assessment report.
    * Digital products.
    * Analog (hard-copy) products.
    * Image and map lineage report.

Figure 13-1   The general steps used to assess the accuracy of remote sensing–derived thematic information.

Unfortunately, even after preprocessing, there is always residual geometric and radiometric error. For example, residual geometric error may cause pixels still to be in their incorrect geographic locations. Similarly, even the best atmospheric correction will not result in a perfect relationship between percent reflectance measured on the ground and percent reflectance measured by an optical remote sensing system for the same geographic area. The residual geometric and radiometric error is passed to subsequent image processing functions.

Sometimes qualitative or quantitative information extraction techniques are based on flawed logic. For example, a hard land-cover classification scheme may not contain classes that are mutually exclusive, exhaustive, and hierarchical. Training sites may be incorrectly labeled during the training

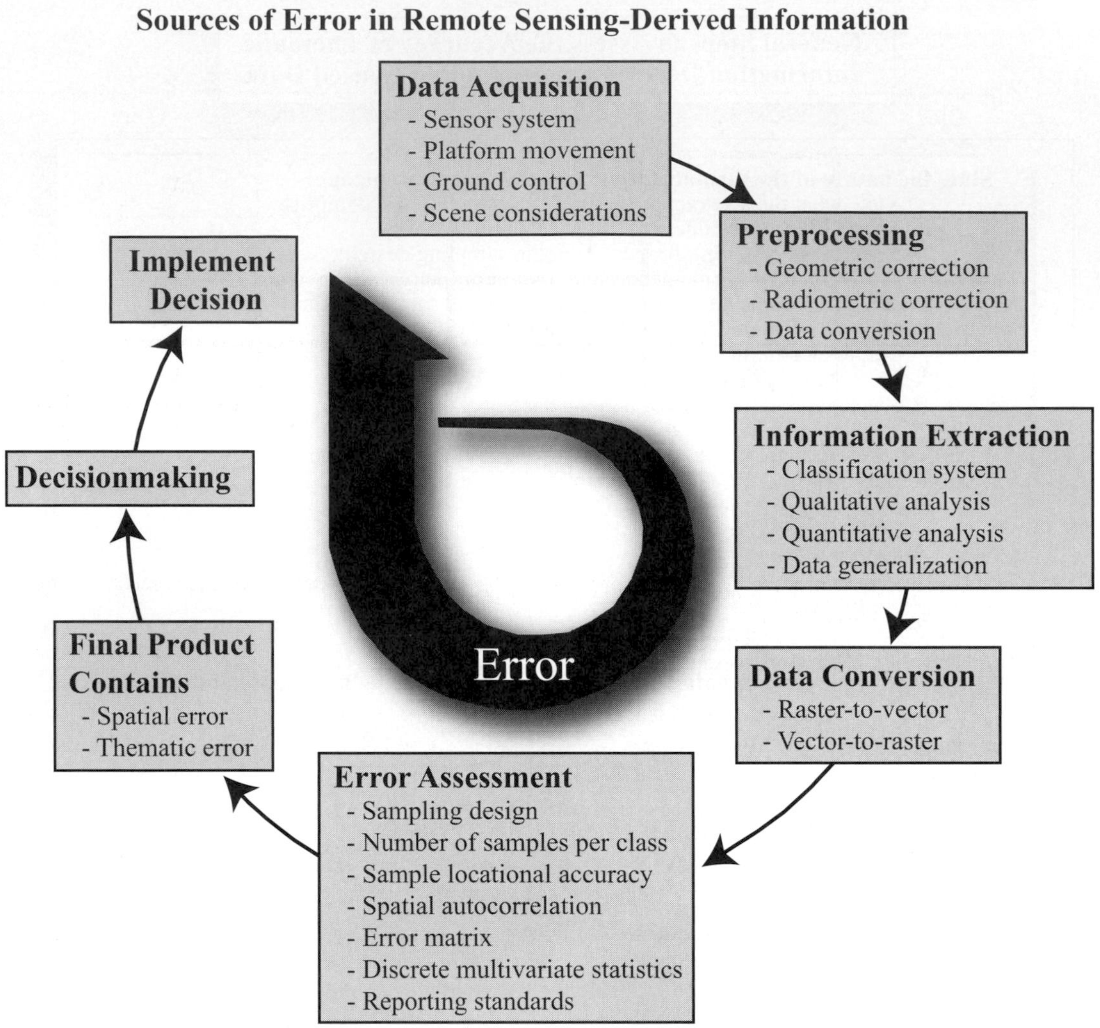

Sources of Error in Remote Sensing-Derived Information

Data Acquisition
- Sensor system
- Platform movement
- Ground control
- Scene considerations

Preprocessing
- Geometric correction
- Radiometric correction
- Data conversion

Information Extraction
- Classification system
- Qualitative analysis
- Quantitative analysis
- Data generalization

Data Conversion
- Raster-to-vector
- Vector-to-raster

Error Assessment
- Sampling design
- Number of samples per class
- Sample locational accuracy
- Spatial autocorrelation
- Error matrix
- Discrete multivariate statistics
- Reporting standards

Final Product Contains
- Spatial error
- Thematic error

Decisionmaking

Implement Decision

Error

Figure 13-2   Remote sensing–derived products such as land-use and land-cover maps contain error. The error accumulates as the remote sensing data are collected and various types of processing take place. An error assessment is necessary to identify the type and amount of error in a remote sensing–derived product (adapted from Lunetta et al., 1991; Bossler et al., 2002).

phase of a supervised classification. Clusters may be incorrectly labeled when performing an unsupervised classification (see Chapter 9). A human interpreter performing visual image interpretation may label a polygon incorrectly.

Thematic information derived from the remotely sensed data may be in raster (grid) or polygon cartographic data structures. Sometimes it is desirable to convert the data from one data structure to another, e.g., raster-to-vector or vector-to-raster. The conversion process may introduce error.

All the error discussed so far accumulates in the remote sensing–derived information (e.g., a land-use or land-cover map). Therefore, it is essential to conduct an error assess-

ment (or accuracy assessment) to place confidence limits on the remote sensing–derived information. Unfortunately, the person conducting the error evaluation could make critical errors regarding the ground reference data sampling design, the number of samples obtained per class, the geometric accuracy of the sample data, and data autocorrelation. Hopefully, a robust multivariate statistical analysis is used so that confidence limits can be specified for the remote sensing–derived thematic information (e.g., hectares of agriculture, forestry).

The final remote sensing–derived product is created when the thematic information is judged to be sufficiently accurate. Unfortunately, the analyst can still introduce error by

Table 13-1.   Characteristics of a typical error matrix consisting of k classes and N ground reference test samples.

		Ground Reference Test Information				
	Class	**1**	**2**	**3**	**k**	**Row total**
Remote Sensing Classification	**1**	$x_{1,1}$	$x_{1,2}$	$x_{1,3}$	$x_{1,k}$	x_{1+}
	2	$x_{2,1}$	$x_{2,2}$	$x_{2,3}$	$x_{2,k}$	x_{2+}
	3	$x_{3,1}$	$x_{3,2}$	$x_{3,3}$	$x_{3,k}$	x_{3+}
	k	$x_{k,1}$	$x_{k,2}$	$x_{k,3}$	$x_{k,k}$	x_{k+}
	Column total	x_{+1}	x_{+2}	x_{+3}	x_{+k}	N

not following standard cartographic procedures for the creation of scale bars and map legends. Metadata may also contain error.

Finally, the people who actually make the decisions often do not understand or appreciate the amount of error that has accumulated in the remote sensing–derived thematic product (Meyer and Werth, 1990). They should be educated about the amount of error in the product and cautioned not to overstate the level of accuracy.

Classification accuracy (or error) assessment was an afterthought rather than an integral part of many remote sensing studies in the 1970s and 1980s. In fact, many studies still simply report a single number (e.g., 85%) to express classification accuracy. Such *non-site*-specific accuracy assessments completely ignore locational accuracy. In other words, only the total amount of a category is considered without regard for its location. A *non-site*-specific accuracy assessment generally yields high accuracy but misleading results when all the errors balance out in a region.

The Error Matrix

To correctly perform a classification accuracy (or error) assessment, it is necessary to systematically compare two sources of information:

1. pixels or polygons in a *remote sensing–derived classification map,* and

2. *ground reference test information* (which may in fact contain error).

The relationship between these two sets of information is commonly summarized in an *error matrix* (sometimes referred to as a contingency table or confusion matrix). Indeed, the error matrix provides the basis on which to both describe classification accuracy and characterize errors, which may help refine the classification or estimates derived from it (Foody, 2002; Wu and Shao, 2002).

An example of a typical error matrix is shown in Table 13-1. The error matrix is used to assess the remote sensing classification accuracy of k classes. The central part of the error matrix is a square array of numbers $k \times k$ in size (e.g., 3×3). The columns of the matrix represent the ground reference test information and the rows correspond to the classification generated from analysis of the remotely sensed data. The intersection of the rows and columns summarize the number of sample units (i.e., pixels, clusters of pixels, or polygons) assigned to a particular category (class) relative to the actual category as verified in the field. The total number of samples examined is N.

The diagonal of the matrix (e.g., $x_{1,1}, x_{2,2}$) summarizes those pixels or polygons that were assigned to the correct class. Every error in the remote sensing classification relative to the ground reference information is summarized in the off-diagonal cells of the matrix (e.g., $x_{1,2}, x_{2,1}, x_{2,3}$). Each error is both an omission from the correct category and a commission to a wrong category. The column and row totals around the margin of the matrix (referred to as *marginals*) are used to compute errors of inclusion (commission errors) and errors of exclusion (omission errors). The outer row and column totals are used to compute producer's and user's accuracy (to be discussed). Some recommend that the error matrix contain proportions rather than individual counts (Stehman and Czaplewski, 1998).

But how do we a) obtain unbiased *ground reference test information* to compare with the remote sensing classification map and fill the error matrix with values, and b) perform the error (or accuracy) assessment? Basically, we must:

- be aware of the problems associated with using training versus ground reference test information;

- determine the total number of samples to be collected for each thematic category;

- design an appropriate sampling scheme (using traditional and/or geostatistical techniques);

- obtain ground reference information at sample locations using a response design; and

- apply appropriate descriptive and multivariate statistics (normal and/or fuzzy) to assess the accuracy of the remote sensing–derived information.

Training versus Ground Reference Test Information

Some image analysts perform an error evaluation based only on the *training pixels* (or training polygons if the study is based on human visual interpretation) used to train or seed a supervised or unsupervised classification algorithm. Unfortunately, the locations of these training sites are usually not random. They are biased by the analyst's a priori knowledge of where certain land-cover types exist in the scene. Because of this bias, an accuracy assessment based on how well the training class data were classified usually results in higher classification accuracies than an accuracy assessment based on ground reference test information (e.g., Muchoney and Strahler, 2002).

The ideal situation is to locate *ground reference test pixels* (or polygons if the classification is based on human visual interpretation) in the study area. These sites are *not* used to train the classification algorithm and therefore represent unbiased reference information. It is possible to collect some ground reference test information prior to the classification, perhaps at the same time as the training data. But the majority of test reference information is often collected after the

classification has been performed using a random sample to collect the appropriate number of unbiased observations per category (to be discussed).

Landscapes often change rapidly. Therefore, it is best to collect both the training and ground reference test information as close to the date of remote sensing data acquisition as possible. This is especially important where the land use or land cover changes rapidly within a single season. For example, agriculture and illicit crops often have relatively short growing seasons or are harvested rapidly and often repeatedly during a season. If a scientist waits too long to obtain the ground reference test information, it may not be possible to tell what was actually in the field the day of the remote sensing data acquisition. This is a serious condition because it becomes very difficult to populate an error matrix with accurate information.

Ideally, the ground reference test data are obtained by visiting the site on the ground and making very careful observations that can then be compared with the remote sensing–derived information for the exact location. Unfortunately, sometimes it is difficult to actually visit all the sites identified in the random sample. Some sites selected in the random sample may be completely inaccessible due to extremely rugged terrain. Others may be inaccessible because private land owners, government agencies, or even criminals (e.g., illegal drug cultivators) deny access.

When this occurs, many scientists obtain higher-spatial-resolution remotely sensed data, interpret it, and use it as a surrogate for ground reference test information (e.g., Morisette et al., 2004). In such cases, the general rule of thumb is that the imagery used to obtain the ground reference test information should be substantially higher in spatial and/or spectral resolution than the imagery used to derive the original thematic information. For example, many studies have used high-spatial-resolution aerial photography (e.g., nominal spatial resolution $<1 \times 1$ m) to extract ground reference test information that is then compared with thematic data produced using Landsat Thematic Mapper imagery (30×30 m). Admittedly, this is not the best solution. However, sometimes it is the only alternative if one wants to populate an error matrix and perform an accuracy (error) assessment.

At some point we must compare the results of the ground reference test information with the remote sensing–derived information at the appropriate location in the thematic map. When we locate a ground reference test unit in the remote sensing–derived thematic map at pixel i, j we do not usually just compare the pixel at that location with the ground reference test data. Rather, we often look at the 8 pixels surrounding pixel i, j and perhaps label the pixel according to the class that has the highest frequency of occurrence (e.g., if there are 3 corn and 6 soybean pixels in the 3×3 window, then we would label the pixel soybean for error matrix evaluation purposes). This procedure is common practice and minimizes the effects of geometric misregistration in the remote sensing–derived product (Jensen et al., 1993).

 Sample Size

The actual number of ground reference test samples to be used to assess the accuracy of individual categories in a remote sensing classification map is a very important consideration. Typically, a design-based inferential framework is adopted in accuracy assessment programs (Stehman, 2000). With such a framework, an appropriately sized sample may be estimated using conventional statistics (Foody, 2002).

Some analysts use an equation based on the *binomial distribution* or the normal approximation to the binomial distribution to compute the required sample size. Others suggest that a *multinomial distribution* be used to determine the sample size because we are usually investigating the accuracy of multiple classes of information on a land-cover map (Congalton and Green, 1999). Alternatively, a model-based inferential framework may be adopted based on geostatistical analysis (which takes into account autocorrelation) to design an efficient sample (Stehman, 2000; Foody, 2002).

Sample Size Based on Binomial Probability Theory

Fitzpatrick-Lins (1981) suggests that the sample size N to be used to assess the accuracy of a land-use classification map be determined from the formula for the binomial probability theory:

$$N = \frac{Z^2(p)(q)}{E^2} \tag{13-1}$$

where p is the expected percent accuracy of the entire map, $q = 100 - p$, E is the allowable error, and $Z = 2$ from the standard normal deviate of 1.96 for the 95% two-sided confidence level. For a sample for which the expected accuracy is 85% at an allowable error of 5%, the number of points necessary for reliable results is:

$$N = \frac{2^2(85)(15)}{5^2} = \text{a minimum of 203 points}.$$

With expected map accuracies of 85% and an acceptable error of 10%, the sample size for a map would be 51:

$$N = \frac{2^2(85)(15)}{10^2} = 51.$$

Thus, the lower the expected percent accuracy (p), and the greater the allowable error (E), the fewer ground reference test samples that need to be collected to evaluate the classification accuracy.

Sample Size Based on Multinomial Distribution

Analysts usually make thematic maps that contain multiple classes (e.g., vegetation, bare soil, water, urban), not just binary information (e.g., vegetation versus everything else). Therefore, some scientists prefer to use equations based on a *multinomial distribution* to determine the sample size necessary to assess classification accuracy. Sample size (N) derived from a multinomial distribution is based on the equation (Tortora, 1978; Congalton and Green, 1999):

$$N = \frac{B\Pi_i(1 - \Pi_i)}{b_i^2} \tag{13-2}$$

where Π_i is the proportion of a population in the ith class out of k classes that has the proportion closest to 50%, b_i is the desired precision (e.g., 5%) for this class, B is the upper $(\alpha/k) \times 100$th percentile of the chi square (χ^2) distribution with 1 degree of freedom, and k is the number of classes.

For example, suppose a land-cover map contains eight classes ($k = 8$) and we know that class Π_i occupies approximately 30% of the map area and that this proportion is closest to 50%. Furthermore, we desire a level of confidence of 95% and a precision (b_i) of 5%. B is determined from the χ^2 table with 1 degree of freedom and $1 - \alpha/k$ as $\chi^2_{(1, 0.99375)} = 7.568$ (Congalton and Green, 1999):

$$1 - \frac{\alpha}{k} = 1 - \frac{0.05}{8} = 0.99375$$

$$N = \frac{7.568(0.30)(1 - 0.30)}{0.05^2}$$

$$N = \frac{1.58928}{0.0025} = 636 \text{ samples}$$

Therefore, in this example 636 samples should be randomly selected to adequately fill the error matrix. Approximately 80 samples per class are required (e.g., $8 \times 80 = 640$).

If we have no idea about the proportion of any of the classes in the land-cover map, then we can use the *worst-case* multinomial distribution algorithm where we assume that one class occupies 50% of the study area (Congalton and Green, 1999):

$$N = \frac{B}{4b^2}. \qquad (13\text{-}3)$$

Holding the precision constant at 5% for all k classes yields:

$$N = \frac{7.568}{4(0.05^2)} = 757 \text{ samples}.$$

Thus, 757 random samples would have to be obtained because we did not have prior knowledge about the true proportion of any of the k classes in this worst-case scenario.

Sometimes a confidence interval of 95% is unrealistic. For example, if we relax the confidence interval from 95% to 85%, as is standard for many land-use and land-cover mapping products, the new $\chi^2_{(1, 0.98125)}$ value for $B = 5.695$. If we maintain the same desired precision (i.e., 5%), then the total number of samples required drops to 478:

$$1 - \frac{\alpha}{k} = 1 - \frac{0.15}{8} = 0.98125$$

$$N = \frac{5.695(0.30)(1 - 0.30)}{0.05^2}$$

$$N = \frac{1.19595}{0.0025} = 478 \text{ samples}.$$

Therefore, approximately 60 samples are required per class (e.g., $8 \times 60 = 480$). The worst-case scenario would require a total of 570 samples (Congalton and Green, 1999):

$$N = \frac{5.695}{4(0.05^2)} = 570 \text{ samples}$$

or approximately 71 samples per class ($8 \times 71 = 568$).

Unfortunately, it is not always possible to obtain such large numbers of random samples. A balance between what is statistically sound and what is practically attainable must be found. Congalton (1991) and Congalton and Green (1999)

suggest that a good rule of thumb is to collect a minimum of 50 samples for each land-cover class in the error matrix. If the area is especially large (e.g., >1 million ha) or the classification has a large number of land-use categories (e.g., >10 classes), the minimum number of samples should be increased to 75 or 100 samples per class. The number of samples can also be adjusted based on the relative importance of that category within the objectives of the project or by the inherent variability within each category. It may be useful to take fewer samples in categories that show little variability, such as water and forest, and increase the sampling in the categories that are more variable, such as agricultural areas. The goal is to obtain an unbiased, representative sample that can be used to populate an error matrix.

 Sampling Design (Scheme)

Scientists cannot possibly visit every pixel or polygon in the remote sensing–derived classification map to assess its accuracy. Therefore, the total sample size (N) and the number of samples required per class (strata) were computed. Now it is necessary to determine the geographic location (x,y) of these samples in the real world so that we can visit them and obtain ground reference test information. *The location of the sample locations must be selected randomly without bias.* Any bias will cause the statistical analysis of the error matrix to over- or underestimate the true accuracy of the thematic map. Therefore, it is necessary to select an appropriate random *sampling design (scheme)*. Each sampling scheme assumes a different sampling model and consequently different variance equations.

There are basically five common sampling designs used to collect ground reference test data for assessing the accuracy of a remote sensing–derived thematic map (Congalton and Green, 1999) (Figure 13-3):

1. random sampling,

2. systematic sampling,

3. stratified random sampling,

4. stratified systematic unaligned sampling, and

5. cluster sampling.

Sampling Methods

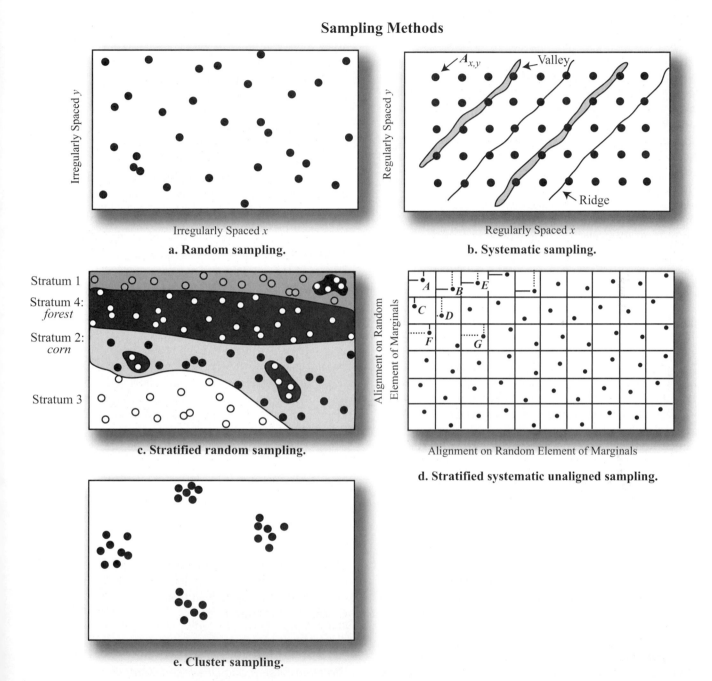

a. Random sampling.

b. Systematic sampling.

c. Stratified random sampling.

d. Stratified systematic unaligned sampling.

e. Cluster sampling.

Figure 13-3 Several geographic sampling methods may be used to infer the statistical characteristics of a population: a) Simple random sampling. b) Systematic sampling that is initiated with random x, y coordinates selected for point A. A biased sample would likely be extracted if the systematic sample were drawn from terrain with ridge and valley topography orientation such as that shown. c) Stratified random sampling. This particular study area has been subdivided into four strata. d) Stratified systematic unaligned sampling. e) Cluster sampling.

Simple Random Sampling

The key to accurately projecting the characteristics of a sample onto a population is to draw the sample with care to ensure that it is truly representative of its population. Simple *random sampling* without replacement always provides adequate estimates of the population parameters, provided the sample size is sufficient (Figure 13-3) (Congalton, 1988). Normally a random number generator is used to identify random x, y coordinates within the study area. Samples located randomly throughout a hypothetical study area are shown in Figure 13-3a. The method ensures that the members of a population are drawn independently with equal probabilities of being selected and that the selection of one sample does not influence the selection of any other sample. An advantage of working with random samples is that an important assumption underlying many statistical techniques is satisfied.

Unfortunately, simple random sampling may undersample small but possibly very important classes (e.g., the small acreage of illicit crops in a large agricultural region) unless the sample size is significantly large. The random samples may also be located in inhospitable or access-denied locations. Finally, some clustering of the random sample points often results and, as a consequence, important spatial properties of the population may be overlooked.

Systematic Sampling

To avoid the shortcomings of random sampling regarding the coverage of an area, a *systematic sampling* plan may be used. This procedure requires that it be possible to work through the study area in some consistent and orderly manner such as along a line of coordinate points. The sampling is normally initiated by the random selection of the starting point x, y coordinates (A) as shown in Figure 13-3b. Unfortunately, if there are periodicities in the data being collected (e.g., the ridge and valley topography of the Appalachian Mountains), the regularly spaced points could hit the same point on a cycle (e.g., a valley) time and again, yielding biased sample estimates of the population. Systematic sampling should be used with caution because it may overestimate the population parameters (Congalton, 1988).

Stratified Random Sampling

Many remote sensing analysts prefer *stratified random sampling* by which a minimum number of samples are selected from each strata (i.e., land-cover category) after the thematic map has been prepared. Stratified random sampling involves two steps. First, the study area is subdivided into land-cover strata based on what is found in the remote sensing–derived thematic map. A strata is created by extracting only pixels (or polygons) associated with a specific class. Some analysts prefer to mask out all unwanted classes to create the strata file. Sample locations are then randomly distributed throughout the geographic strata. For example, all pixels (or polygons in a visual interpretation project) associated with forest were placed in a single stratum (4) shown in Figure 13-3c. A random number generator was then used to allocate a sufficient number of random samples to this forest stratum. Similarly, all pixels (or polygons) associated with corn cultivation were placed in stratum 2 in Figure 13-3c. Random samples were then allocated throughout this stratum. Note that only strata 3 and 4 were completely homogeneous in Figure 13-3c. Strata 1 and 2 had stratum 4 (forest) areas within them. This is quite common. Note that the stratification process resulted in random samples being allocated to the forest islands even though they existed within other strata.

The advantage of stratified random sampling is that all strata (e.g., land-cover classes), no matter how small their proportion of the entire study area, will have samples allocated to them for error evaluation purposes. Without stratification, it is often difficult to find sufficient samples for classes that occupy a very small proportion of the study area. The drawback with stratified random sampling is that one has to wait until the thematic map has been completed in order to allocate samples to the various land-cover strata. The ground reference test information can rarely be obtained on the same day as the remote sensing data acquisition.

Stratified random sampling can also be applied to products derived from remote sensing area-frame sampling as long as the various land-cover field polygon boundaries associated with the individual frames are kept. This allows ground reference test samples to be allocated within each stratum, as discussed.

Stratified Systematic Unaligned Sampling

Stratified systematic unaligned sampling combines randomness and stratification with a systematic interval. It introduces more randomness than just beginning with a random x, y coordinate for the first sample in each stratum. Figure 13-3d demonstrates how a stratified systematic unaligned sample is created. First, point A is selected at random. The x-coordinate of A is then used with a new random y-coordinate to locate *B*, a second random *y*-coordinate to locate *E*, and so on across the top *row* of the stratum under investigation. By a similar process the *y*-coordinate of *A* is used in combination with random *x*-coordinates to locate point C and all successive points in the first *column* of the stratum under investigation. The random *x*-coordinate of *C* and *y*-coordinate of *B* are then used to locate *D*, of *E* and *F* to locate *G*, and so on until all strata have sample elements (Berry, 1962; King, 1969).

Cluster Sampling

Sometimes it is difficult to go into the field to randomly selected locations and obtain the required information. Therefore, some have suggested that several samples can be collected at a single random location. Unfortunately, each pixel in the cluster is not independent of the others. Therefore, Congalton (1988) warns that clusters no larger than 10 pixels and certainly not larger than 25 pixels should be used because of the lack of information added by each pixel beyond these cluster sizes.

Some combination of random and stratified sampling provides the best balance of statistical validity and practical application. Such a system may employ random sampling to collect some assessment data early in the project, and random sampling within strata should be used after the classification has been completed to ensure that enough samples were collected for each category and to minimize periodicity (spatial autocorrelation) in the data (Congalton, 1988). Ideally, the *x, y* locations of the reference test sites are determined using global positioning system (GPS) instruments.

To demonstrate these concepts, consider the 407 reference pixels collected for the Charleston, SC, study based on stratified random sampling after classification. The methods involved making five files, each containing only the pixels of a specific land-cover class (i.e., the classified land-cover pixels in each file were recoded to a value of 1 and the unclassified background to a value of 0). A random number generator was then used to identify random *x, y* coordinates within each stratified file until a sufficient number of points was collected (i.e., at least 50 for each class). The result was a stratified random sample of the five classes. All locations were then visited in the field or evaluated using large-scale orthophotography acquired during the same month as the Landsat TM overpass. Cross-tabulating the information resulted in the creation of the error matrix shown in Table 13-2.

Obtaining Ground Reference Information at Locations Using a Response Design

The actual collection of ground reference samples for use in the accuracy assessment is referred to as the *response design* (Stehman and Czaplewski, 1998). In effect, it is the data collection conducted in response to the sampling design. The response design can be subdivided into an evaluation protocol and a labeling protocol. The *evaluation protocol* refers to selecting the support region on the ground (typically associated with a pixel or polygon) where the ground information will be collected. Atkinson and Curran (1995) define this as the spatial support region consisting of "the size, geometry and orientation of the space on which an observation is defined." Once the location and dimensions of the sampling unit are defined, the *labeling protocol* is initiated and the sampling unit is assigned a hard or fuzzy ground reference label. This ground reference label (e.g., forest) is paired with the remote sensing–derived label (e.g., forest) for assignment in the error matrix.

Evaluation of Error Matrices

After the ground reference test information has been collected from the randomly located sites, the test information is compared pixel by pixel (or polygon by polygon when the remote sensor data are visually interpreted) with the information in the remote sensing–derived classification map. Agreement and disagreement are summarized in the cells of the error matrix. Information in the error matrix may be evaluated using simple descriptive statistics or multivariate analytical statistical techniques.

Descriptive Evaluation of Error Matrices

The *overall accuracy* of the classification map is determined by dividing the total correct pixels (sum of the major diagonal) by the total number of pixels in the error matrix (*N*). Computing the accuracy of individual categories, however, is more complex because the analyst has the choice of dividing the number of correct pixels in the category by the total number of pixels in the corresponding row or column. Traditionally, the total number of correct pixels in a category is divided by the total number of pixels of that category as derived from the reference data (i.e., the column total). This statistic indicates the probability of a reference pixel being correctly classified and is a measure of omission error. This statistic is called the *producer's accuracy* because the producer (the analyst) of the classification is interested in how well a certain area can be classified. If the total number of correct pixels in a category is divided by the total number of pixels that were actually classified in that category, the result is a measure of commission error. This measure, called the *user's accuracy* or reliability, is the probability that a pixel classified on the map actually represents that category on the ground (Story and Congalton, 1986).

Sometimes we are producers of classification maps and sometimes we are users. Therefore, we should always report all three accuracy measures: overall accuracy, producer's accuracy, and user's accuracy, because we never know how the classification will be used (Felix and Binney, 1989). For example, the remote sensing–derived error matrix in Table 13-2 has an overall classification accuracy of 93.86%. However, what if we were primarily interested in the ability to classify just residential land use using Landsat TM data of Charleston, SC? The producer's accuracy for this category was calculated by dividing the total number of correct pixels in the category (70) by the total number of residential pixels as indicated by the reference data (73), yielding 96%, which is quite good. We might conclude that because the overall accuracy of the entire classification was 93.86% and the producer's accuracy of the residential land-use class was 96%, the procedures and Landsat TM data used are adequate for identifying residential land use in this area. Such a conclusion could be a mistake. We should not forget the user's accuracy, which is computed by dividing the total number of correct pixels in the residential category (70) by the total number of pixels classified as residential (88), yielding 80%. In other words, although 96% of the residential pixels were correctly identified as residential, only 80% of the areas called residential are actually residential. A careful evaluation of the error matrix also reveals that there was confusion when discriminating residential land use from commercial and forest land cover. Therefore, although the producer of this map can claim that 96% of the time a geographic area that was residential was identified as such, a user of this map will find that only 80% of the time will an area she or he visits in the field using the map actually be residential. The user may feel that an 80% user's accuracy is unacceptable.

Discrete Multivariate Analytical Techniques Applied to the Error Matrix

Discrete multivariate techniques have been used to statistically evaluate the accuracy of remote sensing–derived classification maps and error matrices since 1983 (Congalton and Mead, 1983; Hudson and Ramm, 1987; Foody, 2002). The techniques are appropriate because remotely sensed

data are discrete rather than continuous and are also binomially or multinomially distributed rather than normally distributed. Statistical techniques based on normal distributions simply do not apply. It is important to remember, however, that there is not a single universally accepted measure of accuracy but instead a variety of indices, each sensitive to different features (Stehman, 1997; Foody, 2002).

It is useful to review several multivariate error evaluation techniques using the error matrix in Table 13-2. First, the raw error matrix can be *normalized* (standardized) by applying an iterative proportional fitting procedure that forces each row and column in the matrix to sum to 1 (not shown). In this way, differences in sample sizes used to generate the matrices are eliminated and individual cell values within the matrix are directly comparable. In addition, because the rows and columns are totaled (i.e., the marginals) as part of the iterative process, the resulting normalized matrix is more indicative of the off-diagonal cell values (i.e., the errors of omission and commission). In other words, all the values in the matrix are iteratively balanced by row and column, thereby incorporating information from that row and column into each cell value. This process then changes the cell values along the major diagonal of the matrix (correct classification), and therefore a normalized overall accuracy can be computed for each matrix by summing the major diagonal and dividing by the total of the entire matrix. Therefore, it may be argued that the normalized overall accuracy is a better representation of accuracy than is the overall accuracy computed from the original matrix because it contains information about the off-diagonal cell values (Congalton, 1991).

Standardized error matrices are valuable for another reason. Consider a situation where analyst 1 uses classification algorithm A and analyst 2 uses classification algorithm B on the same study area to extract the same four classes of information. Analyst A evaluates 250 random locations to derive error matrix A and analyst B evaluates 300 random locations to derive error matrix B. After the two error matrices are standardized, it is possible to directly compare cell values of the two matrices to see which of the two algorithms was better. Therefore, the normalization process provides a convenient way to compare individual cell values between error matrices regardless of the number of samples used to derive the matrix.

Kappa Analysis

Kappa analysis is a discrete multivariate technique of use in accuracy assessment (Congalton and Mead, 1983; Feinstein, 1998; Foody, 2002). The method was introduced to the remote sensing community in 1981 and was first published

Table 13-2. Error matrix of the classification map derived from Landsat Thematic Mapper data of Charleston, SC.

	Reference Data					
	Residential	Commercial	Wetland	Forest	Water	Row total
Residential	70	5	0	13	0	88
Commercial	3	55	0	0	0	58
Wetland	0	0	99	0	0	99
Forest	0	0	4	37	0	41
Water	0	0	0	0	121	121
Column total	73	60	103	50	121	**407**

Overall Accuracy = 382/407 = 93.86%

Producer's Accuracy (omission error)

Residential = 70/73 =   96%   4% omission error
Commercial = 55/60 =   92%   8% omission error
Wetland = 99/103 =   96%   4% omission error
Forest = 37/50 =   74%   26% omission error
Water = 20/22 =   100%   0% omission error

User's Accuracy (commission error)

Residential = 70/88 =   80%   20% commission error
Commercial = 55/58 =   95%   5% commission error
Wetland = 99/99 =   100%   0% commission error
Forest = 37/41 =   90%   10% commission error
Water = 121/121 =   100%   0% commission error

Computation of K_{hat} Coefficient of Agreement

$$\hat{K} = \frac{N \sum_{i=1}^{k} x_{ii} - \sum_{i=1}^{k} (x_{i+} \times x_{+i})}{N^2 - \sum_{i=1}^{k} (x_{i+} \times x_{+i})}$$

where $N = 407$

$$\sum_{i=1}^{k} x_{ii} = (70 + 55 + 99 + 37 + 121) = 382$$

$$\sum_{i=1}^{k} (x_{i+} \times x_{+i}) = (88 \times 73) + (58 \times 60) + (99 \times 103) + (41 \times 50) + (121 \times 121) = 36{,}792$$

therefore $\hat{K} = \dfrac{407(382) - 36792}{407^2 - 36792} = \dfrac{155474 - 36792}{165649 - 36792} = \dfrac{118682}{128857} = 92.1\%$

in a remote sensing journal in 1983 (Congalton, 1981; Congalton et al., 1983).

K_{hat} **Coefficient of Agreement:** Kappa analysis yields a statistic, $\hat{K}$, which is an estimate of Kappa. It is a measure of agreement or accuracy between the remote sensing–derived classification map and the reference data as indicated by a) the major diagonal and b) the chance agreement, which is indicated by the row and column totals (referred to as *marginals*) (Rosenfield and Fitzpatrick-Lins, 1986; Congalton, 1991; Paine and Kiser, 2003). $\hat{K}$ is computed:

$$\hat{K} = \frac{N \sum_{i=1}^{k} x_{ii} - \sum_{i=1}^{k} (x_{i+} \times x_{+i})}{N^2 - \sum_{i=1}^{k} (x_{i+} \times x_{+i})} \quad (13\text{-}4)$$

where k is the number of rows (e.g., land-cover classes) in the matrix, x_{ii} is the number of observations in row i and column i, and x_{i+} and x_{+i} are the marginal totals for row i and column i, respectively, and N is the total number of observations. $\hat{K}$ values >0.80 (i.e., >80%) represent strong agreement or accuracy between the classification map and the ground reference information. $\hat{K}$ values between 0.40 and 0.80 (i.e., 40 to 80%) represent moderate agreement. $\hat{K}$ values <0.40 (i.e., <40%) represent poor agreement (Landis and Koch, 1977).

Congalton and Green (1999) caution that if a standardized error matrix is used that has a great many off-diagonal cells with zeros (i.e., the classification is very good), then the normalized results may disagree with the overall classification accuracy and standard Kappa results.

Charleston, SC, Case Study: The computation of the $\hat{K}$ statistic for the Charleston dataset is summarized in Table 13-2. The overall classification accuracy is 93.86%, and the $\hat{K}$ statistic is 92.1%. The results are different because the two measures incorporated different information. The overall accuracy incorporated only the major diagonal and excluded the omission and commission errors. Conversely, $\hat{K}$ computation incorporated the off-diagonal elements as a product of the row and column marginals. Therefore, depending on the amount of error included in the matrix, these two measures may not agree. Congalton (1991) suggests that overall accuracy, normalized accuracy, and $\hat{K}$ be computed for each matrix to "glean as much information from the error matrix as possible." Computation of the $\hat{K}$ statistic may also be used 1) to determine whether the results presented in the error matrix are significantly better than a random result (i.e., a null hypothesis of $\hat{K} = 0$) or 2) to compare two similar matrices (consisting of identical categories) to determine if they are significantly different.

Savannah River Site, SC, Case Study: It is important to monitor the land cover of clay-capped hazardous waste sites on a repetitive basis to ensure the integrity of the clay-caps (Jensen et al., 2003). This was recently accomplished at the Westinghouse Savannah River Site near Aiken, SC, using DAIS 3715 hyperspectral imagery obtained on July 31, 2002, and a Spectral Angle Mapper classification algorithm.

The accuracy of the classification was determined by first obtaining *in situ* land-cover reference data at the 98 locations shown in **Color Plate 13-1a**. All test sites were inventoried to within $\pm$ 30 cm using survey-grade GPS. The land cover at each of these locations was identified in the classification map (**Color Plate 13-1b**). The classification accuracy results are summarized in Table 13-3. The overall classification accuracy was 89.79% with a $\hat{K}$ statistic of 85.81%.

Conditional K_{hat} Coefficient of Agreement: The conditional coefficient of agreement (K_c) can be used to calculate agreement between the reference and remote sensing–derived data with change agreement eliminated for an individual class for user accuracies using the equation (Congalton and Green, 1999; Paine and Kiser, 2003):

$$\hat{K}_i = \frac{N(x_{ii}) - (x_{i+} \times x_{+i})}{N(x_{i+}) - (x_{i+} \times x_{+i})} \quad (13\text{-}5)$$

where x_{ii} is the number of observations correctly classified for a particular category (summarized in the diagonal of the matrix), x_{i+} and x_{+i} are the marginal totals for row i and column i associated with the category, and N is the total number of observations in the entire error matrix. For example, the conditional K_i coefficient of agreement for the residential land-use class of the Charleston, SC, dataset (Table 13-2) is:

$$\hat{K}_{Resid} = \frac{407(70) - (88 \times 73)}{407(88) - (88 \times 73)} = \frac{28490 - 6424}{35816 - 6424} = 0.75 .$$

This procedure can be applied to each land-cover class of interest.

Procedures such as those discussed allow land-use maps derived from remote sensing to be quantitatively evaluated to determine overall and individual category classification accuracy. Their proper use enhances the credibility of using remote sensing–derived land-cover information.

Fuzzification of the Error Matrix

A hard classification system requires that all classes be mutually exclusive and exhaustive. Furthermore, the classes must usually be hierarchical, meaning that more detailed classes logically can be combined to produce more general but useful classes. Some phenomena in the real world may in certain instances be considered pure or homogeneous (e.g., a monoculture corn field with complete canopy closure). Unfortunately, the world contains many phenomena that are not well mapped using hard classification logic. Sometimes a field or polygon contains multiple materials. These are

Table 13-3. Error matrix of the classification map derived from hyperspectral data of the Mixed Waste Management Facility on the Savannah River Site (Jensen et al., 2003).

	Reference Data				
	Bahia grass	**Centipede**	**Bare soil**	**Concrete**	**Row total**
Bahia grass	31	2	0	0	33
Centipede	6	7	0	2	15
Bare soil	0	0	30	0	30
Concrete	0	0	0	20	20
Column total	37	9	30	22	**98**

Overall Accuracy = 88/98 = 89.79%

Producer's Accuracy (omission error)			**User's Accuracy (commission error)**		
Bahia grass = 31/37 =	84%	16% omission error	Bahia grass = 31/33 =	94%	6% commission error
Centipede = 7/9 =	78%	22% omission error	Centipede = 7/15 =	47%	53% commission error
Bare soil = 30/30 =	100%	0% omission error	Bare soil= 30/30 =	100%	0% commission error
Concrete = 20/22 =	91%	9% omission error	Concrete = 20/20 =	100%	0% commission error

Computation of K_{hat} Coefficient of Agreement

$$\hat{K} = \frac{N \sum_{i=1}^{k} x_{ii} - \sum_{i=1}^{k} (x_{i+} \times x_{+i})}{N^2 - \sum_{i=1}^{k} (x_{i+} \times x_{+i})}$$

where $N = 98$

$$\sum_{i=1}^{k} x_{ii} = (31 + 7 + 30 + 20) = 88$$

$$\sum_{i=1}^{k} (x_{i+} \times x_{+i}) = (33 \times 37) + (15 \times 9) + (30 \times 30) + (20 \times 22) = 2696$$

therefore $\hat{K} = \dfrac{98(88) - 2696}{98^2 - 2696} = \dfrac{8624 - 2696}{9604 - 2696} = \dfrac{5928}{6908} = 85.81\%$

often referred to as endmembers (e.g., corn intercropped with hay, bare soil between rows of corn, shadows, etc.). In this case, per-pixel or polygon class membership may be a matter of degree or percentage rather than assignment to a hard, mutually exclusive class. For example, if a single field or polygon contains 49% corn and 51% hay, should it really be labeled a hay field or should it be called something else? In fact, many natural landscapes often grade gradually into one another at the interface (e.g., it is common to find grass-

land grading into scrub/shrub and then scrub/shrub transitioning into forest).

Fuzzy set theory was developed by Zadeh (1965). Gopal and Woodcock (1994) were the first to suggest that fuzzy logic be used to introduce real-world fuzziness into the classification map accuracy assessment process. Instead of a right or wrong (binary) analysis, map labels are considered partially right (or wrong), generally on a five-category scale. Various

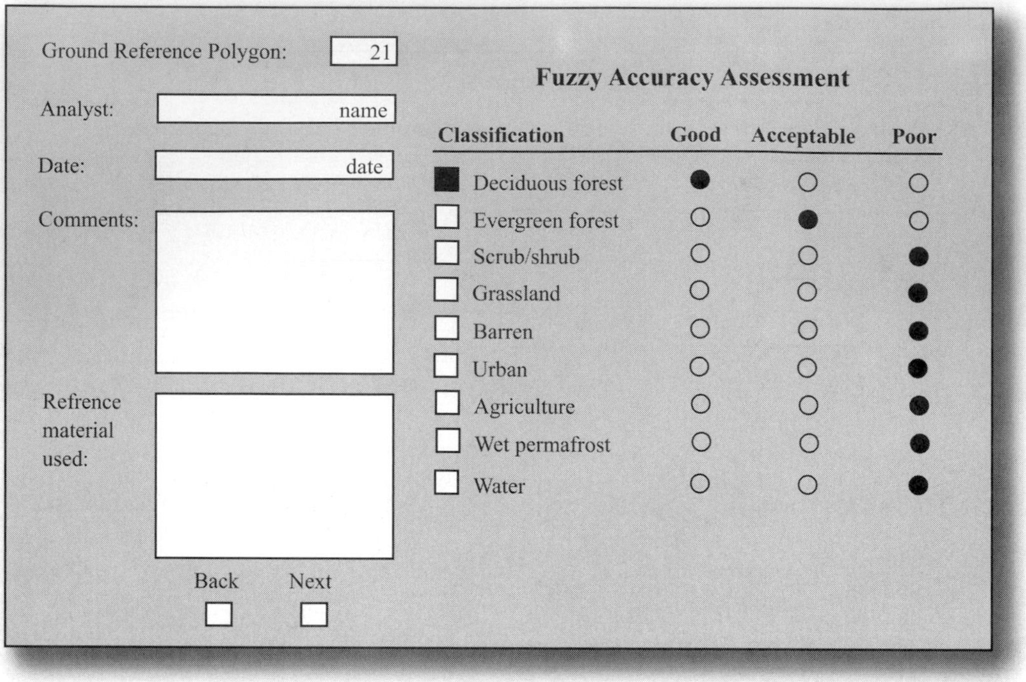

Figure 13-4   A template used for the introduction of fuzzy reference test information in an accuracy assessment of NIMA-sponsored Landsat GeoCover land-cover maps. The ground reference test polygon under consideration (21) has been assigned the deterministic class of deciduous forest. However, the analyst believes it could also be evergreen forest. The analyst is sure it is not one of the other categories. These data can be used to create an error matrix that contains deterministic as well as fuzzy accuracy assessment information (adapted from Green and Congalton, 2003).

statistics are then derived from the fuzzy information, including Max (M), the number of sites with an absolutely right answer (rank of 5); Right (R), the number of sites with a reasonable, good, or absolutely right answer (ranks 3, 4, and 5); and Increase (R–M), which is the difference between the Right and Max functions (Jacobs and Thomas, 2004). These are used to produce more sophisticated fuzzy logic–based accuracy assessment measures.

Congalton and Green (1999) introduced additional methods to incorporate fuzzy logic into an error assessment using a fuzzified error matrix. The first method involves accepting as correct ±1 class along the diagonal in the error matrix. This assumes that the data represent a continuum, e.g., 0% hay, 1–25%, 26–49%, 50–74%, and 75–100%. The marginal column and row totals take this enlargement of the diagonal (i.e., fuzzification) into account, as does the overall classification accuracy.

Another method uses fuzzy logic during the phase when ground reference information is obtained and compared with the remote sensing–derived classified map results. We will

demonstrate this technique using a real-world example that assessed the accuracy of a NIMA-sponsored global Landsat (30 × 30 m) GeoCover land-cover map (Green and Congalton, 2003). It was not practical to visit on the ground all the sites selected in the worldwide stratified random sample. Therefore, NIMA provided high-spatial-resolution national technical means (NTM) imagery of the selected sample sites. These were interpreted and represented the ground reference test information.

Experienced interpreters used a version of the fuzzy logic form shown in Figure 13-4 to enter their ground reference test information. Each ground reference test sample location in the NTM imagery was evaluated for the likelihood of being identified as each of the possible cover types. The analyst first determined the most appropriate label for the sample site, which was then entered in the "classification" column in the form (Figure 13-4). After assigning the deterministic label for the sample, the remaining possible map labels were evaluated as "good," "acceptable," or "poor" candidates for the site's label. For example, a sample site might fall near the classification scheme margin between

Table 13-4.   An error matrix that contains deterministic and fuzzy information used to assess the classification accuracy of NIMA-sponsored Landsat GeoCover thematic map products (adapted from Green and Congalton, 2003).

Reference Data

	D. Forest	E. Forest	Scrub/ Shrub	Grass	Barren	Urban	Ag.	Wet Perma	Water	User's Row Total Deterministic	User's Row Total Fuzzy
Deciduous Forest	48	4/0	2/0	0/1	0/0	0/1	0/1	0/0	0/0	48/56 (85.7%)	54/56 (96.4%)
Evergreen Forest	24/7	17	0/1	0/0	0/0	0/0	0/1	0/0	0/0	17/50 (34%)	41/50 (82%)
Scrub/shrub	0/1	0/1	15	5/1	0/2	0/0	7/15	0/0	0/0	15/57 (26.3%	33/57 (57.9%)
Grass	0/3	0/0	8/1	14	0/0	0/0	18/6	0/0	0/0	14/50 (28%)	40/50 (80%)
Barren	0/0	0/0	0/0	0/0	0	0/0	0/0	0/0	0/0	na	na
Urban	0/1	0/0	0/0	0/0	0/0	20	2/0	0/1	0/0	20/24 (83.3%)	22/24 (91.7%)
Agriculture	0/11	0/1	2/2	3/0	0/1	2/0	29	0/0	0/0	29/51 (56.9%)	36/51 (70.6%)
Wet permafrost	0/0	0/0	0/0	0/0	0/0	0/0	0/0	0	0/0	na	na
Water	0/18	0/3	0/0	0/0	0/0	0/0	1/2	1/0	8	8/33	10/33
Producer's Column Total Deterministic	48/123 (39%)	17/26 (65.4%)	15/31 (48.4%)	14/24 58.3%)	0/3 (0%)	20/22 (90.9%)	29/82 (35.4%)	0/2 (0%)	8/8 (100%)		
Producer's Column Total Fuzzy	78/123 (63.4%)	21/26 (80.8%)	27/31 (87.1%)	22/24 (91.7%)	0/3 (0%)	22/22 (100%)	57/82 (69.5%)	1/2 (50%)	8/8 (100%)		

Overall deterministic accuracy = 151/321 = 47%.
Overall fuzzy accuracy = 236/321 = 73.5%.

evergreen forest and shrub/scrub. In this instance, the analyst might rate evergreen forest as the most appropriate classification, but shrub/scrub as "acceptable." As each site was interpreted, the deterministic and fuzzy assessment reference labels were entered into the error matrix (Table 13-3).

Nondiagonal cells in the error matrix (Table 13-3) contain two tallies that can be used to distinguish class labels that are uncertain or that fall on class margins from class labels that are probably in error. The first number represents sites where the map label matched a "good" or "acceptable" reference label in the fuzzy assessment. Therefore, even though the class label was not considered the most appropriate, it was considered acceptable given the fuzziness of the classification system and the minimal quality of some of the ground reference test data. These sites are considered a "match" for estimating fuzzy assessment accuracy. The second number

in the cell represents those sites whose map class label was considered poor (i.e., an error).

The fuzzy assessment overall accuracy was estimated as the percentage of sites where the "good" or "acceptable" reference label(s) matched the map label. Individual class accuracy was estimated by summing the number of matches for that class's row or column divided by the row or column total. Class accuracy by column represented producer's class accuracy. Class accuracy by row represented user's accuracy.

Overall accuracy for the deterministic components of the matrix was 47% (151/321). The deterministic statistic ignored any variation in the interpretation of reference data and the inherent fuzziness at class boundaries. Including the "good" and "acceptable" ratings, overall fuzzy accuracy was

a. Hypothetical crop polygons.

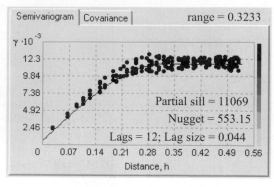

b. Semivariogram

c. Predicted crop polygons

Figure 13-5   a) A hypothetical map containing crop polygons. b) Semivariogram of the distribution of polygons. c) Predicted crop polygons.

74% (236/321). The deterministic Kappa statistic was 37%. It does not account for fuzzy class membership and variation in interpretation of the reference data. From a map user's perspective, individual fuzzy assessment class accuracies ranged from 30.3% (for water) to 96.4% (for deciduous forest). From a producer's perspective, accuracies ranged from 0% (barren) to 100% (water and urban).

Geostatistical Analysis to Assess the Accuracy of Remote Sensing–derived Information

The emphasis in the previous sections on error evaluation is primarily concerned with the accuracy of *thematic* (categorical) information. However, the Earth's surface and remotely sensed images of that surface also have distinct *spatial* properties. Once quantified, these properties can be used for many tasks in remote sensing, including image classification and the sampling of both the image and the ground reference test data (Curran, 1988).

Where there is spatial dependence (i.e., autocorrelation) in a remote sensing–derived land-cover map, the number of sample points required to achieve a given level of confidence might be much less if the study is based on a systematic sample survey instead of a completely random survey (Curran, 1988). Constructing the empirical semivariogram model and evaluating its range and sill may make it possible to identify the critical autocorrelation characteristics of the dataset and design a sampling scheme that ensures that neighboring sample points are as far from one another as is practical for a fixed sample size and area. This type of sampling scheme would minimize the duplication of information that often occurs in random sampling where some sample points are inevitably close (Curran, 1988).

Curran (1988) describes how to test various grid-spacing designs that allow sampling error and sample size to be considered in a spatial context. He suggests that a well-designed systematic survey based on geostatistical analysis can increase the precision over random sampling and decrease the sometimes impossibly large sample sizes associated with random sampling (Curran and Williamson, 1986).

For example, a hypothetical map of remote sensing–derived agricultural crop polygons is shown in Figure 13-5. Its semivariogram and predicted image are also provided (discussed in Chapter 4). The geostatistical spherical kriging process did an excellent job of predicting the original surface. Plus, we now have detailed information about the agricultural field autocorrelation present in the dataset. If this type of information were available over an entire region, it would be possible to identify the optimum sampling frequency that should be used for a) identifying the optimum spatial resolution of the remote sensing system, b) the spacing of individual frames of imagery if the study involves a sampling scheme rather than wall-to-wall mapping, and/or c) allocating ground reference test sites to perform a classification accuracy assessment.

Jacobs and Thomas (2003) used fuzzy set analysis and geospatial statistics (kriging) to create a spatial view of the accuracy of an Arizona GAP Analysis vegetation map.

Image Metadata and Lineage Information for Remote Sensing–derived Products

It is becoming increasingly important to document a) all of the information about the creation of an individual remotely sensed image and b) the various procedures and processes applied to the imagery to create an output product and assess its accuracy.

Individual Image Metadata

Many governments now require that all imagery collected for government use contain detailed metadata (data about data) as to how the images were collected. Therefore, detailed metadata standards have now been developed for raster image datasets by the U.S. Federal Geographic Data Committee (FGDC). This type of information allows the image analyst and decision maker to document the source of the imagery and all of its inherent characteristics. Metadata standards are an essential component of national and global spatial data infrastructures (SDI) (Jensen et al., 2002).

Lineage of Remote Sensing–derived Products

It is not sufficient to collect only image metadata. It is also necessary to carefully document every operation that is performed on the remote sensing data that results in an output product such as a land-use or land-cover map. This is commonly referred to as *lineage documentation*. Unfortunately, manual bookkeeping of the processes used to create a final product is cumbersome and rarely performed. Some digital image processing systems do provide history or audit files to keep track of the iterations and operations performed. However, none of these methods is capable of fulfilling the information requirements of a true lineage report that itemizes the characteristics of image and cartographic sources, the topological relationships among sources, intermediate and final product layers, and a history of the transformations applied to the sources to derive the output products (Lanter, 1991).

Lineage information should be included in *quality assurance reports* associated with every remote sensing–derived thematic product, including:

- a summary of all image and ancillary (e.g., soils map) source materials used to derive the thematic information;

- geoid, datum, and map projection information used to create the thematic information;

- geometric and radiometric transformation parameters;

- the processing steps used to transform the remotely sensed data into information;

- the methods and results of the accuracy assessment;

- the location of the archived original, interim, and final datasets; and

- procedures for accessing the archived information.

Quality assurance is an important part of life today. Image analysts who extract thematic information from remotely sensed data add value and rigor to the product by documenting its lineage.

References

Atkinson, P. M. and P. J. Curran, 1995, "Defining an Optimal Size of Support for Remote Sensing Investigations," *IEEE Transactions Geoscience Remote Sensing*, 33:768–776.

Berry, B. J. L., 1962, "Sampling, Coding, and Storing Flood Plain Data," *Agricultural Handbook*, Washington: U.S. Department of Agriculture, No. 237.

Bossler, J. D., Jensen, J. R., McMaster, R. B. and C. Rizos, 2002, *Manual of Geospatial Science and Technology*, London: Taylor & Francis, 623 p.

Congalton, R. G., 1981, *The Use of Discrete Multivariate Analysis for the Assessment of Landsat Classification Accuracy*, Blacksburg, VA: Virginia Polytechnic Institute and State University, Master's thesis.

Congalton, R. G., 1988, "Using Spatial Autocorrelation Analysis to Explore the Errors in Maps Generated from Remotely Sensed Data," *Photogrammetric Engineering & Remote Sensing*, 54(5):587–592.

Congalton, R. G., 1991, "A Review of Assessing the Accuracy of Classifications of Remotely Sensed Data," *Remote Sensing of Environment*, 37:35–46.

Congalton, R. G. and K. Green, 1999, *Assessing the Accuracy of Remotely Sensed Data: Principles and Practices*, Boca Raton, FL: Lewis Publishers, 137 p.

Congalton, R. G. and R. A. Mead, 1983, "A Quantitative Method to Test for Consistency and Correctness in Photointerpretation," *Photogrammetric Engineering & Remote Sensing*, 49(1):69–74.

Congalton, R. G., Oderwald, R. G. and R. A. Mead, 1983, "Assessing Landsat Classification Accuracy Using Discrete Multivariate

Statistical Techniques," *Photogrammetric Engineering & Remote Sensing*, 49(12):1671–1678.

Curran, P. J., 1988, "The Semivariogram in Remote Sensing: An Introduction," *Remote Sensing of Environment*, 24:493–507.

Curran, P. J. and H. D. Williamson, 1986, "Sample Size for Ground and Remotely Sensed Data," *Remote Sensing of Environment*, 20:31–41.

Feinstein, A. R., 1998, "Kappa Test of Concordance," in *Encyclopedia of Statistical Sciences*, New York: Wiley-Interscience, 2:351–352.

Felix, N. A. and D. L. Binney, 1989, "Accuracy Assessment of a Landsat-assisted Vegetation Map of the Coastal Plain of the Arctic National Wildlife Refuge," *Photogrammetric Engineering & Remote Sensing*, 55(4):475–478.

Fitzpatrick-Lins, K., 1981, "Comparison of Sampling Procedures and Data Analysis for a Land-use and Land-cover Map," *Photogrammetric Engineering & Remote Sensing*, 47(3):343–351.

Fisher, P. F. and S. Pathirana, 1990, "The Evaluation of Fuzzy Membership of Land Cover Classes in the Suburban Zone," *Remote Sensing of Environment*, 34:121–132.

Foody, G. M., 2002, "Status of Land Cover Classification Accuracy Assessment," *Remote Sensing of Environment*, 80:185–201.

Gopal, S. and C. Woodcock, 1994, "Theory and Methods for Accuracy Assessment of Thematic Maps Using Fuzzy Sets," *Photogrammetric Engineering & Remote Sensing*, 60(2):181–188.

Green, K. and R. G. Congalton, 2003, "An Error Matrix Approach to Fuzzy Accuracy Assessment: The NIMA Geocover Project," in Lunetta, R. L. and J. G. Lyons (Eds.), *Geospatial Data Accuracy Assessment,* Las Vegas: U.S. Environmental Protection Agency, Report No. EPA/600/R-03/064, 335 p.

Hudson, W. and C. Ramm, 1987, "Correct Formulation of the Kappa Coefficient of Agreement," *Photogrammetric Engineering & Remote Sensing*, 53(4):421–422.

Jacobs, S. R. and K. A. Thomas, 2003, "Fuzzy Set and Spatial Analysis Techniques for Evaluating Thematic Accuracy of a Land-Cover Map," in Lunetta, R. L. and J. G. Lyons (Eds.), *Geospatial Data Accuracy Assessment,* Las Vegas: U.S. Environmental Protection Agency, Report No. EPA/600/R-03/064, 335 p.

Jensen, J. R., Botchway, K., Brennan-Galvin, E., Johannsen, C., Juma, C., Mabogunje, A., Miller, R., Price, K., Reining, P., Skole,

D., Stancioff, A. and D. R. F. Taylor, 2002, *Down to Earth: Geographic Information for Sustainable Development in Africa*, Washington: National Research Council, 155 p.

Jensen, J. R., Hadley, B. C., Tullis, J. A., Gladden, J., Nelson, S., Riley, S., Filippi, T. and M. Pendergast, 2003, *2002 Hyperspectral Analysis of Hazardous Waste Sites on the Savannah River Site*, Aiken, SC: Westinghouse Savannah River Company, WSRC-TR-2003-0025, 52 p.

Jensen, J. R., Narumalani, S., Weatherbee, O. and H. E. Mackey, 1993, "Measurement of Seasonal and Yearly Cattail and Waterlily Changes Using Multidate SPOT Panchromatic Data," *Photogrammetric Engineering & Remote Sensing*, 59(4):519–525.

Johannsen, C. J., Petersen, G. W., Carter, P. G. and M. T. Morgan, 2003, "Remote Sensing: Changing Natural Resource Management," *Journal of Soil and Water Conservation*, 58(2):42–45.

King, L., 1969, *Statistical Analysis in Geography*, Englewood Cliffs, NJ: Prentice-Hall, 288 p.

Kyriakidis, P. C., Liu, X. and M. F. Goodchild, 2004, "Geostatistical Mapping of Thematic Classification Uncertainty," in Lunetta, R. L. and J. G. Lyons (Eds.), *Geospatial Data Accuracy Assessment,* Las Vegas: U.S. Environmental Protection Agency, Report No. EPA/600/R-03/064, 335 p.

Landis, J. and G. Koch, 1977, "The Measurement of Observer Agreement for Categorical Data," *Biometrics*, 33:159–174.

Lanter, D. P., 1991, "Design of a Lineage-based Meta-database for GIS," *Cartography and Geographic Information Systems*, 18(4):255–261.

Lunetta, R. L. and J. G. Lyons (Eds.), 2003, *Geospatial Data Accuracy Assessment,* Las Vegas: U.S. Environmental Protection Agency, Report No. EPA/600/R-03/064.

Lunetta, R. S., Congalton, R. G., Fenstermaker, L. K., Jensen, J. R., McGwire, K. C. and L. R. Tinney, 1991, "Remote Sensing and Geographic Information System Data Integration: Error Sources and Research Issues," *Photogrammetric Engineering & Remote Sensing,* 57(6):677–687.

Meyer, M. and L. Werth, 1990, "Satellite Data: Management Panacea or Potential Problem?" *Journal of Forestry*, 88(9):10–13.

Morisette, J. T., Privette, J. L., Strahler, A., Mayaux, P. and C. O. Justice, 2003, "Validation of Global Land-Cover Products by the Committee on Earth Observing Satellites," in Lunetta, R. L. and J. G. Lyons (Eds.), *Geospatial Data Accuracy Assessment,* Las

Vegas, NV: U.S. Environmental Protection Agency, Report No. EPA/600/R-03/064, 335 p.

Muchoney, D. M. and A. H. Strahler, 2002, "Pixel- and Site-based Calibration and Validation Methods for Evaluating Supervised Classification of Remotely Sensed Data," *Remote Sensing of Environment*, 81:290–299.

Paine, D. P. and J. D. Kiser, 2003, "Chapter 23: Mapping Accuracy Assessment," *Aerial Photography and Image Interpretation*, 2nd ed., New York: John Wiley & Sons, 465–480.

Rosenfield, G. H. and K. Fitzpatrick-Lins, 1986, "A Coefficient of Agreement as a Measure of Thematic Classification Accuracy," *Photogrammetric Engineering & Remote Sensing*, 52(2):223–227.

Sarndal, E. E. Swensson, B. and J. Wretman, 1992, *Model-Assisted Survey Sampling*, New York: Springer-Verlag.

Stehman, S. V., 1997, "Selecting and Interpreting Measures of Thematic Classification Accuracy," *Remote Sensing of Environment*, 62:77–89.

Stehman, S. V., 2000, "Practical Implications of Design-based Sampling for Thematic Map Accuracy Assessment," *Remote Sensing of Environment*, 72:35–45.

Stehman, S. V., 2001, "Statistical Rigor and Practical Utility in Thematic Map Accuracy Assessment," *Photogrammetric Engineering & Remote Sensing*, 67:727–734.

Stehman, S. V. and R. L. Czaplewski, 1998, "Design and Analysis for Thematic Map Accuracy Assessment: Fundamental Principles," *Remote Sensing of Environment*, 64:331–344.

Story, M. and R. Congalton, 1986, "Accuracy Assessment: A User's Perspective," *Photogrammetric Engineering & Remote Sensing*, 52(3):397–399.

Tortora, R., 1978, "A Note on Sample Size Estimation for Multinomial Populations," *The American Statistician*, 32(3):100–102.

Wu, W. and G. Shao, 2002, "Optimal Combinations of Data, Classifiers, and Sampling Methods for Accurate Characterizations of Deforestation," *Canadian Journal of Remote Sensing*, 28(4):601–609.

Zadeh, L. A., 1965, "Fuzzy Sets," *Information and Control*, 8:338–353.

Index